Springer Collected Works in Mathematics

For further volumes:
http://www.springer.com/series/11104

Hermann Weyl

Gesammelte Abhandlungen IV

Editor

Komaravolu Chandrasekharan

Reprint of the 1968 Edition

Springer

Author
Hermann Weyl
1885 Elmshorn, Germany –
1955 Zürich, Switzerland

Editor
Komaravolu Chandrasekharan
Department of Mathematics
ETH Zürich
Switzerland

ISSN 2194-9875
ISBN 978-3-662-44289-0 (Softcover)
978-3-540-04388-1 (Hardcover)
DOI 10.1007/978-3-662-44290-6
Springer Heidelberg New York Dordrecht London

Library of Congress Control Number: 2012954381

Mathematical Subject Classification (2010): 01-XX, 83-XX

Printed on acid-free paper

Springer is part of Springer Science+Business Media (www.springer.com)

Preface

The name of HERMANN WEYL is enshrined in the history of mathematics. A thinker of exceptional depth, and a creator of ideas, WEYL possessed an intellect which ranged far and wide over the realm of mathematics, and beyond. His mind was sharp and quick, his vision clear and penetrating. Whatever he touched he adorned. His personality was suffused with humanity and compassion, and a keen aesthetic sensibility. Its fullness radiated charm. He was young at heart to the end. By precept and example, he inspired many mathematicians, and influenced their lives. The force of his ideas has affected the course of science. He ranks among the few universalists of our time.

This collection of papers is a tribute to his genius. It is intended as a service to the mathematical community.

Thanks are due to Springer-Verlag for undertaking the publication, and to the Zentenarfonds of the Eidgenössische Technische Hochschule, Zürich, and to its President Dr. J. BURCKHARDT, for a generous subvention. The co-operation of Professor B. ECKMANN has helped the project along.

These papers will no doubt be a source of inspiration to scholars through the ages.

Zürich, May 1968 K. CHANDRASEKHARAN

Preface

The name of HERMANN WEYL is enshrined in the history of mathematics. A thinker of exceptional depth, and a creator of ideas, WEYL possessed an intellect which ranged far and wide over the realm of mathematics, and beyond. His mind was sharp and quick, his vision clear and penetrating. Whatever he touched he adorned. His personality was suffused with humanity and compassion, and a keen aesthetic sensibility. Its fullness radiated charm. He was young at heart to the end. By precept and example, he inspired many mathematicians, and influenced their lives. The force of his ideas has affected the course of science. He ranks among the few universalists of our time.

This collection of papers is a tribute to his genius. It is intended as a service to the mathematical community.

Thanks are due to Springer-Verlag for undertaking the publication, and to the Zentenarfonds of the Eidgenössische Technische Hochschule, Zürich, and to its President Dr. J. BURCKHARDT, for a generous subvention. The co-operation of Professor B. ECKMANN has helped the project along.

These papers will no doubt be a source of inspiration to scholars through the ages.

Zürich, May 1968 K. CHANDRASEKHARAN

Note

These four volumes of papers by HERMANN WEYL contain all those listed in the bibliography given in the *Selecta* HERMANN WEYL, together with four additions. No changes in the text have been made other than those made by the author himself at the time of the publication of his *Selecta*.

An obituary notice by A. WEIL and C. CHEVALLEY, originally published in *l'Enseignement mathématique*, is reproduced at the end, by courtesy of the authors.

The co-operation of the publishers of the various periodicals in which WEYL's work appeared, and particularly of Birkhäuser-Verlag who brought out the *Selecta*, is gratefully acknowledged. The excellent work done by the printer merits a special mention. The frontispiece is from the collection of Mrs. ELLEN WEYL.

Inhaltsverzeichnis Band IV

122.

On the use of indeterminates in the theory of the orthogonal and symplectic groups

American Journal of Mathematics 63, 777—784 (1941)

1. These lines contain a supplement, concerning the use of indeterminates, to my book, *The Classical Groups*, Princeton (1939), which I cite in the following as *CG*. The general infinitesimal operation of the orthogonal group,

$$dx = Sx, \tag{1}$$

is described by a skew-symmetric matrix $S = \| s_{ik} \|$ of which the $n(n-1)/2$ elements s_{ik} $(i < k)$ are indeterminate parameters. It is to be expected that the first main theorem for vector invariants of the orthogonal group will hold if invariance is demanded only with respect to the infinitesimal operations (1). In this form the theorem cannot be affected by the field in which we operate, as long as it is of characteristic zero. In my book I approached the problem of vector invariants for all groups discussed in a systematic way by combining the formal apparatus of Capelli's identities with such non-formal reasoning as is used to prove the geometric theorem of congruence. The topological fact that the proper orthogonal group is a connected manifold shows that for the field K of all real numbers the full group may be replaced by the set of its infinitesimal elements. Actually I employed an algebraic equivalent for this topological argument which, by a somewhat devious procedure, carries the results over to "formal" and "infinitesimal" invariants. I now think it would have been wiser to settle the question of infinitesimal invariants for the orthogonal and the symplectic groups directly by means of another formal identity of Capelli's type which I propose to develop in § 2.

Besides the question of invariants there is connected with any group Γ of linear transformations $A = \| a_{ik} \|$ in a vector space P an ideal $\mathfrak{J}(\Gamma)$. It consists of all polynomials $\Phi(A)$ of the n^2 variables a_{ik} which vanish for each element A of Γ. Whereas invariance under the whole group is secured when it holds for any set of generators of the group, e. g., for the infinitesimal elements, it is essential to postulate the vanishing of Φ for *all* elements A of the group; for it is not true that the vanishing of $\Phi(A_1)$ and $\Phi(A_2)$ implies the vanishing of $\Phi(A_1A_2)$. In proving that $(E-S)/(E+S)$ is a generic zero of the orthogonal ideal, I should therefore have observed that the

* Received May 16, 1941.

expression $(E-S)/(E+S)$ with its $n(n-1)/2$ indeterminates s_{ik} $(i<k)$ constitutes a group, in the sense that the equation

$$(2) \qquad \frac{E-S}{E+S}=\frac{E-S'}{E+S'}\cdot\frac{E-S''}{E+S''}$$

defines a skew-symmetric matrix S in terms of two indeterminate skew-symmetric matrices S' and S''; the parameters of S are rational functions of those of S' and S''. Not only did I fail to emphasize this point, but the wrong formulation of Theorem (5. 3. B) betrays that I overlooked it altogether. The elements

$$(3) \qquad \frac{E-S}{E+S}, \qquad J_n\cdot\frac{E-S}{E+S}$$

constitute a group in the above sense. Hence the hypothesis imposed upon $\Phi(A)$ in that theorem should have required its vanishing for (3) (and not merely for $(E-S)/(E+S)$ and J_n). The corollary immediately following the theorem and Lemma (7. 1. B) ought to be corrected accordingly. (§ 3)

An important link of the investigation is the fact that the set $\Pi_f[(E-S)/(E+S)]$, being a set of orthogonal transformations in the tensor space P_f, is fully reducible. We shall deal with this point without making numerical substitutions for the indeterminates s_{ik} and, in the case of the symplectic group, without recourse to the "unitarian trick." (§ 4) We carry the details through for the symplectic group. The modifications required for the orthogonal group are obvious.

2. The coördinates of a vector x in $n=2\nu$ dimensions will be denoted by

$$x_1,\cdots,x_\nu;\ x_{1'}=x'_1,\cdots,x_{\nu'}=x'_\nu.$$

Indices $i, k,\cdots$ run over the whole range $1,\cdots,\nu,\ 1',\cdots,\nu'$ whereas $\alpha,\beta,\cdots$ run merely over the half range $1,\cdots,\nu$. The skew-product $[xy]$ of two vectors x, y is defined by

$$[xy]=\sum_\alpha (x_\alpha y'_\alpha - x'_\alpha y_\alpha)=\sum_{i,k}\epsilon(ik)x_i y_k.$$

Its matrix $\|\epsilon(ik)\|$ is called I. An infinitesimal operation $dx=Sx$ of the symplectic group has a matrix

$$S=\begin{Vmatrix} s(\alpha\beta), & s(\alpha\beta') \\ s(\alpha'\beta), & s(\alpha'\beta') \end{Vmatrix}$$

for which

$$(4) \qquad s(\alpha'\beta')=-s(\beta\alpha);\quad s(\alpha\beta')=s(\beta\alpha'),\quad s(\alpha'\beta)=s(\beta'\alpha).$$

As its $n(n+1)/2$ parameters s_j we use

$$\text{all } s(\alpha\beta);\quad s(\alpha\beta') \text{ for } \alpha\leqq\beta;\quad s(\alpha'\beta) \text{ for } \alpha\leqq\beta.$$

We refer to these conditions (4) by speaking of an S-matrix. Thus we write the indeterminate S-matrix as $L(s) = \Sigma L_j s_j$.

Let $f(x, y, \cdots)$ be a polynomial homogeneous of a certain degree with respect to the components of each argument vector. Let e be the number of argument vectors and r the degree of f with respect to x. In order to express invariance of f under infinitesimal symplectic transformations, we introduce the matrix $X = x\xi$ with the elements $X(ik) = x_i\xi_k$, composed of a covariant vector x (single column) and a contravariant ξ (single row), and moreover the matrix X' with the elements

$$X'(\alpha\beta) = x_\alpha\xi_\beta - x'_\beta\xi'_\alpha, \qquad X'(\alpha\beta') = x_\alpha\xi'_\beta + x_\beta\xi'_\alpha,$$
$$X'(\alpha'\beta) = x'_\alpha\xi_\beta + x'_\beta\xi_\alpha, \qquad X'(\alpha'\beta') = x'_\alpha\xi'_\beta - x_\beta\xi_\alpha.$$

Incidentally

$$X' = X + IX^*I.$$

Setting $\xi_i = \partial/\partial x_i$ turns $X(ik)$ and $X'(ik)$ into differential operators $d_x(ik)$, $d'_x(ik)$. The infinitesimal invariance of f is expressed by

$$\sum_y d'_y(ik)f = 0,$$

the sum extending over the e argument vectors y of f.

Let us start with the identity

$$\begin{vmatrix} [xy] & (x\eta) \\ (\xi y) & [\xi\eta] \end{vmatrix} = -\operatorname{tr}(XY'). \tag{5}$$

(If one prefers, one may write the trace on the right side as $\frac{1}{2}\operatorname{tr}(X'Y')$, but calculations are a little bit easier with the original form.) Put $\xi = \partial/\partial x$, $\eta = \partial/\partial y$; the polar $D_{yx}f$ and the operator

$$\Theta_{[xy]}f = \sum_a \left(\frac{\partial^2}{\partial x_a \partial y'_a} - \frac{\partial^2}{\partial x'_a \partial y_a}\right) f$$

make their appearance. Both are infinitesimal symplectic invariants if f is. $\Theta_{[xy]}$ plays the same rôle for the symplectic group as the Laplacian operator

$$\Theta_{xx}f = \sum_i \frac{\partial^2 f}{\partial x_i^2}$$

does for the orthogonal group. Handling the operators D and d as if multiplication by x_i and differentiation $\partial/\partial x_i$ were commutable, one would obtain from (5) the equation

$$[xy] \cdot \Theta_{[xy]}f - D_{xy}(D_{yx}f) = -\operatorname{tr}(d_x, d'_y f).$$

What one really gets are the equations

$$[xy] \cdot \Theta_{[xy]} f - D_{xy}(D_{yx} f) + r f = - \operatorname{tr}(d_x, d'_y f)$$

if $y \neq x$, and

$$[xx] \cdot \Theta_{[xx]} f - r(r+n) f = - \operatorname{tr}(d_x, d'_x f)$$

for $y = x$. Under the assumption that f is an infinitesimal symplectic invariant (briefly, an invariant), we thus find

$$\sum_y [xy] \Theta_{[xy]} f - \sum_y{}' D_{xy}(D_{yx} f) = r(r+n+1-e) f \tag{6}$$

where $\sum_y, \sum_y{}'$ designate summations over all e arguments y including or excluding the fixed argument x.

We maintain:

THEOREM I. *Every invariant $f(x, y, \cdots)$ is expressible as a polynomial in terms of the skew-products $[xy], \cdots$ of its arguments $x, y, \cdots$.*

Suppose first $e \leqq n+1$. The polar $D_{yx} f$ for $y \neq x$ and $\Theta_{[xy]} f$ (whether $y = x$ or $y \neq x$) are invariants of lower degree in x than f. Hence the equation (6) proves the theorem by induction with respect to r for any degree r under the hypothesis that it is true for $r = 0$, i. e., for invariants depending only on the $e-1$ arguments $y, \cdots$. Thus induction with respect to e yields the desired result for $e \leqq n+1$. But Capelli's general identity allows us to increase the number of arguments indefinitely once the stage of n arguments is reached. (There is an overlapping inasmuch as the increase of e from n to $n+1$ may be effected either by that identity or by (6). Up to $n+1$ arguments there are no algebraic relations between the skew-products of the arguments.)

For the orthogonal group (6) is replaced by the formula

$$\sum_y (xy) \Theta_{xy} f - \sum_y{}' D_{xy}(D_{yx} f) = r(r+n-1-e) f.$$

It works for $e \leqq n-1$ while Capelli's general identity goes on from $e = n$. The gap $e = n-1 \rightarrow n$ is bridged by Capelli's special identity which brings in the "bracket factor."

3. The relationship

$$A = (E-S)(E+S)^{-1} = (E+S)^{-1}(E-S) \tag{7}$$

between finite and infinitesimal symplectic transformations, A and S, shows that (2) defines an S-matrix in terms of two indeterminate S-matrices S', S'' with the parameters s_j', s_j''. The parameters s_j of S are rational functions $\psi_j(s', s'')$. Writing (2) in the form

$$2(E+S)^{-1}-E=(E+S')^{-1}(E-S')(E-S'')(E+S'')^{-1}$$

one finds at once

$$(E+S)^{-1}=(E+S')^{-1}(E+S'S'')(E+S'')^{-1}$$

or

$$E+S=(E+S'')(E+S'S'')^{-1}(E+S').$$

Hence the functions $\psi_j(s', s'')$ have the determinant

$$\Delta=|E+S'S''|$$

as their common denominator, and

$$|E+S|=|E+S'|\cdot|E+S''|\cdot\Delta^{-1}. \tag{8}$$

We derive the product matrix $\Pi_f A$ in the tensor space P_f from

$$A=(E-S)(E+S)^{-1}, \qquad S=L(s). \tag{9}$$

Evidently it is of the following form

$$\Pi_f A=R(s)/|E+S|^f, \qquad R(s)=\sum_p C_p\,\sigma_p \tag{10}$$

where the numerator $R(s)$ is a polynomial of formal degree nf of the parameters s_j with matric coefficients C_p. Thus σ_p runs over the monomials of degree $\leqq nf$ of the parameters s_j. The same holds for $\Pi^{(f)}(A)$:

$$|E+S|^f\cdot\Pi^{(f)}(A)=\sum_p C_p^{(f)}\sigma_p.$$

We maintain

THEOREM II. *The linear closure of* (10), *i. e., the set* $\mathfrak{C}_f$ *of all linear combinations* $\Sigma\lambda_p C_p$ *with numerical coefficients* λ_p, *is an algebra containing the unit matrix.*

The unit matrix is the coefficient C_0 of the monomial $\sigma_0=1$. Equation (2) entails

$$\Pi_f\left(\frac{E-S}{E+S}\right)=\Pi_f\left(\frac{E-S'}{E+S'}\right)\cdot\Pi_f\left(\frac{E-S''}{E+S''}\right)$$

or, because of (8),

$$\Delta^f\cdot\sum_r C_r\sigma_r=\sum_{p,q} C_pC_q\sigma_p'\sigma_q''.$$

After the substitution of $\psi_j(s', s'')$ for s_j each σ_r turns into a polynomial $\phi_r(s', s'')$ divided by Δ^{nf}. Hence

$$\Delta^{(n-1)f}\cdot\sum_{p,q} C_pC_q\sigma_p'\sigma_q''=\sum_r C_r\phi_r(s', s''). \tag{11}$$

We now apply the following trivial algebraic lemma:

Let $\phi = 1 + \cdots$ be a given polynomial of some variables $x_1, \cdots, x_l$, with the constant term 1. Then the coefficients of an arbitrary polynomial F of degree m may be linearly expressed by the coefficients of $\phi F = G$.

Arranging the terms of F in ascending lexicographic order one obtains recursive linear equations for the unknown coefficients a of F, with the coefficients b of G as the known right members (only terms of G of a degree not exceeding m enter). Denote by F_μ the terms of degree μ in any polynomial $F = F_0 + F_1 + \cdots$. Putting $\phi = 1 - \omega$ and using the power series $(1-\omega)^{-1}$, one finds more explicitly

$$F = \Sigma G_\mu \omega_{\mu_1} \omega_{\mu_2} \cdots$$
$$(\mu \geqq 0;\ \mu_1 \geqq 1, \mu_2 \geqq 1, \cdots;\ \mu + \mu_1 + \mu_2 + \cdots \leqq m).$$

If ϕ has integral coefficients, then the coefficients of the linear combinations expressing the a in terms of the b are likewise integers.

This lemma is of immediate application to the equation (11) in which Δ begins with the constant term 1. The products $\sigma_p' \sigma_q''$ are monomials no two of which are equal, and thus we arrive at equations

$$C_p C_q = \sum_r \gamma^r{}_{pq} C_r$$

with rational integers $\gamma^r{}_{pq}$. They prove our statement. The same equations hold for the matrices $C_p^{(f)}$, and thus the linear combinations $\Sigma \lambda_p C_p^{(f)}$ form an algebra $\mathfrak{C}^{(f)}$. The stage for the whole drama is the field κ of rational numbers (or, as the arithmetician might care to observe, the even narrower ring of rational integers).

4. We go on operating in κ. The next step is

THEOREM III. *The algebra $\mathfrak{C}_f$ in κ is fully reducible.*

Suppose that a subspace $\{a_\lambda\}$ of P_f is spanned by a number of linearly independent vectors

$$a_\lambda = (a_{1\lambda}, \cdots, a_{N\lambda}), \qquad (N = n^f).$$

The orthogonal subspace $\{a_\mu\}$ is spanned by a complete set of linearly independent solutions $x = a_\mu$ of the simultaneous equations

$$\sum_{i=1}^{N} a_{i\lambda} x_i = 0.$$

(The ranges of the two indices λ and μ are disjunct.) In κ the two subspaces have no vector in common except 0 and hence span the whole space; or the matrix U whose columns are the a_λ and a_μ is a non-singular square matrix. By construction the symmetric matrix U^*U decomposes into a (λ, λ)- and a

(μ,μ)-part whereas the rectangles (λ,μ) and (μ,λ) are empty. We now assume the first subspace $\{a_\lambda\}$ to be invariant under $\mathfrak{C}_f$. This circumstance is expressed by an equation

$$R(s)\cdot U = U\cdot Q(s)$$

where $Q(s)$ is reduced in the sense that its (λ,μ)-part is empty. We propose to prove that its (μ,λ)-part is also empty.

To that end, multiply by U^* to the left:

$$U^*R(s)U = (U^*U)Q(s).$$

We then see that the (λ,μ)-part of $U^*R(s)U$ is empty,

$$[U^*R(s)U]_{(\lambda,\mu)} = 0, \tag{12}$$

and that it is sufficient to establish the same fact for the (μ,λ)-part of the same matrix.

If S is an S-matrix, so is its transpose S^*. The following simple linear involutorial substitution of the parameters s_j,

$$s(\alpha\beta)\to s(\beta\alpha),\quad s(\alpha\beta')\to s(\alpha'\beta),\quad s(\alpha'\beta)\to s(\alpha\beta'), \tag{13}$$

changes the indeterminate $S = L(s)$ into S^*, hence A, (9), into A^*, $\Pi_f A$ into $(\Pi_f A)^*$, and since $|E+S| = |E+S^*|$, also $R(s)$ into $R^*(s)$ and $R^*(s)$ back into $R(s)$. Taking the transpose of the equation (12),

$$[U^*R^*(s)U]_{(\mu,\lambda)} = 0,$$

and then carrying out the substitution (13), one arrives at the desired result

$$[U^*R(s)U]_{(\mu,\lambda)} = 0.$$

(13) induces a simple involutorial permutation among the monomials σ_p, $\sigma_p \to \sigma_{p^*}$. Our argument shows that

$$\Sigma C^*_p\,\sigma_p = \Sigma C_p\,\sigma_{p^*} = \Sigma C_{p^*}\,\sigma_p;\quad C^*_p = C_{p^*}.$$

Thus the algebra $\mathfrak{C}_f$ is a set $\mathfrak{C}$ of matrices C which coincides with the set $\mathfrak{C}^*$ of its transposed elements C^*. The method employed yields this general proposition: *For any set $\mathfrak{C}$ of matrices in a real field k with the property $\mathfrak{C}^* = \mathfrak{C}$ reduction results in full reduction if projection modulo the invariant subspace is carried out as projection upon the orthogonal subspace.* This is in line with older investigations by E. Fischer who studied groups $\mathfrak{C}$ of linear transformations in a real field k enjoying the property $\mathfrak{C}^* = \mathfrak{C}$ and showed that their invariants depending on a variable form have a finite integrity basis.[1]

Once in possession of our three theorems we put the same machinery into

[1] *Journal für reine u. ang. Mathematik*, vol. 140 (1911), pp. 48-81.

play as employed on pp. 141-142 of *CG*. We still operate in the field κ. An $f(x, y, \cdots)$ which is invariant under the generic (9) is necessarily an infinitesimal invariant. Apply the hypothesis to tS instead of S and take merely the first power of the parameter t into account. Consequently Theorem I shows that the algebra described as $\mathfrak{A}^{(f)}$ on p. 174 is the commutator algebra of the commutator algebra of $\Pi^{(f)}[(E-S)/(E+S)]$ or of $\mathfrak{C}^{(f)}$. Hence as $\mathfrak{C}^{(f)}$ is an algebra containing the unit matrix (Theorem II) and fully reducible (Theorem III), the algebra $\mathfrak{A}^{(f)}$ is not wider than $\mathfrak{C}^{(f)}$ (R. Brauer's principle). In view of the definition of $\mathfrak{A}^{(f)}$ this fact remains true in any field of characteristic zero. Any polynomial $\Phi(A)$ of degree f depending on the matrix $A = \| a_{ik} \|$ with variable components a_{ik} proceeds from a linear form $\gamma(A^{(f)})$ of an arbitrary bisymmetric $A^{(f)}$ by the substitution $A^{(f)} = \Pi^{(f)}(A)$. If $\Phi(A)$ is annulled by the substitution (9) we must have $\gamma(C_p{}^{(f)}) = 0$, and therefore $\gamma(A^{(f)})$ vanishes for all matrices $A^{(f)}$ of $\mathfrak{A}^{(f)}$. Thus results Theorem (6.3.B) on p. 174 of *CG*. In restating it I propose the following terminology for ideals of polynomials $\phi(x_1, \cdots, x_l)$. Elements $\phi_1, \cdots, \phi_m$ of a given ideal $\mathfrak{J}$, whose degrees are $f_1, \cdots, f_m$ respectively, are said to constitute a *form basis* of $\mathfrak{J}$ if any element ϕ of $\mathfrak{J}$ of degree f may be obtained in the form

$$\phi = h_1\phi_1 + \cdots + h_m\phi_m$$

where h_i is a polynomial of degree $f - f_i$ (which implies, of course, the vanishing of h_i whenever $f - f_i < 0$). By introduction of homogeneous coördinates $\mathfrak{J}$ gives rise to an ideal $\bar{\mathfrak{J}}$ of homogeneous forms $\phi(x_0, x_1, \cdots, x_l)$ such that $\phi(x_0, x_1, \cdots, x_l)$ is in $\bar{\mathfrak{J}}$ if and only if $\phi(1, x_1, \cdots, x_l)$ is in $\mathfrak{J}$. The statement "$\phi_1, \cdots, \phi_m$ constitute a *form basis* of $\mathfrak{J}$" means that the corresponding homogeneous forms $\phi_1, \cdots, \phi_m$ of degrees $f_1, \cdots, f_m$ constitute a *basis* of $\bar{\mathfrak{J}}$.

We arrive at the following proposition concerning the *symplectic ideal* (i. e., the ideal $\mathfrak{J}(\Gamma)$ of the symplectic group Γ) which holds in any field k of characteristic 0:

THEOREM IV. 1) *Let $A = \| a_{ik} \|$ be the matrix with n^2 variable components a_{ik}. The components of the two matrices $A^*IA - I$ and $AIA^* - I$ constitute a form basis, the components of either of them a basis, of the symplectic ideal.*

2) *The symplectic ideal is a prime ideal, and the expression* (9) *a generic zero; i. e., a polynomial $\phi(A)$ vanishes for all symplectic transformations A if it is annulled by the substitution* (9).

123.

Concerning the differential equations of some boundary layer problems

Proceedings of the National Academy of Sciences of the United States of America
27, 578—583 (1941)

I. Mathematically the simplest boundary layer is that forming around a half-plane $y = 0$, $x \geqq 0$ immersed in an incompressible viscous fluid of kinematic viscosity ϵ^2 which flows by with a given constant velocity V in the direction of the x-axis. This two-dimensional steady flow can be described by a stream function $\psi^*(\epsilon;\ x, y)$, and if the viscosity ϵ^2 approaches zero, $\frac{1}{\epsilon}.\psi^*(\epsilon;\ x, \epsilon y)$ will tend to a limit $\psi(x, y)$, the stream function of the boundary layer. In our case ψ is of the form $2x^{1/2}.w(y/2x^{1/2})$ where the function w of the single variable $z = y/2x^{1/2}$ satisfies the non-linear differential equation

$$w''' + 2ww'' = 0 \quad (0 \leqq z < \infty). \tag{1}$$

One has to find a solution w for which $w = w' = 0$ at $z = 0$ and w' approaches V with $z \to \infty$. If $w(z)$ is a solution, so is $\beta.w(\beta z)$ whatever the positive constant β. Let $w = f(z)$ be the solution determined by the initial values $f = f' = 0$, $f'' = 1$. Once we are sure that f' tends to a positive limit a with $z \to \infty$,

$$a = \int_0^\infty f''(z)dz > 0, \tag{2}$$

we may adjust the constant β in such a way,

$$\beta^2 a = V, \tag{3}$$

that the derivative of $w = \beta.f(\beta z)$ approaches V at infinity.

When H. Blasius carried out the integration as a first example of Prandtl's idea of the boundary layer,[1] he made use of the power series for w around $z = 0$ and of a certain asymptotic expression for large values of z, adjusting the constant β so as to make both expressions dovetail in a middle region. It is a fact, although it seems to have escaped notice, that convergence of the power series of f stops somewhere between $z = \sqrt[3]{9} = 2.08$ and $\sqrt[3]{30} = 3.11$. Thus Blasius' method, which has been followed later in many similar cases, can yield results of limited accuracy only, and the mathematician remains in doubt whether the solution f extends over the whole interval $z \geqq 0$ and has the property (2). After rejecting the power series because of its shortcoming, I shall here describe a rapidly converging process of successive and alternating approximations which gives all one may desire and is an excellent tool for actual numerical computation.

The coefficients c_n of

$$f(z) = \Sigma(-1)^n c_n z^{3n+2} \quad (n = 0, 1, \ldots)$$

are determined by the recursive formula

$$(3n + 2)(3n + 1)3n.c_n = 2.\Sigma(3i + 2)(3i + 1)c_i c_k \quad (i + k = n - 1)$$

and $c_0 = {}^1/_2$. Notice that

$$\Sigma_0^{n-1}(3i + 2)(3i + 1) = n(3n^2 - 1)$$

and

$$1/30 \leqq \frac{3n^2 - 1}{3(3n + 2)(3n + 1)} \leqq 1/9 \quad (n = 1, 2, \ldots).$$

Hence the inequalities

$$c_i \leqq \frac{1}{2}\left(\frac{1}{9}\right)^i \text{ and } c_i \geqq \frac{1}{2}\left(\frac{1}{30}\right)^i$$

for $i = 0, \ldots, n - 1$ imply the same for $i = n$. Since they hold for $i = 0$ our statement is proved in the sharp form

$$\frac{1}{2}\left(\frac{1}{30}\right)^n \leqq c_n \leqq \frac{1}{2}\left(\frac{1}{9}\right)^n.$$

As is easily seen, the differential equation $f''' + 2ff'' = 0$ together with the initial conditions $f = f' = 0, f'' = 1$ is equivalent to the following integral equation for $g(z) = f''(z)$:

$$g(z) = \exp(-\textstyle\int_0^z (z - \zeta)^2 g(\zeta) d\zeta) \quad \text{or} \quad g = \Phi\{g\}$$

where the functional operator Φ at the right side has the properties:

$$\Phi\{g\} \geqq 0, \Phi\{g\} \geqq \Phi\{g^*\} \text{ if } g \leqq g^*.$$

Thus we are led to define the successive approximations g_n by

$$g_{n+1} = \Phi\{g_n\} \quad (n = 0, 1, \ldots),$$

starting with $g_0(z) = 0$. The trivial relations $g_1 \geqq g_0 = 0, g_2 \geqq g_0 = 0$ then give rise to two sequences of inequalities:

$$g_0 \leqq g_1, g_1 \geqq g_2, g_2 \leqq g_3, g_3 \geqq g_4, \ldots$$

and

$$g_0 \leqq g_2, g_1 \geqq g_3, g_2 \leqq g_4, g_3 \geqq g_5, \ldots.$$

Hence the g_n perform an alternating pincer movement: all the odd g_n are above the even g_n, the odd g_n move down, the even g_n go up with increasing n. Moreover one readily finds by induction with respect to n:

$$0 \leqq (-1)^n(g_{n+1}(z) - g_n(z)) \leqq \frac{(2z^3)^n}{(3n)!}.$$

Thus in the limit for $n \to \infty$ a solution g results which is approached by the even g_n from below, by the odd g_n from above.

The first approximations are

$$g_0 = 0,\ g_1 = 1,\ g_2 = e^{-1/3\,z^3},$$
$$g_3(z) = \exp(-\int_0^z (z - \zeta)^2 . e^{-1/3\,\zeta^3} d\zeta).$$

For large values of z the exponent in g_3 behaves like a quadratic polynomial,

$$\int_0^z (z - \zeta)^2 e^{-1/3\,\zeta^3} d\zeta \sim a_2 z^2 - 2a'_2 z + a''_2$$

where

$$(a_2, a'_2, a''_2) = \int_0^\infty (1, \zeta, \zeta^2) . e^{-1/3\,\zeta^3} d\zeta$$

and since $g \leqq g_3$ we have

$$g(z) \leqq e^{-kz^2}$$

for sufficiently high z if k is any positive number

$$< a_2 = \int_0^\infty e^{-1/3\,\zeta^3} d\zeta = \sqrt[3]{3} . \Gamma\left(\frac{4}{3}\right).$$

This assures the convergence of

$$\int_0^\infty g(z) dz = \int_0^\infty f''(z) dz = a,$$

and the asymptotic behavior of

$$f(z) = \int_0^z (z - \zeta) g(\zeta) d\zeta$$

is indicated by

$$f(z) \sim \int_0^\infty (z - \zeta) g(\zeta) d\zeta = az - a'.$$

As $g \geqq g_2$ implies $a \geqq a_2$ we find the numerical coefficient $\alpha = a^{-3/2}$ in

$$w''(0) = \beta^3 = \alpha V^{3/2}$$

to be $< \alpha_2 = 0.684$. According to S. Goldstein,[2] $\alpha = 0.664$. Hence the very first approximate value for α which can be derived from our method misses the mark by less than 3%, which makes it likely that $g_3(z)$ is a pretty good approximation of $g(z)$ throughout the entire interval.

II. With slight alterations our process applies to the equation

$$w''' + 2ww'' - w'^2 = 0 \tag{4}$$

which determines the wake behind our flat plate.[3] [The stream function of problem I for large $x \sim l$ assumes a finite value $\varphi(\eta) = \frac{\alpha}{4} V^{3/2}.\eta^2$ in the limit $l \to \infty$ after we make the substitution $y = \eta . l^{1/4}$, which we combine with $x - l = \xi . l^{1/4}$. With the plate ending at the point $x = l$, the stream function

$$\psi(\xi, \eta) = 3\xi^{2/3}.w(\eta/\xi^{1/3}) \quad (\xi \geqq 0)$$

of the wake is determined under the (not too good) assumption that it ties up with $\varphi(\eta)$ for $\eta/\xi^{1/3} \to \infty$.] The boundary conditions are:

$$w = w'' = 0 \text{ at } z = 0 \text{ and } w'' \to \tfrac{\alpha}{6}.V^{3/2} \text{ for } z \to \infty.$$

Let f be the solution with the initial values $f = 0, f' = 1, f'' = 0$. After differentiating (4),

$$f'''' + 2ff''' = 0, \tag{5}$$

we arrive at the following integral equation for $f'' = g$:

$$g(z) = \int_0^z \exp(- z^2 - \int_0^z (z - \zeta)^2 g(\zeta) d\zeta).dz.$$

$g(\infty)$ has a finite positive value, and thus one may adjust the constant δ in the solution $w = \delta . f(\delta z)$ so as to give to $w''(\infty)$ the desired value.

Neither of the equations (1) and (4) contains the independent variable z, and both allow the group of transformations $w(z) \to \beta . w(\beta z)$. Hence, as J. v. Neumann has observed, they are reducible to differential equations of the first order followed by two quadratures. They are thus within reach of Poincaré's general discussion of first-order differential equations. However, it seems certain that our method gives quicker numerical results and, what is more important, it works also in cases of a different type where that reduction is impossible.

III. One such case is the three-dimensional problem of a horizontal flow against a perpendicular wall.[4] Its differential equation

$$w''' + 2ww'' + (V^2 - w'^2) = 0 \tag{6}$$

holds for any value of the viscosity ϵ^2 and not merely in the limit $\epsilon \to 0$. Add the intial conditions $w = w' = 0$ at $z = 0$ and $w' \to V$ for $z \to \infty$. We try to construct a solution f of the equation (5) obtained by differentiation, for which $f = f' = 0, f'' = 1$ for $z = 0$ and the initial value $-a^2$ of f''' is such that f' approaches a finite limit for $z \to \infty$ (which then must equal a). The function $w = \beta . f(\beta z)$ solves our problem if β is chosen by (3). The integral equation for $f'' = g$ now reads

$$g(z) = a^2 . \int_z^\infty e^{-\gamma(\zeta)} d\zeta$$

where

$$\gamma(z) = \int_0^z (z - \zeta)^2 g(\zeta) d\zeta, \; a^2 = 1/\int_0^\infty e^{-\gamma(\zeta)} d\zeta.$$

This time we have to start with $g_0(z) = 1$ and to construct the successive approximations by means of the formulas

$$\gamma_n(z) = \int_0^z (z - \zeta)^2 g_n(\zeta) d\zeta,$$
$$g_{n+1}(z) = \int_z^\infty e^{-\gamma_n(\zeta)} d\zeta / \int_0^\infty e^{-\gamma_n(\zeta)} d\zeta.$$

Readily enough, though by a somewhat more intricate argument, one establishes the alternating pincer movement of the sequence g_n. The question of convergence is of a different character and only convergence of the type of a geometric series is to be expected, because we are now dealing with a real boundary rather than an initial-value problem. I obtain majorizing constants μ_n,

$$0 \leqq (-1)^n (g_n(z) - g_{n-1}(z)) \leqq \mu_n,$$

defined by the recursive equations

$$\mu_1 = 1; \; \mu_n = q_n \cdot \mu_{n-1}, \quad \mu_{n+1} = q_n \cdot \mu_n \quad (n \text{ even})$$

where q_n denotes the constant

$$q_n = \int_0^\infty e^{-\gamma_{n-1}(\zeta)} . \tfrac{1}{3}\zeta^3 d\zeta / \int_0^\infty e^{-\gamma_{n-1}(\zeta)} d\zeta$$
$$= \int_0^\infty g_n(z) . z^2 dz.$$

The second expression shows the alternating pincer movement of the q_n, hence all even $q_n \leqq q_2 = q$. If $q < 1$ our process converges uniformly and better than the geometric series of quotient q. Since $q_1 = 1/3$ our experience with problem I leads us to expect that q is not much larger than $1/3$; a provisional calculation indicates that its value is about 0.41.

Queerly enough, the differential equation of the corresponding two-dimensional problem, first treated by Hiemenz,[5]

$$w''' + 2ww'' + 2(V^2 - w'^2) = 0, \qquad (7)$$

proves less tractable, not so much for want of an appropriate recursive process—which can be set up after twice differentiating (7)—as for want of a proper start $g_0(z)$.

The boundary layer of the two-dimensional flow against an angular bastion of angle $\pi\lambda$ may be determined by conformally mapping the exterior of the angle upon an $(x + iy)$-plane cut along the positive x-axis. Introducing Prandtl's coördinate $y^2/2x^{1/2}$ in that plane, one obtains the differential equation

$$w''' + 2ww'' + 2\lambda(V^2 - w'^2) = 0.$$

Hence the rectangular bastion, $\lambda = {}^1/_2$, leads to Homann's equation (6) which showed itself accessible to our method. I do not know how to deal with values of the constant λ other than $\lambda = 0$ and ${}^1/_2$.

I wish the WPA Mathematical Tables Project could be persuaded to resume computation of tables for the functions f, f', f'' of the problems I, II, III by the systematic procedure here explained.

[1] *Zeitschr. Math. Phys.*, **56,** 1 (1908).

[2] *Proc. Cam. Phil. Soc.*, **26,** 1–30, especially p. 19 (1930). See also Howarth, L., *Proc. Roy. Soc.*, **164A,** 551 (1938), and Goldstein, S., *Modern Developments in Fluid Dynamics,* Oxford, 1938, p. 136.

[3] Goldstein, S., loc. cit.,[2] and *Proc. Roy. Soc.*, **142A,** 545–573 (1933).

[4] Homann, F., *Zeitschr. angew. Math. Mech.*, **16,** 153 (1936).

[5] *Dingler's Polytech. Jour.*, **326,** 321–326 (1911).

124.

Concerning the differential equations of some boundary layer problems. II

Proceedings of the National Academy of Sciences of the United States of America
28, 100—102 (1942)

In a previous note under the same title[1] I have solved by successive approximations the following boundary value problem:

$$(A_\lambda)\begin{cases} w''' + 2ww'' + 2\lambda(k^2 - w'^2) = 0 \text{ for } z \geqq 0; \\ w = w' = 0 \text{ for } z = 0,\ w'(\infty) = k, \end{cases}$$

involving the positive parameter k, for the special values $\lambda = 0$ and $^1/_2$. Convergence of the process for $\lambda = {}^1/_2$ requires a certain constant q to be < 1. The actual computation of q is laborious; however, I wish to report that a comparatively simple estimation yields $q < {}^3/_4$.

For *arbitrary* $\lambda \geqq 0$ I have to take refuge to a much more sophisticated method, that of fixed points of functional transformations, which was inaugurated by Birkhoff and Kellogg in a famous memoir[2] and later formulated in general abstract terms by Schauder and Leray. The experiences with the special cases suggest the following procedure. We eliminate the parameter k from the differential equation by differentiation,

$$w'''' + 2ww''' + 2(1 - 2\lambda)w'w'' = 0,$$

and observe that the function $\kappa.w(\kappa z)$ solves the new equation of fourth order if $w(z)$ does, whatever the positive constant κ. Hence we seek our solution of (A_λ) in the form $w(z) = \kappa.f(\kappa z)$ where

$$\left.\begin{aligned} f'''' + 2ff''' + 2(1 - 2\lambda)f'f'' &= 0 \text{ for } z \geqq 0; \\ f(0) = f'(0) = 0, f''(0) = 1;\ f''(\infty) &= 0. \end{aligned}\right\} \qquad (1)$$

Our line of attack can be explained a little more easily if we replace the infinite by the finite interval $0 \leqq z \leqq a$ and thus the last boundary condition in (1) by $f''(a) = 0$. We then introduce $f'' = g = \varphi$ as the unknown function and start with this set-up:

$$f(z) = \int_0^z (z - \zeta) g(\zeta) d\zeta. \tag{2}$$

$$\left.\begin{aligned} \varphi'' + 2f\varphi' + 2(1 - 2\lambda)f'\varphi &= 0; \\ \varphi(0) = 1, \ \varphi(a) &= 0. \end{aligned}\right\} \tag{3}$$

$$\varphi = g. \tag{4}$$

For an arbitrarily given g we form (2), then solve the linear boundary value problem (3) of familiar type, thus defining the functional operator Φ_λ which carries g into φ, and as the last step (4), we try to determine g as a function whose image coincides with the original,

$$g = \Phi_\lambda\{g\}.$$

The success hinges on the following

LEMMA: *Under the assumption $g(z) \geqq 0$ the linear problem (3) has a unique solution φ; it satisfies the conditions*

$$\varphi \geqq 0, \ \varphi' + 2f\varphi \leqq 0 \tag{5}$$

throughout the interval $0 \leqq z \leqq a$.

To prove this we set

$$p(z) = \exp(2\int_0^z f(\zeta)d\zeta)$$

so that $p(0) = 1$, $p' = 2fp$, and introduce the auxiliary function

$$\varphi_1 = p(\varphi' + 2f\varphi) = (p\varphi)'.$$

The single differential equation for φ is then replaced by the system

$$\left.\begin{aligned} \varphi' + 2f\varphi &= \varphi_1/p, \\ \varphi_1' - 2f\varphi_1 &= 4\lambda p f' \varphi. \end{aligned}\right\} \tag{6}$$

Multiplication by φ_1 and φ, respectively, and addition leads to the formula

$$[\varphi\varphi_1]_b^a = \int_b^a \left(\frac{1}{p}\varphi_1^2 + 4\lambda p f' \varphi^2\right) dz \qquad (b < a). \tag{7}$$

The inequality $g \geqq 0$ which implies $f' \geqq 0$ guarantees the positive definite character of the "Dirichlet integral" at the right side of (7). Our lemma is an easy consequence of this fact.

Thus we have defined the operator Φ_λ for all $g \geqq 0$; the restriction $g \geqq 0$ is harmless because it carries over to the image φ. Since (5) implies $(p\varphi)' \leqq 0$ we have even

$$p\varphi \leqq 1, 0 \leqq \varphi \leqq 1,$$

and may therefore start with the set G of all functions $g(z)$ defined and continuous in the interval $0 \leqq z \leqq a$ for which $0 \leqq g \leqq 1$. This limitation of g has the consequence that

$$0 \leqq f' \leqq z, 0 \leqq f \leqq 1/2z^2 \tag{8}$$

and enables us to ascertain a universal bound for φ', i.e., a number A depending on λ and a only such that

$$0 \leqq -\varphi'(z) \leqq A$$

for $0 \leqq z \leqq a$ and the image $\varphi = \Phi_\lambda\{g\}$ of *every* $g \epsilon G$. First we determine such a universal bound for $\varphi'(0)$. Indeed would φ start off at $z = 0$ with too steep a decline it could not help hitting the ground, $\varphi = 0$, long before z reaches a, considering that the coefficients of its differential equation (3) are bounded by (8). Once $\varphi'(0)$ has been penned in that differential equation does the rest. Hence the images φ of all elements $g \epsilon G$ are equi-continuous in the strict sense that $|z_1 - z_2| \leqq \epsilon$ implies $|\varphi(z_1) - \varphi(z_2)| \leqq A\epsilon$.

We topologize the "space" of all continuous functions of z by interpreting convergence as *uniform* convergence in the interval $0 \leqq z \leqq a$. Then Φ_λ is a continuous operator in G which maps G into the subset G_A consisting of all functions $g \epsilon G$ for which $|z_1 - z_2| \leqq \epsilon$ implies $|g(z_1) - g(z_2)| \leqq A\epsilon$. This suffices for the existence of a "fixed point" $g \epsilon G$ of the operator Φ_λ (see Birkhoff and Kellogg, l.c., Theorem II).

The case of the infinite interval is not essentially more difficult. It turns out that the solution f of (1) which our method yields has the property that $f'' \geqq 0$ and $f''' + 2ff'' \leqq 0$ converge to zero with $z \rightarrow \infty$ at least as strongly as a function of the type $e^{-\gamma z^2} (\gamma > 0)$. Hence

$$\int_0^\infty f''(z)dz = f'(\infty) = \beta$$

converges and we may retrace the step leading from (A_λ) to (1).

THEOREM. *For given positive k the problem (A_λ) has a solution w whose derivative is monotone increasing from 0 to k as z travels from 0 to infinity; the second derivative decreases monotonely from*

$$w''(0) = \alpha k^{3/2} \quad (\alpha = \beta^{-3/2})$$

to zero, approaching zero with $z \rightarrow \infty$ at least as strongly as a function of the type $e^{-\gamma z^2}$.

[1] These PROCEEDINGS, 27, 578–583 (1941).

[2] Birkhoff, G. D., and Kellogg, O. P., *Trans. Am. Math. Soc.*, 23, 96–115 (1922); Schauder, J., *Studia Mathematica*, 2, 171–180 (1930); Leray, J., and Schauder, J., *Ann. Sc. Ec. Norm. Sup.*, 51, 45–78 (1934).

125.

On the differential equations of the simplest boundary-layer problems

Annals of Mathematics 43, 381—407 (1942)

§ 1. The Central Boundary-Value Problem and its Hydrodynamic Interpretation

In the theory of viscous fluids the following non-linear boundary-value problem for a function $w(z)$ of a real variable $z \geqq 0$ involving two constants $k > 0$ and $\lambda \geqq 0$ plays an important part:

$$\left.\begin{aligned} & w''' + 2ww'' + 2\lambda(k^2 - w'^2) = 0 \quad \text{for} \quad z \geqq 0; \\ & w(0) = w'(0) = 0, \quad w'(z) \to k \quad \text{for} \quad z \to \infty. \end{aligned}\right\} \qquad (A_\lambda)$$

We consider λ as a given constant, but k as a variable parameter. A mathematically satisfactory proof of its solvability has never been given, although various numerical devices, including BUSH's differential analyzer, have been set at work on it. We shall here give a complete solution of the problem[1]), first for the two special values $\lambda = 0$ and $\lambda = 1/2$ by a process of alternating approximations, rapidly converging and thus well suited for numerical computations (§§ 2, 4, 5), and then approach the general case (§§ 6, 7) by the method of fixed points of transformations in a functional space,–which is considerably less amenable to calculation. In between (§ 3) the first method will be applied to certain boundary-value problems closely related to (A_0).

There are available two hydrodynamic interpretations of (A_λ). Consider first the steady flow of an incompressible viscous fluid of constant density ϱ and kinematic viscosity ε^2 filling the half $z > 0$ of an m-dimensional Euclidean space with the Cartesian coordinates $x_1, \ldots, x_m$ and the cylindrical coordinates

$$r = (x_1^2 + \cdots + x_{m-1}^2)^{1/2}, \quad z = x_m.$$

If cylindrical symmetry prevails and hence the radial (r) and vertical (z) com-

---

[1]) See the author's preliminary notes in Proc. Nat. Acad. Sci. 27, 578–583 (1941), and 28, 100–102 (1942).

ponents u, v of velocity as well as the pressure ϱp depend on r, z only, then the following differential equations obtain for $z > 0$:

$$\left.\begin{aligned} &\frac{\partial u}{\partial r} + \frac{m-2}{r} u + \frac{\partial v}{\partial z} = 0; \\ &u \frac{\partial u}{\partial r} + v \frac{\partial u}{\partial z} + \frac{\partial p}{\partial r} = \varepsilon^2 \left(\Delta u - \frac{m-2}{r^2} u \right), \\ &u \frac{\partial v}{\partial r} + v \frac{\partial v}{\partial z} + \frac{\partial p}{\partial z} = \varepsilon^2 \Delta v \end{aligned}\right\} \tag{1}$$

where the Laplace operator Δ is defined by

$$\Delta \varphi = \frac{\partial^2 \varphi}{\partial r^2} + \frac{m-2}{r} \frac{\partial \varphi}{\partial r} + \frac{\partial^2 \varphi}{\partial z^2}.$$

These equations are to be combined with the boundary conditions

$$u \to 0, \quad v \to 0 \text{ for } \quad z \to 0.$$

For an ideal fluid, $\varepsilon = 0$, we have this simple solution:

$$u_0(r, z) = \frac{2k}{m-1} r, \quad v_0(r, z) = -2kz,$$

$$p_0 = \text{const} - \frac{1}{2}(u_0^2 + v_0^2) = \text{const} - 2k^2 \left\{ \frac{r^2}{(m-1)^2} + z^2 \right\}$$

arising from the harmonic velocity potential

$$\varphi = k \left(\frac{x_1^2 + \cdots + x_{m-1}^2}{m-1} - x_m^2 \right) = k \left(\frac{r^2}{m-1} - z^2 \right)$$

and involving an arbitrary positive constant k. As is necessary, the vertical (though not the radial) component velocity vanishes along the boundary $z = 0$. The Navier-Stokes equations (1) for the viscous fluid possess a solution of the form

$$u = \frac{2}{m-1} r F'(z), \quad v = -2F(z), \quad p = \text{const} - 2k^2 \left\{ \left(\frac{r}{m-1} \right)^2 + L(z) \right\}$$

which approaches the solution u_0, v_0, p_0 for $z \to \infty$. The first equation (1) is identically satisfied, the second and third yield

$$\varepsilon^2 F''' + 2FF'' + \frac{2}{m-1}(k^2 - F'^2) = 0$$

and

$$k^2 L' = 2FF' + \varepsilon^2 F'' \tag{2}$$

respectively. Setting $F(\varepsilon z) = \varepsilon \cdot w(z)$ we obtain the equations (A_λ) with $\lambda = 1/(m-1)$ from which the viscosity constant has disappeared, so that w is independent of ε. Equation (2) in integrated form gives

$$k^2 \cdot L(\varepsilon z) = \varepsilon^2 \{ w'(z) + w^2(z) \}.$$

Our solution describes approximately the flow of a viscous fluid around an obstacle with a blunt nose in the neighborhood of the forward stagnation point. The cases of physical interest, $m = 2$ and 3, i.e. $\lambda = 1$ and 1/2, have been treated by HIEMENZ and HOMANN respectively[1]).

Let the subscript ε in u_ε, v_ε, p_ε indicate dependence of our flow on the viscosity constant ε. Certainly u_ε, v_ε, p_ε tend to u_0, v_0, p_0 with $\varepsilon \to 0$ in the region $z > 0$, but the convergence cannot be uniform at the boundary because the viscous fluid adheres, the ideal glides along the wall. Hence we have the phenomenon of a boundary layer of thickness $\sim \varepsilon$ in which the velocity rises from 0 at the surface to the external value

$$\bar{u} = u_0(r, 0) = \frac{2k}{m-1} r, \quad \bar{v} = v_0(r, 0) = 0.$$

Indeed

$$u_\varepsilon(r, \varepsilon z), \quad \frac{1}{\varepsilon} \cdot v_\varepsilon(r, \varepsilon z), \quad p_\varepsilon(r, \varepsilon z) \tag{3}$$

tend with $\varepsilon \to 0$ to the values

$$U(r, z) = \frac{2}{m-1} r\, w'(z), \quad V(r, z) = -2\, w(z),$$

$$\bar{p}(r) = p_0(r, 0) = \text{const} - 2\left(\frac{k r}{m-1}\right)^2.$$

[As a matter of fact, the first two quantities (3) are independent of ε, the last differs from $\bar{p}(r)$ by the term $2\varepsilon^2\{w'(z) + w^2(z)\}$ of order ε^2.] According to L. PRANDTL, similar circumstances with regard to convergence for $\varepsilon \to 0$ are to be expected along the surface of any obstacle immersed in a fluid of viscosity ε^2.

We propose to formulate the *two-dimensional* boundary layer problem in terms of *conformal* coordinates ξ_1, ξ_2 which arise from the Cartesian coordinates x_1, x_2 by a conformal transformation. Let u_1, u_2 be the covariant components of velocity with respect to these coordinates ξ_1, ξ_2 and

$$ds^2 = dx_1^2 + dx_2^2 = e(d\xi_1^2 + d\xi_2^2)$$

the square of the line element. The Navier-Stokes equations assume the form

$$\frac{\partial u_1}{\partial \xi_1} + \frac{\partial u_2}{\partial \xi_2} = 0, \tag{4}$$

$$\begin{aligned} \sum_k \frac{\partial u_i}{\partial \xi_k} u_k - \frac{1}{2}\frac{1}{e}\frac{\partial e}{\partial \xi_i} \sum_k u_k^2 + \frac{\partial p}{\partial \xi_i} & \\ = \varepsilon^2 \left\{ \Delta u_i + \frac{1}{e} \sum_k \left(\frac{\partial u_k}{\partial \xi_i} - \frac{\partial u_i}{\partial \xi_k} \right) \frac{\partial e}{\partial \xi_k} \right\} & \end{aligned} \qquad [i, k = 1, 2] \tag{5}$$

where

$$\Delta u_i = \frac{\partial^2 u_i}{\partial \xi_1^2} + \frac{\partial^2 u_i}{\partial \xi_2^2}.$$

---

[1]) HIEMENZ, DINGLER's Polytechn. J. *326*, 321–326 (1911). – F. HOMANN, Z. angew. Math. Mech. *16*, 153 (1936).

We suppose that u_1, u_2, p with $\varepsilon \to 0$ converge to the flow $\mathcal{F}_0$ of an ideal fluid arising from a harmonic potential φ:

$$\overset{0}{u}_i = \frac{\partial \varphi}{\partial \xi_i}, \quad p_0 = \text{const} - \frac{1}{2} \sum_i \overset{0}{u}_i \overset{0}{u}^i = \text{const} - \frac{1}{2e} \sum \overset{0}{u}{}_i^2.$$

Along with φ any multiple $k\varphi$ with a positive constant factor k is equally serviceable, which means that the total strength of the stream may be arbitrarily fixed. Choose ξ_1, ξ_2 so that a multiple $k\zeta$ of $\zeta = \xi_1 + i\xi_2$ is the complex potential of the limiting flow $\mathcal{F}_0$ and let the stream line $\xi_2 = 0$ be the one which forms the boundary. We have good reasons to believe, and this belief is the basis of the boundary-layer theory, that u_1, u_2/ε and p, when expressed in terms of the arguments $\xi = \xi_1$, $\eta = \xi_2/\varepsilon$, tend to limiting functions $U(\xi, \eta)$, $V(\xi, \eta)$, $P(\xi, \eta)$ which satisfy the equations arising from (4), (5) by the same passage to the limit. The second equation (5) then shows that $\partial P/\partial \eta = 0$, and $P(\xi, \eta)$ is therefore independent of η and has the value

$$\bar{p}(\xi) = p_0(\xi, 0) = \text{const} - \frac{k^2}{2\,\bar{e}(\xi)}, \quad \bar{e}(\xi) = e(\xi, 0),$$

throughout the boundary layer. Thereafter the two other equations give

$$\left.\begin{aligned} &\frac{\partial U}{\partial \xi} + \frac{\partial V}{\partial \eta} = 0, \\ &U \frac{\partial U}{\partial \xi} + V \frac{\partial U}{\partial \eta} + h(\xi)\,(k^2 - U^2) = \frac{\partial^2 U}{\partial \eta^2} \end{aligned}\right\} \tag{B}$$

where

$$h(\xi) = \frac{1}{2} \frac{d \log \bar{e}(\xi)}{d\xi}.$$

One has to add the boundary conditions

$$U \to 0, \quad V \to 0 \ \text{ for } \ \eta \to 0 \ \text{ and } \ U \to k \ \text{ for } \ \eta \to \infty. \tag{$\bar{\text{B}}$}$$

A full justification of the basic hypothesis of boundary-layer theory will hardly be possible without changing its differential form as given by these equations into a suitable integral form and without proving the existence of a unique solution of the problem (B, $\bar{\text{B}}$)[1]).

Because of the first equation (B), the flow (U, V) derives from a stream function ψ,

$$U = \frac{\partial \psi}{\partial \eta}, \quad V = -\frac{\partial \psi}{\partial \xi}$$

satisfying the formidable differential equation

$$h(\xi)\left\{k^2 - \left(\frac{\partial \psi}{\partial \eta}\right)^2\right\} + \frac{\partial^2 \psi}{\partial \xi\, \partial \eta} \cdot \frac{\partial \psi}{\partial \eta} - \frac{\partial^2 \psi}{\partial \eta^2} \cdot \frac{\partial \psi}{\partial \xi} = \frac{\partial^3 \psi}{\partial \eta^3} \tag{B_ψ}$$

[1]) Experience shows that in general the assumptions of the theory are fulfilled only along a certain frontal part of the surface of the solid. For the whole theory see S. Goldstein, *Modern Developments in Fluid Dynamics*, vol. I (Oxford, 1938).

and the boundary conditions

$$\psi \to 0, \quad \frac{\partial \psi}{\partial \eta} \to 0 \text{ for } \eta \to 0, \quad \frac{\partial \psi}{\partial \eta} \to k \text{ for } \eta \to \infty. \tag{$\bar{B}_\psi$}$$

Suppose now the obstacle is an angle of $\pi\lambda$ $(0 \leqq \lambda < 2)$ with the origin as vertex and the positive real axis as median. The exterior of the angle is mapped conformally upon the slit $(\xi_1 + i\xi_2)$-plane, the slit extending along the positive real axis, by the analytic function

$$x_1 + i x_2 = \text{const } (\xi_1 + i\xi_2)^{1-\lambda/2},$$

and thus one readily finds

$$h(\xi) = -\frac{\lambda}{2} \cdot \frac{1}{\xi}. \tag{6}$$

The domain for the differential equation (B_ψ) is the quadrant $\xi > 0$, $\eta > 0$. If, more generally, the solid parts the stream symmetrically with a prow of angle $\pi\lambda$ at the origin, then the formula (6) will hold at least approximately in the neighborhood of the forward stagnation point. The problem $(B_\psi, \bar{B}_\psi)$ with this value of $h(\xi)$ is carried into itself by the transformation

$$\xi \to \gamma^2 \cdot \xi, \quad \eta \to \gamma \cdot \eta, \quad \psi \to \gamma \cdot \psi \tag{7}$$

(γ a positive constant). Hence the solution must be of the form

$$\psi(\xi, \eta) = 2\sqrt{\xi} \cdot w\,(\eta/2\sqrt{\xi}), \tag{8}$$

and for the function $w(z)$ one obtains exactly the conditions (A_λ). The case $\lambda = 0$ where the obstacle consists of the half line $y = 0$, $x \geqq 0$ in the (x, y)-plane and the fluid flows by with constant positive velocity k was the first boundary-layer problem to be numerically integrated (BLASIUS, 1907). Arbitrary values of λ have been treated by FALKNER, SKAN, and HARTREE[1]). Of the two hydrodynamic interpretations for (A_λ) which we have described, the second is applicable to all values of λ (at least within the range $0 \leqq \lambda < 2$), the first to the reciprocal integers $\lambda = 1/(m-1)$ only. Both coincide for $\lambda = 1$. We notice in particular that the two-dimensional boundary-layer problem of the rectangular prow is mathematically equivalent to the three-dimensional flow against a straight wall ($\lambda = 1/2$).

§ 2. Solution of Blasius's Problem

Turning to the solution of our problems, we start with the case $\lambda = 0$, which occupies a singular position inasmuch as the parameter k is absent from its differential equation

$$w''' + 2\,w\,w'' = 0. \tag{9}$$

[1]) H. BLASIUS, Z. Math. Phys. *56*, 1 (1908). – V. M. FALKNER and S. W. SKAN, Phil. Mag. *12*, 865 (1931); (British) Aero. Res. Comm. R. & M. 1314. – D. R. HARTREE, Proc. Camb. Phil. Soc. *33*, 223–239 (1937).

For any constant $\varkappa$ the expression $\varkappa \cdot w(\varkappa z)$ is a solution of this equation if $w(z)$ is. Following an argument first advanced by TÖPFER[1]) let $w = f(z)$ be the solution determined by the initial values

$$f(0) = f'(0) = 0, \quad f''(0) = 1.$$

Once we are sure that f extends over the whole interval $0 \leqq z < \infty$ and f' tends to a positive limit β with $z \to \infty$,

$$\beta = \int_0^\infty f''(z)\, dz > 0,$$

we may adjust the constant $\varkappa$ so as to let the derivative of $w = \varkappa \cdot f(\varkappa z)$ approach k at infinity:

$$\varkappa^2 \beta = k, \quad \varkappa = \left(\frac{k}{\beta}\right)^{1/2}. \tag{10}$$

Therefore

$$w''(0) = \varkappa^3 = \alpha k^{3/2}, \quad \alpha = \beta^{-3/2} \tag{11}$$

The value $w''(0)$ is the essential factor in the formula for the skin friction along the immersed plate. Hence skin friction is proportional to the 3/2 power of velocity.

Treat f and f'' in the equation

$$\frac{df''}{dz} + 2 f f'' = 0$$

as two separate functions. Because of the initial condition $f''(0) = 1$ one then obtains

$$f''(z) = \exp\left(-2\int_0^z f(\zeta)\, d\zeta\right).$$

Introduce $f'' = g$ as the unknown function and using the initial values $f(0) = f'(0) = 0$, tie up f, or rather its integral, with g by two successive partial integrations:

$$2\int_0^z f(\zeta)\, d\zeta = \int_0^z (z-\zeta)^2 \cdot f''(\zeta)\, d\zeta.$$

The differential equation plus the initial conditions are thus equivalent to the integral equation

$$g = \Phi\{g\} \tag{12}$$

with the operator

$$\Phi\{g\} = \exp\left(-\int_0^z (z-\zeta)^2 g(\zeta)\, d\zeta\right).$$

Notice the following properties of this operator:

$$\text{(i)} \quad \Phi\{g\} \geqq 0, \quad \text{(ii)} \quad \Phi\{g\} \geqq \Phi\{g^*\} \quad \text{if } g \leqq g^*.$$

[1]) K. TÖPFER, Z. Math. Phys. *60*, 397–398 (1912).

We are led to define a sequence of successive 'approximations' g_n, starting with $g_0(z) = 0$, by

$$g_{n+1} = \Phi\{g_n\} \quad (n = 0, 1, 2, \ldots).$$

The trivial relations $g_1 \geqq g_0 = 0$, $g_2 \geqq g_0 = 0$, implied in (i), give rise by (ii) to two rows of inequalities, namely

$$g_0 \leqq g_1, \quad g_1 \geqq g_2, \quad g_2 \leqq g_3, \quad g_3 \geqq g_4, \quad \ldots \tag{13}$$

and

$$g_0 \leqq g_2, \quad g_1 \geqq g_3, \quad g_2 \leqq g_4, \quad g_3 \geqq g_5, \quad \ldots. \tag{14}$$

The latter may be rearranged as follows:

$$g_0 \leqq g_2 \leqq g_4 \leqq \cdots \quad \text{and} \quad g_1 \geqq g_3 \geqq g_5 \geqq \quad \cdots.$$

In view of (13) these relations prove that the descending sequence of the odd g_n lies above the ascending sequence of the even g_n. Does this 'alternating pincer movement' close in on a uniquely determined limit function $g(z)$?

To answer this question, introduce the abbreviation

$$G(z) = \int_0^z (z - \zeta)^2 \, g(\zeta) \, d\zeta \tag{15}$$

and let

$$0 \leqq g(z) \leqq g^*(z), \quad \Delta g = g^* - g, \quad \Delta\Phi\{g\} = \Phi\{g^*\} - \Phi\{g\}.$$

Since

$$0 \leqq e^{-u} - e^{-v} \leqq v - u \quad \text{if } 0 \leqq u \leqq v$$

we get

$$0 \leqq -\Delta\Phi\{g\} \leqq \Delta G.$$

The increment ΔG arises from $2 \cdot \Delta g$ by thrice integrating from 0 to z. These remarks suffice to establish the inequality

$$|g_{n+1}(z) - g_n(z)| \leqq \frac{(2 z^3)^n}{(3n)!}. \tag{16}$$

Indeed, because $g_0 = 0$, $g_1 = 1$, it holds for $n = 0$, and since threefold integration changes

$$\frac{z^{3n}}{(3n)!} \quad \text{into} \quad \frac{z^{3n+3}}{(3n+3)!}$$

the inequality carries over from n to $n+1$, i.e. from $g_{n+1} - g_n$ to $\Phi\{g_{n+1}\} - \Phi\{g_n\}$. Thus convergence (of the type of the exponential series) is assured by the relation (16), and we obtain a solution

$$g(z) = \lim_{n \to \infty} g_n(z)$$

of (12) which is larger than the even and smaller than the odd g_n.

Uniqueness is established by the remark that *any* solution $g(z)$ of (12) satisfies the inequalities

$$g \geqq g_0, \quad g \leqq g_1, \quad g \geqq g_2, \quad g \leqq g_3, \quad \ldots$$

derived from the trivial one $g \geqq g_0 = 0$ by iterated application of the operator Φ. Thus g is necessarily caught between the tongs of the even and the odd g_n.

Our next concern is the asymptotic behavior of $g(z)$ for $z \to \infty$. Choose any $z_0 > 0$ and set

$$\int_0^{z_0} g_2(\zeta)\, d\zeta = c\ (> 0).$$

As G_2 arises by two-fold integration from $2\int_0^z g_2(\zeta)\, d\zeta$ we get $G_2(z) \geqq c(z - z_0)^2$ and thus

$$g(z) \leqq g_3(z) \leqq e^{-c(z-z_0)^2} \quad \text{for } z \geqq z_0.$$

Consequently

$$\int_0^\infty g(z)\, dz = \beta > 0, \quad \int_0^\infty z\, g(z)\, dz = \beta' > 0$$

converge, the asymptotic behavior of

$$f(z) = \int_0^z (z - \zeta)\, g(\zeta)\, d\zeta$$

is indicated by

$$f(z) \sim \int_0^\infty (z - \zeta)\, g(\zeta)\, d\zeta = \beta z - \beta' \tag{17}$$

and that of $w(z)$ by

$$w(z) \sim k z - \frac{\beta'}{\beta^{1/2}}\, k^{1/2}. \tag{18}$$

As $g(z) \geqq g_2(z) = e^{-z^3/3}$ implies

$$\beta \geqq B = \int_0^\infty e^{-z^3/3}\, dz = 3^{1/3} \cdot \Gamma\left(\frac{4}{3}\right)$$

we find the numerical coefficient α in (11) to be $< 0{\cdot}684$. According to the most reliable computations[1]) $\alpha = 0{\cdot}664$. Hence the very first approximate value for α which can be derived from our method misses the mark by not more than 3 per cent.

Given any positive constant $c < B$ we have seen that, for sufficiently large z,

$$\left(G_n(z) \geqq\right) G_2(z) \geqq c z^2.$$

By making use of this fact, one can sharpen the upper bound in (16) to

$$e^{-c z^2} \cdot \frac{(2 z^3)^n}{(3 n)!}.$$

---

[1]) See TÖPFER, loc. cit. p. 492, footnote [1]), and S. GOLDSTEIN, Proc. Camb. Phil. Soc. *26*, 19–20 (1930).

The maximum value of this function is assumed for $z = (3n/2c)^{1/2}$, and we thus ascertain a constant upper bound μ_n for $|g_{n+1}(z) - g_n(z)|$ in the entire interval $z \geqq 0$ which is essentially of the order

$$\frac{1}{\sqrt{(6\pi n)}} (\mu n)^{-3n/2}, \quad \mu \sim 1{\cdot}8.$$

Such sharper estimates are valuable guides for numerical computation.

Knowing a priori that the positive functions f'', f', f have the upper bounds $1, z, (1/2)\,z^2$ respectively, one could have established the existence (and uniqueness) of the solution f over the entire interval $0 \leqq z < \infty$ within the frame of the classic theory of differential equations. But as those bounds (and other related estimates) are most easily derived from the integral equation (12), I have preferred to carry the construction through on its basis. For the general case (A_λ), $\lambda \neq 0$, I see no other alternative.

Von Neumann pointed out to me that the differential equation (9) of order 3 must be reducible to one of first order (followed by two quadratures) because it permits the group of transformations

$$z \to z + z_0, \quad w(z) \to \varkappa \cdot w(\varkappa z)$$

involving two arbitrary constants z_0 and $\varkappa$, and that thus the problem comes within reach of Poincaré's discussion of first-order differential equations. Setting

$$w = e^{-s}, \quad \frac{dw}{dz} = e^{-2s} \cdot \vartheta(s), \quad -\frac{d\vartheta}{ds} + 2\vartheta = t$$

von Neumann obtains the equation

$$\frac{dt}{d\vartheta} = \frac{t(t + \vartheta + 2)}{\vartheta(2\vartheta - t)}$$

with the initial condition $t \to \infty$ for $\vartheta \to \infty$. After determining $t(\vartheta)$ from this equation, one finds by quadratures s and then z as functions of ϑ from

$$ds = \frac{d\vartheta}{2\vartheta - t}, \quad dz = -\frac{e^s\,ds}{\vartheta} = -e^s \cdot \frac{d\vartheta}{\vartheta(2\vartheta - t)}.$$

§ 3. Generalization. Power Series. Goldstein's Wake Problem

In a trivial manner our method carries over to the equation

$$f^{(\nu+1)} + 2 f f^{(\nu)} = 0 \quad (z \geqq 0) \tag{19}$$

with the initial conditions

$$f = f' = \cdots = f^{(\nu-1)} = 0, \quad f^{(\nu)} = 1 \quad \text{for } z = 0. \tag{20}$$

Here ν may be any positive integer. Setting $f^{(\nu)} = g$ we get the integral equation

$$g(z) = \exp\left(-\frac{2}{\nu!}\int_0^z (z-\zeta)^\nu g(\zeta)\,d\zeta\right),$$

and after defining the alternating sequence $g_n(z)$ accordingly, we find instead of (16)

$$|g_{n+1}(z) - g_n(z)| \leqq \frac{2^n z^{n^*}}{n^*!} \quad [n^* = (\nu+1)\,n].$$

We may even generalize the initial conditions (20) to

$$f(0) = c_0, \ldots, \quad f^{(\nu-1)}(0) = c_{\nu-1}, \quad f^{(\nu)}(0) = 1 \tag{21}$$

with arbitrary constants c_μ. Then our integral equation reads

$$g(z) = \exp\left(-2\,Q(z) - \frac{2}{\nu!}\int_0^z (z-\zeta)^\nu g(\zeta)\,d\zeta\right),$$

where $Q(z)$ is the polynomial

$$Q(z) = \frac{c_0}{1!}\,z + \frac{c_1}{2!}\,z^2 + \cdots + \frac{c_{\nu-1}}{\nu!}\,z^\nu, \tag{22}$$

and convergence follows from the inequality

$$|g_{n+1}(z) - g_n(z)| \leqq A^{n+1} \cdot \frac{2^n z^{n^*}}{n^*!} \left[\leqq A\,\frac{(2\,A\,a^{\nu+1})^n}{n^*!}\right]$$

holding in any interval $0 \leqq z \leqq a$ in which $e^{-2Q(z)} \leqq A$. The solution $g(z)$ satisfies the inequality

$$0 \leqq g(z) \leqq g_1(z) = e^{-2Q(z)}. \tag{23}$$

Let us for a moment return to the simple initial conditions (20). From the lowest case $\nu = 1$ where the solution is an elementary function, namely $f(z) = \tanh(z)$, we learn that we must not expect the Taylor expansion of the solution $f(z)$ around the origin to converge beyond a certain finite limit, which for $\nu = 1$ is reached at the point $z = \pi/2$. I find no indication in the literature that this had been realized in Blasius's case $\nu = 2$. For any ν the coefficients c_n of the power series

$$f(z) = \sum_{n=0}^{\infty} (-1)^n c_n z^{n^*+\nu} \quad [n^* = (\nu+1)\,n]$$

are determined by the recursive equations $c_0 = 1/\nu!$,

$$n^*(n^*+1)\ldots(n^*+\nu)\,c_n = 2\sum (i^*+1)\ldots(i^*+\nu)\,c_i\,c_k \quad (i+k = n-1).$$

Following the same straightforward procedure as in my first note in the Proceedings, we obtain

$$\frac{1}{\nu!}\left\{\frac{2\cdot\nu!}{(2\nu+1)!}\right\}^n \leqq c_n \leqq \frac{1}{\nu!}\left\{\frac{2}{(\nu+1)\cdot(\nu+1)!}\right\}^n$$

and thus for the radius R of convergence the bounds

$$\frac{1}{2}(\nu+1)\cdot(\nu+1)! \leqq R^{\nu+1} \leqq \frac{1}{2}(\nu+1)\ldots(2\nu+1).$$

The essential difference between the flow of a viscous fluid *before* and *behind* an obstacle is clearly exhibited in our problem (A_0) by the fact that no solution w exists if w' is required to assume a *negative* value k for $z\to\infty$.

However, GOLDSTEIN[1]) has treated the wake behind a flat plate under the plausible hypothesis that the flow up to the abscissa $x = l$ is but little modified if the plate, $y = 0$, $x \geqq 0$, ends at this point. We shift the origin of the coordinates to the end of the plate and at the same time enlarge the standard length at the ratio $l^{1/4}:1$, i.e. in our stream function

$$\psi(x, y) = 2\sqrt{x}\cdot w\left(\frac{y}{2\sqrt{x}}\right)$$

we make the substitution

$$x = l + l^{1/4}\cdot\xi, \quad y = l^{1/4}\cdot\eta.$$

We then obtain for $\xi = 0$:

$$\psi = \varphi(\eta) + \cdots, \quad \varphi(\eta) = \frac{1}{4}\alpha\, k^{3/2}\cdot\eta^2.$$

The remainder indicated by the dots tends to zero with $l\to\infty$ and shall be neglected as is permissible for plates of great length. The stream function $\psi(\xi,\eta)$ of the 'wake layer' behind the plate, $\xi > 0$, satisfies the same differential equation as before

$$\frac{\partial^3\psi}{\partial\eta^3} + \frac{\partial\psi}{\partial\xi}\cdot\frac{\partial^2\psi}{\partial\eta^2} - \frac{\partial\psi}{\partial\eta}\cdot\frac{\partial^2\psi}{\partial\xi\,\partial\eta} = 0$$

while symmetry requires $\psi(\xi, -\eta) = -\psi(\xi,\eta)$. Hence under limitation to the half plane $\eta \geqq 0$ the conditions at the fictitious boundary $\eta = 0$ become $\psi = \partial^2\psi/\partial\eta^2 = 0$. We wish to construct that solution which for fixed η and $\xi\to 0$ (or for fixed ξ and $\eta\to\infty$) ties up with our function $\varphi(\eta)$. The problem, including this boundary condition, permits the substitution

$$\xi\to\gamma^3\cdot\xi, \quad \eta\to\gamma\cdot\eta, \quad \psi\to\gamma^2\cdot\psi$$

with an arbitrary constant γ and must thus be of the form

$$\psi(\xi,\eta) = 3\,\xi^{2/3}\cdot w\left(\frac{\eta}{\xi^{1/3}}\right).$$

[1]) S. GOLDSTEIN, Proc. Camb. Phil. Soc. **26**, 18–30 (1930).

For $w(z)$ one readily obtains the differential equation

$$w''' + 2\,w\,w'' - w'^2 = 0. \tag{24}$$

The boundary conditions are: $w = w'' = 0$ at $z = 0$, and

$$w''(z) \to \frac{1}{6}\,\alpha\,k^{3/2} \quad \text{for } z \to \infty. \tag{25}$$

If $w(z)$ is a solution of (24), so is the function $\varkappa \cdot w(\varkappa z)$ involving an arbitrary constant $\varkappa$. Let $w = f(z)$ be the solution of (24) with the initial values $f = 0$, $f' = 1$, $f'' = 0$ for $z = 0$. Then $f'''(0) = 1$ while differentiation changes (24) into

$$w'''' + 2\,w\,w''' = 0. \tag{26}$$

Thus we find ourselves confronted with the case $\nu = 3$; $c_0 = 0$, $c_1 = 1$, $c_2 = 0$ of the general problem (19) + (21) discussed above, and since (23) now reads

$$0 \leqq f'''(z) = g(z) \leqq e^{-z^2},$$

$f''(z)$ tends with $z \to \infty$ to a positive limit

$$\beta^* = \int_0^\infty g(z)\,dz < \sqrt{\frac{\pi}{2}}.$$

Consequently we may adjust the constant $\varkappa^*$ in $w(z) = \varkappa^* \cdot f(\varkappa^* z)$ so as to give $w''(\infty)$ the desired value (25)[1]).

§ 4. Solution of Homann's Problem

Of a more difficult type is the problem (A_λ) for $\lambda \neq 0$, as it involves the parameter k in the boundary conditions as well as in the differential equation itself. Here we are dealing with a real boundary-value problem, which is not reducible, as (A_0) is, to an initial-value problem. Only convergence of the type of a geometric series if any can be expected for the process of successive approximations. By differentiating the differential equation we eliminate the parameter k:

$$w'''' + 2\,w\,w''' + 2\,(1 - 2\,\lambda)\,w'\,w'' = 0. \tag{27}$$

This equation is again invariant under the transformation $w(z) \to \varkappa \cdot w(\varkappa z)$. For $\lambda = 1/2$, the case with which we shall be concerned in the next two sections, we fall back upon the familiar type (26), although the boundary conditions

[1]) A remark by K. Friedrichs to the effect that the assumptions $\nu = 3$, $2\,Q(z) = -z^2$ lead to a wake with back flow caused me to drop the restriction $Q(z) \geqq 0$ for $z \geqq 0$ which the original MS contained (April 4, 1942).

make our problem considerably more intricate than before. Let f denote that solution of (26) for which

$$f(0) = f'(0) = 0, \quad f''(0) = 1, \quad f'''(0) = -\beta^2.$$

It will satisfy the third-order equation

$$f''' + 2ff'' + (\beta^2 - f'^2) = 0,$$

and we expect that, for a certain positive β, the derivative f'' (and f''') will strongly approach 0 with $z \to \infty$. Thus the equation itself forces f' (which is positive throughout the interval) to approach β, and $w = \varkappa \cdot f(\varkappa z)$ will solve our problem if $\varkappa$ is determined by (10).

As before we obtain first

$$f'''(z) = -\beta^2 \cdot \exp\left(-2\int_0^z f(\zeta)\, d\zeta\right)$$

$$= -\beta^2 \cdot \exp\left(-\int_0^z (z-\zeta)^2 f''(\zeta)\, d\zeta\right) = -\beta^2 \cdot e^{-G(z)}$$

and then for $f'' = g$ the equation

$$g(z) = 1 - \beta^2 \cdot \int_0^z e^{-G(\zeta)}\, d\zeta.$$

The constant β^2 is determined by the condition $g(\infty) = 0$, thus

$$\beta^2 = 1 \Big/ \int_0^\infty e^{-G(\zeta)}\, d\zeta.$$

Adhering to the notation (15) we are led to introduce an operator Ψ which produces from any given $g(z)$ the function

$$\Psi\{g\} = \int_z^\infty e^{-G(\zeta)}\, d\zeta \Big/ \int_0^\infty e^{-G(\zeta)}\, d\zeta \tag{28}$$

(provided the integral $\int^\infty$ converges). Evidently

$$0 < \Psi\{g\} \leqq 1,$$

and as we shall presently prove,

$$\Psi\{g\} \geqq \Psi\{g^*\} \quad \text{if } g \leqq g^*. \tag{29}$$

The operator Ψ is applicable to the function $g(z) = 1$ but not to $g(z) = 0$[1]. In order to solve the functional equation

$$g = \Psi\{g\} \tag{30}$$

we therefore construct a sequence of functions $g_n(z)$ by the recursive equation

$$g_{n+1} = \Psi\{g_n\} \quad (n = 1, 2, \ldots)$$

starting with $g_1(z) = 1$, in the hope that the sequence will converge to a solution g of (30). Alternating pincer movement of the g_n is a consequence of (29) and the trivial inequalities $g_2 \leqq g_1$, $g_3 \leqq g_1$.

To prove (29), set as before $g^* - g = \Delta g$. The third derivative of $\Delta G = G^* - G$ is $2 \cdot \Delta g$, and ΔG and its first two derivatives vanish for $z = 0$. Hence $\Delta g \geqq 0$ implies $(\Delta G)''$ to be an increasing function of z, and as it vanishes for $z = 0$ it must be positive throughout. Repeating this argument two more times, we find that $\Delta G(z)$ is an increasing positive function for $z > 0$. Set

$$\int_0^z e^{-G(\zeta)}\, d\zeta = H_1, \quad \int_z^\infty e^{-G(\zeta)}\, d\zeta = H_2,$$

so that

$$\Psi\{g\} = \frac{H_2}{H_1 + H_2}.$$

We then have

$$H_1^* = \int_0^z e^{-G^*(\zeta)}\, d\zeta = \int_0^z e^{-G(\zeta)} \cdot e^{-\Delta G(\zeta)}\, d\zeta \geqq e^{-\Delta G(z)} \cdot H_1,$$

$$H_2^* = \int_z^\infty e^{-G^*(\zeta)}\, d\zeta = \int_z^\infty e^{-G(\zeta)} \cdot e^{-\Delta G(\zeta)}\, d\zeta \leqq e^{-\Delta G(z)} \cdot H_2,$$

or the ratios $\vartheta_1 = H_1^*/H_1$, $\vartheta_2 = H_2^*/H_2$ satisfy the inequalities $(\vartheta_1 \leqq 1, \vartheta_2 \leqq 1)$, $\vartheta_2 \leqq \vartheta_1$. Consequently

$$\frac{H_2^*}{H_1^* + H_2^*} \leqq \frac{H_2}{H_1 + H_2} \quad \text{or} \quad \Psi\{g^*\} \leqq \Psi\{g\}.$$

Again choose a $z_0 > 0$ and set

$$\int_0^{z_0} g_2(\zeta)\, d\zeta = c > 0$$

so that $G_2(z) \geqq c(z - z_0)^2$ and

$$g_3(z) \leqq \text{const} \int_z^\infty e^{-c(\zeta - z_0)^2} \cdot d\zeta \quad \text{for } z \geqq z_0.$$

---

[1]) Application of the operator Ψ becomes unrestricted if one replaces the definition (28) by

$$\Psi\{g\} = \lim_{a \to \infty} \left(\int_z^a \Big/ \int_0^a \right).$$

Then one may start with $g_0(z) = 0$ and find $g_1(z) = 1$. Cf. § 6.

All following g's are smaller than g_3 and hence the same inequality prevails for $g_n(z)$ ($n \geqq 3$) and, provided the limit $\lim_{n\to\infty} g_n(z) = g(z)$ exists, also for $g(z)$. Thus knowing that $g(z) = f''(z)$ converges strongly enough to 0 with $z \to \infty$ we again get the asymptotic formula (17) and (11), (18) for the solution

$$w(z) = \varkappa \cdot f(\varkappa z), \quad \varkappa = \left(\frac{k}{\beta}\right)^{1/2}$$

of our problem ($A_{1/2}$).

In proving *convergence* of the alternating sequence $g_n(z)$ we use the above notations $g(z) \leqq g^*(z)$, Δg, G, etc., and write $H = H_1 + H_2$. Then

$$-\Delta\Psi = \Psi\{g\} - \Psi\{g^*\} = \frac{H_2}{H} - \frac{H_2^*}{H^*}$$
$$= \frac{H_2 - H_2^*}{H} + H_2^* \left(\frac{1}{H} - \frac{1}{H^*}\right) \leqq \frac{H_2 - H_2^*}{H} \leqq \frac{H - H^*}{H}.$$

Thus

$$-\Delta\Psi \leqq \int_0^\infty e^{-G(\zeta)} (1 - e^{-\Delta G(\zeta)})\, d\zeta \Big/ \int_0^\infty e^{-G(\zeta)}\, d\zeta.$$

Suppose we have a constant μ such that

$$0 \leqq \Delta g(z) \leqq \mu.$$

Then

$$1 - e^{-\Delta G(\zeta)} \leqq \Delta G(\zeta) \leqq \mu \cdot \frac{1}{3} \zeta^3$$

and hence

$$0 \leqq -\Delta\Psi \leqq q \cdot \mu$$

where

$$q = \int_0^\infty e^{-G(z)} \cdot \frac{1}{3} z^3\, dz \Big/ \int_0^\infty e^{-G(z)}\, dz.$$

Another expression for the quotient q is

$$\int_0^\infty \Psi\{g\} \cdot z^2\, dz$$

as one verifies by substituting (28) for $\Psi\{g\}$.

In this argument we can choose $g = g_n$ and $g^* = g_{n+1}$ or g_{n-1} for any even $n \geqq 2$ and then we obtain majorizing constants μ_n for all odd and even $n \geqq 1$,

$$|g_{n+1}(z) - g_n(z)| \leqq \mu_n, \tag{31}$$

which are defined by the recursive equations

$$\mu_1 = 1; \quad \mu_n = q_n \cdot \mu_{n-1}, \quad \mu_{n+1} = q_n \cdot \mu_n \quad (n \text{ even}) \tag{32}$$

with

$$q_n = \int_0^\infty e^{-G_n(z)} \cdot \frac{1}{3} z^3\, dz \Big/ \int_0^\infty {}^{-G_n(z)}\, dz$$
$$= \int_0^\infty g_{n+1}(z) \cdot z^2\, dz.$$

The second expression of q_n shows that the constants q_n perform a pincer movement of the same type as the functions $g_n(z)$, and hence all q_n lie between q_1 and q_2. Evaluating by partial integration the integral in the numerator of

$$q_1 = \int_0^\infty e^{-z^3/3} \cdot \frac{1}{3} z^3 \, dz \Big/ \int_0^\infty e^{-z^3/3} \, dz,$$

namely

$$-\frac{1}{3} \int_0^\infty z \cdot de^{-z^3/3},$$

we find $q_1 = 1/3$. By some rough estimates it is proved in § 5 that $q_2 < 0{\cdot}76$; but the value of q_2 is probably not much larger than $q_1 = 0{\cdot}33$. Once we are sure that $q_2 < 1$ we see from (32) and $q_n \leqq q_2$ that the sequence $g_n(z)$ converges at least as strongly as a geometric series of quotient q_2.

Uniqueness is assured since every solution g of the equation $g = \Psi\{g\}$ is necessarily sandwiched in between the odd and even g_n.

§ 5. Proof that $q_2 < 1$

We use the constants

$$A = \int_0^\infty z \cdot e^{-z^3/3} \, dz = 3^{-1/3} \cdot \Gamma\left(\frac{2}{3}\right),$$

$$B = \int_0^\infty e^{-z^3/3} \, dz = 3^{-2/3} \cdot \Gamma\left(\frac{1}{3}\right) = 1{\cdot}288.$$

Their product is

$$A\,B = \frac{1}{3} \Gamma\left(\frac{1}{3}\right) \Gamma\left(\frac{2}{3}\right) = \frac{\pi}{3 \sin(\pi/3)} = \frac{2\pi}{3\sqrt{3}}.$$

q_2 is defined as the quotient

$$\int_0^\infty e^{-G_2(z)} \cdot \frac{1}{3} z^3 \, dz \Big/ \int_0^\infty e^{-G_2(z)} \, dz.$$

Since

$$G_2(z) < G_1(z) = \frac{1}{3} z^3$$

the denominator is greater than B. Let us split the integral of the numerator into the parts

$$\int_0^2 + \int_2^\infty$$

and employ in the first part the initial terms of the power series of $G_2(z)$, in the second part an asymptotic appraisal. We readily find

$$G_2(z) = \frac{1}{3} z^3 - \frac{1}{B} \int_0^z \frac{1}{3} (z - \zeta)^3 \cdot e^{-\zeta^3/3} \, d\zeta \tag{33}$$

and, because $e^{-\zeta^3/3} \leqq 1$,

$$G_2(z) > \frac{1}{3} z^3 - \frac{1}{12 B} z^4$$

and thus, as long as $z \leqq 2$,

$$G_2(z) > c \cdot \frac{1}{12} z^4 \quad \text{with} \quad c = 2 - \frac{1}{B}.$$

Consequently

$$\int_0^2 = \int_0^2 e^{-G_2(z)} \cdot \frac{1}{3} z^3 \, dz < \int_0^2 e^{-cz^4/12} \cdot d\frac{z^4}{12}$$

$$= \frac{1}{c} (1 - e^{-4c/3}) = 0{\cdot}6574.$$

To find an asymptotic estimate write (33) in the form

$$B \cdot G_2(z) = \int_0^\infty \frac{1}{3} \left[z^3 - (z - \zeta)^3 \right] \cdot e^{-\zeta^3/3} \, d\zeta - \frac{1}{3} \int_z^\infty (\zeta - z)^3 \cdot e^{-\zeta^3/3} \, d\zeta.$$

The first term

$$= A z^2 - z + \frac{1}{3} B,$$

the integral of the second term is changed by the substitution $\zeta \to \zeta + z$ into

$$\int_0^\infty \zeta^3 \cdot e^{-(\zeta + z)^3/3} \, d\zeta \leqq e^{-z^3/3} \cdot \int_0^\infty \zeta^3 \cdot e^{-z^2 \zeta} \, d\zeta$$

$$= 6 z^{-8} e^{-z^3/3}.$$

Hence

$$G_2(z) \geqq \frac{A}{B} z^2 - \frac{1}{B} z + \left(\frac{1}{3} - \frac{1}{128 B} e^{-8/3} \right) \quad \text{for } z \geqq 2. \tag{34}$$

Set

$$2 \sqrt{A B} = \frac{1}{b}, \quad \text{i.e.,} \quad (2 b)^4 = \frac{27}{(2 \pi)^2};$$

$$\frac{z}{2 B b} - b = x, \quad \frac{1}{B b} - b = x_0, \quad \frac{1}{3} - b^2 - \frac{1}{128 B} e^{-8/3} = B'$$

so that the right side of (34) equals $x^2 + B'$. Then

$$\int_2^\infty \leqq \int_2^\infty e^{-(x^2 + B')} \cdot \frac{1}{3} z^3 \, dz$$

$$= \frac{1}{3} (2 B b)^4 e^{-B'} \cdot \int_{x_0}^\infty e^{-x^2} (x + b)^3 \, dx.$$

Developing

$$(x+b)^3 = x^3 + 3x^2b + 3xb^2 + b^3$$

one readily finds

$$\int_{x_0}^{\infty} e^{-x^2}(x+b)^3\,dx = \frac{1}{2}e^{-x_0^2}\left(x_0^2 + 1 + 3b\,x_0 + 3b^2\right) + b\left(\frac{3}{2} + b^2\right)\int_{x_0}^{\infty} e^{-x^2}\,dx.$$

But

$$\int_{x_0}^{\infty} e^{-x^2}\,dx = \frac{1}{2}\int_{x_0^2}^{\infty} e^{-t}\cdot\frac{dt}{\sqrt{t}} \leqq \frac{1}{2x_0}\int_{x_0^2}^{\infty} e^{-t}\,dt = \frac{1}{2x_0}\,e^{-x_0^2}.$$

Hence

$$\int_2^{\infty} \leqq \frac{1}{6}(2Bb)^4\,e^{-(B'+x_0^2)}\cdot\left\{x_0^2 + 1 + 3b\,x_0 + 3b^2 + \frac{b}{x_0}\left(\frac{3}{2} + b^2\right)\right\}$$

$$= 0{\cdot}3171.$$

The numerator $\int_0^{\infty}$ of q_2 turns out to be

$$< 0{\cdot}6574 + 0{\cdot}3171 = 0{\cdot}9745$$

and q_2 itself

$$< \frac{0{\cdot}9745}{B} < 0{\cdot}76.$$

§ 6. Set-up for Arbitrary λ

Enriched by the experience gathered in the cases $\lambda = 0$ and $1/2$ we now make bold to attack (A_λ) for arbitrary $\lambda \geqq 0$. We seek the solution in the form $w(z) = \varkappa \cdot f(\varkappa z)$ where

$$\left.\begin{aligned} &f'''' + 2ff''' + 2(1-2\lambda)f'f'' = 0 \quad (\text{for } z \geqq 0);\\ &f(0) = f'(0) = 0,\quad f''(0) = 1;\quad f''(\infty) = 0, \end{aligned}\right\} \tag{35}$$

and introduce $f'' = g = \varphi$ as the unknown function. Hence we start with this set-up:

$$f(z) = \int_0^z (z-\zeta)\,g(\zeta)\,d\zeta. \tag{36}$$

$$\varphi'' + 2f\varphi' + 2(1-2\lambda)f'\varphi = 0, \tag{37}$$

$$\varphi(0) = 1,\quad \varphi(\infty) = 0. \tag{37*}$$

$$\varphi = g. \tag{38}$$

More explicitly: for an arbitrarily given function g we form (36) and then solve the linear boundary value problem (37) + (37*), thus defining the functional operator Φ_λ carrying g into φ; at the last step (38) we ask for a fixed element g of that operator,

$$g = \Phi_\lambda\{g\}. \tag{39}$$

In proving the unique existence of φ we shall fix the precise meaning of the boundary condition $\varphi(\infty) = 0$. (The whole discussion would turn out a bit simpler if we dealt with a finite interval $0 \leqq z \leqq a$ instead.)

Auxiliary Theorem. *If $g(z)$ is any continuous non-negative function and*

$$f'(z) = \int_0^z g(\zeta)\, d\zeta, \quad f(z) = \int_0^z f'(\zeta)\, d\zeta = \int_0^z (z-\zeta)\, g(\zeta)\, d\zeta$$

then (37) *has a unique solution with the properties*

$$\varphi(0) = 1; \quad \varphi(z) \geqq 0, \quad \varphi' + 2f\varphi \leqq 0 \quad (\text{for } z \geqq 0). \tag{40}$$

Proof. Set

$$p(z) = \exp\left(2\int_0^z f(\zeta)\, d\zeta\right)$$

so that $p' = 2fp$ and $p(z) \geqq 1$, and introduce the auxiliary function

$$\varphi_1 = (p\,\varphi)' = p(\varphi' + 2f\varphi).$$

Then

$$\left(\frac{\varphi_1}{p}\right)' = \varphi'' + 2f\varphi' + 2f'\varphi = 4\lambda f'\varphi,$$

and the single equation (37) for φ is replaced by the system

$$\left.\begin{aligned} \varphi' + 2f\varphi &= \frac{1}{p}\cdot\varphi_1, \\ \varphi_1' - 2f\varphi_1 &= 4\lambda p f'\cdot\varphi. \end{aligned}\right\} \tag{41}$$

It defines an infinitesimal linear transformation in two variables φ, φ_1 with the matrix (of vanishing trace)

$$\Theta(z) = \begin{Vmatrix} -2f, & \dfrac{1}{p} \\ 4\lambda p f', & 2f \end{Vmatrix}.$$

Hence the two solutions $(\varphi, \varphi_1) = (\eta, \eta_1)$ and (ϑ, ϑ_1) satisfying the initial conditions

$$\eta = 1, \quad \eta_1 = 0; \quad \vartheta = 0, \quad \vartheta_1 = 1 \quad (\text{for } z = 0)$$

are given by the formula

$$\begin{Vmatrix} \eta, & \vartheta \\ \eta_1, & \vartheta_1 \end{Vmatrix} = \sum_{n=0}^{\infty} \underset{\substack{(0 \leqq z_1 \leqq \cdots \\ \leqq z_n \leqq z)}}{\int \cdots \int} \Theta(z_1) \ldots \Theta(z_n)\, dz_1 \ldots dz_n \tag{42}$$

(where the term $n = 0$ of the series at the right is understood to be the unit matrix). Multiplication of (41) by φ_1, φ respectively, followed by addition and integration, establishes the fundamental relation

$$[\varphi\,\varphi_1]_a^b = \int_a^b \left(\frac{1}{p}\,\varphi_1^2 + 4\,\lambda\,p\,f'\,\varphi^2\right) dz. \tag{43}$$

The fact that $f' \geqq 0$ guarantees the positive definite character of the 'Dirichlet integral' at the right side.

Apply (43) to ϑ:

$$\vartheta\,\vartheta_1 = \int_0^z \left(\frac{1}{p}\,\vartheta_1^2 + 4\,\lambda\,p\,f'\,\vartheta^2\right) dz > 0 \quad \text{for } z > 0.$$

This shows (1) that ϑ_1 never vanishes, therefore never changes sign and, because of $\vartheta_1(0) = 1$, stays positive throughout; and (2) that ϑ has the same sign as ϑ_1. In the same manner we find that $\eta\,\eta_1$, η, η_1 (in this order) are all positive,

$$\vartheta > 0, \quad \vartheta_1 > 0; \quad \eta > 0, \quad \eta_1 > 0 \quad \text{for } z > 0.$$

Next consider a finite interval $0 \leqq z \leqq a$ $(a > 0)$ and determine that solution φ for which $\varphi(a) = 0$, $\varphi_1(a) = -1$. Again we see from

$$\varphi\,\varphi_1 = -\int_z^a \left(\frac{1}{p}\,\varphi_1^2 + 4\,\lambda\,p\,f'\,\varphi^2\right) dz < 0$$

that φ_1 is negative and φ positive throughout the interval $0 \leqq z < a$. In particular, $\varphi(0) > 0$, so that we can divide by $\varphi(0)$ thus constructing a solution $\varphi^{(a)}$ with the boundary values

$$\varphi^{(a)}(0) = 1, \quad \varphi^{(a)}(a) = 0.$$

It satisfies the inequalities

$$\varphi^{(a)} > 0, \quad \varphi_1^{(a)} < 0 \quad \text{for } 0 \leqq z < a.$$

Clearly $\varphi^{(a)}(z)$ is of the form $\eta(z) - l_a \cdot \vartheta(z)$ with a constant l_a for which we find the *positive* value $\eta(a)/\vartheta(a)$. Let $a < b$ and write

$$\varphi^{(b)}(z) = \varphi^{(a)}(z) + (l_a - l_b)\,\vartheta(z).$$

Then

$$l_a - l_b = \frac{\varphi^{(b)}(a)}{\vartheta(a)} > 0,$$

or the positive coefficient l_a decreases with increasing a and thus tends to a limit $l \geqq 0$ for $a \to \infty$. The solution

$$\omega(z) = \eta(z) - l \cdot \vartheta(z)$$

is the one we wish to construct[1]). It has the properties

$$\omega(0) = 1; \quad \omega(z) > 0, \quad \omega_1(z) \leqq 0, \tag{44}$$

and is characterized by the fact that the condition

$$\varphi(z) = \omega(z) - m \cdot \vartheta(z) \geqq 0$$

cannot be satisfied throughout the interval $0 \leqq z < \infty$ for any positive constant m.

It remains to show that no solution φ except this ω satisfies (40). Indeed, according to what has just been stated, any such solution would have to be of the form

$$\varphi(z) = \omega(z) + m \cdot \vartheta(z), \quad m > 0.$$

The required inequality $\varphi_1 \leqq 0$ or $(p\varphi)' \leqq 0$ implies $p\varphi \leqq 1$ for $z \geqq 0$. This remarkable relation prevails in particular for $\varphi = \omega$. On the other hand,

$$\left(\frac{\vartheta_1}{p}\right)' = 4\lambda f' \vartheta \geqq 0,$$

therefore

$$\frac{\vartheta_1}{p} \geqq 1, \quad \vartheta_1 \geqq p;$$

$$(p\vartheta)' = \vartheta_1 \geqq p, \quad p\vartheta \geqq \int_0^z p(\zeta)\, d\zeta \geqq z.$$

The consequent relation $p\varphi \geqq m \cdot z$ is incompatible with $p\varphi \leqq 1$ for positive m.

We have now completely and unambiguously defined the functional operator Φ_λ carrying a given $g(z) \geqq 0$ into the function $\omega(z)$, $\omega = \Phi_\lambda\{g\}$. Since

$$p\,\omega \leqq 1, \quad \text{a fortiori} \quad \omega \leqq 1, \tag{45}$$

our operator Φ_λ obeys the law

$$0 < \Phi_\lambda\{g\} \leqq 1.$$

Hence we can and will restrict ourselves to the set $\mathcal{G}$ of all continuous functions $g(z)$ for which $0 \leqq g \leqq 1$. Were Φ_λ monotone in the sense that $\Phi_\lambda\{g\}$ decreases while g increases, then there would be some hope for successful construction of a fixed point of the operator Φ_λ in the functional space $\mathcal{G}$ by some such alternating process of successive approximation as carried us through in the

[1]) A similar construction in H. Weyl, Nachr. Ges. Wiss. Göttingen *1909*, 39.

special instances $\lambda = 0$ and 1/2. Unfortunately this does not seem to be so, and this calamity forces me to proceed by the general theory concerning fixed points of functional operators which we owe to BIRKHOFF-KELLOGG and SCHAUDER-LERAY[1]). The main point will be to establish 'equi-continuity' for the images $\omega = \Phi_\lambda\{g\}$ of all elements $g \in \mathcal{G}$ and continuity for the operator Φ_λ. The lemmas in the following section are so conceived as to meet this demand.

§ 7. Solving the Problem (A_λ)

In the lemmas 1–4 the function g is supposed to be any element of $\mathcal{G}$ and f, c_0, c_1, c_2, c_3 are numbers not depending on g. The condition $g \leqq 1$ implies $c' \leqq z$.

Lemma 1.

$$0 < -\omega'(0) \leqq c_1.$$

Proof. Denote $-\omega'(0)$ by l and argue as follows:

$$\left(\frac{\omega_1}{p}\right)' = 4\lambda f' \omega \leqq 4\lambda z,$$

$$\frac{\omega_1}{p} \leqq -l + 2\lambda z^2,$$

and then, because $p \geqq 1$ and ω_1 negative,

$$(p\,\omega)' = \omega_1 \leqq \frac{\omega_1}{p} \leqq -l + 2\lambda z^2,$$

$$p\,\omega \leqq 1 - l z + \frac{2}{3}\lambda z^3.$$

Since $\omega(z) > 0$ we get

$$1 - l z + \frac{2}{3}\lambda z^3 > 0 \quad \text{or} \quad l < \frac{1}{z} + \frac{2}{3}\lambda z^2,$$

and taking $z = 1$ we have proved the lemma with $c_1 = 1 + (2/3)\lambda$. However, we exploit our inequality to the full when we choose c_1 as the minimum of the elementary function at the right side of the last inequality, which is given by

$$c_1^3 = \frac{9\lambda}{2}. \tag{46}$$

One could also argue from the equation

$$(p\,\omega')' = 2(2\lambda - 1) f' p\,\omega.$$

---

[1]) G. D. BIRKHOFF and O. D. KELLOGG, Trans. Amer. Math. Soc. *23*, 96–115 (1922). – J. SCHAUDER, Studia Math. *2*, 171–180 (1930). – J. LERAY and J. SCHAUDER, Ann. Sci. Ec. Norm. Sup. *51*, 45–78 (1934).

If $\lambda \leqq 1/2$ then

$$(p\,\omega')' \leqq 0, \quad p\,\omega' \leqq -l,$$

and since $p \leqq e^{z^3/3}$,

$$\omega' \leqq -l \cdot e^{-z^3/3}, \quad 0 \leqq \omega \leqq 1 - l \cdot \int_0^z e^{-\zeta^3/3}\,d\zeta,$$

therefore

$$l \leqq \frac{1}{B} = 0{\cdot}777 \quad \text{with} \quad B = \int_0^\infty e^{-z^3/3}\,dz.$$

If, however, $\lambda \geqq 1/2$, then $f' \leqq z$, $p\,\omega \leqq 1$ yield

$$(p\,\omega')' \leqq 2\,(2\,\lambda - 1)\,z, \quad p\,\omega' \leqq -l + (2\,\lambda - 1)\,z^2$$

and taking the value

$$\int_0^\infty z^2 \cdot e^{-z^3/3}\,dz = \int_0^\infty e^{-z^3/3} \cdot d\left(\frac{1}{3}\,z^3\right) = 1$$

into account, we obtain by the same argument

$$l \leqq \frac{2\,\lambda}{B}.$$

Hence the lemma is satisfied with

$$c_1 = \frac{1}{B} \text{ for } \lambda \leqq \frac{1}{2}, \quad c_1 = \frac{2\,\lambda}{B} \text{ for } \lambda \geqq \frac{1}{2}.$$

For small λ and large λ our first appraisal (46) is better, but in a certain middle range, namely for $0{\cdot}104 \leqq \lambda \leqq 1{\cdot}096$, the second gives a sharper result.

Lemma 2.

$$0 \leqq -p\,\omega' \leqq c_1 + c_2\,z^2, \tag{47}$$

and thus, a fortiori,

$$0 \leqq -\omega' \leqq c_1 + c_2\,z^2. \tag{48}$$

Proof.

$$p\,\omega' = \omega_1 - 2\,p\,f\,\omega \leqq \omega_1 \leqq 0. \tag{49}$$

$$(-p\,\omega')' = 2\,(1 - 2\,\lambda)\,f'\,p\,\omega.$$

Again distinguish the cases $\lambda \geqq 1/2$, $\lambda \leqq 1/2$. In the first case $-p\,\omega'$ decreases and therefore

$$-p\,\omega' \leqq l \leqq c_1.$$

In the second case

$$(-p\,\omega')' \leqq 2\,(1 - 2\,\lambda)\,z,$$

$$-p\,\omega' \leqq l + (1 - 2\,\lambda)\,z^2 \leqq c_1 + (1 - 2\,\lambda)\,z^2.$$

Thus we have established Lemma 2 with

$$c_2 = 0 \quad \text{for } \lambda \geqq \frac{1}{2} \quad c_2 = 1 - 2\,\lambda \quad \text{for } \lambda \leqq \frac{1}{2}.$$

The following two lemmas prepare the way for an asymptotic appraisal of $g(z)$ when z approaches infinity.

Lemma 3.

$$\int_0^1 \omega(z)\, dz \geqq c\ (> 0). \tag{50}$$

Proof. Twice integrating (48) we find the inequalities

$$\omega(z) \geqq 1 - c_1 z - \frac{1}{3} c_2 z^3,$$

$$\int_0^z \omega(\zeta)\, d\zeta \geqq z - \frac{1}{2} c_1 z^2 - \frac{1}{12} c_2 z^4 \geqq z\left(1 - \frac{z}{2 c_3}\right)$$

for $z \leqq 1$ where

$$\frac{1}{c_3} = \max\left(1, c_1 + \frac{1}{6} c_2\right).$$

Thus

$$\int_0^1 \omega(\zeta)\, d\zeta \geqq \int_0^{c_3} \omega(\zeta)\, d\zeta = \frac{1}{2} c_3.$$

This argument is fully exploited by choosing $z = c_0$ as the point where the polynomial $1 - c_1 z - c_2 z^3/3$ changes sign and then computing the area

$$c = \int_0^{c_0} \left(1 - c_1 z - \frac{1}{3} c_2 z^3\right) dz = c_0 \left(1 - \frac{1}{2} c_1 c_0 - \frac{1}{12} c_2 c_0^3\right)$$
$$= \frac{1}{4} c_0 \left(3 - c_0 c_1\right)\ \left(\geqq \frac{1}{2} c_0\right),$$

with the result

$$\int_0^{c_0} \omega(z)\, dz \geqq c.$$

Lemma 4. *If* $\int_0^1 g(z)\, dz \geqq \gamma$ *then*

$$\left.\begin{aligned} &0 < \omega(z) \leqq e^{-\gamma(z-1)^2}, \\ &0 \leqq -(\omega' + 2 f \omega) \leqq (c_1 + c_2 z^2) \cdot e^{-\gamma(z-1)^2} \end{aligned}\right\} \text{ for } z \geqq 1.$$

Proof. The hypothesis implies for $z \geqq 1$:

$$f'(z) \geqq \gamma, \quad f(z) \geqq \gamma(z-1), \quad 2\int_0^z f(\zeta)\, d\zeta \geqq \gamma(z-1)^2,$$

$$p(z) \geqq e^{\gamma(z-1)^2}.$$

Combine this with (45), (47) and (49): $-\omega_1 \leqq -p\,\omega'$.

We topologize the functional space $\mathcal{F}$ consisting of all functions $g = g(z)$ defined and continuous for $z \geqq 0$ by agreeing that a sequence g_n approaches zero, $g_n \to 0$ with $n \to \infty$, if $g_n(z)$ converges to zero *uniformly in each finite interval*; in other words, if to every $\varepsilon > 0$, $z > 0$ one can assign an $N(\varepsilon, z)$ such that $0 \leqq z \leqq z_0$, $n \geqq N(\varepsilon, z_0)$ imply $|g_n(z)| \leqq \varepsilon$. This functional space $\mathcal{F}$ is *complete*, i.e. convergence of a sequence g_n, $g_n - g_m \to 0$ with $n, m \to \infty$, implies convergence to some element g, $g_n - g \to 0$. Our domain $\mathcal{G}$ defined by $0 \leqq g(z) \leqq 1$ is a closed convex part of $\mathcal{F}$. (See Appendix, under 1.)

Let $\delta(\varepsilon, z)$ be any positive function of the variables $\varepsilon > 0$, $z > 0$. The element $g \in \mathcal{G}$ is said to lie in $\mathcal{G}_\delta$ if the inequality $|g(z_1) - g(z_2)| \leqq \varepsilon$ holds whenever

$$0 \leqq z_1, z_2 \leqq z_0 \quad \text{and} \quad |z_1 - z_2| \leqq \delta(\varepsilon, z_0)$$

(equi-continuity of type δ). Clearly $\mathcal{G}_\delta$ is a *compact* subset of $\mathcal{G}$; one has only to consider the values of g for rational arguments, marching them off in Indian file. The same simple argument of interpolation as employed by BIRKHOFF and KELLOGG[1]), in proving their Theorem II (p. 103) yields the following general principle (see Appendix, under 2):

An operator $\Phi\{g\}$ defined and continuous in $\mathcal{G}$ and mapping $\mathcal{G}$ into $\mathcal{G}_\delta$ necessarily has a fixed element $g = \Phi\{g\}$.

According to (48), Lemma 2, our operator Φ_λ maps $\mathcal{G}$ into the subset $\mathcal{G}_\delta$ corresponding to the function

$$\delta(\varepsilon, z) = \frac{\varepsilon}{c_1 + c_2 z^2}.$$

Hence the existence of a solution g of the functional equation $g = \Phi_\lambda\{g\}$ will be proved as soon as we can establish continuity of the operator Φ_λ in $\mathcal{G}$:

Lemma 5. *The image $\omega = \Phi_\lambda\{g\}$ depends continuously on $g \in \mathcal{G}$.*

This fact, which we are now going to prove, is less trivial than it appears, because it implies that ω varies but little when the 'tail' of the function $g \in \mathcal{G}$ (i.e. its values for large values of the argument z) changes arbitrarily. The explicit formula (42) clearly shows that the particular solutions η, η_1 and ϑ, ϑ_1 depend continuously on g; these functions in a finite interval $0 \leqq z \leqq z_0$ do not depend on what $g(z)$ does for $z > z_0$. The salient point is the way in which the constant $l = -\omega'(0)$ in

$$\omega(z) = \eta(z) - l \cdot \vartheta(z)$$

depends on g.

Let $g^{(n)} \in \mathcal{G}$ be a sequence which in the sense of our topology tends to g. Using the notations $\eta^{(n)}$, $\vartheta^{(n)}$, $l^{(n)}$, $\omega^{(n)}$ in an obvious way we have

$$\eta^{(n)} \to \eta, \quad \eta_1^{(n)} \to \eta_1; \qquad \vartheta^{(n)} \to \vartheta, \quad \vartheta_1^{(n)} \to \vartheta_1 \qquad \text{with } n \to \infty.$$

According to Lemma 1, all $l^{(n)}$ lie between 0 and c_1 and thus this sequence has at least one point of condensation l^*. The functions

$$\omega^*(z) = \eta(z) - l^* \cdot \vartheta(z), \quad \omega_1^*(z) = \eta_1(z) - l^* \cdot \vartheta_1(z),$$

---

[1]) G. D. BIRKHOFF and O. D. KELLOGG, loc. cit.

being the limits of a subsequence of $\omega^{(n)}$, $\omega_1^{(n)}$, satisfy the inequalities $\omega^* \geqq 0$, $\omega_1^* \leqq 0$. However, we know that ω is the only solution of this kind, and consequently $l^* = l$. Thus the bounded sequence $l^{(n)}$ has only one condensation point, namely l, and therefore converges to l.

Our proof for the existence of a function $g \in \mathcal{G}$ which is its own image under the operator Φ_λ is now complete. As an image it satisfies the inequalities (49), (50),

$$g' \leqq 0, \quad \int_0^1 g(z)\, dz \geqq c, \tag{51}$$

besides $0 \leqq g \leqq 1$, and as image of a g for which (51) holds, it satisfies the further conditions

$$g(z) \leqq e^{-c(z-1)^2}, \tag{52}$$

$$0 \leqq -(g' + 2 f g) \leqq (c_1 + c_2 z^2) \cdot e^{-c(z-1)^2}$$

for $z \geqq 1$ (Lemma 4). Expressing everything in terms of f we see that $f(0) = f'(0) = 0$, $f''(0) = 1$ and

$$f''' + 2 f f'' - 2 \lambda f'^2 = \text{const.} \tag{53}$$

Moreover f'' is monotone decreasing and f'' as well as $f''' + 2 f f''$ tend to zero with $z \to \infty$ essentially as strongly as e^{-cz^2} Hence the positive integral

$$\int_0^\infty g(z)\, dz = f'(\infty) = \beta$$

converges and for the constant on the right side of (53) we find the value $-2\lambda\beta^2$ so that

$$f''' + 2 f f'' + 2\lambda(\beta^2 - f'^2) = 0.$$

An explicit appraisal of β is obtained from (51) and (52):

$$c < \beta < 1 + \sqrt{\frac{\pi}{2c}}.$$

Putting finally

$$w(z) = \varkappa \cdot f(\varkappa z) \quad \text{with} \quad \varkappa = \left(\frac{k}{\beta}\right)^{1/2}$$

we may formulate our chief result as follows:

Theorem. *For given positive k the problem* (A_λ) *has a solution w whose derivative is monotone increasing from 0 to k as z travels from 0 to infinity; the second derivative decreases monotonely from*

$$w''(0) = \alpha k^{3/2} \quad \left(\alpha = \beta^{-3/2}\right)$$

to zero, approaching zero with $z \to \infty$ at least as strongly as a function of the type $e^{-\gamma z^2}$ ($\gamma > 0$).

So far so good. But I should like to see the aircraft engineer who will apply this method to compute the boundary layer for a given profile of an aerofoil!

We have *not* proved that there is only one solution of the problem (A_λ). One could try to approach the question of uniqueness by studying the continuous variation of the operator Φ_λ with λ[1]). It seems not impossible to attack the general boundary layer equation (B_φ) by the method here developed.

§ 8. Appendix

1. The *bounded* continuous functions g form a normed linear space $\mathscr{F}_0$ in which the norm $\|g\|$ induces our topology if, somewhat artificially, we define the norm by

$$\|g\| = \sum \frac{1}{2^\nu} \{ \max_{\nu-1 \leqq z \leqq \nu} |g(z)| \} \quad (\nu = 1, 2, \ldots).$$

Whereas $\mathscr{F}_0$ is incomplete, the part $\mathscr{G}$ is a complete closed subset of $\mathscr{F}_0$. Hence the 'general principle' on which we base our argument will fit into SCHAUDER'S scheme only if one slightly generalizes his central theorem (Satz 2, on p. 175 of Studia Mathematica 2, 1930) in the following manner. Let $\mathscr{F}_0$ be a normed linear space; a continuous mapping of a complete and closed convex subset $\mathscr{G}$ of $\mathscr{F}_0$ into a compact subset $\mathscr{G}^*$ of $\mathscr{G}$ has a fixed point.

2. In adapting the proof to our conditions, I give it a more constructive twist. For the open interval $0 \leqq z < \infty$ one has to combine ever more refined subdivision with exhaustion. Let therefore n, ν be two positive integers. For any given numbers

$$x_m \quad (m = 0, 1, \ldots, n\nu;\ 0 \leqq x_m \leqq 1)$$

form by *linear interpolation* the function $g(z) = g(z;\ x_0, \ldots, x_{n\nu})$ with the prescribed values

$$g\left(\frac{m}{n}\right) = x_m \quad (m = 0, 1, \ldots, n\nu)$$

in the interval $0 \leqq z \leqq \nu$ and extrapolate it beyond ν by $g(z) = x_{n\nu}$ for $z \geqq \nu$. Denote by x_m^* the values of its image $g^* = \Phi\{g\}$ at the points $z = m/n$ ($m = 0, 1, \ldots, n\nu$). The continuous mapping $x_m \to x_m^*$ of the $(n\nu+1)$-dimensional unit cube $0 \leqq x_0 \leqq 1, \ldots, 0 \leqq x_{n\nu} \leqq 1$ has a fixed point (BROUWER'S Theorem); choose one, let it have the coordinates x_m^0 and set

$$g_{n,\nu}(z) = g(z;\ x_0^0, \ldots, x_{n\nu}^0).$$

[1]) See E. ROTHE, Bull. Amer. Math. Soc. *45*, 606–613 (1939).

The image $g^*_{n,\nu}$ of $g_{n,\nu}$ takes on the same values x^0_m as $g_{n,\nu}$ itself at the points $z = m/n$. Let ν_0 be a positive integer. Since $g^*_{n,\nu} \in \mathcal{G}_\delta$,

$$|g^*_{n,\nu}(z) - x^0_m| \leqq \varepsilon \quad \text{for} \quad \frac{m}{n} \leqq z \leqq \frac{m+1}{n} \quad (m = 0, 1, \ldots, n\nu_0 - 1) \tag{54}$$

provided $\nu \geqq \nu_0$ and $1/n \leqq \delta(\varepsilon, \nu_0)$; in particular $[z = (m+1)/n]$

$$|x^0_{m+1} - x^0_m| \leqq \varepsilon .$$

Hence, because $g_{n,\nu}(z)$ is the linear interpolation of the values x^0_m, the inequality

$$|g_{n,\nu}(z) - x^0_m| \leqq \varepsilon$$

holds under the same conditions as (54) and thus

$$|g^*_{n,\nu}(z) - g_{n,\nu}(z)| \leqq 2\varepsilon \quad \text{for } 0 \leqq z \leqq \nu_0$$

as soon as $\nu \geqq \nu_0$ and $n \geqq 1/\delta(\varepsilon, \nu_0)$. This means that $g^*_{n,\nu} - g_{n,\nu} \to 0$ with n and ν tending to infinity. A subsequence[1]) of the $g^*_{n,\nu} \in \mathcal{G}_\delta$ tends to a limit g, the corresponding subsequence of the $g_{n,\nu}$ to the same limit, and, on account of the continuity of Φ, the relation $g^*_{n,\nu} = \Phi\{g_{n,\nu}\}$ yields $g = \Phi\{g\}$.

---

[1]) By a subsequence of the pairs (n, ν) we mean a sequence (n_i, ν_i) both members of which are monotone: $n_i < n_{i+1}$, $\nu_i < \nu_{i+1}$.

126.

Theory of reduction for arithmetical equivalence. II

Transactions of the American Mathematical Society 51, 203—231 (1942)

1. **Introduction. Lattices over the unit lattice.** Given n linearly independent vectors $\mathfrak{d}_1, \cdots, \mathfrak{d}_n$ in an n-dimensional vector space E^n, the formula

$$\mathfrak{x} = y_1\mathfrak{d}_1 + \cdots + y_n\mathfrak{d}_n \tag{1}$$

yields all vectors of the space E^n or of a lattice $\mathfrak{L}$ in E^n if the coordinates y_i range over all real numbers or all integers, respectively. We take the viewpoint that the lattice $\mathfrak{L}$ is given but the choice of its basis arbitrary. The several bases are connected with one another by unimodular transformations. If $f(\mathfrak{x})$ is a gauge function assigning a "length" $f(\mathfrak{x})$ to each vector $\mathfrak{x}$ the problem of reduction requires normalization of the lattice basis in terms of the given f. A solution is sought for all possible gauge functions or at least for some important class. The most significant class is obtained by running f^2 over all positive quadratic forms.

Following in Dirichlet's and Hermite's footsteps, Minkowski developed such a method of reduction for quadratic forms and established the decisive facts about it. In R1 I approached the same problem in that geometric way which Minkowski had initiated but then abandoned for unknown reasons.

The question may be put in a slightly different form. All linear mappings of E^n carrying $\mathfrak{L}$ into itself carry $f(\mathfrak{x})$ into equivalent gauge functions. The task is to pick out by a universal rule in each class of equivalent gauge functions one particular $f(\mathfrak{x})$ which is called the reduced function of its class. Let $\mathcal{R}_0$, $\mathcal{R}$, $\mathcal{C}$ in the future denote the fields of all rational, real and complex numbers, respectively. Complex numbers are written in the form $\xi = x_0 + x_1 i$ (x_0, x_1 real). It is convenient to insert between the full vector space and the lattice $\mathfrak{L}$, the set E_0^n of all vectors (1) with *rational* coefficients y_i, a set which we describe as an n-dimensional vector space over $\mathcal{R}_0$. Crystallography has found this advisable in distinguishing between the macroscopic and atomistic symmetries of a crystal, and in the theory of algebraic numbers one puts the *field* before the *ring* of its integers.

Let a lattice $\mathfrak{L}$ in E_0^n be given. With respect to any basis $\mathfrak{d}_1, \cdots, \mathfrak{d}_n$ of E_0^n, formula (1), the function $f(\mathfrak{x})$ is represented by a function $g(y_1, \cdots, y_n)$ and the lattice $\mathfrak{L}$ by a "numerical lattice" Λ whose vectors are n-uples $(y_1, \cdots, y_n)$

Presented to the Society, January 1, 1941; received by the editors December 11, 1940.

(1) The first part, which appeared under the same title in these Transactions, vol. 48 (1940), pp. 126–164, is cited as R1.

of rational numbers. (Only if $\mathfrak{d}_1, \cdots, \mathfrak{d}_n$ is a true basis of $\mathfrak{L}$ will Λ be the unit lattice I whose elements are the n-uples of integers.) Hence $f(\mathfrak{x})$ with respect to $\mathfrak{L}$ is represented by g/Λ. All representations g/Λ of $f(\mathfrak{x})/\mathfrak{L}$ are equivalent, i.e., they arise from one another by linear transformations of the coordinates with rational coefficients. In each class of equivalent g/Λ we are to pick one individual, the "reduced" g/Λ. Suppose we have succeeded in doing this by some universal rule. We then have to select, for each Λ that may occur in a reduced g/Λ, a definite basis $\mathfrak{d}_1^*, \cdots, \mathfrak{d}_n^*$ in terms of which $\mathfrak{L}$ is represented by Λ. The equation

$$f^*(\mathfrak{x}) = g(y_1, \cdots, y_n) \quad \text{for} \quad \mathfrak{x} = y_1\mathfrak{d}_1^* + \cdots + y_n\mathfrak{d}_n^*$$

then defines the reduced gauge function f^* in its class. By the first step of reducing g/Λ no essential progress has been made unless the lattices Λ which may occur in a reduced g/Λ are limited to a finite number of possibilities. For only then is the selection of a basis $\mathfrak{d}_1^*, \cdots, \mathfrak{d}_n^*$ for each of these Λ essentially simpler than the original problem.

The Dirichlet-Hermite-Minkowski method of reduction by admitting only bases $\mathfrak{d}_1, \cdots, \mathfrak{d}_n$ of $\mathfrak{L}$ always represents $\mathfrak{L}$ by the one lattice $\Lambda = \mathrm{I}$, the unit lattice. Thus it provides the ideal solution. Minkowski's construction of consecutive shortest distances in the lattice

$$f(\mathfrak{d}_1) = M_1, \cdots, f(\mathfrak{d}_n) = M_n$$

(for which he obtains the inequality $M_1 \cdots M_n V \leqq 2^n$) falls under our more general scheme. That theorem which he describes as indicating a certain "Oekonomie der Strahldistancen" states exactly that there is only an *a priori* limited number of possibilities for Λ with which to count in a reduced g/Λ. In R1 I carried the first method over to those other fields and quasi-fields which have not more than one infinite prime spot, and I found that it works only under the hypothesis that the class number for ideals is 1. Simultaneously Siegel observed that the rougher second method, by which incidentally Minkowski had proved that the class number of positive quadratic forms with *integral* coefficients and a given discriminant is finite, operates without this restrictive hypothesis(2). I add the remark that an argument making no use of the bases of a lattice need not even assume their existence. In an algebraic number field $\mathcal{J}$ we consider any "order" $[\mathcal{J}]$; in general there are several classes of lattices belonging to this order. The theory is limited neither to the principal class nor to the principal order. Following a suggestion by Siegel, P. Humbert generalized the investigation of quadratic forms to an arbitrary algebraic number field $\mathcal{J}$ with several infinite prime spots(2). No doubt the whole problem thereby loses much of its simplicity. But once upon this track one ought to include the quaternions and thus deal also with those noncom-

(2) See P. Humbert, Commentarii Mathematici Helvetici, vol. 12 (1939–1940), pp. 263–306.

mutative division algebras of finite degree over $\mathfrak{R}_0$ for which the concept of infinite prime spots goes through. I resume here the rougher method of reduction with these further generalizations by the same geometric approach as in R1. I am not only interested in the fact that certain numbers are finite; I wish to ascertain reasonably low explicit upper bounds for them. The geometric method yields good results in this regard.

Before concluding this introduction I remind the reader of some simple facts about lattices in E_0^n. A vector $\mathfrak{x}$ in E_0^n is defined as an n-uple $(x_1, \cdots, x_n)$ of rational numbers. The unit vectors $\mathfrak{e}_k = (e_{1k}, \cdots, e_{nk})$ are the columns of the unit matrix $\|e_{ik}\|$. The word lattice means any set of vectors such that $\mathfrak{a}-\mathfrak{b}$ is contained in the set every time $\mathfrak{a}$ and $\mathfrak{b}$ are. We assume that the lattice is n-dimensional, i.e., contains n linearly independent vectors; and discrete, i.e., we require that for any given positive integer q there are not more than a finite number of lattice vectors satisfying the inequalities

$$|x_1| \leqq q, \cdots, |x_n| \leqq q.$$

From now on the term lattice refers only to discrete lattices which have the full dimensionality of their vector space. By a familiar argument one shows that one can find n linearly independent vectors $\mathfrak{l}_1, \cdots, \mathfrak{l}_n$ in a given lattice $\mathfrak{L}$ such that every lattice vector

$$\mathfrak{x} = u_1\mathfrak{l}_1 + \cdots + u_n\mathfrak{l}_n$$

has integral components u_i. By the same construction one adapts the basis $\mathfrak{l}_1, \cdots, \mathfrak{l}_n$ of any lattice Λ containing the unit lattice I to the basis $\mathfrak{e}_1, \cdots, \mathfrak{e}_n$ of I:

$$\begin{aligned} \mathfrak{e}_1 &= c_1\mathfrak{l}_1, \\ \mathfrak{e}_2 &= c_{21}\mathfrak{l}_1 + c_2\mathfrak{l}_2, \\ &\cdots\cdots\cdots\cdots, \\ \mathfrak{e}_n &= c_{n1}\mathfrak{l}_1 + \cdots + c_{n,n-1}\mathfrak{l}_{n-1} + c_n\mathfrak{l}_n. \end{aligned}$$

The integers c_k are positive and the integral skew coefficients c_{ki} $(i<k)$ may be normalized by

$$0 \leqq c_{ki} < c_i \qquad (k = i+1, \cdots, n);$$

then $(\mathfrak{l}_1, \cdots, \mathfrak{l}_n)$ is uniquely determined. The index $j = [\Lambda:\mathrm{I}]$, i.e., the number of vectors in Λ which are incongruent modulo I, equals $c_1 \cdots c_n$. Let $\Lambda^{(k)}$ denote the part of Λ lying in the linear subspace $x_{k+1} = \cdots = x_n = 0$. The index $j_k = [\Lambda^{(k)}:\mathrm{I}^{(k)}]$ equals $c_1 \cdots c_k$. Hence these two lemmas:

LEMMA 1. *j_k is a divisor of j_h for $k<h$.*

LEMMA 2. *The number $h_n(j)$ of different lattices Λ over I of given index $j = [\Lambda:\mathrm{I}]$ is finite.*

Indeed, it equals the sum

$$\sum c_1^{n-1} c_2^{n-2} \cdots c_n^{0}$$

extended over all factorizations $c_1c_2 \cdots c_n = j$ of j. (Incidentally, the numbers $h_n(j)$ for $j=1, 2, \cdots$ have as their generating function the Dirichlet series

$$\sum_{j=1}^{\infty} h_n(j) \cdot j^{-s} = \zeta(s)\zeta(s-1) \cdots \zeta(s-n+1)$$

convergent in the half-plane $\mathfrak{R}s > n$.)

2. **Vector space and lattice over an algebraic field.** Let $\mathfrak{F}$ be any field of finite degree f over $\mathfrak{R}_0$. By carefully putting all factors in their proper places we shall see to it that all arguments and formulas in this and the following two sections remain valid for any division algebra, whether commutative or not, of finite degree over $\mathfrak{R}_0$. We choose a basis $\sigma_1, \cdots, \sigma_f$ of $\mathfrak{F}/\mathfrak{R}_0$ so that each number ξ of $\mathfrak{F}$ is uniquely represented by

$$\xi = x_1\sigma_1 + \cdots + x_f\sigma_f \qquad (x_a \text{ rational}). \tag{2}$$

Any n-uple $(\xi_1, \cdots, \xi_n)$ of numbers ξ_i in $\mathfrak{F}$ is a vector of the n-dimensional vector space E^n over $\mathfrak{F}$. The fundamental operations are addition of two vectors, $\mathfrak{x}+\mathfrak{x}'$, and multiplication $\delta\mathfrak{x}$ of a vector $\mathfrak{x}$ by a number δ (the numerical factor always in front of the vector!). Thus we may write

$$\mathfrak{x} = (\xi_1, \cdots, \xi_n) = \xi_1\mathfrak{e}_1 + \cdots + \xi_n\mathfrak{e}_n.$$

k linearly independent vectors $\mathfrak{d}_1, \cdots, \mathfrak{d}_k$ span a linear subspace $[\mathfrak{d}_1, \cdots, \mathfrak{d}_k]$ consisting of all vectors of the form $\eta_1\mathfrak{d}_1 + \cdots + \eta_k\mathfrak{d}_k$. Any n linearly independent vectors $\mathfrak{d}_1, \cdots, \mathfrak{d}_n$ form a basis of $E^n/\mathfrak{F}$ in terms of which each vector is uniquely expressible as

$$\mathfrak{x} = \eta_1\mathfrak{d}_1 + \cdots + \eta_n\mathfrak{d}_n. \tag{3}$$

The original coordinates ξ_i are connected with the η_i by that nonsingular linear transformation D,

$$\xi_i = \sum_k \eta_k\delta_{ik}, \tag{4}$$

whose matrix $\|\delta_{ik}\|$ has for its columns the vectors $\mathfrak{d}_k = (\delta_{1k}, \cdots, \delta_{nk})$.

Expressing each component ξ_i in terms of the basis σ of $\mathfrak{F}$,

$$\xi_i = x_{i1}\sigma_1 + \cdots + x_{if}\sigma_f,$$

we identify $E^n/\mathfrak{F}$ with the (nf)-dimensional vector space E_0^{nf} over $\mathfrak{R}_0$. The rational numbers x_{ia} are the coordinates of $\mathfrak{x}$ with respect to the basis $\sigma_a\mathfrak{e}_i$. One has to distinguish between linear dependence in $\mathfrak{F}$ and in $\mathfrak{R}_0$.

We suppose we are given a lattice $\mathfrak{L}$ in E_0^{nf}. It will have a basis $\mathfrak{l}_\mu$

($\mu = 1, \cdots, nf$) in terms of which each vector $\mathfrak{x}$ of $\mathfrak{L}$,

$$\mathfrak{x} = \sum_{\mu} u_\mu \mathfrak{l}_\mu \tag{5}$$

has rational *integral* components u_μ. A number δ of $\mathcal{J}$ is said to be a *multiplier* of $\mathfrak{L}$ if the operation $\mathfrak{x} \to \delta\mathfrak{x}$ carries each lattice vector $\mathfrak{x}$ into a lattice vector $\delta\mathfrak{x}$. The multipliers of $\mathfrak{L}$ form an *order* $[\mathcal{J}]$. This assertion is meant to imply the following four properties(³):

1°. The number 1 is in $[\mathcal{J}]$.

2°. $[\mathcal{J}]$ is a ring.

3°. Any given number δ in $\mathcal{J}$ may be multiplied by a positive rational integer m such that $m\delta$ is in $[\mathcal{J}]$.

4°. Each number in $[\mathcal{J}]$ is an integer.

1° and 2° are evident. To prove 3° and 4° we write

$$\delta \mathfrak{l}_\mu = \sum_{\nu} d_{\nu\mu} \mathfrak{l}_\nu \qquad (d_{\mu\nu} \text{ rational}).$$

If δ is any number and m a common denominator of the coefficients $d_{\mu\nu}$, then $m\delta$ is a multiplier. If δ happens to be a multiplier, then the $d_{\mu\nu}$ are rational integers and δ satisfies the equation

$$\left| \delta e_{\nu\mu} - d_{\nu\mu} \right| = 0.$$

In the same manner as for the "principal order" consisting of all integers of $\mathcal{J}$ one proves(⁴) that any order $[\mathcal{J}]$ is a discrete f-dimensional lattice in the f-dimensional vector space $\mathcal{J}/\mathcal{R}_0$, and hence has a basis $\sigma_1, \cdots, \sigma_f$ in terms of which every number ξ of $[\mathcal{J}]$ appears in the form (2) with rational *integral* coefficients x_i.

The transformation D, (4), maps $\mathfrak{L}$ upon a lattice Λ: If $\mathfrak{x} = (\xi_1, \cdots, \xi_n)$ is in $\mathfrak{L}$, then $(\eta_1, \cdots, \eta_n)$ is in Λ, and *vice versa*. We call two lattices *equivalent* and admit them to the same *class* if one is carried into the other by a non-singular transformation D. The lattices Λ of one class express a given lattice $\mathfrak{L}$ in terms of different bases $(\mathfrak{d}_1, \cdots, \mathfrak{d}_n)$ of $E^n/\mathcal{J}$. Obviously two equivalent lattices have the same multipliers.

A lattice $\mathfrak{L}$ is said to belong to the order $[\mathcal{J}]$ if every number of that order is a multiplier of $\mathfrak{L}$. (For $n = 1$ this notion coincides with that of an ideal in $[\mathcal{J}]$, and our classes of lattices with the classes of ideals.) Given an order $[\mathcal{J}]$, the n-uples $(\xi_1, \cdots, \xi_n)$ of numbers ξ_i in $[\mathcal{J}]$ form a lattice I which belongs to the order $[\mathcal{J}]$; we call it the *unit lattice* for $[\mathcal{J}]$. The lattices belonging to a given order $[\mathcal{J}]$ are distributed over a number of classes, the class of I being the principal class.

(³) Notion and name are due to Dedekind. Hilbert in his Zahlbericht introduced the word "ring" for this purpose, but since ring has now acquired a wider meaning I revert, in agreement with such authorities as Artin and Chevalley, to Dedekind's terminology.

(⁴) Cf. H. Weyl, *Algebraic Theory of Numbers*, Princeton, 1940, pp. 145–146.

Let $\sigma_1, \cdots, \sigma_f$ be a basis of $[\mathfrak{J}]$ and $\mathfrak{l}_\mu$ ($\mu=1, \cdots, nf$) a basis of Λ. If Λ contains I the vectors $\sigma_a\mathfrak{e}_k$, which span I, are linear combinations of the $\mathfrak{l}_\mu$ with integral rational coefficients, and their absolute determinant, i.e., the absolute determinant of the transformation connecting the coordinates u_μ with the x_{ka} ($k=1, \cdots, n$; $a=1, \cdots, f$) is the index $j=[\Lambda:\mathrm{I}]$.

Those vectors $(\xi_1, \cdots, \xi_n)$ in Λ for which $\xi_{k+1}= \cdots =\xi_n=0$ form a lattice $\Lambda^{(k)}$ in the k-dimensional space $E^k/\mathfrak{J}$ with the coordinates $\xi_1, \cdots, \xi_k$. Considering Λ as a lattice in E_0^{nf} and using the arrangement

$$x_{11}, \cdots, x_{1f};\ x_{21}, \cdots, x_{2f};\ \cdots$$

of the coordinates in E_0^{nf} one can apply Lemma 1 to kf and $(k+1)f$ instead of k and h and thus one derives a corresponding proposition in $\mathfrak{J}$ instead of $\mathfrak{R}_0$:

LEMMA 3. *In the row of indices*

$$j_k = [\Lambda^{(k)}:\mathrm{I}^{(k)}] \qquad (k = 1, \cdots, n) \tag{6}$$

each number is a divisor of its successor.

The set of vectors $(\xi_1, \cdots, \xi_n)$ in Λ outside $[\mathfrak{e}_1, \cdots, \mathfrak{e}_{k-1}]$, i.e., for which $(\xi_k, \cdots, \xi_n)\neq(0, \cdots, 0)$, will be denoted Λ_k. Thus Λ_k and $\Lambda^{(k-1)}$ are complements in Λ.

We have seen that the number of different lattices Λ over a given lattice I with a given index $[\Lambda:\mathrm{I}]=j$ is finite, namely $h_{nf}(j)$. More exactly, one finds by the same argument that the number $H_f(j_1, \cdots, j_n)$ of different lattices Λ over I with given indices (6) has as its generating function the Dirichlet series of n variables $s_1, \cdots, s_n$:

$$\begin{aligned} \mathrm{Z}_f(s_1 + s_2 + \cdots + s_n - nf)\cdot \mathrm{Z}_f(s_2 + \cdots + s_n - (n-1)f) \cdots \mathrm{Z}_f(s_n - f) \\ = \sum_{j_1, \cdots, j_n} H_f(j_1, \cdots, j_n)\cdot j_1^{-s_1} \cdots j_n^{-s_n} \end{aligned}$$

where

$$\mathrm{Z}_f(s) = \zeta(s + 1) \cdots \zeta(s + f).$$

Hence *a fortiori*:

LEMMA 4. *We have found upper bounds for the number of lattices Λ belonging to a given order $[\mathfrak{J}]$ which contain the unit vectors $\mathfrak{e}_1, \cdots, \mathfrak{e}_n$ and hence the unit lattice* I *for $[\mathfrak{J}]$ and which, moreover, have either a given index $j=[\Lambda:\mathrm{I}]$ or a given row of indices $j_1, \cdots, j_n$.*

3. **Preliminaries about reduction.** Suppose an order $[\mathfrak{J}]$ and a basis $\sigma_1, \cdots, \sigma_f$ of $[\mathfrak{J}]$ to be given. We consider a real-valued function

$$f(\mathfrak{x}) = f(\xi_1, \cdots, \xi_n)$$

which depends on a variable vector $\mathfrak{x}$ in $E^n/\mathfrak{J}$ and is positive except for $\mathfrak{x}=0$, and we assume:

(i_0) For each positive q one can ascertain a positive q' such that the inequality $f(\mathfrak{x}) < q$ implies the nf inequalities

$$(7) \qquad |x_{ia}| < q' \qquad (i = 1, \cdots, n; a = 1, \cdots, f)$$

for the components x_{ia} of the ξ_i.

Let $\mathfrak{L}$ be a lattice belonging to the fixed order $[\mathcal{J}]$. n vectors $\mathfrak{d}_1, \cdots, \mathfrak{d}_n$ of $\mathfrak{L}$ which are linearly independent with respect to $\mathcal{J}$ constitute a semi-basis of $\mathfrak{L}$. Because of the discreteness of $\mathfrak{L}$ there is but a finite number of vectors $\mathfrak{x}$ in $\mathfrak{L}$ satisfying the inequalities (7). Hence Minkowski's construction of consecutive minima of f in $\mathfrak{L}$ is applicable. It yields a semi-basis $\mathfrak{d}_1, \cdots, \mathfrak{d}_n$ of $\mathfrak{L}$ such that

$$f(\mathfrak{x}) \geqq f(\mathfrak{d}_k) = M_k$$

for every vector $\mathfrak{x}$ in $\mathfrak{L}$ outside $[\mathfrak{d}_1, \cdots, \mathfrak{d}_{k-1}]$ (*reduced semi-basis*). Obviously

$$(8) \qquad M_1 \leqq M_2 \leqq \cdots \leqq M_n.$$

The mapping

$$(3) \qquad \mathfrak{x} = \eta_1\mathfrak{d}_1 + \cdots + \eta_n\mathfrak{d}_n \rightarrow (\eta_1, \cdots, \eta_n)$$

carries $f(\mathfrak{x})$ into a function $g(\eta_1, \cdots, \eta_n)$ and $\mathfrak{L}$ into a lattice Λ which contains the unit lattice I for $[\mathcal{J}]$. The function $g(\xi_1, \cdots, \xi_n)$ is *reduced with respect to* Λ, i.e.,

$$(9) \qquad g(\xi_1, \cdots, \xi_n) \geqq g(e_{1k}, \cdots, e_{nk})$$

whenever $(\xi_1, \cdots, \xi_n)$ is in Λ_k.

The M_k are uniquely determined by f and $\mathfrak{L}$; the situation is somewhat less favorable for $\mathfrak{d}_1, \cdots, \mathfrak{d}_n$. Suppose $\mathfrak{d}_1', \cdots, \mathfrak{d}_n'$ is another set constructed according to our prescription. If M_k is actually lower than M_{k+1} then $[\mathfrak{d}_1', \cdots, \mathfrak{d}_k'] = [\mathfrak{d}_1, \cdots, \mathfrak{d}_k]$. (Analogues of Theorems 8 and 9 in R1.)

Being given n real numbers p_k,

$$1 \leqq p_1 \leqq \cdots \leqq p_n,$$

we say that the semi-basis $\mathfrak{d}_1', \cdots, \mathfrak{d}_n'$ of $\mathfrak{L}$ has the property $B(p_1, \cdots, p_n)$ if

$$f(\mathfrak{x}) \geqq \frac{1}{p_k} f(\mathfrak{d}_k')$$

for any vector $\mathfrak{x}$ in $\mathfrak{L}$ outside $[\mathfrak{d}_1', \cdots, \mathfrak{d}_{k-1}']$. Accordingly we ascribe the property $B(p_1, \cdots, p_n)$ to a function $g(\xi_1, \cdots, \xi_n)$ in conjunction with a lattice Λ over I if

$$g(\xi_1, \cdots, \xi_n) \geqq \frac{1}{p_k} \cdot g(e_{1k}, \cdots, e_{nk})$$

whenever $(\xi_1, \cdots, \xi_n)$ is in Λ_k.

If $\mathfrak{d}_k'$ is a semi-basis of $\mathfrak{L}$ with the property $B(p_1, \cdots, p_n)$, then

$$M_k' = f(\mathfrak{d}_k') \leqq p_k M_k.$$

(Analogue of Theorem 8_p.) Indeed, let $\mathfrak{d}_1, \cdots, \mathfrak{d}_n$ be a reduced semi-basis of $\mathfrak{L}$, $f(\mathfrak{d}_k) = M_k$. At least one of the k linearly independent vectors $\mathfrak{d}_1, \cdots, \mathfrak{d}_k$, say $\mathfrak{d}_i$, lies outside $[\mathfrak{d}_1', \cdots, \mathfrak{d}_{k-1}']$; hence

$$f(\mathfrak{d}_i) \geqq \frac{1}{p_k} \cdot f(\mathfrak{d}_k')$$

or

$$M_k' \leqq p_k M_i \leqq p_k M_k.$$

With the same notations I maintain that $[\mathfrak{d}_1', \cdots, \mathfrak{d}_k'] = [\mathfrak{d}_1, \cdots, \mathfrak{d}_k]$ provided $M_{k+1} > p_k M_k$. (Analogue of Theorem 9_p.) Proof: Suppose that one of the vectors $\mathfrak{d}_1', \cdots, \mathfrak{d}_k'$, say $\mathfrak{d}_i'$, is not in $[\mathfrak{d}_1, \cdots, \mathfrak{d}_k]$. Then

$$f(\mathfrak{d}_i') \geqq M_{k+1}.$$

Vice versa, if all k numbers $M_1', \cdots, M_k'$ are less than M_{k+1} then $\mathfrak{d}_1', \cdots, \mathfrak{d}_k'$ lie in $[\mathfrak{d}_1, \cdots, \mathfrak{d}_k]$. The observation that $M_i' \leqq p_i M_i \leqq p_k M_k$ finishes the proof.

The notation of *properly reduced* bases depends on a given multiplicative group U of numbers ϵ in $\mathfrak{J}$ and deals with functions f which satisfy the further condition:

(ii$_0$) $f(\epsilon\mathfrak{x}) = f(\mathfrak{x})$ (ϵ in U).

The semi-basis $\mathfrak{d}_1, \cdots, \mathfrak{d}_n$ of $\mathfrak{L}$ is said to be properly reduced provided the inequality

$$f(\mathfrak{x}) > f(\mathfrak{d}_k)$$

holds with the sign $>$ for any vector $\mathfrak{x}$ of $\mathfrak{L}$ outside $[\mathfrak{d}_1, \cdots, \mathfrak{d}_{k-1}]$ except for the vectors of the special form

$$\mathfrak{x} = \epsilon\mathfrak{d}_k \qquad (\epsilon \text{ in } U).$$

Accordingly $g(\xi_1, \cdots, \xi_n)$ is properly reduced with respect to the lattice Λ over I if

$$g(\xi_1, \cdots, \xi_n) > g(e_{1k}, \cdots, e_{nk})$$

for all vectors $(\xi_1, \cdots, \xi_n)$ in Λ_k except the special vectors

$$\epsilon(e_{1k}, \cdots, e_{nk}) \qquad (\epsilon \text{ in } U).$$

With $\mathfrak{d}_k$ the vectors

$$\mathfrak{d}_k' = \epsilon_k \mathfrak{d}_k \qquad (\epsilon_k \text{ in } U)$$

form a reduced semi-basis of $\mathfrak{L}$ under the sole assumption that they lie in $\mathfrak{L}$.

Because $\mathfrak{x} = \epsilon_k \mathfrak{d}_k$ satisfies

$$f(\mathfrak{x}) = f(\mathfrak{d}_k), \quad \text{a fortiori} \quad f(\mathfrak{x}) \leqq f(\mathfrak{d}_k),$$

there is then, according to (i_0), only a finite number of possibilities for ϵ_k. We set

$$\eta_1 \mathfrak{d}_1 + \cdots + \eta_n \mathfrak{d}_n = \mathfrak{x} = \eta_1' \mathfrak{d}_1' + \cdots + \eta_n' \mathfrak{d}_n'$$

and denote by Λ, Λ' the corresponding images of $\mathfrak{L}$:

$$(\eta_1, \cdots, \eta_n) \text{ in } \Lambda \cdot \rightleftarrows \cdot \mathfrak{x} \text{ in } \mathfrak{L} \cdot \rightleftarrows \cdot (\eta_1', \cdots, \eta_n') \text{ in } \Lambda'.$$

The "special transformation"

(10) $$\eta_k = \eta_k' \epsilon_k \qquad (\epsilon_k \text{ in } U)$$

carries Λ into Λ'. We count in the same *family* any two lattices Λ and Λ' arising from each other by such a special transformation. Given the lattice Λ over I there is only a finite number of special transformations such that the transformed lattice Λ' also contains I. In particular, the group $\{J_\Lambda\}$ of all special transformations J_Λ leaving Λ invariant is finite. If h is its degree, one has $\epsilon_k^h = 1$ ($k = 1, \cdots, n$) for each J_Λ; hence the ϵ_k are roots of unity in $\mathfrak{J}$. The roots of unity in a field $\mathfrak{J}$ form a finite cyclic group; in particular, if $\mathfrak{J}$ has at least one real spot, the only such roots are ± 1. (However, in noncommutative division algebras the group of the roots of unity is, generally speaking, neither Abelian nor finite.)

The simple argument in R1, p. 136, shows:

If $\mathfrak{d}_1, \cdots, \mathfrak{d}_n$ *is a properly reduced semi-basis and* $\mathfrak{d}_1', \cdots, \mathfrak{d}_n'$ *any semi-basis of* $\mathfrak{L}$, *then the sequence of the values* $f(\mathfrak{d}_1), \cdots, f(\mathfrak{d}_n)$ *is lower than* $f(\mathfrak{d}_1'), \cdots, f(\mathfrak{d}_n')$. *If* $\mathfrak{d}_1', \cdots, \mathfrak{d}_n'$ *is reduced and* $\mathfrak{d}_1, \cdots, \mathfrak{d}_n$ *properly reduced, then*

$$\mathfrak{d}_1' = \epsilon_1 \mathfrak{d}_1, \cdots, \mathfrak{d}_n' = \epsilon_n \mathfrak{d}_n \qquad (\epsilon_i \textit{ in } U).$$

4. **Extension to the ground field $\mathfrak{R}$. Minkowski's inequality.** So far the function $f(\mathfrak{x})$ has been defined merely for the vectors in the space $E^n/\mathfrak{J}$. In order to introduce geometry we assign to the variables x_α in (2) arbitrary *real* values:

(2*) $$\xi^* = x_1 \sigma_1 + \cdots + x_f \sigma_f.$$

Sticking to the multiplication table of the basic elements σ_α, we thus extend $\mathfrak{J}/\mathfrak{R}_0$ to a commutative algebra $\mathfrak{J}^*$ over $\mathfrak{R}$. But only in the two cases treated in R1, where $\mathfrak{J}$ is $\mathfrak{R}_0$ itself or an imaginary quadratic field over $\mathfrak{R}_0$, is $\mathfrak{J}^*$ again a field. In general it is not. However, any n-uple $\mathfrak{x}^* = (\xi_1^*, \cdots, \xi_n^*)$ of elements ξ_i^* in $\mathfrak{J}^*$ may be considered as a vector in an (nf)-dimensional vector space E^{nf} over $\mathfrak{R}$ with the real coordinates $x_{i\alpha}$:

$$\xi_i^* = x_{i1}\sigma_1 + \cdots + x_{if}\sigma_f.$$

We now assume $f(\mathfrak{x}^*)$ to be a *gauge function*, i.e., a continuous real-valued function in this space, having the following properties:

(i) $f(\mathfrak{x}^*) > 0$ except for $\mathfrak{x}^* = 0$.
(ii) $f(t\mathfrak{x}^*) = |t| \cdot f(\mathfrak{x}^*)$ for any real factor t.
(iii) $f(\mathfrak{x}_1^* + \mathfrak{x}_2^*) \leqq f(\mathfrak{x}_1^*) + f(\mathfrak{x}_2^*)$.

The gauge body

$$K: \quad f(\xi_1^*, \cdots, \xi_n^*) < 1$$

and also the solid qK defined by $f(\xi_1^*, \cdots, \xi_n^*) < q$ are bounded; hence postulate (i_0) of the previous section is fulfilled. Let V^* be the volume of K computed in terms of the coordinates $x_{i\alpha}$.

Again we fix an order $[\mathfrak{J}]$ and a basis $\sigma_1, \cdots, \sigma_f$ of $[\mathfrak{J}]$. Let Λ be a lattice belonging to this order and containing the unit lattice I for $[\mathfrak{J}]$ and let $f(\xi_1, \cdots, \xi_n)$ be reduced with respect to Λ. The volume of K in terms of the coordinates u_μ as introduced by (5), i.e., measured against the fundamental parallelepiped of Λ, equals $V^* \cdot [\Lambda : \mathrm{I}]$. Hence by the simple argument explained in R1, p. 140, Minkowski's second inequality leads to this formula holding for a gauge function $f(\xi_1^*, \cdots, \xi_n^*)$ which is reduced with respect to Λ:

$$(M_1 \cdots M_n)^f \cdot V^*[\Lambda : \mathrm{I}] \leqq 2^{nf}$$

where

$$M_k = f(e_{1k}, \cdots, e_{nk}).$$

5. **Splitting. The number of reduced lattices is finite.** Up to now everything has worked for a division algebra of degree f over $\mathfrak{R}_0$ just as well as for a field $\mathfrak{J}$. Further progress depends on the structure of $\mathfrak{J}^*$. If $\mathfrak{J}$ is a field, then $\mathfrak{J}^*$ is isomorphic to the direct sum of a number of components $\mathfrak{R}$ and $\mathfrak{C}$. We first study this case.

The decomposition of $\mathfrak{J}^*$ is brought about by conjugation. One knows that $\mathfrak{J}/\mathfrak{R}_0$ has a determining number θ whose powers $1, \theta, \cdots, \theta^{f-1}$ constitute a basis for $\mathfrak{J}$. The number θ satisfies an irreducible equation in $\mathfrak{R}_0$ of degree f. Let θ^α and θ^β, $\bar{\theta}^\beta$ (or in one row: $\theta^{(1)}, \cdots, \theta^{(f)}$) denote its r real and s pairs of complex conjugate roots. They define the f conjugations

$$\xi \to \xi^\alpha; \qquad \xi \to \xi^\beta, \qquad \xi \to \bar{\xi}^\beta$$

each of which projects $\mathfrak{J}$ isomorphically into $\mathfrak{R}$ or $\mathfrak{C}$. We use the notations

$$\xi^\alpha = x^\alpha, \qquad \xi^\beta = x_0^\beta + ix_1^\beta \qquad (\bar{\xi}^\beta = x_0^\beta - ix_1^\beta)$$

and call the $r+s$ numbers ξ^α, ξ^β the *splits*, and the f real numbers x^α; x_0^β, x_1^β

the *splitting coordinates* of ξ. The same applies to any element ξ^* of $\mathfrak{J}^*$. The product $\zeta^* = \xi^*\eta^*$ has the splits

$$\zeta^\alpha = \xi^\alpha\eta^\alpha, \qquad \zeta^\beta = \xi^\beta\eta^\beta.$$

The arithmetician speaks of the different values of the indices α and β as the r real and s imaginary (*infinite prime*) *spots* of $\mathfrak{J}$; for the sake of brevity we often drop the adjectives in parentheses. If a definite arrangement is desired, we write $\alpha = \alpha_1, \cdots, \alpha_r$; $\beta = \beta_1, \cdots, \beta_s$.

The splitting coordinates x^α; x_0^β, x_1^β are connected with the components $x_1, \cdots, x_f$ of ξ^* by the linear substitution

$$\Sigma = \|\sigma_1, \cdots, \sigma_f\| \tag{11}$$

where in the symbol on the right side each term stands for the column of its splitting coordinates (in a definite arrangement). The splitting of $\mathfrak{J}^*$ into r components $\mathfrak{R}$ and s components $\mathfrak{C}$ is established as soon as it is certain that the absolute determinant

$$\Delta = \text{abs.} \left| \sigma_1, \cdots, \sigma_f \right|$$

of the matrix Σ is different from zero. For the particular basis $1, \theta, \cdots, \theta^{f-1}$ one sees that $(-2i)^s \cdot \Delta$ is the Vandermonde determinant of $\theta^{(1)}, \cdots, \theta^{(f)}$, and hence indeed $\Delta \neq 0$. This fact carries over to any basis $\sigma_1, \cdots, \sigma_f$ of $\mathfrak{J}$.

The number in $\mathfrak{J}^*$ with the splitting coordinates x^α; x_0^β, $-x_1^\beta$ is denoted by $\bar{\xi}^*$. As absolute value $|\xi^*|$ we introduce the greatest of the $r+s$ numbers $|\xi^\alpha|$, $|\xi^\beta|$. One could agree on other definitions, but this one is most convenient for our future applications. What usually is called a *unit* in $\mathfrak{J}$ is a number of $\mathfrak{J}$ which is a unity at all *finite* prime spots. None but the infinite prime spots matter for our investigation; hence we take the liberty of using the term "*unit*" for those numbers ϵ of $\mathfrak{J}$ which are unities at all infinite prime spots, i.e., for which the $r+s$ equations $|\epsilon^\alpha| = 1$, $|\epsilon^\beta| = 1$ hold.

For any element δ^* of $\mathfrak{J}^*$ one introduces the real matrix $\|d_{ab}\|$ of the linear operation $\xi^* \to \xi^*\delta^*$ in $\mathfrak{J}^*$:

$$x_a \to \sum_b d_{ab}x_b \qquad \left(\sigma_a\delta^* = \sum_b d_{ba}\sigma_b\right)$$

and its characteristic polynomial

$$\left| te_{ab} - d_{ab} \right| = t^f - d_1 t^{f-1} + \cdots \pm d_f.$$

d_1 and d_f are called trace (tr) and norm (Nm), respectively. In terms of the splitting coordinates our operation of multiplication splits into the transformations

$$x^\alpha \to x^\alpha d^\alpha; \qquad \xi^\beta \to \xi^\beta\delta^\beta,$$

each corresponding to a real or imaginary spot α or β. Of course, $\xi^\beta \to \xi^\beta \delta^\beta$ stands for

$$\overset{\beta}{x}_0 \to \overset{\beta}{x}_0 \overset{\beta}{d}_0 - \overset{\beta}{x}_1 \overset{\beta}{d}_1, \qquad \overset{\beta}{x}_1 \to \overset{\beta}{x}_0 \overset{\beta}{d}_1 + \overset{\beta}{x}_1 \overset{\beta}{d}_0.$$

Hence

$$\operatorname{tr}(\delta^*) = \sum_\alpha \overset{\alpha}{d} + 2 \sum_\beta \overset{\beta}{d}_0,$$

$$\operatorname{Nm}(\delta^*) = \prod_\alpha \overset{\alpha}{d} \cdot \prod_\beta \left\{ (\overset{\beta}{d}_0)^2 + (\overset{\beta}{d}_1)^2 \right\}.$$

If $\delta^* = \delta$ is in $\mathfrak{J}$ the d_{ab} are rational numbers. For a unit ϵ in $\mathfrak{J}$ our formulas show that the determinant $\operatorname{Nm}(\epsilon)$ of the transformation $\xi^* \to \xi^* \epsilon$ is of absolute value 1 and hence as a rational number equal to ± 1.

Considering the trace $\operatorname{tr}(\xi^2)$ one readily verifies that $(2^\delta \Delta)^2$ is rational for any basis $\sigma_1, \cdots, \sigma_f$ and especially a rational integer for a basis of an order $[\mathfrak{J}]$.

The transformation (4) in E^{nf},

$$\xi_i^* = \sum_k \eta_k^* \delta_{ik} \qquad (\delta_{ik} \text{ numbers in } \mathfrak{J}) \tag{4}$$

splits into the components

$$\overset{\alpha}{\xi}_i = \sum_k \overset{\alpha}{\eta}_k \overset{\alpha}{\delta}_{ik}, \qquad \overset{\beta}{\xi}_i = \sum_k \overset{\beta}{\eta}_k \overset{\beta}{\delta}_{ik},$$

each α-component involving n, each β-component $2n$ real variables:

$$\overset{\alpha}{\xi}_k = \overset{\alpha}{x}_k; \qquad \overset{\beta}{\xi}_k = \overset{\beta}{x}_{k0} + i \overset{\beta}{x}_{k1}.$$

How closely can one approximate an element ξ^* of $\mathfrak{J}^*$ by a number γ of our order $[\mathfrak{J}]$ with the basis $\sigma_1, \cdots, \sigma_f$? For an appropriate γ in $[\mathfrak{J}]$ the real components x_a' of $\xi^* - \gamma$,

$$\xi^* - \gamma = x_1' \sigma_1 + \cdots + x_f' \sigma_f,$$

will satisfy the inequalities $|x_a'| \leqq \frac{1}{2}$, and thus

$$|\xi^* - \gamma| \leqq \rho$$

where

$$\rho = \tfrac{1}{2} \cdot \max_{\alpha, \beta} \left(|\overset{\alpha}{\sigma}_1| + \cdots + |\overset{\alpha}{\sigma}_f|, \ |\overset{\beta}{\sigma}_1| + \cdots + |\overset{\beta}{\sigma}_f| \right).$$

The "circles" of radius ρ around all numbers γ of $[\mathfrak{J}]$ cover the whole $\mathfrak{J}^*$. (Such a radius was denoted by the letter r in R1, which now serves a different purpose.)

Let us now return to the situation explained at the end of the previous section and let V be the volume of K computed in terms of the splitting coordinates of $\xi_1^*, \cdots, \xi_n^*$. Then $V = V^*/\Delta^n$. Moreover we observe that

$f(\xi_1^*, \cdots, \xi_k^*, 0, \cdots, 0)$ is reduced with respect to the lattice $\Lambda^{(k)}$, and denoting by V_k the volume of the solid

$$f(\xi_1^*, \cdots, \xi_k^*, 0, \cdots, 0) < 1$$

in E^{kf} computed in terms of the splitting coordinates of $\xi_1^*, \cdots, \xi_k^*$, we obtain these fundamental inequalities for $M_k = f(e_{1k}, \cdots, e_{nk})$:

THEOREM I. *For a reduced* $f(\xi_1, \cdots, \xi_n)/\Lambda$ *one has*

$$(M_1 \cdots M_n)^f \cdot V[\Lambda : \mathrm{I}] \leqq (2^f\Delta)^n, \tag{12}$$

more generally

$$(M_1 \cdots M_k)^f \cdot V_k[\Lambda^{(k)} : \mathrm{I}^{(k)}] \leqq (2^f\Delta)^k. \tag{12_k}$$

At this point we introduce the further assumption:

(ii*) $f(\tau^* \mathfrak{x}^*) \leqq |\tau^*| \cdot f(\mathfrak{x}^*)$ (τ^* any element of $\mathfrak{J}^*$),

and henceforth the term "gauge function" is to be taken in this restricted sense. Following Minkowski's own argument, we then prove

THEOREM II. *For a reduced* $f(\xi_1, \cdots, \xi_n)/\Lambda$ *one always has*

$$j = [\Lambda : \mathrm{I}] \leqq (nf)!\left(\frac{4}{\pi}\right)^{ns} \cdot \left(\frac{\Delta}{f!}\right)^n \tag{13}$$

and more generally

$$j_k = [\Lambda^{(k)} : \mathrm{I}^{(k)}] \leqq (kf)!\left(\frac{4}{\pi}\right)^{ks}\left(\frac{\Delta}{f!}\right)^k \qquad (k = 1, \cdots, n). \tag{13_k}$$

Hence in any class of lattices belonging to the order $[\mathfrak{J}]$ there is always a lattice Λ which contains I and satisfies (13) and (13_k). Together with Lemma 2 this proves[5]:

THEOREM III. *The number of classes of lattices belonging to a given order* $[\mathfrak{J}]$ *is finite.*

---

[5] This theorem is well known. We are concerned only with those lattices Λ over I which are in the class of $\mathfrak{L}$, but as our bounds (13) or the sharper bounds (35) depend on the order rather than on the special class it seemed worth while to mention Theorem III in passing. For a commutative field $\mathfrak{J}$ and its principal order $[\mathfrak{J}]$ E. Steinitz, Mathematische Annalen, vol. 71 (1912), pp. 328–354, and vol. 72 (1912), pp. 297–345, proved that the number for classes of any n is the same as for $n=1$, namely equal to the number of classes of ideals. See also I. Schur, Mathematische Annalen, vol. 71 (1912), pp. 355–367; W. Franz, Journal für die reine und angewandte Mathematik, vol. 171 (1934), pp. 149–161; C. Chevalley, *L'Arithmétique dans les Algèbres de Matrices*, Actualités Scientifiques et Industrielles, no. 323, 1936, in particular Theorems 3, 7 and 8.

(The proposition implies the corresponding one about classes of ideals.) Any gauge function will do for the proof, for instance

$$f(\xi_1^*, \cdots, \xi_n^*) = |\xi_1^*| + \cdots + |\xi_n^*|.$$

We shall soon see that much better upper bounds for the number of classes are obtained by using for f^2 the trace of a positive Hermitian form. However, our present Theorem II goes far beyond Theorem III because it deals with any gauge function f in conjunction with a lattice rather than with lattices alone.

Proof. Observe that the "octahedron"

$$|\xi_1^*| + \cdots + |\xi_n^*| < 1$$

contains no vector of Λ except the zero vector. Hence owing to Minkowski's chief inequality we find this upper bound for its volume W:

$$Wj \leqq (2^f\Delta)^n.$$

Let $(\xi_1, \cdots, \xi_n)$ be a vector in Λ and ξ_k be the last nonvanishing one among its coordinates ξ_i. Then by the definition of reduction

$$f(\xi_1, \cdots, \xi_n) \geqq M_k = f(e_{1k}, \cdots, e_{nk}). \tag{14}$$

On the other hand the assumptions (iii) and (ii*) imply

$$\begin{aligned} f(\xi_1 e_1 + \cdots + \xi_n e_n) &\leqq M_1|\xi_1| + \cdots + M_n|\xi_n| \\ &= M_1|\xi_1| + \cdots + M_k|\xi_k|. \end{aligned} \tag{15}$$

Because of (8) the relations (14) and (15) are incompatible unless

$$|\xi_1| + \cdots + |\xi_n| = |\xi_1| + \cdots + |\xi_k| \geqq 1.$$

We base our computation of W upon the following general remark about gauge functions f in an n-dimensional vector space over $\mathfrak{R}$. If V is the volume of the gauge body $K: f(\mathfrak{x}) < 1$, then the integral $\int$ of e^{-f} over the whole space equals $n!V$. One simply evaluates the integral by decomposing the space into the infinitely thin shells

$$q \leqq f(\mathfrak{x}) < q + dq$$

and thus finds

$$\int = V \cdot \int_0^\infty e^{-q} \cdot nq^{n-1} dq = n!V.$$

Applying this remark to the gauge function $|\xi_1^*| + \cdots + |\xi_n^*|$ in our (nf)-dimensional vector space and to the gauge function $|\xi^*|$ in the f-dimensional space $\mathfrak{J}^*$, one gets this double value for $\int$:

$$(nf)!W = (f!w)^n,$$

w being the volume of the "cylinder" defined by

$$|\mathfrak{x}^*| < 1, \qquad \text{or by } |x^\alpha| < 1, \quad (x_0^\beta)^2 + (x_1^\beta)^2 < 1.$$

Therefore $w = 2^r \pi^s$.

(ii*) entails the property (ii$_0$) of §3, provided U is the group of units in our sense. From now on we shall abide by this convention and interpret the term "properly reduced" accordingly. Then the transformation $\mathfrak{x}^* \to \mathfrak{x}^* \cdot \epsilon$ (ϵ in U) and hence every special transformation (10) has the determinant ± 1 and thus the indices j_k for two lattices Λ and Λ' over I which are in the same family coincide: $j_k = j_k'$ for $k = 1, \cdots, n$.

The values γ^* of a *Hermitian form* in $\mathfrak{J}^*$,

$$\gamma^*(\mathfrak{x}^*) = \sum_{i,k} \xi_i^* \gamma_{ik}^* \bar{\xi}_k^* \qquad (\gamma_{ki}^* = \bar{\gamma}_{ik}^*) \tag{16}$$

are totally real in the sense that $\bar{\gamma}^* = \gamma^*$, or that even the β-splits $\gamma^\beta = g_0^\beta + i g_1^\beta = g^\beta$ of γ^* are real. What such a Hermitian form does is to associate a quadratic form $\{g_{ik}^\alpha\}$ with each real spot α and a Hermitian form $\{\gamma_{ik}^\beta\}$ with every imaginary spot β. The splits of $\gamma^*(\mathfrak{x}^*)$ are

$$g^\alpha = \sum_{i,k} x_i^\alpha g_{ik}^\alpha x_k^\alpha, \qquad g^\beta = \sum_{i,k} \xi_i^\beta \gamma_{ik}^\beta \bar{\xi}_k^\beta \tag{17}$$

where $x_i^\alpha = \xi_i^\alpha$ and ξ_i^β are the splits of ξ_i^*. The form $\gamma^*(\mathfrak{x}^*)$ is said to be positive if each of the r quadratic forms $\{g_{ik}^\alpha\}$ and each of the s Hermitian forms $\{\gamma_{ik}^\beta\}$ is positive definite.

We now apply our theory to the gauge function f introduced by

$$f^2 = \operatorname{tr}(\gamma^*(\mathfrak{x}^*)). \tag{18}$$

In terms of the splits (17) one has

$$f^2 = \sum_\alpha g^\alpha + 2 \sum_\beta g^\beta. \tag{19}$$

The properties (i) to (iii) of §4 are readily verified; (ii*) is also fulfilled because of

$$f^2(\tau^* \mathfrak{x}^*) = \sum_\alpha |\tau^\alpha|^2 g^\alpha + 2 \sum_\beta |\tau^\beta|^2 g^\beta.$$

6. **Quaternion algebra of totally positive norm over a totally real field.** Turning to noncommutative division algebras, we denote by $\mathfrak{Q}$ the quasi-field of quaternions

$$a = a_0 + a_1 i_1 + a_2 i_2 + a_3 i_3$$

whose components a_0, a_1, a_2, a_3 are arbitrary real numbers, and use the notations $\bar{a}$ and $|a|$ in the customary manner:

$$a\bar{a} = |a|^2 = a_0^2 + a_1^2 + a_2^2 + a_3^2.$$

For which of the noncommutative division algebras of finite degree over $\mathfrak{R}_0$ does the concept of infinite prime spots work in a way similar to that in the previous section for fields? I am going to describe one such situation without discussing the question whether or not it is the only one (though, as a matter of fact, it is).

Suppose we are given a field $\mathfrak{E}$ of degree e over $\mathfrak{R}_0$ and two numbers ω_1, ω_2 in $\mathfrak{E}$. We put $\omega_3 = \omega_1\omega_2$ and form the quaternion algebra $\mathfrak{J}$ over $\mathfrak{E}$ whose elements ξ are quadruples $(\xi_0, \xi_1, \xi_2, \xi_3)$ of numbers in $\mathfrak{E}$,

$$\xi = \xi_0 + \xi_1\iota_1 + \xi_2\iota_2 + \xi_3\iota_3, \tag{20}$$

with this multiplication table for the unities $\iota_1, \iota_2, \iota_3$:

$$\iota_1^2 = -\omega_1, \qquad \iota_2^2 = -\omega_2, \qquad \iota_3^2 = -\omega_3;$$

$$\iota_1\iota_2 = -\iota_2\iota_1 = \iota_3, \qquad \iota_3\iota_1 = -\iota_1\iota_3 = \omega_1\iota_2, \qquad \iota_2\iota_3 = -\iota_3\iota_2 = \omega_2\iota_1.$$

The conjugate $\bar{\xi}$ is $\xi_0 - \xi_1\iota_1 - \xi_2\iota_2 - \xi_3\iota_3$ and

$$\xi\bar{\xi} = \xi_0^2 + \omega_1\xi_1^2 + \omega_2\xi_2^2 + \omega_3\xi_3^2.$$

If the equation

$$\xi_0^2 + \omega_1\xi_1^2 + \omega_2\xi_2^2 + \omega_3\xi_3^2 = 0 \tag{21}$$

has no solution $(\xi_0, \xi_1, \xi_2, \xi_3)$ in $\mathfrak{E}$ except (0, 0, 0, 0), then $\mathfrak{J}$ is a division algebra of degree 4 over $\mathfrak{E}$ and of degree $f = 4e$ over $\mathfrak{R}_0$. We assume $\mathfrak{E}$ to be totally real (to have no imaginary infinite prime spot) and ω_1, ω_2 to be totally positive numbers in $\mathfrak{E}$ (i.e., their e conjugates $\omega_1^\alpha, \omega_2^\alpha$ are all positive). Then the quadratic form of the variables $\xi_0, \xi_1, \xi_2, \xi_3$ at the left of (21) is positive definite in each conjugate $\mathfrak{E}^\alpha$ of $\mathfrak{E}$ and hence (21) has no solution except 0. Denoting as before by τ^α the conjugate of any number τ in $\mathfrak{E}$ corresponding to the spot α of $\mathfrak{E}$, we map (20) upon the element

$$\xi^\alpha = \xi_0^\alpha + \xi_1^\alpha(\omega_1^\alpha)^{1/2}\cdot i_1 + \xi_2^\alpha(\omega_2^\alpha)^{1/2}\cdot i_2 + \xi_3^\alpha(\omega_3^\alpha)^{1/2}\cdot i_3 \tag{22}$$

in $\mathfrak{Q}$. This "conjugation" is an isomorphic mapping and defines the "infinite quaternion prime spot" α of $\mathfrak{J}$. (22) are the splits, and the $4e = f$ real numbers

$$x_0^\alpha = \xi_0^\alpha, \qquad x_1^\alpha = \xi_1^\alpha(\omega_1^\alpha)^{1/2}, \qquad x_2^\alpha = \xi_2^\alpha(\omega_2^\alpha)^{1/2}, \qquad x_3^\alpha = \xi_3^\alpha(\omega_3^\alpha)^{1/2}$$

are the "splitting coordinates," of ξ. Application to the elements ξ^*, equation (2*), of $\mathfrak{J}^*$ is immediate.

There is only one thing to settle: The splitting coordinates $x_0^\alpha, x_1^\alpha, x_2^\alpha, x_3^\alpha$ arise from the components x_a by the substitution (11), each σ_a standing for the column of its splitting coordinates. Is its determinant, whose absolute value will again be denoted by Δ, different from zero? To answer the question, let $(\tau_1, \cdots, \tau_e)$ be a basis of $\mathcal{E}$ and set $\Delta_0 = \text{abs.}|\tau_1, \cdots, \tau_e|$. From it we obtain the following basis of $\mathcal{J}$:

$$\tau_a, \qquad \tau_a \iota_1, \qquad \tau_a \iota_2, \qquad \tau_a \iota_3 \qquad\qquad (a = 1, \cdots, e).$$

The Δ of this particular basis is given by

$$\Delta = \prod_{\alpha} (\overset{\alpha}{\omega_1} \overset{\alpha}{\omega_2}) \cdot \overset{4}{\Delta_0} = \text{Nm } \omega_3 \cdot \overset{4}{\Delta_0}.$$

Thus $\Delta \neq 0$ for this and consequently for any basis.

Incidentally Δ is a rational number for any basis of $\mathcal{J}$ and $4^e \cdot \Delta$ a rational integer if $\sigma_1, \cdots, \sigma_f$ is a basis of $[\mathcal{J}]$. The characteristic equation of the multiplication $\xi^* \to \xi^* \cdot \delta$ considered as a linear operation in $\mathcal{J}^*$ is the square of a polynomial (of degree $2e$), and so is the characteristic polynomial of the linear substitution (4) in E^{nf}.

The notion of unit and the absolute value $|\xi^*|$ of any element ξ^* of $\mathcal{J}^*$ are introduced as before. The constant on the right side of (13_k) is to be changed into

$$(kf)! \left(\frac{32}{\pi^2}\right)^{ek} \left(\frac{\Delta}{f!}\right)^k.$$

As gauge functions f we employ in particular those whose square equals

$$\tfrac{1}{4} \operatorname{tr}(\gamma^*) = \sum_{\alpha} g^{\alpha}$$

where γ^* is any positive Hermitian form (16) in $\mathcal{J}^*$.

7. **The theorems of finiteness for quadratic forms.** After so many preliminaries which stake out the ground covered by our investigation, I now come to the core of the matter, which may be explained fairly completely by the simplest example $\mathcal{J} = \mathcal{R}_0$. Here we have only one order $[\mathcal{J}]$ consisting of the ordinary integers $0, \pm 1, \pm 2, \cdots$ and only one class of lattices. For any given lattice Λ over I and any positive quadratic form

$$f^2(\mathfrak{x}) = \sum g_{ik} x_i x_k \qquad\qquad (g_{ki} = g_{ik})$$

the conditions of reduction read:

$$f^2(\mathfrak{x}) \geqq g_{kk} \qquad \text{whenever } \mathfrak{x} = (x_1, \cdots, x_n) \text{ is in } \Lambda_k.$$

Each of them is a linear inequality for the coefficients g_{ij}.

With the notation used in R1 we carry out Jacobi's transformation:

$$f^2 = q_1 z_1^2 + \cdots + q_n z_n^2.$$

The volume V of the ellipsoid $f^2 < 1$ is given by

$$V^2 = \omega_n^2/q_1 \cdots q_n,$$

ω_n being the volume of the n-dimensional sphere. Hence the inequality (12),

$$M_1 \cdots M_n V[\Lambda:\mathrm{I}] \leqq 2^n, \tag{23}$$

turns into

$$g_{11} \cdots g_{nn}[\Lambda:\mathrm{I}]^2 \leqq (2^n/\omega_n)^2 \cdot q_1 \cdots q_n. \tag{24}$$

As Minkowski observed, (23) may be proved much more easily for quadratic forms than for an arbitrary gauge function. By an argument similar to the one employed in proving Theorem II we see that the ellipsoid

$$f'^2 = \frac{q_1}{M_1^2} z_1^2 + \cdots + \frac{q_n}{M_n^2} z_n^2 < 1$$

contains no lattice vector except zero. Hence its volume V' satisfies the inequality

$$V'[\Lambda:\mathrm{I}] \leqq 2^n, \quad \text{and} \quad V' = M_1 \cdots M_n \cdot V.$$

If κ_n is a number such that the part of space covered by impenetrable n-dimensional spheres in any lattice arrangement may never exceed the proportion $\kappa_n:1$ then we can even write $\kappa_n 2^n$ instead of 2^n and thus replace ω_n in (24) by $\pi_n = \omega_n/\kappa_n$. The most primitive choice is $\kappa_n = 1$; however, according to Blichfeldt's ingenious device(6),

$$\kappa_n = (n+2)\cdot 2^{-1-n/2}$$

is a permissible and better value.

Making use of the inequalities

$$g_{ii} \geqq q_i$$

on the left side of (24) we get for the index $j = [\Lambda:\mathrm{I}]$ this upper bound

$$j \leqq 2^n/\pi_n \tag{25}$$

which is a considerable improvement over (13), $j \leqq n!$ For $n = 1, 2, 3$ it yields the result $j = 1$, to which the theory of reduction for binary and ternary forms owes its comparative simplicity.

For similar reasons

$$g_{11} \cdots g_{kk} \cdot j_k^2 \leqq (2^k/\pi_k)^2 \cdot q_1 \cdots q_k, \tag{24$_k$}$$

$$j_k \leqq 2^k/\pi_k \qquad (k = 1, \cdots, n). \tag{25$_k$}$$

Unless the lattice Λ satisfies the n inequalities (25_k) for its indices $j_k = [\Lambda^{(k)}:\mathrm{I}^{(k)}]$ there can be no Λ-reduced forms.

(6) H. F. Blichfeldt, Mathematische Annalen, vol. 101 (1929), pp. 605–608.

Dividing (24_k) by

$$g_{11} \cdots g_{k-1,k-1} \geqq q_1 \cdots q_{k-1}$$

we find that our reduced form satisfies the fundamental relations

$$q_k \geqq \lambda_k g_{kk} \tag{26}$$

where

$$\lambda_k = (j_k \pi_k / 2^k)^2. \tag{27}$$

This lower bound for q_k is much better than the corresponding one holding for the method of reduction studied in R1.

The *first theorem of finiteness* deals with the subset $\Lambda_k(=)$ of Λ_k to which a vector $\mathfrak{x}$ in Λ_k belongs if there exists a Λ-reduced positive quadratic form f^2 satisfying the *equation* $f^2(\mathfrak{x}) = g_{kk}$. *The set* $\Lambda_k(=)$ *is finite.* The proof is as in R1, but the upper bounds arrived at are a good deal lower. The first part of the proof yields the bounds

$$\lambda_i z_i^2 \leqq 1 \qquad (\text{for } i = k, k+1, \cdots, n)$$

where the λ_i are now defined by (27). In the second part one replaces the vector $\mathfrak{x}$ in $\Lambda_k(=)$ by $\mathfrak{x} - \mathfrak{x}_h$ where $\mathfrak{x}_h$ is any vector in $\Lambda^{(h)}$ $(h<k)$ and observes that

$$f^2(\mathfrak{x} - \mathfrak{x}_h) \geqq g_{kk}.$$

This is true in particular if $\mathfrak{x}_h$ is in $\mathrm{I}^{(h)}$, and as in R1 one thus derives the relations

$$\lambda_h z_h^2 \leqq \rho^{2h} \qquad (\rho = 1/2;\ h = 1, \cdots, k-1).$$

Once the discrete lattice Λ is given, the resulting universal upper bounds for $|x_n|, \cdots, |x_1|$ leave only a limited number of possibilities for a vector $\mathfrak{x} = (x_1, \cdots, x_n)$ in Λ.

The *second theorem of finiteness* shall be restated in a more natural and slightly more general form. Let $p \geqq 1$ and $w \geqq 0$ be given. With respect to the lattice Λ over I the positive quadratic form f^2 will be said to have the property $B(p, w)$ provided

$$f^2(\mathfrak{x}) \geqq \frac{1}{p} \cdot f^2(\mathfrak{e}_k) \tag{28}$$

for any vector $\mathfrak{x}$ in Λ_k, and

$$f^2(\mathfrak{e}_k - \mathfrak{x}_h) \geqq f^2(\mathfrak{e}_k) - w \cdot f^2(\mathfrak{e}_h) \tag{28'}$$

for $h<k$ and any vector $\mathfrak{x}_h$ in $\Lambda^{(h)}$. Again, each of these conditions is a linear inequality for the coefficients g_{ij} of f^2. We maintain:

Given two lattices Λ *and* Λ' *over* I, *there is only a limited number of linear* *transformations carrying Λ into Λ' and at the same time capable of carrying an unspecified Λ-reduced form f^2 into an unspecified f'^2 which has the property $B(p, w)$ with respect to Λ'.*

We write the transformation as

$$\mathfrak{x} = \sum_i x_i \mathfrak{e}_i = \sum_i y_i \mathfrak{d}_i:$$

if $(x_1, \cdots, x_n)$ is in Λ, then $(y_1, \cdots, y_n)$ is in Λ', and *vice versa*. In particular, the $\mathfrak{d}_1, \cdots, \mathfrak{d}_n$ are vectors in Λ. (p, $\mathfrak{e}_i$, $\mathfrak{d}_i$ were denoted in R1 by ρ^2, $\mathfrak{b}_i$, $\mathfrak{s}_i$.) More explicitly as has been done in R1, we divide the row of indices $1, \cdots, n$ into a number of sections by means of the subspaces

$$E_k = [\mathfrak{e}_1, \cdots, \mathfrak{e}_k], \qquad E_k' = [\mathfrak{d}_1, \cdots, \mathfrak{d}_k] \qquad (k = 0, 1, \cdots, n).$$

We pick out those $k = l_0, l_1, \cdots, l_v$,

$$0 = l_0 < l_1 < \cdots < l_{v-1} < l_v = n$$

for which $E_k = E_k'$, and divide the range of k into the v sections

$$l_{u-1} < k \leqq l_u \qquad (u = 1, \cdots, v).$$

We then study the possibilities for transformations $(\mathfrak{d}_1, \cdots, \mathfrak{d}_n)$ with given $l_1, \cdots, l_{v-1}$.

By the analogues of Theorems 8_p and 9_p we have

$$g'_{kk} \leqq p g_{kk} \qquad (k = 1, \cdots, n) \tag{29}$$

and moreover

$$g_{i+1,i+1} \leqq p g_{ii} \tag{30}$$

whenever i and $i+1$ are in the same section. Consider a $\mathfrak{d}_k$ of the last section $(l_{v-1} < k \leqq n)$. The first part of the proof in R1 yields for $\mathfrak{x} = \mathfrak{d}_k$ the simple upper bounds

$$\lambda_h z_h^2 \leqq p^{\{k-h\}+1}$$

if h also belongs to the last section, $\{k\}$ denoting 0 or k according as $k \leqq 0$ or $k > 0$. The second part requires a slight modification. Suppose h lies in the uth section $(u < v)$, and set for the moment $l_u = l$. Since $E_l = E_l'$, the vectors in $\Lambda^{(l)}$ are obtained from the expression $y_1\mathfrak{d}_1 + \cdots + y_l\mathfrak{d}_l$ by running $(y_1, \cdots, y_l, 0, \cdots, 0)$ over $\Lambda'^{(l)}$. Hence, according to the postulate (28'):

$$f'^2(\mathfrak{e}_k - \mathfrak{y}') \geqq g'_{kk} - w g'_{ll}$$

for any vector $\mathfrak{y}'$ in $\Lambda'^{(l)}$, or

$$f^2(\mathfrak{d}_k - \mathfrak{x}') \geqq g'_{kk} - w g'_{ll}$$

for any vector $\mathfrak{x}'$ in $\Lambda^{(l)}$, *a fortiori* for any vector in $\Lambda^{(h)}$, *a fortiori* for any vector $\mathfrak{x}'$ in $I^{(h)}$. Following the same argument as in R1, one gets the inequality

$$wg'_{ll} + \rho^2 hg_{hh} \geqq \lambda_h g_{hh} z_h^2 \qquad (\rho = 1/2).$$

But because h and l are in the same section, (30) and (29) lead to

$$g_{ll} \leqq p^{l-h} g_{hh}, \qquad g'_{ll} \leqq pg_{ll} \leqq p^{l-h+1} g_{hh}$$

and thus finally

$$\lambda_h z_h^2 \leqq h\rho^2 + w \cdot p^{l_u - h+1} \qquad (l_{u-1} < h \leqq l_u;\ u = 1, \cdots, v-1).$$

It is clear how the same procedure applies to a k in the lower sections. Denoting the values of the variables $z_1, \cdots, z_n$ for $\mathfrak{x} = \mathfrak{d}_k$ by $z_{1k}, \cdots, z_{nk}$, one finds:

$z_{hk} = 0$ if h is in a higher section than k;

$\lambda_h z_{hk}^2 \leqq p^{|k-h|+1}$ if h and k are in the same section;

$\lambda_h z_{hk}^2 \leqq h\rho^2 + w \cdot p^{l-h+1}$ if h is in a lower section than k which ends with l.

8. **Modifications in arbitrary fields and quasi-fields.** Our next concern is to examine whether any serious modifications of the procedure just described arise in the two general cases of a field and a quaternion algebra over a field. Take the case of the field first. With a positive Hermitian form γ^* in $\mathfrak{J}^*$ we combine its trace f^2:

$$(31) \qquad f^2(\xi_1^*, \cdots, \xi_n^*) = \sum_\alpha \sum_{i,k} g_{ik} \overset{\alpha}{x}_i \overset{\alpha}{x}_k + 2 \sum_\beta \sum_{i,k} \overset{\beta}{\gamma}_{ik} \overset{\beta}{\xi}_i \overset{\beta}{\bar{\xi}}_k \qquad (\overset{\beta}{\xi}_k = \overset{\beta}{x}_{k0} + i\overset{\beta}{x}_{k1}).$$

γ^* is called reduced with respect to Λ if the gauge function f is, i.e., if

$$(32) \qquad f^2(\xi_1, \cdots, \xi_n) \geqq \operatorname{tr}(\gamma_{kk}^*) = M_k^2$$

for any vector $(\xi_1, \cdots, \xi_n)$ in Λ_k. Each part is subjected to its Jacobi transformation:

$$\sum_{i,k} g_{ik} \overset{\alpha}{x}_i \overset{\alpha}{x}_k = \sum_i \overset{\alpha}{q}_i (\overset{\alpha}{z}_i)^2,$$

$$\sum_{i,k} \overset{\beta}{\gamma}_{ik} \overset{\beta}{\xi}_i \overset{\beta}{\bar{\xi}}_k = \sum_i \overset{\beta}{q}_i |\overset{\beta}{\zeta}_i|^2.$$

Besides

$$\operatorname{tr}(q_i) = \sum_\alpha \overset{\alpha}{q}_i + 2 \sum_\beta \overset{\beta}{q}_i, \qquad \operatorname{Nm}(q_i) = \prod_\alpha \overset{\alpha}{q}_i \cdot \prod_\beta (\overset{\beta}{q}_i)^2$$

we introduce the mean value $\langle q_i \rangle$ by

$$f \cdot \langle q_i \rangle = \operatorname{tr}(q_i).$$

In terms of the coordinates x_k^α; x_{k0}^β, x_{k1}^β the volume V of the ellipsoid $f^2(\mathfrak{x}^*) < 1$ is

$$\omega_{nf} \text{ divided by } 2^{ns}\left(\prod_i \operatorname{Nm} q_i\right)^{1/2}.$$

Instead of applying Minkowski's second inequality to the present gauge function we again consider the ellipsoid

$$f'^2(\mathfrak{x}^*) = \operatorname{tr}\left(\sum_i \frac{q_i}{M_i^2}\zeta_i\bar{\zeta}_i\right) < 1$$

which contains no lattice vector except zero, and thus establish the inequality

$$\prod_i (\operatorname{tr} \overset{*}{\gamma}_{ii})^f \cdot [\Lambda:\mathrm{I}]^2 \leqq \frac{(4^{r+s}\Delta^2)^n}{\pi_{nf}^2} \cdot \prod_i \operatorname{Nm} q_i \tag{33}$$

for any reduced γ^*/Λ.

Now enters the only new feature: Making use of the inequality between arithmetic and geometric means in the form

$$\operatorname{Nm} q_i \leqq \langle q_i \rangle^f$$

we infer from

$$\overset{\alpha}{q}_i \leqq \overset{\alpha}{g}_{ii}, \qquad \overset{\beta}{q}_i \leqq \overset{\beta}{\gamma}_{ii}$$

the relation

$$\langle \overset{*}{\gamma}_{ii} \rangle^f \geqq \langle q_i \rangle^f \geqq \operatorname{Nm} q_i,$$

and then (33) yields the following upper bound for $j = [\Lambda:\mathrm{I}]$:

$$j \leqq 1/\mu_n \quad \text{with the abbreviation} \quad \mu_n = \frac{\pi_{nf} \cdot f^{nf/2}}{(2^{r+s}\Delta)^n}.$$

For the same reasons

$$\prod_{i=1}^{k} \langle \overset{*}{\gamma}_{ii} \rangle^f (\mu_k j_k)^2 \leqq \prod_{i=1}^{k} \operatorname{Nm} q_i \tag{34}$$

and hence

$$\mu_k j_k \leqq 1. \tag{35}$$

These estimates are an improved substitute for Theorem II. Combining (34) with

$$\prod_{i=1}^{k-1} \langle \overset{*}{\gamma}_{ii} \rangle^f \geqq \prod_{i=1}^{k-1} \langle q_i \rangle^f \geqq \prod_{i=1}^{k-1} \operatorname{Nm} q_i$$

one gets

$$\mathrm{Nm}\; q_k \geqq (\mu_k j_k)^2 \cdot \langle \gamma^*_{kk} \rangle^f. \tag{36}$$

Not only does this inequality establish a lower bound for the trace of q_k,

$$\mathrm{tr}\; q_k \geqq (\mu_k j_k)^{2/f} \cdot \mathrm{tr}\; \gamma^*_{kk},$$

but it also shows that the geometric mean of the conjugates of q_k is not much smaller than their arithmetic mean. Therefore none of the conjugates can be much smaller than their arithmetic mean. We have a special case of the situation dealt with by the following

LEMMA 5. *Let $f_1, \cdots, f_m$ be positive integers and $u_1, \cdots, u_m; v_1, \cdots, v_m$ two rows of positive numbers. We set*

$$f = f_1 + \cdots + f_m,$$
$$f \cdot \langle u \rangle = f_1 u_1 + \cdots + f_m u_m,$$
$$\mathrm{Nm}\; u = u_1^{f_1} \cdots u_m^{f_m}.$$

If $u_\alpha \leqq v_\alpha$ ($\alpha = 1, \cdots, m$) and

$$\mathrm{Nm}\; u \geqq \mu \cdot \langle v \rangle^f \tag{37}$$

with some constant $\mu \geqq 1$, then

$$u_\alpha \geqq \lambda_\alpha \cdot \langle v \rangle$$

where λ_α depends on μ but not on the u and v.

(In our case r among the weights f_α are 1 and s of them equal 2.)

Proof. In the trivial case $m=1$ one determines λ by

$$\lambda^f = \mu. \tag{38}$$

If $m>1$ we set $u_1 = \lambda \cdot \langle v \rangle$ and assume $\lambda \leqq 1$. Then

$$\mathrm{Nm}\; u = \lambda^{f_1} \langle v \rangle^{f_1} \cdot u_2^{f_2} \cdots u_m^{f_m} \leqq \lambda^{f_1} \langle v \rangle^{f_1} \langle u \rangle_1^{f-f_1}.$$

Here $\langle u \rangle_1$ denotes the arithmetic mean of $u_2, \cdots, u_m$ formed with the weights $f_2, \cdots, f_m$ of sum $f - f_1$:

$$\begin{aligned}(f - f_1) \cdot \langle u \rangle_1 = f_2 u_2 + \cdots + f_m u_m &= f \cdot \langle u \rangle - f_1 u_1 \\ &= f \cdot \langle u \rangle - f_1 \lambda \langle v \rangle \leqq (f - f_1 \lambda) \langle v \rangle.\end{aligned}$$

Therefore

$$\mathrm{Nm}\; u \leqq \mu \cdot \langle v \rangle^f$$

where

(39) $$\mu = \lambda^{f_1}\left(\frac{f - f_1\lambda}{f - f_1}\right)^{f-f_1}.$$

As its logarithmic derivative shows,

(40) $$\frac{d\mu}{\mu} = \frac{f_1(1-\lambda)}{f - f_1\lambda}\cdot\frac{fd\lambda}{\lambda};$$

this function $\mu(\lambda)$ maps the interval $0 \leqq \lambda \leqq 1$ monotonically upon $0 \leqq \mu \leqq 1$ and thus will assume the given value μ ($\leqq 1$) for a certain $\lambda = \lambda_1$ ($\leqq 1$). Thus we cannot have $u_1 < \lambda_1 \cdot \langle v \rangle$ under the condition (37).

(We wish to obtain the best value for the constant λ_1. If one is content with a little less, one may choose λ_1 according to the equation

$$\lambda_1^{f_1}\cdot\left(\frac{f}{f - f_1}\right)^{f-f_1} = \mu,$$

or even, as

$$\left(1 + \frac{f_1}{f - f_1}\right)^{(f-f_1)/f_1} < \mathrm{e} \qquad (= \text{basis of natural logarithms}),$$

$$(\lambda_1 \mathrm{e})^{f_1} = \mu.$$

Incidentally, formula (40) holds good also for the function (38) which rules the trivial case $m = 1, f_1 = f$.)

In this way we ascertain constants λ_k, λ_k' such that

$$\text{each} \quad \overset{\alpha}{q}_k \geqq \lambda_k \cdot \langle \overset{*}{\gamma}_{kk} \rangle \quad \text{and each} \quad \overset{\beta}{q}_k \geqq \lambda_k' \cdot \langle \overset{*}{\gamma}_{kk} \rangle.$$

In case there is only one infinite prime spot, λ_k and λ_k' are determined by the relation

$$(\mu_k j_k)^2 = \lambda_k = \lambda_k'^2$$

in case of several spots by the equations

$$(\mu_k j_k)^2 = \lambda_k\left(\frac{f - \lambda_k}{f - 1}\right)^{f-1} = \lambda_k'^2\left(\frac{f - 2\lambda_k'}{f - 2}\right)^{f-2}$$

together with $\lambda_k \leqq 1$ and $\lambda_k' \leqq 1$.

Similarly for the other case studied in §6, that of a quaternion algebra $\mathfrak{J}$ with totally positive relative norm over a totally real field. The constants μ_k, λ_k in the inequalities

$$\mu_k j_k \leqq 1 \quad \text{and} \quad \overset{\alpha}{q}_k \geqq \lambda_k \cdot \langle \overset{*}{\gamma}_{kk} \rangle$$

are then given by

$$\mu_k = \pi_{4ke}\, \Delta^{-k}(e/4)^{2ke},$$

$$(\mu_k j_k)^{1/2} = \lambda_k \quad \text{or} \quad \lambda_k\left(\frac{e - \lambda_k}{e - 1}\right)^{e-1} \qquad (\lambda_k \leqq 1),$$

according as $e = \frac{1}{4}f$ is 1 or is greater than 1.

After this the proofs for the first and second theorems of finiteness roll along as before.

9. **The pattern of equivalent cells.** The Hermitian forms $\{\gamma_{ik}^*\}$ constitute a linear space of

$$N = f\cdot\tfrac{1}{2}n(n-1) + (r+s)n = f\cdot\tfrac{1}{2}n(n+1) - sn \qquad (\text{field } \mathfrak{J})$$

or

$$N = en(2n-1) \qquad (\text{quasi-field } \mathfrak{J})$$

dimensions, the positive ones a convex cone G in that space. G is an open set; we operate within G throughout. "Form" means any positive Hermitian form.

Let Λ be a lattice over I. A Λ-reduced form γ^* has been characterized by the inequalities

$$(32) \qquad f^2(\xi_1, \cdots, \xi_n) \geqq f^2(e_{1k}, \cdots, e_{nk})$$

holding for $f^2 = \operatorname{tr}(\gamma^*)$ whenever $(\xi_1, \cdots, \xi_n)$ is in Λ_k. For a given vector $(\xi_1, \cdots, \xi_n)$ the equality sign in (32) will hold identically for all Hermitian forms γ^* only if

$$(\xi_1, \cdots, \xi_n) = \epsilon\cdot(e_{1k}, \cdots, e_{nk}) \qquad (\epsilon \text{ a unit}),$$

as follows at once from the expression (31). For any other vector $(\xi_1, \cdots, \xi_n)$ the equation determines a $(N-1)$-dimensional hyperplane in our N-dimensional linear space of forms. This remark justifies our definition of "properly reduced" in terms of the group U of units.

The forms γ^* which are reduced with respect to Λ make up a convex cone G_Λ in G. The properly reduced forms are the inner points of G_Λ; see R1, p. 150. G_Λ may be empty; indeed it will be so unless the indices j_k of Λ satisfy the inequalities (35). Even if it is not empty it may be without inner points. Theorem 10 in R1, together with the first theorem of finiteness, proves:

THEOREM IV. *If G_Λ has inner points, then G_Λ is a convex pyramid defined within G by a limited number of linear inequalities.*

A linear mapping $\mathfrak{x} \to \mathfrak{x}'$ of $E^n/\mathfrak{J}$ upon itself is one satisfying the conditions $(\mathfrak{x}_1 + \mathfrak{x}_2)' = \mathfrak{x}_1' + \mathfrak{x}_2'$ and $(\delta\cdot\mathfrak{x})' = \delta\cdot\mathfrak{x}'$ for any number δ in $\mathfrak{J}$. We also require that $\mathfrak{x} = 0$ is the only vector whose image $\mathfrak{x}'$ equals 0. If $\mathfrak{d}_1, \cdots, \mathfrak{d}_n$ is any basis of $E^n/\mathfrak{J}$ the mapping S may be defined by giving the images $\mathfrak{d}_i' = \mathfrak{d}_i S$ of the $\mathfrak{d}_i$. The mapping S carries a form γ^* into a form γ_S^* according to the equation $\gamma_S^*(\mathfrak{x}S) = \gamma^*(\mathfrak{x})$. An order $[\mathfrak{J}]$ in $\mathfrak{J}$ and a lattice $\mathfrak{L}$ belonging to the order $[\mathfrak{J}]$ are supposed to be given. The linear mappings S which leave $\mathfrak{L}$ invariant are said to form the *modular group*[7]. In terms of a basis $\mathfrak{d}_1, \cdots, \mathfrak{d}_n$ of $E^n/\mathfrak{J}$ the lattice $\mathfrak{L}$ is represented by Λ:

$$\mathfrak{x} = \eta_1\mathfrak{d}_1 + \cdots + \eta_n\mathfrak{d}_n \quad \text{in} \quad \mathfrak{L} \cdot\rightleftarrows\cdot (\eta_1, \cdots, \eta_n) \quad \text{in} \quad \Lambda,$$

and a form $\gamma^*(\mathfrak{x})$ is represented by a form $\Gamma^*(\eta_1, \cdots, \eta_n)$:

$$\gamma^*(\eta_1\mathfrak{d}_1 + \cdots + \eta_n\mathfrak{d}_n) = \Gamma^*(\eta_1, \cdots, \eta_n). \tag{41}$$

The linear mapping defined by $\mathfrak{d}_i \rightarrow \mathfrak{d}_i'$ carries $\eta_1\mathfrak{d}_1 + \cdots + \eta_n\mathfrak{d}_n$ into $\eta_1\mathfrak{d}_1' + \cdots + \eta_n\mathfrak{d}_n'$; hence it leaves $\mathfrak{L}$ invariant and thus belongs to the modular group if and only if $\Lambda = \Lambda'$, Λ and Λ' being the representations of $\mathfrak{L}$ in terms of $\mathfrak{d}_i$ and $\mathfrak{d}_i'$. For a vector $\mathfrak{d}$ in $\mathfrak{L}$ there are not more than a finite number of units ϵ such that $\epsilon\mathfrak{d}$ also is in $\mathfrak{L}$. Indeed, for the splits of $\mathfrak{x} = \epsilon\mathfrak{d}$ one finds

$$|\xi^\alpha| = |\delta^\alpha|, \quad |\xi^\beta| = |\delta^\beta|, \quad \text{a fortiori} \quad |\xi^\alpha| \leqq |\delta^\alpha|, \quad |\xi^\beta| \leqq |\delta^\beta|,$$

which in view of the discrete nature of $\mathfrak{L}$ proves the point.

We want to divide G without gaps and overlappings into domains which are mutually equivalent under the modular group. We shall introduce these cells first as entities which have nothing to do with Hermitian forms, adopting a criterion of identity other than the set-theoretic one. The systematic place for this introduction would have been at the end of §3. Only afterwards shall we explain the meaning of the phrase "a form lies in a cell." Here are the definitions:

A semi-basis $\mathfrak{d}_1, \cdots, \mathfrak{d}_n$ of $\mathfrak{L}$ determines a cell $Z(\mathfrak{d}_1, \cdots, \mathfrak{d}_n)$; the semi-basis $\mathfrak{b}_1, \cdots, \mathfrak{b}_n$ is said to determine the same cell if

$$\mathfrak{b}_i = \epsilon_i\mathfrak{d}_i \qquad (\epsilon_i \text{ units}). \tag{42}$$

Let S be an operation of the modular group. The image Z_S of $Z = Z(\mathfrak{d}_1, \cdots, \mathfrak{d}_n)$ is defined as $Z(\mathfrak{d}_1', \cdots, \mathfrak{d}_n')$ where $\mathfrak{d}_i' = \mathfrak{d}_iS$. (Notice that if Z is written as $Z(\mathfrak{b}_1, \cdots, \mathfrak{b}_n)$, $\mathfrak{b}_i = \epsilon_i\mathfrak{d}_i$, then $Z(\mathfrak{b}_1', \cdots, \mathfrak{b}_n')$ is the same Z_S because $\mathfrak{b}_i' = \epsilon_i\mathfrak{d}_i'$; thus Z_S is independent of the fixation of the defining semi-basis $\mathfrak{d}_1, \cdots, \mathfrak{d}_n$.) Those S of the modular group for which $Z_S = Z$ shall be denoted by J_Z; they form a finite group $\{J_Z\}$. Indeed, for such an $S = J_Z$ one must have

$$\mathfrak{d}_i' = \mathfrak{d}_iS = \sigma_i\mathfrak{d}_i \qquad (\sigma_i \text{ a unit}), \tag{43}$$

and the J_Z are those mappings of the special form (43) which leave $\mathfrak{L}$ invariant. (In terms of another defining semi-basis $\mathfrak{b}_i = \epsilon_i\mathfrak{d}_i$ the same J_Z is expressed by $\mathfrak{b}_i' = \epsilon_i\sigma_i\epsilon_i^{-1}\cdot\mathfrak{b}_i$.) Any operation S of the modular group has the same effect upon Z as J_ZS.

---

(7) If one feels that this term ought to be reserved for the group which is fundamental in the theory of the modules of the theta functions then a new word, say "lattice group," is indicated for our purpose.

In terms of $(\mathfrak{d}_1, \cdots, \mathfrak{d}_n)$ the lattice $\mathfrak{L}$ is represented by an admissible lattice Λ, i.e., by a lattice Λ over I which is equivalent to $\mathfrak{L}$. Hence to the cell $Z=Z(\mathfrak{d}_1, \cdots, \mathfrak{d}_n)$ there corresponds a family of admissible lattices Λ, and the same family to each equivalent cell Z_S. We have a one-to-one correspondence between the classes of equivalent cells on the one hand, and the families of admissible lattices Λ on the other. We distinguish them by different colors. The operations J_Z are represented by the operations J_Λ in terms of the basis $(\mathfrak{d}_1, \cdots, \mathfrak{d}_n)$.

Now we come to the realization of cells as point sets in G. A form γ^* is said to lie in $Z=Z(\mathfrak{d}_1, \cdots, \mathfrak{d}_n)$ if $(\mathfrak{d}_1, \cdots, \mathfrak{d}_n)$ is reduced with respect to γ^*, i.e., if for $f^2=\mathrm{tr}(\gamma^*)$ one has $f^2(\mathfrak{x}) \geqq f^2(\mathfrak{d}_k)$ whenever $\mathfrak{x}$ is in $\mathfrak{L}$ and outside $[\mathfrak{d}_1, \cdots, \mathfrak{d}_{k-1}]$. Because (42) implies

$$f^2(\mathfrak{b}_k) = f^2(\mathfrak{d}_k), \qquad [\mathfrak{b}_1, \cdots, \mathfrak{b}_{k-1}] = [\mathfrak{d}_1, \cdots, \mathfrak{d}_{k-1}],$$

the definition is independent of the fixation of the defining semi-basis $\mathfrak{d}_1, \cdots, \mathfrak{d}_n$. If γ^* lies in $Z=Z(\mathfrak{d}_1, \cdots, \mathfrak{d}_n)$ then the transform Γ^* introduced by (41) lies in G_Λ, and γ_S^* lies in Z_S.

The fact that there always exists a reduced semi-basis for a given γ^* and the concluding sentence of §3 can now be stated thus:

(a) Every point γ^* lies in at least one cell Z.

(b) An inner point of a cell Z cannot lie in a cell Z' unless Z' is the same as Z (or briefly: different cells have no inner points in common).

The fact (a) will of course not be altered by suppressing all *empty* cells and their colors. Thus we have to look only for those admissible Λ whose indices satisfy the conditions (35); and this brings the colors down to a limited number. Will (a) still prevail after suppressing all cells without inner points and their colors? The answer is affirmative because there is no inner clustering of cells in G. This is a consequence of the second theorem of finiteness, which now takes on the following form. Let $\mathfrak{a}_1, \cdots, \mathfrak{a}_n$ be a semi-basis of $\mathfrak{L}$, $p \geqq 1$ and $w \geqq 0$. The form γ^* is said to lie in $Z(\mathfrak{a}_1, \cdots, \mathfrak{a}_n | p, w)$ if

$$f^2(\mathfrak{x}) \geqq \frac{1}{p} \cdot f^2(\mathfrak{a}_k)$$

whenever $\mathfrak{x}$ is in $\mathfrak{L}$ and outside $[\mathfrak{a}_1, \cdots, \mathfrak{a}_{k-1}]$, and if, moreover,

$$f^2(\mathfrak{a}_k - \mathfrak{x}_h) \geqq f^2(\mathfrak{a}_k) - w \cdot f^2(\mathfrak{a}_h)$$

whenever $h<k$ and $\mathfrak{x}_h$ is in $\mathfrak{L}$ and $[\mathfrak{a}_1, \cdots, \mathfrak{a}_h]$.

THEOREM V. *There is only a finite number of operations S of the modular group such that the image Z_S into which a given cell Z is thrown by S will have points in common with the domain $Z(\mathfrak{a}_1, \cdots, \mathfrak{a}_n | p, w)$.*

Application to $p=1$, $w=0$ proves in particular that a cell borders on not more than a finite number of other cells. And since $Z(\mathfrak{a}_1, \cdots, \mathfrak{a}_n | p, w)$ sweeps over the whole G if p and w increase to infinity we are sure that the cells cluster around no point in the interior of G ("the modular group is properly discontinuous in G"). We therefore definitely admit only those colors whose cells are N-dimensional solids, i.e., have inner points. In our summary we talk of them as point sets in G.

THEOREM VI. (a) *G is divided into a pattern of cells, each cell bearing a color out of a finite palette of colors. The cells cover G without gaps and overlappings. Each cell is a solid convex pyramid (in G). The mappings of the modular group leave this design, including its coloring, invariant. Any two cells of the same color can be carried one into the other by an operation of the modular group.*

(b) *Given a point in G and a cell Z one can assign a neighborhood $\mathfrak{N}$ to the point such that there is only a limited number of operations S of the modular group for which the image Z_S penetrates into $\mathfrak{N}$.*

(c) *The operations of the modular group which carry a cell into itself form a finite subgroup. This group of linear operations in the vector space $E^n/\mathfrak{J}$ is equivalent (in $\mathfrak{J}$) to a group whose elements are of the special form*

$$\xi_1 \to \xi_1\epsilon_1, \cdots, \xi_n \to \xi_n\epsilon_n \qquad (\epsilon_i \text{ units}).$$

(Of course, in view of statement (c) the statement (b) could have been replaced by the simpler one that only a finite number of cells penetrate into $\mathfrak{N}$.)

We form a *nucleus* by selecting one cell Z_c of each color c. All cells adjacent to the nuclear cells form a *wreath* around the nucleus. Here the word "adjacent" may be interpreted either in the wide sense of "having a point in common," or in the narrower sense of "having a wall of $N-1$ dimensions in common."

THEOREM VII. *Determine for each cell Z_c' of color c in the wreath an operation S_c' of the modular group which maps the nuclear cell Z_c of color c into Z_c'. The S_c' thus selected, together with the operations J_c of the modular group which carry Z_c into itself, generate the whole group if all colors c are taken into account.*

Were it not for the groups $\{J_\Lambda\}$ the nucleus would form a fundamental domain. As it is, one has first to replace in our construction each G_Λ by a part G_Λ' which in G_Λ is a fundamental domain for the finite group of special transformations

$$J_\Lambda: \quad \xi_k \to \xi_k\epsilon_k \qquad (\epsilon_k \text{ a unit})$$

carrying G_Λ into itself. The effect of J_Λ upon the coefficient γ_{ik}^* is described by

$$\overset{*}{\gamma}_{ik} \to \epsilon_i \overset{*}{\gamma}_{ik} \bar{\epsilon}_k.$$

If in one split α the transformation of the variable $\gamma_{ik}^\alpha = \Xi$,

$$\Xi \to \overset{\alpha}{\epsilon}_i \,\Xi\, \overset{\alpha}{\bar{\epsilon}}_k = \overset{\alpha}{\epsilon}_i \,\Xi (\overset{\alpha}{\epsilon}_k)^{-1},$$

is the identity, then the same is true of every split. Hence it is sufficient to consider one split α only, and after choosing it we write simply $\gamma^{\alpha}_{ik}=\gamma_{ik}$, $\epsilon^{\alpha}_{i}=\epsilon_{i}$. If the transformation $\Xi\rightarrow\epsilon_1\Xi\epsilon_2^{-1}$ is the identity, one must have $\epsilon_1=\epsilon_2$ as the specialization $\Xi=1$ shows, and $\Xi\rightarrow\epsilon_2\Xi\epsilon_1^{-1}$ is also the identity. Moreover, if $\Xi\rightarrow\epsilon_1\Xi\epsilon_2^{-1}$ and $\Xi\rightarrow\epsilon_2\Xi\epsilon_3^{-1}$ are identities, then $\Xi\rightarrow\epsilon_1\Xi\epsilon_3^{-1}$ is. Consequently we may well limit ourselves first to the coefficients γ_{ik} $(i<k)$ on one side of the diagonal, and then more particularly to

$$\Xi_1 = \gamma_{12},\ \Xi_2 = \gamma_{23},\ \cdots,\ \Xi_{n-1} = \gamma_{n-1,n}.$$

Let us at once consider the most disagreeable case, that of a quaternion quasi-field $\mathcal{J}$ as described in §6.

The group $\{J_A\}$ induces a group of transformations of the type

$$\Xi \rightarrow \epsilon_1 \Xi \bar{\epsilon}_2 \qquad (|\epsilon_1| = |\epsilon_2| = 1)$$

for $\gamma_{12}=\Xi_1=\Xi$. This is a finite group of orthogonal transformations J in the space of the four components X_0, X_1, X_2, X_3 of the variable quaternion Ξ. Denote by ΞJ the transform of Ξ by J. The simplest way of ascertaining a fundamental domain for this group $\{J\}$ is as follows: One chooses a point $\Xi=\mathrm{A}$ which differs from all its transforms $\mathrm{A}J$ $(J\neq\text{identity})$. The fundamental domain consists of all points Ξ which are nearer to the center A than to the other equivalent centers $\mathrm{A}J$ and is thus characterized by the inequalities

$$\Xi\cdot\overline{\mathrm{A} - \mathrm{A}J} + (\mathrm{A} - \mathrm{A}J)\cdot\bar{\Xi} \geqq 0.$$

Fortunately these are linear inequalities, namely of the form

$$a_0X_0 + a_1X_1 + a_2X_2 + a_3X_3 \geqq 0$$

(a_0, a_1, a_2, a_3 being the components of $\mathrm{A}-\mathrm{A}J$). After having done this we limit ourselves to those operations J_A which leave Ξ_1 unchanged. They form a subgroup and we study its influence upon $\Xi_2, \cdots, \Xi_{n-1}$. The next step would consist in singling out Ξ_2. By induction we thus obtain a finite number of subsidiary linear inequalities each concerned with the four components of one of the variables $\gamma_{12}, \cdots, \gamma_{n-1,n}$ only, and by them we define the fundamental domain G_A' in G_A for the group $\{J_A\}$.

I set little store by this whittling down of G_A to G_A'. It seems less artificial to operate with the whole cells Z; in doing so one has to keep in mind that the modular group in its influence upon Z matters only modulo $\{J_Z\}$.

127.

On geometry of numbers

Proceedings of the London Mathematical Society 47, 268—289 (1942)

1. Minkowski stated his fundamental theorem concerning convex solids in relationship to a lattice as follows. Let the lattice in the n-dimensional space of vectors $\mathfrak{x} = (x_1, \ldots, x_n)$ consist of all vectors $\mathfrak{a}$ with integral components, and let $f(\mathfrak{x})$ be any "gauge function", *i.e.* any continuous function in vector space enjoying the following properties:

$$f(\mathfrak{x}) > 0 \text{ except for the origin } \mathfrak{x} = \mathfrak{o} = (0, \ldots, 0);$$

$$f(t\mathfrak{x}) = |t| . f(\mathfrak{x}) \text{ for any real factor } t;$$

$$f(\mathfrak{x} + \mathfrak{x}') \leqslant f(\mathfrak{x}) + f(\mathfrak{x}').$$

Then $f(\mathfrak{x}) < 1$ defines an (open) convex solid $\mathfrak{K}$ with $\mathfrak{o}$ as its centre. V being the volume of $\mathfrak{K}$ and $2M$ the minimum of $f(\mathfrak{a})$ for all lattice vectors $\mathfrak{a} \neq \mathfrak{o}$, Minkowski's universal inequality is

$$M^n V \leqslant 1. \tag{1.1}$$

If $\mathfrak{S}$ is any set in vector space, we can place it round a given point a of the n-dimensional affine point space; the ensuing point set $\mathfrak{S}_a$ consists of all points x such that the vector $\mathfrak{x} = \overrightarrow{ax}$ is in $\mathfrak{S}$. Minkowski's argument is based on the simple remark that the convex solid

$$\mathfrak{K}(q): \; f(\mathfrak{x}) < q,$$

homothetic with $\mathfrak{K}$, when placed round each of the lattice points a, causes no overlapping so long as $q \leqslant M$, and hence up to that limit its volume $q^n V$ stays less than or equal to 1. With q growing beyond M, the solids of the lattice begin to overlap. Studying the portion of the space covered by them, Minkowski arrives at the sharper inequality†

$$M_1 \dots M_n V \leqslant 1, \tag{1.2}$$

where the numbers M_k may be described thus: A bubble of shape $\mathfrak{K}$ while being blown up describes the series of figures $\mathfrak{K}(q)$ with increasing q; $q = 2M = 2M_1$ is the first moment when a lattice point $\mathfrak{d}_1 \neq \mathfrak{o}$ enters into the bubble, $2M_k$ that stage at which for the first time a lattice point $\mathfrak{d}_k$ enters off the $(k-1)$-dimensional linear manifold $[\mathfrak{d}_1, \dots, \mathfrak{d}_{k-1}]$ spanned by $\mathfrak{d}_1, \dots, \mathfrak{d}_{k-1}$. Clearly $M_1 \leqslant M_2 \leqslant \dots \leqslant M_n$. In other words, $\mathfrak{a} = \mathfrak{d}_k$ is that one among the lattice vectors $\mathfrak{a}$ outside the manifold $[\mathfrak{d}_1, \dots, \mathfrak{d}_{k-1}]$ for which $f(\mathfrak{a})$ assumes the least possible value $2M_k$. So long as $q \leqslant 2M_k$ the body $\mathfrak{K}(q)$ contains less than k linearly independent lattice vectors, while for $q > 2M_k$ their number is at least k.

The vectors

$$\pm \mathfrak{d}_1/2M_1, \quad \pm \mathfrak{d}_2/2M_2, \quad \dots, \quad \pm \mathfrak{d}_n/2M_n,$$

and hence their convex closure, are contained in $\mathfrak{K}$. If $P = |\mathfrak{d}_1 \dots \mathfrak{d}_n|$ is the volume of the parallelotope spanned by the lattice vectors $\mathfrak{d}_1, \dots, \mathfrak{d}_n$, the volume of that closure is

$$\frac{2^n}{n!} \frac{P}{2M_1 \dots 2M_2} = \frac{P}{n!} \frac{1}{M_1 \dots M_n}.$$

Therefore
$$\frac{P}{n!} \leqslant M_1 \dots M_n V \leqslant 1,$$

and thus the positive integer $P \leqslant n!$.

This note consists of two parts, the first stating Minkowski's problem in terms of characteristic functions ϕ which are capable of the two "truth values" 0 and 1 only, and then generalizing it to probability functions that move in the range $0 \leqslant \phi \leqslant 1$. Overlapping is not excluded, and is taken care of by using Boolean rather than ordinary sums. The substance of the argument remains unaltered, yet the essential points come out more clearly.

If P were equal to 1, the inequality (1.2) would tell an important fact about reduction under arithmetic equivalence, namely about normalization of the basis of our lattice in terms of a given convex solid $\mathfrak{K}$. As it stands, it points in a slightly different direction. However, by an almost trivial twist, one can make it tally with the problem of reduction. This is done in the second part, which concludes with some remarks about the special case of the reduction of quadratic forms.

---

† *Geometrie der Zahlen* (Leipzig, 1896, 1910), 211–218.

[*Added* January, 1940. Prof. L. J. Mordell, to whom I sent a copy of the MS., kindly pointed out to me two papers which I had overlooked; one by H. Davenport, *Quart. J. of Math.*, 10 (1939), 119–121, bears closely on Part I by containing another proof for the inequality (1.2), while the other by K. Mahler, *Quart. J. of Math.*, 9 (1938), 259–262, actually anticipates the main result of Part II, namely Theorem V. I am still surprised that the discovery of this fact in 1938 by Dr. Mahler, and of the simple remark on which it is based, had to wait for forty-two years after Minkowski established his inequality (1.2).]

I.

2. A set in the affine space of points $x = (x_1, \ldots, x_n)$ may be described by its characteristic function $\phi(x)$ which is equal to 1 inside the set and to 0 outside the set. The characteristic function ϕ for the union of two sets with the characteristic functions ϕ_1, ϕ_2 is their "Boolean" sum

$$\phi = \phi_1 + \phi_2 - \phi_1 \phi_2,$$

which reduces to the ordinary sum only if the sets do not overlap. Any number a in the interval $0 \leqslant a \leqslant 1$ is said to be a *probability*. If a and b are probabilities of two statistically independent events A and B, then their Boolean sum

$$c = a \vee b = a + b - ab = 1 - (1-a)(1-b),$$

which again satisfies the inequality $0 \leqslant c \leqslant 1$, is the probability of the event "A or B". The Boolean sum is commutative and associative. $a' \leqslant a$ implies $(a' \vee b) \leqslant (a \vee b)$, as the expression $a \vee b = a(1-b) + b$ at once shows; and consequently

$$a' \leqslant a, \quad b' \leqslant b \quad \text{imply} \quad (a' \vee b') \leqslant (a \vee b). \tag{2.1}$$

The Boolean sum of a finite number of probabilities $a_1, \ldots, a_h$ is

$$\underset{i}{\mathrm{S}}\, a_i = a_1 \vee a_2 \vee \ldots \vee a_h = 1 - \prod_{i=1}^{h} (1 - a_i)$$

$$= \sum_{(i)} a_i - \sum_{(i,k)} a_i a_k + \sum_{(i,k,l)} a_i a_k a_l - \ldots$$

$$= \frac{1}{1!} \sum_{i} a_i - \frac{1}{2!} \sum_{i,k} a_i a_k + \frac{1}{3!} \sum_{i,k,l} a_i a_k a_l - \ldots.$$

Summation extends to all throws of 1 or 2 or 3 or ... different indices from the set 1, ..., h, in the second line without regard to their order, in the third line with regard to their order. Brackets as in (i, k, l) indicate the first kind of summation.

A function $\phi(x)$ in our space which satisfies the condition $0 \leqslant \phi \leqslant 1$ is called a probability function.

Let $\phi(x)$ now be an (integrable) probability function vanishing outside a finite region $\mathfrak{H}$ of the x-space which may be fixed as a parallelotope:

$$A_i < x_i < B_i \quad (i = 1, \ldots, n).$$

The lattice translation $\mathfrak{a} = (a_1, \ldots, a_n)$ with integral components a_i,

$$x_i' = x_i + a_i,$$

carries $\phi(x)$ into a function

$$\phi(x, \mathfrak{a}) = \phi(x_1 - a_1, \ldots, x_n - a_n).$$

In these circumstances, the Boolean sum

$$\psi(x) = \underset{\mathfrak{a}}{\mathrm{S}}\, \phi(x, \mathfrak{a})$$

extending to all lattice translations $\mathfrak{a}$ has a meaning, since for each x only a limited number of its terms do not disappear. $\psi(x)$ is a probability function with period 1 in each of the variables $x_1, \ldots, x_n$. We form the integral

$$J[\phi] = \int_E \psi(x)\, dx \quad [dx = dx_1\, dx_2 \ldots dx_n]$$

extending over the fundamental mesh of the lattice

$$E: \quad 0 < x_1 < 1, \ldots, 0 < x_n < 1.$$

Obviously $0 \leqslant \psi(x) \leqslant 1$ entails

(2.2) $$0 \leqslant J[\phi] \leqslant 1.$$

Moreover, for two probability functions ϕ_2 and $\phi_1 \leqslant \phi_2$ we have $\psi_1 \leqslant \psi_2$, because of the law of monotony (2.1) for Boolean sums, and hence

(2.3) $$J[\phi_1] \leqslant J[\phi_2].$$

The explicit expression for ψ is

$$\psi(x) = 1 - \prod_{\mathfrak{a}} \{1 - \phi(x, \mathfrak{a})\}$$

$$= \frac{1}{1!} \sum_{\mathfrak{a}} \phi(x, \mathfrak{a}) - \frac{1}{2!} \sum_{\mathfrak{a}, \mathfrak{b}} \phi(x, \mathfrak{a})\, \phi(x, \mathfrak{b}) + \frac{1}{3!} \sum_{\mathfrak{a}, \mathfrak{b}, \mathfrak{c}} \phi(x, \mathfrak{a})\, \phi(x, \mathfrak{b})\, \phi(x, \mathfrak{c}) - \ldots$$

[$\mathfrak{a}, \mathfrak{b}, \mathfrak{c}, \ldots$ lattice vectors].

In substituting $\mathfrak{a}+\mathfrak{b}, \mathfrak{a}+\mathfrak{c}, \ldots$ for $\mathfrak{b}, \mathfrak{c}, \ldots$ we see that $\psi(x)$ may be written as an ordinary sum

$$\psi(x)=\sum_{\mathfrak{a}} \phi^*(x, \mathfrak{a}),$$

where

$$\phi^*(x)=\phi(x)\left\{\frac{1}{1!}-\frac{1}{2!} \sum_{\mathfrak{b}} \phi(x, \mathfrak{b})+\frac{1}{3!} \sum_{\mathfrak{b}, \mathfrak{c}} \phi(x, \mathfrak{b}) \phi(x, \mathfrak{c})-\ldots\right\}$$

[$\mathfrak{b}, \mathfrak{c}, \ldots$, lattice vectors $\neq \mathfrak{o}$].

This trick works because the lattice translations form a *group*. When we return to "combinations" without regard to order, the second factor on the right in the last formula is

$$\phi^{**}(x)=1-\tfrac{1}{2} \sum_{(\mathfrak{a})} \phi(x, \mathfrak{a})+\tfrac{1}{3} \sum_{(\mathfrak{a}, \mathfrak{b})} \phi(x, \mathfrak{a}) \phi(x, \mathfrak{b})-\ldots. \tag{2.4}$$

[$\mathfrak{a}, \mathfrak{b}, \ldots$ lattice vectors $\neq \mathfrak{o}$].

Take a real parameter t ranging from 0 to 1, and introduce the product

$$\prod_{\mathfrak{a} \neq \mathfrak{o}}\{1-t \phi(x, \mathfrak{a})\}=1-t \sum_{(\mathfrak{a})} \phi(x, \mathfrak{a})+t^2 \sum_{(\mathfrak{a}, \mathfrak{b})} \phi(x, \mathfrak{a}) \phi(x, \mathfrak{b})-\ldots.$$

Its integration from $t=0$ to 1 leads to (2.4). Hence our final definitions are:

$$\phi^*(x)=\phi(x) \cdot \phi^{**}(x), \quad \phi^{**}(x)=\int_0^1 \prod_{\mathfrak{a} \neq \mathfrak{o}}\{1-t \phi(x, \mathfrak{a})\} d t ; \tag{2.5}$$

$$J[\phi]=\int_{\infty} \phi^*(x) d x. \tag{2.6}$$

∞ indicates that the integral extends over the whole x-space. Notice that $0<\phi^{**}(x) \leqslant 1$ and $\phi_1^{**} \geqslant \phi_2^{**}$ if $\phi_1 \leqslant \phi_2$. (2.6) proves $J[\phi]$ to be invariant under translation, namely

$$J[\phi']=J[\phi] \quad \text{for} \quad \phi'(x)=\phi(x-x^0). \tag{2.7}$$

If $0 \leqslant \phi_1 \leqslant \phi$ and $\phi-\phi_1 \leqslant \epsilon$, we find that

$$\phi^*(x)=\phi(x) \cdot \phi^{**}(x) \leqslant \phi(x) \cdot \phi_1^{**}(x)$$

and thus, from (2.6), derive the result

$$J[\phi]-J[\phi_1] \leqslant \int_{\mathfrak{H}}\{\phi(x)-\phi_1(x)\} \phi_1^{**}(x) d x \leqslant H \epsilon, \tag{2.8}$$

where H is the volume of $\mathfrak{H}$. We summarize:

THEOREM I. *The integral* (2.6) *is less than or equal to* 1 *for any probability function* ϕ *which vanishes outside a finite region.* $J[\phi]$ *is invariant under translation, and*

$$J[\phi_1] \leqslant J[\phi_2] \quad \textit{if} \quad \phi_1(x) \leqslant \phi_2(x).$$

Specilization of the theorem for the characteristic function ϕ of an (open) set $\mathfrak{S}$ of definite Jordan volume V yields the simple Minkowski-Blichfeldt-Scherrer theorem: If $\mathfrak{S}_0$ has no point in common with the sets $\mathfrak{S}_a$ placed round the other lattice points $a \neq 0$, or, what is the same thing: if $\mathfrak{S}_0$ never contains both end points x and $x+a$ of a lattice vector $a \neq 0$, then its volume is less than or equal to 1. I consider Theorem I to be the natural general form of this principle, and our proof is nothing but a neat analytical form of the simplest argument by which that principle has been proved†.

[Our affine x-space could be replaced by any differentiable manifold U of points x which carries an infinitesimal volume measure, and the group of lattice translations by any group of differentiable volume-preserving one-to-one transformations $x \to x^s$ of U which

(1) is discrete in the strictest sense, so that the set of points x^s, equivalent to any given point x, never has a condensation point,

(2) has a compact fundamental domain; *i.e.*, the manifold, arising from U by identification of equivalent points, is supposed to be compact.]

3. With Siegel‡ we introduce the Fourier coefficients of the periodic function $\psi(x)$:

$$J_{\mathfrak{l}} = \int_E \psi(x) \,.\, e(l_1 x_1 + \ldots + l_n x_n)\, dx$$

corresponding to the various lattice-vectors $\mathfrak{l} = (l_1, \ldots, l_n)$ in the dual space, with $e(x)$ as an abbreviation for $e^{2\pi i x}$, and formulate the Parseval equation

$$\int_E \big(\psi(x)\big)^2 dx = \sum_{\mathfrak{l}} |J_{\mathfrak{l}}|^2.$$

The zero coefficient J_0 is our $J = J[\phi]$. Making use of the equation

$$\int_E \psi(x) \,.\, \omega(x)\, dx = \int_\infty \phi^*(x) \,.\, \omega(x)\, dx$$

† G. Hajós, *Act. Litt. Sci., Szeged*, 6 (1934), 224–225. H. F. Blichfeldt, *Trans. American Math. Soc.*, 15 (1914), 227–235. W. Scherrer, *Math. Annalen*, 86 (1922), 99–107.

‡ *Acta Math.*, 65 (1935), 307–323.

which holds for any periodic function $\omega(x)$, we get

$$J_{\mathfrak{l}} = \int_\infty \phi^*(x) \, . \, e(l_1 x_1 + \ldots + l_n x_n)\, dx \tag{3.1}$$

and

$$\int_E \big(\psi(x)\big)^2 dx = \int_\infty \phi^*(x) \, . \, \psi(x)\, dx.$$

We find that

$$\phi^* \psi = \phi^* \Big\{1 - \prod_{\mathfrak{a}} \Big(1 - \phi(x, \mathfrak{a})\Big)\Big\}$$

$$= \phi^* - \phi^*(1-\phi) \, . \prod_{\mathfrak{a} \neq \mathfrak{o}} \{1 - \phi(x, \mathfrak{a})\}$$

and thus

$$\phi^* \psi = \phi^* - \phi(1-\phi)\, \phi_* \, \phi^{**} \quad \text{with}$$
$$(3.2) \qquad \phi_*(x) = \prod_{\mathfrak{a} \neq \mathfrak{o}} \{1 - \phi(x, \mathfrak{a})\}.$$

THEOREM II. *If* ϕ_* *and* ϕ^{**} *are defined by* (3.2), (2.4) *and* $J_{\mathfrak{l}}$ *denotes the integral* (3.1), *then the following equation holds for* $J = J[\phi]$:

$$J(1-J) = \int_\infty \phi(1-\phi) \, . \, \phi_* \, \phi^{**} \, dx + \sum_{\mathfrak{l} \neq \mathfrak{o}} |J_{\mathfrak{l}}|^2.$$

It shows very clearly how $0 \leqslant \phi \leqslant 1$ leads to $0 \leqslant J \leqslant 1$.

[This argument works on any differentiable manifold U on which we have a transitive group Γ of differentiable one-to-one transformations, the lattice consisting of a discrete subgroup of Γ whose fundamental domain is compact. The harmonics replacing $e(l_1 x_1 + \ldots + l_n x_n)$ are obtained by a slight modification of the process described by the author, *Annals of Math.*, 35 (1934), 486–499.]

4. We now introduce an assumption of *convexity* by supposing $\{\phi(x) \geqslant r\}_x$, *i.e.* the locus $\mathfrak{F}_r$ of the points x where $\phi(x) \geqslant r$, to be a convex set for any $r > 0$. The relations $\phi(x) \geqslant r$ and $\phi(x') \geqslant r$ then imply

$$\phi\Big(tx + (1-t)\, x'\Big) \geqslant r \tag{4.1}$$

for any number t between 0 and 1. *A function* ϕ *having this property in our n-dimensional space preserves it on any linear subspace.* The "terrace" $\{r \leqslant \phi(x) < r'\}_x$ between the levels r and $r' > r$ has a Jordan volume because it is the difference between the two convex bodies $\mathfrak{F}_{r'}$ and $\mathfrak{F}_r$. Let

$$0 = r_0 < r_1 < r_2 < \ldots < r_h = 1 \tag{4.2}$$

be any division of the interval $0 \leqslant r \leqslant 1$ into subintervals of lengths $r_{i+1} - r_i \leqslant \epsilon$, and let $\phi_\Delta(x)$, $\phi^\Delta(x)$ be the corresponding step functions which approximate ϕ from below and above with an error less than or equal to ϵ. They vanish outside $\mathfrak{H}$ and are defined inside $\mathfrak{H}$ by

$$\phi_\Delta(x) = r_i \text{ and } \phi^\Delta(x) = r_{i+1} \text{ if } r_i \leqslant \phi(x) < r_{i+1}.$$

The Riemann integrals of ϕ_Δ and ϕ^Δ clearly differ by not more than $H\epsilon$; this proves the Riemann integrability of ϕ. Moreover, by (2.3) and (2.8),

$$J[\phi_\Delta] \leqslant J[\phi] \leqslant J[\phi^\Delta] \quad\text{and}\quad J[\phi^\Delta] - J[\phi_\Delta] \leqslant H\epsilon,$$

and these inequalities could also serve to define $J[\phi]$ by mean of the Riemann integrals $J[\phi_\Delta]$ and $J[\phi^\Delta]$ of the approximate terraced functions.

We repeat our assumptions (A) about ϕ:

(i) $0 \leqslant \phi(x) \leqslant 1$;

(ii) $\phi(x) = 0$ outside a finite region $\mathfrak{H}$;

(iii) the sets $\mathfrak{F}_r$: $\{\phi(x) \geqslant r\}_x$ are convex.

Minkowski's second step deals with the process of dilation as defined by

$$\phi_q(x_1, \ldots, x_n) = \phi\left(\frac{x_1}{q}, \ldots, \frac{x_n}{q}\right) \tag{4.3}$$

and, properly generalized, leads to

Theorem III. *Under the assumptions* (A),

$$J[\phi_q] \geqslant J[\phi] \textit{ for } q \geqslant 1. \tag{4.4}$$

Proof. Suppose that we have a point x^0 where $\phi(x)$ takes on its maximum value. If x is any point and $\phi(x) = r$, we then have $\phi(x^0) \geqslant r$, and, by (4.1),

$$\phi\left(\frac{1}{q}x + \left(1 - \frac{1}{q}\right)x^0\right) \geqslant r = \phi(x) \text{ for any } q \geqslant 1. \tag{4.5}$$

For the moment denote the function on the left-hand side by $\phi_q'(x)$. Formula (4.5) entails

$$J[\phi_q'] \geqslant J[\phi],$$

and thus, since ϕ_q' arises from ϕ_q by a translation,

$$J[\phi_q] = J[\phi_q'] \geqslant J[\phi].$$

Since we have neither assumed $\phi(x)$ to be continuous nor the sets $\mathfrak{F}_r$ to be closed, the existence of a maximum is doubtful. But the following slight modification will do. Let r_i in the finite sequence (4.2) be the highest level for which $\mathfrak{F}_{r_i}$ is not empty. We cut off the top of the mountain at the altitude r_i, *i.e.* we form

$$\hat{\phi}(x) = \min\bigl(r_i, \phi(x)\bigr)$$

and choose x^0 as a point in $\mathfrak{F}_{r_i}$. Then we get

$$\phi_q'(x) \geqslant \hat{\phi}(x) \quad (q \geqslant 1)$$

instead of (4.5), and hence

$$J[\phi_q] \geqslant J[\hat{\phi}].$$

However, according to (2.8)

$$J[\hat{\phi}] \geqslant J[\phi] - H\epsilon,$$

and by driving ϵ towards 0 the inequality (4.4) is re-established. (Instead of the decapitated mountain, we might have used the terraced mountain ϕ_Δ.)

Since
$$(\phi_{q'})_q = \phi_{q'q},$$

we may claim Theorem III to assert that

$$J[\phi_q] = J(q)$$

is a monotonically increasing function of the parameter q of dilation.

5. Our next move is to modify the procedure by taking into consideration only the lattice vectors

$$(5.1) \qquad (a_1, \ldots, a_k, 0, \ldots, 0) \quad (a_i \text{ integers}, \ k \geqslant 1)$$

in the linear subspace

$$x_{k+1} = \ldots = x_n = 0.$$

$\phi^* = \phi^{(k)}$ is now defined by the last equation (2.5) with the product ranging over the vectors $\mathfrak{a} \neq \mathfrak{o}$ of this k-dimensional lattice, and

$$J^{(k)}[\phi] = \int_\infty \phi^{(k)}(x)\, dx.$$

The integration is then conveniently carried out in two steps, first with respect to $x_1, \ldots, x_k$ for any constant $x_{k+1}, \ldots, x_n$, and secondly with respect to $x_{k+1}, \ldots, x_n$. Thus we find that

$$\phi_1 \leqslant \phi_2 \quad \text{implies} \quad J^{(k)}[\phi_1] \leqslant J^{(k)}[\phi_2],$$

while the inequality

$$J^{(k)}[\phi] \leqslant J^{(k)}[\phi_1] + H\epsilon,$$

holding under the condition $0 \leqslant \phi - \phi_1 \leqslant \epsilon$, follows by the same argument as (2.8), without a break in integration.

Like ϕ itself, the function

$$\Phi(x_1 \ldots x_k) = \phi(x_1 \ldots x_k, x_{k+1}^0 \ldots x_n^0)$$

of k variables satisfies the assumptions (A), and therefore

$$J[\Phi_q] \geqslant J[\Phi] \quad \text{for} \quad q \geqslant 1. \tag{5.2}$$

Notice that $\Phi_q(x_1 \ldots x_k) = \bar{\phi}_q(x_1 \ldots x_k, x_{k+1}^0 \ldots x_n^0)$

with $\bar{\phi}_q(x) = \phi\left(\dfrac{x_1}{q}, \ldots, \dfrac{x_k}{q}, x_{k+1}, \ldots, x_n\right).$

Thus, by integrating (5.2) with respect to $x_{k+1}, \ldots, x_n$, we get

$$J^{(k)}[\bar{\phi}_q] \geqslant J^{(k)}[\phi].$$

The substitution

$$\frac{x_{k+1}}{q} = x'_{k+1}, \quad \ldots, \quad \frac{x_n}{q} = x'_n \tag{5.3}$$

performed on the last $n-k$ variables results in the equation

$$J^{(k)}[\phi_q] = q^{n-k} \,.\, J^{(k)}[\bar{\phi}_q].$$

Hence $J^{(k)}[\phi_q] \geqslant q^{n-k} \,.\, J^{(k)}[\phi] \quad \text{for} \quad q \geqslant 1.$

For our Riemann integrals the performance of the integration in two steps $x_1 \ldots x_k | x_{k+1} \ldots x_n$ and of the substitution (5.3) offers no difficulty. But maybe it is preferable first to argue for the step function ϕ_Δ, thus avoiding the difficulty about the maximum of ϕ mentioned above, and then to combine the ensuing inequality

$$J^{(k)}[\phi_q] \geqslant q^{n-k} \,.\, J^{(k)}[\phi_\Delta]$$

with $J^{(k)}[\phi_\Delta] \geqslant J^{(k)}[\phi] - H\epsilon.$

For $k=0$ we define $\phi^{(0)}=\phi$ and therefore

$$J^{(0)}[\phi_q]=q^n.\int_\infty \phi(x)\,dx.$$

Any k-dimensional discrete lattice $\mathfrak{L}_k$ may serve here instead of (5.1), since the first k coordinate vectors $\mathfrak{e}_1, \ldots, \mathfrak{e}_k$ may be so chosen as to span the pre-assigned lattice (that is, so that the lattice consists of the vectors $a_1\mathfrak{e}_1+\ldots+a_k\mathfrak{e}_k$ with integral components a_i).

THEOREM IV. *By means of a k-dimensional lattice* $\mathfrak{L}_k$ *we define*

$$\phi^{(k)}(x)=\phi(x).\int_0^1 \prod_{\mathfrak{a}}\{1-t\phi(x,\mathfrak{a})\}\,dt,$$

the product extending to all vectors $\mathfrak{a}\neq 0$ *of the lattice* $\mathfrak{L}_k$, *and*

$$J^{(k)}[\phi]=\int_\infty \phi^{(k)}(x)\,dx, \qquad J_k(q)=J^{(k)}[\phi_q].$$

Then $J_k(q)/q^{n-k}$ *is an increasing function of* q. *In particular,*

$$J_0(q)/q^n=\int_\infty \phi(x)\,dx.$$

The application to the situation considered by Minkowski is immediate. ϕ is now the characteristic function of the convex solid $\mathfrak{K}$: $f(x)<1$. Let $\mathfrak{L}_k$ consist of the lattice vectors in $[\mathfrak{b}_1, \ldots, \mathfrak{b}_k]$. So long as $q\leqslant M_{k+1}$, we have

$$\phi_q^{(k)}=\phi_q{}^*, \quad J_k(q)=J(q).$$

Indeed, the solids $\mathfrak{K}(q)$ and $\mathfrak{K}_a(q)$ do not overlap if a is a lattice point outside $[\mathfrak{g}_1, \ldots, \mathfrak{g}_k]$ and $q\leqslant M_{k+1}$. We thus find that

$$J(q)=q^n.V \quad \text{for} \quad q\leqslant M_1,$$

in particular

$$J(M_1)=M_1{}^n.V.$$

Then

$$J(q)\geqslant\left(\frac{q}{M_1}\right)^{n-1}.J(M_1)=M_1q^{n-1}.V \quad \text{for} \quad M_1\leqslant q\leqslant M_2,$$

and hence

$$J(M_2)\geqslant M_1M_2^{n-1}.V.$$

Continuing in the same manner, we get

$$J(q)\geqslant M_1\ldots M_{k-1}q^{n-k+1}.V \quad \text{for} \quad M_{k-1}\leqslant q\leqslant M_k$$
$$(k=1, \ldots, n)$$

and ultimately

$$J(q) \geqslant M_1 \dots M_n . V \quad \text{for} \quad q \geqslant M_n.$$

Since $J(q)$ always remains less than or equal to 1, Minkowski's inequality (1.2) is proved.

II.

6. Contrary to the procedure in I, yielding the vectors $\mathfrak{d}_k$ and ratios of dilation $2M_k$, the *problem of reduction* requires the determination of vectors $\mathfrak{e}_k$ and ratios N_k according to the following inductive rule:

Let $\mathfrak{a}$ range over all lattice vectors outside the $(k-1)$-dimensional manifold $[\mathfrak{e}_1 \dots \mathfrak{e}_{k-1}]$ such that every lattice vector in the k-dimensional manifold $[\mathfrak{e}_1 \dots \mathfrak{e}_{k-1}\,\mathfrak{a}]$,

$$\mathfrak{x}_{k-1} + x\mathfrak{a} \quad \text{with } \mathfrak{x}_{k-1} \text{ in } [\mathfrak{e}_1 \dots \mathfrak{e}_{k-1}],$$

has an integral $\mathfrak{a}$-component x. $\mathfrak{a} = \mathfrak{e}_k$ is a vector within this range $\mathfrak{R}_k$ for which $f(\mathfrak{a})$ takes on the least possible value N_k.

As the first step, we may take $\mathfrak{e}_1 = \mathfrak{d}_1$, so that $N_1 = 2M_1$. By induction we readily conclude that $\mathfrak{e}_1, \dots, \mathfrak{e}_k$ span (*i.e.* form a basis of) the whole lattice in the manifold $[\mathfrak{e}_1 \dots \mathfrak{e}_k]$; hence $\mathfrak{e}_1, \dots, \mathfrak{e}_n$ span the whole n-dimensional lattice. Relative to this coordinate system, the lattice vectors

$$\mathfrak{x} = x_1\mathfrak{e}_1 + \dots + x_n\mathfrak{e}_n = (x_1, \dots, x_n)$$

are again described by integral components x_i, and our reduction is equivalent to the following conditions:

$$f(x_1, x_2, \dots, x_n) \geqslant f(\delta_1^k, \dots, \delta_n^k) = N_k \tag{6.1}$$

which hold provided that $x_1, x_2, \dots, x_n$ are integers and $x_k, \dots, x_n$ are without common divisor. The δ_i^k are the Kronecker δ's and thus $\mathfrak{e}_k = (\delta_r^k, \dots, \delta_n^k)$, the k-th unit vector.

This characterization (6.1) obviously entails the inequalities

$$N_1 \leqslant N_2 \leqslant \dots \leqslant N_n.$$

We have normalized the basis of our n-dimensional lattice relative to the convex solid by means of the n sets (6.1) of infinitely many inequalities, and this is what the problem of reduction demands. The question arises whether one can establish a universal upper bound for $N_1 \dots N_n V$. I now prove

Theorem V. *If a reduced lattice basis $\mathfrak{e}_k$ is so chosen as to satisfy the conditions* (6.1), *then*

$$N_1 \dots N_n V \leqslant \mu_n, \tag{6.2}$$

where

$$\mu_n = 2^n \cdot (\tfrac{3}{2})^{\frac{1}{2}(n-1)(n-2)}. \tag{6.3}$$

Proof. One of the vectors $\mathfrak{d}_1, \ldots, \mathfrak{d}_k$, say $\mathfrak{d}$, lies outside the linear manifold

$$E_{k-1} = [\mathfrak{e}_1 \ldots \mathfrak{e}_{k-1}].$$

In

$$E_k^* = [\mathfrak{e}_1 \ldots \mathfrak{e}_{k-1}\, \mathfrak{d}]$$

we can find a lattice vector $\mathfrak{e}^*$ which, together with $\mathfrak{e}_1, \ldots, \mathfrak{e}_{k-1}$, spans the whole lattice in E_k^*. Out of the finite number of lattice vectors of the form

$$x_1 \mathfrak{e}_1 + \ldots + x_{k-1} \mathfrak{e}_{k-1} + y\mathfrak{d} \tag{6.4}$$

with

$$-\tfrac{1}{2} < x_i \leqslant \tfrac{1}{2} \quad (i = 1, \ldots, k-1), \quad 0 < y \leqslant 1,$$

select one, say $\mathfrak{e}^*$, with the lowest y. Since $\mathfrak{d}$ itself must be an integral combination of $\mathfrak{e}_1, \ldots, \mathfrak{e}_{k-1}, \mathfrak{e}^*$, this y is the reciprocal $1/e$ of a positive integer e. If $e = 1$, $\mathfrak{e}^*$ coincides with $\mathfrak{d}$. Our $\mathfrak{e}^*$ is in $\mathfrak{R}_k$ and thus in the competition for the lowest value N_k of $f(\mathfrak{a})$. But, since $f(\mathfrak{d})$, as one of the numbers

$$2M_1, \ldots, 2M_k,$$

is certainly less than or equal to $2M_k$, we infer from these remarks and (6.4) that

$$f(\mathfrak{e}^*) \leqslant 2M_k \quad \text{if} \quad e = 1,$$

$$f(\mathfrak{e}^*) \leqslant \frac{2M_k}{e} + \tfrac{1}{4}(N_1 + \ldots + N_{k-1}) \leqslant M_k + \tfrac{1}{4}(N_1 + \ldots + N_{k-1}) \quad \text{if} \quad e \geqslant 2.$$

Hence N_k cannot exceed the larger of the two numbers

$$M_k + \tfrac{1}{4}(N_1 + \ldots + N_{k-1}) \quad \text{and} \quad 2M_k. \tag{6.5}$$

This allows us to establish universal inequalities of the kind

$$N_i \leqslant 2\theta_i M_i. \tag{6.6}$$

Indeed (6.6) is true for $i = 1$ with $\theta_1 = 1$. Once it holds good for

$$i = 1, \ldots, k-1,$$

the bound (6.5) for N_k yields a similar inequality for $i = k$ with θ_k as the greater of the two numbers

$$1 \quad \text{and} \quad \tfrac{1}{2}(1 + \theta_1 + \ldots + \theta_{k-1}).$$

This determines the factors θ_k by recursion, and we readily verify that

$$\theta_1 = 1, \quad \theta_k = (\tfrac{3}{2})^{k-2} \quad \text{for} \quad k \geqslant 2.$$

Therefore (1.2) implies

$$N_1 \dots N_n V \leqslant 2^n . \theta_1 \dots \theta_n = \mu_n.$$

It would be an interesting problem to ascertain the lowest constant μ_n for which the relation (6.2) holds.

Fuller exploitation of the method yields the following

GENERALIZED THEOREM V. *Let* $q_1, \dots, q_n$ *be given positive numbers greater than or equal to* 1. *If the inequality*

$$f(x_1, x_2, \dots, x_n) \geqslant \frac{1}{q_k} . f(\delta_1{}^k, \dots, \delta_n{}^k)$$

holds whenever $x_1, \dots, x_n$ *are integers and* $x_k, \dots, x_n$ *have no common factor* ($k = 1, \dots, n$), *then the values*

$$N_k = f(\delta_1{}^k, \dots, \delta_n{}^k)$$

satisfy the relations $\qquad q_k N_{k+1} \geqslant N_k \quad (k = 1, \dots, n-1)$

and

$$N_1 \dots N_n V \leqslant 2q_1 \dots q_n (1+q_1)^{n-1} \cdot (1+\tfrac{1}{2}q_2)^{n-2} \dots (1+\tfrac{1}{2}q_{n-1})^1.$$

The proof is practically the same as before. The recursive equations for the θ_k are now

$$\theta_1 = q_1 \quad \text{and} \quad \frac{2\theta_k}{q_k} = 1 + \theta_r + \dots + \theta_{k-1} \quad (k \geqslant 2)$$

i.e.

$$\frac{2\theta_2}{q_2} = 1 + q_1, \quad \frac{2\theta_k}{q_k} + \theta_k = \frac{2\theta_{k+1}}{q_{k+1}} \qquad (k \geqslant 2),$$

with the solution

$$\theta_k = \tfrac{1}{2} q_k (1+q_1) \cdot (1+\tfrac{1}{2}q_2) \dots (1+\tfrac{1}{2}q_{k-1}) \quad (k \geqslant 2).$$

7. Here is another similar proposition:

THEOREM VI. *Put*

$$h_\nu = \nu^2 - \tfrac{1}{2}\nu - 1 \quad (\nu = 1, 2, \dots)$$

and

$$\rho_n = 2^{2\nu-n} . (2\nu+2)(2\nu+3) \dots (\nu+n+1) \quad \textit{for} \quad h_\nu \leqslant n \leqslant h_{\nu+1}.$$

We can determine a lattice basis $\mathfrak{e}_1{}^*, \dots, \mathfrak{e}_n{}^*$ *such that the distances* $f(\mathfrak{e}_k{}^*) = M_k{}^*$ *satisfy the inequality*

(7.1) $$M_1{}^* \dots M_n{}^* V \leqslant \rho_n.$$

This constant ρ_n is much lower than the one in Theorem V. Its increase with n is best judged by the recurrence formulae

$$\rho_n = \frac{n+\nu+1}{2}\,\rho_{n-1} \quad \text{if} \quad 1+h_\nu \leqslant n \leqslant h_{\nu+1},$$

and
$$\rho_n = \frac{(n+\nu)(n+\nu+1)}{2(n+\nu)+1}\,\rho_{n-1} \quad \text{for odd } \nu \text{ and } n = h_\nu + \tfrac{1}{2}.$$

Proof. Out of the vector basis $\mathfrak{d}_1, \ldots, \mathfrak{d}_n$, we construct a lattice basis $\mathfrak{e}_1^*, \ldots, \mathfrak{e}_n^*$ according to the recurrent equations

$$\mathfrak{e}_k^* = \frac{1}{e_k}\,\mathfrak{d}_k + a_{k,\,k-1}\,\mathfrak{d}_{k-1} + \ldots + a_{k,\,1}\,\mathfrak{d}_1 \tag{7.2}$$

with the conditions

$$-\frac{1}{2e_i} < a_{k,\,i} \leqslant \frac{1}{2e_i} \quad (i<k),$$

the e_k being positive integers. If $e_k = 1$, all the coefficients $a_{k,\,k-1}, \ldots, a_{k,\,1}$ vanish, and $f(\mathfrak{e}_k^*) = 2M_k$. However, if $e_k \geqslant 2$,

$$M_k^* = f(\mathfrak{e}_k^*) \leqslant \frac{2}{e_k}\,M_k + \sum_{i=1}^{k-1} M_i/e_i \leqslant \eta_k M_k$$

with
$$\eta_k = \frac{2}{e_k} + \left(\frac{1}{e_{k-1}} + \ldots + \frac{1}{e_1}\right) \quad (e_k \geqslant 2).$$

To cover both cases we put $\eta_k = 2$ for $e_k = 1$. From (1.2) we now get

$$M_1^* \ldots M_n^*\, V \leqslant \eta_1 \ldots \eta_n,$$

and it remains to discuss the product $\eta_1 \ldots \eta_n$.

We put

$$\delta_k = 2 \text{ or } \delta_k = 1 \quad \text{according as} \quad e_k = 1 \text{ or } e_k \geqslant 2;$$

$$\eta_k' = 2 \text{ if } \delta_k = 2, \quad \text{and} \quad \eta_k' = \delta_k + \tfrac{1}{2}(\delta_{k-1} + \ldots + \delta_1) \text{ if } \delta_k = 1.$$

Then
$$\eta_k \leqslant \eta_k' \quad \text{and} \quad \eta_1 \ldots \eta_n \leqslant \eta_1' \ldots \eta_n'.$$

Among the 2^n values of the product

$$\eta_1' \ldots \eta_n' \tag{7.3}$$

resulting from the 2^n possible sequences $\delta_1, \ldots, \delta_n$ of 1's and 2's, we have to ascertain the largest. An interchange of two consecutive figures 1 and 2 in this sequence whereby

$$\delta_k = 1,\ \delta_{k+1} = 2 \quad \text{is changed into} \quad \delta_k = 2,\ \delta_{k+1} = 1$$

(all other figures remaining unaltered) affects merely the two factors $\eta_k' \eta_{k+1}'$ in (7.3), turning their values

$$[1+\tfrac{1}{2}(\delta_{k-1}+\ldots+\delta_1)].2 \quad \text{into} \quad 2.[2+\tfrac{1}{2}(\delta_{k-1}+\ldots+\delta_1)],$$

and so increasing (7.3). For a given number ν of 2's in the sequence, the maximum value is thus attained when they come first ($\delta_1, \ldots, \delta_\nu$), and then (7.3) is equal to

$$2^\nu \cdot \frac{2\nu+2}{2} \cdot \frac{2\nu+3}{2} \cdots \frac{2\nu+1+\mu}{2} \qquad (\mu = n-\nu)$$

i.e.

$$2^{2\nu-n} \cdot (2\nu+2)(2\nu+3) \ldots (\nu+1+n) = w_\nu.$$

Let ν pass through the values $\nu = 0, 1, \ldots, n$. Since

$$w_\nu = \frac{4(\nu+1+n)}{2\nu(2\nu+1)} \cdot w_{\nu-1},$$

w_ν is on the upgrade ($w_\nu > w_{\nu-1}$ or $w_\nu \geqslant w_{\nu-1}$) so long as

$$\nu(2\nu+1) < 2(\nu+1+n) \quad \text{or} \quad \nu(2\nu+1) \leqslant 2(\nu+1+n),$$

and then goes down. Hence the maximum is attained for the last ν for which $\nu(\nu-\frac{1}{2})$ does not exceed $n+1$.

Here is a table for the lowest values of ρ_n:

$n =$ 1,	2,	3,	4,	5,	6,	7,	....	
$\rho_n =$ 2,	4,	12,	42,	168,	756,	3960,	...,	

Higher bounds, for instance

$$\rho_n = (n+1)! \quad \text{or} \quad \rho_n = 2 \cdot n!,$$

can even more easily be obtained. Again we have the question of the best value of ρ_n in the inequality (7.1).

8. A positive-definite *quadratic form*

(8.1) $$f(x) = \Sigma g_{ik} x_i x_k \quad (i, k = 1, \ldots, n)$$

is said to be *reduced* if it satisfies the inequalities

(8.2) $$f(x) \geqslant g_{kk} \quad \text{for every } x \text{ in } \mathfrak{R}_k \quad \text{and} \quad k = 1, \ldots, n,$$

where $\mathfrak{R}_k$ is the set of all lattice points $x = (x_1, \ldots, x_n)$ for which $x_k, \ldots, x_n$ are without common divisor. While for every positive form of discriminant $D = \det(g_{ik})$,

$$g_{11} \cdots g_{nn} \geqslant D,$$

we have found that for a reduced form

$$\lambda_n g_{11} \cdots g_{nn} \leqslant D \quad \text{with} \quad \lambda_n = (\omega_n/\mu_n)^2. \tag{8.3}$$

Here ω_n denotes the volume of the n-dimensional unit sphere, most handily described by the recurrent relation

$$\omega_n = \frac{2\pi}{n} \cdot \omega_{n-2} \quad \text{together with} \quad \omega_0 = 1,\ \omega_1 = 2.$$

Indeed, with $\sqrt{f}$ as gauge function, the convex body $\mathfrak{K}:\ f < 1$, becomes an ellipsoid of volume $\omega_n/\sqrt{D}$, as is readily proved by linear transformation of $f(x)$ into the normal form $x_1^2 + \ldots + x_n^2$.

The linear substitutions S of the variables x_i with integral coefficients s_{ik} of determinant $|s_{ik}| = \pm 1$ (or only those of determinant 1) form a group $\{S\}$. Two quadratic forms f and Sf which arise from each other by a substitution of the group are said to be *equivalent*. Every positive form is equivalent to a reduced form.

All quadratic forms $f = \{g_{ik}\}$ with real coefficients $g_{ik} = g_{ki}$ form an $N = \frac{1}{2}n(n+1)$-dimensional linear space R, the positive ones an open convex region G in R, the reduced forms a convex subset Z of G which is closed relative to G. Both G and Z are cones and may, therefore, be considered as sets in the $(N-1)$-dimensional sphere of rays in R issuing from the origin. A convex cone H is a set having these three properties:

(i) $f = 0$ is not in H;

(ii) if f is in H, so is tf, where t is any positive factor;

(iii) with f and f' in H, the sum $f + f'$ is in H.

We operate throughout in G; all topological words referring to subcones of G are meant relative to G (not R).

Each of the inequalities (8.2) for the variables g_{ij} is of the form

$$\sum_{i,j} a_{ij} g_{ij} \geqslant 0 \tag{8.4}$$

with integral coefficients a_{ij}. We expressly exclude from the set $\mathfrak{Z}$ of inequalities (8.2) those for which all coefficients a_{ij} vanish, namely (8.2) with $x = \pm(\delta_1^k, \ldots, \delta_n^k)$.

Each individual inequality (8.4) of $\mathfrak{Z}$ defines a half-space bounded by a plane

$$\sum_{i,j} a_{ij} g_{ij} = 0. \tag{8.5}$$

A substitution S carries the central cell Z into an equivalent cell Z_S which is also convex. The equivalent cells jointly cover the whole space G. Z_J coincides with Z for any of the 2^n substitutions

$$J: \; x_1 \to \pm x_1, \; \ldots, \; x_n \to \pm x_n,$$

and S and SJ have the same effect upon Z.

From (8.3) Minkowski deduced two important theorems of finiteness:

I. *A finite number of the inequalities* $\mathfrak{Z}$ *suffices to define the cell* Z.

II. *There is only a finite number of substitutions* S *capable of carrying a point of* Z *into a point of* Z (worded more precisely, the demand on the element S of $\{S\}$ which limits it to a finite set is to the effect that there should exist two points f, f' in Z such that $f' = Sf$).

Minkowski has a straightforward algebraic proof of II, in § 7 of his paper "Diskontinuitätsbereich für arithmetische Äquivalenz"†. His proof of I is involved because he establishes I simultaneously with an inequality of the nature of (8.3) by a complicated induction. This entanglement was dissolved in a joint paper by L. Bieberbach and I. Schur‡ (oh tempi passati!). They improve on I by breaking it up into the two theorems:

I*. *There is only a finite number of lattice points* x *in* $\mathfrak{R}_k$ *for which there exists a reduced form* $f_0 = \{g^0_{ij}\}$ *satisfying the equation*

$$f_0(x) = g^0_{kk}. \tag{8.6}$$

I**. *If we substitute for* x *in succession each of the lattice points assigned by Theorem* I*, *then the corresponding inequalities*

$$f(x) \geqslant g_{kk} \tag{8.7}$$

for $k = 1, \ldots, n$ *completely close in the cell* Z *in* G.

While the proof of I* again is a straightforward algebraic affair based on (8.3), I** depends on the topological argument that a straight line joining a point inside to a point outside a convex region must cross the border. I explain the Bieberbach-Schur argument for I** in this simplified form.

---

† *Journal für Math.*, 129 (1905), 220–274; *Gesammelte Abhandlungen* 2 (Leipzig, 1911), 53–100.

‡ *Sitzungsber. Berlin Akad.*, (1928), 519–523; (1929), 508.

I say that those points f (in G) for which all the inequalities $\mathfrak{Z}$ hold with the sign $>$ rather than $\geqslant$ form the *core* of Z. Any diagonal form

$$g_1 x_1^2 + \dots + g_n x_n^2 \quad \text{with} \quad 0 < g_1 < g_2 < \dots < g_n$$

evidently belongs to the core of Z. So do the inner points of Z. I maintain that conversely every point of the core is an inner point of Z. Indeed if f is a positive form, U_ϵ a sufficiently small neighbourhood of f and A any positive number, then the points x of the x-space such that $f_\epsilon(x) \leqslant A$ holds for at least one f_ϵ in U_ϵ belong to a limited region. Therefore all lattice points x with a finite number of exceptions satisfy the inequality $f_\epsilon(x) > A$ for every f_ϵ in U_ϵ. This follows readily from Jacobi's step transformation of f_ϵ into a square sum. Hence we have a finite subset $\mathfrak{Z}^0$ of $\mathfrak{Z}$ such that every form f_ϵ in U_ϵ satisfies all inequalities of $\mathfrak{Z}-\mathfrak{Z}^0$. Provided that the inequalities $\mathfrak{Z}^0$ hold for f with the $>$ sign, we may shrink U_ϵ so as to have the inequalities $\mathfrak{Z}^0$ satisfied throughout U_ϵ. This establishes I** by the topological argument mentioned above.

Let f_0 again be a point on the boundary of Z and let x range over the finite number of lattice points x in $\mathfrak{R}_k$ which satisfy (8.6). Studying the convex wedge defined by the corresponding simultaneous inequalities (8.7), we find† that those inequalities $\mathfrak{Z}$ suffice which are satisfied with the $=$ sign for $N-1$ linearly independent points f in Z. They are at the same time indispensable, and the corresponding equations describe the $(N-1)$-dimensional planes which contain the faces of the pyramidal cell Z in G.

Another way of constructing these faces would be by determining the common boundaries of Z with the adjacent equivalent cells Z_S. But Z and Z_S have a point f_0 in common only if both f_0 and $S^{-1}f_0$ are in Z. Hence the finiteness of the number of plane faces must also be a consequence of II, so that I ought to be deducible from II. Let us examine the situation more carefully.

Only the substitutions J are capable of transforming a point in the core of Z into a point of Z. Hence the two convex cells Z and Z_S have no inner point in common provided that S is not a J. The points which they have in common, if any, form a convex cone W in a linear submanifold of $N-1$ or ... or 1 dimensions. If $N-1$ is the correct number of dimensions we speak of a wall separating Z from Z_S. Choose f_0 as an inner point of W in its $(N-1)$-dimensional plane. It satisfies one of the inequalities $\mathfrak{Z}$, say (8.4), with the sign $=$ while every point of W satisfies it with the $\geqslant$ sign. Therefore the whole W clearly lies in the plane (8.5), and thus I is again established, *provided that we can show that every point f on the boundary of Z*

† For an elementary treatment see H. Weyl, *Comm. Math. Helvet.*, 7 (1934–5), 290–306.

lies on a wall separating Z from an adjacent cell Z_S. We deduce this from the fact, important in itself, that the equivalent cells cluster only towards the boundary of G.

On closer examination Minkowski's proof of II reveals that there is not more than a finite number of S capable of carrying a point of Z into a point of $G(\rho, \mu)$. Here ρ is any number greater than 1 and μ is any positive number. If D_k denotes the determinant of the form $f(x_1 \dots x_k 0 \dots 0)$, the subcone $G(\rho, \mu)$ of G is defined by the following simultaneous inequalities:

$$\left.\begin{aligned} &D_k \geqslant \frac{\lambda_k}{\rho} g_{11} \cdots g_{kk}, \\ &g_{kk} \geqslant \frac{1}{\rho} g_{k-1,k-1} \quad [g_{00}=0], \\ &f(x_1, \dots, x_{k-1}, 1, 0, \dots, 0) \geqslant g_{kk} - \mu g_{11} \end{aligned}\right\} \qquad (k=1, \dots, n)$$

for any integers $x_1, \dots, x_{k-1}$.

All points of Z are *inner* points of $G(\rho, \mu)$. This is more than sufficient to settle our question. Choose a definite $\rho > 1$ and $\mu > 0$. Let f be a point on the boundary of Z and let $f_1, \dots, f_\nu, \dots$ be a sequence of points outside Z approaching f. We may assume that they lie in $G(\rho, \mu)$. A certain substitution S_ν^{-1} of $\{S\}$ will carry f_ν into a point $f_\nu' = S_\nu^{-1} f_\nu$ of Z. In consequence of our statement about $G(\rho, \mu)$, a certain S recurs infinitely often among the S_ν. This S is no J. The points f_ν of the corresponding selected sequence all lie in Z_S, and so does their limit f. Consequently any point f on the boundary of Z belongs to the common boundary of Z and an equivalent Z_S. The domain $G(\rho, \mu)$ increases when ρ and μ increase, and with $\rho \to \infty$, $\mu \to \infty$ exhausts the whole G. At every stage $G(\rho, \mu)$ has points in common with only a finite number of cells Z_S. An elementary argument like the one applied before shows that among the common boundaries $Z|Z_S$ (which exist in finite number only) none but the "walls", the $(N-1)$-dimensional ones, matter. The substitutions S which carry Z into cells Z_S bordering on Z along a wall, together with the J, generate the entire group $\{S\}$.

The Generalized Theorem V proves that all our statements about $G(\rho, \mu)$ remain true if we define it by the following *linear* inequalities:

$$f(x_1, \dots, x_n) \geqslant \frac{1}{\rho} g_{kk}$$

whenever the x_i are integers and $x_k, \dots, x_n$ have no common factor,

$$f(x_1, \dots, x_n) \geqslant g_{kk} - \mu g_{11}$$

whenever the x_i are integers and $(x_k, x_{k+1}, \ldots, x_n) = (1, 0, \ldots, 0)$. Incidentally the estimates may be improved considerably if we carry out the comparison of the coefficients g_{kk}, g'_{kk} of two equivalent forms f, f' in Z and $G(\rho, \mu)$ respectively for any convex bodies rather than ellipsoids by the procedure of § 6.

Throughout we have studied Z relative to G; this is really the more natural standpoint. However, the results I and I** are true even relative to the space R of all quadratic forms. Z is not closed in R; there may be boundary points f of Z relative to R which are not positive forms. I maintain that for such a form the equation $g_{11} = 0$ holds. Since f is non-negative, it can be brought into the form

$$\xi_1^2 + \ldots + \xi_m^2 \quad (m < n)$$

by a transformation with real coefficients. We can determine a lattice point $x \neq 0$ for which the m linear forms $\xi_1, \ldots, \xi_m$ are in absolute value less than any preassigned positive number ϵ, either by Minkowski's theorem (1.1) for parallelotopes, or by a simple application of Dirichlet's principle of the distribution of $\nu+1$ objects over ν boxes. Hence $g_{11} > 0$ would contradict the inequality $f(x) \geqslant g_{11}$ which holds for all lattice points $x \neq 0$, and Z relative to R is defined by the inequalities described under I* and I** together with $g_{11} \geqslant 0$. But even the addition of $g_{11} \geqslant 0$ becomes superfluous if $n \geqslant 2$. The substitution

$$x_1 = 1, \quad x_2 = \pm 1, \quad x_3 = \ldots = x_n = 0$$

yields the inequality

$$g_{11} \pm 2g_{12} + g_{22} \geqslant g_{22}$$

for any reduced form, or $2|g_{12}| \leqslant g_{11}$. The same holds for $g_{13}, \ldots, g_{1n}$. Hence the boundary form f even satisfies the n equations

$$(8.8) \qquad g_{11} = g_{12} = \ldots = g_{1n} = 0.$$

The fact that Z borders on the plane $g_{11} = 0$ only along this linear variety (8.8) of n dimensions less, proves our point.

Minkowski's investigations set out from the reduction of quadratic forms. Owing to some "unexpected difficulties" the second half of his book *Geometrie der Zahlen* never appeared, and fourteen years after the discovery of the basic idea of the geometry of numbers he published his theory of reduction in a strictly arithmetical form, without a hint of geometric approach. What was the snag he struck? After having taken the main hurdle with the general inequality (1.2), did he fail to notice the simple remark that led us on to (6.2), or was he unable to unravel the inductive kink by which in his final publication the establishment of the inequality (6.2) is tied to the selection of a finite number of inequalities $\mathfrak{Z}$? It is now idle to speculate. But of him it might be said as of Saul that he went out to look after his father's asses and found a kingdom.

For quadratic forms there is a simple algebraic method of deriving (1.2) from (1.1). Minkowski describes it in *Geometrie der Zahlen*, § 51; and very likely this suggested to him the general inequality (1.2). Hence every improvement of the constant γ_n in the inequality for positive forms

$$M^n \leqslant \gamma_n D$$

in which M is the minimum of $f(x)$ for all lattice points $x \neq 0$ entails a proportional improvement in (1.2) and for the constants μ_n and ρ_n in the estimates (6.2), (7.1) for ellipsoids. Our value was $\gamma_n = (2^n/\omega_n)^2$. Blichfeldt† succeeded in improving it by the factor

$$(1+\tfrac{1}{2}n)^2 \,.\, 2^{-n}.$$

This leads to the following λ_n:

$$1/\lambda_n = 2^n(1+\tfrac{1}{2}n)^2\,\omega_n^{-2} \,.\, (\tfrac{9}{4})^{\frac{1}{2}(n-1)(n-2)}. \tag{8.9}$$

Remak recently obtained a better estimate‡ which differs from (8.9) essentially by having $\frac{5}{4}$ instead of $\frac{9}{4}$. However, the chief interest of our constant μ_n in (6.2) lies in its validity for all convex solids whatsoever.

Theorem VI together with Blichfeldt's improved γ_n shows that every positive form has an equivalent one f satisfying the relation

$$\lambda^*{}_n g_{11} \cdots g_{nn} \leqslant D,$$

where

$$1/\lambda_n{}^* = (1+\tfrac{1}{2}n)^2\, 2^{-n}\, \omega_n^{-2}\, \rho_n{}^2.$$

Such an estimate is best suited to prove that there is only a finite number of classes of equivalent positive forms with integral coefficients.

The Institute for Advanced Study,
Princeton, N.J., U.S.A.

---

† *Trans. American Math. Soc.*, 15 (1914), 227–235; *cf.* R. Remak, *Math. Zeitschrift*, 26 (1927), 694–699, and Blichfeldt, *Math. Annalen*, 101 (1929), 605–608.

‡ *Compositio Math.*, 5 (1938), 368–391.

128.

Elementary note on prime number problems of Vinogradoff's type (R. D. James and H. Weyl)

American Journal of Mathematics 64, 539—552 (1942)

1. Stating the problem. Vinogradoff's asymptotic formula slightly amended. By his ingenious methods Vinogradoff succeeded in establishing an asymptotic formula for the number of decompositions of a large integer N into three primes, $N = p_1 + p_2 + p_3$. When one follows the exposition in his paper, *Recueil Mathématique*, vol. 2 (44) (1937), pp. 179-195, one finds that he introduces another number N_1 (p. 192), and writing k instead of N_1 one sees that he actually determines the solutions of the equation $p_1 + p_2 + p_3 = k$ by primes $p_i \leqq N$. In this form the problem admits of an immediate generalization[1]: Let $r \geqq 3$ non-vanishing integers $a_1, \cdots, a_r$ and r positive real numbers $\beta_1, \cdots, \beta_r$ be given once for all. We ask for the number $A(N, k)$ of representations of an integer k in the form

$$k = a_1 p_1 + \cdots + a_r p_r \tag{1}$$

by means of primes

$$p_1 \leqq \beta_1 N, \cdots, p_r \leqq \beta_r N. \tag{2}$$

The problem thus stated makes sense even if some of the coefficients a_i are negative. The integer k and the real number $N \geqq 2$ are considered as variables. The letter C stands throughout for a positive constant, not always the same, which is independent of N and k; if other variables enter, indices attached to C will indicate on which of these variables C depends. The main result is a formula

$$A(N, k) = R(N, k) \cdot \mathfrak{S}(k) + (N^{r-1}/n^r) \cdot \rho \tag{I}$$

in which the relative error ρ, with $N \rightarrow \infty$ and uniformly with respect to k, tends to zero faster than any negative power n^{-g} of $n = \log N$. The factor

* Received June 19, 1941.

[1] J. G. van der Corput has published a number of results concerning this problem, *C. R. Ac. Sci. Paris*, vol. 205 (1937), pp. 479-481 and 591-592; *Nederlandsche Akademie van Wetenschappen, Proceedings*, vol. 41 (1938), p. 80. He even imposes upon the primes p_i congruence restrictions modulo a given integer. But as he mentions existence theorems only, a full account of the asymptotic situation might help "to put the house in order."

$R(N,k)$ takes account of the "quantitative" aspect of k (infinite prime spot) in so far as $R(N,x)$ is defined for arbitrary real values of x, whereas the singular series $\mathfrak{S}(k)$ is an arithmetical function of k alone. For any positive integer q we set

$$q_i = q/(q,a_i), \qquad q^* = q/(q,k),$$

and by means of the Möbius and Euler functions $\mu(q)$ and $\phi(q)$ form

(3) $\chi(q) = (\mu(q_1)/\phi(q_1)) \cdots (\mu(q_r)/\phi(q_r)) \cdot (\mu(q^*)/\phi(q^*)) \cdot \phi(q).$

Then

$$\mathfrak{S}(k) = \mathfrak{S}(k; a_1, \cdots, a_r) = \sum_{q=1}^{\infty} \chi(q). \tag{4}$$

We notice that $\mathfrak{S}(k; a_1, \cdots, a_r)$ depends symmetrically on $k; a_1, \cdots, a_r$, although it should not be forgotten that k may take the value 0. The quantitative factor R may be given the following form. Using the abbreviation $e^{2\pi i x} = e(x)$ we introduce the following functions of a real variable z:

$$J_i(z) = \int_{\sqrt{N}}^{\beta_i N} (e(a_i z x)/\log x)\, dx \qquad (i = 1, \cdots, r). \tag{5}$$

Then $R(N,x)$ is the Fourier transform of their product,

$$R(N,x) = \int_{-\infty}^{+\infty} J_1(z) \cdots J_r(z) \cdot e(-xz)\, dz. \tag{6}$$

The proof depends on Vinogradoff's sums extending over the primes p,

$$S(\eta) = S(N,\eta) = \sum_{p \leqq N} e(\eta p), \qquad S_i(\eta) = S(\beta_i N, a_i \eta),$$

and starts with the exact formula

$$A(N,k) = \int_0^1 S_1(\eta) \cdots S_r(\eta) e(-k\eta)\, d\eta. \tag{7}$$

Following Vinogradoff's exposition step by step we call attention to a few points where we deviate from him. [1] The lower limit in (5) or in the typical function

$$J(z) = \int_{\sqrt{N}}^{\beta N} (e(zx)/\log x)\, dx$$

has been set at $\sqrt{N}$ rather than at 2. Looking at the context where this function makes its first appearance, one realizes that the results are not affected by this change. One immediate advantage is that the estimates

$$|J(z)| \leqq 2\beta N/n, \qquad |J(z)| \leqq 2/(\pi n \,|\, z \,|) \tag{8}$$

hold for *all* z, hence

(9) $$|J_i(z)| \leqq C \cdot \delta(z)$$

throughout where

$$\delta(z) = \begin{cases} N/n \text{ for } |z| \leqq N^{-1}, \\ 1/(n|z|) \text{ for } |z| \geqq N^{-1}. \end{cases}$$

A more important advantage will presently become evident. [2] The integration in (6) which Vinogradoff restricts within the bounds $\pm n^{3h}/N$ has been extended to $\pm \infty$. Because of the second inequality (8) the value of R is not changed thereby by more than $CN^{r-1}n^{-3h(r-1)}$. It is exactly this step which makes the main term in the formula (I) independent of Vinogradoff's exponent h and gives a result from which h has vanished altogether. [3] It is not hard to see why in the expression of $\chi(q)$ the numbers $q_1, \cdots, q_r$ occur. A rational point is of the form $\eta = m/q$ where m is an element of the set G_q of all $\phi(q)$ prime residues mod q. In the reduced form the fraction $a_1\eta = a_1 m/q$ has the denominator q_1 and hence $e(a_1\eta p)$ depends only on the residue class mod q_1 to which p belongs. The following elementary arithmetical observations are to the point. q' being a divisor of q, $q = q'd$, the number of elements m of G_q which are congruent mod q' to a given element m' of $G_{q'}$ has the same value v for each m'. Clearly $v \leqq d$ and $v = \phi(q)/\phi(q')$. In case a'/q' is the reduced form of a fraction a/q we therefore have $v \leqq (a, q)$, *a fortiori*

(10) $$\phi(q) \leqq |a| \cdot \phi(q')$$

if $a \neq 0$, and the sum

$$\sum_{m \in G_q} e(am/q) = v \cdot \sum_{m' \in G_{q'}} e(a'm'/q') = (\phi(q)/\phi(q')) \cdot \mu(q').$$

The last remark finds application for $a = -k$. Because of (10), $\mathfrak{S}(k)$ is majorized by the convergent sum

$$|a_1 \cdots a_r| \cdot \sum_q (\phi(q))^{-r-1}.$$

[4] One crucial point in the estimate for the contribution to the integral (7) of part II of the circumference $0 \leqq \eta \leqq 1$ is the fact that

$$\int_0^1 |S(\eta)|^2 d\eta = \pi(N) \leqq N.$$

An equally good result is obtained in our case where all the factors $S_i(\eta)$ are different, by observing that $\int_0^1 |S_1(\eta)| \cdot |S_2(\eta)| \, d\eta$ does not exceed the square root of

$$\int_0^1 |S_1(\eta)|^2 d\eta \cdot \int_0^1 |S_2(\eta)|^2 d\eta = \pi(\beta_1 N) \cdot \pi(\beta_2 N) \leqq \beta_1\beta_2 \cdot N^2.$$

By (9) we realize that

$$| R(N, x) | \leqq C \cdot N^{r-1}/n^r.$$

Hence we set

$$R(N, x) = (N^{r-1}/n^r) \cdot R^*(N, x)$$

and then find not only $| R^*(N, x) | \leqq C$ but also

$$\text{(II)} \qquad | R^*(N, x) - R^*(N, x') | \leqq C \cdot (| x - x' | / N),$$

which indicates that $R^*(N, x)$ varies very slowly over wide stretches of the variable x if N is large. Of course C now means constant with respect to N and x (or N and x/N). It seems reasonable to measure x against N. Let us therefore write $R^*(N, xN) = R_N(x)$. Then (6) and the inequality (9) prove that the first $r-2$ derivatives of $R_N(x)$ with respect to x are continuous and satisfy relations

$$\text{(11)} \qquad | R_N^{(\mu)}(x) | \leqq C \qquad (\mu = 0, 1, \cdots, r-2).$$

We have thus arrived at the following two propositions:

Theorem I. *A formula* (I) *holds in which the singular series* $\mathfrak{S}(k)$ *is defined by* (4) *and the relative error* $\rho = \rho(N, k)$ *satisfies an inequality*

$$| \rho(N, k) | \leqq C_g \cdot n^{-g}$$

however large the integral exponent $g \geqq 0$.

Theorem II. $R(N, x)$ *is defined for all real* x, *and if we set*

$$R(N, xN) = (N^{r-1}/n^r) \cdot R_N(x),$$

then

$$| R_N(x) | \leqq C, \qquad | R_N(x) - R_N(x') | \leqq C \cdot | x - x' |.$$

2. Analysis of the quantitative factor. In this section we give a simple interpretation of $R(N, x)$ which leads to an asymptotic expansion in terms of powers of n^{-1}. Since k is not representable in the form (1) under the conditions (2) unless k/N lies between the limits

$$A = \sum_{i=1}^{r} \tfrac{1}{2}\beta_i(a_i - | a_i |) \qquad \text{and} \qquad A' = \sum_{i=1}^{r} \tfrac{1}{2}\beta_i(a_i + | a_i |)$$

we must expect $R_N(x)$ to be non-negative for all values of x and to vanish outside the finite interval $A \leqq x \leqq A'$. Our new representation will put these facts in evidence.

We introduce a measure $d\sigma$ for infinitesimal areas $d\sigma$ in any of the parallel planes $E(x)$,

$$a_1x_1 + \cdots + a_rx_r = \text{const.} = x$$

of the $(x_1,\cdots,x_r)$-space, such that the volume of a (straight or oblique) cylinder erected over $d\sigma$ and bounded by the planes $E(x)$ and $E(x+dx)$ is given by $d\sigma\cdot dx$. The expression

$$R(N,x)=\int_{-\infty}^{+\infty}\left\{\int\cdots\int_{B(N)}\frac{e(z(a_1x_1+\cdots+a_rx_r))}{\log x_1\cdots\log x_r}\,dx_1\cdots dx_r\right\}e(-xz)\,dz$$

where $B(N)$ denotes the block

$$\sqrt{N}\leqq x_1\leqq\beta_1N,\quad\cdots,\qquad\sqrt{N}\leqq x_r\leqq\beta_rN$$

suggests forming the integral

$$F(x)=\int(1/\log x_1)\cdots(1/\log x_r)\,d\sigma \tag{12}$$

over the intersection of the plane $E(x)$ with the block $B(N)$. Then

$$R(N,x)=\int_{-\infty}^{+\infty}\left\{\int_{-\infty}^{+\infty}e(zx')F(x')\,dx'\right\}e(-xz)\,dz,$$

and by Fourier's theorem,

$$R(N,x)=F(x).$$

It is evident that $F(x)$ vanishes outside the interval $AN\leqq x\leqq A'N$. We shall show in the next section that it has a continuous derivative, and thus justify our application of Fourier's theorem. We formulate our result as follows: [2]

THEOREM III. *Idealize the r primes p_i as statistically independent continuous quantities x_i such that $\phi_i(x)dx$,*

$$\phi_i(x)=1/\log x\quad\textit{for}\quad\sqrt{N}\leqq x\leqq\beta_iN,\qquad 0\quad\textit{elsewhere,} \tag{13}$$

is the probability of x_i assuming values between x and $x+dx$, then $R(N,x)\cdot dx$ is the probability that $a_1x_1+\cdots+a_rx_r$ lies between x and $x+dx$.

This is a familiar device in the theory of probability: The link betwwen the densities of the probability distributions of r independent variables x_i and a linear combination of them is expressed as simple multiplication in terms of the Fourier transforms of the densities.

---

[2] Even before Vinogradoff made his fundamental discovery enabling him to prove the 3-term Goldbach theorem without such generalized Riemann hypotheses as Hardy-Littlewood had used, H. Rademacher gave this "probability expression" of the quantitative factor occuring in the asymptotic formula for the r-term Goldbach problem. *Math. Zeitschr.*, vol. 25 (1926), p. 628. Van der Corput obtained similar integrals in a number of related problems, mostly of non-linear character. *Nederlandsche Akademie van Wetenschappen, Proceedings*, vol. 41 (1938), pp. 25-26; 97-107; 344-349; *Math. Annalen*, vol. 116 (1938), pp. 1-50.

Using the notation $\xi = -\log x$ we may write our formula in this way:

$$R_N(x) = \int \frac{n}{n-\xi_1} \cdots \frac{n}{n-\xi_r}\, d\sigma, \tag{14}$$

the integral extending over the intersection of $E(x)$ with the block

$$B_N: \qquad N^{-1/2} \leqq x_1 \leqq \beta_1, \cdots, N^{-1/2} \leqq x_r \leqq \beta_r.$$

The asymptotic series for $R_N(x)$ is obtained by expanding the individual factors of the integrand as power series in n^{-1}:

$$\frac{n}{n-\xi} = \sum_{i=0}^{\infty} \xi^i / n^i,$$

$$\frac{n}{n-\xi_1} \cdots \frac{n}{n-\xi_r} = \sum_{i=0}^{\infty} (1/n^i) \cdot p_i(\xi_1, \cdots, \xi_r) \tag{15}$$

where

$$p_i(\xi_1, \cdots, \xi_r) = \sum \xi_1^{i_1} \cdots \xi_r^{i_r}$$
$$(i_1 \geqq 0, \cdots, i_r \geqq 0, \qquad i_1 + \cdots + i_r = i).$$

In the individual terms we extend the integration down to $x_i = 0$ instead of stopping at $N^{-1/2}$; the error to be expected will not be appreciably higher than $O(N^{-1/2})$. We therefore form the integrals

$$\gamma_i = \gamma_i(x) = \int_{E'(x)} p_i(\xi_1, \cdots, \xi_r) \cdot d\sigma \tag{16}$$

extending over the intersection $E'(x)$ of the plane $E(x)$ with the block

$$B: \qquad 0 \leqq x_1 \leqq \beta_1, \cdots, \qquad 0 \leqq x_r \leqq \beta_r,$$

and then arrive at

THEOREM IV. *$R_N(x)$ has the asymptotic expansion*

$$R_N(x) \sim \gamma_0 + \gamma_1 n^{-1} + \gamma_2 n^{-2} + \cdots \tag{17}$$

in the sense that the absolute error committed by breaking off after the g-th term $\gamma_{g-1} n^{-(g-1)}$ is less than $C_g n^{-g}$, the constant C_g being independent of N and x.

For $A(N, k)$ we find the asymptotic formula [3]

$$A(N, k) \sim (N^{r-1}/n^r) \mathfrak{S}(k) (\gamma_0(\kappa) + \gamma_1(\kappa) n^{-1} + \cdots), \tag{18}$$

[3] Vinogradoff has merely the first term of this asymptotic expansion, and the error he gives is $O(\log n/n)$ instead of $O(1/n)$. In the paper cited above, he follows an intricate and indirect approach. Much simpler is the method given on pp. 55-57 of his paper in *Travaux de l'Institut Mathématique de Tbilissi*, vol. 3 (1938). But he uses the formula (6) instead of our (12).

uniformly valid for all k; the letter κ stands for k/N. The thoroughly elementary proof follows in the next section. $R_N(x)$ and $\gamma_i(x)$ vanish outside the interval $A \leqq x \leqq A'$, $R_N(x)$ is non-negative throughout, and the same holds for $\gamma_i(x)$ provided $\beta_1 \leqq 1, \cdots, \beta_r \leqq 1$.

The first coefficient γ_0 is of particular interest. It equals the area of the $(r-1)$-dimensional convex polygon $E'(x)$. Using the method of convolution explained in the next section, we obtain by induction with respect to r this fairly explicit result: Put $\alpha_i = a_i\beta_i$ and for any function $f(x)$ denote the difference $f(x) - f(x - \alpha_i)$ by $\Delta_i f(x)$. Introduce

$$U(x) = \begin{cases} 0 & \text{for} \quad x \leqq 0, \\ \dfrac{x^{r-1}}{(r-1)!} & \text{for} \quad x \geqq 0. \end{cases}$$

Then

$$\gamma_0(x) = \frac{1}{a_1 \cdots a_r} \Delta_1 \cdots \Delta_r U(x).$$

We add a few obvious observations concerning the question under which circumstances some of the inequalities (2) are redundant.

If all $a_1, \cdots, a_r$ are positive, one will ask for the number of representations of a large integer N in the form

$$N = a_1 p_1 + \cdots + a_r p_r$$

by means of r primes p_i, without introducing any upper bounds for the p_i. However, one may add the conditions

$$p_1 \leqq N/a_1, \cdots, p_r \leqq N/a_r$$

without altering the problem. Hence it falls under our scheme. The number $A(N)$ of representations equals $(N^{r-1}/n^r) R_N \cdot \mathfrak{S}(N)$ with a relative error of the type ρ mentioned in Theorem I, and for R_N and the coefficients γ_i of its asymptotic expansion we have expressions which arise from (14) and (16), respectively, by setting $x = 1$ and replacing the blocks B_N and B by the "infinite" blocks

$$B_N^{(0)}: \quad x_1 \geqq N^{-1/2}, \cdots, x_r \geqq N^{-1/2},$$
$$B^{(0)}: \quad x_1 \geqq 0, \cdots, x_r \geqq 0$$

respectively. In particular,

$$\gamma_0 = \frac{1}{a_1 \cdots a_r} \cdot \frac{1}{(r-1)!}$$

If $a_1, \cdots, a_s$ are positive and $a_{s+1}, \cdots, a_r$ negative, then the equation (1) together with the $r - s$ inequalities

$$(2_s) \qquad p_{s+1} \leqq \beta_{s+1}N, \cdots, p_r \leqq \beta_r N$$

results in the following bounds for $p_1, \cdots, p_s$:

$$p_i \leqq \frac{x_0 - \alpha_{s+1} - \cdots - \alpha_r}{a_i} N \qquad (i = 1, \cdots, s)$$

as long as $k \leqq x_0 N$. Hence the number $A(N, k)$ of solutions of (1) in primes satisfying the inequalities (2_s) is asymptotically given by a formula (18) where $\gamma_i(x)$ is the integral (16) extending over the intersection of $E(x)$ with the partially infinite block

$$x_1 \geqq 0, \cdots, x_s \geqq 0, \qquad 0 \leqq x_{s+1} \leqq \beta_{s+1}, \cdots, \qquad 0 \leqq x_r \leqq \beta_r.$$

Its validity, however, is uniform only for $\kappa \leqq x_0$ if x_0 is any preassigned number. In particular,

$$\gamma_0(x) = \frac{1}{a_1 \cdots a_r} \cdot \Delta_{s+1} \cdots \Delta_r U(x).$$

Examples: 1) $p_1 + p_2 - p_3 = k, \qquad p_3 \leqq N.$

$\gamma_0(x) = \frac{1}{2} + x$	$\frac{1}{2}(1+x)^2$	0
for $x \geqq 0$	$-1 \leqq x \leqq 0$	$x \leqq -1$

2) $\qquad p_1 + p_2 - p_3 = k; \qquad p_1 \leqq N, \quad p_2 \leqq N.$

Write the equation in the form $p_3 - p_1 - p_2 = -k$.

$\gamma_0(x) = 1$	$1 - \frac{1}{2}x^2$	$\frac{1}{2}(2+x)^2$	0
for $x \geqq 0$	$-1 \leqq x \leqq 0$	$-2 \leqq x \leqq -1$	$x \leqq -2$

3. Proofs of Theorems III and IV. We have to investigate the integral

$$F(x) = \int_{E(x)} \phi_1(x_1) \cdots \phi_r(x_r) \cdot d\sigma,$$

where the ϕ_i are defined by (13). Let us, more generally, discuss the case of arbitrary densities ϕ_i each vanishing outside a finite interval. Using the x_1-axis as the direction in which we erect a cylinder over the area $d\sigma$ on $E(x)$ we obtain

$$(x) = \frac{1}{|a_1|} \cdot \int_{-\infty}^{+\infty} \cdots \int \phi_1\left(\frac{x - a_2x_2 - \cdots - a_rx_r}{a_1}\right) \cdot \phi_2(x_2) \cdots \phi_r(x_r)\, dx_2 \cdots d$$

Write

$$\phi^{[i]}(x) = \frac{1}{|a_i|} \cdot \phi\left(\frac{x}{a_i}\right)$$

and introduce the linear, commutative and distributive operation o of convolution by

$$f_1 \circ f_2 = \int_{-\infty}^{+\infty} f_1(x-y) f_2(y)\,dy = \int_{-\infty}^{+\infty} f_2(x-y) f_1(y)\,dy.$$

Then

$$F = \phi_1^{[1]} \circ \phi_2^{[2]} \circ \cdots \circ \phi_r^{[r]}.$$

The notation $\times \phi_1 \times \phi_2 \times \cdots \times \phi_r$ will also be used for this convolution. But then a permutation of the factors must be accompanied by the same permutation of the numbers $a_1, \cdots, a_r$ on which the definition of the $\times$ multiplication is based.

Let f_1, f_2 denote functions vanishing outside a finite interval.

Lemma 1. *If f_1, f_2 are piecewise continuous, then $f = f_1 \circ f_2$ is continuous. $|f| \leqq |f_1| \circ |f_2|$, and $|f_1(x)| \leqq M_1$ implies $|f(x)| \leqq M_1 \cdot \int |f_2|\,dx$.*

Suppose f_1 continuous and f_2 of bounded variation, and set

$$g = (f_1, df_2) = \int_{-\infty}^{+\infty} f_1(x-y) \cdot df_2(y).$$

If

$$\int_{-\infty}^{x} f_1 dx = F_1(x), \qquad \int_{-\infty}^{x} g dx = G(x),$$

then

$$G = (F_1, df_2) = f_1 \circ f_2.$$

Hence:

Lemma 2. *If f_1 is continuous and f_2 of bounded variation, then $g = (f_1, df_2)$ is continuous, and $|f_1(x)| \leqq M_1$ implies $|g(x)| \leqq M_1 \cdot \int |df_2|$. The derivative of $f = f_1 \circ f_2$ is g, and if f_1 has a continuous derivative f'_1 then $g' = (f'_1, df_2)$.*

From these lemmas we deduce that our $F = f_1 \circ \cdots \circ f_r$ whose factors are $f_i = \phi_i^{[i]}$, ϕ_i being defined by (13), has the continuous derivative

$$F' = (f_1 \circ \cdots \circ f_{r-1}), df_r.$$

One may even continue derivation up to the order $r-2$:

$$F^{(r-2)} = (f_1 \circ f_2),\ df_3, \cdots, df_r.$$

The $(r-1)$-th derivative may still be written in a similar form but it has discontinuities at the points

$$x_1 = a_1\rho_1 + \cdots + a_r\rho_r \quad . \qquad (\rho_i = \delta_i \text{ or } \delta'_i) \tag{19}$$

where $\delta_i = \sqrt{N}$, $\delta'_i = \beta_i N$ are the discontinuities of $\phi_i(x)$. One may readily write down the jumps of $F^{(r-1)}$ at the points (19) in terms of the jumps $\Delta_i(\rho_i)$ of the individual functions ϕ_i. As a matter of fact, our $R = F$ is piecewise *analytic*, the pieces being sewed together at the seams (19) in such a fashion that the function itself and its derivatives up to the order $r-2$ go over continuously. We may also confirm (11) by this method, even including $\mu = r-1$.

The proof of Theorem IV would hardly need any explanation were $\xi = -\log x$ bounded in the intervals in which it is used. This being not the case,[4] we feel obliged to go into details. First determine the remainder P_g of the expansion (15) left over when one breaks off after the g-th term. Applying to each factor $n/(n-\xi_i)$ in succession the finite geometric progression broken off after a suitable term, with its well-known remainder, we find that P_g equals the sum of all terms of the following form

$$n^{-g}\cdot\xi_1^{i_1}\cdots\xi_{s-1}^{i_{s-1}}\cdot\xi_s^h\frac{n}{n-\xi_s}\cdot\frac{n}{n-\xi_{s+1}}\cdots\frac{n}{n-\xi_r}$$

where $1\leqq s\leqq r$, $i_1+\cdots+i_{s-1}=i<g$ and $h=g-i$. An additional feature is the extension of the integration down to the point $x=0$ instead of $N^{-1/2}$ in the terms of the approximating sum. Therefore the following situation prevails. We denote by $l^*(x)$ the typical function equaling

$$n/(n-\xi)\ \text{for}\ N^{-1/2}\leqq x\leqq\beta,\qquad 0\ \text{elsewhere,}$$

and by $l(x)$ any function

$$\xi^g\ \text{for}\ 0\leqq x\leqq\beta,\qquad 0\ \text{elsewhere,}$$

with an integral exponent $g\geqq 0$. This particular $l(x)$ is said to be of order g. Moreover we introduce a function $L(x)$ of order $g\geqq 1$ which equals

$$\xi^g n/(n-\xi)\ \text{for}\ N^{-1/2}\leqq x\leqq\beta,\qquad 0\ \text{elsewhere,}$$

and a function $\Lambda(x)$ of order $g\geqq 0$ equaling

$$-\xi^g/n^g\ \text{for}\ 0\leqq x\leqq N^{-1/2},\qquad 0\ \text{elsewhere.}$$

Then the difference

$$R_N(x)-(\gamma_0(x)+\cdots+\gamma_{g-1}(x)\cdot n^{-(g-1)})$$

appears as composed of terms

$$n^{-g}(\times l_1\times\cdots\times l_{s-1}\times L_s\times l^*_{s+1}\times\cdots\times l^*_r) \tag{20}$$

and

$$n^{-i}(\times l_1\times\cdots\times l_{s-1}\times\Lambda_s\times l^*_{s+1}\times\cdots\times l^*_r). \tag{21}$$

The sum of the orders of $l_1,\cdots,l_{s-1}$ is any $i<g$ and the order of L_s in (20) is $g-i$ whereas the order of Λ_s in (21) is any number $<g-i$.

---

[4] It would be the case, though, if we replaced (2) by conditions $\beta'_i N\leqq p_i\leqq\beta_i N$ with *positive* β'_i. If the differences $\beta_i-\beta'_i$ are small, this amounts to an approximate fixation of the ratio $p_1:p_2:\cdots:p_r$. Cf. van der Corput, l. c.[1].

We now reap our reward for having raised the lower limit in the integral J from Vinogradoff's 2 to $\sqrt{N}$: l^* and Λ vanish to the left of the point $N^{-1/2}$ and for $x \geqq N^{-1/2}$ we have $n/(n-\xi) \leqq 2$. Consequently only these two facts have to be established:

(22) $$\times |\cdot l_1| \times \cdots \times |l_r| \leqq C$$

and

(23) $$\times |l_1| \times \cdots \times |l_{s-1}| \times |\Lambda_s| \times |l_{s+1}| \times \cdots \times |l_r| \leqq C \cdot N^{-1/2}.$$

($l_{s+1}, \cdots, l_r$ are l's of order 0; but it is superfluous to make use of this particular circumstance.) The first inequality holds for $r \geqq 2$. If it is true for $r = 2$ it follows by induction from Lemma 1 as soon as we make sure that

$$\int | l(x) | \, dx \leqq C.$$

But this inequality is trivial in view of the explicit expression

(24) $$\int_0^x (\log x)^g dx = x\{(\log x)^g - g(\log x)^{g-1} + \cdots \pm g!\}.$$

For $r = 2$ we have to show that

$$\int_0^x |(\log y/a_1)^g \cdot (\log (x-y)/a_2)^h| \, dy$$

is bounded as long as the parameter $x > 0$ stays bounded. The substitution $y = x \cdot \eta$ carries our integral over into a form

$$x \cdot \int_0^1 |(\log x/a_1 + \log \eta)^g \cdot (\log x/a_2 + \log (1-\eta))^h| \, d\eta$$

from which this fact follows immediately. (The essential point is that the integral tends to 0 with $x \to 0$.)

In (23) we push the factor Λ_s to the rear and apply the result (22) to $r-1$ instead of r. Formula (24) gives

$$\int |\Lambda| \, dx \leqq n^{-g} \cdot \int_0^{N^{-1/2}} |\log x|^g dx \leqq C \cdot N^{-1/2},$$

and hence by Lemma 1 the inequality (23) is established.

This ends our proof of Theorem IV. Moreover we learn from the inequality (22) that the coefficients $\gamma_i(x)$ are bounded functions of x; it would be quite easy to prove their continuity.

4. The singular series. The investigation of $\mathfrak{S}$ follows the familiar Hardy-Littlewood pattern. Since $\chi(q)$ has the multiplicative property $\chi(q \cdot q') = \chi(q) \cdot \chi(q')$ for any co-prime q, q', the sum $\mathfrak{S}$ is the product over all primes p of

$$1 + \chi(p) + \chi(p^2) + \cdots.$$

A simple computation yields

THEOREM V. *If d is the greatest common divisor of the numbers*

$$k;\ a_1, \cdots, a_r \tag{25}$$

then

$$\mathfrak{S}(k; a_1, \cdots a_r) = d \cdot \mathfrak{S}(k/d; a_1/d, \cdots, a_r/d). \tag{26}$$

Assuming (25) *without common divisor, we distribute the prime numbers p over $r+1$ classes, p being in the class Γ_t $(t = 0, 1, \cdots, r)$ if $t+1$ of the numbers* (25) *are prime to p. Then*

$$\mathfrak{S} = \prod_{t=0}^{r} \prod_{p \in \Gamma_t} \left\{ 1 - \frac{(-1)^t}{(p-1)^t} \right\}. \tag{27}$$

A formula like (26) was to be expected because the equation (1) coincides with

$$(a_1/d) \cdot p_1 + \cdots + (a_r/d) \cdot p_r = k/d.$$

It was also to be expected, as (27) indeed proves, that $\mathfrak{S}$ vanishes if there are prime numbers p which go into r of the numbers (25) (but not into all of them), or if

$$a_1 + \cdots + a_r \not\equiv k \qquad (\text{mod } 2).$$

In all other cases we deduce from (27):

$$\mathfrak{S} \geqq 2 \cdot \prod_{\nu=1}^{\infty} \{1 - 1/4\nu^2\} = 4/\pi.$$

5. Prime powers. For any given integral exponent $l \geqq 1$ we may study this set of relations

$$\begin{aligned} &a_1 x_1^{\,l} + \cdots + a_r x_r^{\,l} = k, \\ &x_1^{\,l} \leqq \beta_1 N, \cdots, x_r^{\,l} \leqq \beta_r N. \end{aligned} \tag{28}$$

and may ask for solutions in non-negative *integers,* or more specfically in positive *primes* x_i. Let the numbers of solutions of these Waring's and Vinogradoff's problems be $A^0(N, k)$ and $A(N, k)$ respectively. If necessary we add the number r of terms as a subscript, or even give the full signature

$$\left(r_l;\ \begin{matrix} a_1, \cdots, a_r \\ \beta_1, \cdots, \beta_r \end{matrix} \right)$$

Put

$$\sigma_0(a/q) = (1/q) \cdot \sum_t e(at^l/q), \qquad \sigma(a/q) = (1/\phi(q)) \cdot \sum_t e(at^l/q)$$

where the first sum extends over *all* residues t mod q while the second sum is restricted to the reduced residues t mod q. Neither sum changes if one replaces the fraction a/q by its reduced form a'/q'. Hence we may well say that they depend on the fraction rather than on a and q separately. The singular series

$$\mathfrak{S}_0(k) = \sum_q \chi_0(q) \qquad \text{and} \qquad \mathfrak{S}(k) = \sum_q \chi(q)$$

in Waring's and Vinogradoff's problems arise from these "characters"

$$\chi_0(q) = \sum_{m \in G_q} \sigma_0(a_1 m/q) \cdots \sigma_0(a_r m/q) \cdot e(-k/q),$$
$$\chi(q) = \sum_{m \in G_q} \sigma(a_1 m/q) \cdots \sigma(a_r m/q) \cdot e(-k/q),$$

respectively. The quantitative factors R_0 and R in the asymptotic formulas

$$(29) \qquad A^0(N,k) \sim \mathfrak{S}_0(k) R_0(N,k), \qquad A(N,k) \sim \mathfrak{S}(k) R(N,k)$$

are

$$l^{-r} \int (x_1 \cdots x_r)^{\lambda-1} d\sigma \qquad \text{and} \qquad \int \frac{(x_1 \cdots x_r)^{\lambda-1} d\sigma}{\log x_1 \cdots \log x_r} \qquad (\lambda = 1/l)$$

respectively. The first integral extends over the intersection of $E(x)$ with the block

$$0 \leqq x_1 \leqq \beta_1 N, \quad \cdots, \qquad 0 \leqq x_r \leqq \beta_r N,$$

the second over the intersection of the same plane with the block

$$\sqrt{N} \leqq x_1 \leqq \beta_1 N, \quad \cdots, \qquad \sqrt{N} \leqq x_r \leqq \beta_r N.$$

The first integral equals $N^{\lambda r - 1} \cdot R^0(k/N)$ where $R^0(x)$ is the same integral extending over $E'(x)$. (In computing these expressions we have made the substitution $x_i{}^l \to x_i$. The asymptotic development of $R(N,x)$ may be obtained in the same way as for $l = 1$.) We think the parallel is fairly instructive: the difference in the singular series is due to the fact that the residues of primes mod q are necessarily relatively prime to q (apart from the "few" primes going into q) whereas integers leave all residues mod q. The difference in the quantitative factor comes from the fact that 1 is the density with which the integers x are distributed over the x-axis while $1/\log x$ is the density of the primes (a fact which has to be combined with the *uniform* equidistribution over the respective residues modulo any integer q).

More irregular problems that can be tackled at all will conform to the same scheme.

The main terms of the asymptotic formulas (29) are of the order of magnitude $N^{\lambda r - 1}$ and $N^{\lambda r - 1}/n^r$ respectively. Of course we may claim their validity only for a sufficiently large number r of terms. The limit is set by the simpler of the two problems. Indeed, the efficiency of the method depends on two circumstances: 1) the contribution to (7) of part II of the circumference $0 \leqq \eta \leqq 1$ must be negligible; 2) the singular series must converge sufficiently well. The first is the main point. For our present prime number problem we have to set

$$S_i(\eta) = \sum_p e(a_i \eta p^l) \qquad (p^l \leqq \beta_i N).$$

Suppose that by analyzing Waring's problem we have ascertained an even number $s = 2u \geqq 4$ such that

$$A_s{}^0(N, 0) \leqq C \cdot N^{\lambda_s - 1} n^K, \tag{30}$$

the integral exponent $K \geqq 0$ being independent of N. This is somewhat less than the assumption of the asymptotic formula of Waring's problem for $r = s$. It is then maintained that the asymptotic formula for Vinogradoff's problem holds good for $r \geqq 1 + s$, in the sense that the relative error ρ is of the type mentioned in Theorem I. Following Vinogradoff [5] we use the relation

$$\int_0^1 | S_1(\eta) \cdots S_u(\eta) |^2 \, d\eta = A_{2u}(N, 0) \leqq A_{2u}{}^0(N, 0)$$

where A refers to the signature

$$\left(s = 2u; \begin{matrix} a_1, \cdots, a_u, & -a_1, \cdots, -a_u \\ \beta_1, \cdots, \beta_u, & \beta_1, \cdots, \beta_u \end{matrix} \right),$$

and then by Schwarz's inequality and (30) we find the relation

$$\int_0^1 | S_1(\eta) \cdots S_u(\eta) | \cdot | S_{u+1}(\eta) \cdots S_{2u}(\eta) | \, d\eta \leqq C \cdot N^{\lambda_s - 1} n^K$$

which forms the base for settling the first point

The accessory point, namely convergence of the singular series, is taken care of by the following estimate for σ due to Loo-keng Hua [6]:

$$| \sigma(a/q) | \leqq C_\epsilon \cdot q^{-\frac{1}{2} + \epsilon} \tag{31}$$

for $(a, q) = 1$ where ϵ is any positive number $< \frac{1}{2}$ and C_ϵ depends neither on a nor q. The requirement of "sufficiently good" convergence of the series $\Sigma\, \chi(q)$ means that it is majorized by a series $C \cdot \Sigma q^{-\alpha}$ with some exponent $\alpha > 1$. Because of (31) this is obviously the case as soon as $r \geqq 5$.

University of Saskatchewan, Saskatoon, Saskatchewan,
and
Institute for Advanced Study, Princeton, N. J.

[5] See Vinogradoff, l.c.[3].

[6] *Math. Zeitschr.*, vol. 44 (1938), pp. 335-346, Lemma 1. 3.

129.

On the theory of analytic curves

(H. Weyl and J. Weyl)

Proceeding of the National Academy of Sciences of the United States of America
28, 417—421 (1942)

In a paper of great importance and beauty,[1] L. V. Ahlfors has simplified and vastly expanded the theory of meromorphic curves inaugurated by the authors of this note.[2] In the meantime the younger of us had generalized the first and second main theorems of our theory to arbitrary *analytic curves* whose parameter $\mathfrak{p}$ varies over a given Riemann surface $\mathfrak{F}$.[3] The following observations will show how by a proper adaptation of Ahlfors' method this more general theory may be brought up to the same degree of completion as the special case of the meromorphic curves.

As in reference 3 let G denote any compact region of $\mathfrak{F}$ which surrounds a given nucleus G_0, and $\bar{G}$ its complement. We speak of the *condenser* G whose outer conductor is $\bar{G}$ and whose inner conductor G_0 carries unit charge. Let $\varphi = \varphi_G$ be the potential, which vanishes in $\bar{G}$ and assumes a constant value R in G_0. To be precise, the space of the condenser is $G^* - G_0^*$ where G_0^*, G^* are defined by $\varphi = R$, $\varphi > 0$, respectively; for the sake of simplicity we identify G_0^*, G^* with G_0, G. The constant R is called the potential (or the reciprocal capacity) of the condenser. Ahlfors' formula (19) for the order function T_l is generalized as follows. Let $x(\mathfrak{p})$ be our curve and t a local uniformizing parameter at any point of the Riemann surface. Form

$$X_t^l = [x, dx/dt, \ldots, d^{l-1}x/dt^{l-1}],$$

$$S_t^l = 2\,|X_t^{l-1}|^2 \cdot |X_t^{l+1}|^2/|X_t^l|^4.$$

The integral element $S_t\, dt\, d\bar{t}$ is invariant; if one uses any analytic differential dz instead of dt one gets an S_z connected with S_t by the equation

$$S_z\,|dz/dt|^2 = S_t.$$

We arrive at this expression for T_l,

$$T_l = \int\!\!\int \varphi S_t^l \cdot dt\, d\bar{t}, \qquad (1)$$

the integral extending over the entire Riemann surface (or over G only).

Our basic idea is to decompose the space of the condenser into thin layers by the equi-potential lines $\varphi = \varphi_1$. For a fixed value $\varphi_1 = R - r$ the inequality $\varphi_1 \leqq \varphi \leqq R$ defines a region G_r, and the condenser G_r has the potential $R - \varphi_1 = r$. The harmonic function φ is the real part of an analytic function $\varphi - i\sigma = f$ in $G - G_0$ which, to be sure, is not single-valued; but the differential df is, and this is all that counts. Use df instead of dt in that part of the integral (1) which extends over $G - G_0$. Omitting the index l we form the integral

$$Q(r) = \int S_f \cdot d\sigma \quad (d\sigma \geqq 0)$$

along the line C_r: $\varphi = R - r$. (Here the differentiation d/df in S_f really amounts to $d/d\sigma$.) The flux through the line, $\int d\sigma$, is the total charge 1 (not 2π; being mathematicians, we use the Heaviside units). The formula (1) now reads

$$T = A_0 R + \int_0^R \varphi \cdot Q(R - \varphi) \cdot d\varphi$$
$$= A_0 R + \int_0^R (R - r) \cdot Q(r)\, dr,$$

where A_0 is the integral of $S_l\, dt\, \overline{dt}$ over the nucleus G_0 and hence independent of G. Applying the formula to the condenser G_r we obtain for $T(G_r)$:

$$T(r) = A_0 r + \int_0^r (r - \rho)\, Q(\rho)\, d\rho.$$

This proves at once that $T(r)$ is a (positive increasing) *convex* function of the variable r and that

$$d^2T/dr^2 = Q(r). \tag{2}$$

The second main theorem has been formulated in reference 3 in two different forms, first on the basis of a given "meromorphic" function z on $\mathfrak{F}$, and then by means of the intrinsic non-Euclidean metric on $\mathfrak{F}$ to which the theory of uniformization leads. In the first form the meromorphic character of the *differential* dz *in* G is sufficient, and it is not necessary to use the same dz for each domain G. The really important relation arises from the choice $dz = df$. This differential is not defined inside G_0. Hence one must apply the fundamental formula to $G - G_0$, and the separate treatment of G_0 brings in a topological moment. In the notation used loc. cit., we arrive at the *Plücker formulas for analytic curves*

$$V_l + (T_{l+1} - 2T_l + T_{l-1}) = \Omega_l(r) \,\big|_0^R + 2\pi\eta$$

in which the compensating term $\Omega_l(r)$ is the integral

$$\Omega_l(r) = {}^1/_2 \int \log S_f^l \cdot d\sigma \quad (d\sigma \geqq 0)$$

taken along the line C_r and

$$\eta = \Sigma_{\mathfrak{w}}\varphi(\mathfrak{w}) - \nu_0 R.$$

The last sum extends over the "critical points" $\mathfrak{w}$, i.e., the zeros of df inside $G - G_0$ (or, what is the same, the zeros of the electric field strength $-\operatorname{grad}\varphi$). ν_0 is an integer, namely the Euler characteristic of G_0. It is clear that η depends on G, but neither on the index l nor on the curve $x(\mathfrak{p})$. It corresponds to the term $2p - 2$ in Plücker's formulas for an algebraic curve of genus p. The value of η is little influenced by whether or not one includes in the sum critical points $\mathfrak{w}$ near the outer wall of the condenser where $\varphi = 0$. But it seems to violate the law of continuity at the inner wall. However, according to M. Morse's now classical relation,

$$\Sigma_{\mathfrak{w}} 1 = \nu_0 + \nu,$$

where $-\nu$ is the Euler characteristic of G. Hence one can write

$$\eta = -\Sigma_{\mathfrak{w}}(R - \varphi(\mathfrak{w})) + \nu R,$$

a formula in which the critical points near the *inner* wall $\varphi = R$ count very little. Notice the inequality

$$-\nu_0 R \leqq \eta \leqq \nu R.$$

Again dropping the index l we have

$$2\Omega(r) \leqq \log Q(r).$$

The potential $R(G)$ of the condenser G increases if G is enlarged, and hence converges either to ∞ or to a finite limit R_0 under exhaustion of $\mathfrak{F}$ by G (case of infinite or finite total potential). Choose a number $\kappa > 1$. Considering that $T'(r) \geqq A_0$ and thus $T(r) \geqq A_0$ as soon as $r \geqq 1$ we find in familiar fashion from (2) that an inequality

$$\log Q(r) > \kappa^2 \cdot \log T(r) + (\kappa + 1) \log C$$

with a given positive constant C cannot hold throughout a subinterval of $1 \leqq r \leqq R$ of length B/C where

$$2\int_{A_0}^{\infty} y^{-\kappa}\, dy = 2A_0{}^{-(\kappa-1)}/(\kappa - 1) = B$$

is independent of G. Hence *in the case of infinite total potential we see that there will be values r in a boundary strip $R(G) - \beta \leqq r \leqq R(G)$ of preassigned width β for which the inequality*

$$2\Omega(r) \leqq \kappa^2 . \log T(r) + (\kappa + 1) \log B/\beta$$

holds, as soon as $R(G) \geqq \beta + 1$. With a given constant $b > 1/2$ this fact prevents a relation like

$$\Omega(G) \geqq b \cdot \log T(G)$$

from holding for all sufficiently large domains G; but it says a good deal more about the behavior of $\Omega(G)$ with respect to $\log T(G)$.

In the *case of finite total potential* R_0 the function φ_G tends to a limit φ with the exhaustion of $\mathfrak{M}$ by G, and it is then natural to use only the regions $G_r(0 \leqq r < R_0)$ defined by $R_0 - r \leqq \varphi \leqq R_0$. *Almost everywhere*, i.e., with the exception of an r-set over which the integral of $(R_0 - r)^{-1}$ is finite, *the inequality*

$$2\Omega(r) < \kappa^2 \log T(r) + (\kappa + 1) \log (R_0 - r)^{-1}$$

will hold.

In establishing the defect relations by Ahlfors' procedure, let us stick to our use of the symbol ω, as explained for the meromorphic case by formula (5.2) of reference 2; it differs from Ahlfors' usage by an arbitrary positive factor K which he adds to the function θ. In handling arbitrary analytic curves one had better abstain from Ahlfors' trick of tying up the exponent $\alpha < 1$ with the outer circle R of the annular condenser by the relation $1 - \alpha = 1/T(R)$. We include the cases of *non-general position* as in reference 3, section 7, by attaching non-negative weights $\lambda(E^h)$ to the several h-spreads E^h of the given finite set. Ahlfors' method works best for the lowest case of a finite set of 1-spreads or "points" c. Set

$$m_l(r;\ c) = \int \log \frac{|X^l| \cdot |c|}{|[X^l,\ c]|} \cdot d\sigma,$$

the integral running over the curve C_r. Suppose that the weights $\lambda(c)$ attached to the points c of the given finite set will load no l-spread by more than 1. Then

$$\Sigma_c\lambda(c)\left\{\alpha m_l(r;\ c) - m_{l-1}(r;\ c)\right\} + \Omega_l(r) = {}^1/_2\omega(T_l).$$

In particular for $l = 1$ the first term of the left member may be written as

$$\alpha \cdot \Sigma_c m_1(r;\ c)$$

under the assumption that the points c are *distinct*. It is remarkable that no further restriction on the position of the points c is required. The cases $h > 1$ in which $[X^l, E^h]$ takes the place of $[X^l, c]$ can also be treated with fixed exponents. Ahlfors' second set of defect relations arises from application to the dual curve of the first set thus obtained.

A more detailed account will probably be published in a planned monograph on analytic curves in the *Annals of Mathematics Studies*.

[1] Ahlfors, L. V., "The Theory of Meromorphic Curves," *Acta Societatis Scientiarum Fennicae*, Nova Series A, Tom. III, No. 4 (1941).

[2] Weyl, H. and J., *Ann. Math.*, **39**, 516–538 (1938).

[3] Weyl, J., *Ibid.*, **42**, 371–408 (1941).

130.

On Hodge's theory of harmonic integrals

Annals of Mathematics 44, 1—6 (1943)

The attempt which HODGE made in Chapter III of his beautiful book[1]) to establish *the existence of harmonic integrals with preassigned periods* has not been entirely successful because the proof is partly based on a false statement (p. 136) concerning the behavior of the solution of a non-homogeneous integral equation when the spectrum parameter approaches an eigen value. In a Princeton seminar on the subject, BOHNENBLUST pointed out that counter examples are readily available even for linear equations with a finite number of unknowns. For instance the equation $\lambda x + A x = c$ with

$$A = \begin{pmatrix} 0 & 1 \\ 0 & 0 \end{pmatrix}, \quad c = \begin{pmatrix} 1 \\ 0 \end{pmatrix}, \quad x = \begin{pmatrix} x_1 \\ x_2 \end{pmatrix}$$

is solvable for $\lambda = 0$ (x_1 arbitrary, $x_2 = 1$) and yet the solution for $\lambda \neq 0$,

$$x_1 = \frac{1}{\lambda}, \quad x_2 = 0$$

does not tend to a limit with $\lambda \to 0$.

In his book HODGE uses the *parametrix method* first developed for a single elliptic differential equation by LEVI and HILBERT[2]). Building on the formal foundations laid by HODGE, I will show here how the argument can be made conclusive. HILBERT'S procedure served me as a model.

Let n be the dimensionality of our Riemannian manifold. I denote by $*u$, Du the dual form and the derivative of any (linear differential) form u and use the abbreviation Δ for the operator $D*D$. For two forms u, v of rank $p, n-p$ respectively (v, u) designates the integral of the product $v \cdot u$ over the whole manifold. $(*u, u)$ is positive unless $u = 0$. An immediate consequence is

Lemma 1. *$\Delta u = 0$ implies $Du = 0$.*

Indeed $(D*Du, u) = 0$ leads to $(*Du, Du) = 0$, hence $Du = 0$.

---

[1]) W. V. D. HODGE, *The Theory and Applications of Harmonic Integrals* (Cambridge, 1941). See also Proc. London Math. Soc. [2] *41*, 483–496 (1936), where HODGE ascribes the idea of using HILBERT'S parametrix method to H. KNESER. I find it hard to judge whether a previous proof along different lines (Proc. London Math. Soc. [2] *38*, 72 (1933)) is complete, or rather how much effort is needed to make it complete. For the Euclidean case, see W. V. D. HODGE, Proc. London Math. Soc. [2] *36*, 257 (1932), and H. WEYL, Duke Math. J. *7*, 411–444 (1940).

[2]) E. E. LEVI, Mem. Soc. ital. Sci. [3a] *16* (1909); D. HILBERT, *Grundzüge einer allgemeinen Theorie der linearen Integralgleichungen* (Leipzig, 1912), pp. 223–231.

In the following, f, u, φ, η are forms of rank p and g, v, ψ, ϑ forms of rank $n-p$. The rank p is fixed; no induction with respect to p takes place. The goal is to prove the following

Theorem I. *For any given null form* g, $g \sim 0$, *the equation*

$$\Delta u = g \tag{1}$$

has a solution u.

I copy HODGE's two basic formulas (3) and (4) on pp. 132, 133 of his book, replacing $p-1$ by p and using the abbreviation $1/\gamma = (-1)^{np}(n-2)\alpha_n$. Let K be the operator with the kernel $\gamma \cdot K_p(x, y)$ which carries any form $u(x)$ into $\gamma \int K_p(x, y) \cdot u(y)$, and K' its transpose. The 'parametrix' operators Q, P with the kernels $\gamma \cdot \omega_p(x, y)$ and $\gamma \cdot \omega_{p-1}(x, y)$ are symmetric,

$$(Q v, g) = (Q g, v).$$

Finally, I set $DPD = \Pi$. HODGE's formulas read

$$K u - u = Q\Delta u + (-1)^n \Pi * u, \tag{2}$$

$$K' v - v = \Delta Q v + (-1)^n * \Pi v. \tag{2'}$$

The solutions of the equations

$$K u - u = 0, \quad K' v - v = 0$$

will be called the eigen forms of the kernels K and K' (*scilicet* 'for the eigen value 1'). We try to solve our problem by means of the non-homogeneous integral equation suggested by (2),

$$K u - u = Q g. \tag{E}$$

It is essential to study this equation not only for null forms g *but in a wider set* $\mathcal{G}$; the success of the method depends on the proper choice of that linear space $\mathcal{G}$. Here is my definition:

g belongs to $\mathcal{G}$ whenever PDg is closed,

$$DPDg = \Pi g = 0.$$

Every form of the type

$$f = \Pi v \quad (v \text{ arbitrary})$$

is said to belong to $\mathcal{F}$. Evidently $\mathcal{G}$ contains all *closed* forms g whereas all elements f of $\mathcal{F}$ are *null* forms. $\mathcal{F}$ and $\mathcal{G}$ are orthogonal:

Lemma 2. $(g, f) = 0$ *for* $g \in \mathcal{G}$, $f \in \mathcal{F}$.

Indeed, if PDg is closed, then

$$(PDg, Dv) = 0 = (Dg, PDv),$$

an equation which may at once be changed into

$$(g, DPDv) = 0.$$

I take over HODGE's Lemma I on p. 142:

Lemma 3. *If ψ is any eigen form of K' then $Q\psi$ is closed.*

For the sake of completeness I repeat the simple proof. Equation (2′) yields for $\xi = Q\psi$:

$$\Delta\xi = (-1)^{n-1} * \Pi\,\psi, \tag{3}$$

hence $D*\Delta\xi = D*D*D\xi = 0$ and then by double application of Lemma 1[1]),

$$D*D\xi = 0, \quad D\xi = 0.$$

Incidentally we learn from (3) and the intermediate equation $\Delta\xi = 0$ that $\Pi\psi = 0$, or that *the eigen forms ψ of K' lie in $\mathcal{G}$.*

We analyze the eigen forms of K and K' as follows. Within the linear space of all eigen forms $\bar\varphi$ of K we consider the subspace $\mathfrak{f}$ of the *closed* eigen forms φ and choose our basis

$$\varphi_1, \ldots, \varphi_l, \quad \bar\varphi_1, \ldots, \bar\varphi_m$$

for all eigen forms accordingly, i.e. $\varphi_1, \ldots, \varphi_l$ span $\mathfrak{f}$. Equation (2) yields

$$Q\Delta\bar\varphi = (-1)^{n-1}\Pi * \bar\varphi. \tag{4}$$

This proves on the one hand that each closed eigen form φ of K satisfies the condition $\Pi*\varphi = 0$,

Lemma 4. $*\varphi \in \mathcal{G}$ *for every* $\varphi \in \mathfrak{f}$.

It shows on the other hand that $\bar\psi = \Delta\bar\varphi$ satisfies the conditions

$$\Delta Q\bar\psi = 0, \quad \Pi\bar\psi = 0$$

because the operators $\Delta\Pi$ and $\Pi\Delta$ annihilate. It then follows from (2′) that $\bar\psi$ is an eigen form of K'. The m forms $D\bar\varphi_1, \ldots, D\bar\varphi_m$ are linearly independent by construction, and hence by Lemma 1 the same is true for the forms

$$\bar\psi_1 = \Delta\bar\varphi_1, \ldots, \quad \bar\psi_m = \Delta\bar\varphi_m.$$

The transposed kernel K' has the same number $l+m$ of linearly independent eigen forms as K. We determine a basis

$$\bar\psi_1, \ldots, \bar\psi_m; \quad \psi_1, \ldots, \psi_l \tag{5}$$

of which the $\bar\psi$'s are a part.

The integral equation (E) is solvable if and only if

$$(Qg, \psi) = 0 = (g, Q\psi)$$

for every eigen form ψ of K', or with the notation $\xi = Q\psi$, if

$$(g, \xi) = 0. \tag{6}$$

[1]) One differentiation may be saved here by applying the formula $(Ds, Dt) = 0$ holding for any two forms s, t with continuous first derivatives of rank $p-1$ and $n-p-1$ (see H. WEYL, Duke Math. J. 7, 426) 1940); these SELECTA p. 469) to $s = PD\psi$ and $t = *D\xi$ with the result $(\Pi\psi, \Delta\xi) = 0 = (*\Delta\xi, \Delta\xi)$ whence $\Delta\xi = 0 = \Pi\psi$.

Let us say that ψ is *of the first kind* when $\xi = Q\psi \in \mathcal{F}$. The forms $\bar{\psi}_1, \ldots, \bar{\psi}_m$ are of the first kind, on account of the equation (4). We choose our basis (5) so that

$$\bar{\psi}_1, \ldots, \bar{\psi}_m; \quad \psi_1, \ldots, \psi_\nu$$

span the linear manifold of all eigen forms of K' of the first kind. By Lemma 2 the relation (6) holds good for *any* $g \in \mathcal{G}$ in case ψ is of the first kind, and thus the $m+l$ conditions (6) reduce to the last $l-\nu$ of them,

$$(g, Q\psi_{\nu+1}) = 0, \ldots, \quad (g, Q\psi_l) = 0. \tag{7}$$

Let $\mathcal{G}_1$ denote the set of those forms $g \in \mathcal{G}$ which satisfy the conditions (7). *We have found that under the assumption $g \in \mathcal{G}_1$ the integral equation* (E) *has a solution u.*

For this solution u we obtain from (2):

$$Q(g - \Delta u) = (-1)^n \cdot \Pi * u, \tag{8}$$

hence $\Delta Q(g - \Delta u) = 0$. Combining this with $\Pi(g - \Delta u) = 0$ and applying (2′) to $v = g - \Delta u$ one finds

$$g - \Delta u = \psi = \bar{c}_1 \bar{\psi}_1 + \cdots + \bar{c}_m \bar{\psi}_m + c_1 \psi_1 + \cdots + c_l \psi_l$$

to be an eigen form of K'. More precisely, because of (8), $Q\psi \in \mathcal{F}$, ψ is an eigen form of the first kind, which forces $c_{\nu+1}, \ldots, c_l$ to vanish. Writing u for $u + \bar{c}_1\bar{\varphi}_1 + \cdots + \bar{c}_m\bar{\varphi}_m$ we arrive at the following

Intermediary Proposition. *For any $g \in \mathcal{G}_1$ there exists a form u and ν constants $c_1, \ldots, c_\nu$ such that*

$$g - \Delta u = c_1 \psi_1 + \cdots + c_\nu \psi_\nu. \tag{9}$$

We know from Lemma 4 that the dual form $*\varphi$ of any element φ of $\mathfrak{f}$ lies in $\mathcal{G}$. That subspace of $\mathfrak{f}$ the elements φ of which satisfy the conditions

$$(*\varphi, Q\psi_{\nu+1}) = 0, \ldots, (*\varphi, Q\psi_l) = 0$$

is of a dimensionality $\mu \geqq \nu$. Let the basis $\varphi_1, \ldots, \varphi_l$ of $\mathfrak{f}$ be so chosen that $\varphi_1, \ldots, \varphi_\mu$ span this subspace. From (9) we obtain for the ν unknowns c_β the μ linear equations

$$\sum_\beta H_{\alpha\beta} \cdot c_\beta = (g, \varphi_\alpha) \quad (\alpha = 1, \ldots, \mu; \beta = 1, \ldots, \nu) \tag{10}$$

where

$$H_{\alpha\beta} = (\psi_\beta, \varphi_\alpha).$$

I maintain:

Lemma 5. *$\|H_{\alpha\beta}\|$ is a non-singular square matrix.*

Once this is established we have reached the goal. For then the ν conditions

$$(g, \varphi_\alpha) = 0 \quad (\alpha = 1, \ldots, \nu)$$

imply $c_\alpha = 0$ whereby (9) reduces to $g - \Delta u = 0$. In other words, if $g \in \mathcal{G}$ satisfies the relations

$$(g, Q\psi_{\nu+1}) = 0, \ldots, (g, Q\psi_l) = 0; \quad (g, \varphi_1) = 0, \ldots, (g, \varphi_\nu) = 0 \tag{11}$$

then the equation (1) is solvable. A null form g fulfills all our requirements, because the φ_i and $Q\psi_i$ are closed, the first by construction, the others by Lemma 3.

Proof of Lemma 5. We have found the equations (10) to be solvable if $g \in \mathcal{G}_1$. For

$$\varphi = a_1 \varphi_1 + \cdots + a_\mu \varphi_\mu \tag{12}$$

the integral $(*\varphi, \varphi)$ is a positive definite quadratic form of $a_1, \ldots, a_\mu$. Hence we can determine the coefficients a_i in (12) so as to assign arbitrary values b_α to the integrals

$$(*\varphi, \varphi_\alpha) \quad (\alpha = 1, \ldots, \mu).$$

But $g = *\varphi \in \mathcal{G}_1$. Hence we see that the equations

$$\sum_\beta H_{\alpha\beta} c_\beta = b_\alpha \quad (\alpha = 1, \ldots, \mu; \beta = 1, \ldots, \nu)$$

have a solution c_β for arbitrary b_α. In view of $\mu \geqq \nu$ this statement is equivalent to our lemma.

In proving Theorem I we actually showed that the equation $\Delta u = g$ is solvable if $g \in \mathcal{G}$ satisfies the conditions (11). Hence each such g is a null form, and the linear space $\mathcal{G}$ is of finite dimensionality $\leqq l$ modulo the space of null forms. As $\mathcal{G}$ contains all closed forms of rank $n-p$, we find *a fortiori* that the number R'_{n-p} of linearly independent closed forms of rank $n-p$ modulo null is finite and $\leqq l$. The conditions (11) are of the type $(g, f) = 0$ where f runs over certain specified closed forms of rank p. Consider the 'inner product' (g, f) of any two closed forms g, f of rank $n-p$ and p respectively; the factors matter only modulo null. Our proof implies this further fact:

Theorem II. *If the inner product* (g, f) *vanishes for a given closed* g *and all closed* f, *then* $g \sim 0$.

It is of course also true that the product cannot vanish for a given closed f and all closed g unless $f \sim 0$. Both facts together give the duality law

$$R'_{n-p} = R'_p. \tag{13}$$

Theorem II has nothing to do with any Riemannian metric. DE RHAM's second theorem follows at once from it by means of the expression of the product (g, f) in terms of the periods of g and f (HODGE, p. 85, last line), but it is essentially simpler since it deals with closed forms only, and not with forms and cycles. Its proof on an arbitrary manifold should be correspondingly easier.

The following proposition is equivalent to Theorem I for the rank $p-1$ instead of p:

Theorem III. *For any form f there exists a uniquely determined $\eta \sim f$ such that $*\eta$ is closed. If f be closed, then η is harmonic.*

Indeed, set $f = Dt + \eta$, t being of rank $p-1$. The requirement $D*\eta = 0$ leads to the equation $D*Dt = D*f$ which is solvable by Theorem I.

The new proposition shows at once that for any rank p the space of closed forms modulo null may be identified with the space of harmonic forms. This makes the equation (13) particularly lucid because $*u$ is harmonic if u is and vice versa. The same proposition provides another proof for Theorem II, because one has merely to substitute $*\eta$ for ϑ in order to see that the vanishing of the inner product (η, ϑ) of a fixed harmonic form η with every harmonic ϑ implies $\eta = 0$. The observation (Hodge, p. 139) that on account of (2) the harmonic p-forms are eigen forms of K again proves the inequality $R'_p \leqq l$.

The link with the homology theory of cycles is established by de Rham's first theorem stating that a p-cycle C is homologous zero if the integral of every closed p-form f over C vanishes.

131.

Obituary: David Hilbert 1862—1943

Obituary Notices of Fellows of the Royal Society 4, 547—553 (1944)
American Philosophical Society Year Book 387—395 (1944)

At the beginning of this year, February 14, died in Göttingen, Germany, David Hilbert, upon whom the world looked during the last decades as the greatest of the living mathematicians. At the age of eighty-one he succumbed to a compound fracture of the thigh brought about by a domestic accident.

Hilbert was born on January 23, 1862, at the city of Königsberg in East Prussia. He descended from a family which had long been settled there and had brought forth a series of physicians and judges. During his entire life he preserved uncorrupted the Baltic accent of his home. For a long time Hilbert remained faithfully attached to the town of his forbears, and it was well deserved that in his late years it bestowed its honorary citizenship upon him. It was the University of Königsberg at which he studied, where in 1884 he received his doctor's degree, and where in 1886 he was admitted as Privatdozent; there, moreover, he was appointed Ausserordentlicher Professor, in 1892, succeeding his teacher and friend Adolf Hurwitz, and in the following year advanced to a full professorship. The continuity of this Königsberg period was interrupted only by a semester's studies at Erlangen, and by a scholar's journey undertaken during the year before his habilitation, which brought him to Felix Klein at Leipzig and to Paris where he was attracted mainly to Ch. Hermite. It was on Klein's initiative that Hilbert was called to Göttingen in 1895; there he remained until the end of his life. He was retired in the year 1930.

In 1932 he was elected to foreign membership in the American Philosophical Society.

Beginning in his student years at Königsberg a close friendship tied him to Hermann Minkowski, his junior by two years, and it was with deep satisfaction that in 1902 he succeeded in drawing Minkowski also to Göttingen. Only too early did the close collaboration of the two friends find its end with Minkowski's death in 1909. Hilbert and Minkowski were the real heroes of the great and brilliant period which mathematics experienced during the first decade of this century in Göttingen, unforgettable to those who lived through it. Klein ruled over it like a distant god, "divus Felix," from above the clouds; his high time of mathematical productivity lay behind him. Among the authors of the great number of valuable dissertations which in these fruitful years were written under Hilbert's guidance we find many Anglo-Saxon names, names of men who subsequently have played a considerable role in the development of American mathematics. The physical set-up within which this unbridled scientific life unfolded was quite modest. Not until many years after the first world war, after Felix Klein had gone and Richard Courant had succeeded him, towards the end of the sadly brief period of the German Republic, did Klein's dream of the Mathematical Institute at Göttingen come true. But soon the Nazi storm broke and those who had laid the plans and who taught there besides Hilbert were scattered over the earth, and the years after 1933 became for Hilbert years of ever deepening tragic loneliness.

Hilbert was of slight build. Above the small lower face with its goatee there rose the dome of a powerful, in later years bald, skull. He was physically agile, a tireless walker, a good skater, and a passionate gardener. Until 1925 he was of firm health. Then he fell ill of pernicious anemia. Yet this illness only temporarily paralyzed his restless activity in teaching and research. He was among the first with whom the liver treatment, inaugurated by G. R. Minot at Harvard, proved successful; undoubtedly it saved Hilbert's life at that time.

Hilbert's research left an indelible imprint on practically all branches of mathematical science. Yet in rather strictly separated successive periods he gave himself over with impassioned exclusiveness to but a single subject at a time. Perhaps his deepest investigations are those on the theory of number fields. His monumental report on the "Theorie der algebraischen Zahlkörper," which he submitted to the Deutsche Mathematiker-Vereinigung, is dated as of the year 1897, and as far as I know Hilbert did not publish another paper in this field after 1899. The methodical unity of mathematics was for him a matter of belief and experience. It appeared to him essential that—in the face of the manifold interrelations and for the sake of the fertility of research—the productive mathematician should make himself at home in all fields. I quote his own words: "The question is forced upon us whether mathematics is once to face what other sciences have long ago experienced, namely to fall apart into subdivisions whose representatives are hardly able to understand each other and whose connections for this reason will become ever looser. I neither believe nor wish this to happen; the science of mathematics as I see it is an indivisible whole, an organism whose ability to survive rests on the connection between its parts." Also theoretical physics was drawn by Hilbert into the domain of his research; during a whole decade beginning in 1912 it stood at the center of his interest. Great, fruitful problems appear to him as the life nerve of mathematics. "Just as every human enterprise prosecutes final aims," says he, "so mathematical research needs problems. Their solution steals the force of the investigator." Famous is Hilbert's lecture at the International Congress of Mathematicians at Paris in 1900 where he tries to feel out the immediate future of mathematics by posing twenty-three unsolved problems; they have indeed, as we can state today in retrospect, played an eminent role in the development of mathematics during the subsequent forty-three years. A characteristic feature of Hilbert's method is a peculiarly direct attack on problems, unfettered by any algorithms; he always goes back to the questions in their original simplicity. When it is a matter of transferring the theory of linear equations from a finite to an infinite number of unknowns he begins by getting rid of the calculatory tool of determinants. A truly great example of far reaching significance is his mastery of Dirichlet's principle which, originally springing from mathematical physics, provided Riemann with the foundation of his theory of algebraic functions and Abelian integrals, but which subsequently had fallen a victim of Weierstrass's pitiless criticism. Hilbert salvaged it in its entirety. The whole finely wrought apparatus of Calculus of Variations was here consciously set aside. We only need to mention the names R.

Courant and M. Morse to indicate what role this direct method of Calculus of Variations was destined to play in recent times. It seems to me that with Hilbert the mastering of single concrete problems and the forming of general abstract concepts are balanced in a particularly fortunate manner. He came out of a time in which the algorithm had played a more extensive part, and therefore he stressed rather strongly a conceptual procedure; but in the meantime our advance in this direction has been so uninhibited and with so little concern for a growth of the problematics in depth that many of us have begun to fear for the mathematical substance. In Hilbert simplicity and rigor go hand in hand. The growing demand for rigor, imposed by the critical reflections of the nineteenth century upon those parts of mathematics which operate in the continuum, was felt by most investigators as a heavy yoke that made their steps dragging and awkward. Full of longing and with bad conscience they gazed back upon Euler's era of happy-go-lucky analysis. With Hilbert rigor figures no longer as enemy, but as promoter of simplicity. Yet the secret of Hilbert's creative force is not plumbed by any of these remarks. A further element of it, I feel, was his sensitivity in registering hints which revealed to him general relations while solving special problems. This is most magnificently exemplified by the way along which, during his number-theoretical period, he was led to the enunciation of his general theorems on class fields and the general law of reciprocity.

In a few words we shall now recall Hilbert's most important achievements. In the years 1888–92 he proved the fundamental finiteness theorems of the *theory of invariants* for the full projective group. His method, though yielding a proof for the existence of a finite basis for the invariants, does not enable one actually to construct it in a concrete individual case. Hence the exclamation by the great algorithmician P. Gordan, at the appearance of Hilbert's paper: "This is not mathematics; this is theology!" It reveals an antithesis which reaches down to the very roots of mathematics. Hilbert, however, in further penetrating investigations, furnished the means for a finite execution of the construction.

His papers on the theory of invariants had the unexpected effect of withering, as it were overnight, a discipline which so far had stood in full bloom. Its central problems he had finished once and for all. Entirely different was his effect on the *theory of number fields,* which he took up in the years 1892–98. It is a great pleasure to watch how, step by step in a succession of papers ascending from the special to the general, the adequate concepts and methods are evolved and the essential connections come to light. These papers proved of extraordinary fertility for the future. On the purely number-theoretical side I mention the names of Furtwängler, Takagi, Artin, Hasse, Chevalley, and on the number-and-function theoretical one, those of Fuëter and Hecke.

During the subsequent period, 1898–1902, the *foundations of geometry* are nearest to Hilbert's heart, and he is seized by the idea of axiomatics. The soil was well prepared, especially by the Italian school of geometers. Yet it was as if over a landscape, wherein but a few men with a superb sense of orientation had found their way in murky twilight, the sun had risen all at once. Clear and clean-cut we find stated the axiomatic conception according to which geometry is a hypothetical deductive system; it depends on the "implicit definitions" of the concepts of spatial objects and relations which the axioms contain, and not on a description of their intuitive content. A complete and natural system of geometric axioms is set up. They are required to satisfy the logical demands of consistency, independence, and completeness, and by means of quite a few peculiar geometries, constructed *ad hoc*, the proof of independence is furnished in detail. The general ideas appear to us today almost banal, but in these examples Hilbert unfolds his typical wealth of invention. While in this fashion the geometric concepts become formalized, the logical ones function as before in their intuitive significance. The further step where logic too succumbs to formalization, thus giving rise to a purely symbolic mathematics—a step upon which Hilbert already pondered at this epoch, as a paper read to the International Congress of 1904 proves, and which is inevitable for the ultimate justification of the role played by the infinite in mathematics—was systematically followed up by Hilbert during the final years of his mathematical productivity, from 1922 on. In contrast to L. E. J. Brouwer's intuitionism, which finds itself forced to abandon major parts of historical mathematics as untenable, Hilbert attempts to save the holdings of mathematics in their entirety by proving its formalism free of contradiction. Admittedly the question of truth is thus shifted into the question of consistency. To a limited extent the latter has been established by Hilbert himself in collaboration with P. Bernays, by J. von Neumann, and G. Gentzen. In recent times, however, the entire enterprise has become questionable on account of K. Gödel's surprising discoveries. While Brouwer has made clear to us to what extent the intuitively certain falls short of the mathematically provable, Gödel shows conversely to what extent the intuitively certain goes beyond what (in an arbitrary but fixed formalism) is capable of mathematical proof. The question for the ultimate foundations and the ultimate meaning of mathematics remains open; we do not know in which direction it will find its final solution nor even whether a final objective answer can be expected at all. ''Mathematizing'' may well be a creative activity of man, like language or music, of primary originality, whose historical decisions defy complete objective rationalization.

A chance occasion, a lecture in 1901 by the Swedish mathematician, E. Holmgren, in Hilbert's seminar dealing with the, then but recently published, now classical paper of Fredholm's on *integral equations,* provided the impulse which started Hilbert on his investigations on this subject that absorbed his attention until 1912. Fredholm had limited himself to setting up the analogue of the theory of linear equations, while Hilbert recognizes that the analogue of the transformation onto principal axes of quadratic forms yields the theory of the eigenvalues and eigenfunctions for the vibration problems of physics. He develops the parallel between integral equations and sum equations in infinitely many unknowns, and subsequently proceeds to push ahead from the spectral theory of ''completely continuous'' to the much more general one of ''bounded'' quadratic forms. Today these things present themselves to us in the framework of a general theory of Hilbert space. Astonishing indeed is the variety of interesting applications which integral equations find in the most diverse branches of mathematics and physics. I mention Hilbert's own solution of Riemann's problem of monodromy for linear differential equations, a far reaching generalization of the existence theorem for algebraic functions on a preassigned Riemann surface, and his treatment of the kinetic theory of gases, also the completeness theorem for the representations of a continuous compact group, and finally in recent times the construction of harmonic integrals on an arbitrary Riemannian manifold, successfully accomplished by the use of Hilbertian means. Thus only under Hilbert's hands did the full fertility of Fredholm's great idea unfold. But it was also due to his influence that the theory of integral equations became a world-wide fad in mathematics for a considerable length of time, producing an enormous literature for the most part of rather ephemeral value. It was not merit but a favor of fortune when, beginning in 1923 (Heisenberg, Schrödinger) the spectral theory in Hilbert space was discovered to be the adequate mathematical instrument of quantum physics. This later impulse led to a re-examination of the entire complex of problems with refined means (J. von Neumann, M. Stone, and others).

The integral equations are followed by Hilbert's physical period. Significant though it was for Hilbert's rounded personality as a scientist, it produced a lesser harvest than the purely mathematical ones, and may here be passed over. I shall mention instead two single, somewhat isolated, accomplishments that were to have a great effect: his vindication of Dirichlet's principle; and his proof of a famous century-old conjecture of Waring's, carrying the statement that every integer can be written as a sum of four squares over from squares to arbitrary powers. The physical period is finally succeeded by the last one, already mentioned above, in the course of which Hilbert gives an entirely new turn to the question concerning the foundation and the truth content of mathematics itself. A fruit of Hilbert's pedagogical activity during this period is the charming book by him and Cohn-Vossen, *Anschauliche Geometrie.*

This summary, though far from being complete, may suffice to indicate the universality and depth of Hilbert's mathematical work. He has impressed the seal of his spirit upon a whole era of mathematics. And yet I do not believe that his research work alone accounts for the brilliance that eradiated from him, nor for his tremendous influence. Gauss and Riemann, to mention two other Göttingers, were greater mathematicians than Hilbert, and yet their immediate effect upon their contemporaries was undoubtedly smaller. Part of this is certainly due to the changing conditions of time, but the character of the men is probably more decisive. Hilbert's was a nature filled with the zest of living, seeking the intercourse of other people, and delighting in the exchange of scientific ideas. He had his own free manner of learning and teaching. His comprehensive mathematical knowledge he acquired not so much from lectures as in conversations with Minkowski and Hurwitz. "On innumerable walks, at times undertaken day by day," he tells in his obituary on Hurwitz, "we browsed in the course of eight years through every corner of mathematical science." And as he had learned from Hurwitz, so he taught in later years his own pupils—on far flung walks through the woods surrounding Göttingen or, on rainy days, as peripatetics, in his covered garden walk. His optimism, his spiritual passion, and his unshakable faith in the value of science were irresistibly contagious. He says: "The conviction of the solvability of each and every mathematical problem spurs us on during our work; we hear within ourselves the steady call: there is the problem; search for the solution. You can find it by sheer thinking, for in mathematics there is no *ignorabimus.*" His enthusiasm did get along with criticism, but not with scepticism. The snobbish attitude of pretended indifference, of "merely fooling around with things" or even of playful cynicism, did not exist in his circle. Hilbert was enormously industrious; he liked to quote Lichtenberg's saying: "Genius is industry." Yet for all this there was light and laughter around him. Under the influence of his dominating power of suggestion one readily considered important whatever he did; his vision and experience inspired confidence in the fruitfulness of the hints he dropped. It is moreover decisive that he was not merely a scientist but a scientific personality, and therefore capable not only of teaching the technique of his science but also of being a spiritual leader. Although not committing himself to one of the established epistemological or metaphysical doctrines, he was a philosopher in that he was concerned with the life of the idea as it realizes itself among men and as an indivisible whole; he had the force to evoke it, he felt responsible for it in his own sphere, and measured his individual scientific efforts against it. Last, not least, also the environment helped. A university such as Göttingen, in the halcyon days before 1914, was particularly favorable for the development of a living scientific school. Once a band of disciples had gathered around Hilbert, intent upon research and little worried by the chore of teaching, it was but natural that in joint competitive aspiration of related aims each should stimulate the other; there was no need that everything come from the master.

His homeland and America were those among all countries to feel Hilbert's impact most thoroughly. His influence upon American mathematics was not restricted to his immediate pupils. Thus, for instance, the Hilbert of the foundations of geometry had a profound effect on E. H. Moore and O. Veblen; the Hilbert of integral equations on George D. Birkhoff.

A picture of Hilbert's personality should also touch upon his attitude regarding the great powers in the lives of men: social and political organization, art, religion, morals and manners, family, friendship, love. Suffice it to say here that he was singularly free from all national and racial prejudices, that in all questions, political, social, or spiritual, he stood forever on the side of freedom, frequently in isolated opposition against the compact majority of his environment. Unforgotten by all those present remains the unanimous and prolonged applause which greeted him in 1928 at Bologna, the first International Congress of Mathematicians at which, following a lengthy struggle, the Germans were once more admitted. It was a telling expression of veneration for the great mathematician whom everyone knew to have risen from a severe illness, but at the same time an expression of respect for the independent attitude, "au dessus de la mêlée," from which he had not wavered during the world conflict. With veneration, gratitude, and love his memory will be preserved beyond the gates of death by many a mathematician in this country and abroad.

132.

David Hilbert and his mathematical work

Bulletin of the American Mathematical Society 50, 612—654 (1944)
Portugiesische Übersetzung: Boletin da Sociedade de Matemática de São Paulo 1, 76—104 (1946) und 2, 37—60 (1947)

A great master of mathematics passed away when David Hilbert died in Göttingen on February the 14th, 1943, at the age of eighty-one. In retrospect it seems to us that the era of mathematics upon which he impressed the seal of his spirit and which is now sinking below the horizon achieved a more perfect balance than prevailed before and after, between the mastering of single concrete problems and the formation of general abstract concepts. Hilbert's own work contributed not a little to bringing about this happy equilibrium, and the direction in which we have since proceeded can in many instances be traced back to his impulses. No mathematician of equal stature has risen from our generation.

America owes him much. Many young mathematicians from this country, who later played a considerable role in the development of American mathematics, migrated to Göttingen between 1900 and 1914 to study under Hilbert. But the influence of his problems, his viewpoints, his methods, spread far beyond the circle of those who were directly inspired by his teaching.

Hilbert was singularly free from national and racial prejudices; in all public questions, be they political, social or spiritual, he stood forever on the side of freedom, frequently in isolated opposition against the compact majority of his environment. He kept his head clear and was not afraid to swim against the current, even amidst the violent passions aroused by the first world war that swept so many other scientists off their feet. It was not mere chance that when the Nazis "purged" the German universities in 1933 their hand fell most heavily on the Hilbert school and that Hilbert's most intimate collaborators left Germany either voluntarily or under the pressure of Nazi persecution. He himself was too old, and stayed behind; but the years after 1933 became for him years of ever deepening tragic loneliness.

It was another Germany in which he was born on January 23, 1862, and grew up. Königsberg, the eastern outpost of Prussia, the city of Kant, was his home town. Contrary to the habit of most German students who used to wander from university to university, Hilbert studied at home, and it was in his home university that he climbed the first rungs of the academic ladder, becoming Privatdozent and in due time ausserordentlicher Professor. During his entire life he preserved uncorrupted the characteristic Baltic accent. His reputation as a leading algebraist was well established when on Felix Klein's initiative he was offered a full professorship at Göttingen in 1895. From then on until the end of his life Hilbert remained in Göttingen. He was retired in 1930.

When one inquires into the dominant influences acting upon Hilbert in his formative years one is puzzled by the peculiarly ambivalent character of his relationship to Kronecker: dependent on him, he rebels against him. Kronecker's work is undoubtedly of paramount importance for Hilbert in his algebraic period. But the old gentleman in Berlin, so it seemed to Hilbert, used his power and authority to stretch mathematics upon the Procrustean bed of arbitrary philosophical principles and to suppress such developments as did not conform: Kronecker insisted that existence theorems should be proved by explicit construction, in terms of integers, while Hilbert was an early champion of Georg Cantor's general set-theoretic ideas. Personal reasons added to the bitter feeling.[1] A late echo of this old feud is the polemic against Brouwer's intuitionism with which the sexagenarian Hilbert opens his first article on "Neubegründung der Mathematik" (1922): Hilbert's slashing blows are aimed at Kronecker's ghost whom he sees rising from his grave. But inescapable ambivalence even here —while he fights him he follows him: reasoning along strictly intuitionistic lines is found necessary by him to safeguard non-intuitionistic mathematics.

More decisive than any other influence for the young Hilbert at Königsberg was his friendship with Adolf Hurwitz and Minkowski. He got his thorough mathematical training less from lectures, teachers or books, than from conversation. "During innumerable walks, at times undertaken day after day," writes Hilbert in his obituary on Hurwitz, "we roamed in these eight years through all the corners of mathematical science, and Hurwitz with his extensive, firmly grounded and well ordered knowledge was for us always the leader." Closer and of a very intimate character was Hilbert's lifelong friendship with Minkowski. The Königsberg circle was broken up when Hurwitz in 1892 left for Zürich, soon to be followed by Minkowski. Hilbert first became Hurwitz's successor in Königsberg and then moved on to Göttingen. The year 1902 saw him and Minkowski reunited in Göttingen where a new chair of mathematics had been created for Minkowski. The two friends became the heroes of the great and brilliant period which our science experienced during the

[1] How Georg Cantor himself in his excitability suffered from Kronecker's opposition is shown by his violent outbursts in letters to Mittag-Leffler; see A. Schoenflies, *Die Krisis in Cantors mathematischem Schaffen*, Acta Math. vol. 50 (1928) pp. 1–23.

following decade in Göttingen, unforgettable to those who lived through it. Klein, for whom mathematical research had ceased to be the central interest, ruled over it as a distant but benevolent god in the clouds. Too soon was this happy constellation dissolved by Minkowski's sudden death in 1909. In a memorial address before the Göttingen Gesellschaft der Wissenschaften, Hilbert spoke thus about his friend: "Our science, which we loved above everything, had brought us together. It appeared to us as a flowering garden. In this garden there are beaten paths where one may look around at leisure and enjoy oneself without effort, especially at the side of a congenial companion. But we also liked to seek out hidden trails and discovered many a novel view, beautiful to behold, so we thought, and when we pointed them out to one another our joy was perfect."

I quote these words not only as testimony of a friendship of rare depth and fecundity that was based on common scientific interest, but also because I seem to hear in them from afar the sweet flute of the Pied Piper that Hilbert was, seducing so many rats to follow him into the deep river of mathematics. If examples are wanted let me tell my own story. I came to Göttingen as a country lad of eighteen, having chosen that university mainly because the director of my high school happened to be a cousin of Hilbert's and had given me a letter of recommendation to him. In the fullness of my innocence and ignorance I made bold to take the course Hilbert had announced for that term, on the notion of number and the quadrature of the circle. Most of it went straight over my head. But the doors of a new world swung open for me, and I had not sat long at Hilbert's feet before the resolution formed itself in my young heart that I must by all means read and study whatever this man had written. And after the first year I went home with Hilbert's *Zahlbericht* under my arm, and during the summer vacation I worked my way through it—without any previous knowledge of elementary number theory or Galois theory. These were the happiest months of my life, whose shine, across years burdened with our common share of doubt and failure, still comforts my soul.

The impact of a scientist on his epoch is not directly proportional to the scientific weight of his research. To be sure, Hilbert's mathematical work is of great depth and universality, and yet his tremendous influence is not accounted for by it alone. Gauss and Riemann, to mention two other Göttingers, are certainly of no lesser stature than Hilbert, but they made little stir among their contemporaries and no "school" of devoted followers formed around them. No doubt this is due in part to the changing conditions of time, but the character of the men is probably more decisive. A taste for solitude, even obscurity, is in no way irreconcilable with great creative gifts. But Hilbert's was a nature filled with the zest of living, seeking intercourse with other people, above all with younger scientists, and delighting in the exchange of ideas. Just as he had learned from Hurwitz, so he taught his own pupils, at least those in whom he took a deeper personal interest: on far-flung walks through the woods surrounding Göttingen or, on rainy days, as "peripatetics" in his covered garden walk. His optimism, his spiritual passion, his unshakable faith in the supreme value of science, and his firm confidence in the power of reason to find simple and clear answers to simple and clear questions were irresistibly contagious. If Kant through critique and analysis arrived at the principle of the supremacy of practical reason, Hilbert incorporated, as it were, the supremacy of pure reason—sometimes with laughing arrogance (*arrogancia* in the Spanish sense), sometimes with the ingratiating smile of intellect's spoiled child, but most of the time with the seriousness of a man who believes and must believe in what is the essence of his own life. His enthusiasm was compatible with critical acumen, but not with scepticism. The snobbish attitude of pretended indifference, of "merely fooling around with things," or even of playful cynicism, were unknown in his circle. You had better think twice before you uttered a lie or an empty phrase to him: his directness could be something to be afraid of. He was enormously industrious and liked to quote Lichtenberg's saying: "Genius is industry." Yet for all this there was light and laughter around him. He had great suggestive power; it sometimes lifted even mediocre minds high above their natural level to astonishing, though isolated achievements. I do not remember which mathematician once said to him: "You have forced us all to consider important those problems which you considered important." His vision and experience inspired confidence in the fruitfulness of the hints he dropped. He did not hide his light under a bushel. In his papers one encounters not infrequently utterances of pride in a beautiful or unexpected result, and in his legitimate satisfaction he sometimes did not give to his predecessors on whose ideas he built all the credit they deserved. The problems of mathematics are not isolated problems in a vacuum; there pulses in them the life of ideas which realize themselves *in concreto* through our human endeavors in our historical existence, but forming an indissoluble whole transcend any particular science. Hilbert had the power to evoke this life; against it he measured his individual scientific efforts and felt responsible for it in his own sphere. In this sense he was a philosopher, not in the sense of adhering to one of the established epistemological or metaphysical doctrines. Does not in such personal qualities of the academic teacher, rather than in any objectivities or universally accepted metaphysics, lie the answer to Hutchins's quest for a true *universitas literarum*?

Were it my aim to give a full picture of Hilbert's personality I should have to touch upon his attitude regarding the great powers in the lives of men: social and political organization, art, religion, morals and manners, family, friendship, love, and I should also probably have to indicate some of the shadows cast by so much light. I wanted merely to sketch the mathematical side of his personality in an attempt to explain, however incompletely, the peculiar charm and the enormous influence which he exerted. In appraising the latter one must not overlook the environmental factor. A German university in a small town like Göttingen, especially in the halcyon days before 1914, was a favorable milieu for the development of a scientific school. The high social prestige of the professors and everything connected with the university created an atmosphere the like of which has hardly ever existed in America. And once a band of disciples had gathered around Hilbert, intent on research and little worried by the chore of teaching, how could they fail to stimulate one another! We have seen the same thing happening here at Princeton during the first years of the Institute for Advanced Study; there is a kind of snowball effect in the formation of such condensation points of scientific research.

Before giving a more detailed account of Hilbert's work, it remains to characterize in a few words the peculiarly Hilbertian brand of mathematical thinking. It is reflected in his literary style which is one of great *lucidity*. It is as if you were on a swift walk through a sunny open landscape; you look freely around, demarcation lines and connecting roads are pointed out to you, before you must brace yourself to climb the hill; then the path goes straight up, no ambling around, no detours. His style has not the terseness of many of our modern authors in mathematics, which is based on the assumption that printer's labor and paper are costly but the reader's effort and time are not. In carrying out a complete induction Hilbert finds time to develop the first two steps before formulating the general conclusion from n to $n+1$. How many examples illustrate the fundamental theorems of his algebraic papers—examples not constructed ad hoc, but genuine ones worth being studied for their own sake!

In Hilbert's approach to mathematics, *simplicity* and *rigor* go hand in hand. The generation before him, nay even most analysts of his time, felt the growing demand for rigor imposed upon analysis by the critique of the 19th century, which culminated in Weierstrass, as a heavy yoke that made their steps dragging and awkward. Hilbert did much to change that attitude. In his famous address, *Mathematische Probleme*, delivered before the Paris Congress in 1900, he stresses the importance of great concrete fruitful *problems*. "As long as a branch of science," says he, "affords an abundance of problems, it is full of life; want of problems means death or cessation of independent development. Just as every human enterprise prosecutes final aims, so mathematical research needs problems. Their solution steels the force of the investigator; thus he discovers new methods and viewpoints and widens his horizon." "One who without a definite problem before his eyes searches for methods, will probably search in vain." The *methodical unity of mathematics* was for him a matter of belief and experience. Again I quote his own words: "The question is forced upon us whether mathematics is once to face what other sciences long ago experienced, namely the falling apart into subdivisions whose representatives are hardly able to understand each other and whose connections for this reason will become ever looser. I neither wish nor believe it. The science of mathematics as I see it is an indivisible whole, an organism whose ability to survive rests on the connection between its parts." A characteristic feature of Hilbert's method is a peculiarly *direct attack* on problems, unfettered by algorithms; he always goes back to the questions in their original simplicity. An outstanding example is his salvage of Dirichlet's principle which had fallen a victim of Weierstrass's criticism, but his work abounds in similar examples. His strength, equally disdainful of the convulsion of Herculean efforts and of surprising tricks and ruses, is combined with an uncompromising *purity*.

Hilbert helped the reviewer of his work greatly by seeing to it that it is rather neatly cut into different periods during each of which he was almost exclusively occupied with one particular set of problems. If he was engrossed in integral equations, integral equations seemed everything; dropping a subject, he dropped it for good and turned to something else. It was in this characteristic way that he achieved universality. I discern five main periods: i. Theory of invariants (1885–1893). ii. Theory of algebraic number fields (1893–1898). iii. Foundations, (a) of geometry (1898–1902), (b) of mathematics in general (1922–1930). iv. Integral equations (1902–1912). v. Physics (1910–1922). The headings are a little more specific than they ought to be. Not all of Hilbert's algebraic achievements are directly related to invariants. His papers on calculus of variations are here lumped together with those on integral equations. And of course there are some overlappings and a few stray children who break the rules of time, the most astonishing his proof of Waring's theorem in 1909.

His Paris address on "Mathematical problems" quoted above straddles all fields of our science. Trying to unveil what the future would hold in store for us, he posed and discussed twenty-three unsolved problems which have indeed, as we can now state in retrospect, played an important role during the following forty odd years. A mathematician who had solved one of them thereby passed on to the honors class of the mathematical community.

LITERATURE

Hilbert's *Gesammelte Abhandlungen* were published in 3 volumes by J. Springer, Berlin, 1932–35. This edition contains his *Zahlbericht*, but not his two books:

Grundlagen der Geometrie, 7th ed., Leipzig, 1930.

Grundzüge einer allgemeinen Theorie der linearen Integralgleichungen, Leipzig and Berlin, 1912.

Hilbert is co-author of the following works:

R. Courant and D. Hilbert, *Methoden der mathematischen Physik*, Berlin, vol. 1, 2d ed., 1931, vol. 2, 1937.

D. Hilbert and W. Ackermann, *Grundzüge der theoretischen Logik*, Berlin, 1928.

D. Hilbert and S. Cohn-Vossen, *Anschauliche Geometrie*, Berlin, 1932.

D. Hilbert and P. Bernays, *Grundlagen der Mathematik*, Berlin, vol. 1, 1934, vol. 2, 1939.

The Collected Papers contain articles by B. L. van der Waerden, H. Hasse, A. Schmidt, P. Bernays, and E. Hellinger, on Hilbert's work in algebra, in number theory, on the foundations of geometry and arithmetics, and on integral equations. These articles trace the development after Hilbert, giving ample references. The reader may also consult a number of Die Naturwissenschaften vol. 10 (1922) pp. 65–104, dedicated to Hilbert, which surveys his work prior to 1922, and an article by L. Bieberbach, *Ueber den Einfluss von Hilberts Pariser Vortrag über "Mathematische Probleme" auf die Entwicklung der Mathematik in den letzten dreissig Jahren*, Die Naturwissenschaften vol. 18 (1930) pp. 1101–1111. O. Blumenthal wrote a life of Hilbert (Collected Papers, vol. 3, pp. 388–429).

I omit all quotations of literature covered by these articles.

THEORY OF INVARIANTS

The classical theory of invariants deals with polynomials $J = J(x_1 \cdots x_n)$ depending on the coefficients $x_1, \cdots, x_n$ of one or several ground forms of a given number of arguments $\eta_1, \cdots, \eta_g$. Any linear substitution s of determinant 1 of the g arguments induces a certain linear transformation $U(s)$ of the variable coefficients

$x_1, \cdots, x_n$, $x \to x' = U(s)x$, whereby $J = J(x_1 \cdots x_n)$ changes into a new form $J(x_1' \cdots x_n') = J^s(x_1 \cdots x_n)$. J is an invariant if $J^s = J$ for every s. (The restriction to *unimodular* transformations s enables one to avoid the more involved concept of *relative* invariants and to remove the restriction to homogeneous polynomials, with the convenient consequence that the invariants form a *ring*.) The classical problem is a special case of the general problem of invariants in which s ranges over an arbitrary given abstract group Γ and $s \to U(s)$ is any representation of that group (that is, a law according to which every element s of Γ induces a linear transformation $U(s)$ of the n variables $x_1, \cdots, x_n$ such that the composition of group elements is reflected in composition of the induced transformations). The development before Hilbert had led up to two main theorems, which however had been proved in very special cases only. The first states that *the invariants have a finite integrity basis*, or that we can pick a finite number among them, say $i_1, \cdots, i_m$, such that every invariant J is expressible as a polynomial in $i_1, \cdots, i_m$. An identical relation between the basic invariants $i_1, \cdots, i_m$ is a polynomial $F(z_1 \cdots z_m)$ of m independent variables $z_1, \cdots, z_m$ which vanishes identically by virtue of the substitution

$$z_1 = i_1(x_1 \cdots x_n), \cdots, z_m = i_m(x_1 \cdots x_n).$$

The second main theorem asserts that *the relations have a finite ideal basis*, or that one can pick a finite number among them, say $F_1, \cdots, F_h$, such that every relation F is expressible in the form

$$F = Q_1F_1 + \cdots + Q_hF_h, \tag{1}$$

the Q_i being polynomials of the variables $z_1, \cdots, z_m$.

I venture the guess that Hilbert first succeeded in proving the *second* theorem. The relations F form a subset within the ring $k[z_1 \cdots z_m]$ of all polynomials of $z_1, \cdots, z_m$ the coefficients of which lie in a given field k. When Hilbert found his simple proof he could not fail to notice that it applied to any set of polynomials Σ whatsoever and he thus discovered one of the most fundamental theorems of algebra, which was instrumental in ushering in our modern abstract approach, namely:

(A) *Every subset Σ of the polynomial ring $k[z_1 \cdots z_m]$ has a finite ideal basis.*

Is it bad metaphysics to add that his proof turned out so simple *because* the proposition holds in this generality? The proof proceeds by the adjoining of one variable z_i after the other, the individual step being taken care of by the statement: If a given *ring* r satisfies the condition (P): that every subset of r has a finite ideal basis, then the ring $r[z]$ of polynomials of a single variable z with coefficients in r satisfies the same condition (P). Once this is established one gets not only (A) but also an arithmetic refinement discussed by Hilbert in which the field k of rational numbers is replaced by the ring of rational integers.

The subset Σ of relations to which Hilbert applies his theorem (A) is itself an *ideal*, and thus the ideal $\{F_1, \cdots, F_h\}$, that is, the totality of all elements of the form (1), $Q_i \varepsilon k[z_1 \cdots z_m]$, not only contains, but coincides with, Σ. The proof, however, works even if Σ is not an ideal, and yields at one stroke (1) the enveloping ideal $\{\Sigma\}$ of Σ and (2) the reduction of that ideal to a finite basis, $\{\Sigma\} = \{F_1, \cdots, F_h\}$.

Construction of a full set of relations $F_1, \cdots, F_h$ would finish the investigation of the algebraic structure of the ring of invariants were it true that any relation F can be represented in the form (1) *in one way only*. But since, generally speaking, this is not so, we must ask for the "vectors of polynomials" $M = (M_1, \cdots, M_h)$ for which $M_1F_1 + \cdots + M_hF_h$ vanishes identically in z (syzygy of first order). These linear relations M between $F_1, \cdots, F_h$ again form an ideal to which Theorem (A) is applicable, the basis of the M thus obtained giving rise to syzygies of the second order. To the first two main theorems Hilbert adds a third to the effect that if redundance is avoided, the chain of syzygies breaks off after at most m steps.

All this hangs in the air unless we can establish the *first main theorem*, which is of an altogether different character because it asks for an integrity, not an ideal basis. Discussing invariants we operate in the ring $k_x = k[x_1 \cdots x_n]$ of polynomials of $x_1, \cdots, x_n$ in a given field k. Hilbert applies his Theorem (A) to the totality $\mathfrak{J}$ of all *invariants J for which* $J(0, \cdots, 0) = 0$ (a subring of k_x which, by the way, is not an ideal!) and thus determines an ideal basis $i_1, \cdots, i_m$ of $\mathfrak{J}$. Each of the invariants $i = i_r$ may be decomposed into a sum $i = i^{(1)} + i^{(2)} + \cdots$ of homogeneous forms of degree 1, 2, $\cdots$, and as the summands are themselves invariants, the i_r may be assumed to be homogeneous forms of degrees $\nu_r \geq 1$. Hilbert then claims that the $i_1, \cdots, i_m$ constitute an integrity basis for all invariants. I use a finite group Γ consisting of N elements s (although this case of the general problem of invariants was never envisaged by Hilbert himself) in order to illustrate the idea by which the transition is made. Every invariant J is representable in the form

$$J = c + L_1 i_1 + \cdots + L_m i_m \qquad (L_r \varepsilon k_x) \tag{2}$$

where c is the constant $J(0)$. If J is of degree ν one may lop off in L_r all terms of higher degree than $\nu - \nu_r$ without destroying the equation. If it were possible by some process to change the coefficients L in (2) into invariants, the desired result would follow by induction with respect to the degree of J. In the case of a finite group such a process is readily found: the process of averaging. The linear transformation $U(s)$ of the variables $x_1, \cdots, x_n$ induced by s carries (2) into

$$J = c + \overset{s}{L_1} \cdot i_1 + \cdots + \overset{s}{L_m} \cdot i_m.$$

Summation over s and subsequent division by the number N yields the relation

$$J = c + L_1^* i_1 + \cdots + L_m^* i_m$$

where

$$L_r^* = \frac{1}{N} \cdot \sum_s \overset{s}{L_r}.$$

It is of the same nature as (2), except for the decisive fact that according to their formation the new coefficients L^* are invariants.[2]

Actually Hilbert had to do, not with a finite group but with the classical problem in which the group Γ consists of all linear transformations s of g variables $\eta_1, \cdots, \eta_g$, and instead of the averaging process he had to resort to a differentiation process invented by Cayley, the so-called Cayley Ω-process, which he skillfully adapted for his end. (It is essential in Cayley's process that the g^2 components of the matrix s are independent variables; instead of the absolutely invariant polynomials J one has to consider relatively invariant homogeneous *forms* each of which has a definite degree and weight.)

Hilbert's theorem (A) is the foundation stone of the general theory of *algebraic manifolds*. Let us now think of k more specifically as the field of all complex numbers. It seems natural to define an algebraic manifold in the space of n coordinates $x_1, \cdots, x_n$ by a number of simultaneous algebraic equations $f_1 = 0, \cdots, f_h = 0$ ($f_i \varepsilon k_x$). According to Theorem (A), nothing would be gained by admitting an infinite number of equations. Let us denote by $Z(f_1, \cdots, f_h)$ the set of points $x = (x_1, \cdots, x_n)$ where $f_1, \cdots, f_h$ and hence all elements of the ideal $\mathfrak{J} = \{f_1, \cdots, f_h\}$ vanish simultaneously. g vanishes on $Z(f_1, \cdots, f_h)$ whenever $g \varepsilon \{f_1, \cdots, f_h\}$, but the converse is not generally true. For

[2] The example of finite groups is used here as an illustration only. Indeed, a direct elementary proof of the first main theorem for finite groups that makes no use of Hilbert's principle (A) has been given by E. Noether, Math. Ann. vol. 77 (1916) p. 89. In dividing by N we have assumed the field k to be of characteristic zero.

instance x_1 vanishes wherever x_1^3 does, and yet x_1 is not of the form $x_1^3 \cdot q(x_1 \cdots x_n)$. The language of the algebraic geometers distinguishes here between the simple plane $x_1=0$ and the triple plane, although the point set is the same in both cases. Hence what they actually mean by an algebraic manifold is the polynomial ideal and not the point set of its zeros. But even if one cannot expect that every polynomial g vanishing on $Z(f_1, \cdots, f_h)=Z(\mathcal{J})$ is contained in the ideal $\mathcal{J}=\{f_1, \cdots, f_h\}$ one hopes that at least some *power* of g will be. Hilbert's "*Nullstellensatz*" states that this is true, at least if k is the field of complex numbers. It holds in an arbitrary coefficient field k provided one admits points x the coordinates x_i of which are taken from *k or any algebraic extension of k*. Clearly this Nullstellensatz goes to the root of the very concept of algebraic manifolds.[3]

Actually Hilbert conceived it as a tool for the investigation of invariants. As we are now dealing with the full linear group let us consider homogeneous invariants only and drop the adjective homogeneous. Exclude the constants (the invariants of degree 0). Suppose we have ascertained μ non-constant invariants $J_1, \cdots, J_\mu$ such that every non-constant invariant vanishes wherever they vanish simultaneously. An ideal basis of the set $\mathfrak{J}$ of all non-constant invariants certainly meets the demand, but a system $J_1, \cdots, J_\mu$ may be had much more cheaply. Indeed, by a beautiful combination of ideas Hilbert proves that if for a given point $x=x^0$ there exists at all an invariant which neither vanishes for $x=x^0$ nor reduces to a constant, then there exists such an invariant whose weight does not exceed a certain a priori limit W (for example, $W=9n(3n+1)^8$ for a ternary ground form of order n). Hence the $J_1, \cdots, J_\mu$ may be chosen from the invariants of weight not greater than W, and they thus come within the grasp of explicit algebraic construction.

When Hilbert published his proof for the existence of a finite ideal basis, Gordan the formalist, at that time looked upon as the king of invariants, cried out: "This is not mathematics, it is theology!" Hilbert remonstrated then, as he did all his life, against the disparagement of existential arguments as "theology," but we see how, by digging deeper, he was able to meet Gordan's constructive demands. By combining the Nullstellensatz with the Cayley process he further showed that every invariant J is an integral *algebraic* (though not an integral *rational*) function of $J_1, \cdots, J_\mu$, satisfying an equation

[3] B. L. van der Waerden's book *Moderne Algebra*, vol. 2, 2nd ed., 1940, gives on pp. 1–72 an excellent account of the general algebraic concepts and facts with which we are here concerned.

$$J^e + G_1 J^{e-1} + \cdots + G_e = 0$$

in which the G's are polynomials of $J_1, \cdots, J_\mu$. Hence it must be possible by suitable algebraic extensions to pass from $J_1, \cdots, J_\mu$ to a full integrity basis. From there on familiar algebraic patterns such as were developed by Kronecker and as are amenable to explicit construction may be followed.

After the formal investigations from Cayley and Sylvester to Gordan, Hilbert inaugurated a new epoch in the theory of invariants. Indeed, by discovering new ideas and introducing new powerful methods he not only brought the subject up to the new level set for algebra by Kronecker and Dedekind, but made such a thorough job of it that he all but finished it, at least as far as the full linear group is concerned. With justifiable pride he concludes his paper, *Ueber die vollen Invariantensysteme*, with the words: "Thus I believe the most important goals of the theory of the function fields generated by invariants have been attained," and therewith quits the scene.[4]

Of later developments which took place after Hilbert quit, two main lines seem to me the most important: (1) The averaging process, which we applied above to finite groups, carries over to continuous compact groups. By this transcendental process of integration over the group manifold, Adolf Hurwitz treated the real orthogonal group. The method has been of great fertility. The simple remark that invariants for the real orthogonal group are *eo ipso* also invariant under the full complex orthogonal group indicates how the results can be transferred even to non-compact groups, in particular, as it turns out, to all semi-simple Lie groups. (2) Today the theory of invariants for arbitrary groups has taken its natural place within the frame of the theory of representations of groups by linear substitutions, a development which owes its greatest impulse to G. Frobenius.

Although the first main theorem has been proved for wide classes of groups Γ we do not yet know whether it holds for every group. Such attempts as have been made to establish it in this generality were soon discovered to have failed. A promising line for an algebraic attack is outlined in item 14 of Hilbert's Paris list of Mathematical Problems.

Having dwelt in such detail on Hilbert's theory of invariants, we must be brief with regard to his other, more isolated, contributions to algebra. The first paper in which the young algebraist showed his real

[4] I recommend to the reader's attention a brief résumé of his invariant-theoretic work which Hilbert himself wrote for the International Mathematical Congress held at Chicago in conjunction with the World Fair in 1893; Collected Papers, vol. 2, item 23.

mettle concerns the conditions under which a form with real coefficients is representable as a sum of squares of such forms, in particular with the question whether the obviously necessary condition that the form be positive for all real values of its arguments is sufficient. By ingenious continuity arguments and algebraic constructions Hilbert finds three special cases for which the answer is affirmative, among them of course the positive definite quadratic forms, but counter-examples for all other cases. Similar methods recur in two papers dealing with the attractive problem of the maximum number of real ovals of an algebraic curve or surface and their mutual position. Hilbert conjectured that, irrespective of the number of variables, every *rational function* with real (or rational) coefficients is a sum of squares of such functions provided its values are positive for real values of the arguments; and in his *Grundlagen der Geometrie* he pointed out the role of this fact for the geometric constructions with ruler and "Eichmass." Later O. Veblen conceived, as the basis of the distinction between positive and negative in any field, the axiom that no square sum equals zero. Independently of him, E. Artin and O. Schreier developed a detailed theory of such "real fields," and by means of it Artin succeeded in proving Hilbert's conjecture.[5]

In passing I mention Hilbert's irreducibility theorem according to which one may substitute in an irreducible polynomial suitable integers for all of the variables but one without destroying the irreducibility of the polynomial, and his paper on the solution of the equation of ninth degree by functions with a minimum number of arguments. They became points of departure for much recent algebraic work (E. Noether, N. Tschebotareff and others). Finally, it ought to be recorded that on the foundations laid by Hilbert a detailed theory of polynomial ideals was erected by E. Lasker and F. S. Macaulay which in turn gave rise to Emmy Noether's general axiomatic theory of ideals. Thus in the field of algebra, as in all other fields, Hilbert's conceptions proved of great consequence for the further development.

Algebraic number fields

When Hilbert, after finishing off the invariants, turned to the theory of *algebraic number fields*, the ground had been laid by Dirichlet's analysis of the group of units more than forty years before, and by Kummer's, Dedekind's and Kronecker's introduction of ideal divisors. The theory deals with an algebraic field κ over the field k

[5] O. Veblen, Trans. Amer. Math. Soc. vol. 7 (1906) pp. 197–199. E. Artin and O. Schreier, Abh. Math. Sem. Hamburgischen Univ. vol. 5 (1926) pp. 85–99; E. Artin, ibid. pp. 100–115.

of rational numbers. One of the most important general results beyond the foundations had been discovered by Dedekind, who showed that the rational prime divisors of the discriminant of κ are at the same time those primes whose ideal prime factors in κ are not all distinct (ramified primes). l being a rational prime, the adjunction to κ of the lth root of a number α in κ yields a relative cyclic field $K=\kappa(\alpha^{1/l})$ of degree l over κ, *provided* κ *contains the lth root of unity* $\zeta=e^{2\pi i/l}$ (and according to Lagrange, the most general relative cyclic field of degree l over κ is obtained in this fashion). It may be said that it was this circumstance which forced Kummer, as he tried to prove Fermat's theorem of the impossibility of the equation $\alpha^l+\beta^l=\gamma^l$, to pass from the rational ground field k to the cyclotomic field $\kappa_l=k(\zeta)$ and then to conceive his ideal numbers in κ_l and to investigate whether the number of classes of equivalent ideal numbers in κ_l is prime to l. Hilbert sharpened his tools in resuming Kummer's study of the relative cyclic fields of degree l over κ_l, which he christened "Kummer fields."

His own first important contribution was a theory of *relative Galois fields* K over a given algebraic number field κ. His main concern is the relation of the Galois group Γ of K/κ to the way in which the prime ideals of κ decompose in K. Given a prime ideal $\mathfrak{P}$ in K of relative degree f, those substitutions s of Γ for which $s\mathfrak{P}=\mathfrak{P}$ form the splitting group. As always in Galois theory one constructs the corresponding subfield of K/κ (splitting field), to which a number of K belongs if it is invariant under all substitutions of the splitting group. The substitutions t which carry every integer A in K into one, tA, that is congruent to A mod $\mathfrak{P}$ form an invariant subgroup of the splitting group of index f, called the inertial group, and the corresponding field (inertial field) is sandwiched in between the splitting field and K. Let $\mathfrak{p}$ be the prime ideal in κ into which $\mathfrak{P}$ goes, and $\mathfrak{P}^e$ the exact power of $\mathfrak{P}$ by which $\mathfrak{p}$ is divisible. I indicate the nature of Hilbert's results by the following central theorem of his: In the splitting field of $\mathfrak{P}$ the prime ideal $\mathfrak{p}$ in κ splits off the prime factor $\mathfrak{p}^*=\mathfrak{P}^e$ of degree 1 (therefore the name!); in passing from the splitting to the inertial field $\mathfrak{p}^*$ stays prime but its degree increases to f; in passing from the inertial to the full field K, $\mathfrak{p}^*$ breaks up into e equal prime factors $\mathfrak{P}$ of the same degree f. For later application I add the following remarks. If $\mathfrak{P}$ goes into $\mathfrak{p}$ in the first power only, $e=1$ (which is necessarily so provided $\mathfrak{p}$ is not a divisor of the relative discriminant of K/κ), then the inertial group consists of the identity only. In that case the theory of Galois's strictly finite fields shows that the splitting group is cyclic of order f and that its elements $1, s, s^2, \cdots, s^{f-1}$ are uniquely determined by the congruences

$$sA \equiv A^P, \qquad s^2A \equiv A^{P^2}, \cdots \qquad (\text{mod } \mathfrak{P})$$

holding for every integer A. Here P is the number of residues in κ modulo $\mathfrak{p}$ and thus P^f the number of residues in K modulo $\mathfrak{P}$. Today we call $s=\sigma(\mathfrak{P})$ the Frobenius substitution of $\mathfrak{P}$; it is of paramount importance that one particular generating substitution of the splitting group may thus be distinguished among all others. One readily sees that for any substitution u of the Galois group $\sigma(u\mathfrak{P})=u^{-1}\cdot\sigma(\mathfrak{P})\cdot u$. Thus if the Galois field K/κ is *Abelian*, the substitution $\sigma(\mathfrak{P})=\sigma(u\mathfrak{P})$ depends on $\mathfrak{p}$ only and may be denoted by $(\frac{K}{\mathfrak{p}})$.

In 1893 the Deutsche Mathematiker-Vereinigung asked Hilbert and Minkowski to submit within two years a report on number theory. Minkowski dropped out after a while. Hilbert's monumental report *Die Theorie der algebraischen Zahlkörper* appeared in the Jahresberichte of 1896 (the preface is dated April 1897). What Hilbert accomplished is infinitely more than the Vereinigung could have expected. Indeed, his report is a jewel of mathematical literature. Even today, after almost fifty years, a study of this book is indispensable for anybody who wishes to master the theory of algebraic numbers. Filling the gaps by a number of original investigations, Hilbert welded the theory into an imposing unified body. The proofs of all known theorems he weighed carefully before he decided in favor of those "the principles of which are capable of generalization and the most useful for further research." But before such a selection could be made that "further research" had to be carried out! Meticulous care was given to the notations, with the result that they have been universally adopted (including, to the American printer's dismay, the German letters for ideals!) He greatly simplified Kummer's theory, which rested on very complicated calculations, and he introduced those concepts and proved a number of those theorems in which we see today the foundations of a general theory of relative Abelian fields. The most important concept is the norm residue symbol, a pivotal theorem on relative cyclic fields, his famous Satz 90 (Collected Works, vol. I, p. 149). From the preface in which he describes the general character of number theory, and the topics covered by his report in particular, let me quote one paragraph:

"The theory of number fields is an edifice of rare beauty and harmony. The most richly executed part of this building, as it appears to me, is the theory of Abelian fields which Kummer by his work on the higher laws of reciprocity, and Kronecker by his investigations on the complex multiplication of elliptic functions, have opened up to us. The deep glimpses into the theory which the work of these two mathematicians affords reveals at the same time that there still lies an abundance of priceless treasures hidden in this domain, beckoning as a rich reward to the explorer who knows the value of such treasures and with love pursues the art to win them."

Hilbert himself was the miner who during the following two years brought to light much of the hidden ore. The analogy with the corresponding problems in the realm of algebraic functions of one variable where Riemann's powerful instruments of topology and Abelian integrals are available was for him a guiding principle throughout (cf. his remarks in item 12 of his Paris Problems). It is a great pleasure to watch how, step by step, advancing from the special to the general, Hilbert evolves the adequate concepts and methods, and the essential conclusions emerge. I mention his great paper dealing with the relative quadratic fields, and his last and most important *Ueber die Theorie der relativ Abelschen Zahlkörper*. On the basis of the examples he carried through in detail, he conceived as by divination and formulated the basic facts about the so-called class fields. Whereas Hilbert's work on invariants was an end, his work on algebraic numbers was a beginning. Most of the labor of such number theorists of the last decades, as Furtwängler, Takagi, Hasse, Artin, Chevalley, has been devoted to proving the results anticipated by Hilbert. By deriving from the ζ-function the existence of certain auxiliary prime ideals, Hilbert had leaned heavily on transcendental arguments. The subsequent development has gradually eliminated these transcendental methods and shown that though they are the fitting and powerful tool for the investigation of the distribution of prime ideals they are alien to the problem of class fields. In attempting to describe the main issues I shall not ignore the progress and simplification due to this later development.

Hilbert's theory of norm residues is based on the following discoveries of his own: (1) he conceived the basic idea and defined the norm residue symbol for all non-exceptional prime spots; (2) he realized the necessity of introducing infinite prime spots; (3) he formulated the general law of reciprocity in terms of the norm symbol; (4) he saw that by means of that law one can extend the definition of the norm symbol to the exceptional prime spots where the really interesting things happen.—It was an essential progress when E. Artin later (5) replaced the roots of unity by the elements of the Galois group as values of the residue symbol. In sketching Hilbert's problems I shall make use of this idea of Artin's and also of the abbreviating language of (6) Hensel's p-adic numbers and (7) Chevalley's *idèles*.[6]

As everybody knows an integer a indivisible by the prime $p\neq 2$ is said to be a quadratic residue if the congruence $x^2\equiv a \pmod{p}$ is solvable. Gauss introduced the symbol $(\frac{a}{p})$ which has the value $+1$ or -1 according to whether a is a quadratic residue or non-residue mod p, and observed that it is a character, $(\frac{a}{p})\cdot(\frac{a'}{p})=(\frac{aa'}{p})$. Indeed, the p residues modulo p—as whose representatives one may take $0, 1, \cdots, (p-1)$—form a *field*, and after exclusion of 0 a *group* in which the quadratic residues form a subgroup of index 2. Let $K=k(b^{1/2})$ be a quadratic field which arises from the rational ground field k by adjunction of the square root of the rational number b. An integer $a\neq 0$ is called by Hilbert *a p-adic norm* in K if modulo any given power of p it is congruent to the norm of a suitable integer in K. He sets $(\frac{a,\,K}{p})=+1$ if a is p-adic norm, -1 in the opposite case, and finds that this p-adic norm symbol again is a character. The investigation of numbers modulo arbitrarily high powers of p was systematized by K. Hensel in the form of his p-adic numbers, and I repeat Hilbert's definition in this language: "The rational number $a\neq 0$, or more generally the p-adic number $a_p\neq 0$, is a p-adic norm in K if the equation

$$a_p = \mathrm{Nm}\,(x + yb^{1/2}) = x^2 - by^2$$

has a p-adic solution $x=x_p$, $y=y_p$; the norm symbol (a_p, K) equals $+1$ or -1 according to whether or not a_p is (p-adic) norm in K." The p-adic numbers form a field $k(p)$ and after exclusion of 0 a multiplicative group G_p in which, according to Hilbert's result, the p-adic norms in K form a subgroup of index 2 or 1. The cyclic nature of the factor group is the salient point. One easily finds that the p-adic squares form a subgroup G_p^2 of index 4 if $p\neq 2$, of index 8 if $p=2$, and thus the factor group G_p/G_p^2 is not cyclic and could not be described by a single character. Of course every p-adic square is a p-adic norm in K. Both steps, the substitution of K-norms for squares and the passage from the modulus p to arbitrarily high powers of p; the first step amounting to a relaxation, the second to a sharpening of Gauss's condition for quadratic residues; are equally significant for the success of Hilbert's definition.

Every p-adic number $a_p\neq 0$ is of the form $p^h\cdot e_p$ where e_p is a p-adic unit, and thus a_p is of a definite *order* h (at p). An ordinary rational number a coincides with a definite p-adic number $I_p(a)=a_p$. Here I_p symbolizes a homomorphic projection of k into $k(p)$:

$$I_p(a + a') = I_p(a) + I_p(a'), \qquad I_p(aa') = I_p(a)\cdot I_p(a').$$

[6] The latest account of the theory is C. Chevalley's paper *La théorie du corps de classes*, Ann. of Math. vol. 41 (1940) pp. 394–418.

The character $\left(\frac{a, K}{p}\right)$ is identical with $(I_p(a), K)$.

We come to Hilbert's second discovery: he realized that simple laws will not result unless one adds to the "finite prime spots" p one infinite prime spot q. By definition the q-adic numbers are the real numbers and $I_q(a)$ is the real number with which the rational number a coincides. Hence the real number a_q is a q-adic norm in K if the equation $a_q = x^2 - by^2$ has a solution in real numbers x, y. Clearly if $b > 0$ or K is real, this is the case for every a_q; if however $b < 0$ or K is imaginary, only the positive numbers a_q are q-adic norms. Hence

$$(a_q, K) = 1 \text{ if } K \text{ real}; \qquad (a_q, K) = \operatorname{sgn} a_q \text{ if } K \text{ imaginary}.$$

The fact that the norm symbol is a character is thus much more easily verified for the infinite prime spot than for the finite ones.

Hilbert's third observation is to the effect that Gauss's reciprocity law with its two supplements may be condensed into the elegant formula

$$(3) \qquad \prod_p (I_p(a), K) = \prod_p \left(\frac{a, K}{p}\right) = 1,$$

the product extending over the infinite and all finite prime spots p. There is no difficulty in forming this product because almost all factors (that is, all factors with but a finite number of exceptions) equal unity. Indeed, if the prime p does not go into the discriminant of K, then $(a_p, K) = 1$ for every p-adic unit a_p. Formula (3) is the first real vindication for the norm residue idea, which must have given Hilbert the assurance that the higher reciprocity laws had to be formulated in terms of norm residues.

A given rational number a assigns to every prime spot p a p-adic number $a_p = I_p(a)$. On which features of this assignment does one rely in forming the product (3)? The obvious answer is given by Chevalley's notion of *idèle*: an idèle $\mathbf{a}$ is a function assigning to every prime spot p a p-adic number $a_p \neq 0$ which is a p-adic unit for almost all prime spots p. The idèles form a multiplicative group J_k. By virtue of the assignment $a_p = I_p(a)$ every rational number $a \neq 0$ gives rise to an idèle, called the *principal idèle a*. With the idèles $\mathbf{a}$ at our disposal we might as well return to the notation $\left(\frac{\mathbf{a}, K}{p}\right)$ for (a_p, K). The formula

$$(4) \qquad \phi_K(\mathbf{a}) = (\mathbf{a}, K) = \prod_p (a_p, K) = \prod_p \left(\frac{\mathbf{a}, K}{p}\right)$$

defines a character ϕ_K, the norm character, in the group J_k of all idèles. The reciprocity law in Hilbert's form (3) maintains that

$$(a, K) = 1 \tag{5}$$

if $\boldsymbol{a}$ is principal. By the very definition of the norm symbol (a_p, K) the same equation holds if $\boldsymbol{a}$ is a norm in K, that is, if a_p is a p-adic norm in K for every prime spot p. Two idèles $\boldsymbol{a}, \boldsymbol{a}'$ are said to be equivalent, $\boldsymbol{a} \sim \boldsymbol{a}'$, if their quotient $\boldsymbol{a}'\boldsymbol{a}^{-1}$ is principal. Let us denote by $\mathrm{Nm} J_K$ the group of all idèles which are equivalent to norms in K. Then (5) holds for all idèles $\boldsymbol{a}$ of $\mathrm{Nm} J_K$; it would be good to know that it holds for no other idèles, or, in other words, that $\mathrm{Nm} J_K$ is a subgroup of J_k of index 2.

The stage is now reached where the experiences gathered for a quadratic field K over the rational ground field k may be generalized to any relative Abelian field K over a given algebraic number field $\kappa = k(\theta)$. First a word about the infinite prime spots of κ. The defining equation $f(\theta) = 0$, an irreducible equation in k of some degree m, has m distinct roots $\theta', \theta'', \cdots, \theta^{(m)}$ in the continuum of complex numbers. Suppose that r of them are real, say $\theta', \cdots, \theta^{(r)}$. Then each element α of κ has its r real conjugates $\alpha', \cdots, \alpha^{(r)}$, and $\alpha^{(t)}$ arises from α by a homomorphic projection $I^{(t)}$ of κ into the field of all real numbers,

$$\alpha \to \alpha^{(t)} = I^{(t)}(\alpha) \qquad (t = 1, \cdots, r).$$

We therefore speak of r real infinite prime spots $\mathfrak{q}', \cdots, \mathfrak{q}^{(r)}$ with the corresponding projections $I' = I_{\mathfrak{q}'}, \cdots, I^{(r)}$; the fields $\kappa(\mathfrak{q}'), \cdots, \kappa(\mathfrak{q}^{(r)})$ are identical with the field of all real numbers. Thus α is an nth $\mathfrak{q}'$-adic power if the equation $\alpha' = \xi'^n$ has a real solution ξ'. One sees that this imposes a condition only if n is even, and then requires α' to be positive. (In the complex domain the equation is always solvable whether n be even or odd, and that is the reason why we ignore the complex infinite prime spots altogether.)

The finite prime spots are the prime ideals $\mathfrak{p}$ of κ. In studying a Galois field K/κ of relative degree n we first exclude the ramified ideals $\mathfrak{p}$ which go into the relative discriminant of K/κ. An unramified ideal $\mathfrak{p}$ of κ factors in K into a number g of distinct prime ideals $\mathfrak{P}_1 \cdots \mathfrak{P}_g$ of relative degree f, $fg = n$. It is easily seen that a $\mathfrak{p}$-adic number $\alpha_\mathfrak{p} \neq 0$ is a $\mathfrak{p}$-adic norm in K if and only if its order (at $\mathfrak{p}$) is a multiple of f. In particular, the $\mathfrak{p}$-adic units are norms. Thus we encounter a situation which is essentially simpler than the one taken care of by Gauss's quadratic residue symbol: the norm character of $\alpha_\mathfrak{p}$ depends only on the order i at $\mathfrak{p}$ of $\alpha_\mathfrak{p}$. It is now clear how to proceed: we choose a primitive fth root of unity ζ and define $(\alpha_\mathfrak{p}, K) = \zeta^i$ if $\alpha_\mathfrak{p}$ is of order i. This function of $\alpha_\mathfrak{p} \neq 0$ is a character which assumes the value 1 for the norms and the norms only. But here is the rub: there is no algebraic property distinguishing the several primitive fth roots of unity from one another. Thus the choice of ζ among them remains arbitrary. One could put up with this if one dealt with one prime ideal only. But when one has to take all prime spots simultaneously into account, as is necessary in forming products of the type (4), then the arbitrariness involved in the choice of ζ for each $\mathfrak{p}$ will destroy all hope of obtaining a simple reciprocity law like (5). I shall forego describing the devices by which Eisenstein, Kummer, Hilbert, extricated themselves from this entanglement. By far the best solution was found by Artin: if K/κ is Abelian, then the Frobenius substitution $(\frac{K}{\mathfrak{p}})$ is uniquely determined by K and $\mathfrak{p}$ and is an element of order f of the Galois group Γ of K/κ. Let this element of the Galois group replace ζ in our final definition of the $\mathfrak{p}$-adic norm symbol:

$$(6) \qquad (\alpha_{\mathfrak{p}}, K) = \left(\frac{\boldsymbol{\alpha}, K}{\mathfrak{p}}\right) = \left(\frac{K}{\mathfrak{p}}\right)^i \text{ if } \alpha_{\mathfrak{p}} \text{ is of order } i \text{ at } \mathfrak{p}.$$

We could now form for any idèle $\boldsymbol{\alpha}$ the product

$$\prod_{\mathfrak{p}} (\alpha_{\mathfrak{p}}, K) = \prod_{\mathfrak{p}} \left(\frac{\boldsymbol{\alpha}, K}{\mathfrak{p}}\right) = (\boldsymbol{\alpha}, K)$$

extending over all finite and (real) infinite prime spots $\mathfrak{p}$ and formulate the reciprocity law asserting that $(\boldsymbol{\alpha}, K) = 1$ for any principal idèle $\boldsymbol{\alpha}$—had we not excluded certain exceptional prime spots in our definition of $(\alpha_{\mathfrak{p}}, K)$, namely the infinite prime spots and the ramified prime ideals. In the special case he investigated Kummer had succeeded in obtaining the correct value of $(\alpha_{\mathfrak{p}}, K)$ for the exceptional $\mathfrak{p}$ by extremely complicated calculations. Hilbert's fourth discovery is a simple and ingenious method of circumventing this formidable obstacle which blocked the road to further progress. Let us first restrict ourselves to idèles $\boldsymbol{\alpha}$ which are nth powers at our exceptional prime spots; in other words, we assume that the equation $\alpha_{\mathfrak{p}} = \xi_{\mathfrak{p}}^n$ is solvable for the $\mathfrak{p}$-adic values $\alpha_{\mathfrak{p}}$ of $\boldsymbol{\alpha}$ at this finite number of prime spots. There is no difficulty in defining $(\boldsymbol{\alpha}, K)$ under this restriction:

$$(\boldsymbol{\alpha}, K) = \prod_{\mathfrak{p}}{}'(\alpha_{\mathfrak{p}}, K),$$

the product extending, as indicated by the accent, over the non-exceptional prime spots only, for which we know what $(\alpha_{\mathfrak{p}}, K)$ means. Under the same restriction we prove (with Artin) the reciprocity law

$$(7) \qquad (\boldsymbol{\alpha}, K) = 1 \text{ if } \boldsymbol{\alpha} \text{ is principal,}$$

and observe that by its very definition $(\boldsymbol{\alpha}, K) = 1$ if $\boldsymbol{\alpha}$ is norm. We now return to an *arbitrary* idèle $\boldsymbol{\alpha}$. It is easily shown that there exists an equivalent idèle $\boldsymbol{\alpha}^*\sim\boldsymbol{\alpha}$ which is an nth power at all exceptional prime spots, but of course there will be plenty of them. However, the restricted law of reciprocity insures that

$$(\boldsymbol{\alpha}^*, K) = \prod_{\mathfrak{p}}{}'(\alpha_{\mathfrak{p}}^*, K)$$

has the same value for every one of the $\boldsymbol{\alpha}^*$'s, and it is this value which we now denote by $(\boldsymbol{\alpha}, K)$. This definition adopted, the reciprocity law (7) and the statement that $(\boldsymbol{\alpha}, K)=1$ for every norm $\boldsymbol{\alpha}$ follow at once *without restriction*. Thus the reciprocity law itself is made the tool for getting the exceptional prime spots under control!

Once $(\boldsymbol{\alpha}, K)$ is known for every idèle $\boldsymbol{\alpha}$ we can compute $(\alpha_{\mathfrak{p}}, K)$ for a given prime spot $\mathfrak{p}$ and a given $\mathfrak{p}$-adic number $\alpha_{\mathfrak{p}}\neq 0$ as the value of $(\boldsymbol{\alpha}, K)$ for that "primary" idèle, also denoted by $\alpha_{\mathfrak{p}}$, which equals $\alpha_{\mathfrak{p}}$ at $\mathfrak{p}$ and 1 at any other prime spot. (The idèle $\boldsymbol{\alpha}$ is the product of its primary components, $\boldsymbol{\alpha}=\prod_{\mathfrak{p}}\alpha_{\mathfrak{p}}$.) One expects the following two propositions to hold:

I. $(\alpha_{\mathfrak{p}}, K)=1$ *if and only if* $\alpha_{\mathfrak{p}}$ *is a* $\mathfrak{p}$*-adic norm.*

II. *Given a prime ideal* $\mathfrak{p}$, $(\alpha_{\mathfrak{p}}, K)=1$ *for every* $\mathfrak{p}$*-adic unit* $\alpha_{\mathfrak{p}}$ *if and only if* $\mathfrak{p}$ *is unramified.*

The direct parts of I and II:

(I_0) $\alpha_{\mathfrak{p}}=$norm implies $(\alpha_{\mathfrak{p}}, K)=1$;

(II_0) $\mathfrak{p}$ unramified implies $(\alpha_{\mathfrak{p}}, K)=1$ for every $\mathfrak{p}$-adic unit $\alpha_{\mathfrak{p}}$, were settled above. The converse statement of I_0 is trivial for the non-exceptional prime spots; but owing to the indirect definition of the norm symbol for the exceptional prime spots, the proofs of the converse of I_0 for the exceptional spots and of the converse of II_0 are rather intricate. From II we learn that for none of the ramified prime ideals $\mathfrak{p}$ does the norm character of $\alpha_{\mathfrak{p}}$ depend on the order of $\alpha_{\mathfrak{p}}$ only: this simple feature which makes the definition (6) possible is limited to non-ramified $\mathfrak{p}$. One would also expect:

III. *If the principal idèle* α *is an idèle norm in* K, *then the number* α *is norm of a number in* K.

This is true for *cyclic* fields K/κ, but in general not for Abelian fields.

Let us again denote by $\mathrm{Nm}J_K$ the subgroup in J_κ of the idèles which are equivalent to norms. Then the norm symbol $\phi_K(\boldsymbol{\alpha})=(\boldsymbol{\alpha}, K)$ determines a homomorphic mapping of the factor group $J_\kappa/\mathrm{Nm}J_K$ into the Galois group of K/κ. One would expect that this mapping is one-to-one:

IV. *By means of the norm symbol the factor group* $J_\kappa/\mathrm{Nm}J_K$ *is isomorphically mapped onto the Galois group of* K/κ.

I, II, III$_c$ (the subscript c indicating restriction to cyclical fields) and IV are the main propositions of what one might call the ***norm theory of relative Abelian fields***. They refer to a ***given*** field K/κ.

There is a second part of the theory, the *class field theory proper*, which is concerned with the manner in which *all possible* relative Abelian fields K over κ are reflected in the structure of the group J_κ of idèles in κ. Each such field K determines, as we have seen, a subgroup $\mathrm{Nm}J_K$ of J_κ of finite index. The question arises *which* subgroups J_κ^* of J_κ are generated in this way by Abelian fields K/κ. Clearly the following conditions are necessary:

(1) Every principal idèle is in J_κ^*.

(2) There is a natural number n such that every nth power of an idèle is in J_κ^*.

(3) There is a finite set S of prime spots such that α is in J_κ^* provided α is a unit at every prime spot and equals 1 at the prime spots of S.

The main theorem concerning class fields states that these conditions are also sufficient.

V. *Given a subgroup* J_κ^* *of* J_κ *fulfilling the above three conditions* (and therefore, as one readily verifies, of finite index), *there exists a uniquely determined Abelian field* K/κ *such that* $J_\kappa^* = \mathrm{Nm}J_K$.

We divide the idèles of κ into *classes* by throwing two idèles into the same class if their quotient is in J_κ^*. Then J_κ/J_κ^* is the class group and K is called the corresponding *class field*. The most important example results if one lets J_κ^* consist of the *unit idèles* α whose values $\alpha_\mathfrak{p}$ are $\mathfrak{p}$-adic units at every prime spot $\mathfrak{p}$.[7] Then the classes may be described as the familiar classes of *ideals*: two ideals are put in the same class if their quotient is a principal ideal (α) springing from a number α positive at all real infinite prime spots. The corresponding class field K, the so-called absolute class field, is of relative discriminant 1, and the largest unramified Abelian field over κ (Theorem II). Its degree n over κ is the class number of ideals, its Galois group isomorphic to the class group of ideals in κ (Theorem IV). f being the least power of $\mathfrak{p}$ which lies in the principal class, $\mathfrak{p}$ decomposes into n/f distinct prime ideals in K, each of relative degree f. This last statement does nothing but repeat the norm definition of the class field. Hence the way in which $\mathfrak{p}$ factors in K depends only on the class to which $\mathfrak{p}$ belongs. The easiest way of extending the theory from the case with no ramified prime ideals, which was preponderant in Hilbert's thought,

---

[7] At the (real) infinite prime spots the positive numbers are considered the units.

to Takagi's ramified case is by substituting idèles for ideals. Hilbert also stated that every ideal in κ becomes a principal ideal in the absolute class field. It is today possible to show that this is so, by arguments, however, which are far from being fully understood, because this question transcends the domain of Abelian fields.

As was stated above, Hilbert did not prove these theorems in their full generality, but taking his departure from Gauss's theory of genera in quadratic fields and Kummer's investigations he worked his way gradually up from the simplest cases, developing as he went along the necessary machinery of new concepts and propositions about them until he could survey the whole landscape of class fields. We cannot attempt here to give an idea of the highly involved proofs. The completion of the work he left to his successors. The day is probably still far off when we shall have a theory of relative *Galois* number fields of comparable completeness.

Kronecker had shown, and Hilbert found a simpler proof for the fact, that Abelian fields over the rational ground field k are necessarily subfields of the cyclotomic fields, and are thus obtained from the transcendental function $e^{2\pi i x}$ by substituting rational values for the argument x. For Abelian fields over an imaginary quadratic field the so-called complex multiplication of the elliptic and modular functions plays a similar role ("Kronecker's Jugendtraum"). While Heinrich Weber following in Kronecker's footsteps, and R. Fuëter under Hilbert's guidance, made this dream come true, Hilbert himself began to play with modular functions of several variables which are defined by means of algebraic number fields, and to study their arithmetical implications. He never published these investigations, but O. Blumenthal, and later E. Hecke, used his notes and developed his ideas. The results are provocative, but still far from complete. It is indicative of the fertility of Hilbert's mind at this most productive period of his life that he handed over to his pupils a complex of problems of such fascination as that of the relation between number theory and modular functions.[8]

There remain to be mentioned a particularly simple proof of the transcendence of e and π with which Hilbert opened the series of his arithmetical papers, and the 1909 paper settling Waring's century-old conjecture. I should classify the latter paper among his most original ones, but we can forego considering it more closely because a decade later Hardy and Littlewood found a different approach which yields

[8] R. Fuëter, *Singuläre Moduln und complexe Multiplication*, 2 vols., Leipzig, 1924, 1927; cf. also H. Hasse, J. Reine Angew. Math. vol. 157 (1927) pp. 115–139. O. Blumenthal, Math. Ann. vol. 56 (1903) pp. 509–548, vol. 58 (1904) pp. 497–527. E. Hecke, Math. Ann. vol. 71 (1912) pp. 1–37, vol. 74 (1913) pp. 465–510.

asymptotic formulas for the number of representations, and it is the Hardy-Littlewood "circle method" which has given rise in recent times to a considerable literature on this and related subjects.[9]

Axiomatics

There could not have been a more complete break than the one dividing Hilbert's last paper on the theory of number fields from his classical book, *Grundlagen der Geometrie*, published in 1899. Its only forerunner is a note of the year 1895 on the straight line as the shortest way. But O. Blumenthal records that as early as 1891 Hilbert, discussing a paper on the role of Desargues's and Pappus's theorems read by H. Wiener at a mathematical meeting, made a remark which contains the axiomatic standpoint in a nutshell: "It must be possible to replace in all geometric statements the words *point*, *line*, *plane*, by *table*, *chair*, *mug*."

The Greeks had conceived of geometry as a deductive science which proceeds by purely logical processes once the few axioms have been established. Both Euclid and Hilbert carry out this program. However, Euclid's list of axioms was still far from being complete; Hilbert's list is complete and there are no gaps in the deductions. Euclid tried to give a descriptive definition of the basic spatial objects and relations with which the axioms deal; Hilbert abstains from such an attempt. All that we must know about those basic concepts is contained in the axioms. The axioms are, as it were, their implicit (necessarily incomplete) definitions. Euclid believed the axioms to be evident; his concern is the real space of the physical universe. But in the deductive system of geometry the evidence, even the truth of the axioms, is irrelevant; they figure rather as hypotheses of which one sets out to develop the logical consequences. Indeed there are many different material interpretations of the basic concepts for which the axioms become true. For instance, the axioms of n-dimensional Euclidean vector geometry hold if a distribution of direct current in a given electric circuit, the n branches of which connect in certain branching points, is called a vector, and Joule's heat produced per unit time by the current is considered the square of the vector's length. In building up geometry on the foundation of its axioms one will attempt to economize as much as possible and thus illuminate the role of the several groups of axioms. Arranged in their natural

[9] It must suffice here to quote the first paper in this line: G. H. Hardy and J. E. Littlewood, Quart. J. Math. vol. 48 (1919) pp. 272–293, and its latest successor which carries Waring's theorem over to arbitrary algebraic fields: C. L. Siegel, Amer. J. Math. vol. 66 (1944) pp. 122–136.

hierarchy they are the axioms of incidence, order, congruence, parallelism, and continuity. For instance, if the theory of geometric proportions or of the areas of polygons can be established without resorting to the axioms of continuity, this ought to be done.

In all this, though the execution shows the hand of a master, Hilbert is not unique. An outstanding figure among his predecessors is M. Pasch, who had indeed traveled a long way from Euclid when he brought to light the hidden axioms of order and with methodical clarity carried out the deductive program for projective geometry (1882). Others in Germany (F. Schur) and a flourishing school of Italian geometers (Peano, Veronese) had taken up the discussion. With respect to the economy of concepts, Hilbert is more conservative than the Italians: quite deliberately he clings to the Euclidean tradition with its three classes of undefined elements, points, lines, planes, and its relations of incidence, order and congruence of segments and angles. This gives his book a peculiar charm, as if one looked into a face thoroughly familiar and yet sublimely transfigured.

It is one thing to build up geometry on sure foundations, another to inquire into the logical structure of the edifice thus erected. If I am not mistaken, Hilbert is the first who moves freely on this higher "metageometric" level: systematically he studies the mutual independence of his axioms and settles the question of independence from certain limited groups of axioms for some of the most fundamental geometric theorems. His method is the *construction of models*: the model is shown to disagree with one and to satisfy all other axioms; hence the one cannot be a consequence of the others. One outstanding example of this method had been known for a considerable time, the Cayley-Klein model of non-Euclidean geometry. For Veronese's non-Archimedean geometry Levi-Civita (shortly before Hilbert) had constructed a satisfactory arithmetical model. The question of *consistency* is closely related to that of independence. The general ideas appear to us almost banal today, so thoroughgoing has been their influence upon our mathematical thinking. Hilbert stated them in clear and unmistakable language, and embodied them in a work that is like a crystal: an unbreakable whole with many facets. Its artistic qualities have undoubtedly contributed to its success as a masterpiece of science.

In the construction of his models Hilbert displays an amazing wealth of invention. The most interesting examples seem to me the one by which he shows that Desargues's theorem does not follow from the plane incidence axioms, but that the plane incidence axioms combined with Desargues's theorem enable one to embed the plane in a higher dimensional space in which all incidence axioms hold; and then the other example by which he decides whether the Archimedean axiom of continuity is necessary to restore the full congruence axioms after having curtailed them by the exclusion of reflections.

What is the building material for the models? Klein's model of non-Euclidean geometry could be interpreted as showing that he who accepts Euclidean geometry with its points and lines, and so on, can by mere change of nomenclature also get the non-Euclidean geometry. Klein himself preferred another interpretation in terms of projective space. However, Descartes's analytic geometry had long provided a more general and satisfactory answer, of which Riemann, Klein and many others must have been aware: All that we need for our construction is the field of real *numbers*. Hence any contradiction in Euclidean geometry must show up as a contradiction in the arithmetical axioms on which our operations with real numbers are based. Nobody had said that quite clearly before Hilbert. He formulates a complete and simple set of axioms for real numbers. The system of arithmetical axioms will have its exchangeable parts just as the geometric system has. From a purely algebraic standpoint the most important axioms are those characterizing a (commutative or non-commutative) *field*. Any such abstract number field may serve as a basis for the construction of corresponding geometries. *Vice versa*, one may introduce numbers and their operations in terms of a space satisfying certain axioms; Hilbert's Desarguesian *Streckenrechnung* is a fine example. In general this reverse process is the more difficult one. The Chicago school under E. H. Moore took up Hilbert's investigations, and in particular O. Veblen did much to reveal the perfect correspondence between the projective spaces obeying a set of simple incidence axioms (and no axioms of order), and the abstractly defined number fields.[10]

What the question of independence literally asks is to make sure that a certain proposition cannot be deduced from other propositions. It seems to require that we make the propositions, rather than the things of which they speak, the object of our investigation, and that as a preliminary we fully analyze the logical mechanism of deduction. The method of models is a wonderful trick to avoid that sort of logical investigations. It pays, however, a heavy price for thus shirking the fundamental issue: it merely reduces everything to the question of consistency for the arithmetical axioms, which is left unanswered. In

[10] Among later contributions to this question I mention W. Schwan, *Streckenrechnung und Gruppentheorie*, Math. Zeit. vol. 3 (1919) pp. 11–28. A complete bibliography of geometric axiomatics since Hilbert would probably cover many pages. I refrain from citing a list of names.

the same manner *completeness*, which literally means that every general proposition about the objects with which the axioms deal can be decided by inference from the axioms, is replaced by *categoricity* (Veblen), which asserts that any imaginable model is isomorphic to the one model by which consistency is established. In this sense Hilbert proves that there is but "one," the Cartesian geometry, which fulfills all his axioms. Only in the case of G. Fano's and O. Veblen's finite projective spaces, for example, of the projective plane consisting of seven points, the model is a purely combinatorial scheme, and the questions of consistency, independence and completeness can be answered in the absolute sense. Hilbert never seems to have thought of illustrating his conception of the axiomatic method by purely combinatorial schemes, and yet they provide by far the simplest examples.

An approach to the foundations of geometry entirely different from the one followed in his book is pursued by Hilbert in a paper which is one of the earliest documents of set-theoretic topology. From the standpoint of mechanics the central task which geometry ought to perform is that of describing the mobility of a solid. This was the viewpoint of Helmholtz, who succeeded in characterizing the group of motions in Euclidean space by a few simple axioms. The question had been taken up by Sophus Lie in the light of his general theory of continuous groups. Lie's theory depends on certain assumptions of differentiability; to get rid of them is one of Hilbert's Paris Problems. In the paper just mentioned he *does* get rid of them as far as Helmholtz's problem in the plane is concerned. The proof is difficult and laborious; naturally continuity is now the foundation—and not the keystone of the building as it had been in his Grundlagen book. Other authors, R. L. Moore, N. J. Lennes, W. Süss, B. v. Kérékjarto, carried the problem further along these topological lines. A half-personal reminiscence may be of interest. Hilbert defines a two-dimensional manifold by means of neighborhoods, and requires that a class of "admissible" one-to-one mappings of a neighborhood upon Jordan domains in an x, y-plane be designated, any two of which are connected by continuous transformations. When I gave a course on Riemann surfaces at Göttingen in 1912, I consulted Hilbert's paper and noticed that the neighborhoods themselves could be used to characterize that class. The ensuing definition was given its final touch by F. Hausdorff; the Hausdorff axioms have become a byword in topology.[11] (However, when it comes to explaining what a *differenti-*

[11] A parallel development, with E. H. Moore as the chief prompter, must have taken place in this country. As I have to write from memory mainly, it is inevitable that my account should be colored by the local Göttingen tradition.

able manifold is, we are to this day bound to Hilbert's roundabout way; cf. Veblen and Whitehead, *The foundations of differential geometry*, Cambridge, 1932.)

The fundamental issue of an *absolute proof of consistency* for the axioms which should include the whole of mathematical analysis, nay even Cantor's set theory in its wildest generality, remained in Hilbert's mind, as a paper read before the International Congress at Heidelberg in 1904 testifies. It shows him on the way, but still far from the goal. Then came the time when integral equations and later physics became his all-absorbing interest. One hears a loud rumbling of the old problem in his Zürich address, *Axiomatisches Denken*, of 1917. Meanwhile the difficulties concerning the foundations of mathematics had reached a critical stage, and the situation cried for repair. Under the impact of undeniable antinomies in set theory, Dedekind and Frege had revoked their own work on the nature of numbers and arithmetical propositions, Bertrand Russell had pointed out the hierarchy of types which, unless one decides to "reduce" them by sheer force, undermine the arithmetical theory of the continuum; and finally L. E. J. Brouwer by his intuitionism had opened our eyes and made us see how far generally accepted mathematics goes beyond such statements as can claim real meaning and truth founded on evidence. I regret that in his opposition to Brouwer, Hilbert never openly acknowledged the profound debt which he, as well as all other mathematicians, owe Brouwer for this revelation.

Hilbert was not willing to make the heavy sacrifices which Brouwer's standpoint demanded, and he saw, at least in outline, a way by which the cruel mutilation could be avoided. At the same time he was alarmed by signs of wavering loyalty within the ranks of mathematicians some of whom openly sided with Brouwer. My own article on the *Grundlagenkrise* in Math. Zeit. vol. 10 (1921), written in the excitement of the first postwar years in Europe, is indicative of the mood. Thus Hilbert returns to the problem of foundations in earnest. He is convinced that complete certainty can be restored without "committing treason to our science." There is anger and determination in his voice when he proposes "die Grundlagenfragen einfürallemal aus der Welt zu schaffen." "Forbidding a mathematician to make use of the principle of excluded middle," says he, "is like forbidding an astronomer his telescope or a boxer the use of his fists."

Hilbert realized that the mathematical statements themselves could not be made the subject of a mathematical investigation whose aim is to answer the question of their consistency in its primitive sense, lest they be first reduced to mere *formulas*. Algebraic formulas like $a+b=b+a$ are the most familiar examples. The process of deduction by which formulas previously obtained give rise to new formulas must be described without reference to any meaning of the formulas. The deduction starts from certain initial formulas, the axioms, which must be written out explicitly. Whereas in his *Grundlagen der Geometrie* the meaning of the geometric terms had become irrelevant, but the meaning of logical terms, as "and," "not," "if then," had still to be understood, now every trace of meaning is obliterated. As a consequence, logical symbols like $\rightarrow$ in $a \rightarrow b$, read: a implies b, enter into the formulas. Hilbert fully agrees with Brouwer in that the great majority of mathematical propositions are not "real" ones conveying a definite meaning verifiable in the light of evidence. But he insists that the non-real, the "ideal propositions" are indispensable in order to give our mathematical system "completeness." Thus he parries Brouwer, who had asked us to give up what is meaningless, by relinquishing the pretension of meaning altogether, and what he tries to establish is not *truth* of the individual mathematical proposition, but *consistency* of the system. The game of deduction when played according to rules, he maintains, will never lead to the formula $0 \neq 0$. In this sense, and in this sense only, he promises to salvage our cherished classical mathematics in its entirety.

For those who accuse him of degrading mathematics to a mere game he points first to the introduction of ideal elements for the sake of completeness as a common method in all mathematics—for example, of the ideal points outside an accessible portion of space, without which space would be incomplete—; secondly, to the neighboring science of physics where likewise not the individual statement is verifiable by experiment, but in principle only the system as a whole can be confronted with experience.

But how to make sure that the "game of deduction" never leads to a contradiction? Shall we prove this by the same mathematical method the validity of which stands in question, namely by deduction from axioms? This would clearly involve a regress *ad infinitum*. It must have been hard on Hilbert, the axiomatist, to acknowledge that the insight of consistency is rather to be attained by *intuitive reasoning* which is based on evidence and not on axioms. But after all, it is not surprising that ultimately the mind's seeing eye must come in. Already in communicating the rules of the game we must count on understanding. The game is played in silence, but the rules must be *told* and any reasoning about it, in particular about its consistency, communicated by *words*. Incidentally, in describing the indispensable intuitive basis for his *Beweistheorie* Hilbert shows himself an accomplished master of that, alas, so ambiguous medium of communication, language. With regard to what he accepts as evident in this "metamathematicial" reasoning, Hilbert is more papal than the pope, more exacting than either Kronecker or Brouwer. But it cannot be helped that our reasoning in following a hypothetic sequence of formulas leading up to the formula $0 \neq 0$ is carried on in hypothetic generality and uses that type of evidence which a formalist would be tempted to brand as application of the principle of complete induction. Elementary arithmetics can be founded on such intuitive reasoning as Hilbert himself describes, but we need the formal apparatus of variables and "quantifiers" to invest the infinite with the all important part that it plays in higher mathematics. Hence Hilbert prefers to make a clear cut: he becomes strict formalist in mathematics, strict intuitionist in metamathematics.

It is perhaps possible to indicate briefly how Hilbert's formalism restores the *principle of the excluded middle* which was the main target of Brouwer's criticism. Consider the infinite sequence of numbers $0, 1, 2, \cdots$. Any property A of numbers (for example, "being prime") may be represented by a propositional function $A(x)$ ("x is prime"), from which a definite proposition $A(b)$ arises by substituting a concrete number b for the variable x ("6 is prime"). Accepting the principle which Brouwer denies and to which Hilbert wishes to hold on, that (i) either there exists a number x for which $A(x)$ holds, or (ii) $A(x)$ holds for no x, we can find a "representative" r for the property A, a number such that $A(b)$ implies $A(r)$ whatever the number b, $A(b) \rightarrow A(r)$. Indeed, in the alternative (i) we choose r as one of the numbers x for which $A(x)$ holds, in the alternative (ii) at random. Thus Aristides is the representative of honesty; for, as the Athenians said, if there is any honest man it is Aristides. Assuming that we know the representative we can decide the question whether there is an honest man or whether all are dishonest by merely looking at *him*: if he is dishonest everybody is. In the realm of numbers we may even make the choice of the representative unique, in case (i) choosing $x = r$ as the *least* number for which $A(x)$ holds, and $r = 0$ in the opposite case (ii). Then r arises from A by a certain operator ρ_x, $r = \rho_x A(x)$, applicable to every imaginable property A. A propositional function may contain other variables $y, z, \cdots$ besides x. Therefore it is necessary to attach an index x to ρ, just as in integrating one must indicate with respect to which variable one integrates. ρ_x eliminates the variable x; for instance $\rho_x A(x, y)$ is a function of y alone. The word quantifier is in use for this sort of operator. Hence we write our axiom as follows:

$$A(b) \rightarrow A(\rho_x A(x)). \tag{8}$$

It is immaterial whether we fix the representative in the unique manner described above; our specific rule would not fit anyhow unless x ranges over the numbers 0, 1, 2, $\cdots$. Instead we imagine a quantifier ρ_x of universal applicability which, as it were, selects the representative for us. Zermelo's axiom of choice is thus woven into the principle of the excluded middle. It is a bold step; but the bolder the better, as long as it can be shown that we keep within the bounds of consistency!

In the formalism, propositional functions are replaced by *formulas* the handling of which must be described without reference to their meaning. In general, variables $x, y, \cdots$ will occur among the symbols of a formula $\mathfrak{A}$. We say that the symbol ρ_x *binds* the variable x in the formula $\mathfrak{A}$ which follows the symbol[12] and that x occurs *free* in a formula wherever it is not bound by a quantifier with index x. $x, y, \rightarrow, \rho_x$ are symbols entering into the formulas; the German letters are no such symbols, but are used for communication. It is more natural to describe our critical axiom (8) as a rule for the formation of axioms. It says: take any formula $\mathfrak{A}$ in which only the variable x occurs free, and any formula $\mathfrak{b}$ without free variables, and by means of them build the formula

$$\mathfrak{A}(\mathfrak{b}) \rightarrow \mathfrak{A}(\rho_x \mathfrak{A}). \tag{9}$$

Here $\mathfrak{A}(\mathfrak{b})$ stands for the formula derived from $\mathfrak{A}$ by putting in the entire formula $\mathfrak{b}$ for the variable x *wherever x occurs free.*

In this way formulas may be *obtained* as *axioms* according to certain rules. *Deduction* proceeds by the rule of syllogism: From two formulas $\mathfrak{a}$ and $\mathfrak{a} \rightarrow \mathfrak{b}$ *previously obtained*, in the second of which the first formula reappears at the left of the symbol $\rightarrow$, one *obtains* the formula $\mathfrak{b}$.

How does Hilbert propose to show that the game of deduction will never lead to the formula $0 \neq 0$? Here is the basic idea of his procedure. As long as one deals with "finite" formulas only, formulas from which the quantifiers $\rho_x, \rho_y, \cdots$ are absent, one can decide whether they are true or false by merely looking at them. With the entrance of ρ such a descriptive valuation of formulas becomes impossible: evidence ceases to work. But a concretely given deduction is a sequence of

[12] If we wish the rule that ρ_x binds x in all that comes after to be taken literally, we must write $\mathfrak{a} \rightarrow \mathfrak{b}$ in the form $\rightarrow \{ {}^{\mathfrak{a}}_{\mathfrak{b}}$. The formulas will then look like genealogical trees.

formulas in which only a limited number of instances of the axiomatic rule (9) will turn up. Let us assume that the only quantifier which occurs is ρ_x and wherever it occurs it is followed by the *same finite* formula $\mathfrak{A}$, so that the instances of (9) are of the form

$$\mathfrak{A}(\mathfrak{b}_1) \to \mathfrak{A}(\rho_x\mathfrak{A}), \cdots, \mathfrak{A}(\mathfrak{b}_h) \to \mathfrak{A}(\rho_x\mathfrak{A}). \tag{10}$$

Assume, moreover, $\mathfrak{b}_1, \cdots, \mathfrak{b}_h$ to be finite. We then carry out a *reduction*, replacing $\rho_x\mathfrak{A}$ by a certain finite $\mathfrak{r}$ wherever it occurs as part of a formula in our sequence. In particular, the formulas (10) will change into

$$\mathfrak{A}(\mathfrak{b}_1) \to \mathfrak{A}(\mathfrak{r}), \cdots, \mathfrak{A}(\mathfrak{b}_h) \to \mathfrak{A}(\mathfrak{r}). \tag{11}$$

We now see how to choose $\mathfrak{r}$: if by examining the finite formulas $\mathfrak{A}(\mathfrak{b}_1), \cdots, \mathfrak{A}(\mathfrak{b}_h)$ one after the other, we find one that is true, say $\mathfrak{A}(\mathfrak{b}_3)$, then we take $\mathfrak{b}_3$ for $\mathfrak{r}$. If every one of them turns out to be false, we choose $\mathfrak{r}$ at random. Then the h reduced formulas (11) are "true" and our hypothesis that the deduction leads to the false formula $0\neq 0$ is carried *ad absurdum*. The salient point is that a concretely given deduction makes use of a limited number of explicitly exhibited individuals $\mathfrak{b}_1, \cdots, \mathfrak{b}_h$ only. If we make a wrong choice, for example, by choosing Alcibiades rather than Aristides as the representative of incorruptibility, our mistake will do no harm as long as the few people (out of the infinite Athenian crowd) with whom we actually deal are all corruptible.

A slightly more complicated case arises when we permit the $\mathfrak{b}_1, \cdots, \mathfrak{b}_h$ to contain ρ_x, but always followed by the same $\mathfrak{A}$. Then we first make a *tentative* reduction replacing $\rho_x\mathfrak{A}$ by the number 0, say. The formulas $\mathfrak{b}_1, \cdots, \mathfrak{b}_h$ are thus changed into reduced finite formulas $\mathfrak{b}_1^0, \cdots, \mathfrak{b}_h^0$ and (10) into

$$\mathfrak{A}(\overset{0}{\mathfrak{b}_1}) \to \mathfrak{A}(0), \cdots, \mathfrak{A}(\overset{0}{\mathfrak{b}_h}) \to \mathfrak{A}(0).$$

This reduction will do unless $\mathfrak{A}(0)$ is false and at the same time one of the $\mathfrak{A}(\mathfrak{b}_1^0), \cdots, \mathfrak{A}(\mathfrak{b}_h^0)$, say $\mathfrak{A}(\mathfrak{b}_3^0)$, is true. But then we have in $\mathfrak{b}_3^0$ a perfectly legitimate representative of $\mathfrak{A}$, and a second reduction which replaces $\rho_x\mathfrak{A}$ by $\mathfrak{b}_3^0$ will work out all right.

However, this is only a modest beginning of the complications awaiting us. Quantifiers $\rho_x, \rho_y, \cdots$ with different variables and applied to different formulas will be piled one upon the other. We make a tentative reduction; it will go wrong in certain places and from that failure we learn how to correct it. But the corrected reduction will probably go wrong at other places. We seem to be driven around in a vicious circle, and the problem is to direct our consecutive corrections in such a manner as to obtain assurance that finally a reduction will result that makes good at all places in our given sequence of formulas. Nothing has contributed more to revealing the circle-like character of the usual transfinite arguments of mathematics than these attempts to make sure of consistency in spite of all circles.

The symbolism for the formalization of mathematics as well as the general layout and first steps of the proof of consistency are due to Hilbert himself. The program was further advanced by younger collaborators, P. Bernays, W. Ackermann, J. von Neumann. The last two proved the consistency of "arithmetics," of that part in which the dangerous axiom about the conversion of predicates into sets is not yet admitted. A gap remained which seemed harmless at the time, but already detailed plans were drawn up for the invasion of analysis. Then came a catastrophe: assuming that consistency is established, K. Gödel showed how to construct arithmetical propositions which are evidently true and yet not deducible within the formalism. His method applies to Hilbert's as well as any other not too limited formalism. Of the two fields, the field of formulas obtainable in Hilbert's formalism and the field of real propositions that are evidently true, neither contains the other (provided consistency of the formalism can be made evident). Obviously *completeness* of a formalism in the absolute sense in which Hilbert had envisaged it was now out of the question. When G. Gentzen later closed the gap in the consistency proof for arithmetics, which Göbel's discovery had revealed to be serious indeed, he succeeded in doing so only by substantially lowering Hilbert's standard of evidence.[13] The boundary line of what is intuitively trustworthy once more became vague. As all hands were needed to defend the homeland of arithmetics, the invasion of analysis never came off, to say nothing of general set theory.

This is where the problem now stands; no final solution is in sight. But whatever the future may bring, there is no doubt that Brouwer and Hilbert raised the problem of the foundations of mathematics to a new level. A return to the standpoint of Russell-Whitehead's *Principia Mathematica* is unthinkable.

Hilbert is the champion of axiomatics. The axiomatic attitude seemed to him one of universal significance, not only for mathematics, but for all sciences. His investigations in the field of physics are conceived in the axiomatic spirit. In his lectures he liked to illustrate the method by examples taken from biology, economics, and so on. The modern epistemological interpretation of science has been profoundly influenced by him. Sometimes when he praised the axiomatic method

[13] G. Gentzen, Math. Ann. vol. 112 (1936) pp. 493–565.

he seemed to imply that it was destined to obliterate completely the constructive or genetic method. I am certain that, at least in later life, this was not his true opinion. For whereas he deals with the primary mathematical objects by means of the axioms of his symbolic system, the formulas are constructed in the most explicit and finite manner. In recent times the axiomatic method has spread from the roots to all branches of the mathematical tree. Algebra, for one, is permeated from top to bottom by the axiomatic spirit. One may describe the role of axioms here as the subservient one of fixing the range of variables entering into the explicit constructions. But it would not be too difficult to retouch the picture so as to make the axioms appear as the masters. An impartial attitude will do justice to both sides; not a little of the attractiveness of modern mathematical research is due to a happy blending of axiomatic and genetic procedures.

Integral Equations

Between the two periods during which Hilbert's efforts were concentrated on the foundations, first of geometry, then of mathematics in general, there lie twenty long years devoted to analysis and physics.

In the winter of 1900–1901 the Swedish mathematician E. Holmgren reported in Hilbert's seminar on Fredholm's first publications on integral equations, and it seems that Hilbert caught fire at once. The subject has a long and tortuous history, beginning with Daniel Bernoulli. The mathematicians' efforts to solve the (mechanical, acoustical, optical, electromagnetical) problem of the oscillations of a continuum and the related boundary value problems of potential theory span a period of two centuries. Fourier's *Théorie analytique de la chaleur* (1822) is a landmark. H. A. Schwarz proved for the first time (1885) the existence of a proper oscillation in two and more dimensions by constructing the fundamental frequency of a membrane. The last decade of the nineteenth century saw Poincaré on his way to the development of powerful function-theoretic methods; C. Neumann and he came to grips with the harmonic boundary problem; Volterra studied that type of integral equations which now bears his name, and for linear equations with infinitely many unknowns Helge von Koch developed the infinite determinants. Most scientific discoveries are made when "their time is fulfilled"; sometimes, but seldom, a genius lifts the veil decades earlier than could have been expected. Fredholm's discovery has always seemed to me one that was long overdue when it came. What could be more natural than the idea that a set of linear equations connected with a discrete set of mass points gives way to an integral equation when one passes to the limit of a continuum? But the fact that in the simpler cases a differential rather than an integral equation results in the limit riveted the mathematicians' attention for two hundred years on differential equations!

It must be said, however, that the simplicity of Fredholm's results is due to the particular form of his equation, on which it was hard to hit without the guidance of the problems of mathematical physics to which he applied it:

$$x(s) - \int_0^1 K(s, t)x(t)dt = f(s) \qquad (0 \leqq s \leqq 1).$$

Indeed the linear operator which in the left member operates on the unknown function x producing a given f, $(E-K)x=f$, consists of two parts, the identity E and the integral operator K, which in a certain sense is weak compared to E. Fredholm proved that for this type of integral equation the two main facts about n linear equations with the same number n of unknowns hold: (1) The homogeneous equation $[f(s)=0]$ has a finite number of linearly independent solutions $x(s)=\phi_1(s), \cdots, \phi_h(s)$, and the homogeneous equation with the transposed kernel $K'(s, t)=K(t, s)$ has the same number of solutions, $\psi_1(s), \cdots, \psi_h(s)$. (2) The nonhomogeneous equation is solvable if and only if the given f satisfies the h linear conditions

$$\int_0^1 f(s)\psi_i(s)ds = 0 \qquad (i = 1, \cdots, h).$$

Following an artifice used by Poincaré, Fredholm introduces a parameter λ replacing K by λK and obtains a solution in the form familiar from finite linear equations, namely as a quotient of two determinants of H. v. Koch's type, either of which is an entire function of the parameter λ.

Hilbert saw two things: (1) after having constructed Green's function K for a given region G and for the potential equation $\Delta u=0$ by means of a Fredholm equation on the boundary, the differential equation of the oscillating membrane $\Delta\phi+\lambda\phi=0$ changes into a homogeneous integral equation

$$\phi(s) - \lambda \int K(s, t)\phi(t)dt = 0$$

with the symmetric kernel K, $K(t, s)=K(s, t)$ (in which the parameter λ is no longer artificial but of the very essence of the problem); (2) the problem of ascertaining the "eigen values" λ and "eigen functions" $\phi(s)$ of this integral equation is the analogue for integrals of the transformation of a quadratic form of n variables onto principal axes. Hence the corresponding theorem for the quadratic integral form

$$\int_0^1 \int_0^1 K(s, t)x(s)x(t)dsdt \tag{12}$$

with an *arbitrary symmetric kernel* K must provide the general foundation for the theory of oscillations of a continuous medium. If others saw the same, Hilbert saw it at least that much more clearly that he bent all his energy on proving that proposition, and he succeeded by the same direct method which about 1730 Bernoulli had applied to the oscillations of a string: passage to the limit from the algebraic problem. In carrying out the limiting process he had to make use of the Koch-Fredholm determinant. He finds that there is a sequence of eigen values $\lambda_1, \lambda_2, \cdots$ tending to infinity, $\lambda_n \to \infty$ for $n \to \infty$, and an orthonormal set of corresponding eigen functions $\phi_n(s)$,

$$\phi_n(s) - \lambda_n \int_0^1 K(s, t)\phi_n(t)dt = 0,$$

$$\int_0^1 \phi_m(s)\phi_n(s)ds = \delta_{mn},$$

such that

$$\int_0^1 \int_0^1 K(s, t)x(s)x(t)dsdt = \sum \xi_n^2/\lambda_n,$$

ξ_n being the Fourier coefficient $\int_0^1 x(s)\phi_n(s)ds$. The theory implies that every function of the form

$$y(s) = \int_0^1 K(s, t)x(t)dt$$

may be expanded into a uniformly convergent Fourier series in terms of the eigen functions ϕ_n,

$$y(s) = \sum \eta_n\phi_n(s), \qquad \eta_n = \int_0^1 y(s)\phi_n(s)ds.$$

Hilbert's passage to the limit is laborious. Soon afterwards E. Schmidt in a Göttingen thesis found a simpler and more constructive proof for these results by adapting H. A. Schwarz's method invented twenty years before to the needs of integral equations.

From finite forms the road leads either to integrals or to infinite series. Therefore Hilbert considered the same problem of orthogonal transformation of a given quadratic form

$$\sum K_{mn} x_m x_n \tag{13}$$

into a form of the special type

$$\kappa_1 \xi_1^2 + \kappa_2 \xi_2^2 + \cdots \qquad (\kappa_n = 1/\lambda_n \to 0) \tag{14}$$

also for infinitely many (real) variables $(x_1, x_2, \cdots)$ or vectors x in a space of a denumerable infinity of dimensions. Only such vectors are admitted as have a finite length $|x|$,

$$|x|^2 = x_1^2 + x_2^2 + \cdots;$$

they constitute what we now call the Hilbert space. The advantage of Hilbert space over the "space" of all continuous functions $x(s)$ lies in a certain property of completeness, and due to this property one can establish "complete continuity" as the necessary and sufficient condition for the transformability of a given quadratic form K, (13), into (14), by following an argument well known in the algebraic case: one determines $\kappa_1, \kappa_2, \cdots$ as the consecutive maxima of K on the "sphere" $|x|^2 = 1$.

As suggested by the theorem concerning a quadratic integral form, the link between the space of functions $x(s)$ and the Hilbert space of vectors $(x_1, x_2, \cdots)$ is provided by an arbitrary *complete* orthonormal system $u_1(s), u_2(s), \cdots$ and expressed by the equations

$$x_n = \int_0^1 x(s) u_n(s) ds.$$

Bessel's inequality states that the square sum of the Fourier coefficients x_n is less than or equal to the square integral of $x(s)$. The relation of completeness, first introduced by A. Hurwitz and studied in detail by W. Stekloff, requires that in this inequality the *equality* sign prevail. Thus the theorem on quadratic forms of infinitely many variables at once gives the corresponding results about the eigen values and eigen functions of symmetric kernels $K(s, t)$—or would do so if one could count on the uniform convergence of $\sum x_n u_n(s)$ for any given vector $(x_1, x_2, \cdots)$ in Hilbert space. In the special case of an eigen vector of that quadratic form (13) which corresponds to the integral form (12),

$$x_n = \lambda \sum_m K_{nm} x_m,$$

Hilbert settles this point by forming the uniformly convergent series

$$\lambda \sum_m x_m \int_0^1 K(s, t) u_m(t) dt$$

which indeed yields a continuous function $\phi(s)$ with the nth Fourier coefficient

$$\lambda \sum K_{nm} x_m = x_n,$$

and thus obtains the eigen function of $K(s, t)$ for the eigen value λ. Soon afterwards, under the stimulus of Hilbert's investigations. E. Fischer and F. Riesz proved their well known theorem that the space of all functions $x(s)$ the square of which has a finite Lebesgue integral enjoys the same property of completeness as Hilbert space, and hence one is mapped isomorphically upon the other in a one-to-one fashion by means of a complete orthonormal system $u_n(s)$. I mention these details because the historic order of events may have fallen into oblivion with many of our younger mathematicians, for whom Hilbert space has assumed that abstract connotation which no longer distinguishes between the two realizations, the square integrable functions $x(s)$ and the square summable sequences $(x_1, x_2, \cdots)$. I think Hilbert was wise to keep within the bounds of continuous functions when there was no actual need for introducing Lebesgue's general concepts.

Perhaps Hilbert's greatest accomplishment in the field of integral equations is his extension of the theory of spectral decomposition from the completely continuous to the so-called *bounded* quadratic forms. He finds that then the point spectrum will in general have condensation points and a continuous spectrum will appear beside the point spectrum. Again he proceeds by directly carrying out the transition to the limit, letting the number of variables $x_1, x_2, \cdots$ increase *ad infinitum*. Again, not long afterwards, simpler proofs for his results were found.

While thus advancing the boundaries of the general theory, he did not lose sight of the ordinary and partial differential equations from which it had sprung. Simultaneously with the young Italian mathematician Eugenio Elia Levi he developed the parametrix method as a bridge between differential and integral equations. For a given elliptic differential operator Δ^* of the second order, the parametrix $K(s, t)$ is a sort of qualitative approximation of Green's function, depending like the latter on an argument point s and a parameter point t. It is supposed to possess the right kind of singularity for $s=t$ so that the nonhomogeneous equation $\Delta^* u = f$ for

$$u = K\rho, \qquad u(s) = \int K(s, t)\rho(t)dt$$

gives rise to the integral equation $\rho + L\rho = f$ for the density ρ, with a kernel $L(s, t) = \Delta_s^* K(s, t)$ regular enough at $s = t$ for Fredholm's theory to be applicable. It is important to give up the assumption that K satisfies the equation $\Delta^* K = 0$, because in general a fundamental solution will not be known for the given differential operator Δ^*. In order not to be bothered by boundary conditions, Hilbert assumes the domain of integration to be a *compact* manifold, like the surface of a sphere, and finds that the method works if the parametrix, besides having the right kind of singularity, is symmetric with respect to argument and parameter.

What has been said should be enough to make clear that in the terrain of analysis a rich vein of gold had been struck, comparatively easy to exploit and not soon to be exhausted. The linear equations of infinitely many unknowns had to be investigated further (E. Schmidt, F. Riesz, O. Toeplitz, E. Hellinger, and others); the continuous spectrum and its appearance in integral equations with "singular" kernels awaited closer analysis (E. Hellinger, T. Carleman); ordinary differential equations, with regular or singular boundaries, of second or of higher order, received their due share of attention (A. Kneser, E. Hilb, G. D. Birkhoff, M. Bôcher, J. D. Tamarkin, and many others).[14] It became possible to develop such asymptotic laws for the distribution of eigen values as were required by the thermodynamics of radiation (H. Weyl, R. Courant). Expansions in terms of orthogonal functions were studied independently of their origin in differential or integral equations. New light fell upon Stieltjes's continued fractions and the problem of momentum. The most ambitious began to attack nonlinear integral equations. A large international school of young mathematicians gathered around Hilbert and integral equations became the fashion of the day, not only in Germany, but also in France where great masters like E. Picard and Goursat paid their tributes, in Italy and on this side of the Atlantic. Many good papers were written, and many mediocre ones. But the total effect was an appreciable change in the aspect of analysis.

Remarkable are the applications of integral equations outside the field for which they were invented. Among them I mention the following three: (1) Riemann's problem of determining n analytic func-

[14] For later literature and systems of differential equations see Axel Schur, Math. Ann. vol. 82 (1921) pp. 213–239; G. A. Bliss, Trans. Amer. Math. Soc. vol. 28 (1926) pp. 561–584; W. T. Reid, ibid. vol. 44 (1938) pp. 508–521.

tions $f_1(z), \cdots, f_n(z)$, regular except at a finite number of points, which by analytic continuation around these points suffer preassigned linear transformations. The problem was solved by Hilbert himself, and subsequently in a simpler and more complete form by J. Plemelj. (A very special case of it is the existence of algebraic functions on a Riemann surface if that surface is given as a covering surface of the complex z-plane.) Investigations by G. D. Birkhoff on matrices of analytic functions lie in the same line. (2) Proof for the completeness of the irreducible representations of a compact continuous group. This is an indispensable tool for the approach to the general theory of invariants by means of Adolf Hurwitz's integration method, and with its refinements and generalizations plays an important role in modern group-theoretic research, including H. Bohr's theory of almost periodic functions.[15] Contact is thus made with Hilbert's old friend, the theory of invariants. (3) Quite recently Hilbert's parametrix method has served to establish the central existence theorem in W. V. D. Hodge's theory of harmonic integrals in compact Riemannian spaces.[16]

The story would be dramatic enough had it ended here. But then a sort of miracle happened: the spectrum theory in Hilbert space was discovered to be the adequate mathematical instrument of the new quantum physics inaugurated by Heisenberg and Schrödinger in 1923. This latter impulse led to a reexamination of the entire complex of problems with refined means (J. von Neumann, A. Wintner, M. H. Stone, K. Friedrichs). As J. von Neumann was Hilbert's collaborator toward the close of that epoch when his interest was divided between quantum physics and foundations, the historic continuity with Hilbert's own scientific activities is unbroken, even for this later phase. What has become of the theory of abstract spaces and their linear operators in our times lies beyond the bounds of this report.

A picture of Hilbert's "analytic" period would be incomplete without mentioning a second motif, *calculus of variations*, which crossed the dominating one of integral equations. The "theorem of independence" with which he concludes his Paris survey of mathematical problems (1900) is an important contribution to the formal apparatus of that calculus. But of much greater consequence was his audacious direct assault on the functional maxima and minima problems. The whole finely wrought machinery of the calculus of variations is here

---

[15] H. Weyl and F. Peter, Math. Ann. vol. 97 (1927) pp. 737–755. A. Haar, Ann. of Math. vol. 34 (1933) pp. 147–169. J. von Neumann, Trans. Amer. Math. Soc. vol. 36 (1934) pp. 445–492. Cf. also L. Pontrjagin, *Topological groups*, Princeton, 1939.

[16] W. V. D. Hodge, *The theory and applications of harmonic integrals*, Cambridge, 1941. H. Weyl, Ann. of Math. vol. 44 (1943) pp. 1–6.

consciously set aside. He proposes instead to construct the minimizing function as the limit of a sequence of functions for which the value of the integral under investigation tends to its minimum value. The classical example is Dirichlet's integral in a two-dimensional region G,

$$D[u] = \iint_G \left\{\left(\frac{\partial u}{\partial x}\right)^2 + \left(\frac{\partial u}{\partial y}\right)^2\right\} dxdy.$$

Admitted are all functions u with continuous derivatives which have given boundary values. d being the lower limit of $D[u]$ for admissible u, one can ascertain a sequence of admissible functions u_n such that $D[u_n] \to d$ with $n \to \infty$. One cannot expect the u_n themselves to converge; rather they have to be prepared for convergence by the smoothing process of integration. As the limit function will be harmonic and the value of the harmonic function at any point P equals its mean value over any circle K around P, it seems best to replace $u_n(P)$ by its mean value in K, with the expectation that this mean value will converge toward a number $u(P)$ which is independent of the circle and in its dependence on P solves the minimum problem. Besides integration Hilbert uses the process of sifting a suitable subsequence from the u_n before passing to the limit. Owing to the simple inequality

$$\{D[u_m - u_n]\}^{1/2} \leqq \{D[u_m] - d\}^{1/2} + \{D[u_m] - d\}^{1/2}$$

discovered by S. Zaremba this second step is unnecessary.

Hilbert's method is even better suited for problems in which the boundary does not figure so prominently as in the boundary value problems. By a slight modification one is able to include point singularities, and Hilbert thus solved the fundamental problem for flows on Riemann surfaces, providing thereby the necessary foundation for Riemann's own approach to the theory of Abelian integrals, and he further showed that Poincaré's and Koebe's fundamental theorems on uniformization could be established in the same way. We should be much better off in number theory if methods were known which are as powerful for the construction of relative Abelian and Galois fields over given algebraic number fields as the Riemann-Hilbert transcendental method proves to be for the analogous problems in the fields of algebraic functions! Its wide application in the theory of conformal mapping and of minimal surfaces is revealed by the work of the man who was Hilbert's closest collaborator in the direction of mathematical affairs at Göttingen for many years, Richard Courant.[17]

[17] A book by Courant on the Dirichlet principle is in preparation.

Of a more indirect character, but of considerable vigor, is the influence of Hilbert's ideas upon the whole trend of the modern development of the calculus of variations; in Europe Carathéodory, Lebesgue, Tonelli could be mentioned among others, in this country the chain reaches from O. Bolza's early to M. Morse's most recent work.

Physics

Already before Minkowski's death in 1909, Hilbert had begun a systematic study of theoretical physics, in close collaboration with his friend who had always kept in touch with the neighboring science. Minkowski's work on relativity theory was the first fruit of these joint studies. Hilbert continued them through the years, and between 1910 and 1930 often lectured and conducted seminars on topics of physics. He greatly enjoyed this widening of his horizon and his contact with physicists, whom he could meet on their own ground. The harvest however can hardly be compared with his achievements in pure mathematics. The maze of experimental facts which the physicist has to take into account is too manifold, their expansion too fast, and their aspect and relative weight too changeable for the axiomatic method to find a firm enough foothold, except in the thoroughly consolidated parts of our physical knowledge. Men like Einstein or Niels Bohr grope their way in the dark toward their conceptions of general relativity or atomic structure by another type of experience and imagination than those of the mathematician, although no doubt mathematics is an essential ingredient. Thus Hilbert's vast plans in physics never matured.

But his application of integral equations to kinetic gas theory and to the elementary theory of radiation were notable contributions. In particular, his asymptotic solution of Maxwell-Boltzmann's fundamental equation in kinetic gas theory, which is an integral equation of the second order, clearly separated the two layers of phenomenological physical laws to which the theory leads; it has been carried out in more detail by the physicists and applied to several concrete problems. In his investigations on general relativity Hilbert combined Einstein's theory of gravitation with G. Mie's program of pure field physics. For the development of the theory of general relativity at that stage, Einstein's more sober procedure, which did not couple the theory with Mie's highly speculative program, proved the more fertile. Hilbert's endeavors must be looked upon as a forerunner of a unified field theory of gravitation and electromagnetism. However, there was still much too much arbitrariness involved in Hilbert's Hamiltonian function; subsequent attempts (by Weyl, Eddington,

Einstein himself, and others) aimed to reduce it. Hopes in the Hilbert circle ran high at that time; the dream of a universal law accounting both for the structure of the cosmos as a whole, and of all the atomic nuclei, seemed near fulfillment. But the problem of a unified field theory stands to this day as an unsolved problem; it is almost certain that a satisfactory solution will have to include the material waves (the Schrödinger-Dirac ψ for the electron, and similar field quantities for the other nuclear particles) besides gravitation and electromagnetism, and that its mathematical frame will not be a simple enlargement of that of Einstein's now classical theory of gravitation.

Hilbert was not only a great scholar, but also a great teacher. Witnesses are his many pupils and assistants, whom he taught the handicraft of mathematical research by letting them share in his own work and its overflow, and then his lectures, the notes of many of which have found their way from Göttingen into public and private mathematical libraries. They covered an extremely wide range. The book he published with S. Cohn-Vossen on *Anschauliche Geometrie* is an outgrowth of his teaching activities. Going over the impressive list attached to his Collected Papers (vol. 3, p. 430) one is struck by the considerable number of courses on general topics like "Knowledge and Thinking," "On the Infinite," "Nature and Mathematics." His speech was fairly fluent, not as hesitant as Minkowski's, and far from monotonous. He had no difficulty in finding the pregnant words, and liked to emphasize short pivotal phrases by repeating them several times. On the whole, his lectures were a faithful reflection of his spirit; direct, intense; how could they fail to be inspiring?

133.

Concerning a classical problem in the theory of singular points of ordinary differential equations

Actas de la Academia Nacional de Ciencias Exactas, Físicas y Naturales de Lima
7, 21—60 (1944)

1. *Introduction: Aim, Hypotheses, Qualitative Propositions*

In connection with some hydrodynamical investigations I recently had to make use of the classical theory of singular points of an ordinary differential equation of first order. The older form of the theory, going back to the days of Briot and Bouquet, depends on the assumption of analyticity. However, later treatments, by Bendixson and above all by Perron, have relaxed the conditions enormously.[1] In particular, Perron goes as far as one can possibly go, but his arguments are subtle and to a large extent of non-constructive character. I wish to return here to the problem in its parametric form

$$\frac{dx}{dt} = F(x, y), \qquad \frac{dy}{dt} = G(x, y) \tag{1}$$

[1] I. Bendixson, Acta Mathematica *24*, 1901, 1-88.
O. Perron, Math. Zeitschr. *15*, 1922, 121-146; *16*, 1923, 273 - 295.

under assumptions sufficiently wide for all applications, yet so devised as to make possible a direct and straightforward approach that leads to remarkably sharp results. Theorem IX is the most striking instance. The stress lies on the explicit and precise estimates. The constructions we give are of a nature that alone satisfies the practical needs: they can actually be carried through with any degree of accuracy that is wanted; i.e. the error in the results will be less than $\epsilon = 10^{-n}$, n being a preassigned integer, provided the data are known with an uncertainty not exceeding a corresponding amount δ. Every limit equation is supported by an explicit estimate of the error.

The right members of the equations (1) are supposed to be real functions of the real variables x, y defined in the neighborhood of, and vanishing at the origin O:

$$F(0,0) = 0, \qquad G(0,0) = 0.$$

In first approximation the infinitesimal transformation of the x, y -plane described by (1) is linear,

$$F(x,y) \sim m_{11}x + m_{12}y, \qquad G(x,y) \sim m_{21}x + m_{22}y.$$

We assume that the eigenvalues, the roots $\rho = k, l$ of the secular equation

$$\begin{vmatrix} \rho + m_{11}, & m_{12} \\ m_{21}, & \rho + m_{22} \end{vmatrix} = 0,$$

are real and distinct,

$$(m_{11} - m_{22})^2 + 4\, m_{12}\, m_{21} > 0\,. \tag{2}$$

After a suitable affine change of coordinates we may then suppose that $F(x, y) \sim -kx$, $G(x, y) \sim -ly$. We therefore set

$$F(x, y) = -kx + f(x, y), \qquad G(x, y) = -ly + g(x, y)\,.$$

Let us use the abbreviation p for the point (x, y) and $|p|$ for $\max(|x|, |y|)$. (Distance from the origin will be measured consistently by this number and not by $\sqrt{x^2+y^2}$). We stay within a neighborhood $\mathfrak{N}_0$ of the origin, $|p| \leqq R_0$, fixed once for all, in which the functions $f(p)$, $g(p)$ are defined,

$p(t) = (x(t), y(t))$ is a *solution*, say in the interval $0 \leqq t \leqq T$, if $|x(t)|$, $|y(t)| \leqq R_0$ and

$$\frac{dx}{dt} = -kx + f(x, y)\,, \qquad \frac{dy}{dt} = -ly + g(x, y) \tag{3}$$

for $0 \leqq t \leqq T$. Our essential hypothesis concerning f and g is a *Lipschitz condition*

$$|f(p) - f(p_*)|\,, \quad |g(p) - g(p_*)| \leqq \chi(r) \cdot |p - p_*| \tag{4}$$

which holds whenever p and p_* are both $\leqq r\,(\leqq R_0)$, $\chi(r)$ denoting a given monotone increasing function in the interval

$0 \leqq r \leqq R_0$ which vanishes for $r = 0$, $\chi(0) = 0$, so strongly that the integral

$$\int_0^{R_0} \chi(r) . \frac{dr}{r} = \lim_{r \to 0} \int_r^{R_0} \chi(r) . \frac{dr}{r} \tag{5}$$

converges. We are particularly interested in the simple function $\chi(r) = C r^{\delta}$ involving two positive constants C and δ. Since $f(0,0) = 0$, $g(0,0) = 0$, (4) implies

$$|f(p)|, \quad |g(p)| \leqq \chi(|p|) . |p| . \tag{6}$$

In view of the application mentioned above, two cases will hold our main interest:

(A_+) : *Both eigenvalues are positive*, $0 < l < k$;

(A_-) : *The eigenvalues are of opposite sign*, $l < 0$, $k > 0$.

In either instance the larger eigenvalue k is positive and hence both (A_+) and (A_-) may be subsumed under

$$(A) : \quad l \leqq k, \quad k > 0 .$$

In terms of the matrix $\| m_{ij} \|$ case (A_+) is characterized by (2) and

$$m_{11} m_{22} - m_{12} m_{21} > 0 , \quad m_{11} + m_{22} > 0 ,$$

case (A_-) by (2) and

$$m_{11}\, m_{22} - m_{12}\, m_{21} < 0\,.$$

Here are some of the more conspicuous facts concerning the solutions of our differential equations under these conditions.

THEOREM I. *In case* (A_+) *any solution which comes sufficiently near the origin can be continued to arbitrarily high values of the independent variable* t *and converges toward the origin for* $t \to +\infty$. *More precisely,* $e^{lt} \cdot x(t)$ *tends to zero for* $t \to +\infty$ *and* $e^{lt} \cdot y(t)$ *to a definite limit* b.

THEOREM II. *Given a number* a, *there exists, under the hypothesis* (A), *a uniquely determined solution obeying the asymptotic law*

$$(7) \qquad e^{kt} \cdot x(t) \to a\,, \qquad e^{kt} \cdot y(t) \to 0 \qquad (\text{for } t \to +\infty)\,.$$

THEOREM III. *In the subcase* (A_+) *the solutions thus obtained are the only ones for which the limit* b *in Theorem I vanishes. In the subcase* (A_-) *they are the only ones which go into the origin for* $t \to \infty$.

A translation $t \to t + t_1$ of the independent variable carries a solution into a solution. If we do not distinguish between "equivalent" solutions arising from one another in this way, and disregard the trivial solution $x(t) = y(t) = 0$, then we can normalize the preassigned number in Theorem II by $a = \pm 1$, thus obtaining exactly *two* solutions, which approach the origin from opposite sides along the x-axis.

2. *Quantitative Explication of Theorem I*

For any two points (x, y), (x_*, y_*) in $\mathcal{H}_0$ we write

$$\Delta x = x - x_*, \qquad \Delta f = f(x, y) - f(x_*, y_*),$$

for any sequence of points (x_n, y_n)

$$\Delta x_n = x_n - x_{n-1}, \quad \Delta f_n = f(x_n, y_n) - f(x_{n-1}, y_{n-1}) = f_n - f_{n-1}.$$

Thus

$$x_n = \Sigma_{\nu=0}^{n} \Delta x_\nu$$

provided we agree to set

$$x_{-1} = y_{-1} = 0, \tag{8}$$

a convention which automatically implies $f_{-1} = g_{-1} = 0$. If (x, y), (x_*, y_*), (x_n, y_n) depend on a parameter t we may or may not put that argument in evidence.

After writing our differential equations (3) in the form

$$\frac{d}{dt}(e^{kt} x) = e^{kt} f(x, y), \qquad \frac{d}{dt}(e^{lt} y) = e^{lt} g(x, y)$$

we can weld them together with the initial conditions $x(0)=x^0$, $y(0)=y^0$ in the integral equations

$$\text{(9)}\qquad \begin{cases} e^{kt}\, x(t) = x^0 + \int_0^t e^{k\tau} f(x(\tau),\ y(\tau))\, d\tau\,, \\ e^{lt}\, y(t) = y^0 + \int_0^t e^{l\tau} g(x(\tau),\ y(\tau))\, d\tau\,. \end{cases}$$

For the difference of two solutions $p(t)$, $p_*(t)$ we obtain

$$\text{(10)}\qquad \begin{cases} e^{kt}.\Delta x(t) = \Delta x^0 + \int_0^t e^{k\tau}\, \Delta f(\tau)\, d\tau\,, \\ e^{lt}.\Delta y(t) = \Delta y^0 + \int_0^t e^{l\tau}\, \Delta g(\tau)\, d\tau\,. \end{cases}$$

This section is dedicated to the case

$$(\mathrm{A}_+):\qquad 0 < l < k\,.$$

In constructing a solution $(x(t),\ y(t))$ with preassigned initial values x^0, y^0 we shall give the classical method of successive approximations such a turn as to yield the solution and a precise estimate at one stroke for the entire infinite interval $t \geqq 0$. It seems reasonable to define the chain of successive approximations $x_n(t)$, $y_n(t)$ by the recurrent integral equations

(11)
$$\begin{cases} e^{kt} x_n(t) = x^0 + \int_0^t e^{k\tau} f(x_{n-1}(\tau), \; y_{n-1}(\tau)) \, d\tau , \\ e^{lt} y_n(t) = y^0 + \int_0^t e^{l\tau} g(x_{n-1}(\tau), \; y_{n-1}(\tau)) \, d\tau , \end{cases} \qquad (n=0,1,2,\ldots)$$

fixing the start by (8). The zero approximation then is

$$x_0(t) = x^0 \, e^{-kt} , \qquad y_0(t) = y^0 \, e^{-lt} ,$$

and for the differences we obtain

(12)
$$\begin{cases} e^{kt} . \Delta \, x_{n+1}(t) = \int_0^t e^{k\tau} . \Delta \, f_n(\tau) \; d\tau , \\ e^{lt} . \Delta \, y_{n+1}(t) = \int_0^t e^{l\tau} . \Delta \, g_n(\tau) \; d\tau . \end{cases}$$

Concerning the initial point $p^0 = (x^0, y^0)$ we assume it to lie in a neighborhood of the origin

$$\mathfrak{N} = \mathfrak{N}(r_0) , \qquad |p^0| \leqq r_0 \qquad (0 < r_0 \leqq R_0) ,$$

so small that

$$\chi(r_0) = l\,\vartheta_0 < l\,.$$

Then convergence can be established by the following estimate

$$(13) \qquad e^{lt}\,.\,|\Delta p_\nu(t)| \leqq |p^0|\,.\,\frac{(\varphi(t))^\nu}{\nu!} \qquad (\nu = 0, 1, 2, \ldots);$$

the function $\varphi(t)$ will presently be determined within the bounds $0 \leqq \varphi(t) \leqq lt$ (for $t \geqq 0$). The inductive proof is based on the inequalities implied in (12):

$$(14) \qquad \begin{cases} e^{lt}\,|\Delta y_{n+1}(t)| \leqq \int_0^t e^{l\tau}\,|\Delta g_n(\tau)|\;d\tau\,, \\ e^{lt}\,|\Delta x_{n+1}(t)| \leqq \int_0^t e^{l\tau}\,|\Delta f_n(\tau)|\;d\tau\,. \end{cases}$$

The right to substitute lt for kt in the inequality for Δx_{n+1} derives from the fact that, owing to $l \leqq k$,

$$e^{k(\tau-t)} \leqq e^{l(\tau-t)} \quad \text{for } \tau \leqq t \quad \text{(in particular for } \tau = 0).$$

(13) is correct for $\nu = 0$; suppose it to hold for $\nu = 0, 1, \ldots, n$.

Then

$$|p_\nu(t)| \leqq |p^\circ| \cdot e^{-lt+\varphi(t)} \leqq r_0 \cdot e^{-lt+\varphi(t)} \leqq r_0 \leqq R_0 \qquad (\nu=0,1,\ldots, n) .$$

Applying the Lipschitz condition (4) to $p_{n-1}(t)$ and $p_n(t)$ with

$$r = r(t) = r_0 \cdot e^{\varphi(t)-lt}$$

we obtain

$$e^{lt} \cdot |\Delta p_{n+1}(t)| \leqq \int_0^t \chi(r(\tau)) \cdot |\Delta p_n(\tau)| \, d\tau .$$

Using now (13) for $\nu = n$ we shall clearly succeed in obtaining the same relation for $\nu = n+1$ provided

(15) $$\chi(r(t)) = \frac{d\varphi}{dt} .$$

Because of $\frac{1}{r}\frac{dr}{dt} = \frac{d\varphi}{dt} - l$ this gives

(16) $$\frac{dr}{r} \cdot \frac{1}{l-\chi(r)} = -\,dt$$

and thus the following rule emanates: for $0 < r \leqq r_0$ form the function

$$(16') \qquad t = t(r) = \int_0^{r_0} \frac{dr}{r} \cdot \frac{1}{l - \chi(r)}$$

which increases monotonely with decreasing r and tends to ∞ as $r \to 0$, and its inverse

$$r = r_0 \cdot \rho = r_0 \cdot \rho(t) ,$$

ρ decreasing monotonely from $\rho(0) = 1$ to 0 as t runs from 0 to ∞ (Since $\chi(r) \leqq \chi(r_0) < l$ the denominator in (16′) causes no trouble.) $\varphi(t)$ is defined as

$$(17) \qquad lt + \log \rho = \int_r^{r_0} \frac{dr}{r} \cdot \frac{\chi(r)}{l - \chi(r)} .$$

The expression itself shows that $\varphi(t) \geqq 0$, but also that $\varphi(t) \leqq lt$ since $\rho(t)$ does not exceed 1. Another obvious estimate is

$$lt(1 - \vartheta_0) \leqq \log \frac{1}{\rho} \quad \text{or} \quad \rho \leqq e^{-lt(1-\vartheta_0)} .$$

The inductive proof of the inequalities (13) thus finished, we see that $p_n(\tau)$ uniformly converges to a limit $p(\tau)$ for $0 \leqq \tau \leqq t$, and hence the integral equations (9) result for $p(t)$ from the recurrent integral equations (11) for the approximating $p_n(t)$. At the same time the following sharp estimate is obtained

$$(18) \qquad |p(t)| \leqq |p^0| \cdot \rho(t)$$

Up to now we have used merely the monotone character of $\chi(r)$. However, the assumption (5) implies (and is actually equivalent to) the fact that $\rho(t)$ tends to zero with $t \to \infty$ as strongly as e^{-lt}. Indeed, under the hypothesis (5) one concludes from (17) that $lt + \log \rho$ tends monotonely to the limit

$$m = m(r_0) = \int_0^{r_0} \frac{dr}{r} \cdot \frac{\chi(r)}{l - \chi(r)} \quad .$$

Set $M = M(r_0) = e^m$. The following three appraisals for $\rho(t)$ have been found :

$$(19) \quad \rho(t) \leqq 1, \quad \rho(t) \leqq e^{-lt(1-\vartheta_0)}, \quad \rho(t) \leqq M e^{-lt} \quad (\text{for } t \geqq 0).$$

We summarize :

Theorem IV. *In the case* (A_+) *we have constructed a definite solution existing in the entire infinite interval* $t \geqq 0$ *and there satisfying the inequality* (18), *under the sole assumption that the initial point* (x^0, y^0) *lies in a neighborhood* $\mathfrak{N}(r_0)$ *for which*

$\chi(r_0) = l\vartheta_0 < l$. *The function* $\frac{r}{r_0} = \rho(t)$ *is obtained by the inversion of the relation*

$$(16') \qquad \int_r^{r_0} \frac{dr}{r(l-\chi(r))} = t \qquad (0 < r \leqq r_0) .$$

It satisfies the inequalities (19).

For $\chi(r) = C r^\delta$ the differential relation (15) permits immediate integration and we find

$$\rho(t) = \{ \vartheta_0 + (1-\vartheta_0)\, e^{\delta l t} \}^{-\delta'} \quad \text{where} \quad \delta' = \frac{1}{\delta} .$$

Verify that $r = r_0 \rho$ satisfies (16) ! The value of M is $(1-\vartheta_0)^{-\delta'}$.

Uniqueness for the solutions of (3) is established by a classical argument to the extent that coincidence of two solutions $p(t)$, $p_*(t)$ for $t = 0$ implies their coincidence in any interval $0 \leqq t \leqq T$ in which they both exist. It is desirable, however, to transform this qualitative statement into a quantitative one asserting that the difference $p(t) - p_*(t)$ is small in a specified way throughout the interval of existence if it is small at the beginning $t = 0$. Here we get the following exact and remarkably sharp result:

THEOREM V. *In the case* (A_+) *any two solutions* $p(t)$ *and* $p_*(t)$ *which both start in the neighborhood* $\mathfrak{N}(r_0)$ *of the origin,* $\chi(r_0) = l\vartheta_0 < l$, *satisfy the relation*

$$|\Delta p(t)| \leqq |\Delta p^0| \cdot \rho(t) \qquad \text{for } t \geqq 0 . \tag{20}$$

(18) is the special case $p_*(t) = 0$ of (20). This theorem is proved by the following simple

Lemma 1. Let $q(t) \geqq 0$ and $u(t)$ bet two continuous functions in an interval $0 \leqq t \leqq T$ satisfying the relation

$$u(t) \leqq \int_0^t q(\tau)\, u(\tau)\, d\tau . \tag{21}$$

Then $u(t) \leqq 0$.

Proof. Form $\int_0^t q(\tau)\, d\tau = Q(t)$ so that (21) becomes

$$u(t) \leqq \int_0^t u(\tau) \cdot dQ(\tau) .$$

Choose a positive constant k which is an upper bound for $u(t)$ in the interval $0 \leqq t \leqq T$. Induction with respect to $n = 0, 1, 2, \ldots$ yields the inequalities

$$u(t) \leqq k \cdot \frac{(Q(t))^n}{n!} \qquad (0 \leqq t \leqq T).$$

Passage to the limit $n \to \infty$ proves the lemma.

As the equations (12) imply the inequalities (14) so do the difference relations (10) lead to the integral inequalities

$$e^{lt} \cdot |\Delta x(t)| \leqq |\Delta x^{\circ}| + \int_0^t e^{l\tau} |\Delta f(\tau)| \, d\tau ,$$

$$e^{lt} \cdot |\Delta y(t)| \leqq |\Delta y^{\circ}| + \int_0^t e^{l\tau} |\Delta g(\tau)| \, d\tau$$

and from there by means of the Lipschitz condition to

$$e^{lt} |\Delta p(t)| \leqq |\Delta p^{\circ}| + \int_0^t \chi(r(\tau)) \cdot e^{l\tau} |\Delta p(\tau)| \, d\tau .$$

But

$$e^{\varphi(t)} = 1 + \int_0^t \chi(r(\tau)) \, e^{\varphi(\tau)} \, d\tau .$$

Thus

$$u(t) = e^{lt}\,|\Delta p(t)| - |\Delta p^{\circ}|\,e^{\varphi(t)}$$

satisfies the homogeneous integral inequality (21) with $q(t) = \chi(r(t))$, and Lemma 1 yields the desired result

$$e^{lt}\,|\Delta p(t)| - |\Delta p^{\circ}|\,e^{\varphi(t)} \leqq 0 \quad \text{or} \quad |\Delta p(t)| \leqq |\Delta p^{\circ}| \cdot \rho(t)\,.$$

It ought to be mentioned that $m(r_0)$ can be made as small as one wishes by choosing r_0 (and the neighborhood) $\mathfrak{N}(r_0)$ to which p° is restricted) sufficiently small. This observation is better exploited by substituting an arbitrary initial moment $t_1 \geqq 0$ for $t_0 = 0$. Set $r(t_1) = r_1$. The equations

$$\int_r^{r_0} \frac{dr}{r} \cdot \frac{1}{l - \chi(r)} = t\,, \qquad \int_{r_1}^{r_0} \frac{dr}{r} \cdot \frac{1}{l - \chi(r)} = t_1$$

give by subtraction

$$\int_r^{r_1} \frac{dr}{r} \cdot \frac{1}{l - \chi(r)} = t - t_1 \qquad (t \geqq t_1)\,.$$

Consequently the relation (20) or $|\Delta p| \leqq |\Delta p^{\circ}| \cdot \frac{r}{r_0}$ carried over to the new initial moment t_1 asserts that

$$|\triangle p(t)| \leqq |\triangle p(t_1)| \cdot \frac{r}{r_1} \quad \text{or} \quad \frac{|\triangle p(t)|}{\rho(t)} \leqq \frac{|\triangle p(t_1)|}{\rho(t_1)} \quad (\text{for } t \geqq t_1)$$

thus proving

THEOREM VI.

$$\frac{|\triangle p(t)|}{\rho(t)}, \quad \textit{in particular} \quad \frac{|p(t)|}{\rho(t)}$$

is a monotone–decreasing function for $t \geqq 0$.

Omitting the zero terms from the sums $\Sigma_{\nu=0}^{n} \triangle x_\nu$, $\Sigma_{\nu=0}^{n} \triangle y_\nu$, which we majorized by

$$|p^0|\, e^{-lt}\, \Sigma \frac{(\varphi(t))^\nu}{\nu!}$$

we get

$$|x(t) - x^0 e^{-kt}|, \qquad |y(t) - y^0 e^{-lt}|$$

$$\leqq |p^0|\, e^{-lt} (e^{\varphi(t)} - 1) = |p^0| \left\{ \frac{r}{r_0} - e^{-lt} \right\}.$$

Again, replace $t_0 = 0$ by any $t_1 \geqq 0$. The result is an upper bound for

(22) $\quad |e^{lt}\, y(t) - e^{lt_1}\, y(t_1)|$ and $|e^{lt}\, x(t) - e^{kt_1}\, x(t_1) \cdot e^{-(k-l)\,t}|$;

namely

$$\frac{|p(t_1)|}{r_1}(r\, e^{lt} - r_1\, e^{lt_1}) \leqq \frac{|p^0|}{r_0}(r\, e^{lt} - r_1\, e^{lt_1}) =$$

(23) $|p^0| \left\{ exp \int_r^{r_0} - exp \int_{r_1}^{r_0} \right\}$ with the integrand $\dfrac{\chi(r) \cdot dr}{r(l - \chi(r))} = \chi(r) \cdot dt.$

Hence:

THEOREM VII. *For* $t \geqq t_1$ *the differences* (22) *are less than* (23) *or* $|p^0|$ *times*

$$exp \int_0^t \chi(r_0\, \rho(\tau))\, d\tau - exp \int_0^{t_1} \chi(r_0\, \rho(\tau))\, d\tau .$$

Given any $\epsilon > 0$, one may determine t_1 so that (23) does not exceed ϵ for $t \geqq t_r$. Then

$$|y(t)\, e^{lt} - y(t_1)\, e^{lt_1}| \leqq \epsilon$$

for $t \geqq t_1$, consequently $y(t) \cdot e^{lt}$ has a limit b and the error $|e^{lt} \cdot y(t) - b|$ does not exceed

$$\text{(24)} \qquad exp \int_0^\infty \chi(r_0\, \rho(\tau))\, d\tau - exp \int_0^t \chi(r_0\, \rho(\tau))\, d\tau .$$

For $x(t)$ the conclusion is modified by the presence in the subtrahend of the factor $e^{-(k-l)t}$ which tends to zero with $t \rightarrow \infty$. Hence the subtrahend will be less than ϵ from a certain t on, and from that t on $e^{lt} \cdot |x(t)|$ is less than 2ϵ, consequently $e^{lt} \cdot x(t) \rightarrow 0$ with $t \rightarrow +\infty$. Thus Theorem VII implies the asymptotic statement in Theorem I all qualitative assertions of which have now been implemented by quantitative estimates.

One could wish to have a simpler expression for the error (24) even at the expense of losing some of its sharpness. First, it is clear that (24) is less than

$$\text{(25)} \qquad \mathrm{M} \int_t^\infty \chi(r_0\, \rho(\tau))\, d\tau .$$

This could have been deduced directly from the formula

$$e^{lt}\, y(t) = \left(y^0 + \int_0^\infty \right) - \int_t^\infty = b - \int_t^\infty ,$$

the integrand of which, $e^{l\tau} . g(x(\tau), y(\tau))$, is of modulus

$$\leqq \chi(r_0\, \rho(\tau)) \cdot |p^0| \; e^{l\tau} \rho(\tau) \leqq |p^0| \cdot \mathrm{M}\, \chi(r_0\, \rho(\tau)) .$$

But having used the appraisal $\rho(\tau) \leqq M e^{-l\tau}$ outside the function sign χ we may use it as well inside and thus replace (25) by the higher

$$|p^{\circ}| \cdot M \int_{t}^{\infty} \chi(r_{\circ} M e^{-l\tau}) \, d\tau \, .$$

For $\chi(r) = C r^{\delta}$ this is

$$|p^{\circ}| M \cdot \frac{l\vartheta_{\circ}}{1-\vartheta_{\circ}} \int_{t}^{\infty} e^{-\delta l \tau} \, d\tau = |p^{\circ}| \cdot \frac{M \vartheta_{\circ}}{\delta(1-\vartheta_{\circ})} e^{-\delta l t} \, .$$

Hence the error is of the order $e^{-\delta l t}$, and the final result takes on the simple and satisfactory form

$$(26_y) \quad |y(t) - b e^{-lt}| \leqq |p^{\circ}| \frac{M \vartheta_{\circ}}{\delta(1-\vartheta_{\circ})} e^{-l' t} \qquad \text{with } l' = (1+\delta)\, l \, .$$

According the formula for $x(t)$ a similar treatment in which the factor $e^{k(\tau - t)}$ in the integrand is retained and not replaced by the larger $e^{l(\tau - t)}$ we find

$$|x(t) - x^{\circ} e^{-kt}| \leqq |p^{\circ}| M e^{-kt} \int_{0}^{t} e^{(k-l)\tau} \chi(r_{\circ} M e^{-l\tau}) \, d\tau \, .$$

For $\chi(r) = C\, r^{\delta}$ the right member has the value

$$|p^{\circ}|\, M \frac{l\,\vartheta_{o}}{1-\vartheta_{o}} \cdot e^{-kt} \int_{0}^{t} e^{(k-l')\tau}\, d\tau ,$$

the factor after the dot equalling

$$t \cdot \int_{0}^{1} exp\,\{-l'\,t \cdot \mu - k\,t\,(1-\mu)\}\; d\mu = \frac{e^{-l't} - e^{-kt}}{k-l'} .$$

The first expression gives an upper bound

$$t\, e^{-Lt} , \qquad L = min\,(l', k)$$

which is not upset by the "small denominator" $k-l'$. The ensuing inequality

$$(26_x) \qquad |x(t)| \leqq |p^{\circ}| \left\{ 1 + \frac{M\,\vartheta_{o}}{1-\vartheta_{o}}\, l\,t \right\} e^{-Lt}$$

is as good a result as one can expect to hold, because, speaking in the language of oscillations rather than of damping, k and $(1+\delta)\, l$ are the lowest frequencies $> l$ which threaten resonance.

3. *The Standard Solution*

In this section we place ourselves under the general assumptions (A): $l \leqq k$, $k > 0$ and, given a number a, we want to construct a solution satisfying the conditions (7) for $t \to +\infty$. To that end we have to look for a solution of the integral equations

$$
(27) \qquad \begin{cases} x(t) = a e^{-kt} - \int_t^{\infty} e^{k(\tau-t)} \cdot f(x(\tau), y(\tau))\, d\tau , \\ y(t) = - \int_t^{\infty} e^{l(\tau-t)} \cdot g(x(\tau), y(x(\tau))\, d\tau . \end{cases}
$$

Notice that the limit 0 previously appearing in the integrals is now replaced by ∞. The construction itself will take care of the convergence of these integrals. Because of $l \leqq k$ or $\beta = \frac{l}{k} \leqq 1$ we have $e^{l(\tau-t)} \leqq e^{k(\tau-t)}$ for $\tau \geqq t$. Granted that our equations have a unique solution, it is readily seen that it will depend on a and t only in the combination $s = a e^{-kt}$. (It is not without importance to include the possibility $a = 0$.) Writing $x = \xi(s)$, $y = \eta(s)$ and denoting the vector $(\xi(s), \eta(s))$ by $\pi(s)$ we thus claim that the following equations have a unique solution

$$
(28) \qquad \begin{cases} \xi(s) = s - k^{-1} \int_0^s \frac{s}{\sigma} f(\xi(\sigma), \eta(\sigma)) \frac{d\sigma}{\sigma} , \\ \eta(s) = - k^{-1} \int_0^s \left(\frac{s}{\sigma}\right)^{\beta} g(\xi(\sigma), \eta(\sigma)) \frac{d\sigma}{\sigma} , \end{cases}
$$

and in constructing that solution we shall find that

$$\xi(s) = s + o(s), \qquad \eta(s) = o(s) \qquad \text{for } s \to 0.$$

Any constant a being given,

$$x = \xi(a e^{-kt}) = X_a(t), \qquad y = \eta(a e^{-kt}) = Y_a(t) \tag{29}$$

will then be a solution of our original differential equations possessing the required asymptotic properties for $t \to = \infty$.

The scheme for the computation of the successive approximations $\xi_n(s)$, $\eta_n(s)$ is obvious: starting with $\xi_{-1}(s) = \eta_{-1}(s) = 0$ we set

$$\xi_n(s) = s - k^{-1} \int_0^s \frac{s}{\sigma} f(\xi_{n-1}(\sigma), \eta_{n-1}(\sigma)) \frac{d\sigma}{\sigma},$$

$$\eta_n(s) = - k^{-1} \int_0^s \left(\frac{s}{\sigma}\right)^{\beta} g(\xi_{n-1}(\sigma), \eta_{n-1}(\sigma)) \frac{d\sigma}{\sigma}$$

At the first step the zero approximation $\xi_0(s) = s$, $\eta_0(s) = 0$ is obtained. Let $s > 0$. On the basis of the assumption $\beta \leqq 1$ the following inequalities are derived in thc same way as (14):

$$\left\{ \begin{aligned} |\Delta \xi_{n+1}(s)| &\leqq k^{-1} \int_0^s \frac{s}{\sigma} \Delta f_n(\sigma) \frac{d\sigma}{\sigma}, \\ |\Delta \eta_{n+1}(s)| &\leqq k^{-1} \int_0^s \frac{s}{\sigma} \Delta g_n(\sigma) \frac{d\sigma}{\sigma}. \end{aligned} \right. \tag{30}$$

It is maintained that they imply

$$(31)\qquad |\Delta \pi_\nu(s)| \leqq s \cdot \frac{(\psi(s))^\nu}{\nu!} \qquad \text{for } \nu = 0,1,2,\ldots$$

where the positive function $\psi(s)$ is still to be determined. The procedure follows the same road as before. The differential equation for $\psi(s)$ turns out to be

$$k^{-1} \cdot \chi(s\, e^{\psi(s)}) = s \cdot \frac{d\psi}{ds} .$$

Introduction of $R = R(s) = s\, e^{\psi(s)}$ changes it into

$$s \cdot \frac{dR}{R} = \{1 + k^{-1} \chi(R)\}\, ds \qquad \text{or} \qquad \frac{dR}{R} \cdot \frac{1}{1 + k^{-1}\chi(R)} = \frac{ds}{s} .$$

Except that t is measured in the exponential scale s, the only difference over against (16) is the factor $+k^{-1}$ having taken the place of $-l^{-1}$ in the denominator $1 + k^{-1}\chi(R)$. In order to carry out the quadrature in agreement with the asymptotic condition $\psi(s) \to 0$ for $s \to 0$ we subtract $\frac{dR}{R} = d \log R$ and thus obtain

$$(32)\qquad \log R - \int_0^R \frac{\chi(R) \cdot dR}{R(k + \chi(R))} = \log s .$$

As its derivative shows, the left member is a monotone increasing function of R mapping the interval $0 < R \leqq R_0$ onto a certain interval $0 < s \leqq s_0$ of the s-axis. The inverse function shall be denoted by $R = R(s)$ $(0 < s \leqq s_0)$ while $\psi(s)$ is defined as $\log \frac{R}{s}$. Two circumstances are essential for the success of the method: (1) $R(s)$ must remain $\leqq R_0$; this is taken care of by limiting s to the interval $0 < s \leqq s_0$, $\log s_0$ being the value of (32) for $R = R_0$. (2) $\psi(s)$ or $\log \frac{R}{s}$ must be positive; this is implied in the defining formula (32).

Once (31) is proved, the existence of a limit function $\pi(s)$ is guaranteed and the approximating as well as the limit function satisfies the inequality

$$(33) \qquad |\pi_n(s)|, \quad |\pi(s)| \leqq s \cdot e^{\psi(s)} = R(s).$$

A little care is required in establishing the integral equations (28) for the limit functions $\xi(s)$, $\eta(s)$. To that end observe that $|\Delta \pi_n(\sigma)|$ has the upper bound $\sigma \cdot \frac{(\psi(s_0))^n}{n!}$ for $0 < \sigma \leqq s_0$ and consequently

$$|\Delta f_n(\sigma)|, \quad |\Delta g_n(\sigma)| \leqq \chi(R(\sigma))\,\sigma\,\frac{(\psi(s_0))^n}{n!}.$$

Observe moreover that the integral

$$k^{-1}\int_0^s \chi(R(\sigma))\,\sigma \cdot \frac{d\sigma}{\sigma^2} = \int_0^R \frac{\chi(R)\,dR}{R(k+\chi(R))}$$

converges at the lower limit $\sigma = 0$ (or $R = 0$). For this reason summation and integration may be interchanged in

$$\int_0^s \Sigma_n \,\Delta f_n(\sigma) \cdot \frac{d\sigma}{\sigma^2}$$

and the corresponding expression for g.

The asymptotic law obeyed by the solution for $s \to 0$ is here more easily ascertained than in § 2, because it has here been used to select the proper solution while in § 2 initial conditions for $t = 0$ served as the selective principle. Lopping off the zero approximation $\xi_0(s) = s$, $\eta_0(s) = 0$ we find

$$(34) \qquad \left|\frac{\xi(s)}{s} - 1\right|, \quad \left|\frac{\eta(s)}{s}\right| \leqq e^{\psi(s)} - 1 = \frac{R}{s} - 1,$$

the definition (32) showing that $\frac{R}{s} \to 1$ with $s \to 0$.

We add the negative side of the s-interval, $0 > s \geqq -s_0$, operating there with the same function of comparison $R(-s) = R(s)$, and summarize:

THEOREM VIII. *The differential equations*

$$(35) \qquad s\frac{dx}{ds} = -x + k^{-1}f(x, y), \qquad s\frac{dy}{ds} = -\beta y + k^{-1}g(x, y)$$

have a solution $x = \xi(s)$, $y = \eta(s)$ *in the interval* $|s| \leqq s_0$ *the behavior of which at* $s = 0$ *is described by the limit relations* $\frac{\xi(s)}{s} \to 1$, $\frac{\eta(s)}{s} \to 0$ *for* $s \to 0$. *More precisely it obeys the laws* (33) *and* (34) *where the function* $R = R(s)$ *is determined by inversion of the equation* (32) *and* $\log s_0$ *is the value of the function* $\log s$ *of* R, *eq.* (32), *for* $R = R_0$.

For $\chi(r) = Cr^{\delta}$ one readily finds (or verifies)

$$e^{-\delta\psi(s)} = 1 - \left(\frac{C}{k}\right)s^{\delta}, \qquad \frac{R}{s} = \left\{1 - \left(\frac{C}{k}\right)s^{\delta}\right\}^{-\delta'},$$

$$\frac{dR}{ds} = \left(\frac{R}{s}\right)^{\delta+1}.$$

s_0 is obtained from

$$C s_0^{\delta} = \frac{k\,\theta_0}{k+\theta_0}\,, \qquad \theta_0 = C R_0^{\delta}\,.$$

The differential equations (35) give rise to the following estimates concerning the derivatives $\frac{d\xi}{ds} = \xi'(s)$, $\frac{d\eta}{ds} = \eta'(s)$ in the neighborhood of $s = 0$:

$$(36_x) \qquad |\xi'(s) - 1| \leqq \left(\frac{R}{s} - 1\right) + k^{-1}\,\chi(R)\cdot\frac{R}{s}\,,$$

$$(36_y) \qquad |\eta'(s)| \leqq |\beta|\left(\frac{R}{s} - 1\right) + k^{-1}\chi(R)\cdot\frac{R}{s}\,.$$

The right members tend to zero for $s \to 0$. Consequently $\xi'(s)$, $\eta'(s)$ are continuous even at $s=0$, assuming there the values $\xi'(0) = 1$, $\eta'(0) = 0$. The right member of (36_x) equals $\frac{dR}{ds} - 1$. Hence we may replace (36_x) and (36_y) by the following relations of simpler appearance

$$(37) \quad |\xi'(s) - 1| \leqq \frac{dR}{ds} - 1, \qquad |\eta'(s)| \leqq \gamma\left(\frac{dR}{ds} - 1\right),$$

γ being the larger of the two numbers 1, $|\beta|$.

$\xi'(s)$ being positive in the neighborhood of $s = 0$, the function $\xi(s)$ will be monotone-increasing in a certain interval $|s| \leqq s_1$ $(0 < s_1 \leqq s_0)$ and thus have an inverse $s = \sigma(x)$ existing in the interval $|x| \leqq R_1$ and satisfying there the inequality $|\sigma(x)| \leqq s_1$ as well as the equation $\xi(\sigma(x)) = x$, provided R_1 is chosen in accordance with $0 < R_1 \leqq \xi(s_1)$. Apart from its parametrization the trajectory $x = \xi(s)$, $y = \eta(s)$ is then representable in the simple form

$$y = \Phi(x) = \eta(\sigma(x)) \,. \tag{38}$$

$\Phi(x)$ has a continuous derivative $\Phi'(x)$ which vanishes for $x = 0$. The differential equations (35) for $\xi(s)$ and $\eta(s)$ show that the relation

$$\Phi'(x)\{k\,x - f(x, y_1)\} = l\,y_1 - g(x, y_1) \tag{39}$$

holds by virtue of the substitution $y_1 = \Phi(x)$ for $|x| \leqq R_1$.

For the purpose of future application we are interested in choosing R_1 so small that we are sure the inequality

$$|\Phi'(x)| \leqq 1 \qquad \text{and thus} \quad |\Phi(x)| \leqq |x|$$

will hold for $|x| \leqq R_1$. It is *a priori* clear that this can be done, but the explicit estimates which we took care to establish enable us without further ado to ascertain a fairly favorable value of R_1. Indeed, it suffices to see to it that

$$\frac{dR}{ds} - 1 \leqq \frac{1}{(1+\gamma)} \qquad \text{for} \quad s = s_1 \leqq s_0$$

and choose

$$R_1 = s_1 \cdot \frac{\gamma}{1+\gamma} \left(\geqq \frac{1}{2} s_1 \right),$$

the latter choice being vindicated by the inequalities

$$\left| \frac{\xi(s_1)}{s_1} - 1 \right| \leqq \frac{R(s_1)}{s_1} - 1 \leqq \left(\frac{dR}{ds} \right)_{s_1} - 1 \leqq \frac{1}{1+\gamma}, \quad \frac{\xi(s_1)}{s_1} \geqq \frac{\gamma}{1+\gamma}.$$

4. *Comparison with the Standard Solution*

The existence theorems IV and VIII are not restricted to the case of *two* unknown functions $x(t)$, $y(t)$. For any number of unknowns they refer to the lowest and the highest eigenvalues and are dependent on the assumption that these eigenvalues, the lowest in the one case, the highest in the other case, are positive. But what follows now is limited to the two-function problem. We resume the case (A_+) in which the lower eigenvalue l is positive.

a being a given constant, the substitution $s = a \cdot e^{-kt}$ transforms the parametrized trajectory

$$\Lambda: \quad x = \xi(s), \quad y = \eta(s)$$

constructed in the proof of Theorem VIII into a solution (29) of our original differential equations (3). Each of these "exceptional" solutions exists from a certain value of t on. Apart from the trivial solution $x(t) = y(t) = 0$ each is equivalent to one of the two solutions (X_+, Y_+), (X_-, Y_-) which correspond to the values $a = +1$ and -1 respectively. Since $\xi'(0) = 1$, $\eta'(0) = 0$ the trajectory Λ passes through the origin in the direction of the x-axis. It is stated in Theorem III that for all non-exceptional solutions $x(t)$, $y(t)$ the limit b of $e^{lt}\, y(t)$ for $t \to \infty$ is different from zero, which means that all their trajectories run into the origin with $t \to \infty$ in the direction of the y-axis. We have to prove that if $b = 0$ then the solution under consideration coincides with one of the exceptional solutions $X_a(t)$, $Y_a(t)$. However, we shall replace this qualitative by a quantitative statement, saying that if b is small, the solution differs from a certain exceptional solution by an amount which is small of the same order of magnitude as b. Let us explain this in detail!

Our task is to study the solution $\mathfrak{L}: p(t) = (x(t), y(t))$ with the initial value $p^0 = (x^0, y^0)$. As will be remembered, the existence of this solution is guaranteed for $t \geqq 0$, provided $|p^0| \leqq r_0$ and r_0 so chosen that

$$r_0 \leqq R_1, \quad \chi(r_0) = l\vartheta_0 < l;$$

it satisfies the inequality

$$(18) \qquad |p(t)| \leqq |p^0| \cdot \rho(t).$$

Moreover we know that $e^{lt} \cdot y(t)$ has a limit $b = b(\mathfrak{L})$ for $t \to \infty$. The finer details of the behavior of $x(t)$ described by Theorem VII and subsequent remarks including formula (26_x) are now irrelevant; it suffices to use the estimate $|p(t)| \leqq |p^0| M e^{-lt} = O(e^{-lt})$ implied in (18). The assumption $r_0 \leqq R_1$ enables us to establish correspondence between the trajectories $\mathfrak{L}$ and Λ by dropping perpendiculars: the point p and p_1 on $\mathfrak{L}$ and Λ with the abscissa $x(t)$ have the ordinates $y(t)$ and $y_1(t) = \Phi(x(t))$ respectively. This works because $x(t)$ lies within the range of the argument x in the function $\Phi(x)$,

$$|x(t)| \leqq |p(t)| \leqq |p^0| \leqq r_0 \leqq R_1.$$

In particular, the point $s = s^0$ on Λ which corresponds to the point $t = 0$ on $\mathfrak{L}$ is determined by $\xi(s^0) = x(0) = x^0$ or $s^0 = \sigma(x^0)$.

The landmarks of our further course are staked out by the following propositions:

Lemma 2.

$$|y(t)-y_1(t)| \cdot \rho^2(t)\, e^{3lt}$$

is a monotone-increasing function for $t \geqq 0$.

Lemma 3.

$$|y(0)-y_1(0)| \leqq M^2 \cdot |b(\mathfrak{L})|\,.$$

THEOREM IX. *Suppose* $0<l\leqq k$. *Given a point* $p^{\circ} \in \mathfrak{N}(r_0)$,

a constant a *may be so fixed that the solution* $\mathfrak{L}$: $p(t)=(x(t),\ y(t))$ *with the initial point* $p(0) = p^{\circ}$ *stays as near to the exceptional solution* $(X_a(t),\ Y_a(t))$ *for* $t>0$ *as indicated by the inequalities*

$$|x(t)-X_a(t)|, \qquad |y(t)-Y_a(t)| \leqq M^3 \cdot |b(\mathfrak{L})|\, e^{-lt}\,.$$

This is as sharp a result as one could possibly expect: for small b the upper bound is small of the order of b, and an additional factor e^{-lt} takes care of the exponential decrease to zero with $t \to +\infty$. The decisive step is the proof of Lemma 2 wich proceeds as follows:

Denote the difference $y(t)-y_1(t)$ by $\Delta_1 y(t) = z(t)$ and accordingly

$$\Delta_1 f(t) = f(x(t), y(t)) - f(x(t), y_1(t)) .$$

$z(t)$ measures the vertical distance of the two trajectories $\mathfrak{S}$ and Λ in terms of the parameter t of the first. By subtracting the two equations for $y(t)$ and $y_1(t)$,

$$\frac{dy}{dt} = -l\,y(t) + g(x(t), y(t)) ,$$

$$\frac{dy_1}{dt} = \Phi'(x(t))\,\{-kx(t) + f(x(t), y(t))\}$$

one gets

$$\frac{d}{dt}(\Delta_1 y) = \{-ly + g(x, y)\} - \Phi'(x)\,\{-kx + f(x, y)\}.$$

and again subtracting (39),

$$0 = \{-ly_1 + g(x, y_1)\} - \Phi'(x)\,\{-kx + f(x, y_1)\} ,$$

a relation for the difference $z = \Delta_1 y$ results,

(41) $$\frac{dz}{dt} = -lz + S,$$

in which

$$S(t) = \Delta_1 g(t) - \Phi'(x(t))\, \Delta_1 f(t)\ .$$

The upper bound (18) holds good not only for $p = (x, y)$ but also for $p_1(x, y_1)$ because from $|p|$ it carries over to $|x|$ and from there to $|y_1| = |\Phi(x)| \leqq |x|$. In view of these upper bounds the Lipschitz condition yields*

$$|\Delta_1 f|,\quad |\Delta_1 g| \leqq \chi(r) \cdot |\Delta_1 y(t)| \qquad \text{with} \qquad r = r_0 \cdot \rho(t)$$

and considering that $|\Phi'(x(t))| \leqq 1$,

$$|S(t)| \leqq 2\chi(r(t)) \cdot |\Delta_1 y(t)|\ .$$

With the help of a similar appraisal of the remainder term $(\Delta f, \Delta g)$ in

$$\frac{d}{dt}(\Delta x) = -k \cdot \Delta x + \Delta f, \qquad \frac{d}{dt}(\Delta y) = -l \cdot \Delta y + \Delta g$$

---

* Warning: It is not advisable to form the function $\dfrac{\{f(x, y) - f(x, \Phi(x))\}}{\{y - \Phi(x)\}}$ which, though bounded, will cease to exist along the trajectory A.

we previously obtained an upper bound of the two-component vector $\Delta p = (\Delta x, \Delta y)$ in terms of its *initial* value Δp° (see Theorem V). Here we apply it to ascertain an upper bound for $\Delta_1 y$ in terms of its *asymptotic* value for $t \to \infty$. But this is possible only when we are dealing with a scalar $\Delta_1 y$ rather than a vector.

Multiply (41) by z,

$$\frac{1}{2}\frac{dz^2}{dt} = -l z^2 + z\,S$$

and then infer from

$$|z\,S| \leqq 2\chi(r) \cdot z^2$$

the inequality

$$\frac{1}{2}\frac{dz^2}{dt} \geqq -(l + 2\chi(r))\, z^2 . \tag{42}$$

Now $l + 2\chi(r)$ is the derivative of $lt + 2\varphi(t)$, hence

$$U = e^{lt+2\varphi} = \rho^2 e^{8lt}$$

satisfies the equation

$$\frac{dU}{dt} = (l + 2\chi(r))\, U \quad \text{or} \quad \frac{1}{2}\frac{dU^2}{dt} = (l + 2\chi(r))\, U^2 . \tag{43}$$

(42) and (43) imply

$$\frac{1}{2}\frac{d}{dt}(z^2 U^2) \geqq 0 ,$$

hence $|z|U = |z|\rho^2 e^{3lt}$ is monotone–increasing as was claimed by Lemma 2.

The limit of $|z|U$ for $t \to \infty$ is easily determined. Indeed

$$y_1(t) = \Phi(x(t)) = o(|x(t)|) = o(e^{-lt}) \qquad (\text{for } t \to \infty) ,$$

hence

$$lim_{t\to\infty} \{y(t) - y_1(t)\}\, e^{lt} = b \qquad \text{and thus}$$

$$lim_{t\to\infty}\, z(t)\, U(t) = b\, M^2 .$$

Because $|z|U$ is increasing, its value for $t = 0$ cannot exceed its limit for $t \to \infty$: this is the content of Lemma 3.

As the next step we fix the constant a so that $X_a(0)$ becomes equal to $x(0) = x^\circ$, $a = \sigma(x^\circ)$, and thus $Y_a(0) = \Phi(X_a(0)) = y_1(0)$. Introduce the differences

$$\Delta x(t) = x(t) - X_a(t) , \qquad \Delta y(t) = y(t) - Y_a(t) .$$

Their initial values for $t = 0$ are 0 and $z(0) = y(0) - y_1(0)$ respectively. Consequently Theorem V yields

$$| \Delta p(t) | \leqq | z(0) | \cdot \rho(t) \leqq | z(0) | \cdot M e^{-lt} .$$

Combination with Lemma 3 leads to Theorem IX.

Notice once more that by proper choice of r_0 we can bring $M = M(r_0)$ as near to unity as we wish.

The method for proving *the second half of Theorem III* is similar, though somewhat simpler. Now l is negative. Choose $r_0 > 0$ so that

$$\chi(r_0) < \frac{1}{2} |l| \quad \text{and} \quad r_0 \leqq R_1 .$$

Suppose we have a solution $\mathfrak{L}$: $p(t) = (x(t), y(t))$ approaching the origin as t tends to $+\infty$. We may ascertain a t_0 such that $p(t)$ lies in the neighborhood $\mathfrak{N} = \mathfrak{N}(r_0)$ of the origin for $t \geqq t_0$. The above construction of projecting $\mathfrak{L}$ perpendicularly on to $\mathbf{A}$ can then be carried out, and the resulting $p_1(t) = (x(t), y_1(t))$ will also stay in $\mathfrak{N}$ for $t \geqq t_0$. We then have

$$|\Delta_1 f|, \ |\Delta_1 g| \leqq \chi(r_0) \cdot |z|$$

and after setting $\dfrac{2\chi(r_0)}{|l|} = \vartheta < 1$ the remainder S in the relation

$$\frac{dz}{dt} = -lx + S = |l|\, z + S$$

is of modulus $\leqq 2\chi(r_0) \cdot |z| = \vartheta\, |l| \cdot |z|$. Hence

$$\frac{1}{2}\,\frac{dz^2}{dt} \geqq (1-\vartheta)\, |l|\, z^2\,,$$

which means that the vertical distance $|p-p_1| = |z|$ of $\mathfrak{L}$ and Λ increases at least exponentially with time:

(44) $$|z(t)| \geqq |z(t_0)| \cdot e^{(1-\vartheta)|l|(t-t_0)} \qquad \text{for } t \geqq t_0\,.$$

This is in obvious disagreement with $z(t) \to 0$ for $t \to +\infty$ unless

(45) $$z(t_0) = 0\,.$$

Choose a by

$$a\, e^{-kt_0} = \sigma(x(t_0)) \qquad \text{or} \qquad \xi(a\, e^{-kt_0}) = x(t_0)\,.$$

Then (45) states that the two solutions $(x(t), y(t))$ and $(X_a(t), Y_a(t))$ have the same initial values for $t = t_0$. As they both exist throughout the entire infinite interval $t \geqq t_0$ (and are of modulus $\leqq r_0$), the classical argument for uniqueness (based on a special case of Lemma 1) proves that they coincide over the same range of t.

Also this part of Theorem III can be given a quantitative turn. Suppose a solution $\mathfrak{L}$: $p(t) = (x(t), y(t))$ is known to remain in the neighborhood $\mathfrak{N}$ of the origin during a very long time interval, say for $0 \leqq t \leqq T$. Fix a constant a in the manner described above; $\Delta x(t) = x(t) - X_a(t)$ then vanishes for $t = 0$ whereas $\Delta y(0) = \{y(t) - Y_a(t)\}_{t=0}$ is the vertical distance $y(0) - y_1(0)$. Our assumption implies that the vertical distance of $\mathfrak{L}$ and Λ does not exceed $2 r_0$ after the lapse of time T. But then (44) requires that $|\Delta p|$ did not exceed the extremely small amount $2 r_0 \cdot e^{-(1-\vartheta)|l|T}$ at the beginning $t = 0$:

Theorem X. *Suppose* $l < 0$, $k > 0$. *Given a solution* $(x(t), y(t))$ *that remains in the neighborhood* $\mathfrak{N} = \mathfrak{N}(r_0)$ *of the origin during the time interval* $0 \leqq t \leqq T$, *we may fix* a *so that the difference* $\Delta p(t) = (x(t) - X_a(t), y(t) - Y_a(t))$ *at the beginning* $t=0$ *is of modulus* $\leqq 2 r_0 \cdot e^{-(1-\vartheta)|l|T}$.

134.

Comparison of a degenerate form of Einstein's with Birkhoff's theory of gravitation

Proceedings of the National Academy of Sciences of the United States of America
30, 205—210 (1944)

Whereas electric charge is not universally proportional to the inertial mass of bodies their gravity is. This fundamental fact, supported both by daily experience and the most refined experiments, led Einstein to the conception that inertia and gravitation are one (principle of equivalence) and thus to his theory of general relativity. The main reason for my and many others' belief in that theory is the radical explanation it affords for the fact just mentioned. Any theory which breaks up the unity of inertia and gravitation, as Birkhoff's recent theory of gravitation in a flat world[1] does, throws us back into the position before Einstein where we had to accept the identity of mass and weight without understanding it. Is there any reason for such a withdrawal?

One way of breaking the bond between inertia and gravitation is suggested by the study of weak gravitational fields in Einstein's theory: One lets the g_{ik} differ but infinitesimally from constant values δ_{ik}, $g_{ik} = \delta_{ik} + \epsilon h_{ik}$ (δ_{ik} = inertia, h_{ik} = gravitation), throws away all higher powers of the constant factor ϵ except the first, writes h_{ik} for ϵh_{ik} and adopts the equations thus obtained as strictly valid. In a review[2] of Birkhoff's paper I had tried to illustrate his scheme by analogy with this "degenerate Einstein theory." In a recent note[3] A. Barajas claims to disprove the validity of this comparison. But he misinterprets my intention, and his calculations are not entirely correct. I hope to clarify the situation by the following fuller statement of the case, in which I compare the complete and the "degenerate" Einstein theories (E, D) with the theory of Birkhoff (B).

In terms of a suitable parameter σ the equations of the world trajectory of a particle have the common form

$$\frac{d^2x_i}{d\sigma^2} = -\gamma\cdot\Gamma^i_{pq}u^p u^q \qquad (u^i = dx_i/d\sigma), \tag{1}$$

which is invariant with respect to the substitution $\sigma \to k\sigma$. The components Γ^i_{pq} of the gravitational field are in the respective theories defined by

$$\Gamma_{i,\,pq} = g_{ij}\Gamma^{j}_{pq} = \frac{1}{2}\left(\frac{\partial g_{ip}}{\partial x_q} + \frac{\partial g_{iq}}{\partial x_p} - \frac{\partial g_{pq}}{\partial x_i}\right), \tag{2E}$$

$$\Gamma_{i,\,pq} = \delta_{ij}\Gamma^{j}_{pq} = \frac{1}{2}\left(\frac{\partial h_{ip}}{\partial x_q} + \frac{\partial h_{iq}}{\partial x_p}\right) - \frac{\partial h_{pq}}{\partial x_i}, \tag{2B}$$

$$\Gamma_{i,\,pq} = \delta_{ij}\Gamma^{j}_{pq} = \frac{1}{2}\left(\frac{\partial h_{ip}}{\partial x_q} + \frac{\partial h_{iq}}{\partial x_p} - \frac{\partial h_{pq}}{\partial x_i}\right). \tag{2D}$$

The constant factor γ, specific for the individual particle, is the ratio gravity/mass. Once the cut performed by which D arises from E, a combination $\delta_{ik} + \mu h_{ik}$ with this or that constant μ has as much and as little significance in D as in B. In B and D there is no more reason for γ to be independent of the particle than one can expect the ratio η = charge/mass to be the same for all particles moving in an electromagnetic field according to the equations

$$\frac{d^2x_i}{d\sigma^2} = \eta \cdot F^{i}_{q}u^{q} \qquad \left(F_{ik} = \frac{\partial \varphi_k}{\partial x_i} - \frac{\partial \varphi_i}{\partial x_k}\right).$$

We use the familiar notations

$$ds^2 = g_{ik}dx_i dx_k \text{ in E,}$$

$$ds^2 = \delta_{ik}dx_i dx_k = dx_0^2 - (dx_1^2 + dx_2^2 + dx_3^2) \text{ in B and D.} \tag{3}$$

From (2E) and (2B) there follows the relation

$$\left(\frac{ds}{d\sigma}\right)^2 = \text{const.} \tag{4E, B}$$

Birkhoff's choice of the Γ^{i}_{pq} was, in fact, suggested by the postulated relation (4B). No such relation (4) holds in D. [One could remedy this by a suitable transformation of the parameter σ before letting $\epsilon \to 0$; the recipe for the construction of D is not entirely unambiguous! But following Mr. Barajas I shall stick to our present interpretation of D.]

What I had in mind when I said that B and D are "essentially of the same nature" (not "much the same," as Mr. Barajas quotes me) were the following points: (1) Both operate in a flat world and are linear with respect to h_{ik}. (2) Both have a *tensor* h_{ik} as gravitational potential, and the ponderomotoric force exerted by the gravitational field depends *quadratically* on u^i. One should not forget that before Einstein a gravitational potential with 10 components was an unheard-of thing; this complication was forced upon him by the principles of general relativity rather than by any observed facts about the dependence of the force of gravitation on velocity. Notice also the formal similarity of the expressions (2B) and (2D). (3) In both theories inertia and gravitation are separate entities; hence no account can be given for the identity of weight and mass, while the equation $\gamma = 1$ is a necessity in E—but pointing to these analogies I did not for a moment suggest that B and D are identical in their numerical consequences!

In a theory of discrete particles the static centrally symmetric solutions of the homogeneous field equations are decisive. With such calibration of the distance $r = (x_1^2 + x_2^2 + x_3^2)^{1/2}$ from the center as renders Einstein's metric 3-space a conformal map of Euclidean space this solution in E is given by

$$ds^2 = \left(\frac{1 - m/2r}{1 + m/2r}\right)^2 \cdot dx_0^2 - (1 + m/2r)^4(dx_1^2 + dx_2^2 + dx_3^2). \quad \text{(5E)}$$

It contains only *one* constant m. In the limit for small m/r one obtains

$$h_{00} = -2m/r, \qquad h_{0\alpha} = 0, \qquad h_{\alpha\beta} = (2m/r)\delta_{\alpha\beta} \quad [\alpha, \beta = 1, 2, 3]. \text{(5')}$$

The homogeneous field equations assumed by Birkhoff are

$$\Box h_{ik} = 0 \qquad \text{where } \Box \varphi = \delta^{pq} \frac{\partial^2 \varphi}{\partial x_p \partial x_q}. \qquad (6)$$

Their most general static centrally symmetric solution involves 3 arbitrary constants a, b, l:

$$\begin{aligned} h_{00} &= -2a/r, \qquad h_{0\alpha} = 0; \\ h_{\alpha\beta} &= (2b/r)\delta_{\alpha\beta} + l(r^{-3}\delta_{\alpha\beta} + 3r^{-5}x_\alpha x_\beta). \end{aligned} \qquad \text{(5B)}$$

From the present standpoint this is a serious disadvantage of B. (5′) is contained in (5B) as the special case $a = b$, $l = 0$. To have a common basis for the computation of the consequences of (2B) and (2D) I assume the h_{ik} to be given by (5B) with $l = 0$, and set $\gamma = 1$ for all particles.

Then one finds for the period of revolution T of a planet around the sun and for the advance ω of its perihelion in the three competing theories the approximate formulae

$$\begin{aligned} &T^2 \sim 4\pi^2 r_0^3/M, \qquad \omega \sim 2\pi j/r_0(1 - e^2); \\ &M_E = m, \qquad M_B = 2a, \qquad M_D = a; \\ &j_E = 3m, \qquad j_B = 2a + 4b, \qquad j_D = 2a + 3b. \end{aligned}$$

$r_0(1 - e)$ and $r_0(1 + e)$ are the limits between which the distance of the planet from the center (sun) varies. In carrying out the calculation for D, Mr. Barajas changes at a certain step b into $-b$ and erroneously makes use of the relation (4B) which does not hold in D. In our notations, and with the dot used as an abbreviation of $d/d\sigma$, the correct equations in D read

$$\ddot{x}_0 = -\frac{2a}{r^2}\dot{x}_0\dot{r}$$

$$\ddot{x}_\alpha = -\frac{x_\alpha}{r^3}\{a\dot{x}_0^2 + b(\dot{x}_1^2 + \dot{x}_2^2 + \dot{x}_3^2)\} + \frac{2b}{r^2}\dot{x}_\alpha\dot{r},$$

from which, after proper normalization of the constant factor k in σ and of the invariant plane, follow the integrals:

$$\dot{x}_0 = e^{2a/r},$$

$$x_3 = 0, \qquad x_1\dot{x}_2 - x_2\dot{x}_1 = C\cdot e^{-2b/r},$$

$$\dot{x}_1^2 + \dot{x}_2^2 = \frac{a}{2a+b}\,(e^{4a/r} - e^{-2b/r}) + C_0\cdot e^{-2b/r}.$$

For the deflection 2δ of a light ray whose asymptote has the distance r_0 from the center one finds

$$\delta \sim i/r_0; \qquad i_E = 2m, \qquad i_B = 2(a+b), \qquad i_D = a+b,$$

if one assumes that the light ray follows a trajectory which satisfies the condition $(ds/d\sigma)^2 = 0$ in E and B and the same condition *at infinity* in D. It is true that the special choice $a = b = m/2$ reduces the *three* values of M, j, i in B to those in E, because the linear relation $M + j = 2i$ satisfied in E holds also in B, whereas $M + j = 3i$ in D.

However, the hypothesis on which our computation of δ is based is of a very dubious nature in B and D. In truth the propagation of light is determined by Maxwell's electromagnetic field equations, from which the concept of light ray emerges by letting the frequency of the light waves tend to infinity. In Einstein's general relativity theory the g_{ik} affect Maxwell's equations in such a way that the light rays thus defined coincide automatically with the nul geodesics; and the deflection $\delta = 2m/r_0$ results. But in B with its flat world there is no necessity for modifying Maxwell's equations, and hence the light rays are straight lines ($i_B = 0$) and do not coincide with the nul trajectories—unless the discrepancy is removed by some artificial modification of the electromagnetic equations which has to be constructed *ad hoc.*

Finally we discuss the relative shift $\Delta\lambda/\lambda$ of wave lengths λ which is observed by a spectroscopist who compares the light emitted from two atoms of the same nature. In Einstein's theory there is no doubt about the principles on which the computation of $\Delta\lambda/\lambda$ rests: the constitution of the atom determines the frequencies of its oscillations measured in terms of its *proper time s*, propagation takes place in spherical waves according to the electromagnetic equations. For the static field (5E), two resting atoms at fixed distances r and a resting observer one thus obtains $\Delta\lambda/\lambda \sim m\cdot\Delta(1/r)$. Desperate steps must be taken to escape the conclusion $\Delta\lambda = 0$ in B. If I do not misunderstand him, Mr. Barajas interprets $dx_0/d\sigma$ as the energy of a photon traveling along its nul trajectory from one to the other atom and, by grafting Planck's quantum law on his classical scheme, assumes that the frequency of the photon varies in proportion to this its energy! In this way (to which the quantum theory of light lends little support) he manages to obtain a red shift $\Delta\lambda/\lambda \sim 2a\cdot\Delta(1/r)$.

It is well known that in E the formula (4E) is nothing but a special case of the law that the scalar product of two vectors v^i, w^i at a common world point P does not change if both undergo parallel displacement to an infinitely near point:

$$dv^i = -\Gamma^i_{pq} v^p dx_q, \qquad dw^i = -\Gamma^i_{pq} w^p dx_q$$

imply $d(v^i w_i) = 0$. Perpendicular vectors remain perpendicular. Among other applications of the latter fact I mention Fokker's computation of the geodesic precession of the Earth's axis,[4] for which, assuming the orbit to be circular, he finds the value $3\pi m/r_0$. How would this phenomenon be treated in B? Birkhoff's rule (2B) naturally fails to produce a law of constancy of the same generality; and yet the special case where v and w are both parallel to the displacement is made the decisive postulate.

When one passes from discrete particles to continuous distribution of charge, energy and momentum and thus from the homogeneous to non-homogeneous field equations, the divergences between the theories E and B become more pronounced than their analogies, for the simple fact that Birkhoff's theory is based on the duality of field and matter in a form which physics has abandoned during the last 40 years. In his scheme mechanical equations appear independently besides field equations, and matter is represented by a charge vector s^i and an energy-momentum tensor T^{ik} of the type

$$s^i = \rho u^i, \qquad T^{ik} = \mu u^i u^k \tag{7}$$

which involve a velocity field u^i (hydrodynamical models of atoms!). In a steady development physics has come to the conclusion that the mechanical equations, the differential conservation laws for charge, energy and momentum, follow from the field equations, and in formulating its exact field laws it has learned to dispense with quantities like (7) which are tied to the idea of a "fluid," a continuously distributed moving substance. It is true that quantum physics by its statistical interpretation of the field equations gives rise to a certain complementary duality of "wave" and "particle"; but this is an altogether different affair.

A consistent linear field theory of gravitation is well suited to illuminate the problem of energy and momentum; it might also help, though this is much more problematic, in pointing the way to the correct unification of gravitation and electromagnetism. I propose, therefore, to continue this discussion at another place.

[1] Birkhoff, G. D., these PROCEEDINGS, **29**, 231 (1943).

[2] *Mathematical Reviews*, **4**, 285 (1943).

[3] These PROCEEDINGS, **30**, 54 (1944).

[4] *Versl. K. Akad. Wetensch. Amsterdam*, **29**, p. 611; Schouten, J. A., *ibid.*, p. 1150; Weyl, H., *Raum Zeit Materie*, 5th ed., 1923, p. 320.

135.

How far can one get with a linear field theory of gravitation in flat space-time?

American Journal of Mathematics 66, 591—604 (1944)

Introduction and Summary. G. D. Birkhoff's attempt to establish a linear field theory of gravitation within the frame of special relativity [1] makes it desirable to probe the potentialities and limitations of such a theory in more general terms. In thus continuing a discussion begun at another place [2] I find that the differential operators at one's disposal form a 5 dimensional linear manifold. But the requirement that the field equations imply the law of conservation of energy and momentum in the simple form $\partial T_i{}^k/\partial x_k = 0$ limit these ∞^5 possibilities to ∞^2, which, however, reduce easily to two cases, a regular one (L) and a singular one (L'). The regular case (L) is nothing but Einstein's theory of weak fields. Resembling very closely Maxwell's theory of the electromagnetic field, it satisfies a principle of gauge invariance involving 4 arbitrary functions, and although its gravitational field exerts no force on matter, it is well suited to illustrate the role of energy and momentum, charge and mass in the interplay between matter and field. It might also help, though this is much more problematic, in pointing the way to a more satisfactory unification of gravitation and electricity than we at present possess. Birkhoff follows the opposite way: by avoiding rather than adopting the ∞^2 special operators mentioned above, his "dualistic" theory (B) destroys the bond between mechanical and field equations, which is such a decisive feature in Einstein's theory.

1. Maxwell's theory of the electromagnetic field and the monistic linear theory of gravitation (*L*). Gauge invariance. Within the frame of special relativity and its metric ground form

$$ds^2 = \delta_{ik} dx_i dx_k = dx_0^2 - (dx_1^2 + dx_2^2 + dx_3^2)$$

an electromagnetic field is described by a skew tensor

$$f_{ik} = \partial\phi_k/\partial x_i - \partial\phi_i/\partial x_k$$

derived from a vector potential ϕ_i and satisfies Maxwell's equations

---

* Received August 9, 1944.

[1] *Proceedings of the National Academy of Sciences*, vol. 29 (1943), p. 231.

[2] *Proceedings of the National Academy of Sciences*, vol. 30 (1944), p. 205.

(1) $$\partial f^{ki}/\partial x_k = s^i \quad \text{or } D_i\phi = \Box\phi_i - \partial\phi'/\partial x_i = s_i$$

where s^i is the density-flow of electric charge and

$$\phi' = \partial\phi^i/\partial x_i, \qquad \Box\phi = \delta^{pq}(\partial^2\phi/\partial x_p\partial x_q).$$

The equations do not change if one substitutes

(2) $$\phi^*_i = \phi_i - \partial\lambda/\partial x_i \quad \text{for} \quad \phi_i,$$

λ being an arbitrary function of the coördinates (“*gauge invariance*”), and they imply the differential conservation law of electric charge:

(3) $$\partial s^i/\partial x_i = 0.$$

As is easily verified, there are only two ways in which one may form a vector field by linear combination of the second derivatives of a given vector field ϕ_i, namely,

$$\Box\phi_i \text{ and } \partial\phi'/\partial x_i \qquad (\phi' = \partial\phi^p/\partial x_p).$$

The only linear combination $D_i\phi$ of these two vector fields which satisfies the identity $(\partial/\partial x_i)(D^i\phi) = 0$ is the one occurring in (1),

$$D_i\phi = \Box\phi_i - \partial\phi'/\partial x_i.$$

Herein lies a sort of mathematical justification for Maxwell's equations.

Taking from Einstein's theory of gravitation the hint that gravitation is represented by a symmetric tensor potential h_{ik}, but trying to emulate the linear character of Maxwell's theory of the electromagnetic field, one could ask oneself what symmetric tensors $\bar{D}_{ik}h$ can be constructed by linear combination from the second derivatives of h_{ik}. The answer is that there are 5 such expressions, namely

(4) $$\Box h_{ik}, \quad \partial h'_i/\partial x_k + \partial h'_k/\partial x_i, \quad h''\delta_{ik}, \quad \partial^2 h/\partial x_i\partial x_k, \quad \Box h\cdot\delta_{ik}$$

where

$$h = h_p{}^p, \qquad h'_i = \partial h_i{}^p/\partial x_p, \qquad h'' = \partial^2 h^{pq}/\partial x_p\partial x_q.$$

With any linear combination $\bar{D}_{ik}h$ of these 5 expressions one could set up the field equations of gravitation

(5) $$\bar{D}_{ik}h = T_{ik}$$

the right member of which is the energy-momentum tensor T_{ik}. In analogy to the situation encountered in Maxwell's theory one may ask further for which linear combinations $\bar{D}_{ik}$ the identity

$$(\partial/\partial x_k)(\bar{D}_i^{\cdot k}h) = 0$$

will hold, and one finds that this is the case if, and only if, $\bar{D}_{ik}h$ is of the form

(6) $\alpha\{\Box h_{ik} - (\partial h'_i/\partial x_k + \partial h'_k/\partial x_i) + h''\delta_{ik}\} + \beta\{\partial^2 h/\partial x_i\partial x_k - \Box h\cdot\delta_{ik}\}$,

α and β being arbitrary constants. In this case the field equations ($\bar{5}$) entail the differential conservation law of energy and momentum

(7) $$\partial T_i^{\ k}/\partial x_k = 0.$$

With two constants a, b ($a \neq 0$, $a \neq 4b$) we can make the substitution

$$h_{ik} \rightarrow a\cdot h_{ik} - b\cdot h\delta_{ik}$$

and thereby reduce α, β to the values 1, 1, provided $\alpha \neq 0$, $\alpha \neq 2\beta$. Hence, disregarding these singular values, we may assume as our field equations

(5) $$D_{ik}h \equiv \{\Box h_{ik} - (\partial h'_i/\partial x_k + \partial h'_k/\partial x_i) + h''\delta_{ik}\} + \{\partial^2 h/\partial x_i\partial x_k - \Box h\cdot\delta_{ik}\} = T_{ik}.$$

$D_{ik}h$ remains unchanged if h_{ik} is replaced by

(8) $$h^*_{ik} = h_{ik} + (\partial\xi_i/\partial x_k + \partial\xi_k/\partial x_i)$$

where ξ_i is an arbitrary vector field. Hence we have the same type of correlation between gauge invariance and conservation law for the gravitational field as for the electromagnetic field, and it is reasonable to consider as physically equivalent any two tensor fields h, h^* which are related by (8).

The linear theory of gravitation (L) in a flat world at which one thus arrives with a certain mathematical necessity is nothing else but Einstein's theory for weak fields. Indeed, on replacing Einstein's g_{ik} by $\delta_{ik} + 2\kappa\cdot h_{ik}$ and then neglecting higher powers of the gravitational constant κ, one obtains (5), and the property of gauge invariance (8) reflects the invariance of Einstein's equations with respect to arbitrary coördinate transformations.[3]

By proper normalization of the arbitrary function λ in (2) one may impose the condition $\phi' = 0$ upon the ϕ_i, thus giving Maxwell's equations a form often used by H. A. Lorentz:

(9) $$\Box\phi_i = s_i, \qquad \partial\phi^i/\partial x_i = 0.$$

---

[3] Cf. A. Einstein, *Sitzungsber. Preuss. Ak. Wiss.* (1916), p. 688 (and 1918, p. 154).

In the same manner one can choose the ξ_i in (8) so that $\gamma_{ik} = h_{ik} - \frac{1}{2}h \cdot \delta_{ik}$ satisfies the equations

(10) $$\partial \gamma_i{}^k / \partial x_k = 0 \text{ and}$$

(11) $$\Box \gamma_{ik} = T_{ik}.$$

In one important respect gauge invariance works differently for electromagnetic and gravitational fields: If one splits the tensor of derivatives $\phi_{k,i} = \partial \phi_k / \partial x_i$ into a skew and a symmetric part,

$$\phi_{k,i} = \tfrac{1}{2}(\phi_{k,i} - \phi_{i,k}) + \tfrac{1}{2}(\phi_{k,i} + \phi_{i,k}),$$

the first part is not affected by a gauge transformation whereas the second can locally be transformed into zero. In the gravitational case *all* derivatives $\partial h_{ik}/\partial x_p$ can locally be transformed into zero. Hence we may construct, according to Faraday and Maxwell, an energy-momentum tensor L_{ik} of the electromagnetic field,

(12) $$L_i{}^k = f_{ip} f^{pk} - \tfrac{1}{2}\delta_i{}^k (ff), \qquad (ff) = \tfrac{1}{2} f_{pq} f^{qp},$$

depending quadratically on the gauge invariant field components

$$f_{ik} = \phi_{k,i} - \phi_{i,k},$$

but no tensor G_{ik} depending quadratically on the derivatives $\partial h_{ik}/\partial x_p$ exists, if gauge invariance is required, other than the trivial $G_{ik} \equiv 0$.

2. Particles as centers of force, and the charge vector and energy-momentum tensor of a continuous cloud of substance. Conceiving a resting particle as a center of force, let us determine the *static centrally symmetric solutions* of our homogeneous field equations (1) and (5) ($s^i = 0$, $T_{ik} = 0$). One easily verifies that *in the sense of equivalence* the most general such solution is given by the equations

(13) $$\phi_0 = e/4\pi r, \qquad \phi_i = 0 \text{ for } i \neq 0;$$

(14) $$\gamma_{00} = m/4\pi r, \qquad \gamma_{ik} = 0 \text{ for } (i, k) \neq (0, 0),$$

r being the distance from the center. As was to be hoped, it involves but two constants, *charge* e and *mass* m. The center itself appears as a singularity in the field. Indeed ϕ_0 and the factor ϕ in $\phi_\alpha = \phi x_\alpha$ [$\alpha = 1, 2, 3$] must be functions of r alone, and the relations

$$\Delta\phi_0 = 0, \qquad \partial\phi_\alpha/\partial x_\alpha = 0 \qquad\qquad [\alpha = 1, 2, 3]$$

implied in (9) then yield

$$\phi_0 = a/r, \qquad \phi = b/r^3, \qquad \phi_\alpha = -(\partial/\partial x_\alpha)(b/r).$$

Substitution of $\phi_\alpha - \partial\lambda/\partial x_\alpha$ for ϕ_α with $\lambda = - b/r$ changes ϕ_α into zero. In the same manner (14) is obtained from the equations (10 & 11).

A continuous cloud of "charged dust" can be characterized by its velocity field u^i ($u_i u^i = 1$) and the rest densities μ, ρ of mass and charge. It is well known that its equations of motion and the differential conservation laws of mass and charge result if one sets $s^i = \rho u^i$ in Maxwell's equations and lets T_i^k in (7) consist of the Faraday-Maxwell field part (12) and the kinetic part $\mu u_i u^k$:

$$\partial(\rho u^i)/\partial x_i = 0, \qquad \partial(\mu u^i)/\partial x_i = 0;$$
$$\mu du_i/ds = \rho \cdot f_{ip} u^p.$$

Since the motion of the individual dust particle is determined by $dx_i/ds = u^i$ we have written d/ds for $u^k \partial/\partial x_k$. In this manner Faraday explained by his electromagnetic tensions (flow of momentum) the fact that the *active* charge which generates an electric field is at the same time the *passive* charge on which a given field acts. At its present stage our theory (L) accounts for the force which an electromagnetic field exerts upon matter, but the gravitational field remains a powerless shadow. From the standpoint of Einstein's theory this is as it should be, because the gravitational force arises only when one continues the approximation beyond the linear stage. We pointed out above that no remedy for this defect may be found in a gauge invariant gravitational energy-momentum tensor. However, the theory (L) explains why active gravity, represented by the scalar factor μ in the kinetic term $\mu u_i u_k$ as it appears in the right member T_{ik} of the gravitational equations (5), is at the same time inertial mass: this is simply another expression of the fact that the mechanical equations (7) are a consequence of those field equations.

We have seen that even in empty space the field part of energy and momentum must not be ignored, and thus a particle should be described by the static centrally symmetric solution of the equations

$$(15) \qquad D_i\phi = 0, \qquad D_{ik}h - L_{ik} = 0$$

(of which the second set is no longer strictly linear!). Again we find, after proper gauge normalization,

$$\phi_0 = e/4\pi r, \qquad \phi_1 = \phi_2 = \phi_3 = 0, \tag{13}$$

and then

$$\left\{\begin{array}{ll} \gamma_{00} = m/4\pi r - \frac{1}{4}(e/4\pi r)^2, & \gamma_{0\alpha} = 0, \\ \gamma_{\alpha\beta} = -(e/4\pi r)^2 \cdot (x_\alpha x_\beta/4r^2) & \end{array}\right. \qquad [\alpha, \beta = 1, 2, 3]. \tag{14_e}$$

As before, two characteristic constants e and m appear. *At distances much larger than the "radius" $e^2/4\pi m$ of the particle the gravitational influence of charge becomes negligible compared with that of mass.*

3. The singular case. In normalizing the operator (6) by $\alpha = \beta = 1$ we had to exclude the cases $\alpha = 0$, $\beta = 1$ and $\alpha = 1$, $\beta = 1/2$. The first is clearly without interest because it deals with a field described by a scalar h rather than a tensor h_{ik}. But the differential operator (6), D'_{ik}, corresponding to the values $\alpha = 1$, $\beta = 1/2$ and the attendant field equations

$$D'_{ik}h = T_{ik} \tag{5'}$$

deserve a moment's attention. $D'_{ik}h$ remains unchanged if h_{ik} is replaced by

$$h^*_{ik} = h_{ik} + \eta\delta_{ik} + (\partial\xi_i/\partial x_k + \partial\xi_k/\partial x_i)$$

where the 5 functions η, ξ_i are subject to the one restriction $\partial\xi^i/\partial x_i = 0$. By proper gauge normalization one may reduce the field equations (5′) to the form

$$\partial h_i{}^k/\partial x_k = 0, \tag{10'}$$

$$\Box h_{ik} + \tfrac{1}{2}(\partial^2 h/\partial x_i \partial x_k - \Box h \cdot \delta_{ik}) = T_{ik}. \tag{11'}$$

The static centrally symmetric solution of the homogeneous equations ($T_{ik} = 0$) is the following counterpart to (14):

$$h_{00} = 0, \quad h_{0\alpha} = 0, \quad h_{\alpha\beta} = (m'/4\pi r)(\delta_{\alpha\beta} - x_\alpha x_\beta/r^2) \quad [\alpha, \beta = 1, 2, 3]$$

The same electric part as in (14_e) may be superimposed. It seems remarkable that besides (L) this possibility (L') exists.

4. Derivation of the mechanical laws wihtout hypotheses about the inner structure of particles. In principle the idea of substance had already been overcome by Newton's dynamical interpretation of Nature. His particles are centers of force, the inertial mass is a dynamic coefficient and not, as the scholastic definition pretends, quantity of substance. Boscovich, Ampère and others took the extreme view that the centers of force are points without extension. Modern atomistic physics has raised the discrete structure of matter above all doubt. Although it does not forbid us to picture the elementary particles as something of continuous extension, one must admit that, so far, speculations about their "interior" have never borne fruit. Indeed we can explain the laws of reaction of particles with the continuous field without committing ourselves to any hypotheses concerning their inner structure, simply *by describing a particle through the surrounding "local" field.* I proceed to illustrate this fundamental point first by Maxwell's equations and then by our linear theory (L).

A particle describes a narrow channel in the 4 dimensional world. The only assumption concerning the electromagnetic potential ϕ_i we make is that outside this channel Maxwell's homogeneous equations

$$\partial f^{ki}/\partial x_k = 0 \tag{16}$$

are satisfied. By arbitrary continuous extension we fill the channel with a *fictitious field* ϕ_i and then *define* s^i by (1). The relation (3) is a consequence of this definition, and (16) asserts that s^i vanishes outside the channel. Let S_t denote the plane $x_0 = \text{const.} = t$, S^*_t the portion of S_t inside the channel, Ω the surface of the channel and Ω_t the intersection of Ω with S_t (or the boundary of S^*_t). The surface Ω_t surrounds the particle in the 3-space S_t. Integrating (3) over S_t we find

$$de/dt = 0 \quad \text{for} \quad e = \iiint_{S^*_t} s^0 dx_1 dx_2 dx_3;$$

hence e does not vary in time. More generally, it can be stated that the vector field s^i sends the same flow e through any 3 dimensional surface crossing the channel. Application of this fact to two different cross sections S_t confirms the above result; application to two cross sections $x_0 = \text{const.}$ and $x^*_0 = \text{const.}$ corresponding to two different admissible coördinate systems x and x^* (which are linked by a Lorentz transformation) proves e to be an *invariant*. Finally we must show that it is independent of the fictitious "filling." But according to the definition of s^0,

$$e = \iiint_{S^*_t} (\partial f^{01}/\partial x_1 + \partial f^{02}/\partial x_2 + \partial f^{03}/\partial x_3)\, dx_1 dx_2 dx_3$$

is the flow of the electric field (f^{01}, f^{02}, f^{03}) through Ω_t and hence is completely determined by the *real* field on Ω. For this introduction of the charge e it does not matter whether the particle is an actual singularity of the field or covers a (small) region where the known laws in empty space are suspended (and unknown laws take their place). If the field surrounding the particle is described by (13) then the flow e of the electric field through Ω_t is the constant designated by the same letter in (13). Approximately one can ascribe a world direction u^i to the channel, and it is clear that, if numerous particles of nearly the same velocity u^i, each with its charge e, are encountered in a macroscopic "volume element" of space, their effect can macroscopically be accounted for by a convective current ρu^i.

Faute de mieux, H. A. Lorentz and H. Poincaré used this expression also for the infinitesimal volume elements of an electron, and the question arose by what cohesive forces the charges of the several parts of an electron are held together against their electrostatic repulsion. Compared with this primitive viewpoint (which was elaborated in considerable detail by M. Abraham) G. Mie's field theory of paricles,[4] which expressed the current s^i in terms of the same fundamental quantities, namely ϕ_i, as the field itself, signified an enormous progress. But also this theory, in spite of some highly attractive features, the great hopes it once raised and its development by men like D. Hilbert, M. Born and others, has remained in the limbo of speculative physics. The sober non-committal attitude here described was the third stage in the history of our problem. [A fourth has been opened by quantum physics: Following in Schrödinger's footsteps, Dirac expressed s^i in terms of the 4 spinor components of the electronic field ψ. This is a simple extension of the scheme of field physics, which in itself is as natural as the appearance of the Maxwellian L_{ik} in the gravitational field equations (15). However an entirely new feature, statistical interpretation based on quantization of the field laws, "creates" in quantum physics the discrete particles. The singularities to which this process of quantization gives rise constitute a difficulty at least as serious in quantum as in classical physics.]

Let us return to the classical standpoint and proceed from electricity to gravitation. After bridging the channel by a fictitious field h_{ik} we integrate the identities

$$(\partial/\partial x_k)(D_i{}^k h) = 0$$

over a cross section S^*_t of the channel, thus obtaining the mechanical equations

$$dJ_i/dt = P_i \tag{16}$$

in which

$$J_i = \int\int\int_{S^*_t} D_i{}^0 h \cdot dx_1 dx_2 dx_3$$

and $-P_i$ is the flow of the vector field $(D_i{}^1 h, D_i{}^2 h, D_i{}^3 h)$ on S_t through Ω_t.

---

[4] *Ann. d. Phys.*, vols. 37, 39, 40 (1912/13).

By its definition P_i does not depend on the fictitious filling, and from this fact and (16) it follows that the same is true for J_i. Indeed define $J_i^{(1)}$ by a filling 1, $J_i^{(2)}$ by a filling 2, consider two distinct cross sections S_1, S_2, $t = t_1$ and $t = t_2$, and construct a filling 3 that coincides with 1 in the neighborhood of S_1, with 2 in the neighborhood of S_2. Applying (16) to these three fillings and recalling that P_i remains unaffected one finds

$$J_i^{(1)}(t_2) - J_i^{(1)}(t_1) = \int_{t_1}^{t_2} P_i dt, \qquad J_i^{(2)}(t_2) - J_i^{(2)}(t_1) = \int_{t_1}^{t_2} P_i dt,$$

$$J_i^{(3)}(t_2) - J_i^{(3)}(t_1) = J_i^{(2)}(t_2) - J_i^{(1)}(t_1) = \int_{t_1}^{t_2} P_i dt;$$

hence

$$J_i^{(1)}(t) = J_i^{(2)}(t) \quad \text{for} \quad t = t_1 \text{ and } t_2.$$

When dealing with an *isolated* system we can assume that $D_{ik}h$ vanishes outside the channel; then $P_i = 0$. Let us choose an arbitrary constant contravariant vector l^i and form the vector field $q^k = l^i \cdot D_i^k h$, which satisfies the equation $\partial q^k / \partial x_k = 0$ and under our assumption vanishes outside the channel. The argument previously applied to s^k proves that

$$\int\int\int_{S^*_t} q^0 dx_1 dx_2 dx_3 = l^i J_i$$

is constant in time and an invariant. Hence J_i *are the components of a covariant vector.* In this way we introduce the energy-momentum vector J of an isolated particle and obtain the conservation law

$$J_i = \text{const.} \tag{17}$$

For the *static* field (14) one may compute J_i by means of a *static* filling. Then $J_1 = J_2 = J_3 = 0$ and J_0 is the integral of

$$D_0{}^0 h = -\Delta\gamma_{00} + \partial^2\gamma^{\alpha\beta}/\partial x_\alpha \partial x_\beta \qquad [\alpha, \beta = 1, 2, 3]$$

over a sphere S^*_0 around the center, hence the flow through its surface Ω_0 of the spatial vector

$$-\{\partial\gamma_{00}/\partial x_\alpha + \partial\gamma_\alpha{}^\beta/\partial x_\beta\}.$$

But this flow may be computed from the *real* field and thus turns out to be radial and of strength $m/4\pi \cdot 1/r^2$; consequently $J_0 = m$.

Since J_i is a covariant vector, our result $J_0 = m$, $J_1 = J_2 = J_3 = 0$ carries over from a resting isolated particle to one moving in the direction u^i:

(18) $$J_i = mu_i.$$

For a particle interacting with other particles we can not assume that $D_{ik}h$ vanishes outside the channel, and the conservation law (17) must be replaced by the mechanical equations (16). We might call P external force and J energy-momentum; both, as we have seen, are independent of the filling, but there is no reason why J should be a vector. We get beyond this general scheme by an approximate evaluation of P and J, based on the field equations (15) which hold outside Ω and the character of the local field surrounding the particle. Computation of J_0 for the static centrally symmetric field (14_e) by the same method as for the special case $e = 0$ yields

$$J_0 = m - \tfrac{1}{2} \cdot (e^2/4\pi a),$$

provided Ω_0 is the sphere of radius a. Notice that $J_0(a)$ tends to $-\infty$ and not to zero with $a \to 0$. The energy between two spheres of different radii a has the correct value of the electric field energy $(e^2/8\pi)[1/a]$; *nevertheless the total energy* ($a \to \infty$) *is not infinite but* m.

The electric field will be a superposition of the local fields generated by the several particles. In terms of a suitable system of coördinates in which the particle under consideration momentarily (for $t = 0$) rests we shall, therefore, have a field $F_{ik} + f_{ik}$ on $\Omega_0 = \Omega_{t=0}$ where

$$(f_{01}, f_{02}, f_{03}) = (e/4\pi r^3)(x_1, x_2, x_3), \qquad f_{12} = f_{23} = f_{31} = 0,$$

while F_{ik} is practically constant, i.e. varies on Ω_0 essentially less than f_{ik} (though it may well be stronger than f_{ik}). A familiar calculation then gives for the flow of

$$-(D_i^1 h, D_i^2 h, D_i^3 h) = -(L_i^1, L_i^2, L_i^3)$$

the value $P_i = eF_{i0}$.

Were f_{ik} the total electric field we could assume that the (local) gravitational field surrounding the particle, for $t = 0$ and outside Ω_0, is given by (14_e), and we should obtain

(19) $$J_0 = m, \qquad J_1 = J_2 = J_3 = 0,$$

provided *the radius a of the sphere Ω_0 is large in comparison with the radius $e^2/4\pi m$ of the particle.* We fix Ω_0 in this manner: it is at this point that the necessity for keeping away from the particle arises. The equations (14) will still hold with sufficient accuracy on and outside Ω_0 if not only e^2/a^4 but also *the energy of the "outer" field ΣF_{ik}^2 on Ω_0 is small compared to m/a^3.*

Cut the channel by two cross sections $x_0 = \text{const.}$, $x^*_0 = \text{const.}$, belonging to two different coördinate systems x, x^* and going through a common point inside the channel. Let l again be an arbitrary constant contravariant vector with the components l^i in the one, l^{*i} in the other coördinate system. The difference of the respective integrals l^iJ_i, $l^{*i}J^*_i$ is the flow of

$$(l^i \cdot D_i{}^1h,\ l^i \cdot D_i{}^2h,\ l^i \cdot D_i{}^3h) = (l^iL_i{}^1,\ l^iL_i{}^2,\ l^iL_i{}^3)$$

through the part of the channel surface Ω between these two cross cuts, and hence, under the above assumptions, of a lower order of magnitude than m. *With this approximation J_i is a covariant vector,* and thus the formula (18) becomes applicable not only for the cross section $t = 0$ where the particle rests momentarily, but for any cross section $x_0 = t = \text{const.}$

Of course, (16) has to be interpreted in integral fashion,

$$[J_i] = J_i(t_1) - J_i(0) = \int_0^{t_1} P_i dt,$$

and here we may set, with sufficient approximation, $J_i(t) = m(t)u_i(t)$. The equation itself shows that an appreciable change of J_i, one that is comparable with m, can be expected only after a lapse of time t_1 of order $m/e\,|F|$, which is large in comparison with the radius a of Ω_0: Our assumptions imply that J_i or m and u^i change but slowly (*quasi-stationary motion*).

But with these precautions in mind, the differential equation

$$(d/dt)(mu_i) = eF_{i0} \tag{20}$$

may now be claimed as holding for $t = 0$. The component $i = 0$ gives $dm/dt = 0$; hence *the mass m stays constant.* By a known simple technique (20) is changed into its invariant form

$$m\,du_i/ds = e \cdot F_{ip}u^p$$

which will hold along the entire channel. The deduction indicates clearly the hypotheses to which the approximate validity of this Lorentz equation of motion of a particle is bound.[5] We now understand why quantities of the type $s^i = \rho u^i$, $T_i{}^k = \mu u_i u^k$ can account in a rough manner for the interaction between field and a cloud of charged dust in which near particles have nearly the same velocity.

---

[5] I have repeated here for the linear theory an argument which I first developed within the frame of general relativity in the 4th and in more detail in the 5th edition of my book "Raum Zeit Materie"; see the latter edition, Berlin 1923, pp. 277-286. The purely gravitational case was treated with the greatest care in a more recent paper by A. Einstein, L. Infeld and B. Hoffmann, *Annals of Mathematics,* vol. 39 (1938), pp. 65-100.

5. Vague suggestions about a future unification of gravitation and electromagnetism. In spite of such achievements nobody will believe in the sufficiency of the linear theory (L). For, as we said above, its gravitational field is a shadow without power. The fundamental fact that *passive gravity and inertial mass* always coincide appears to me convincing proof that *general relativity* is the only remedy for this shortcoming. But thereby the gravitational constant κ enters the picture, and one knows that the ratio of the electric and gravitational radii of an electron, $(e^2/m) : \kappa m = e^2/\kappa m^2$, is a pure number of the order of magnitude 10^{40}. This circumstance and Mach's old idea that the plane of the Foucault pendulum is carried around by the stars in their daily revolution, point to a construction in which the gravitational force is bound to the totality of masses in the universe. Our present theory, Maxwell + Einstein, with its inorganic juxtaposition of electromagnetism and gravitation, cannot be the last word. Such juxtaposition may be tolerable for the linear approximation (L) but not in the final generally relativistic theory. Transition from (L) with its flat world to general relativity should raise both, not only the gravitational, but also the electromagnetic part, above the linear level and, as it changes the gauge transformations of the former into non-linear transformations of coördinates, something similar ought to happen to the gauge transformations of the ϕ_i.

After adding Dirac's 4 spin components of the electronic field ψ to the fundamental field quantities ϕ_i, h_{ik} the electric gauge invariance [6] states that the field equations do not change under the substitution of

$$e^{i\lambda} \cdot \psi, \qquad \phi_k - \frac{h}{e}\frac{\partial \lambda}{\partial x_k} \quad \text{for} \quad \psi, \phi_k$$

(h = Planck's quantum of action): the process of "covariant derivation" of ψ is defined by $\partial/\partial x_k + (ie/h)\phi_k$. Thus the electromagnetic field ϕ_i appears as a sort of appendage of the ψ-field. It is natural to expect the h_{ik} to be appended in a similar manner to quantities associated with other elementary particles. Thus incompleteness of our present theory on the linear level, a *premature* transition to general relativity, might have their share in blocking the view towards a satisfactory unification. For these reasons a linear theory of gravitation like (L), though necessarily preliminary in character, may still deserve the physicist's attention.

---

[6] This principle of gauge invariance is analogous to one by which the author in 1918 made the first attempt at a unification of electromagnetism and gravitation. He has long since realized that it does not connect electricity and gravitation (ϕ_i and g_{ik}), as he then believed, but the electric with the electronic field (ϕ_i with ψ). In this form, in which the exponent of the gauge factor $e^{i\lambda}$ is pure imaginary and not real, it expresses well established atomistic facts, and the connecting coefficient, h/e, is a known atomistic and not an unknown cosmologic constant.

6. A free paraphrase of Birkhoff's recent linear theory of gravitation (*B*). The linear theory (*B*), however, is essentially different from (*L*). It seems to me characteristic for Birkhoff's conception that he uses the kinetic quantities $s^i = \rho u^i$, $T_i^k = \mu u_i u^k$ not only for a macroscopic description of matter, but a late follower of Lord Kelvin, even for the construction of fluid models of atoms, and that he preserves the duality of field and matter also in the form of mechanical equations which do not follow from the field equations. In contrast to this "dualistic" scheme Einstein's theory and its linear approximation (*L*) are "monistic."

Since Birkhoff wishes to avoid the fact that mechanical equations such as (7) follow from the field equations, he must choose for the left side $\tilde{D}_{ik}h$ of his linear field equations ($\tilde{5}$) any combination of the 5 tensors (4) which is *not* of the special form (6). He picks, somewhat arbitrarily, $\Box h_{ik}$ or rather $\Box h_{ik} - \frac{1}{2}\Box h \cdot \delta_{ik}$; but it seems wiser not to commit oneself too early. He is then at liberty to add to the left member of (7) a term representing the action of the gravitational field on matter. Assuming that force to be quadratic in u^i, as in Einstein's theory, he writes

$$(\partial/\partial x_k)(\mu u_i u^k + L_i^k) + \Gamma_{i,pq} u^p u^q = 0 \tag{21}$$

and finds

$$\Gamma_{i,pq} = (\sigma/2)(\partial h_{ip}/\partial x_q + \partial h_{iq}/\partial x_p - 2\partial h_{pq}/\partial x_i) \tag{22}$$

as the mathematically simplest expression by which the differential law of conservation of mass

$$(\partial/\partial x_k)(\mu u^k) = 0$$

is upheld. (In Einstein's theory one has instead $\Gamma_{i,pq} = -\frac{\mu}{2}\frac{\partial g_{pq}}{\partial x_i}$, provided μ and L_i^k denote the scalar and tensorial *densities*, not scalar and tensor.) But since no theory in which inertia and gravitation are separate entities can explain the universal proportionality of passive gravity and inertial mass, there is no reason why the scalar field σ should be the same as μ (instead one might expect that for a substance *of given chemical constitution* μ and σ are connected by some equation of state $F(\mu, \sigma) = 0$). However, just as Maxwell's L_i^k accounts for the identity of active and passive charge, one can hope in this theory to establish the identity of active and passive gravity by a gravitational energy tensor. For that purpose it is necessary to assume $\sigma u_i u_k$ rather than $\mu u_i u_k + L_{ik}$ as the right member T_{ik} of the field equations ($\tilde{5}$),

$$\tilde{D}_{ik}h = \sigma u_i u_k,$$

and one will try to construct a symmetric tensor G_{ik} which is quadratic in the derivatives $\partial h_{pq}/\partial x_i$ such that the following identity holds:

$$(23) \qquad \partial G_i^k/\partial x_k = (\tfrac{1}{2}\partial h_{ip}/\partial x_q + \tfrac{1}{2}\partial h_{iq}/\partial x_p - \partial h_{pq}/\partial x_i) \cdot \tilde{D}^{pq}h.$$

Then (21) would indeed assume the form of a differential law of conservation of energy and momentum:

$$(24) \qquad (\partial/\partial x_k)(\mu u_i u^k + L_i^k + G_i^k) = 0.$$

There are 16 linearly independent tensors G_{ik} of this sort, and I have checked whether for any linear combination of them a relation like (23) can hold; the result was negative. This applies in particular to the field equations which Birkhoff adopts:

$$\tilde{D}_{ik}h \equiv \Box h_{ik} - \tfrac{1}{2}\Box h \cdot \delta_{ik} = \sigma u_i u_k$$

(and which he interprets in a slightly different manner in terms of a fluid of peculiar nature). It may, therefore, be said that Birkhoff *sacrifices the conservation law of energy and momentum to that of mass.*

That it is possible to develop a theory of dualistic type in which the conservation law for energy-momentum holds is proved by a certain interpretation of the " degenerate Einstein theory " (D) which I had used to illustrate (B): One starts with the field equations of (L) in the normalized form (10 & 11), sets $T_{ik} = \sigma u_i u_k$, *throws away the supplementary conditions* (10) in order to make room for an extra term in the mechanical equations (7) and finally replaces the latter not by (21), but by

$$(\partial/\partial x_k)(\mu u_i u^k + L_i^k) - \frac{\sigma}{2}\frac{\partial h_{pq}}{\partial x_i} u^p u^q = 0.$$

Of course, mass is not conservative in this set-up; one finds instead

$$\partial(\mu u^k)/\partial x_k = (\sigma/6)(\partial h_{pq}/\partial x_r + \partial h_{qr}/\partial x_p + \partial h_{rp}/\partial x_q) u^p u^q u^r.$$

But the conservation laws for energy and momentum (24) hold if one defines the gravitational energy tensor G_{ik} by

$$G_i^k = -H_i^k + \tfrac{1}{2}\delta_i^k \cdot H \qquad (H = H_p^p)$$

where

$$H_{ik} = \frac{1}{2}\frac{\partial h_{pq}}{\partial x_i}\frac{\partial h^{pq}}{\partial x_k} - \frac{1}{4}\frac{\partial h}{\partial x_i}\frac{\partial h}{\partial x_k}.$$

But it is not my intention to propagandize this or any other dualistic theory of gravitation!

136.

Fundamental domains for lattice groups in division algebras. I, II

I: Festschrift zum 60. Geburtstag von Prof. Dr. A. Speiser, Orell Füssli Zürich, 218—232 (1945)
II: Commentarii mathematici Helvetici 17, 283—306 (1944/45)

In a series of publications[1]) which continue the great tradition of GAUSS, DIRICHLET, HERMITE, MINKOWSKI and which I am sure will be admired by generations of mathematicians to come, SIEGEL has vastly added to our knowledge in the theory of definite and indefinite quadratic forms and of arithmetical reduction. The purpose of the present article, conceived as a commentary on the all too laconic Part III of his paper on Discontinuous Groups, will be fulfilled if it makes SIEGEL's results concerning the group of units in a simple order more accessible to the average reader. Because of its importance the subject seems to me worthy of a fuller treatment.

We do not know for what reasons MINKOWSKI, when he developed the theory of reduction for quadratic forms with real coefficients and n integral variables[2]), abandoned his own number-geometric methods, although he had invented them with that very goal in mind. That his methods work quite satisfactorily was evinced by a paper of mine of more recent date[3]). At the same time, and under SIEGEL's stimulus, the theory was extended to arbitrary commutative orders by the late HUMBERT; a bold, though partially abortive attack on the non-commutative case, aiming in a slightly different direction,

---

[1]) C. L. SIEGEL, *Über die analytische Theorie der quadratischen Formen*, Ann. Math. *36*, 527–606 (1935); *37*, 230–263 (1936). *Über die Zetafunktionen indefiniter quadratischer Formen*, Math. Z. *43*, 682–708 (1938); *44*, 398–426 (1939). *Einheiten quadratischer Formen*, Abh. Math. Sem. Hansischen Univ. *13*, 209–239 (1940). *Equivalence of Quadratic Forms*, Amer. J. Math. *63*, 658–680 (1941). *Contributions to the Theory of the Dirichlet L-Series and the Epstein Zeta-Functions*, Ann. Math. *44*, 143–172 (1943). *On the Theory of Indefinite Quadratic Forms*, Ann. Math. *45*, 577–622 (1944). *The Average Measure of Quadratic Forms with Given Determinant and Signature*, Ann. Math. *45*, 667–685 (1944). – *Discontinuous Groups*, Ann. Math. *44*, 674–689 (1943). – *Einführung in die Theorie der Modulfunktionen n-ten Grades*, Math. Ann. *116*, 617–657 (1939). *Note on Automorphic Functions of Several Variables*, Ann. Math. *43*, 613–616 (1942). *Symplectic Geometry*, Amer. J. Math. *65*, 1–86 (1943).

[2]) H. MINKOWSKI, *Diskontinuitätsbereich für arithmetische Äquivalenz*, Gesammelte Abhandlungen II, *1911*, 53–100. Cf. also L. BIEBERBACH und I. SCHUR, Sitz.-Ber. Preuss. Akad. Wiss. *1928*, 510–535; *1929*, 508. – R. REMAK, Compos. Math. *5*, 368–391 (1938).

[3]) H. WEYL, Transact. Amer. Math. Soc. *48*, 126–164 (1940).

was made by EICHLER[1]). SIEGEL finished the job. His success can be ascribed to three principal ideas: (1) adoption of a modified process of reduction which MINKOWSKI had used only for quadratic forms with integral coefficients, and by which the limitation to fields of class number 1 is overcome; (2) foundation of that process on the *trace* of a quadratic form; (3) correct definition of a positive quadratic form in a semi-simple algebra over the real field. As will become clear in our account, the only fact relevant for this purpose is the existence of an anti-automorphic involution in such an algebra.

A problem of central significance in MINKOWSKI's work has thus been solved in its true generality by a suitable adaptation of his methods. In writing this article, which stresses the methodological viewpoint and tries to summarize the whole development, and in dedicating it to ANDREAS SPEISER, I cannot help feeling that I am still acting on the same commission which brought SPEISER and me together some thirty-five years ago, when we collaborated in editing MINKOWSKI's Collected Papers.

A. PRELIMINARIES

§ 1. Lemmas on Matrices

We base the treatment of quadratic forms in a division algebra on a number of elementary lemmas concerning matrices. Let $X = \| x_{ik} \|$ and Y be any g-rowed real square matrices, whereas G is real and symmetric, $G = G'$, and x denotes a vector, i.e. a column of g real numbers. The facts that the quadratic form $x'G\,x = G[x]$ is $\geqq 0$ for all x or that $G[x] > 0$ for all $x \neq 0$ will be expressed by $G \geqq 0$, $G > 0$ respectively.

1.1. $\operatorname{tr}(X'X) > 0$ except if $X = 0$.

We sometimes use the notation $\operatorname{tr}(X'X) = (\operatorname{abs} X)^2$.

1.2. $\operatorname{abs}(X + Y) \leqq \operatorname{abs} X + \operatorname{abs} Y$, $\operatorname{abs}(XY) \leqq \operatorname{abs} X \cdot \operatorname{abs} Y$.

1.3. $G \geqq 0$ implies $\operatorname{tr}(X'G\,X) \geqq 0$,
$G > 0$ implies $\operatorname{tr}(X'G\,X) > 0$ except for $X = 0$.

Proof.

$$\operatorname{tr}(X'G\,X) = \sum_i \Big(\sum_{k,l} x_{ki}\, g_{kl}\, x_{li}\Big). \qquad (1)$$

The ith summand is $\geqq 0$ if $G \geqq 0$, and in case $G > 0$ it is actually > 0 unless $x_{1i} = \cdots = x_{gi} = 0$.

Let $r' \leqq \cdots \leqq r^{(g)}$ be the eigenvalues of G, $r' = r$ the lowest, $r^{(g)} = r^*$ the highest. By an orthogonal transformation $G[x]$ can be brought into the diagonal form $r'x_1^2 + \cdots + r^{(g)}x_g^2$. r and r^* are minimum and maximum of the quadratic form $G[x]$ under the condition $E[x] = x'x = 1$. All $r^{(i)}$ are positive if $G > 0$.

[1]) P. HUMBERT, Comm. Math. Helv. *12*, 263–306 (1939/40). [Cf. also H. WEYL, Transact. Amer. Math. Soc. *51*, 203–231 (1942).] – M. EICHLER, Comm. Math. Helv. *11*, 253–272 (1938/39).

1.4. tr $(X'GX)$ lies between $r \cdot \operatorname{tr}(X'X)$ and $r^* \cdot \operatorname{tr}(X'X)$.

Proof. For the ith term in the sum (1) one has

$$r \sum_k x_{ki}^2 \leqq \sum_{k,l} x_{ki}\, g_{kl}\, x_{li} \leqq r^* \cdot \sum_k x_{ki}^2 .$$

Sum over i.

1.5. Let $G > 0$. Then $|G| \leqq (g^{-1} \cdot \operatorname{tr} G)^g$.

Proof. $|G| = r' \ldots r^{(g)}$, $\operatorname{tr} G = r' + \cdots + r^{(g)}$; $r^{(i)} > 0$.

1.6. Let $G > 0$ (and $g \geqq 2$). Suppose we have a positive constant $c \leqq 1$ such that $|G| \geqq c \cdot (g^{-1} \cdot \operatorname{tr} G)^g$. Then

$$\frac{r^*}{r} \leqq \frac{1 + \sqrt{1-c}}{1 - \sqrt{1-c}} = e. \tag{2}$$

Proof. We have to show: Given $g \geqq 2$ positive numbers

$$(r =)\, r' \leqq \cdots \leqq r^{(g)}\, (= r^*) \quad \text{with the average} \quad \langle r \rangle = \frac{1}{g}\,(r' + \cdots + r^{(g)}),$$

the inequality $r' \ldots r^{(g)} \geqq c \langle r \rangle^g$ implies $r^*/r \leqq e$. One may assume $\langle r \rangle = 1$. Then

$$r\, r^*\, r'' \ldots r^{(g-1)} \geqq c. \tag{3}$$

The g numbers $(r + r^*)/2$, $(r + r^*)/2$, r'', ..., $r^{(g-1)}$ have the arithmetical mean $\langle r \rangle = 1$. Hence their geometric mean cannot exceed unity:

$$\left(\frac{r + r^*}{2}\right)^2 r'' \ldots r^{(g-1)} \leqq 1. \tag{4}$$

(3) and (4) imply

$$\frac{4\, r\, r^*}{(r + r^*)^2} \geqq c, \quad \left(\frac{r^* - r}{r^* + r}\right)^2 \leqq 1 - c,$$

$$r^* \,(1 - \sqrt{1-c}) \leqq r\,(1 + \sqrt{1-c}).$$

1.7. Let $G > 0$ and suppose we have a number t and a positive constant $c \leqq 1$ such that

$$g^{-1} \cdot \operatorname{tr} G \leqq t, \quad |G| \geqq c\, t^g.$$

Then we can ascertain two positive numbers b and $B \leqq g$ which depend on c only such that

$$b\, t \leqq r^{(i)} \leqq B\, t. \tag{5}$$

We have to derive the inequalities (5) from the fact that the g positive numbers $r^{(i)}$ satisfy the relations

$$\langle r \rangle \leqq t, \quad r' \ldots r^{(g)} \geqq c\, t^g.$$

In doing so we may assume $t = 1$. The assumption $x\, r^{(2)} \ldots r^{(g)} \geqq c$, in which x stands for r' implies

$$x \left(\frac{r'' + \cdots + r^{(g)}}{g-1} \right)^{g-1} \geqq c.$$

Because of

$$r'' + \cdots + r^{(g)} = g\langle r \rangle - x \leqq g - x,$$

this implies

$$f(x) \equiv x \left(\frac{g-x}{g-1} \right)^{g-1} \geqq c.$$

The logarithmic derivative

$$\frac{1}{f} \frac{df}{dx} = \frac{g}{x} \frac{1-x}{g-x}$$

shows that $f(x)$ increases monotonely from 0 to 1 and then decreases monotonely from 1 to 0, while x travels first from 0 to 1 and then from 1 to g. Therefore $f(x) \geqq c$ requires x to lie between the two roots $x = b$ and $x = B$ the equation $f(x) = c$ has between 0 and g; $b \leqq 1$, $B \geqq 1$. (For $g = 1$ one must set $b = c$, $B = 1$.)

§ 2. Division Algebra. Vectors, Linear Transformations. Lattice

Let $\mathfrak{F}$ be a division algebra of rank g over the field k of rational numbers. o, ε are the zero and unit elements in $\mathfrak{F}$. With every element α of $\mathfrak{F}$ one associates the mapping

$$(\alpha):\ \xi \to \eta = \alpha\, \xi.$$

This 'regular representation' is faithful; indeed, if α and β are distinct, the mappings (α) and (β) carry ε into two distinct elements. With respect to a given basis $\omega_1, \ldots, \omega_g$ every element

$$\xi = x_1\, \omega_1 + \cdots + x_g\, \omega_g$$

is expressed by the column x of its (rational) coordinates x_i (column representing ξ), and the mapping (α): $\eta = \alpha\, \xi$ is expressed by a matrix $A = \| a_{ik} \|$ (matrix representing α):

$$y_i = \sum_k a_{ik}\, x_k, \quad y = A\, x,$$

where $\alpha\, \omega_i = \Sigma_k\, a_{ki}\, \omega_k$. The matrix algebra consisting of the matrices A thus associated with the elements α of $\mathfrak{F}$ will be denoted by $(\mathfrak{F})$. A change of basis by any non singular linear transformation U changes A into $U^{-1} A U$, hence does not affect the characteristic polynomial $|t\, E - A|$; in particular

$$\operatorname{tr} A = \sum_i a_{ii}, \quad |A| = \det a_{ik}$$

will be designated as $\operatorname{tr} \alpha$ and $\operatorname{Nm} \alpha$.

A *vector* $\mathfrak{x}$ in the n-dimensional vector space $S^n/\mathfrak{F}$ over $\mathfrak{F}$ is a column of n elements $(\xi_1, \ldots, \xi_n)$ of $\mathfrak{F}$. Addition $\mathfrak{x}+\mathfrak{y}$ and multiplication $\mathfrak{x}\,\alpha$ by an element α of $\mathfrak{F}$ are defined in the obvious manner. $\mathfrak{o}$ is the vector $(o, \ldots, o)$ and $\mathfrak{e}_\nu$ the νth column of the unit matrix $E = \|\varepsilon_{\mu\nu}\|$ of degree n:

$$\varepsilon_{\mu\nu} = \varepsilon \quad \text{for} \quad \mu = \nu, \qquad = o \quad \text{for} \quad \mu \neq \nu.$$

Then $\mathfrak{x} = \mathfrak{e}_1\xi_1 + \cdots + \mathfrak{e}_n\xi_n$. Any m vectors $\mathfrak{d}_1, \ldots, \mathfrak{d}_m$ are linearly independent (in $\mathfrak{F}$) if $\mathfrak{d}_1\eta_1 + \cdots + \mathfrak{d}_m\eta_n = \mathfrak{o}$ implies $\eta_1 = \cdots = \eta_m = o$; all vectors of the form $\mathfrak{x} = \mathfrak{d}_1\eta_1 + \cdots + \mathfrak{d}_m\eta_m$ then constitute the m-dimensional subspace $[\mathfrak{d}_1, \ldots, \mathfrak{d}_m]$ with the basis $\mathfrak{d}_1, \ldots, \mathfrak{d}_m$. (For $m = 0$ the subspace consists of the vector $\mathfrak{o}$ only.)

If we replace each component ξ_μ of the vector $\mathfrak{x}$ by the column x_μ of its coordinates $x_{\mu 1}, \ldots, x_{\mu g}$ with respect to a given basis $\omega_1, \ldots, \omega_g$ of $\mathfrak{F}$, we obtain a vector of the $(N = ng)$-dimensional vector space S^N/k over k, which we also denote by $\mathfrak{x}$.

Let n vectors $\mathfrak{d}_1, \ldots, \mathfrak{d}_n$ be given; $\mathfrak{d}_\nu = \mathfrak{e}_1\delta_{1\nu} + \cdots + \mathfrak{e}_n\delta_{n\nu}$. The equation

$$\mathfrak{x} = \mathfrak{d}_1\,\eta_1 + \cdots + \mathfrak{d}_n\,\eta_n \tag{6}$$

breaks up into the n component equations

$$\xi_\mu = \sum_\nu \delta_{\mu\nu}\eta_\nu \tag{7}$$

and these are equivalent to

$$x_\mu = \sum_\nu D_{\mu\nu}y_\nu$$

where $D_{\mu\nu}$ is the matrix representing $\delta_{\mu\nu}$. Hence $\mathfrak{d}_1, \ldots, \mathfrak{d}_n$ are linearly independent if and only if the matrix

$$D = \begin{Vmatrix} D_{11}, & \ldots, & D_{1n} \\ \ldots & \ldots & \ldots \\ D_{n1}, & \ldots, & D_{nn} \end{Vmatrix} \tag{8}$$

of degree N is of non-vanishing determinant, and then *every* vector $\mathfrak{x}$ is expressible in the form (6). As, in particular, $\mathfrak{e}_1, \ldots, \mathfrak{e}_n$ are thus expressible,

$$\mathfrak{e}_\nu = \mathfrak{d}_1\theta_{1\nu} + \cdots + \mathfrak{d}_n\theta_{n\nu},$$

the linear transformation (7), $\Delta = \|\delta_{\mu\nu}\|$, has an inverse $\Delta^{-1} = \|\theta_{\mu\nu}\|$.

A linear mapping $\mathfrak{x} \to \mathfrak{y} = \mathfrak{x}^s$ is one for which

$$(\mathfrak{x} + \mathfrak{x}_1)^s = \mathfrak{x}^s + \mathfrak{x}_1^s, \quad (\mathfrak{x}\,\alpha)^s = \mathfrak{x}^s\,\alpha,$$

whatever the vectors $\mathfrak{x}$, $\mathfrak{x}_1$ and the element α in $\mathfrak{F}$. After interchanging ξ and η, (7) may also be interpreted as the expression of such a linear mapping.

It is non-singular and has an inverse if $\mathfrak{x} = \mathfrak{o}$ is the only vector mapped into $\mathfrak{y} = \mathfrak{o}$.

By extension of the rational field k to the *field K of all real numbers* $\mathfrak{F}$ becomes an algebra $\mathfrak{F}_K$ of rank g over K. All that has been said about $\mathfrak{F}$ remains valid in $\mathfrak{F}_K$. However $\mathfrak{F}_K$ will, in general, have ceased to be a division algebra. Indeed, only if $|D| = \mathrm{Nm}\,\delta \neq 0$, has the element δ of $\mathfrak{F}_K$ a reciprocal δ^{-1} and the equation $\xi = \delta\,\eta$ for η the solution $\eta = \delta^{-1}\,\xi$. (This is merely the special case $n = 1$ of our remarks concerning linear substitutions.) In $S^n/\mathfrak{F}$ the following principle of the step-by-step construction of a basis of a linear subspace holds: *If $m-1$ vectors $\mathfrak{d}_1, \ldots, \mathfrak{d}_{m-1}$ are linearly independent and $\mathfrak{d}$ does not lie in $[\mathfrak{d}_1, \ldots, \mathfrak{d}_{m-1}]$ then $\mathfrak{d}_1, \ldots, \mathfrak{d}_{m-1}, \mathfrak{d}$ are linearly independent.* It goes without saying that this principle breaks down in $S^n/\mathfrak{F}_K$.

In terms of N vectors $\mathfrak{a}_1, \ldots, \mathfrak{a}_N$ of $S^n/\mathfrak{F}$ which are linearly independent in k, every vector $\mathfrak{x}$ of $S^n/\mathfrak{F}$ may be expressed as a linear combination $u_1\mathfrak{a}_1 + \cdots + u_N\mathfrak{a}_N$ with rational coefficients u_p. The $\mathfrak{x}$ with integral u_p form a *lattice* $\mathfrak{A}$, 'spanned' by the vectors $\mathfrak{a}_1, \ldots, \mathfrak{a}_N$. Let $\mathfrak{d}_1, \ldots, \mathfrak{d}_n$ be any n vectors of $\mathfrak{A}$ which are linearly independent in $\mathfrak{F}$: we speak of them as constituting a *semi-basis* of $\mathfrak{A}$ (in $\mathfrak{F}$). (6) will lie in $\mathfrak{A}$ whenever $(\eta_1, \ldots, \eta_n)$ lies in a certain lattice $\mathfrak{L}$, the 'representation of $\mathfrak{A}$ in terms of the semi-basis $\mathfrak{d}_1, \ldots, \mathfrak{d}_n$'. As $\mathfrak{d}_\nu$ is represented by $\mathfrak{e}_\nu$, $\mathfrak{L}$ contains the unit vectors $\mathfrak{e}_\nu$. Any two semi-bases $\mathfrak{d}_1, \ldots, \mathfrak{d}_n$ and $\mathfrak{d}_1^*, \ldots, \mathfrak{d}_n^*$ of $\mathfrak{A}$ lead to two representations $\mathfrak{L}$ and $\mathfrak{L}^*$ of $\mathfrak{A}$ which are carried into one another by a linear substitution $\eta_\mu^* = \sum_\nu \alpha_{\mu\nu}\,\eta_\nu$ and are in this sense equivalent or belong to the same class. Hence all possible representations $\mathfrak{L}$ of $\mathfrak{A}$ in terms of semi-bases satisfy two requirements: they belong to the same class and they contain the unit vectors $\mathfrak{e}_\nu$ ('admissible lattices' $\mathfrak{L}$).

The element β of $\mathfrak{F}$ is called a *multiplier* of $\mathfrak{A}$ if $\mathfrak{x}\beta$ lies in $\mathfrak{A}$ whenever $\mathfrak{x}$ does. The multipliers of a given lattice form an order $\{\mathfrak{F}\}$. The multipliers of $\mathfrak{A}$ are identical with those of the representing lattice $\mathfrak{L}$; even without regard to $\mathfrak{A}$ it is obvious that equivalent lattices $\mathfrak{L}$ and $\mathfrak{L}^*$ have the same multipliers. With $\mathfrak{e}_1, \ldots, \mathfrak{e}_n$ every vector

$$\mathfrak{x} = \mathfrak{e}_1\,\xi_1 + \cdots + \mathfrak{e}_n\,\xi_n = (\xi_1, \ldots, \xi_n)$$

the components ξ_μ of which are in $\{\mathfrak{F}\}$ lies in $\mathfrak{L}$. All these vectors form a lattice $\mathfrak{J}$, the 'unit lattice' of $\{\mathfrak{F}\}$. If $\omega_1, \ldots, \omega_g$ is a minimal basis of $\{\mathfrak{F}\}$, $\mathfrak{J}$ is spanned by the N vectors $\mathfrak{e}_\mu\omega_i$ ($\mu = 1, \ldots, n$; $i = 1, \ldots, g$), and if each component $\xi_\mu = x_{\mu 1}\omega_1 + \cdots + x_{\mu g}\,\omega_g$ is replaced by the column x_μ of its coordinates $(x_{\mu 1}, \ldots, x_{\mu g})$, then $\mathfrak{J}$ consists of those vectors $\mathfrak{x}$ for which all N coordinates $x_{\mu i}$ are rational integers. The index $j = [\mathfrak{L}:\mathfrak{J}]$ is the number of vectors in $\mathfrak{L}$ which are incongruent mod $\mathfrak{J}$. On limiting ourselves to the vectors of the form

$$(\xi_1, \ldots, \xi_m, o, \ldots, o)$$

in $\mathfrak{L}$ and $\mathfrak{J}$ we obtain two (mg)-dimensional lattices $\mathfrak{L}_m \supset \mathfrak{J}_m$ and a corresponding index $j_m = [\mathfrak{L}_m:\mathfrak{J}_m]$. In the series $j_1, \ldots, j_n$ each term is a divisor of the subsequent one, and the last coincides with j. There is only a finite number of lattices $\mathfrak{L} \supset \mathfrak{J}$ with a given index j, *a fortiori* with a given series of indices $j_1, \ldots, j_n$. That number is further limited by the restriction to lattices $\mathfrak{L}$ of a given class.

Let $f(\mathfrak{x})$ be a gauge function in MINKOWSKI's sense, so that $f(\mathfrak{x}) < 1$ defines a symmetric convex body in $S^n/\mathfrak{F}_K$. Denote by V its volume computed in terms of the above described coordinates $x_{\mu i}$. The fundamental parallelotope of $\mathfrak{J}$,

$$0 \leqq x_{\mu i} < 1 \quad (\mu = 1, \ldots, n;\, i = 1, \ldots, g)$$

has the volume 1 in these coordinates, hence that of $\mathfrak{L}$ the volume $j^{-1} = [\mathfrak{L}:\mathfrak{J}]^{-1}$. Assume now that the convex body contains no vector $\mathfrak{x}$ of $\mathfrak{L}$ besides the origin, in other words that $\mathfrak{x} = \mathfrak{o}$ is the only solution of $f(\mathfrak{x}) < 1$ in $\mathfrak{L}$. Then the basic principle of MINKOWSKI's Geometry of Numbers yields the inequality $2^{-N} \cdot V \leqq j^{-1}$ or

$$V[\mathfrak{L}:\mathfrak{J}] \leqq 2^{ng}. \tag{9}$$

A linear mapping s, $\mathfrak{x} \to \mathfrak{x}^s$, is called a *lattice transformation* if it maps the lattice $\mathfrak{A}$ in one-to-one fashion into itself. The '*lattice group*' of all such transformations is the main object of our study. s carries a semi-basis $\mathfrak{d}_1, \ldots, \mathfrak{d}_n$ of $\mathfrak{A}$ into a semi-basis $\mathfrak{d}_1^s, \ldots, \mathfrak{d}_n^s$; *vice versa* a semi-basis $\mathfrak{d}_1^*, \ldots, \mathfrak{d}_n^*$ arises from $\mathfrak{d}_1, \ldots, \mathfrak{d}_n$ by a lattice substitution s, $\mathfrak{d}_\mu^* = \mathfrak{d}_\mu^s$, if, and only if, the representations $\mathfrak{L}$ and $\mathfrak{L}^*$ of $\mathfrak{A}$ in terms of the $\mathfrak{d}$ and the $\mathfrak{d}^*$ coincide.

B. THE BASIC CONCEPT OF QUADRATIC FORM

§ 3. Conjugate Complex

The decomposition of the matrix algebra $(\mathfrak{F}_K)$ yields a certain basis $\omega_1^0, \ldots, \omega_g^0$—we call it *normal basis*—such that with every matrix A in $(\mathfrak{F}_K)$ the transpose A' also belongs to $(\mathfrak{F}_K)$. To prove this, use for a moment the following notations: For an abstract algebra $\mathfrak{a}$ let $\boldsymbol{M}_h\, \mathfrak{a}$ be the algebra of all h-rowed matrices $\|\alpha_{ik}\|$, the coefficients α_{ik} of which are elements of $\mathfrak{a}$. For a matrix algebra $\mathfrak{A}$ let $\boldsymbol{M}_h\, \mathfrak{A}$ and $\boldsymbol{I}_h\, \mathfrak{A}$ be the algebras consisting of the matrices

$$\begin{Vmatrix} A_{11}, & A_{12}, & \ldots, & A_{1h} \\ A_{21}, & A_{22}, & \ldots, & A_{2h} \\ \multicolumn{4}{c}{\dotfill} \\ A_{h1}, & A_{h2}, & \ldots, & A_{hh} \end{Vmatrix} (A_{ik} \in \mathfrak{A}), \quad \begin{Vmatrix} A, & 0, & \ldots, & 0 \\ 0, & A, & \ldots, & 0 \\ \multicolumn{4}{c}{\dotfill} \\ 0, & 0, & \ldots, & A \end{Vmatrix} (A \in \mathfrak{A},\ h \text{ rows})$$

while $\mathfrak{A} + \mathfrak{B}$ consists of the matrices

$$\begin{Vmatrix} A & 0 \\ 0 & B \end{Vmatrix} \quad (A \in \mathfrak{A},\ B \in \mathfrak{B}).$$

After extension to K, $\mathfrak{F}_K$ is still a semi-simple algebra and hence, according to WEDDERBURN, of the form

$$M_{h_1}\, \mathfrak{F}_1 + M_{h_2}\, \mathfrak{F}_2 + \cdots$$

where $\mathfrak{F}_1, \mathfrak{F}_2, \ldots$ are division algebras over K and the $+$ sign indicates a direct sum. Therefore the regular representations $(\mathfrak{F}_K)$ and $(\mathfrak{F}_1)$, $(\mathfrak{F}_2)$, ... are connected by the relation

$$(\mathfrak{F}_K) = I_{h_1} M_{h_1} (\mathfrak{F}_1) + I_{h_2} M_{h_2} (\mathfrak{F}_2) + \cdots. \tag{10}$$

This decomposition refers to an appropriate basis. There are only three possibilities for $\mathfrak{F}_p$ $(p = 1, 2, \ldots)$: it is either the field K itself, basis 1; or the complex field over K, basis 1, i; or the quaternion algebra, basis 1, i, j, k in the usual notation. In the three cases the generic matrix of $(\mathfrak{F}_p)$ is

$$a, \quad \begin{Vmatrix} a_0, & -a_1 \\ a_1, & a_0 \end{Vmatrix}, \quad \begin{Vmatrix} a_0, & -a_1, & -a_2, & -a_3 \\ a_1, & a_0, & -a_3, & a_2 \\ a_2, & a_3, & a_0, & -a_1 \\ a_3, & -a_2, & a_1, & a_0 \end{Vmatrix}.$$

Hence A_p' lies in $(\mathfrak{F}_p)$ if A_p does, and then (10) shows at once that A' lies in $(\mathfrak{F}_K)$ if A does.

Our result means two things: (1) Every element α of $\mathfrak{F}_K$ has a *conjugate* $\bar{\alpha}$ which arises from it by an anti-automorphic involution; i.e. the conjugate of $\bar{\alpha}$ is α and, x being any real number, the conjugates of $x\alpha$, $\alpha + \beta$, $\alpha\beta$ are $x\bar{\alpha}$, $\bar{\alpha} + \bar{\beta}$, $\bar{\beta}\bar{\alpha}$. (2) If A is the matrix representing α in terms of the normal basis, then $\bar{\alpha}$ is represented by A'. Since A' has the same trace and determinant as A, one finds at once

$$\operatorname{tr} \bar{\alpha} = \operatorname{tr} \alpha, \quad \operatorname{Nm} \bar{\alpha} = \operatorname{Nm} \alpha.$$

Because we keep this involution $\xi \to \bar{\xi}$ and the normal basis ω_i^0 fixed, we need not discuss how far they are uniquely determined.

The transition $\xi \to \bar{\xi}$ is a linear operation and hence in terms of any basis $\omega_1, \ldots, \omega_g$ expressed by a matrix

$$J = \|j_{kl}\|, \quad \bar{x} = J x \quad \text{where} \quad \bar{\omega}_k = \sum_l j_{lk}\, \omega_l. \tag{11}$$

As J^2 is unity the determinant $|J| = \pm 1$.

In the following we shall have to make use of two bases of $\mathfrak{F}_K$ only: the fixed normal basis $\omega_1^0, \ldots, \omega_g^0$, and a fixed minimal basis $\omega_1, \ldots, \omega_g$ of the order $\{\mathfrak{F}\}$. For the remainder of this section and for the next section our argument refers to the normal basis, and small Greek letters denote elements of $\mathfrak{F}_K$.

α is *symmetric* if $\bar{\alpha} = \alpha$, *skew* if $\bar{\alpha} = -\alpha$. The symmetric and the skew α form linear subspaces of $\mathfrak{F}_K$, the dimensionalities of which we denote by g^+ and g^-. Since every element α is the unique sum of a symmetric and a skew element,

$$\alpha = \frac{1}{2}(\alpha + \bar{\alpha}) + \frac{1}{2}(\alpha - \bar{\alpha}),$$

$g^+ + g^- = g$. g^+ is at least 1 since ε is symmetric.

$\gamma > 0$ and $\gamma \geqq 0$ indicate for a symmetric γ that the representing symmetric matrix $G(= G')$ is > 0 or $\geqq 0$ respectively.

$\operatorname{tr}(\bar{\xi}\eta) = x' T_0 y$ is a symmetric bilinear form because the conjugate of $\bar{\xi}\eta$ is $\bar{\eta}\xi$.

Lemma 3. 1. The quadratic form $T_0[x] = x' T_0 x = \operatorname{tr}(\bar{\xi}\xi)$ is positive.

This follows at once from 1.1. We write $\operatorname{tr}(\bar{\xi}\xi) = (\operatorname{abs}\xi)^2$ and suppose γ to be symmetric. The other lemmas in § 1 then give rise to the following propositions:

Lemma 3.2. $\operatorname{abs}(\xi + \eta) \leqq \operatorname{abs}\xi + \operatorname{abs}\eta$, $\operatorname{abs}(\xi\eta) \leqq \operatorname{abs}\xi \cdot \operatorname{abs}\eta$.

Lemma 3.3. $\gamma \geqq 0$ implies $\operatorname{tr}(\bar{\xi}\gamma\xi) \geqq 0$; $\gamma > 0$ implies $\operatorname{tr}(\bar{\xi}\gamma\xi) > 0$ except for $\xi = 0$.

Let $r' \leqq \cdots \leqq r^{(g)}$ be the eigenvalues of the symmetric matrix G representing γ, $r = r'$ the lowest, $r^* = r^{(g)}$ the highest.

Lemma 3.4. $\operatorname{tr}(\bar{\xi}\gamma\xi)$ lies between $r \cdot \operatorname{tr}(\bar{\xi}\xi)$ and $r^* \cdot \operatorname{tr}(\bar{\xi}\xi)$.

Lemma 3.5. Let $\gamma > 0$. Then $\operatorname{Nm}\gamma \leqq (g^{-1} \cdot \operatorname{tr}\gamma)^g$.

Lemma 3.6. Let $\gamma > 0$ and suppose we have a positive constant $c \leqq 1$ such that

$$\operatorname{Nm}\gamma \geqq c\,(g^{-1} \cdot \operatorname{tr}\gamma)^g.$$

Then $r^*/r \leqq e$ where $e = e(c) = (1 + \sqrt{1 - c})/(1 - \sqrt{1 - c})$.

Lemma 3.7. Let $\gamma > 0$ and suppose we have a number t and a positive constant $c \leqq 1$ such that

$$\operatorname{tr}\gamma \leqq g\,t, \quad \operatorname{Nm}\gamma \geqq c\,t^g.$$

Then

$$b\,t \leqq r^{(i)} \leqq B\,t$$

where $b = b(c)$ and $B = B(c)$ are determined as described in 1.7.

In the interest of reasonably low estimates we add the following remark. Let f^2 be the rank of the division algebra $\mathfrak{F}$ over its center. The decomposition of the matric algebra $(\mathfrak{F})$ in the field of all complex numbers shows that its generic element is a string of f equal matrices A diagonally arranged,

$$\left\| \begin{matrix} A & 0 & \ldots & 0 \\ 0 & A & \ldots & 0 \\ \ldots & \ldots & \ldots & \ldots \\ 0 & 0 & \ldots & A \end{matrix} \right\|.$$

Hence the sequence of the eigenvalues $r', \ldots, r^{(g)}$ of the matrix G representing a symmetric γ consists of h sections each containing f equal numbers, $g = f\,h$. If this is taken into account, one may replace c by $c^{1/f}$ and g by h in the definitions of the constants e, b and B entering into Lemmas 3.6 and 3.7.

§ 4. Quadratic Forms. Jacobi Transformation

A *quadratic form* (in $\mathfrak{F}_K$) is defined by an n-rowed square matrix

$$\Gamma = \left\| \begin{matrix} \gamma_{11}, \ldots, \gamma_{1n} \\ \cdots\cdots\cdots \\ \gamma_{n1}, \ldots, \gamma_{nn} \end{matrix} \right\| \tag{12}$$

of elements $\gamma_{\mu\nu}$ of $\mathfrak{F}_K$ which satisfy the symmetric conditions

$$\gamma_{\nu\mu} = \bar{\gamma}_{\mu\nu}.$$

These Γ form a linear space H of

$$g_n = \frac{n(n-1)}{2}\, g + n\, g^+ = \frac{n(n+1)}{2}\, g^+ + \frac{n(n-1)}{2}\, g^- \tag{13}$$

dimensions. The value of a given quadratic form Γ for a given vector $\mathfrak{x} = (\xi_1, \ldots, \xi_n)$,

$$\Gamma[\mathfrak{x}] = \sum_{\mu,\nu} \bar{\xi}_\mu \gamma_{\mu\nu} \xi_\nu \tag{14}$$

is a symmetric element of $\mathfrak{F}_K$. By a linear substitution

$$\Delta : \mathfrak{x} = \mathfrak{d}_1\eta_1 + \cdots + \mathfrak{d}_n\eta_n, \quad \xi_\mu = \sum_\nu \delta_{\mu\nu}\,\eta_\nu \tag{6}$$

the form (12) passes into another form $\bar{\Delta}'\Gamma\Delta$. Γ is said to be *positive* if the symmetric matrix of degree N

$$\boldsymbol{G} = \left\| \begin{matrix} G_{11}, \ldots, G_{1n} \\ \cdots\cdots\cdots \\ G_{n1}, \ldots, G_{nn} \end{matrix} \right\| > 0,$$

$G_{\mu\nu}$ being the matrix of degree g representing $\gamma_{\mu\nu}$ in terms of the normal basis ω_i^0. (This definition is suggested by the fact that it is the non-vanishing of the determinant (8) which characterizes a non-singular linear substitution.) In other words we assume that for any n columns $x_\mu = (x_{\mu 1}, \ldots, x_{\mu g})$ of real numbers $x_{\mu i}$ ($\mu = 1, \ldots, n$; $i = 1, \ldots, g$)

$$\boldsymbol{G}[\mathfrak{x}] \equiv \sum_{\mu,\nu} x'_\mu\, G_{\mu\nu}\, x_\nu > 0 \tag{15}$$

except if $(x_1, \ldots, x_n) = (0, \ldots, 0)$. The positive Γ form an open convex cone H^+ in H which is certainly not empty since the unit matrix $\|\varepsilon_{\mu\nu}\|$ lies in it.

It would not do to define the positive character of the form Γ by the condition $\Gamma[\mathfrak{x}] > 0$ for $\mathfrak{x} \neq \mathfrak{o}$. For then there would exist no positive forms at all (unless $\mathfrak{F}_K$ is, like $\mathfrak{F}$ itself, a division algebra). Indeed the inequality $\bar{\xi}\gamma\xi > 0$ implies $\text{Nm}\,\gamma \cdot (\text{Nm}\,\xi)^2 > 0$, hence $\text{Nm}\,\xi \neq 0$, and thus the relation $\bar{\xi}\gamma\xi > 0$ will never be satisfied for elements $\xi \neq 0$ of $\mathfrak{F}_K$ the norm of which vanishes.

Suppose Γ is positive in our sense. Then necessarily $G_{11} > 0$, hence $|G_{11}| \neq 0$ and $\varkappa_1 = \gamma_{11} > 0$ has a reciprocal γ_{11}^{-1}. Write

$$\begin{aligned} \delta_{12} &= \gamma_{11}^{-1}\gamma_{12}, \ldots, \delta_{1n} = \gamma_{11}^{-1}\gamma_{1n}, \\ \zeta_1 &= \xi_1 + \delta_{12}\xi_2 + \cdots + \delta_{1n}\xi_n, \\ z_1 &= x_1 + D_{12}x_2 + \cdots + D_{1n}x_n. \end{aligned}$$

We then have the parallel formulas

$$\Gamma[\mathfrak{x}] = \bar{\zeta}_1\varkappa_1\zeta_1 + \sum_{\mu,\nu\,2}^{n} \bar{\xi}_\mu \gamma_{\mu\nu}^{(2)} \xi_\nu,$$

$$\mathbf{G}[\mathfrak{x}] = z_1' K_1 z_1 + \sum_{\mu,\nu\,2}^{n} x_\mu' G_{\mu\nu}^{(2)} x_\nu.$$

This is the first step of JACOBI's transformation. The $(n-1)$-rowed matrix

$$\Gamma^{(2)} = \|\gamma_{\mu\nu}^{(2)}\| \quad (\mu, \nu = 2, \ldots, n)$$

is symmetric in the same sense as Γ is. If one chooses $x_2, \ldots, x_n$ arbitrarily but determines x_1 by

$$z_1 = x_1 + D_{12}x_2 + \cdots + D_{1n}x_n = 0$$

one sees that the new form

$$\sum_{\mu,\nu=2}^{n} x_\mu' G_{\mu\nu}^{(2)} x_\nu$$

is positive and hence by our definition $\Gamma^{(2)} > 0$. This makes the inductive continuance of JACOBI's process possible, and we end up with a formula

$$\Gamma[\mathfrak{x}] = \bar{\zeta}_1\varkappa_1\zeta_1 + \cdots + \bar{\zeta}_n\varkappa_n\zeta_n \tag{16}$$

where each $\varkappa$ is symmetric and positive and

$$\zeta_\mu = \xi_\mu + \sum_{\nu>\mu} \delta_{\mu\nu}\xi_\nu. \tag{17}$$

Note the parallel formulae

$$\begin{aligned} \mathbf{G}[\mathfrak{x}] &= z_1' K_1 z_1 + \cdots + z_n' K_n z_n, \\ z_\mu &= x_\mu + \sum_{\nu>\mu} D_{\mu\nu} x_\nu. \end{aligned}$$

If Nm Γ designates the determinant $|\boldsymbol{G}|$, we therefore have

$$|\boldsymbol{G}| = |K_1| \dots |K_n| \quad \text{or} \quad \text{Nm}\, \Gamma = \text{Nm}\, \varkappa_1 \dots \text{Nm}\, \varkappa_n.$$

It follows from (16) that $\Gamma[\mathfrak{x}] \geqq 0$. But more important for us is the fact that the number

$$t_\Gamma[\mathfrak{x}] = \text{tr}\,(\Gamma[\mathfrak{x}]) = \sum_{\mu,\nu} \text{tr}\,(\bar{\xi}_\mu \gamma_{\mu\nu} \xi_\nu) \tag{18}$$

is positive except for $\mathfrak{x} = \mathfrak{o}$, as the same expression (16) and Lemma 3.3 show. $t_\Gamma[\mathfrak{x}]$ is thus a positive quadratic form in the ordinary sense in the vector space $S^n/\mathfrak{F}_K$ or S^N/K. *It is on this form and not on* $\mathbf{G}[\mathfrak{x}]$ *that we must base the theory of reduction.*

The relation

$$\gamma_{mm} = \varkappa_m + \sum_{\mu<m} \bar{\delta}_{\mu m} \varkappa_\mu \delta_{\mu m}$$

proves by Lemma 3.3 that

$$\text{tr}\, \varkappa_m \leqq \text{tr}\, \gamma_{mm}. \tag{19}$$

All positive quadratic forms Γ in $\mathfrak{F}_K$ are equivalent to the unit form E:

Lemma 4.1. Γ is positive if and only if it is expressible as $\bar{A}'A$ by means of a non-singular

$$A = \begin{Vmatrix} \alpha_{11}, & \dots, & \alpha_{1n} \\ \dots & \dots & \dots \\ \alpha_{n1}, & \dots, & \alpha_{nn} \end{Vmatrix}.$$

(For $n = 1$: $\gamma > 0$ if and only if $\gamma = \bar{\alpha}\alpha$, Nm $\alpha \neq 0$.)

Proof. It is clear that $\bar{A}'A = \Gamma$ is positive provided A is non-singular; for then $\boldsymbol{G} = \boldsymbol{A}'\boldsymbol{A}$ where

$$\boldsymbol{A} = \begin{Vmatrix} A_{11}, & \dots, & A_{1n} \\ \dots & \dots & \dots \\ A_{n1}, & \dots, & A_{nn} \end{Vmatrix}. \tag{20}$$

Vice versa, let Γ be positive and denote the (positive) eigenvalues of $\boldsymbol{G}$ by $r', \dots, r^{(N)}$ and their diagonal matrix by $\boldsymbol{R}$. Then there exists an orthogonal $\boldsymbol{U}$ such that

$$\boldsymbol{G} = \boldsymbol{U}'\boldsymbol{R}\boldsymbol{U} = \boldsymbol{U}^{-1}\boldsymbol{R}\boldsymbol{U}.$$

Let $\boldsymbol{R}^{1/2}$ be the diagonal matrix of the positive square roots $r'^{1/2}, r''^{1/2}, \dots$ and

$$\boldsymbol{A} = \boldsymbol{U}'\boldsymbol{R}^{1/2}\boldsymbol{U} = \boldsymbol{U}^{-1}\boldsymbol{R}^{1/2}\boldsymbol{U}.$$

Then $\boldsymbol{A}$ is symmetric and

$$\boldsymbol{G} = \boldsymbol{A}^2 = \boldsymbol{A}'\boldsymbol{A}.$$

Suppose the sequence $r', r'', \dots$ contains m distinct numbers $r_1, r_2, \dots$ LAGRANGE's formula of interpolation gives a polynomial $f(x)$ of formal degree $m-1$ with real coefficients such that

$$\cdot f(r_1) = r_1^{1/2}, \quad f(r_2) = r_2^{1/2}, \dots$$

But then $f(\boldsymbol{R}) = \boldsymbol{R}^{1/2}$, $f(\boldsymbol{G}) = \boldsymbol{A}$. The matrix $A = f(\Gamma)$ solves our problem.

The lemma shows that the question whether Γ is symmetric and positive can be decided once the conjugation $\xi \to \bar{\xi}$ is given; the answer does not depend on the choice of the normal basis in terms of which this process appears as transposition $X \to X'$. We can now finally dismiss the normal basis, and from here on it is the *minimal basis* $\omega_1, \dots, \omega_g$ of $\{\mathfrak{F}\}$ by which we determine the representing column x and matrix X of any element ξ of $\mathfrak{F}$ or $\mathfrak{F}_K$.

C. AN OLD STORY RETOLD (WITH SOME MINOR ADAPTATIONS)

§ 5. Minkowski's Fundamental Inequality

Express the symmetric bilinear form

$$\operatorname{tr}(\bar{\xi}\,\eta) = x'\, T_0\, y$$

in terms of that basis. We compute the determinant d of its coefficients $t^0_{ik} = \operatorname{tr}(\bar{\omega}_i\, \omega_k)$ as follows. If J represents the operation $\xi \to \bar{\xi}$ in terms of the basis ω_i, (11), then

$$\operatorname{tr}(\bar{\omega}_i\, \omega_k) = \sum_l j_{li} \operatorname{tr}(\omega_l\, \omega_k)$$

or

$$\|\operatorname{tr}(\bar{\omega}_i\, \omega_k)\| = J' \cdot \|\operatorname{tr}(\omega_i\, \omega_k)\|,$$

hence

$$d = \pm\, |\operatorname{tr}(\omega_i\, \omega_k)|.$$

d is necessarily positive. $\operatorname{tr}(\omega_i\, \omega_k)$ are the coefficients of the bilinear form $\operatorname{tr}(\xi\,\eta)$, which is also symmetric because $\operatorname{tr}(XY)$ depends symmetrically on the two matrices X, Y. Thus we find that d is the absolute value of the determinant $|\operatorname{tr}(\omega_i\, \omega_k)|$. This absolute value is independent of the choice of the minimal basis $\omega_1, \dots, \omega_g$ of $\{\mathfrak{F}\}$ and is therefore known as the *discriminant* of $\{\mathfrak{F}\}$. Computation of $\operatorname{tr}(\omega_i\, \omega_k)$ by means of the basis ω_l itself shows that these coefficients and therefore d are rational integers. The non-degeneracy of the symmetric bilinear form $\operatorname{tr}(\xi\,\eta)$ implied by $d \neq 0$ is an important fact which concerns the division algebra $\mathfrak{F}$ over k (though our proof passes through $\mathfrak{F}_K$ by means of the conjugation $\xi \to \bar{\xi}$).

Lemma 5.1. The determinant d of the quadratic form

$$\operatorname{tr}(\bar{\xi}\,\xi) = x'\,T_0\,x = T_0[x]$$

equals the discriminant of $\{\mathfrak{F}\}$ and is a positive rational integer.

Since the conjugate of $\bar{\xi}\,\gamma\,\eta$ is $\bar{\eta}\,\gamma\,\xi$, provided γ is symmetric, $\operatorname{tr}(\bar{\xi}\,\gamma\,\eta) = x'T\,y$ is a symmetric bilinear form. Its coefficients are $t_{ik} = \operatorname{tr}(\bar{\omega}_i\,\gamma\,\omega_k)$. We have

$$\gamma\,\omega_k = \sum_l g_{lk}\,\omega_l,$$

$$\operatorname{tr}(\bar{\omega}_i\,\gamma\,\omega_k) = \sum_l \operatorname{tr}(\bar{\omega}_i\,\omega_l)\cdot g_{lk},$$

$$|\operatorname{tr}(\bar{\omega}_i\,\gamma\,\omega_k)| = |\operatorname{tr}(\bar{\omega}_i\,\omega_k)|\cdot|G|.$$

Thus:

Lemma 5.2. The determinant of the quadratic form $\operatorname{tr}(\bar{\xi}\,\gamma\,\xi) = T_\gamma[x]$ equals $d\cdot\operatorname{Nm}\gamma$.

In terms of the fixed minimal basis $\omega_1, \ldots, \omega_g$ of the order $\{\mathfrak{F}\}$ we express each component ξ_μ of a vector $\mathfrak{x} = (\xi_1, \ldots, \xi_n)$ by the column of its coordinates $x_{\mu i}$, $\xi_\mu = \Sigma_i\, x_{\mu i}\,\omega_i$, and now use the $N = ng$ quantities $x_{\mu i}$ ($\mu = 1, \ldots, n$; $i = 1, \ldots, g$) as coordinates of $\mathfrak{x}$; they follow one another in the order $\mu i = 11, \ldots, 1g; 21, \ldots, 2g; \ldots$ The JACOBI transformation (17) then appears as a linear transformation

$$z_\mu = x_\mu + \sum_{\nu>\mu} D_{\mu\nu} x_\nu \tag{21}$$

which connects the coordinates $z_{\mu i}$ with $x_{\mu i}$ and has the triangular matrix

$$D = \begin{Vmatrix} E, & D_{12}, & \ldots, & D_{1n} \\ 0, & E\;\;, & \ldots, & D_{2n} \\ \cdots & \cdots & \cdots & \cdots \\ 0, & 0\;\;, & \ldots, & E \end{Vmatrix}. \tag{22}$$

Hence (16) and Lemma 5.2 prove that the determinant of the quadratic form $\mathfrak{t}[\mathfrak{x}] = \mathfrak{t}_\Gamma[\mathfrak{x}]$ of the variables $x_{\mu i}$ equals

$$d^n\cdot\operatorname{Nm}\varkappa_1\ldots\operatorname{Nm}\varkappa_n.$$

A lattice $\mathfrak{A}$ is given, and $\{\mathfrak{F}\}$ with the minimal basis $\omega_1, \ldots, \omega_g$ is the order of its multipliers. Since for any positive t there is only a finite number of lattice vectors $\mathfrak{x}$ for which $\mathfrak{t}[\mathfrak{x}] \leqq t$ we can construct the successive minima of $\mathfrak{t}[\mathfrak{x}]$ as follows: Among all lattice vectors $\mathfrak{x} \neq \mathfrak{o}$ the minimum t_1 of $\mathfrak{t}[\mathfrak{x}]$ is attained for $\mathfrak{x} = \mathfrak{d}_1$; among all lattice vectors not in $[\mathfrak{d}_1]$ the minimum t_2 of $\mathfrak{t}[\mathfrak{x}]$ is attained for $\mathfrak{x} = \mathfrak{d}_2$; etc. For any m, $1 \leqq m \leqq n$, let $\mathfrak{x}$ range over all lattice vectors not in $[\mathfrak{d}_1, \ldots, \mathfrak{d}_{m-1}]$; $\mathfrak{t}[\mathfrak{x}]$ then assumes its minimum t_m for a certain $\mathfrak{x} = \mathfrak{d}_m$. The $\mathfrak{d}_1, \ldots, \mathfrak{d}_n$ thus obtained by induction constitute a semi-basis for $\mathfrak{A}$ such that

$$\mathfrak{t}[\mathfrak{x}] \geqq \mathfrak{t}[\mathfrak{d}_m]$$

whenever $\mathfrak{x}$ is in $\mathfrak{A}$ but not in $[\mathfrak{d}_1, \ldots, \mathfrak{d}_{m-1}]$; $m = 1, \ldots, n$. Let us express this by saying that $\mathfrak{d}_1, \ldots, \mathfrak{d}_n$ is a *reduced semi-basis of* $\mathfrak{A}$ *with respect to* Γ; we have proved the existence of such a basis. The consecutive minima $t_m = t[\mathfrak{d}_m]$ increase with the index m, $0 < t_1 \leqq t_2 \leqq \cdots \leqq t_n$.

By carrying out the transformation (6) $\mathfrak{A}$ is turned into a lattice $\mathfrak{L}$ containing the unit vectors $\mathfrak{e}_\mu$, the form Γ into a form of the variables η. Denoting the new form by Γ again, and the variables by ξ instead of η, we are facing the following situation: $\mathfrak{L}$ is a given admissible lattice and, for any $m = 1, \ldots, n$,

$$t_\Gamma[\mathfrak{x}] \geqq t_\Gamma[\mathfrak{e}_m] \tag{23}$$

whenever $\mathfrak{x} = (\xi_1, \ldots, \xi_n)$ is in $\mathfrak{L}$ and $(\xi_m, \ldots, \xi_n) \neq (o, \ldots, o)$. Let us say under these circumstances that Γ is an *$\mathfrak{L}$-reduced form*. Set again $t[\mathfrak{e}_m] = t_m$ and observe that the N-dimensional 'sphere' defined by

$$f^2(\mathfrak{x}) = t_1^{-1} \cdot \operatorname{tr}(\bar{\zeta}_1 \varkappa_1 \zeta_1) + \cdots + t_n^{-1} \cdot \operatorname{tr}(\bar{\zeta}_n \varkappa_n \zeta_n) < 1 \tag{24}$$

contains no lattice vector $\mathfrak{x} \neq \mathfrak{o}$, provided Γ is reduced[1]). Indeed, let ξ_m be the last non-vanishing component of the non-vanishing lattice vector $\mathfrak{x}$. We have

$$f^2(\mathfrak{x}) = \sum_{\nu=1}^{m} t_\nu^{-1} \cdot \operatorname{tr}(\bar{\zeta}_\nu \varkappa_\nu \zeta_\nu) \geqq t_m^{-1} \cdot \sum_{\nu=1}^{m} \operatorname{tr}(\bar{\zeta}_\nu \varkappa_\nu \zeta_\nu) = t_m^{-1} \cdot t[\mathfrak{x}],$$

but by (23) $t[\mathfrak{x}] \geqq t_m$. The 'length' $f(\mathfrak{x})$ is a gauge function. If v_n is the volume of the n-dimensional unit sphere, $v_n = \{\Gamma(1/2)\}^n / \Gamma(1+n/2)$ then the volume of the sphere (24) equals

$$v_{ng} \cdot \left\{ \frac{t_1^g \ldots t_n^g}{d^n \operatorname{Nm} \varkappa_1 \ldots \operatorname{Nm} \varkappa_n} \right\}^{1/2},$$

and hence the inequality (9) yields

$$4^{-ng} v_{ng}^2 d^{-n} t_1^g \ldots t_n^g [\mathfrak{L} : \mathfrak{J}]^2 \leqq \operatorname{Nm} \varkappa_1 \ldots \operatorname{Nm} \varkappa_n.$$

If one knows a number $\tau_n \leqq 1$ such that congruent non-overlapping spheres in a lattice arrangement cannot occupy more than the part τ_n of the total n-dimensional space, then v_N may here be replaced by the larger constant $\pi_N = v_N/\tau_N$. BLICHFELDT has shown that, for instance, $\tau_n = (n+2) \cdot 2^{-(n+2)/2}$ is a legitimate choice[2]). Setting

$$\left(\frac{g}{4}\right)^{ng} \pi_{ng}^2 d^{-n} = A_n,$$

---

[1]) In this way MINKOWSKI himself proceeded for quadratic forms; see *Geometrie der Zahlen* (1896), pp. 196–199. About his general inequality $S_1 \ldots S_n V \leqq 2^n$ ibid., pp. 211–219; compare H. DAVENPORT, Quart. J. Math. *10*, 119–121 (1939). – H. WEYL, Proc. London Math. Soc. [2] *47*, 270–279 (1942).

[2]) H. F. BLICHFELDT, Math. Ann. *101*, 605–608 (1929).

$[\mathfrak{L}:\mathfrak{J}] = j = j_n$, $c_n = A_n j_n^2$, we have arrived at the following fundamental inequality:

$$c_n \cdot \prod_{\nu=1}^{n} (g^{-1} \operatorname{tr} \gamma_{\nu\nu})^g \leqq \prod_{\nu=1}^{n} \operatorname{Nm} \varkappa_\nu . \tag{25_n}$$

The docked form arising from the reduced $\Gamma[\mathfrak{x}]$ by setting

$$\xi_{m+1} = \cdots = \xi_n = 0$$

is an $\mathfrak{L}_m$-reduced form of the vector $(\xi_1, \ldots, \xi_m)$. Hence a similar inequality holds for every $m = 1, \ldots, n$:

$$c_m \cdot \prod_{\nu=1}^{m} (g^{-1} \operatorname{tr} \gamma_{\nu\nu})^g \leqq \prod_{\nu=1}^{m} \operatorname{Nm} \varkappa_\nu \tag{25_m}$$

where $c_m = A_m j_m^2$. From Lemma 3.5 and (19) we learn that

$$(g^{-1} \cdot \operatorname{tr} \gamma_{\nu\nu})^g \geqq (g^{-1} \cdot \operatorname{tr} \varkappa_\nu)^g \geqq \operatorname{Nm} \varkappa_\nu .$$

Applying this in (25_m) for $\nu = 1, \ldots, m$ we obtain the important inequality

$$c_m = A_m j_m^2 \leqq 1 \quad (m = 1, \ldots, n); \tag{26}$$

if, however, we apply it for $\nu = 1, \ldots, m-1$ only, we find

$$\operatorname{Nm} \varkappa_m \geqq c_m (g^{-1} \operatorname{tr} \gamma_{mm})^g , \tag{27}$$

a fortiori, in view of (19)

$$\operatorname{Nm} \varkappa_m \geqq c_m (g^{-1} \operatorname{tr} \varkappa_m)^g . \tag{27'}$$

From (26) upper bounds depending on $\{\mathfrak{F}\}$ only result for the indices $j_1, \ldots, j_n$. As there is but a finite number of lattices $\mathfrak{L}$ over $\mathfrak{J}$ with given indices, we have thus proved the

First Theorem of Finiteness. *There exist $\mathfrak{L}$-reduced forms for not more than a finite number of admissible lattices $\mathfrak{L}$.*

(27) gives occasion to apply Lemmas 3.6 and 3.7 by identifying $\varkappa_m$ and $g^{-1} \cdot \operatorname{tr} \gamma_{mm} = t_m/g$ with the quantities γ and t of the lemmas. Let therefore r_m and r_m^* be the least and the largest eigenvalue of the positive symmetric matrix K_m and set $e(c_m) = e_m$, $b(c_m) = b_m$, $B(c_m) = B_m$. Then

$$r_m \geqq \frac{b_m t_m}{g}, \quad r_m^* \leqq \frac{B_m t_m}{g} \quad (\text{for } g \geqq 1); \tag{28}$$

$$\frac{r_m^*}{r_m} \leqq e_m \quad (\text{for } g \geqq 2). \tag{29}$$

In the following we denote by M constants depending on $\mathfrak{L}$ only, not always the same. b_m, B_m, e_m are of this nature.

By Lemma 3.4

$$\operatorname{tr}(\bar{\zeta}_m \varkappa_m \zeta_m) \geqq r_m \cdot \operatorname{tr}(\bar{\zeta}_m \zeta_m) \geqq g^{-1} \cdot b_m t_m \cdot \operatorname{tr}(\bar{\zeta}_m \zeta_m) \tag{30}$$

and

$$\operatorname{tr}(\bar{\zeta}_m \varkappa_m \zeta_m) \leqq g^{-1} \cdot B_m t_m \cdot \operatorname{tr}(\bar{\zeta}_m \zeta_m). \tag{31}$$

§ 6. The Pyramid of Reduced Forms

Our next goal is two-fold: we shall derive upper bounds M for abs $\delta_{\mu\nu}$ $(\mu < \nu)$ and show that the 'cell' Z_0 of the $\mathfrak{L}$-reduced forms Γ is defined within H^+ by a *finite* number of linear inequalities and hence is a convex pyramid.

Let m be one of the numbers $1, \ldots, n$, Γ an $\mathfrak{L}$-reduced form, $t = t_\Gamma$, and $\mathfrak{x} = (\xi_1, \ldots, \xi_n)$ a lattice vector for which

$$(\xi_m, \ldots, \xi_n) \neq (o, \ldots, o) \tag{32}$$

and

$$t[\mathfrak{x}] = t_m \tag{33}$$

(equality, not inequality). We maintain that under these circumstances upper bounds M may be ascertained for all abs ζ_ν. Indeed, our equation (33) reads

$$\sum_{\nu=1}^{n} \operatorname{tr}(\bar{\zeta}_\nu \varkappa_\nu \zeta_\nu) = t_m,$$

hence

$$\operatorname{tr}(\bar{\zeta}_\nu \varkappa_\nu \zeta_\nu) \leqq t_m \leqq t_\nu \quad \text{for} \quad \nu \geqq m.$$

By (30)

$$\operatorname{tr}(\bar{\zeta}_\nu \varkappa_\nu \zeta_\nu) \geqq g^{-1} b_\nu t_\nu \cdot \operatorname{tr}(\bar{\zeta}_\nu \zeta_\nu).$$

Consequently

$$b_\nu \cdot \operatorname{tr}(\bar{\zeta}_\nu \zeta_\nu) \leqq g \quad (\text{for } \nu = m, \ldots, n). \tag{34}$$

Similar bounds for $\nu < m$ depend on the fact that Γ is reduced. Any element of $\mathfrak{F}_K$ is congruent mod $\{\mathfrak{F}\}$ to a 'reduced' element

$$\xi = x_1 \omega_1 + \cdots + x_g \omega_g,$$

i.e. one for which $-1/2 < x_i \leqq 1/2$. Let ϱ^2 be an upper bound of $\operatorname{tr}(\bar{\xi}\xi) = T_0[x]$ in the unit cube $-1/2 \leqq x_i \leqq 1/2$. If $\mathfrak{F}_K$ is endowed with a metric by the positive form $T_0[x]$, then the g-dimensional linear space $\mathfrak{F}_K$ is completely covered by circles of radius ϱ around all elements of $\{\mathfrak{F}\}$. Let $\nu < m$. The lattice vector $\mathfrak{x}$ is changed into another lattice vector $\mathfrak{x}^0 = \mathfrak{x} - \mathfrak{a}$ by subtracting a vector of the form $\mathfrak{a} = (\alpha_1, \ldots, \alpha_\nu, o, \ldots, o)$, the components $\alpha_1, \ldots, \alpha_\nu$

of which lie in $\{\mathfrak{F}\}$. The components $\xi_{\nu+1}, \ldots, \xi_n$ are not affected, $\xi_\mu^0 = \xi_\mu$ for $\mu > \nu$; hence

$$(\xi_m^0, \ldots, \xi_n^0) = (\xi_m, \ldots, \xi_n) \neq (o, \ldots, o).$$

Because Γ is reduced, we then must have

$$\mathfrak{t}[\mathfrak{x}^0] \geqq t_m = \mathfrak{t}[\mathfrak{x}],$$

and this gives

$$\sum_{\mu=1}^{n} \operatorname{tr}(\bar{\zeta}_\mu \varkappa_\mu \zeta_\mu) \leqq \sum_{\mu=1}^{n} \operatorname{tr}(\bar{\zeta}_\mu^0 \varkappa_\mu \zeta_\mu^0).$$

But the terms in the left and right sums coincide for $\mu > \nu$. Therefore

$$\sum_{\mu=1}^{\nu} \operatorname{tr}(\bar{\zeta}_\mu \varkappa_\mu \zeta_\mu) \leqq \sum_{\mu=1}^{\nu} \operatorname{tr}(\bar{\zeta}_\mu^0 \varkappa_\mu \zeta_\mu^0). \tag{35}$$

One may choose $\alpha_\nu, \ldots, \alpha_1$ in $\{\mathfrak{F}\}$ one after the other so that $\zeta_\nu^0, \ldots, \zeta_1^0$ are reduced mod $\{\mathfrak{F}\}$. Then

$$\operatorname{tr}(\bar{\zeta}_\mu^0 \varkappa_\mu \zeta_\mu^0) \leqq g^{-1} B_\mu t_\mu \cdot \operatorname{tr}(\bar{\zeta}_\mu^0 \zeta_\mu^0) \leqq g^{-1} B_\mu t_\mu \varrho^2 \quad (\mu \leqq \nu),$$

and consequently the right member of (35) does not exceed

$$g^{-1} \varrho^2 (B_1 t_1 + \cdots + B_\nu t_\nu) \leqq g^{-1} \varrho^2 (B_1 + \cdots + B_\nu) t_\nu.$$

In the left member we retain only the last term

$$\operatorname{tr}(\bar{\zeta}_\nu \varkappa_\nu \zeta_\nu) \geqq g^{-1} b_\nu t_\nu \cdot \operatorname{tr}(\bar{\zeta}_\nu \zeta_\nu).$$

Hence

$$b_\nu \cdot \operatorname{tr}(\bar{\zeta}_\nu \zeta_\nu) \leqq \varrho^2 (B_1 + \cdots + B_\nu) \quad \text{for} \quad \nu = 1, \ldots, m-1. \tag{34*}$$

Given the reduced Γ, the equation (33) holds in particular for $\mathfrak{x} = \mathfrak{e}_m$; therefore by (34*)

$$\operatorname{abs}^2 \delta_{\nu\mu} \leqq \frac{\varrho^2 (B_1 + \cdots + B_\nu)}{b_\nu} \quad (\nu < \mu). \tag{36}$$

In view of Lemma 3.2 and the recursion formulas

$$\xi_n = \zeta_n,$$
$$\xi_{n-1} + \delta_{n-1,n} \xi_n = \zeta_{n-1},$$
$$\cdots\cdots\cdots\cdots\cdots\cdots$$

our upper bounds M for abs ζ_ν and abs $\delta_{\nu\mu}$ as given by (34), (34*), (36) entail similar bounds for abs ξ_ν,

$$\operatorname{abs}^2 \xi_\nu \leqq M.$$

Representing ξ_ν by its column $x_\nu = (x_{\nu 1}, \ldots, x_{\nu g})$ in terms of the basis $\omega_1, \ldots, \omega_g$ we thus find

$$T_0[x_\nu] \leqq M,$$

and since $T_0[x]$ is a positive quadratic form completely determined by $\{\mathfrak{F}\}$ we obtain upper bounds M for all the coordinates $x_{\nu i}$ of $\mathfrak{x}$,

$$|x_{\nu i}| \leqq M. \tag{37}$$

$\mathfrak{J}$ is a subgroup of index j of the additive Abelian group $\mathfrak{L}$; hence the multiple j of any vector of $\mathfrak{L}$ lies in $\mathfrak{J}$, or in other words the $x_{\nu i}$ are integers divided by j. Thus (37) leaves only a finite number of possibilities for $\mathfrak{x}$, independent of the special reduced form Γ.

For a more careful formulation of this result let m again be one of the numbers $1, \ldots, n$. We call a vector $\mathfrak{x} = (\xi_1, \ldots, \xi_n)$ in $\mathfrak{L}$ an essential lattice vector of rank m if $(\xi_m, \ldots, \xi_n) \neq (o, \ldots, o)$ and if *there exists* an $\mathfrak{L}$-reduced form Γ such that

$$t_\Gamma[\mathfrak{x}] = t_\Gamma[\mathfrak{e}_m].$$

Then we have proved that there is only a finite number of essential lattice vectors $\mathfrak{x}$ of rank m.

The cell Z_0 of the reduced forms Γ is defined within H^+ by an infinite number of linear inequalities $L(\Gamma) \geqq 0$: for each $m \leqq n$ and each lattice vector $\mathfrak{x}$ satisfying (32) we have such an L, namely

$$L(\Gamma) = t_\Gamma[\mathfrak{x}] - t_\Gamma[\mathfrak{e}_m] = \sum_{\mu,\nu} \operatorname{tr}(\bar{\xi}_\mu \gamma_{\mu\nu} \xi_\nu) - \operatorname{tr} \gamma_{mm} \geqq 0. \tag{38}$$

We speak of the positive forms Γ as *points* in H^+. Let Γ_0 be a point in Z_0, Γ outside Z_0. Then there is at least one inequality L which is not satisfied by Γ, $L(\Gamma) < 0$. But as there is only a finite number of lattice vectors $\mathfrak{x}$ for which $t_\Gamma[\mathfrak{x}] \leqq t_\Gamma[\mathfrak{e}_1]$ or $\leqq t_\Gamma[\mathfrak{e}_2]$ or $\ldots$, there is not more than a finite number of inequalities $L = L_1, \ldots, L_h$ which are violated by Γ: the planes L_p 'separate' Γ_0 from Γ. Traveling from Γ_0 to Γ along the straight segment $\overrightarrow{\Gamma_0 \Gamma}$ the variable point $u\Gamma_0 + (1-u)\Gamma$ $(0 \leqq u < 1)$ will cross the plane L_p at $u = u_p$, $u_p L_p(\Gamma_0) + (1 - u_p) L_p(\Gamma) = 0$. Let u_1 be the least of the numbers $u_1, \ldots, u_h$, and $u_1 \Gamma_0 + (1 - u_1)\Gamma$ the corresponding point Γ_1. Then Γ_1 is obviously reduced, but satisfies the equation $L_1(\Gamma_1) = 0$. Hence L_1 is essential in the sense that there exists a reduced form Γ_1 for which the equation $L_1(\Gamma_1) = 0$ holds. However, $L_1(\Gamma) < 0$. Thus we have shown: If Γ violates any of the inequalities L it violates in particular an essential L. Or formulated in the positive way: a point Γ satisfying the essential among the inequalities L, satisfies them all. But we know that there is only a finite number of essential inequalities L (which correspond to what we called above the essential lattice vectors $\mathfrak{x}$ of ranks $m = 1, \ldots, n$). This finishes the proof of the

Second Theorem of Finiteness. *The cell Z_0 of $\mathfrak{L}$-reduced forms is defined within the space H^+ of all positive forms by means of a finite number of linear inequalities; in this sense it is a convex pyramid.*

(The proof makes use of the assumption that Z_0 is not empty. Of course an empty Z_0 may also be defined by a finite number of linear inequalities, but one must not choose them from among the inequalities L.)

§ 7. The Pattern of Cells

The following geometric terminology suggests itself. Any semi-basis $\mathfrak{d}_1, \ldots, \mathfrak{d}_n$ of $\mathfrak{A}$ determines a *cell* $Z = Z(\mathfrak{d}_1, \ldots, \mathfrak{d}_n)$; the point Γ *lies in the cell* if $\mathfrak{d}_1, \ldots, \mathfrak{d}_n$ is reduced with respect to Γ, i.e. if

$$t_\Gamma[\mathfrak{x}] \geqq t_\Gamma[\mathfrak{d}_m]$$

for all vectors $\mathfrak{x}$ that are in $\mathfrak{A}$ but not in $[\mathfrak{d}_1, \ldots, \mathfrak{d}_{m-1}]$ $(m = 1, \ldots, n)$. Because there exists a reduced semi-basis of $\mathfrak{A}$ with respect to any given positive form Γ, each point Γ lies in at least one cell Z: *the cells cover H^+ without gaps.* To each cell $Z(\mathfrak{d}_1, \ldots, \mathfrak{d}_n)$ there corresponds an admissible lattice $\mathfrak{L}$, namely the representation of $\mathfrak{A}$ in terms of $\mathfrak{d}_1, \ldots, \mathfrak{d}_n$; the same lattice to two cells if and only if they arise from each other by a lattice substitution s,

$$Z = Z(\mathfrak{d}_1 \ldots, \mathfrak{d}_n), \quad Z^s = Z(\mathfrak{d}_1,, \ldots, \mathfrak{d}_n^s).$$

s carries Γ into the form Γ^s defined by

$$\Gamma^s[\mathfrak{x}^s] = \Gamma[\mathfrak{x}] \quad \text{or} \quad \Gamma^s[\mathfrak{x}] = \Gamma[\mathfrak{x}^{s^{-1}}];$$

Γ^s lies in Z^s when Γ lies in Z. We distinguish the different admissible lattices $\mathfrak{L}$ by different colors and paint the cells accordingly. Provided we omit the empty cells (the lattices $\mathfrak{L}$ for which there are no $\mathfrak{L}$-reduced forms) only a finite number of colors is needed. A lattice transformation s carries this pattern of cells including its coloring into itself. We know the non-empty cells to be convex pyramids.

There will, however, be overlappings: a point Γ may belong to a number of distinct cells. Indeed, the two cells $Z(\mathfrak{d}_1, \ldots, \mathfrak{d}_n)$ and $Z(\mathfrak{d}_1\alpha_1, \ldots, \mathfrak{d}_n\alpha_n)$ completely cover one another if the α_μ are unitary factors. Here an element α of $\mathfrak{F}$ (or of $\mathfrak{F}_K$) is said to be unitary if

$$\bar{\alpha}\,\alpha = \varepsilon.$$

The norm of a unitary element α equals ± 1 and its reciprocal $\alpha^{-1} = \bar{\alpha}$. Hence $\operatorname{tr}(\bar{\alpha}\,\tau\,\alpha) = \operatorname{tr}\tau$ for every element τ of $\mathfrak{F}_K$; in particular the value (14) is not changed by passing from the argument $\mathfrak{x}$ to $\mathfrak{x}\,\alpha$, whatever the quadratic form $\Gamma = \|\gamma_{\mu\nu}\|$. The unitary elements form a group.

As above, we consider the cells as entities *sui generis*, not as point sets; but we identify certain cells according to the rule: *Any two semi-bases* $\mathfrak{d}_1, \ldots, \mathfrak{d}_n$ *and* $\mathfrak{c}_1, \ldots, \mathfrak{c}_n$ belong to the same family or *determine the same cell if* $\mathfrak{c}_\mu = \mathfrak{d}_\mu \alpha_\mu$ *where* α_μ *is unitary*. In this case $\mathfrak{c}_1, \ldots, \mathfrak{c}_n$ is a reduced basis with respect to the quadratic form Γ whenever $\mathfrak{d}_1, \ldots, \mathfrak{d}_n$ is; hence whether or not a point Γ lies in a cell Z does not depend on the basis, $\mathfrak{d}_1, \ldots, \mathfrak{d}_n$ or $\mathfrak{c}_1, \ldots, \mathfrak{c}_n$, by which Z is defined. It is also true that the image Z^s of Z by a given lattice substitution s is independent of the defining basis because $\mathfrak{c}_\mu = \mathfrak{d}_\mu \alpha_\mu$ implies $\mathfrak{c}_\mu^s = \mathfrak{d}_\mu^s \alpha_\mu$. Two admissible lattices $\mathfrak{L}$ and $\mathfrak{L}^*$ are said to belong to the same family if they arise from each other by a 'special substitution'

$$\xi_\mu = \alpha_\mu \xi_\mu^* \quad (\alpha_\mu \text{ unitary}). \tag{39}$$

There is a one-to-one correspondence between the classes of equivalent cells and the families of lattices $\mathfrak{L}$, and we now paint all lattices $\mathfrak{L}$ of the same family and all cells of the corresponding class with the same color. Given an admissible $\mathfrak{L}$, the number of special substitutions (39) we have to reckon with is *a priori* limited. Indeed, since $\mathfrak{e}_\mu$ is contained in $\mathfrak{L}^*$, the vectors

$$(\alpha_1, o, \ldots, o), \quad (o, \alpha_2, \ldots, o), \ldots, (o, o, \ldots, \alpha_n)$$

must be in $\mathfrak{L}$, and the inequalities

$$\operatorname{tr}(\bar{\alpha}_\mu \alpha_\mu) \leqq g$$

will leave but a finite number of possibilities open for them. This shows two things: (1) a family of admissible lattices contains only a finite number of members; (2) *the group of special substitutions carrying* $\mathfrak{L}$ *into itself is finite.* If one writes out the substitution

$$\xi_1 = \alpha_1 \xi_1^*, \ldots, \xi_m = \alpha_m \xi_m^*$$

in terms of the coordinates $x_{\mu i}$ ($\mu = 1, \ldots, m$; $i = 1, \ldots, g$) of $\xi_1, \ldots, \xi_m$ its determinant equals $\mathrm{Nm}\,\alpha_1 \ldots \mathrm{Nm}\,\alpha_m = \pm 1$. Hence the fundamental parallelotopes of $\mathfrak{L}_m$ and $\mathfrak{L}_m^*$ have the same (mg)-dimensional volume provided $\mathfrak{L}$ and $\mathfrak{L}^*$ belong to the same family. Thus two lattices of the same family have the same row of indices $j_1, \ldots, j_n$.

$\mathfrak{d}_1, \ldots, \mathfrak{d}_n$ is said to be a *properly reduced* semi-basis of $\mathfrak{A}$ with respect to Γ, and Γ is said to *belong to the core* of $Z(\mathfrak{d}_1, \ldots, \mathfrak{d}_n)$ if

$$\boldsymbol{t}_\Gamma[\mathfrak{x}] > \boldsymbol{t}_\Gamma[\mathfrak{d}_m]$$

for all vectors $\mathfrak{x}$ in $\mathfrak{A}$ which are not in $[\mathfrak{d}_1, \ldots, \mathfrak{d}_{m-1}]$ and not of the special form $\mathfrak{d}_m \alpha$ (α unitary) ($m = 1, \ldots, n$). One proves at once: Is $\mathfrak{d}_1, \ldots, \mathfrak{d}_n$ a properly reduced and $\mathfrak{d}_1^*, \ldots, \mathfrak{d}_n^*$ a reduced semi-basis with respect to Γ, then $\mathfrak{d}_\mu^* = \mathfrak{d}_\mu \alpha_\mu$, the α_μ being unitary. Or: *A point belonging to the core of a cell lies in no other cell.*

Does this mean that there are no overlappings? Considering a cell $Z = Z(\mathfrak{d}_1, \ldots, \mathfrak{d}_n)$ as the set of points Γ lying in Z we must show that any *inner* point of Z belongs to its core.

Again, we make the substitution

$$\mathfrak{x} = \mathfrak{d}_1 \eta_1 + \cdots + \mathfrak{d}_n \eta_n \tag{40}$$

and afterwards write ξ for η, with the effect that $\mathfrak{e}_\mu$, $\mathfrak{L}$, Z_0 take the place of $\mathfrak{d}_\mu$, $\mathfrak{A}$, $Z(\mathfrak{d}_1, \ldots, \mathfrak{d}_n)$. For each $m = 1, \ldots, n$ and each vector $\mathfrak{x}$ in $\mathfrak{L}$ outside $[\mathfrak{e}_1, \ldots, \mathfrak{e}_{m-1}]$ which is not of the form $\mathfrak{e}_m \alpha$ (α unitary) we set up the linear form

$$L(\Gamma) = \sum_{\mu, \nu} \operatorname{tr} (\bar{\xi}_\mu \gamma_{\mu\nu} \xi_\nu) - \operatorname{tr} (\gamma_{mm}). \tag{41}$$

If it is sure that none of these L vanishes identically, then an inner point Γ of Z_0 necessarily satisfies the strict inequalities $L(\Gamma) > 0$ and hence belongs to the core of Z_0. Thus we must prove that, given a vector $\mathfrak{x}$ of Z, (41) vanishes identically in Γ only if $\mathfrak{x} = \mathfrak{e}_m \alpha$, α unitary. It suffices to do this for $m = 1$. Let γ be any symmetric element of $\mathfrak{F}_K$. Choosing for Γ in succession the n diagonal matrices $\|\gamma_{\mu\nu}\|$ with the elements

$$\{\gamma, o, \ldots, o\}, \quad \{o, \varepsilon, o, \ldots, o\}, \quad \ldots, \quad \{o, o, \ldots, \varepsilon\}$$

along the diagonal we deduce from the identity $L(\Gamma) \equiv 0$:

$$\operatorname{tr} (\bar{\xi}_1 \gamma \xi_1) = \operatorname{tr} \gamma, \quad \operatorname{tr} (\bar{\xi}_2 \xi_2) = 0, \quad \ldots, \quad \operatorname{tr} (\bar{\xi}_n \xi_n) = 0,$$

therefore $\xi_2 = \cdots = \xi_n = o$, $\mathfrak{x} = \mathfrak{e}_1 \alpha$, and $\xi_1 = \alpha$ satisfies the equation $\operatorname{tr}(\bar{\alpha}\gamma\alpha) = \operatorname{tr}\gamma$ for every symmetric γ. Specialize further by setting first $\gamma = \varepsilon$ and then $\gamma = \alpha\bar{\alpha}$:

$$\operatorname{tr} \varepsilon = g, \quad \operatorname{tr} (\bar{\alpha} \alpha) = g, \quad \operatorname{tr} (\bar{\alpha} \alpha \bar{\alpha} \alpha) = \operatorname{tr} (\alpha \bar{\alpha}) = \operatorname{tr} (\bar{\alpha} \alpha) = g.$$

Consequently the trace of the square of $\beta = \bar{\beta} = \bar{\alpha}\alpha - \varepsilon$ is zero; but $\operatorname{tr}(\bar{\beta}\beta) = 0$ implies $\beta = o$, hence α is unitary.

Thus there is no overlapping inasmuch as no inner point of one cell lies in any other cell of our pattern. In the next section we shall show that there is no clustering of cells inside H^+. For this reason we still have a covering of H^+ without gaps even when we retain only the cells with inner points; these are solid pyramids.

An assembly of cells $Z = Z(\mathfrak{d}_1, \ldots, \mathfrak{d}_n)$ in which each color is represented by one member would constitute a fundamental domain for the lattice group if the group $\mathfrak{g}_Z$ of lattice transformations carrying Z into itself consisted of the identity only. As this will not be so, generally speaking, we first have to whittle down Z to a fundamental domain within Z of the finite group $\mathfrak{g}_Z$. In view of our rule of identification for cells a lattice transformation s carrying

$Z(\mathfrak{d}_1, \ldots, \mathfrak{d}_n)$ into itself must be of the form $\mathfrak{d}_\mu \to \mathfrak{d}_\mu^s = \mathfrak{d}_\mu \alpha_\mu$ where the factors α_μ are unitary. It transforms the vector $\mathfrak{d}_1 \eta_1 + \cdots + \mathfrak{d}_n \eta_n$ into

$$\mathfrak{d}_1^s \eta_1 + \cdots + \mathfrak{d}_n^s \eta_n = \mathfrak{d}_1 \eta_1^s + \cdots + \mathfrak{d}_n \eta_n^s$$

where $\eta_\mu^s = \alpha_\mu \eta_\mu$. Hence after the substitution (40) which replaces $\mathfrak{A}$ and $Z(\mathfrak{d}_1, \ldots, \mathfrak{d}_n)$ by $\mathfrak{L}$ and Z_0 respectively, the group $\mathfrak{g}_Z$ is made up of those special transformations

$$\{\alpha_1, \ldots, \alpha_n\}: \quad \xi_\mu^s = \alpha_\mu \xi_\mu \quad (\alpha_\mu \text{ unitary})$$

which carry $\mathfrak{L}$ into itself. They induce a group $\mathfrak{g}$ of linear transformations in the space H of quadratic forms $\Gamma = \|\gamma_{\mu\nu}\|$:

$$\Gamma \to \Gamma^s, \quad \gamma_{\mu\nu} \to \bar{\alpha}_\mu^{-1} \gamma_{\mu\nu} \alpha_\nu^{-1} = \alpha_\mu \gamma_{\mu\nu} \bar{\alpha}_\nu. \tag{42}$$

Denote by $\mathfrak{g}_2$ the invariant subgroup of $\mathfrak{g}_1 = \mathfrak{g}$ the elements of which leave γ_{12} unchanged, by $\mathfrak{g}_3$ the invariant subgroup of $\mathfrak{g}_2$ whose elements leave γ_{13} unchanged, ..., by $\mathfrak{g}_n$ the invariant subgroup of $\mathfrak{g}_{n-1}$ the elements of which leave γ_{1n} unchanged. $\mathfrak{g}_n$ consists of the identity only. Indeed, for any element $\{\alpha_1, \alpha_2, \alpha_3, \ldots, \alpha_n\}$ of $\mathfrak{g}_2$ the substitution $\xi \to \alpha_1 \xi \bar{\alpha}_2$ is the identity, therefore $\alpha_1 \bar{\alpha}_2 = \varepsilon$ or $\alpha_1 = \alpha_2$. Hence all elements of $\mathfrak{g}_n$ are of the form $\{\alpha, \ldots, \alpha\}$ and $\xi \to \alpha \xi \bar{\alpha}$ is the identity. But then (42) is the identity.–In its influence upon γ_{12} the group $\mathfrak{g}_1$ is actually $\mathfrak{g}_1/\mathfrak{g}_2$. We endow the g-dimensional space of the variable 'point' $\xi = \gamma_{12}$ with a metric by means of the positive form tr $(\bar{\xi} \xi)$. The operations of $\mathfrak{g}_1/\mathfrak{g}_2$ are linear metric-preserving ('orthogonal') mappings of the ξ-space. In familiar fashion we construct a fundamental domain for this finite group as follows. We choose a point $\xi = \pi_0$ which is carried into $h+1$ distinct points $\pi_0, \pi_1, \ldots, \pi_h$ by the $h+1$ operations of $\mathfrak{g}_1/\mathfrak{g}_2$ and set up the h linear inequalities expressing that the variable point ξ lies at least as near to π_0 as to $\pi_1, \ldots, \pi_h$:

$$l^{(r)}(\xi) \equiv \operatorname{tr}\{(\bar{\pi}_0 - \bar{\pi}_r)\, \xi\} \geqslant 0 \quad (r = 1, \ldots, h).$$

By adding these h inequalities $l(\gamma_{12}) \geqq 0$ to the ones $L(\Gamma) \geqq 0$ defining Z_0, we obtain a convex part $Z_0^{(2)}$ of Z which is invariant under the group $\mathfrak{g}_2$ but whose $h+1$ images generated by the operations of $\mathfrak{g}_1/\mathfrak{g}_2$ cover Z_0 without gaps and overlappings. We then carry out the same construction for γ_{13} with respect to the group $\mathfrak{g}_2/\mathfrak{g}_3, \ldots,$ for γ_{1n} with respect to the group $\mathfrak{g}_{n-1}/\mathfrak{g}_n = \mathfrak{g}_{n-1}$. Thus by a number of linear inequalities

$$l_2^{(r_2)}(\gamma_{12}) \geqq 0, \ldots, l_2^{(r_n)}(\gamma_{1n}) \geqq 0 \quad (1 \leqq r_2 \leqq h_2, \ldots, 1 \leqq r_n \leqq h_n),$$

each concerning only one coefficient of Γ, a convex part $Z_0^\bullet$ of Z_0 is constructed, the images of which by the mappings s of $\mathfrak{g}$ cover Z_0 without gaps and overlappings. Denote the corresponding part of $Z(\mathfrak{d}_1, \ldots, \mathfrak{d}_n)$ by $Z^\bullet(\mathfrak{d}_1, \ldots, \mathfrak{d}_n)$:

an assembly of such $Z^{\bullet}(\mathfrak{d}_1, \ldots, \mathfrak{d}_n)$ in which each color is represented by one member constitutes a fundamental domain for the lattice group. (In the case $n = 1$, one has of course to proceed in the same manner in the g^+-dimensional space of a symmetric variable $\gamma = \gamma_{11}$ rather than in the g-dimensional spaces of $\gamma_{12}, \ldots, \gamma_{1n}$.)

§ 8. The Third Theorem of Finiteness

To make sure that the pattern of cells shows no inner clustering in H^+ it is not sufficient, as MINKOWSKI seems to have believed, to prove that each cell borders on not more than a finite number of neighbors. Rather, one has to introduce a variable subregion H_t of H^+ depending on a real parameter $t > 0$ in such manner that it grows as t increases and sweeps over the whole region H^+ as t tends to infinity, and then to prove that there is only a finite number of lattice substitutions s carrying a given cell $Z = Z(\mathfrak{d}_1, \ldots, \mathfrak{d}_n)$ into cells Z^s which have points in common with H_t.

In analyzing MINKOWSKI's proof I came to adopt the following definition of the expanding subregion. Given $p \geqq 1$, $w > 0$ and a semi-basis $\mathfrak{c}_1, \ldots, \mathfrak{c}_n$ of $\mathfrak{A}$, we say that the positive form Γ lies in

$$Z(\mathfrak{c}_1, \ldots, \mathfrak{c}_n | p, w)$$

if

$$t_\Gamma[\mathfrak{x}] \geqq p^{-1} \cdot t_\Gamma[\mathfrak{c}_m]$$

for every $m = 1, \ldots, n$ and every vector $\mathfrak{x}$ that is in $\mathfrak{A}$ but not in $[\mathfrak{c}_1, \ldots, \mathfrak{c}_{m-1}]$, and if moreover

$$t_\Gamma[\mathfrak{c}_m - \mathfrak{y}] \geqq t_\Gamma[\mathfrak{c}_m] - w \cdot t_\Gamma[\mathfrak{c}_\mu]$$

for $\mu < m$ whenever $\mathfrak{y}$ is in $\mathfrak{A}$ and in $[\mathfrak{c}_1, \ldots, \mathfrak{c}_\mu]$. While p and w increase to infinity, the set $Z(\mathfrak{c}_1, \ldots, \mathfrak{c}_n | p, w)$ grows, and any given point Γ of H^+ will finally come to lie in it. Instead of p and w one could of course introduce a lot more parameters, but could also reduce the two parameters to one, for instance by setting $p = \exp w$; it makes little difference either way.

The Theorem of Discontinuity. *There is only a finite number of lattice substitutions s carrying a given cell $Z = Z(\mathfrak{d}_1, \ldots, \mathfrak{d}_n)$ into cells Z^s that have points in common with*

$$H_{p,w} = Z(\mathfrak{c}_1, \ldots, \mathfrak{c}_n | p, w) : Z^s \cap H_{p,w} \neq 0.$$

From the beginning we may assume $\mathfrak{d}_\mu = \mathfrak{e}_\mu$. Then $\mathfrak{A}$ coincides with $\mathfrak{L}$ and $Z(\mathfrak{d}_1, \ldots, \mathfrak{d}_n)$ with Z_0. The image Z_0^s has points in common with $H_{p,w}$ if $Z_0 \cap H_{p,w}^{s^{-1}} \neq 0$. Let s^{-1} carry $\mathfrak{c}_1, \ldots, \mathfrak{c}_n$ into $\mathfrak{e}_1^*, \ldots, \mathfrak{e}_n^*$, and Γ be a common point of Z_0 and $Z(\mathfrak{e}_1^*, \ldots, \mathfrak{e}_n^* | p, w)$. Then the $\mathfrak{e}_\mu^*$ are vectors in $\mathfrak{L}$. Write $t_\Gamma[\mathfrak{e}_m] = t_m$, $t_\Gamma[\mathfrak{e}_m^*] = t_m^*$. The form Γ is $\mathfrak{L}$-reduced whereas

$$t_\Gamma[\mathfrak{x}] \geqq p^{-1} \cdot t_\Gamma[\mathfrak{e}_m^*] \tag{43}$$

whenever $\mathfrak{x}$ is in $\mathfrak{L}$ and outside $[\mathfrak{e}_1^*, \ldots, \mathfrak{e}_{m-1}^*]$, and

$$t_\Gamma[\mathfrak{e}_m^* - \mathfrak{y}] \geqq t_\Gamma[\mathfrak{e}_m^*] - w \cdot t_\Gamma[\mathfrak{e}_\mu^*]$$

whenever $\mathfrak{y}$ is in $\mathfrak{L}$ and in $[\mathfrak{e}_1^*, \ldots, \mathfrak{e}^*]$ $(\mu < m \leqq n)$. From these facts bounds M which do not depend on Γ are to be derived for the vectors $\mathfrak{e}_m^*$. We omit the subscript Γ in t_Γ. Two lemmas point the way:

Lemma 8.1.

$$t_m^* \leqq p\, t_m. \tag{44}$$

Proof. At least one of the m vectors $\mathfrak{e}_1, \ldots, \mathfrak{e}_m$, say $\mathfrak{e}_\mu$, lies outside $[\mathfrak{e}_1^*, \ldots, \mathfrak{e}_{m-1}^*]$. Then by (43)

$$t[\mathfrak{e}_\mu] \geqq p^{-1} \cdot t[\mathfrak{e}_m^*] \quad \text{or} \quad t_\mu \geqq p^{-1} t_m^*,$$

and since $t_1 \leqq t_2 \leqq \cdots \leqq t_m$ *a fortiori* $t_m \geqq p^{-1} t_m^*$.

Lemma 8.2. If the two spaces $S^m = [\mathfrak{e}_1, \ldots, \mathfrak{e}_m]$ and $S_*^m = [\mathfrak{e}_1^*, \ldots, \mathfrak{e}_m^*]$ do not coincide, then

$$t_{m+1} \leqq p\, t_m. \tag{45}$$

Indeed, if $\mathfrak{e}_\mu^*$ $(\mu < m)$ lies outside $[\mathfrak{e}_1, \ldots, \mathfrak{e}_m]$, then by the definition of reduction $t_\mu^* = t[\mathfrak{e}_\mu^*] \geqq t_{m+1}$. Therefore by (44), $p\, t_\mu \geqq t_{m+1}$ and *a fortiori* $p\, t_m \geqq t_{m+1}$.

The second lemma suggests introduction of those numbers

$$m = l_0, l_1, \ldots, l_v \quad (0 = l_0 < l_1 < \cdots < l_v = n)$$

for which $S^m = S_*^m$ and to divide the row $m = 1, 2, \ldots, n$ into the sections

$$l_0 < m \leqq l_1, \quad l_1 < m \leqq l_2, \ldots, \quad l_{v-1} < m \leqq l_v. \tag{46}$$

The inequality (45) holds if m and $m+1$ belong to the same section; hence

$$t_\mu \leqq p^{\mu-\nu}\, t_\nu \tag{47}$$

for $\mu > \nu$ provided ν and μ are in the same section.

From now on the proof follows closely the line of Minkowski's proof of the first theorem of finiteness. Again we use Jacobi's transformation for Γ and the notations that go with it. We consider the possibilities for $\mathfrak{e}_1^*, \ldots, \mathfrak{e}_n^*$ which correspond to a definite partition (46) into sections. For a given m let λ be any of the numbers $l_1, \ldots, l_{v-1}$ which is less than m. Because of (44) and $[\mathfrak{e}_1^*, \ldots, \mathfrak{e}_\lambda^*] = [\mathfrak{e}_1, \ldots, \mathfrak{e}_\lambda]$ the vector $\mathfrak{x} = \mathfrak{e}_m^*$ of $\mathfrak{L}$ satisfies the following inequalities

$$t[\mathfrak{x}] \leqq p\, t_m,$$

$$t[\mathfrak{x} - \mathfrak{a}] \geqq t[\mathfrak{x}] - w\, p\, t_\lambda \tag{48}$$

whenever $\mathfrak{a}$ is in $\mathfrak{L}$ and $[\mathfrak{e}_1, \ldots, \mathfrak{e}_\lambda]$. We maintain that they are reconcilable only with a finite number of possibilities for $\mathfrak{x}$. As in § 6 the first inequality yields

$$b_\nu \cdot \operatorname{tr}(\bar{\zeta}_\nu\, \zeta_\nu) \leqq p\, g \quad \text{for} \quad \nu \geqq m. \tag{49}$$

Owing to (47), the same argument still works for $\nu < m$ provided ν is in the same section as m, with the result

$$b_\nu \cdot \mathrm{tr}\,(\bar{\zeta}_\nu\, \zeta_\nu) \leqq p^{m-\nu+1} \cdot g. \tag{49*}$$

If, however, ν is in a lower section than m, $l_{u-1} < \nu \leqq l_u = \lambda < m$, we apply the inequality (48),

$$\mathfrak{t}[\mathfrak{x}] - \mathfrak{t}[\mathfrak{x} - \mathfrak{a}] \leqq w\, p\, t_\lambda \tag{50}$$

to a vector $\mathfrak{a} = \mathfrak{e}_1\alpha_1 + \cdots + \mathfrak{e}_\nu\alpha_\nu$ the components $\alpha_1, \ldots, \alpha_\nu$ of which belong to the order $\{\mathfrak{F}\}$. Setting $\mathfrak{x}^0 = \mathfrak{x} - \mathfrak{a}$ we may ascertain $\alpha_\nu, \ldots, \alpha_1$ so that $\zeta_\nu^0, \ldots, \zeta_1^0$ become reduced mod $\{\mathfrak{F}\}$, and then (50) gives

$$t_\nu\, b_\nu \cdot \mathrm{tr}\,(\bar{\zeta}_\nu\, \zeta_\nu) \leqq \sum_{\mu=1}^{\nu} t_\mu\, B_\mu \cdot \mathrm{tr}\,(\bar{\zeta}_\mu^0\, \zeta_\mu^0) + w\, g\, p\, t_\lambda,$$

$$b_\nu \cdot \mathrm{tr}\,(\bar{\zeta}_\nu\, \zeta_\nu) \leqq \varrho^2(B_1 + \cdots + B_\nu) + w\, g\, p^{\lambda-\nu+1}. \tag{49**}$$

These bounds M for all $\mathrm{abs}^2\, \zeta_\nu$ lead by means of (36) and Lemma 3.2 to similar bounds for all $\mathrm{abs}^2\, \xi_\nu$. Since $\mathfrak{x} = \mathfrak{e}_m^*$ lies in S^l, provided the section to which m belongs ends with l, one might add to (49) the remark that ζ_ν and ξ_ν vanish for $\nu > l$.

D. SIEGEL'S RESULTS

§ 9. The Jacobi Transform of t_Γ

SIEGEL follows another procedure[1]). He carries through the JACOBI transformation of the quadratic form $t_\Gamma[\mathfrak{x}]$ of the N variables $x_{\mu i}$. By means of the substitution (21) we transformed it into

$$T_1[z_1] + \cdots + T_n[z_n]$$

where $T_m[z] = \mathrm{tr}\,(\bar{\zeta}\, \varkappa_m\, \zeta)$. Setting

$$\begin{Vmatrix} T_1 & & & \\ & T_2 & & \\ & & \ddots & \\ & & & T_n \end{Vmatrix} = \boldsymbol{T}$$

besides (22) we have

$$\boldsymbol{D}'\,\boldsymbol{T}\boldsymbol{D} = \boldsymbol{t}_\Gamma$$

Upper bounds M that depend on $\mathfrak{L}$ only were found for $\mathrm{abs}^2\, \delta_{\mu\nu}$; do they imply upper bounds of the same nature for the coefficients of the representing matrices $D_{\mu\nu}$?

[1]) C. L. SIEGEL, Ann. Math. *44*, 687 (1943).

To ask the question is to answer it. Let ξ be an element of $\mathfrak{F}_K$ such that $\operatorname{abs}^2 \xi = T_0[x] \leqq M$. Write down the multiplication table of the basis ω_i:

$$\omega_i \, \omega_k = \sum_l e_{lk}^{(i)} \omega_l .$$

Then the matrices representing $\omega_1, \ldots, \omega_g$ in terms of this basis are $\| e_{kl}^{(1)} \|, \ldots, \| e_{kl}^{(g)} \|$, hence $\xi = x_1 \omega_1 + \cdots + x_g \omega_g$ is represented by

$$\| x_{ik} \| \quad \text{where} \quad x_{ik} = e_{ik}^{(1)} x_1 + \cdots + e_{ik}^{(g)} x_g .$$

$T_0[x] \leqq M$ yields upper bounds M for the absolute values of the coordinates $x_1, \ldots, x_g$ and thus for the absolute values of x_{ik}. However, this is not the most direct proof. Denote for a moment by $\| x_{ik}^0 \|, \| x_{ik} \|$ the two matrices representing ξ in terms of the normal basis ω_i^0 and the basis ω_i respectively, and by $U = \| u_{ik} \|$, $U^{-1} = \| \tilde{u}_{ik} \|$ the transformation that leads from one to the other. Then

$$x_{ik} = \sum_{j,h} \tilde{u}_{ij} x_{jh}^0 \, u_{hk} ,$$

$$x_{ik}^2 \leqq \sum_j \tilde{u}_{ij}^2 \cdot \sum_{j,h} (x_{jh}^0)^2 \cdot \sum_h u_{hk}^2 .$$

Here

$$\sum_{j,h} (x_{jh}^0)^2 = \operatorname{tr} (\bar{\xi} \, \xi) \leqq M .$$

and therefore

$$x_{ik}^2 \leqq \tilde{u}_i^0 \, M \, u_k^0$$

where $u_i^0, \tilde{u}_i^0$ are the elements in the diagonal of $U' U$ and $(U' U)^{-1}$.

To complete our task we have to perform the JACOBI transformation on each of the forms $T = T_m$. We compare them with the form T_0 which is independent of Γ. Thus we are dealing with two positive quadratic forms $T[x]$ and $T_0[x]$ of g variables x and carry out their JACOBI transformation,

$$T = D' \, Q \, D = Q[D] \quad \text{and} \quad T_0 = Q_0[D_0] .$$

Q, Q_0 are diagonal matrices with the positive terms $q_1, \ldots, q_g; q_1^0, \ldots, q_g^0$ along the diagonal while D and D_0 are triangular,

$$D = \left\| \begin{matrix} 1, & d_{12}, & \ldots, & d_{1n} \\ & 1, & \ldots, & d_{2n} \\ & \cdots & \cdots & \\ & & & 1 \end{matrix} \right\| , \quad D_0 = \left\| \begin{matrix} 1, & d_{12}^0, & \ldots, & d_{1n}^0 \\ & 1, & \ldots, & d_{2n}^0 \\ & \cdots & \cdots & \\ & & & 1 \end{matrix} \right\| .$$

Introduce the triangular $C = D \, D_0^{-1} = \| c_{ik} \|$.

Lemma 9.1. Suppose we have two positive constants r, r^* such that

$$r\, T_0[x] \leqq T[x] \leqq r^*\, T[_0x].$$

Then

$$r\, q_i^0 \leqq q_i \leqq r^*\, q_i^0$$

and

$$c_{ik}^2 \leqq \left(\frac{r^*}{r} - 1\right) \cdot \frac{q_k^0}{q_i^0} \quad (i < k).$$

[This is a quantitative reinforcement of the qualitative statement that a form T determines its Jacobi transformation uniquely: $r^* = r = 1$ implies $Q = Q_0$, $(C = E)$, $D = D_0$.]

Proof. Instead of $T \leqq r^* T_0$ or $Q[D] \leqq r^* Q_0[D_0]$ one may write $Q[C] \leqq r^* Q_0$. This inequality for the matrices $Q[C]$ and Q_0 implies the corresponding inequalities for the elements in their diagonal:

$$q_k + \sum_{i<k} q_i\, c_{ik}^2 \leqq r^*\, q_k^0. \tag{51}$$

Therefore $q_k \leqq r^* q_k^0$. Interchanging T and T_0 one finds in the same way $q_k^0 \leqq q_k/r$. By taking this result into account (51) yields

$$\sum_{i<k} q_i^0\, c_{ik}^2 \leqq \left(\frac{r^*}{r} - 1\right) q_k^0.$$

Apply the lemma to our forms

$$T_m = Q_m[D_m].$$

Returning to the notations of § 5, (28), (29), we have indeed

$$r_m\, T_0[x] \leqq T_m[x] \leqq r_m^*\, T_0[x]$$

and therefore obtain for the coefficients $c_{ik}^{(m)}$ of the triangular matrix $C_m = D_m D_0^{-1}$ the upper bounds

$$(c_{ik}^{(m)})^2 \leqq (e_m - 1) \cdot \frac{q_k^0}{q_i^0} \quad (i < k). \tag{52}$$

Similar such bounds, which depend on $\mathfrak{L}$ only, follow then for $D_m = C_m D_0$ and ultimately for the triangular matrix

$$\left\| \begin{matrix} D_1 & & & \\ & D_2 & & \\ & & \ddots & \\ & & & D_n \end{matrix} \right\| \cdot \left\| \begin{matrix} E, & D_{12}, & \ldots, & D_{1n} \\ & E, & \ldots, & D_{2n} \\ & \ldots & \ldots & \ldots \\ & & & E \end{matrix} \right\| = \tilde{D}$$

which effects the transformation of $\boldsymbol{t}_\Gamma$ into the diagonal matrix

$$\boldsymbol{Q} = \begin{Vmatrix} Q_1 & & & \\ & Q_2 & & \\ & & \ddots & \\ & & & Q_n \end{Vmatrix}, \quad \boldsymbol{t}_\Gamma = \boldsymbol{Q}[\tilde{\boldsymbol{D}}].$$

Denote the diagonal elements

$$q_{11}, \ldots, q_{1g} \mid q_{21}, \ldots, q_{2g} \mid \cdots \mid q_{n1}, \ldots, q_{ng}$$

of $\boldsymbol{Q}$ in this order by $q_1, \ldots, q_N$. For any two consecutive ones which do not jump the partitions $|$, like $q_{\nu i}$ and $q_{\nu, i+1}$ $(i = 1, \ldots, g-1)$, we find by Lemma 9.1

$$\frac{q_{\nu, i}}{q_{\nu, i+1}} \leqq \frac{r_\nu^*}{r_\nu} \cdot \frac{q_i^0}{q_{i+1}^0},$$

hence by Lemma 3.6 or formula (29)

$$\frac{q_{\nu, i}}{q_{\nu\ i+1}} \leqq e_\nu \cdot \frac{q_i^0}{q_{i+1}^0}. \tag{53}$$

In order to cross a partition, for instance from $q_{\nu g}$ to $q_{\nu+1,1}$ we appeal to (28) and Lemma 9.1:

$$q_{\nu g} \leqq r_\nu^* \, q_g^0 \leqq g^{-1} \, B_\nu \, t_\nu \cdot q_g^0,$$
$$q_{\nu+1,1} \geqq r_{\nu+1} \, q_1^0 \geqq g^{-1} \, b_{\nu+1} \, t_{\nu+1} \cdot q_1^0.$$

But $t_{\nu+1} \geqq t_\nu$; therefore

$$\frac{q_{\nu g}}{q_{\nu+1,1}} \leqq \frac{B_\nu}{b_{\nu+1}} \cdot \frac{q_g^0}{q_1^0}. \tag{53*}$$

We thus obtain bounds M for all the quotients

$$\frac{q_K}{q_{K+1}} \qquad (K = 1, \ldots, N-1)$$

and all the coefficients d_{KL} of $\tilde{\boldsymbol{D}}$. Any positive quadratic form

$$\tilde{\boldsymbol{G}}[\mathfrak{x}] = \sum_{K, L=1}^{N} g_{KL} \, x_K \, x_L$$

of N variables $x_1, \ldots, x_N$ determines uniquely its Jacobi transformation

$$z_K = x_K + \sum_{L>K} d_{KL} \, x_L,$$
$$\tilde{\boldsymbol{G}}[\mathfrak{x}] = \sum_K q_K \, z_K^2.$$

Given a positive number t, let us say that $\tilde{\boldsymbol{G}}$ belongs to the set $\mathfrak{R}_t$ if

$$\frac{q_K}{q_{K+1}} \leqq t \quad \text{and} \quad d_{KL}^2 \leqq t \quad (K < L).$$

We have constructed a number a which depends on $\mathfrak{L}$ only, such that for every $\mathfrak{L}$-reduced form Γ the corresponding $\boldsymbol{t}_\Gamma$ belongs to $\mathfrak{R}_a$.

The quadratic forms of N real variables $x_1, \ldots, x_N$ with real coefficients form a linear space $\mathfrak{R}$ in which the positive ones form an open convex cone $\mathfrak{R}^+$. In SIEGEL's conception the theorem of discontinuity deals with this space $\mathfrak{R}$ of dimensionality $N(N+1)/2$ rather than with the space of quadratic forms Γ in $\mathfrak{F}_K$ of dimensionality g_n. It is clear that with t increasing to infinity, $\mathfrak{R}_t$ will exhaust $\mathfrak{R}^+$, and the set H_t of all positive Γ for which the corresponding $\boldsymbol{t}_\Gamma$ lies in $\mathfrak{R}_t$ will exhaust H^+. A lattice transformation

$$s: (\xi_1, \ldots, \xi_n) \to (\xi_1^s, \ldots, \xi_n^s), \quad \xi_\mu^s = \sum_\nu \alpha_{\mu\nu} \xi_\nu \tag{54}$$

when expressed in terms of the N coordinates $x_{\mu i}$ appears as the linear transformation $\boldsymbol{A}$, (20); it has the property that the coefficients both of $\boldsymbol{A}$ and its inverse $\boldsymbol{A}^{-1}$ are rational numbers with the common denominator j. A general principle of SIEGEL's[1]) asserts that, given $a > 0$ and $t > a$, there exists but a finite number of transformations $\boldsymbol{A}$ of this character which carry $\mathfrak{R}_a$ into sets that have points in common with $\mathfrak{R}_t$. This principle, which is a very powerful tool in all investigations concerning quadratic forms, including the indefinite ones, permits him to transfer the problem of the discontinuity of the lattice group from H to $\mathfrak{R}$. Compared to MINKOWSKI's approach which the previous section followed in its outline, this method has the disadvantage of yielding undesirably high estimates for the number of such images Z_0^s of Z_0 as may be expected to have points in common with H_t. But it recommends itself by the generality of the underlying principle.

The lattice group consists of certain linear substitutions (54) and is, therefore, contained in the continuous group W of all non-singular linear substitutions $A = \|\alpha_{\mu\nu}\|$, $\alpha_{\mu\nu} \in \mathfrak{F}_K$. Consider continuous representations of W which become discontinuous under restriction of the variable element s of W to the lattice group, in the sense that in the representation space no set of points which are equivalent under this group has an accumulation point. As the Theorem of Discontinuity in either of its two forms proves, the representation $\Gamma \to \Gamma^s = \Gamma[A^{-1}]$ in the space H^+ of all positive quadratic forms Γ is of this discontinuous nature. SIEGEL devotes the major part of his paper[2]) to developing a general principle from which it follows that among all representations of such nature our $\Gamma \to \Gamma^s$ is the 'most compact' and therefore of least dimension. I shall not deal here with this side of the problem of reduction.

§ 10. Volume of the Fundamental Domain

Let us consider *that portion of the pyramid Z_0 of the $\mathfrak{L}$-reduced positive $\Gamma = \|\gamma_{\mu\nu}\|$ whose points Γ satisfy the condition*

$$\mathrm{Nm}\, \Gamma \leq 1. \tag{55}$$

[1]) C. L. SIEGEL, Abh. Math. Sem. Hansischen Univ. *13*, 217 (1940), Satz 3.

[2]) C. L. SIEGEL, *Discontinuous Groups*, Ann. Math. *44*, 674–684 (1943). Cf. also M. EICHLER, Comm. Math. Helv. *11*, 253–272 (1938/39).

Using the JACOBI transformation of the coefficients $\gamma_{\mu\nu}$ into $\varkappa_\mu$, $\delta_{\mu\nu}$ $(\mu < \nu)$ SIEGEL proved[1]) that this portion of Z_0 *has a finite volume* V *in the* g_n*-dimensional linear space* H. I describe here an alternative procedure which operates directly with the $\gamma_{\mu\nu}$ and leads to simpler estimates.

Dealing first with the lateral $\gamma_{\mu\nu}$ $(\mu < \nu)$ set for a moment $-\gamma_{\mu\mu}^{-1}\gamma_{\mu\nu} = \beta_{\mu\nu}$ $(\mu < \nu)$ and take $\beta_{25} = \beta$ as an example. Write $\gamma_{\mu\mu} = \gamma_\mu$. Choose any element ξ of $\{\mathfrak{F}\}$ and apply (38) to the lattice vector $\mathfrak{x} = (\xi_1, \ldots, \xi_n)$ of which all components ξ_μ vanish, except $\xi_2 = \xi$, $\xi_5 = \varepsilon$. One finds

$$\operatorname{tr}(\bar{\xi}\gamma_{22}\xi + \bar{\xi}\gamma_{25} + \gamma_{52}\xi) \geqq 0 \quad (\gamma_{52} = \bar{\gamma}_{25})$$

or

$$\operatorname{tr}\{(\bar{\beta} - \bar{\xi})\gamma_2(\beta - \xi)\} \geqq \operatorname{tr}(\bar{\beta}\gamma_2\beta). \tag{56}$$

Determine ξ in $\{\mathfrak{F}\}$ so that $\beta - \xi = \beta_0$ is reduced mod$\{\mathfrak{F}\}$. We make use of the upper bound furnished by Lemma 3.5, but replace the largest eigenvalue ν^* by the sum $\operatorname{tr}\gamma$ of all eigenvalues:

$$\operatorname{tr}(\bar{\beta}_0\gamma_2\beta_0) \leqq \operatorname{tr}\gamma_2 \cdot \operatorname{tr}(\bar{\beta}_0\beta_0) \leqq \varrho^2 \cdot \operatorname{tr}\gamma_2.$$

Hence (56) implies

$$\operatorname{tr}(\bar{\beta}_{\mu\nu}\gamma_\mu\beta_{\mu\nu}) \leqq \varrho^2 \cdot \operatorname{tr}\gamma_\mu \quad (\mu = 2,\ \nu = 5),$$

or more generally, for $\mu < \nu$,

$$\operatorname{tr}(\bar{\gamma}_{\mu\nu}\gamma_\mu^{-1}\gamma_{\mu\nu}) \leqq \varrho^2 \cdot \operatorname{tr}\gamma_\mu. \tag{57}$$

For a fixed γ_μ we extend the integration with respect to the g real coefficients of the variable $\gamma_{\mu\nu}$ over the entire ellipsoid (57). Since by Lemma 5.2 $d \cdot (\operatorname{Nm}\gamma_\mu)^{-1}$ is the determinant of the quadratic form

$$\operatorname{tr}(\bar{\xi}\gamma_\mu^{-1}\xi) = T_\mu^*[x]$$

we find for the volume V_μ of this ellipsoid

$$V_\mu^2 = v_g^2\, d^{-1} \operatorname{Nm}\gamma_\mu \cdot \varrho^{2g} (\operatorname{tr}\gamma_\mu)$$

or, because of $\operatorname{Nm}\gamma_\mu \leqq (g^{-1} \cdot \operatorname{tr}\gamma_\mu)^g$,

$$V_\mu \leqq k(\operatorname{tr}\gamma_\mu)^g = k\, t_\mu^g$$

where

$$k = v_g\, \varrho^g\, (g^g \cdot d)^{-1/2}.$$

In view of (25_n) and (55) it remains to integrate

$$k^{n(n-1)/2}\,(t_1^{n-1}\, t_2^{n-2} \ldots t_n^0)^g$$

---

1) C. L. SIEGEL, *Discontinuous Groups*, Ann. Math. **44**, 688 (1943).

with respect to the $n g^+$ coordinates of the variable symmetric positive $\gamma_1, \ldots, \gamma_n$ over the region described by

$$\left.\begin{array}{l} 0 < t_1 \leqq t_2 \leqq \cdots \leqq t_n, \\ t_1 \ldots t_n \leqq M_0^n \quad (M_0 = g \cdot c_n^{-1/ng}) \end{array}\right\}. \tag{58}$$

Let v_+/g^+ be the volume in the space of the g^+ coefficients of an arbitrary symmetric element γ of the bounded portion described by $\gamma > 0$, $\operatorname{tr}\gamma \leqq 1$. Then the volume of the infinitely thin shell

$$\gamma > 0, \quad t \leqq \operatorname{tr}\gamma \leqq t + dt$$

is $v_+ t^{g^+-1} dt$, and we obtain as an upper bound for V the integral of

$$k^{n(n-1)/2} v_+^n t_1^{(n-1)g + (g^+-1)} \ldots t_n^{g^+-1}$$

extended with respect to the real variables $t_1, \ldots, t_n$ over (58). It is only at this last quite elementary step that we are dealing with a non-bounded domain. (58) implies

$$\begin{array}{ll} t_1 \ldots t_{n-1} t_n & \leqq M_0^n, \\ t_1 \ldots t_{n-2} t_{n-1}^2 & \leqq M_0^n, \\ t_1 \ldots t_{n-3} t_{n-2}^3 & \leqq M_0^n, \quad t_\mu > 0. \\ \ldots\ldots\ldots\ldots\ldots\ldots & \\ t_1^n & \leqq M_0^n, \end{array}$$

Integration over the larger domain described by these inequalities may be carried out step by step, first with respect to t_n,

$$0 < t_n \leqq M_0^n t_1^{-1} \ldots t_{n-1}^{-1},$$

then with respect to t_{n-1},

$$0 < t_{n-1} \leqq M_0^{n/2} t_1^{-1/2} \ldots t_{n-2}^{-1/2}$$

..., with the result

$$V \leqq \frac{2^{n-1}}{n!\, g^+ g^{n-1}} k^{n(n-1)/2} v_+^n M_0^{gn}.$$

For the rational case, $g = 1$, this upper limit is

$$\frac{2^{n-1}}{n!} c_n^{-(n+1)/2}.$$

By following existing models for $g = 1$ one can find explicit expressions for the volume of the fundamental domain.

§ 11. The Group of Units in an Order of a Simple Algebra

From the fundamental domain of the lattice group and its images one obtains at once a fundamental domain for any subgroup of the lattice group of finite index. *This remark settles the problem for the group of units in any order of a simple algebra over k.*

According to WEDDERBURN, the elements A of such an algebra $\mathfrak{F}_n$ consist of the matrices

$$\begin{Vmatrix} \alpha_{11}, \ldots, \alpha_{1n} \\ \ldots\ldots\ldots\ldots \\ \alpha_{n1}, \ldots, \alpha_{nn} \end{Vmatrix}$$

formed by means of arbitrary elements $\alpha_{\mu\nu}$ of a division algebra $\mathfrak{F}$. Let again g be the rank of $\mathfrak{F}$ and set $N = ng$. $Nn = n^2 g$ is the rank of $\mathfrak{F}_n$. As previously, we interpret A as a linear mapping of the vector space $S^n/\mathfrak{F}$. Let $\{\mathfrak{F}_n\}$ be an order in $\mathfrak{F}_n$ and $\Omega_1, \ldots, \Omega_{Nn}$ a minimal basis of $\{\mathfrak{F}_n\}$. The first columns of all elements A of $\{\mathfrak{F}\}$ form a vector lattice $\mathfrak{A}$; the element A of $\{\mathfrak{F}_n\}$ is a mapping that carries every vector of the lattice $\mathfrak{A}$ into a vector of $\mathfrak{A}$. The *units* in $\{\mathfrak{F}_n\}$ are those elements A of $\{\mathfrak{F}_n\}$ which are one-to-one mappings of $\mathfrak{A}$ into itself. We now consider *all* elements B of $\mathfrak{F}_n$ which, interpreted as mappings, carry lattice vectors into lattice vectors; they form an order $\{\mathfrak{F}_n\}^* \supset \{\mathfrak{F}_n\}$, the units of which are our old lattice transformations s. We maintain that the group of units of $\{\mathfrak{F}_n\}$ is a subgroup *of finite index* within the group of units of $\{\mathfrak{F}_n\}^*$.

Indeed, express $\Omega_1, \ldots, \Omega_{Nn}$ in terms of a minimal basis $\Omega_1^*, \ldots, \Omega_{Nn}^*$ of $\{\mathfrak{F}_n\}^*$. The coefficients are rational integers; denote the absolute value of its determinant by h. Then the multiple $h\,B$ of any element B of $\{\mathfrak{F}_n\}^*$ is an element A of $\{\mathfrak{F}_n\}$. Let B_1, B_2 be two units in $\{\mathfrak{F}_n\}^*$, the difference of which is of the form $h\,B$, B in $\{\mathfrak{F}_n\}^*$. Then

$$B_2\,B_1^{-1} - E = h \cdot B\,B_1^{-1} = h\,B_3$$

or

$$B_2\,B_1^{-1} = A \quad \text{where} \quad A = E + h\,B_3.$$

Here B_3 lies in $\{\mathfrak{F}_n\}^*$, hence $h\,B_3$, A in $\{\mathfrak{F}_n\}$. Consequently the index whose finiteness we claim cannot exceed h^{Nn}.

137.

Encomium (Wolfgang Pauli)

Science 103, 216—218 (1946)

IT IS DIFFICULT TO IMAGINE what the history of physics would have been without the influence of Pauli during the last twenty-odd years. As another Nobel laureate recently expressed it, "Pauli for many years has been the conscience and criterion of truth for a large part of the community of theoretical physicists." Thus, there is complete unanimity the world over that Pauli has amply deserved the recognition now accorded his work by the Royal Swedish Academy, to whose hands, by Nobel's will, the distribution of the Nobel Prize for Physics is entrusted. . . .

I think it is very fortunate that through the accident of Nobel's birth the lot to bestow this highest international honor for scientific achievement fell to Sweden, one of the Scandinavian countries. Indeed, these countries march in the vanguard of civilization; nowhere on this planet has man come nearer to the fulfillment of his dream of a happy and free life, with justice and equal opportunity for all, where good prevails over evil, and beauty and truth can shine and are loved. In physics and mathematics in particular the Scandinavian countries have, during the last decades, contributed more than their share to the advancement of our knowledge. It is enough to mention one name, that of Niels Bohr, who has exerted the most extraordinary influence upon the development of physics and on the whole generation of younger physicists in the last thirty years. Pauli himself is his disciple.

The impression prevails that it has been harder for a theoretical than for an experimental physicist to win the Nobel laurels. One obvious reason is that it is more difficult to assess at an early stage the importance of a theoretical discovery. When modern quantum physics came into being around 1925, one often spoke of it as boys' physics—"Knabenphysik." Indeed, at the time neither Heisenberg, nor Dirac, nor Pauli were over twenty-five (de Broglie and Schrödinger were somewhat older). It is gratifying that now all the boys who enacted this great scientific drama have been crowned by the Swedish Academy.

Born and educated in Vienna, Wolfgang Pauli started his scientific career in Munich under Arnold Sommerfeld. Perhaps I am among the first with whom he established scientific contacts, for the first papers he published dealt with a unified field theory of gravitation and electromagnetism which I had propounded in 1918. He dealt with it in a truly Paulinean fashion—namely, he dealt it a pernicious blow. Pauli's article on relativity theory, written in these years for the ***Mathematical Encyclopaedia,*** is a mature and masterly work which shows the author in full command of both the mathematical and physical aspects of the subject; and yet it was the work of a young man of twenty. After having earned his Ph.D. in Munich, Pauli migrated to Göttingen, since the time of Gauss a center of mathematical and physical research, where Max Born and James Franck were teaching at that time, and from there he went to Copenhagen and came under Niels Bohr's strong formative influence. From 1923 to 1928 he was *Dozent* at the University of Hamburg, and since then he has occupied a chair for theoretical physics at the Eidgenössische Technische Hochschule at Zürich. The year 1935–1936 he spent as a visiting professor at our Institute. In 1940, immediately after the invasion of Denmark and Norway by the Nazis, when it was clear that all the other neutral European countries were in danger of being overrun by the swastika, the Institute made an effort to bring Niels Bohr and Pauli to this country. Bohr considered it his patriotic duty to stay in Copenhagen, but we were lucky enough to get Pauli. I hope he does not regret that he came, even though Switzerland was spared the fate of being invaded by Hitler's hordes.

Let me now cast a quick glance over Pauli's principal achievements in physics—although a mathematician is hardly entitled to speak about them with authority. We mathematicians feel near to Pauli since he is distinguished among physicists by his highly developed organ for mathematics. Even so, he is a physicist; for he has to a high degree what makes the physicist: the genuine interest in the experimental facts in all their puzzling complexity. His accurate, instinctive estimate of the relative weight of relevant experimental facts has been an unfailing guide for him in his theoretical investigations. Pauli combines in an exemplary way physical insight and mathematical skill.

As I have already mentioned, Pauli began his work in the sign of relativity theory. Although he later returned to this theory on one or two occasions, his main work, by which he should be judged as a creative physicist, is in quantum physics. Here it is natural to distinguish the periods before and after the Heisenberg-Schrödinger break-through to a consistent quantum theory of the atom in 1925. In the time before this dramatic event one had to operate with Niels Bohr's models and a compromise that Bohr vaguely formulated as the Principle of Correspondence, and to find one's way through the maze of spectroscopic facts more by divination ("Schnauze" is Pauli's word for it) than by theory. It is remarkable that in this period Pauli scored some of his greatest successes. For instance, he saw that the so-called hyperfine structure of spectral lines is to be ascribed to a quantum character of the nucleus rather than to the electronic shell of the atom. But above all, his investigations concerning the Zeeman effect gradually led him to the discovery of the exclusion principle, according to which no two electrons may be in the same quantum state. This was a very bold conception. The exclusion principle, strange and incomprehensible as it is from the standpoint of classical physics, is decisive for an understanding of the periodic system of chemical elements. It is a lasting achievement, which will hardly be affected by any future changes of our physical theories.

One would expect that in stable equilibrium each electron revolving around the nucleus occupies the lowest possible energy level, according to the Planck-Bohr quantum rule. Instead, when we run over the chemical elements in their natural order, we find that only the first two electrons, in hydrogen and helium, are bound in this lowest state. Then a sort of saturation seems to be reached. The next element, lithium, has only one valence electron. As the spectra show, the eight electrons from lithium to neon are all bound in the next higher level, and with neon again a closed shell which can admit no further electrons, seems to have been completed. It is these fundamental facts which Pauli's exclusion principle explains. In developing it he had to overcome an accessory difficulty. When he started his work the quantum state of an electron was characterized by three quantum numbers. But that led to shells of 1, 4, 9 . . . electrons instead of 2, 8, 18 . . . , as we find in nature. He accounted for this "duplicity" by a fourth quantum number of the electron. Shortly afterward, Goudsmit and Uhlenbeck suggested that this quantum number had its origin in an angular momentum, a spin of the electron. Again it was Pauli who, once the foundations of the new quantum mechanics were established, first succeeded in describing correctly the nature of this momentum, which is radically different from that of a spinning top and not to be accounted for by classical concepts.

The Pauli principle reveals a general mysterious property of the electron, the importance of which is by no means limited to spectroscopy.. Pauli himself applied it to the statistics of particles in a degenerate gas, thus explaining the paramagnetic properties of such gases. A paper on the paramagnetism of metals laid the foundations for the quantum mechanical theory of electrons in metals. A step of great consequences, for which Dirac's quantum theory of radiation had paved the way, was taken in a joint investigation by Pauli and Heisenberg on the quantization of the field equations: thereby wave mechanics passed from the theory of a single particle to that of the interaction of an indefinite number of particles. Pauli's studies of the intimate relationship between spin and statistical behavior of particles naturally led him to investigate the dynamics of the meson. The meson is now a generally accepted particle in nuclear physics. Of a more dubious character seems an invention of Pauli's, the most elusive of all elementary particles, which he dubbed neutrino, and others call Paulino. It is a particle without charge and mass, which nevertheless seems to be indispensable if the laws of conservation of energy and angular momentum are to be safeguarded. Here are question marks for the future.

My brief account is far from complete. I have not mentioned Pauli's two great articles on quantum theory, written for the *Handbuch der Physik* in 1925 and 1933. Enormous, but difficult to assess, is the influence Pauli has exerted by correspondence and discussion. In view of the discontinuous leaps by which theoretical physics develops, the stream of his scientific production has been remarkably steady. Indeed, when I compare the theoretical physicist with the mathematician I find that the former has a much harder lot. If the mathematician cannot solve a problem, he modifies it until he can solve it; no impenetrable reality limits the freedom of his imagination. So he is liable to succumb to Peer Gynt's temptation: " 'Go around,' said the crooked." Not so the physicist. He has to face the hard facts of nature. The problem of the atom must be solved straightforwardly; otherwise, no further progress is possible. Therefore, theorectical physics has affluent periods when, after persistent efforts, a new stage of theoretical interpretation has been reached, as was the case, for instance, in 1925; then there is all of a sudden plenty of highly satisfactory work for the theorist. But this alternates with stagnant periods where nothing else seems possible than to wait patiently for the slow accumulation of new facts by the experimentalists—facts which refuse to fall into any recognizable theoretical pattern. I have the greatest admiration for the courage and ingenuity with which Pauli has met this intriguing situation.

Another tension tells on the theoretical physicist—that between pure science and applications. He is a theorist and thereby committed to the contemplative life and its ideals. As Dilthey once said, "das vom Eigenleben unabhängige Glück des Sehens" is one of the most primitive and basic blessings of our existence. True, the physicist's contemplation is not a purely passive attitude—it is creative construction, but construction in symbols, resembling the creative work of the musician. On the other hand, science, since it discloses reality, is applicable to reality. Thus it is called upon to serve for the benefit and malefit of mankind. Its technical applications are used to make man's life more comfortable and more miserable, to build and to destroy. To what extent shall and can the theorist take responsibility for the practical consequences of his discoveries? What a beautiful theoretical edifice is quantum physics—and what a terrible thing is the atomic bomb! When they helped to develop the latter, did the physicists do nothing but their duty as citizens of a country engaged in total war, or did they prostitute their science? I think the experience of the last years has shown that there is little danger that the call of national duty will not be heeded by the scientists when the life of the nation is at stake, but that there is great danger indeed that in the fight for the basic values of our existence we may lose these values themselves; that the relentless pursuit of science—strange antinomy!—may imperil its very foundations in man's life. Pauli has all his life been deeply interested in philosophy. The wisdom of the Chinese sages seems to have a special appeal for him. No wonder that his sympathies are with those who are not willing to sacrifice the spiritual for the secular, and who are not willing to accept efficiency as the ultimate criterion. . . .

138.

Mathematics and logic. A brief survey serving as a preface to a review of "The Philosophy of Bertrand Russell"

The American Mathematical Monthly 53, 2—13 (1946)

1. Reduction of mathematics to set theory: the logical apparatus. The reduction of mathematics to set-theory was the achievement of the epoch of Dedekind, Frege and Cantor, roughly between 1870 and 1895. As to the basic notion of set (to which that of function is essentially equivalent) there are two conflicting views: a set is considered either a collection of things (Cantor), or synonymous with a property (attribute, predicate) of things. In the latter case "x is a member of the set γ," in formula $x \epsilon \gamma$, means nothing but that x has the property γ. The property of being red or being odd is certainly prior to the set of all red bodies or of all odd numbers. On the other hand, if with regard to a bag of potatoes or a curve drawn by pencil on paper, the property of a potato to be in the bag or of a point to lie on the curve is introduced, then the set (or a more concrete structure representing the set) is prior to the property. Whatever the epistemological significance of this distinction, it leaves the mathematician cool, since for any property γ we may speak of the set γ of all elements which have the property, and with respect to a given set γ we may speak of the property to be a member of γ. When he adopts the term *set* in preference to *property* the mathematician indicates his intention to consider co-extensive properties as identical, two properties α and β being co-extensive if every element that has the property α has also the property β and *vice versa* (set = "Begriffsumfang," Frege). Thus he will identify red and round in spite of their different "meanings" if every red body in the world happens to be round and *vice versa*.

The property π of "being prime" is represented by the propositional function $P(x)$, read "x is prime," with an argument x the range of significance of which is circumscribed by the concept "number." (The natural numbers $1, 2, 3, \cdots$ shall simply be called numbers; other numbers will be specified as rational, real, *etc.*) Indeed, understanding of the (false) proposition "6 is prime" requires that one understand what it means for *any* number x to be prime. Hence the proposition $P(6)$ arises from the propositional function $P(x)$ by the substitution $x=6$. Besides properties we must consider binary, ternary, $\cdots$, relations, represented by propositional functions of $2, 3, \cdots$ arguments.

Although the mathematician need care little whether the language of properties or sets is used, he cannot afford to ignore another distinction sometimes confused with it: the distinction between what is considered as *given* and what he *constructs* from the given by the iterated combination of certain explicitly described constructive processes. For instance, in the axiomatic setup of elementary geometry one considers as given three categories of objects—points, lines (=straight lines) and planes, and a few relations between these objects (as "point *lies on* plane"). More complicated relations must be "defined," *i.e.*, logically constructed from these primitive relations. In that intuitive theory of natural numbers (arithmetic) which is truly basic for all mathematics, even the objects are not given but constructed from the first number 1 by iterating one process, addition of 1; while all arithmetical relations are logically constructed from the one basic relation thus established: $y=x+1$, "x is followed by y." On the other hand, in a phenomenology of nature one would have to deal not only with categories of objects, as "bodies" or "events," but also with whole categories of properties which are prior to all construction, *e.g.* with the continuum of color qualities.

Logical construction of propositional functions from other propositional functions consists in the combined iterated application of a few elementary operations. Among them are the primitive logical operators $\sim$ (not), $\cap$ (and), $\cup$ (or), and the two quantifiers $(\exists x)$, "there exists an x such that," and $(\forall x)$, "for all x." For instance, from two propositional functions $S(x)$, $T(x)$ we form

$$\sim S(x), \qquad S(x) \cap T(x), \qquad S(x) \cup T(x), \qquad (\exists x)S(x), \qquad (\forall x)S(x).$$

The quantifiers carry an argument x as index and "kill" that argument in the propositional function following the quantifier, just as the substitution of an individual number, say $x=6$, does. The arithmetic operations $+$ and $\times$ are primarily applied to numbers and from there carry over to functions, while the process of integration with respect to a variable x by its very nature refers to a function $f(x)$. Just so, the logical operations $\sim$, $\cap$, $\cup$ deal primarily with propositions, while $(\exists x)$, $(\forall x)$ refer to propositional functions involving a variable x. The operator $\cup$ is primitive in the sense that the truth value (true or false) of $a \cup b$ depends only on the truth values of a and b. The same holds for $\sim$ and $\cap$. It is convenient to add the primitive operator of implication for which I use Hilbert's symbol $\rightarrow$: $a \rightarrow b$ is false if a is true and b false, but true for the three other combinations: a true, b true; a false, b true; a false, b false. Propositions without argument result when all variables have been eliminated by substitution of explicitly given individuals or by quantifiers. The construction of arithmetic propositions and propositional functions constitutes their "meanings."

In introducing properties of numbers we presuppose that we know what we mean by "*any* number"; we shall say that the *category* of the elements to which the argument in the propositional function under consideration refers must be given. We are assuming that this category is a closed realm of things existing in themselves, or, as we shall briefly say, is *existential*, when we ask with respect to a given property γ of its elements whether there *exists* an element of the property γ, with the expectation that, whatever the property γ, the question has a definite meaning and that there either exists such an element or every one of its elements has the opposite property $\sim\gamma$. In number theory or in elementary geometry we assume the numbers or the points, lines, planes, to constitute existential categories in this sense. However, we envisage only single individual properties and relations, never anything like the category of "all possible properties of numbers." This situation changes radically with the set-theoretic approach.

There we are forced to consider properties of numbers x as *objects* ξ of a new type to which the numbers stand in the relation $x \in \xi$. The proposition "6 is prime" is now looked upon as arising from the binary relation $x \in \xi$ by substituting 6 for x, and the property π of being prime for the argument ξ. The "copula" ϵ corresponds to the word "is" in the spoken sentence "6 is prime." Any propositional function $P(x)$, like "x is prime," gives rise to a corresponding property $\pi = [x]P(x)$ (the property of being prime) such that $P(x)$ is equivalent to $x \in \pi$. The operator $[x]$ which effects transition from the propositional function to the corresponding property or set kills the argument x. For the sake of uniformity of notation we shall write henceforth $\epsilon(x; \xi)$ instead of $x \in \xi$. In the same way a binary propositional function $P(x, y)$ defines a relation $\pi = [x, y]P(x, y)$ and $\epsilon(xy; \pi)$ expresses the same as $P(x, y)$, namely that x and y are in this relation π.

2. Two examples. Let our further reflections be guided by two typical examples taken from Dedekind's set-theoretic analysis of the two decisive steps in the building up of mathematics: his analysis of the sequence of numbers (in "*Was sind und was sollen die Zahlen*" 1887) and of the continuum of real numbers (in "*Stetigkeit und Irrationalzahlen*" 1872). Our critical attention will be kept more alert by using Frege's terminology of properties rather than that of sets.

I. A property α of numbers is said to be hereditary if for any number x that has the property α, the follower $x+1$ also has it. Dedekind defines: A number b is less than a if there *exists* a hereditary property that a has but b has not.

Here it is not only supposed that we know what we mean by *any* property, but we refer to the totality of all possible properties. In applying the quantifiers to properties of numbers as well as to numbers, it is absolutely imperative to look upon properties as secondary objects related to our primary objects, the numbers, by the copula relation ϵ. Heredity is even a property of properties. To be consistent we have to imagine objects of type 1 (the numbers), of type 2 (the properties of numbers), type 3, $\cdots$, and the fundamental relation $\epsilon(x_n; x_{n+1})$ connects a variable x_n of type n with one x_{n+1} of type $n+1$. Let $\mathrm{I}(\xi)$ denote the proposition that ξ is hereditary. The definition of this propositional function whose Greek argument refers to the category "properties (or sets) of numbers" is as follows:

$$\mathrm{I}(\xi) = (\forall x)\{\epsilon(x; \xi) \rightarrow \epsilon(x+1; \xi)\},$$

and Dedekind's definition of $x<y$:

$$(x < y) = (\exists \xi)\{\mathrm{I}(\xi) \cap \epsilon(y; \xi) \cap \sim \epsilon(x; \xi)\}.$$

II. Dedekind, as Eudoxos had done more than 2000 years before him, characterizes a non-negative real number α by the set of all positive rational numbers (fractions) $x > \alpha$. But for him any arbitrarily constructed (non-empty) set α of fractions (satisfying a certain condition, namely that along with any fraction b it contains every fraction $>b$) creates a corresponding real number α. The real number α is but a *façon de parler* for this set α. A set α consists of all fractions x satisfying a certain propositional function $A(x)$; $\alpha = [x]A(x)$. Let $\mathrm{I}(\xi)$ be a propositional function whose argument ξ refers to properties of fractions. The (greatest) lower bound $\gamma = [x]C(x)$ of a set of non-negative real numbers ξ can then be obtained as the *join* of all sets ξ for which $\mathrm{I}(\xi)$ holds:

$$C(x) = (\exists \xi)\{\mathrm{I}(\xi) \cap (x \in \xi)\}. \tag{1}$$

In this way Dedekind *proves* that any set of non-negative real numbers has a lower bound. Again the quantifier $(\exists \xi)$ is applied to "all possible properties of fractions."

3. Levels or no levels? The constructive and the axiomatic standpoints. But now let us pause to think what we have been doing. Properties of fractions are constructed by combined iterated application of a number of elementary logical operations $O_1, O_2, \cdots, O_h$. Let us call any property α thus obtained a *constructible* property, or a property of level 1. We can then interpret $(\exists \xi)$ in the definition (1) as "There exists a *constructible* property ξ," and this application of the quantifier is legitimate provided we admit its applicability to natural numbers. Indeed, the different manners in which one can form finite sequences with iteration out of h symbols $O_1, \cdots, O_h$ is not essentially more complicated than the possible finite sequences of one symbol 1. But the property $\gamma = [x]C(x)$ defined by (1) is certainly not identical in its meaning with any of the properties of level 1 because it is defined *in terms of the totality of all properties of level* 1. It is therefore a property of higher level 2. Nevertheless it may be *coextensive* with a property of level 1, and this is exactly what Russell's "axiom of reducibility" claims. But if the properties are constructed there is no room for an axiom here; it is a question which ought to be decided on the ground of the construction; and in our case that is a hopeless business. On the other hand, the edifice of our classical analysis collapses if we have to admit different levels of real numbers such that a real number is of level $l+1$ when it is defined in terms of the totality of real numbers of level l. If we wish to save Dedekind's proof we must abandon the constructive standpoint and assume that there is *given*, independently of all construction, an existential category "properties" or "second-type objects" (among which the constructible ones form but a small part) such that the following axiom holds, replacing the definition (1): For a given third-type object i there exists a second-type object γ such that

$$x \in \gamma \cdot \equiv \cdot (\exists \xi)\{(x \in \xi) \cap (\xi \in \mathrm{i})\}.$$

Here the arguments x and ξ refer to objects of first and second types, and $\equiv$ means "coextensive." That is a bold, an almost fantastic axiom; there is little justification for it in the real world in which we live, and none at all in the evidence on which our mind bases its constructions. With the assumption that properties constitute an existential category of given objects we return from Dedekind, who wanted to *construct* the real numbers out of the rational ones, to Eudoxos, for whom they were *given* by the points on a line; and instead of proving the existence of the lower bound on the ground of a definition of real numbers, we accept it as an axiom.

In reflecting on the source of the antinomies which had shown up at the fringe of Cantor's general set-theory, Russell realized the necessity of distinguishing between the several levels [1]. No doubt, by this fundamental insight, which he expressed somewhat loosely by his vicious circle principle: "No totality can contain members defined in terms of itself," he cured the disease but, as shown by the Dedekind example, also imperiled the very life of the patient. Classical analysis, the mathematics of real variables as we know it and as it is applied in geometry and physics, has simply no use for a continuum of real numbers of different levels. With his axiom of reducibility Russell therefore abandoned the road of logical analysis and turned from the constructive to the existential-axiomatic standpoint,—a complete volte-face.* After thus abolishing the several levels of properties, he still has the hierarchy of types: primary objects, their properties, properties of their properties, *etc.* And he finds that this alone will stop the known antinomies. But in the resulting system mathematics is no longer founded on logic, but on a sort of logician's paradise, a universe endowed with an "ultimate furniture" of rather complex structure and governed by quite a number of sweeping axioms of closure. The motives are clear, but belief in this transcendental world taxes the strength of our faith hardly less than the doctrines of the early Fathers of the Church or of the scholastic philosophers of the Middle Ages.

4. The Russell universe. Let us describe this structure in a little more detail. A few primary categories of elements are given; they are the ranges of significance for the lowest types of arguments. In a relation, each of the n ($=1$ or 2 or 3 or $\cdots$) arguments $x_1, \cdots, x_n$ refers to a certain "type" $k_1, \cdots, k_n$; the relation itself is of a type $k^* = \{k_1, \cdots, k_n\}$, determined by $k_1, \cdots, k_n$, that stands higher than any of the constituents $k_1, \cdots, k_n$. Draw a diagram representing k^* by a dot and $k_1, \cdots, k_n$ by a row of dots below k^* joined with it by straight lines, as you would depict a man k^* and his descendants in a geneological tree. Descending from k^* to its n constituents, and from them to their constituents, *etc.*, one obtains a "topological tree" in which each *end point* is associated with one of the primary categories: this diagram describes the type k^*. The fundamental relation is $\epsilon(x_1 \cdots x_n; x^*)$ where $x_1, \cdots, x_n$ refer to given types $k_1, \cdots, k_n$ and x^* to $k^* = \{k_1, \cdots, k_n\}$. Existential categories of elements are supposed to be given, one for each possible type (including the lowest, the primary types). We mentioned at the beginning with what data axiomatic elementary geometry operates, or a phenomenology of nature may have to operate; our present "Russell universe" U is seen to be incomparably richer.

* I know very well that this is at odds with Russell's own interpretation; he in the course of time became more and more inclined to visualize sets as "logical fictions." "Though," as Gödel adds, "perhaps the word fiction need not necessarily mean that these things do not exist, but only that we have no direct perceptions of them." In the second edition of Volume I of the Principia Mathematica, an attempt is made to prove independently of the axiom of reducibility that at least all levels of *natural numbers* can be reduced to the five lowest. But as Gödel observes, the proof is far from being conclusive. [2]

We give a few of the more obvious axioms on which its theory is to be erected. The universal normal form for propositional functions involving n variables $x_1, \cdots, x_n$ of given types $k_1, \cdots, k_n$ is $\epsilon(x_1 \cdots x_n; a^*)$. Indeed, such a relation is itself an element a^* of type $k^* = \{k_1, \cdots, k_n\}$. The relation of identity $(x=y)$ between elements of type k is itself an element of type $\{k, k\}$; call it $I = I_k$. Similarly, let $E = E_{k_1 \cdots k_n}$ be the ϵ relation with its $n+1$ arguments of types $k_1, \cdots, k_n, k^* = \{k_1, \cdots, k_n\}$. The existence of these special elements must be stipulated explicitly:

Axiom 1. There is an element $I = I_k$ of type $\{k, k\}$ such that $\epsilon(xy; I)$ holds if and only if the elements x, y of type k are identical. There is an element $E = E_{k_1 \cdots k_n}$ of type $K = \{k_1, \cdots, k_n, k^*\}$ such that $\epsilon(x_1 \cdots x_n x^*; E)$ is coextensive with $\epsilon(x_1 \cdots x_n; x^*)$, the variables $x_1, \cdots, x_n$ and x^* ranging over their respective categories $k_1, \cdots, k_n$ and $k^* = \{k_1, \cdots, k_n\}$.

The composite property "red or round" is no longer constructed from the descriptive properties "red," "round," but belongs with them to the category of properties given prior to all construction. Its existence must be guaranteed by one of the simpler axioms of closure.

Axiom 2. Given an element a^* and an element b^* of type $k^* = \{k_1, \cdots, k_n\}$ there exists an element c^* of the same type such that $\epsilon(x_1 \cdots x_n; c^*)$ is coextensive with

$$\epsilon(x_1 \cdots x_n; a^*) \cup \epsilon(x_1 \cdots x_n; b^*)$$

(each x_i varying over its category k_i).

Substitution of a definite element b for a variable is taken care of by the

Axiom 3. Given an element a^* of type $k^* = \{k_1, \cdots, k_n\}$ and an element b_n of type k_n, there exists an element c of type $k = \{k_1, \cdots, k_{n-1}\}$ such that $\epsilon(x_1 \cdots x_{n-1}; c)$ is coextensive with $\epsilon(x_1 \cdots x_{n-1} b_n; a^*)$ (if $x_1, \cdots, x_{n-1}$ vary over the categories $k_1, \cdots, k_{n-1}$).

Elimination of a variable x_n by the corresponding quantifier $(\exists x_n)$ changes a relation a^* of type k^* into a relation a of type k:

Axiom 4. Given an element a^* of type $k^* = \{k_1, \cdots, k_n\}$, there exists an element a of type $k = \{k_1, \cdots, k_{n-1}\}$ such that $\epsilon(x_1 \cdots x_{n-1}; a)$ is coextensive with $(\exists x_n)\epsilon(x_1 \cdots x_{n-1} x_n; a^*)$.

These are only a few typical axioms that indicate the direction. The reader must not look for them in the *Principia Mathematica*, which are conceived in a different style. But our system U embodies the same ideas in a form that seems to me both natural in itself and advantageous for a comparison with other systems W, Z presently to be discussed.

Our axioms serve as a basis for deduction in the same way as, for instance, the axioms of geometry; deduction takes place by that sort of logic on which one is used to rely in geometry or analysis, including the free use of "there exists" and "all" with reference to any fixed type in the hierarchy of types and the elements of the corresponding categories. While these categories, as well as the basic relation ϵ, are considered as undefined, the logical terms like "not" $\sim$, "if then" $\rightarrow$, "there is" $(\exists x)$, *etc.*, have to be understood in their meaning and do not form part of the axiomatic system: the formalism of symbolic logic is merely used for the sake of conciseness.

If the *Principia Mathematica* set out to base mathematics on pure logic the result, as we now see, is quite different: an axiomatic world system has taken the place of logic. Its very structure, the hierarchy of types, cannot be described without resort to the intuitive concept of iteration. To develop, in Dedekind-Frege fashion, a theory of natural numbers from this system is therefore an enterprise of doubtful value.

5. A constructive compromise. Realizing the highly transcendental character of the axiomatic universe from which this system deduces mathematics, one wonders whether it is not possible to stick, in spite of everything, to the constructive standpoint, which seems so much more natural to the mathematician. We accept the hierarchy of types; but we assume only one category of primary objects, the numbers; and one basic binary relation between numbers, namely "x is followed by y." All other relations of the various types are explicitly constructed, the quantifiers $(\exists x)$ and $(\forall x)$ being applied only to numbers and not to arguments of higher type. No axioms are postulated. What we can get in this way constitutes the ground level, or level 1. One could build over it a second level containing relations which are constructed by applying the quantifiers to the totality of relations of this or that type constructible on the first level, and proceed in the same manner from the second to a third level, *etc.* One would obtain a "ramified hierarchy" of types and levels. But in this way, as we have said before, nothing resembling our classical Calculus will result. The temptation to pass beyond the first level of construction must be resisted; instead, one should try to make the range of constructible relations as wide as possible *by enlarging the stock of basic operations*. It is *a priori* clear that *iteration* in some form must find a place among these irreducible principles of construction—contrary to the Dedekind-Frege program.

We begin again at the beginning. Let $R(x, \xi)$ be a binary propositional function the two arguments x, ξ of which are of types k, κ respectively; for instance the relation "x is less than ξ" between numbers. We can then speak of the property of a number "to be less than ξ" (or of the set of all numbers less than ξ). This is obviously a property depending on ξ. In general terms we can form $[x]R(x, \xi) = r^*(\xi)$, which is an element of $\{k\}$ depending on ξ such that $R(x, \xi)$ is coextensive with $\epsilon(x; r^*(\xi))$. If in addition to $R(x, \xi)$ we have a propositional function $S(x^*)$ whose argument is of type $k^* = \{k\}$, we can form $T(\xi) = S(r^*(\xi))$, or more explicitly,

$$T(\xi) = S([x]R(x, \xi)).$$

This is the *process of substitution* generating $T(\xi)$ from $R(x, \xi)$ and $S(x^*)$.

Take now the particular situation that $\kappa = \{k\}$. Then argument and value of the function $r^*(\xi) = [x]R(x, \xi)$ are of the same type κ, and whenever that happens *iteration* becomes possible. Thus we define by complete induction a relation $T(n, \xi)$ in which the argument n refers to the primary category of numbers, as follows:

$$T(1, \xi) = S(\xi);$$
$$T(n + 1, \xi) = T(n, r^*(\xi)) \qquad (n = 1, 2, \cdots).$$

Adding the operation of substitution and iteration, as illustrated by these examples, to the other elementary logic operations, but without applying the quantifiers to anything else than numbers, the writer was able (in *Das Kontinuum*, 1918) to build up in a purely constructive way, and without axioms, a fair part of classical analysis, including for instance Cauchy's criterion of convergence for infinite sequences of real numbers.* In this system iteration plays the role which in set-theory was played by the uninhibited application of quantifiers. Our construction honestly draws the consequences of Russell's logical insight into the tower of levels, which Dedekind had ignored inadvertently and Russell himself, afraid of its consequences, razed to the ground by his axiom of reducibility. Considering their common origin the axiomatic system U as outlined in section 4 and this constructive approach are remarkably different. But even here we have adhered to the belief that "there is" and "all" make sense when applied to natural numbers: in addition to logic we rely on this existential creed and the idea of iteration.

6. Brouwer's intuitionistic mathematics. Essentially more radical and a further step toward pure constructivism is Brouwer's intuitionistic mathematics [3]. Brouwer made it clear, as I think beyond any doubt, that there is no evidence supporting the belief in the existential character of the totality of all natural numbers, and hence the principle of excluded middle in the form "Either there is a number of the given property γ, or all numbers have the property $\sim\gamma$" is without foundation. The first part of the sentence is an *abstract* from some statement of fact in the form: The number thus and thus constructed has the property γ. The second part is one of *hypothetic* generality, asserting something only if $\cdots$; *viz.*, *if* you are actually given a number, you may be sure that it has the property $\sim\gamma$. The sequence of numbers which grows beyond any stage already reached by passing to the next number, is a manifold of possibilities open towards infinity; it remains forever in the status of creation, but is not a closed realm of things existing in themselves. That we blindly converted one into the other is the true source of our difficulties, including the antinomies—a source of more fundamental nature than Russell's vicious circle principle indicated. Brouwer opened our eyes and made us see how far classical mathematics, nourished by a belief in the "absolute" that transcends all human possibilities of realization, goes beyond such statements as can claim real meaning and truth founded on evidence. According to his view and reading of history, classical logic was abstracted from the mathematics of finite sets and

* But of course the theorem of the lower bound of an arbitrary set of non-negative real numbers could not be upheld.

their subsets. (The word finite is here to be taken in the precise sense that the members of such a set are explicitly exhibited one by one.) Forgetful of this limited origin, one afterwards mistook that logic for something above and prior to all mathematics, and finally applied it, without justification, to the mathematics of infinite sets. This is the Fall and original sin of set-theory, for which it is justly punished by the antinomies. Not that such contradictions showed up is surprising, but that they showed up at such a late stage of the game!

Thanks to the notion of "Wahlfolge," that is a sequence *in statu nascendi* in which one number after the other is *freely chosen* rather than determined by law, Brouwer's treatment of real variables is in the closest harmony with the intuitive nature of the continuum; this is one of the most attractive features of his theory. But on the whole, Brouwer's mathematics is less simple and much more limited in power than our familiar "existential" mathematics. It is for this reason that the vast majority of mathematicians hesitate to go along with his radical reform.

7. The Zermelo brand of axiomatics; sets and classes. From this excursion to the left wing of the "constructionists" we return to the universe U with its hierarchy of types. Once one has committed oneself to the existential or axiomatic viewpoint, can one not go forth in the same direction and even erase all differences of types, taking only such precautions as are absolutely necessary to avoid the known contradictions? This is what Zermelo did in his *Untersuchungen über die Grundlagen der Mengenlehre*, 1908 [4]. His axioms deal with only one (existential) category of objects called elements or sets, and one basic relation $x \in y$, "x is member of y." But he is forced to give up the principle that any well-defined property γ determines an element c such that $x \in c$ whenever the element x has the property γ and *vice versa*. Properties are used by him merely to cut out subsets *from a given set*. Hence his axiom of selection: "Given a well-defined property γ and an element a, there is an element a' such that $x \in a'$ if and only if x is member of a and has at the same time the property γ." The notion of a well-defined property which enters into it is somewhat vague. But we know that we can make it precise by constructing properties by iterated combined application of some elementary constructive processes. Instead of saying that x has the property γ, let us say that x is a member of the class γ, $x \in \gamma$. We thus distinguish between elements or sets on the one hand, classes on the other, and formulate the axioms in terms of two undefined categories of objects, elements and classes. Since we postulate that two elements a, b are identical in case $x \in a$ and $x \in b$ are coextensive, and since each element a is associated with the class α of all elements x satisfying the condition $x \in a$, we are justified in identifying a with that class α. Then *every element is a class* and the axioms deal with one undefined fundamental relation $x \in \xi$, "the element x is member of the class ξ," which has absorbed Zermelo's relation $x \in y$ between elements. The principles for the construction of properties are replaced by corresponding axioms for classes; *e.g.*, given two classes α and β, there exists a class γ

such that the statement $(x \in \alpha) \cup (x \in \beta)$ about an arbitrary element x is coextensive with $x \in \gamma$.

Since the axiom of selection can only generate smaller sets out of a given set, we need some vehicle that carries us in the opposite direction. Therefore two axioms are added guaranteeing the existence of the set of all subsets of a given set and the join of a given set of sets. It is essential that they be limited to sets = elements and do not apply to classes.

With the introduction of classes, which is due to Fraenkel, von Neumann, Bernays, and others, the axioms assume the same self-sustaining character as, for instance, the axioms of geometry; no longer do such general notions as "any well-defined property" penetrate into the axiomatic system from the outside. A complete table of axioms for this system, Z as we shall call it, is to be found on the first pages of Gödel's monograph, *Consistency of the Continuum Hypothesis*, Annals of Mathematics Studies No. 3, Princeton, 1940 [5]. Even before the turn of the century Cantor himself had moved in the same direction by distinguishing "consistent classes" = sets and inconsistent classes [6]. Not the hierarchy of types, but the non-admission among the decent "sets" of such classes as are too "big," averts here the disaster of the familiar antinomies.

One might object to a system like Z on the ground that it does not rest on a real insight into the causes of the antinomies but patches up Cantor's original conception by a minimum of concessions necessary to avoid the known contradictions. Indeed, we have no assurance of the consistency of Z, except from the empirical fact that so far no contradictions have resulted from it. But we are in no better position toward Russell's universe U. And Z has its great advantages over U: it is of an essentially simpler structure and seems to be the most adequate basis for what is actually done in present-day mathematics. In particular, the "existential" Dedekind-Frege theory of numbers can be derived from it (Zermelo), and Gödel was able to show (*l.c.*) that Zermelo's far-reaching axiom of choice in a very sharp form is consistent with the other axioms of Z

8. Complete formalization and the question of consistency. Pessimistic conclusions. A new turn in the axiomatization of mathematics of paramount importance was inaugurated by Hilbert's "Beweistheorie" (since 1922) [7]. Hilbert sets out *to prove* (not the truth, but) *the consistency of mathematics*. He realizes that to that end mathematics and logic must first be completely formalized: all statements are to be replaced by formulas in which now also the logical operators $\sim$, $\cap$, $(\exists x)$ *etc.*, appear as undefined symbols. Thus formalized logic is absorbed into formalized mathematics.* The formulas have no meaning. A mathematical demonstration is a concrete sequence of formulas in which a formula is derived from the preceding ones according to certain rules comprehensible without recourse to any meaning of the formulas—just as in a game of chess each position is derived from the preceding one by a move obeying certain

* In this regard the ground was well prepared for Hilbert by the Principia Mathematica. For the sake of comparison I mention one other completely formalized system, that of Quine. [8]

rules. *Consistency*, the fact that no such game of deduction may end with the formula $\sim(1=1)$, must be proved by intuitive reasoning about the formulas which rests on evidence rather than on axioms, and respects throughout the limits of evidence as disclosed by Brouwer. But in this thinking about demonstrations, in following a hypothetic sequence of formulas leading up to the end formula $\sim(1=1)$ our mind cannot help using that type of evidence in which the possibility of iteration is founded. In the axiomatization of mathematics Hilbert is forced to proceed with more restraint than Zermelo: if he is too liberal with his axioms he will lose all chance of ever proving their consistency; he is guided by an at least vaguely preconceived plan for such a proof. It is for this reason that he finds it advisable, for instance, to distinguish various levels of variables.

Hilbert's formulas are concrete structures consisting of concrete symbols; the order in which the symbols follow one another in a formula, and also their identity in the same or different formulas, must be recognizable irrespective of little variations in their execution. Handling these symbols, we move on the same level of understanding as guides our daily life in our relationship to such tools as hammer or table or chair. Hilbert sees in it the most important prelogical foundation of mathematics, in fact of all science. But in addition his axioms of mathematics and the intuition of iteration of which the metamathematical non-axiomatic reasoning about mathematics makes use are other extralogical ingredients of his system.

Our brief survey may be summarized in a little diagram in which the constructive tendency increases toward the left, the axiomatic toward the right, and also the relative "depth" of the foundations is indicated. Frege and after him Russell had hoped (1) to develop the theory of natural numbers on a sure basis without recourse to the intuitive idea of indefinite iteration, and (2) to make mathematics a part of logic. We have now seen that none of the systems

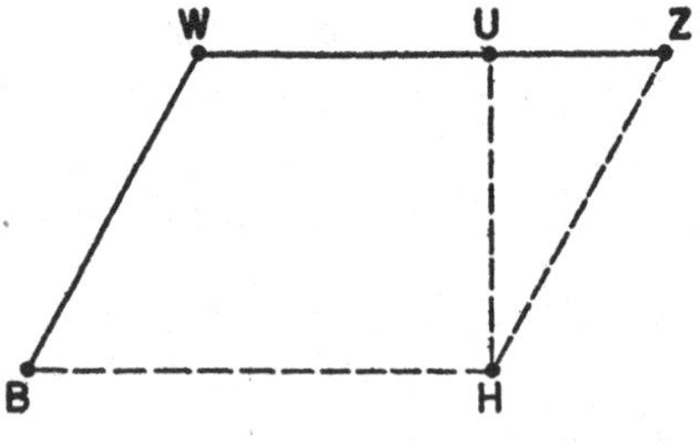

B = Brouwer
H = Hilbert
W = Weyl

discussed gives any hope of accomplishing (2), the subjugation of mathematics by logic. In *U* and *Z* elaborate systems of axioms form the basis, in *W*, *B*, and to a lesser degree also in *U*, the intuitive idea of iteration is indispensable. The extra-logical foundations of Hilbert's theory have just been described. The only system which in a sense can claim to reach the goal (1) is *Z*. But even there the theory of numbers does not rest on logic alone, but on a highly transcendental system of axioms (the belief in whose consistency is supported by empirical facts, but not by reasons). Poincaré has proved right in his defense of mathematical induction as an indispensable and irreducible tool of mathematical reasoning.

It is likely that all mathematicians ultimately would have accepted Hilbert's approach had he been able to carry it out successfully. The first steps were inspiring and promising. But then Gödel dealt it a terrific blow (1931), from which it has not yet recovered. Gödel enumerated the symbols, formulas, and sequences of formulas in Hilbert's formalism in a certain way, and thus transformed the assertion of consistency into an arithmetic proposition. He could show that this proposition can neither be proved nor disproved within the formalism [9]. This can mean only two things: either the reasoning by which a proof of consistency is given must contain some argument that has no formal counterpart within the system, *i.e.*, we have not succeeded in completely formalizing the procedure of mathematical induction; or hope for a strictly "finitistic" proof of consistency must be given up altogether. When G. Gentzen finally succeeded in proving the consistency of arithmetic [10] he trespassed those limits indeed by claiming as evident a type of reasoning that penetrates into Cantor's "second class of ordinal numbers."

From this history one thing should be clear: we are less certain than ever about the ultimate foundations of (logic and) mathematics. Like everybody and everything in the world today, we have our "crisis." We have had it for nearly fifty years. Outwardly it does not seem to hamper our daily work, and yet I for one confess that it has had a considerable practical influence on my mathematical life: it directed my interests to fields I considered relatively "safe," and has been a constant drain on the enthusiasm and determination with which I pursued my research work. This experience is probably shared by other mathematicians who are not indifferent to what their scientific endeavors mean in the context of man's whole caring and knowing, suffering and creative existence in the world.

REFERENCES

1. Mathematical Logic as Based on the Theory of Type, Am. Jour. Math., v. 30, 1908, pp. 222–262. B. Russell and A. N. Whitehead, Principia Mathematica, 3 vols., Cambridge, 1910–13; 2nd ed. of vol. I, 1935.

2. Gödel, Russell's Mathematical Logic, in The Philosophy of Bertrand Russell, pp. 127, 145, 146.

3. Brouwer's thesis Over de grondslagen der wiskunde appeared in 1907. For a list of his papers on the subject see A. Church's Bibliography of Symbolic Logic, Jour. Symb. Logic, v. 1, 1936, pp. 121–218.

4. Math. Ann., v. 65, 1908, pp. 261–281.

5. See also P. Bernays, Jour. Symb. Logic, v. 2, 1937, pp. 65–77 and the references given there.

6. Cf. G. Cantor, Gesammelte Abhandlungen, ed. E. Zermelo, 1932, pp. 443–451 (correspondence Cantor-Dedekind).

7. Vol. 3 of Hilbert's Collected Papers, 1937, and Hilbert-Bernays, Grundlagen der Mathematik, 2 vols. 1934 and 1939.

8. Am. Math. Monthly, v. 44, 1937, pp. 70–80.

9. Monatsh. Math. Phys., v. 38, 1931, pp. 173–198.

10. Math. Ann., v. 112, 1936, pp. 493–565.

139.

Comment on a paper by Levinson

American Journal of Mathematics 68, 7—12 (1946)

The following lines, though adding little to the substance of Professor Levinson's paper, may help to shed some light on his interesting result. This result is not limited to linear equations; it holds for such perturbations as are majorized by linear perturbations. Levinson passes from a given solution $\mathfrak{x}$ of the complete equation to a corresponding one $\mathfrak{z}$ of the approximate equation, $\mathfrak{x} \to \mathfrak{z}$. We add the inverse process $\mathfrak{z} \to \mathfrak{x}$: transition from the unperturbed to the perturbed phenomenon.

A system of differential equations for n functions $x_i(t)$ of the real variable t may be looked upon as a single differential equation for the vector (column) $\mathfrak{x}$ with the components x_i, and assuming the right member to consist of a linear part $A\mathfrak{x}$ with a constant coefficient matrix $A = \| a_{ik} \|$ and a perturbation $\mathfrak{v}$; we may write the equation in the form

$$(1) \qquad d\mathfrak{x}/dt = A\mathfrak{x} + \mathfrak{v}, \qquad \mathfrak{v}(t) = \mathfrak{f}(t, \mathfrak{x}(t)),$$

$\mathfrak{f}(t, \mathfrak{x})$ being a given vector function of t and a variable vector $\mathfrak{x}$. Form $U(t) = e^{At}$, the one-parameter group of linear transformations generated by the infinitesimal A:

$$dU/dt = AU = UA, \quad U(0) = \text{unit matrix } E; \quad U(t-\tau) = U(t)U^{-1}(\tau),$$

and define the absolute values $\| \mathfrak{x} \|$, $\| A \|$ of vectors and matrices by

$$\| \mathfrak{x} \|^2 = \Sigma \mid x_i \mid^2, \qquad \| A \|^2 = \Sigma \mid a_{ik} \mid^2.$$

The equation (1) may be written

$$(d/dt)(U^{-1}\mathfrak{x}) = U^{-1}\mathfrak{v}$$

("variation of constants"), whence follows

$$(2) \qquad \mathfrak{x}(t) = \mathfrak{z}(t) + U(t) \int_0^t U^{-1}(\tau)\mathfrak{v}(\tau)\, d\tau,$$

$$(3) \qquad \mathfrak{z}(t) = U(t)\mathfrak{a} \qquad (\mathfrak{a} = \text{const.})$$

---

* Received December 5, 1945.

being a solution of

$$d\mathfrak{z}/dt = A\mathfrak{z}. \tag{4}$$

It can also be verified directly that if $\mathfrak{x}$ is a solution of (1), then the $\mathfrak{z}$ defined by (2) or

$$\mathfrak{z}(t) = \mathfrak{x}(t) - U(t)\int_0^t U^{-1}(\tau)\mathfrak{v}(\tau)\,d\tau = \mathfrak{x}(t) - \int_0^t U(t-\tau)\mathfrak{v}(\tau)\,d\tau \tag{5}$$

is a solution of $d\mathfrak{z}/dt = A\mathfrak{z}$. This is, in general form, the lemma from which Mr. Levinson's investigation starts.

Let us now assume that

$$\| \mathfrak{f}(t,\mathfrak{x}) \| \leq \| \mathfrak{x} \| \cdot g(t). \tag{I}$$

If the perturbation is linear, $\mathfrak{f}(t,\mathfrak{x}) = G(t)\mathfrak{x}$, then (I) holds with $g(t) = \| G(t) \|$.

LEMMA. *Suppose*

$$\| \mathfrak{z}(t) \| \leq a, \qquad \| U(t) \| \leq c \quad \textit{for} \quad 0 \leq t \leq t_1$$

and set

$$c\int_0^t g(\tau)\,d\tau = h(t).$$

Then the equation (5) *implies*

$$\| \mathfrak{x}(t) \| \leq a \cdot e^{h(t)} \qquad (0 \leq t \leq t_1).$$

Proof. For $x(t) = \| \mathfrak{x}(t) \|$ one obtains, as a consequence of assumption (I), the integral inequality

$$x(t) \leq a + c\int_0^t x(\tau)g(\tau)\,d\tau = a + \int_0^t x(\tau)\cdot dh(\tau);$$

hence for $y(t) = x(t) - a\cdot e^{h(t)}$ the inequality

$$y(t) \leq \int_0^t y(\tau)\cdot dh(\tau).$$

Choose a constant b such that $y(t) \leq b$ for $0 \leq t \leq t_1$. The inequalities

$$y(t) \leq b\cdot (h(t))^n/n! \qquad (n = 0, 1, 2, \cdots;\ 0 \leq t \leq t_1)$$

then follow one after the other by induction. In the limit for $n \to \infty$ one gets $y(t) \leq 0$.

Let us now introduce Levinson's two hypotheses:

(II) convergence of $q = \int_0^\infty g(\tau)\,d\tau$;

(III) boundedness of $U(t)$ for $t \to \infty$, $\| U(t) \| \leq c$ for $t \geq 0$.

Then of course every solution $\mathfrak{z}$, (3), of (4) is bounded, $\| \mathfrak{z}(t) \| \leq a = c \| \mathfrak{a} \|$. Our lemma yields at once

$$\| \mathfrak{x}(t) \| \leq a^* = a \cdot e^{cq} \quad (\text{for } t \geq 0), \tag{6}$$

and hence the

THEOREM. *Under the hypotheses* (I), (II), (III) *every solution of* (1) *is bounded for* $t \geq 0$.

Instead of (6) Levinson uses the rougher estimate

$$\| \mathfrak{x}(t) \| \leq a/(1-\theta) \qquad (0 \leq t \leq t_1;\ \theta = h(t_1)),$$

which he establishes in a more indirect way. Its validity is limited to intervals $0t_1$ for which $\theta < 1$. Nonetheless, the theorem can be derived from it: one simply replaces the lower limit of integration 0 by a t_0 so chosen that

$$\theta = c \int_{t_0}^\infty g(\tau)\,d\tau < 1.$$

In a suitable coordinate system $U(t)$ breaks up into blocks (elementary divisors) of the form

$$\left\| \begin{matrix} e^{\lambda t} \cdot t_0, & 0, & \cdots & 0 \\ e^{\lambda t} \cdot t_1, & e^{\lambda t} \cdot t_0, & \cdots & 0 \\ \cdot & \cdot & \cdots & \cdot \\ \cdot & \cdot & \cdots & \cdot \\ e^{\lambda t} \cdot t_{m-1}, & e^{\lambda t} \cdot t_{m-2}, & \cdots & e^{\lambda t} \cdot t_0 \end{matrix} \right\| \qquad [\ t_i = (1/i!)\,t^i\ ].$$

Hence hypothesis (III) implies that either $\Re\lambda < 0$ or $\Re\lambda = 0$; but in the latter case the block must consist of one row only $(m = 1)$. After uniting all elementary divisors of the first and second kind into V_1, V_2 respectively, one gets a decomposition $U = \left\| \begin{matrix} V_1 & 0 \\ 0 & V_2 \end{matrix} \right\|$ in which $V_1(t) \to 0$ for $t \to \infty$ and $V_2(t)$ is bounded not only for $t \to +\infty$, but also for $t \to -\infty$. Using the corresponding decomposition of the unit matrix

$$E = \left\| \begin{matrix} E_1 & 0 \\ 0 & 0 \end{matrix} \right\| + \left\| \begin{matrix} 0 & 0 \\ 0 & E_2 \end{matrix} \right\| = I_1 + I_2$$

into two complementary idempotents I_1, I_2, one can write

$$U = UI_1 + UI_2 = U_1 + U_2.$$

The idempotents I_1 and I_2 commute with A (and hence with U). In this form the description is independent of the coordinate system.

$$U_1(t) \to 0 \text{ for } t \to \infty, \qquad \| U_2(t) \| \leq c_2 \text{ for } t \leq 0.$$

Every vector $\mathfrak{x}$ may be split according to $I_1\mathfrak{x} + I_2\mathfrak{x} = \mathfrak{x}_1 + \mathfrak{x}_2$. For a solution $\mathfrak{z}$ of (4) the first part $\mathfrak{z}_1$ is damped, the second $\mathfrak{z}_2$ a pure sinusoidal oscillation.

Let us now decompose U in (5) into $U_1 + U_2$ and in the second part replace the lower limit of integration 0 by ∞:

$$\mathfrak{x}(t) - \int_0^t U_1(t-\tau)\mathfrak{v}(\tau)\,d\tau + \int_t^\infty U_2(t-\tau)\mathfrak{v}(\tau)\,d\tau = \mathfrak{z}(t). \tag{7}$$

Since $\mathfrak{x}(t)$ is bounded for $t \geq 0$, (6), and

$$\| \mathfrak{v}(\tau) \| = \| \mathfrak{f}(\tau, \mathfrak{x}(\tau)) \| \leq \| \mathfrak{x}(\tau) \| \, g(\tau) \leq a^* g(\tau) \text{ for } \tau \geq 0,$$
$$\| U_2(t-\tau)\mathfrak{v}(\tau) \| \leq c_2 a^* g(\tau) \text{ for } \tau \geq t \geq 0,$$

the integral extending to infinity converges by virtue of hypothesis (III). Observe that

$$U_2(t-\tau) = U(t-\tau)I_2 = U(t)U(-\tau)I_2 = U(t)U_2(-\tau).$$

The fact that $\mathfrak{z}(t)$, (7), satisfies (4) provided $\mathfrak{x}$ is a solution of (1) is not affected by the shift of the limit of integration in the U_2-part, as one sees either by direct verification or (as Levinson does) by the simple transformation

$$U(t) \int_0^t U_2(-\tau)\mathfrak{v}(\tau)\,d\tau = U(t)\mathfrak{a}_2 - U(t) \int_t^\infty U_2(-\tau)\mathfrak{v}(\tau)\,d\tau$$

where $\mathfrak{a}_2 = \int_0^\infty U_2(-\tau)\mathfrak{v}(\tau)\,d\tau$. Therefore formula (7) associates with every solution $\mathfrak{x}$ of (1) a unique solution $\mathfrak{z}$ of (4), $\mathfrak{x} \to \mathfrak{z}$. Multiplication by I_1, I_2 splits (7) into the two equations

$$\mathfrak{x}_1(t) - \int_0^t U_1(t-\tau)\mathfrak{v}(\tau)\,d\tau = \mathfrak{z}_1(t);$$
$$\mathfrak{x}_2(t) + \int_t^\infty U_2(t-\tau)\mathfrak{v}(\tau)\,d\tau = \mathfrak{z}_2(t),$$

and thus one arrives at

LEVINSON'S THEOREM. *For every solution $\mathfrak{x}$ of* (1) *the part $\mathfrak{x}_1(t)$ tends to* 0 *with $t \to \infty$, and there exists a solution $\mathfrak{z}$ of* (4) *such that*

(8) $\mathfrak{x}_1(t) - \mathfrak{z}_1(t) = 0$ *for* $t = 0$, $\mathfrak{x}_2(t) - \mathfrak{z}_2(t) \to 0$ *for* $t \to \infty$.

Only the statement of the first sentence remains to be proved:

$$\int_0^t U_1(t-\tau)\mathfrak{v}(\tau)\,d\tau \to 0 \text{ for } t \to \infty.$$

Split $\int_0^t$ into $\int_0^{t/2} + \int_{t/2}^t$ and, ϵ being a given positive number, assume that

$$\| U_1(t) \| \leq c_1 \text{ for } t \geq 0; \qquad \int_0^\infty g(\tau)\,d\tau = q;$$

$$\| U_1(t) \| \leq \epsilon \text{ for } t \geq t_\epsilon \ (\geq 0), \qquad \int_{t_\epsilon}^\infty g(\tau)\,d\tau \leq \epsilon.$$

Then as soon as $t \geq 2t_\epsilon$, the absolute value of the first part is less than or equal to $\epsilon \int_0^{t/2} \| \mathfrak{v}(\tau) \|\,d\tau \leq a^* q\epsilon$, that of the second part less than or equal to $c_1 \int_{t/2}^\infty \| \mathfrak{v}(\tau) \|\,d\tau \leq a^* c_1 \epsilon$.

In studying the inverse process, the transition $\mathfrak{z} \to \mathfrak{x}$, we replace (I) by the stronger Lipschitz condition

(I*) $\mathfrak{f}(t, 0) = 0, \qquad \| \mathfrak{f}(t, \mathfrak{x}) - \mathfrak{f}(t, \mathfrak{x}^*) \| \leq \| \mathfrak{x} - \mathfrak{x}^* \| \cdot g(t),$

which again is fulfilled in the linear case $\mathfrak{f}(t, \mathfrak{x}) = G(t)\mathfrak{x}$. For a given $\mathfrak{z}$ the integral equation (5) for $\mathfrak{x}$ may then be solved by successive approximations just as in the classical case $A = 0$. Adding Levinson's hypotheses (II) and (III) we turn at once to the more complicated integral equation (7) and show that it has a unique solution $\mathfrak{x}(t)$ provided

$$\theta = (c_1 + c_2) \int_0^\infty g(\tau)\,d\tau < 1.$$

Proof. Define the successive approximations $\mathfrak{x}_n$ by $\mathfrak{x}_0 = \mathfrak{z}$,

(9) $$\mathfrak{x}_{n+1}(t) = \mathfrak{z}(t) + \int_0^t U_1(t-\tau)\mathfrak{v}_n(\tau)\,d\tau - \int_t^\infty U_2(t-\tau)\mathfrak{v}_n(\tau)\,d\tau$$

$(n = 0, 1, 2, \cdots)$ where $\mathfrak{v}_n(\tau) = \mathfrak{f}(\tau, \mathfrak{x}_n(\tau))$. Since (I*) implies

$$\| \mathfrak{v}_n(t) - \mathfrak{v}_{n-1}(t) \| \leq \| \mathfrak{x}_n(t) - \mathfrak{x}_{n-1}(t) \| \cdot g(t)$$

(even for $n=0$ if one sets $\mathfrak{x}_{-1}(t)=0$), the inequality $\|\mathfrak{x}_n(t)-\mathfrak{x}_{n-1}(t)\| \leq a_n$ leads, by means of (9), to

$$\|\mathfrak{x}_{n+1}(t)-\mathfrak{x}_n(t)\| \leq \theta a_n.$$

Hence if $\|\mathfrak{z}(t)\| \leq a$ for $t \geq 0$, then

$$\|\mathfrak{x}_n(t)-\mathfrak{x}_{n-1}(t)\| \leq a\theta^n,$$

and the series $\mathfrak{x}_0+(\mathfrak{x}_1-\mathfrak{x}_0)+(\mathfrak{x}_2-\mathfrak{x}_1)+\cdots$ converges at least as well as the geometric series with the quotient θ. Uniqueness is established by an argument of the same type. I like to arrange it as follows. Suppose a function $\mathfrak{x}(t)$ satisfying the equation (7) is given. Then one finds for the difference $\Delta_n\mathfrak{x}=\mathfrak{x}-\mathfrak{x}_{n-1}$:

$$\Delta_{n+1}\mathfrak{x}(t)=\int_0^t U_1(t-\tau)\cdot\Delta_n\mathfrak{v}(\tau)\,d\tau-\int_t^\infty U_2(t-\tau)\cdot\Delta_n\mathfrak{v}(\tau)\,d\tau$$

where $\Delta_n\mathfrak{v}=\mathfrak{v}-\mathfrak{v}_{n-1}$. If $\|\mathfrak{x}(t)\| \leq a^*$ the last equation yields by induction

$$\|\Delta_n\mathfrak{x}(t)\| \leq a^*\theta^n \qquad (n=0,1,2,\cdots)$$

which proves the uniform convergence of $\mathfrak{x}_n(t)$ toward the given $\mathfrak{x}(t)$. We summarize:

THEOREM. *Under the hypotheses* (I*), (II), (III) *we effect the splitting* $U=U_1+U_2$ *described above. Let*

$$\|U_1(t)\| \leq c_1 \text{ for } t \geq 0, \qquad \|U_2(t)\| \leq c_2 \text{ for } t \leq 0.$$

If t_0 *be so near to infinity that*

$$\theta=(c_1+c_2)\int_{t_0}^\infty g(\tau)\,d\tau<1$$

then we have established a one-to-one correspondence between the solutions $\mathfrak{x}$ *of* (1) *and* $\mathfrak{z}$ *of* (4) *such that*

$$\mathfrak{x}_1(t)-\mathfrak{z}_1(t)=0 \text{ for } t=t_0, \qquad \mathfrak{x}_2(t)-\mathfrak{z}_2(t)\to 0 \text{ for } t\to\infty.$$

Contribution 140 "A remark on the coupling of gravitation and electron (Actas de la Academia Nacional de Ciencias Exactas, Físicas y Naturales de Lima 11, 1—17 (1948))" was not taken in, as, due to typographical errors, it is incomprehensible.

141.

A remark on the coupling of gravitation and electron

The Physical Review 77, 699—701 (1950) (abgekürzte Fassung von 140)

In Einstein's theory of gravitation one may consider either the metric potentials g_{ik}, or both the metric potentials and the components of the affine connection, as independent quantities (metric *vs.* mixed theory) in the coupling of gravitational and electronic fields. This makes a difference for the ensuing equations, but one that can be easily straightened out by a slight change of the Lagrangian; see the Theorem at the end of the paper.

ON the level of classical field theory the welding together of Maxwell's electromagnetic, Dirac's electronic and Einstein's gravitational fields (**F**, **E** and **G**) affords no serious difficulty—whatever interpretation quantum physics may impose upon the resulting field equations, and however unsatisfactory a Lagrangian may be that is made up by addition of four separate parts, one for **F**, two for **E**, and one for **G**. The formulation of Dirac's theory of the electron in the frame of general relativity has to its credit one feature which should be appreciated even by the atomic physicist who feels safe in ignoring the role of gravitation in the building-up of the elementary particles: its explanation of the quantum mechanical principle of "gauge invariance" that connects Dirac's ψ with the electromagnetic potentials.

In contrast to Einstein's original "metric" conception in terms of the g_{pq} there was later developed, by Eddington, by Einstein himself, and recently by Schrödinger, an affine field theory operating with the components Γ_{pq}^r of an affine connection. But in 1925 Einstein also advocated a "mixed" formulation by means of a Lagrangian in which both the g_{pq} and the Γ_{pq}^r are taken as basic field quantities and submitted to independent arbitrary infinitesimal variations.[1] In certain respects this seems to be the most natural procedure. But even when the electromagnetic field is taken into account by adding the Lagrangian characteristic for Maxwell's theory, the resulting equations turn out to be the same as in Einstein's purely metric theory. This ceases to be true for the interaction between gravitation and electron; yet, as I shall show here, coincidence can be reestablished by adding to the Lagrangian one simple term, of a structure not dissimilar to that of the Dirac mass term.

A Cartesian frame E in a four-dimensional orthogonal vector space consists of four linearly independent vectors $\mathbf{e}(\alpha)$ ($\alpha=0, 1, 2, 3$), the transition between any two such ("equally admissible") frames being effected by an orthogonal transformation (rotation). Any vector $\mathbf{v}$ is uniquely representable relative to E in the form $\sum_\alpha v(\alpha)\mathbf{e}(\alpha)$ by its "*ortho-components*" $\mathbf{v}(\alpha)$. The square of the length of the vector $\mathbf{v}$ is the invariant $\sum_\alpha v^2(\alpha)$. We assume the reality conditions characteristic for the Lorentz-Minkowski vector geometry: $\mathbf{e}(1), \mathbf{e}(2), \mathbf{e}(3)$ are real space-like vectors while $\mathbf{e}(0)/i$ is a time-like vector pointing toward the future. An infinitesimal Lorentz rotation $d\mathbf{e}(\alpha)=\sum_\beta do(\alpha\beta)\cdot\mathbf{e}(\beta)$ is defined by a skew-symmetric matrix $\|do(\alpha\beta)\|$ of which the components $\alpha, \beta=1, 2, 3$ are real and the components $\alpha=0$, $\beta=1, 2, 3$ pure imaginary. With respect to the Lorentz frame E the wave function of the electron has four complex components $\psi_1, \psi_2; \psi_3, \psi_4$ which we arrange in a *column* ψ. Let $\tilde{\psi}$ denote the *row* of the conjugate-complex numbers $\bar{\psi}_1, \bar{\psi}_2; \bar{\psi}_3, \bar{\psi}_4$ and form the following Hermitian expressions $s(\alpha)=\tilde{\psi}S(\alpha)\psi$, $sp(\alpha)=\tilde{\psi}Sp(\alpha)\psi$:

$s(\alpha)$, $sp(\alpha)$	α
$i(\bar{\psi}_1\psi_1+\bar{\psi}_2\psi_2)\pm i(\bar{\psi}_3\psi_3+\bar{\psi}_4\psi_4)$	0
$(\bar{\psi}_1\psi_1-\bar{\psi}_2\psi_2)\mp(\bar{\psi}_3\psi_3-\bar{\psi}_4\psi_4)$	1
$(\bar{\psi}_1\psi_2+\bar{\psi}_2\psi_1)\mp(\bar{\psi}_3\psi_4+\bar{\psi}_4\psi_3)$	2
$i(\bar{\psi}_2\psi_1-\bar{\psi}_1\psi_2)\mp i(\bar{\psi}_4\psi_3-\bar{\psi}_3\psi_4)$	3

where the upper signs refer to $s(\alpha)$, the lower to $sp(\alpha)$. The ψ-components transform under the influence of a Lorentz rotation of the frame so as to make $s(\alpha)$ the ortho-components of an invariant vector (charge-current vector). $sp(\alpha)$ are then the components of a pseudo-vector (the spin, which changes sign if left and right in 3-space are interchanged). We make use also of the bilinear expressions $\tilde{\psi}S(\alpha)\chi$ in which χ is a quantity transforming cogrediently with ψ. The components ψ are determined but for an arbitrary factor $e^{i\lambda}$, the phase λ of which is a real constant.

The metric field at a given point P in the real four-dimensional world is described by assigning to P a Lorentz-frame $E=E(P)$, but does not change if E is submitted to an arbitrary Lorentz rotation. Let the world around P be referred to (real) coordinates x^p ($p=0, 1, 2, 3$). The prototype of a "vector at P" is the line element joining the point P, coordinates x^p, with an infinitely near point P', coordinates x^p+dx^p. The coordinate increments dx^p are the *contravariant components* of this vector. The contravariant components $e^p(\alpha)$ of the four vectors $\mathbf{e}(\alpha)$ that constitute the frame $E(P)$ describe analytically the metric field at P in terms of the coordinates x^p. The laws of nature are

* This is an abbreviated version of a note that appeared under the same title in Actas de la Academía Nacional de Ciencias de Lima **11** (1948), but was made incomprehensible by numerous misprints.

[1] Sitzungsber., Preuss. Akad. der Wissensch. (1925), p. 414.

invariant (1) with respect to arbitrary transformations of the coordinates x^p and (2) with respect to arbitrary Lorentz rotations of the frames $E(P)$ at all points P (the rotation of the frame $E(P)$ at P depending in an arbitrary manner on the point P). Let $\|e_p(\alpha)\|$ denote the matrix reciprocal to $\|e^p(\alpha)\|$,

$$\textstyle\sum_\alpha e^p(\alpha)\cdot e_q(\alpha)=\delta_q{}^p,\quad \sum_p e^p(\alpha)\cdot e_p(\beta)=\delta(\alpha\beta).$$

Here $\|\delta_q{}^p\|$ and $\|\delta(\alpha\beta)\|$ stand for the unit matrix. Let ϵ^{-1} be the absolute value of the determinant of the $e^p(\alpha)$. $\epsilon\cdot dx^0dx^1dx^2dx^3=\epsilon\cdot dx$ is the invariant volume element of the world. The contravariant, the ortho- and the covariant components v^p, $v(\alpha)$, v_p of a (real) vector $\mathbf{v}$ are connected by the relations

$$\begin{aligned} v^p&=e^p(\alpha)\cdot v(\alpha) \quad\text{or}\quad v(\alpha)=e_p(\alpha)\cdot v^p,\\ v_p&=e_p(\alpha)\cdot v(\alpha) \quad\text{or}\quad v(\alpha)=e^p(\alpha)\cdot v_p \end{aligned} \tag{1}$$

(in the writing-down of which the well-known convention about the omission of summation signs has been adopted). Multiplication by ϵ changes a scalar into a scalar density, a vector into a vector density, etc. We indicate this process by transmuting an italic into the corresponding German letter.

Infinitesimal parallel displacement from P to P' carries the frame $\mathbf{e}(\alpha)=\mathbf{e}(\alpha; P)$ at P into a frame at P' that proceeds from the local frame $\mathbf{e}(\alpha; P')$ at P' by a certain infinitesimal rotation

$$d\Omega=\|do(\alpha\beta)\|,\quad do(\alpha\beta)=o_p(\alpha\beta)\cdot dx^p.$$

We thus come to describe the metric field and the affine connection by the 16+24 quantities $e^p(\alpha)$ and $o_p(\alpha\beta)\{=-o_p(\beta\alpha)\}$.[2] The Riemann curvature tensor is given by

$$R_{pq}(\alpha\beta)=\left\{\frac{\partial o_q(\alpha\beta)}{\partial x^p}-\frac{\partial o_p(\alpha\beta)}{\partial x^q}\right\} +\{o_p(\alpha\gamma)o_q(\gamma\beta)-o_q(\alpha\gamma)o_p(\gamma\beta)\},$$

and hence the scalar curvature equals

$$R=e^p(\alpha)e^q(\beta)R_{pq}(\alpha\beta).$$

The integral $\int\mathfrak{R}\cdot dx$ of $\mathfrak{R}=\epsilon\cdot R$ is an invariant; following Einstein we adopt $\mathfrak{R}$ as the Lagrangian of the gravitational field.

Transcription of Dirac's theory of the electron into general relativity yields for the Lagrangian $\mathfrak{L}+m_0\mathfrak{l}$ of the electronic field the sum of two terms; the principal term $\mathfrak{L}$ is responsible for the most decisive general features of quantum mechanics, the accessory term $m_0\mathfrak{l}$ contains the mass m_0 of the electron as a constant factor:

$$L=\frac{1}{i}\left\{\bar\psi e^p(\alpha)S(\alpha)\frac{\partial\psi}{\partial x^p}-\frac{\partial\bar\psi}{\partial x^p}e^p(\alpha)S(\alpha)\psi\right\} -\frac{1}{i}\sum e^p(\alpha)o_p(\beta\gamma)sp(\delta), \tag{2}$$

with the sum $\sum$ extending over all even permutations $\alpha\beta\gamma\delta$ of 0123;

$$l=(\bar\psi_3\psi_1+\bar\psi_4\psi_2)+(\bar\psi_1\psi_3+\bar\psi_2\psi_4). \tag{3}$$

In general relativity the phase λ of the arbitrary factor $e^{i\lambda}$ of the ψ's is an arbitrary function of the position P. The infinitesimal transformation $d\psi=dZ\cdot\psi$ of the ψ's that is induced by an infinitesimal rotation $d\Omega=\|do(\alpha\beta)\|$ of the local frame, therefore, contains an indeterminate additive term $i\cdot df\cdot\psi$ with a pure imaginary scalar factor $i\cdot df$. Hence in a full description of the affine connection there will occur, in addition to the matrix $d\Omega$, also a scalar $df=f_p dx^p$ depending linearly on the dx^p. The derivative $\partial\psi/\partial x^p$ in L has to be replaced by $\partial\psi/\partial x^p+if_p\psi$, where f_p are the covariant components of a vector field, and all laws must be invariant with respect to the "gauge transformation" (or should one rather say "phase transformation"?)

$$\psi\to e^{i\lambda}\cdot\psi,\quad f_p\to f_p-\frac{\partial\lambda}{\partial x^p}.$$

Experience shows that f_p accounts for the effect of the electromagnetic field on the electron and is to be identified with the electromagnetic potential φ_p, measured in a suitable atomic unit, $f_p=(e/\hbar c)\cdot\varphi_p$. Form the field strength $f_{pq}=\partial f_q/\partial x^p-\partial f_p/\partial x^q$. The generation of the electromagnetic field by the electronic charge-current $s(\alpha)$ is accounted for by adding the electromagnetic term

$$\mathfrak{M}=f_{pq}\mathfrak{f}^{pq} \tag{4}$$

to the total Lagrangian.

In the Riemann-Einstein metric geometry the metric quantities $e^p(\alpha)$ determine the affine connection, i.e. the quantities $o_p(\alpha\beta)$, in the following manner. From the Poisson brackets

$$\frac{\partial e^p(\alpha)}{\partial x^q}\cdot e^q(\beta)-\frac{\partial e^p(\beta)}{\partial x^q}\cdot e^q(\alpha)=[e(\alpha),e(\beta)]^p,$$

and in accordance with the rules (1) for the juggling of indices

$$o(\gamma;\alpha\beta)=e^p(\gamma)\cdot o_p(\alpha\beta),$$
$$[e(\alpha),e(\beta)](\gamma)=e_p(\gamma)\cdot[e(\alpha),e(\beta)]^p.$$

Then

$$o(\alpha;\beta\gamma)+o(\beta;\gamma\alpha)=[e(\alpha),e(\beta)](\gamma)$$

or more explicitly

$$2o(\gamma;\alpha\beta)=[e(\gamma),e(\alpha)](\beta) +[e(\beta),e(\gamma)](\alpha)-[e(\alpha),e(\beta)](\gamma). \tag{5}$$

In a quantity like L which involves the $o_p(\alpha\beta)$ one may replace the latter by their expressions $o_p(\alpha\beta)\{e\}$ in terms of the $e^p(\alpha)$ as derived from (5). Let the resulting expression be denoted by L^*, the asterisk indicating that $o_p(\alpha\beta)\{e\}$ has been substituted for $o_p(\alpha\beta)$. In particular R^* is Riemann's scalar curvature

[2] See for this whole analytic treatment: H. Weyl, Zeits. f. Physik 56, 330 (1929).

written as an algebraic combination of the $e^p(\alpha)$, their first and second derivatives.

For the moment let us ignore the electromagnetic potentials f_p as well as the Dirac mass term, and let κ denote Einstein's gravitational constant multiplied by $\hbar/c$. Then Dirac's and Einstein's equations in the metric theory require the variation of

$$\int (\mathfrak{R}^*+\kappa\mathfrak{L}^*)dx$$

to vanish for such arbitrary infinitesimal variations $\delta\psi$ and $\delta e^p(\alpha)$ as vanish outside a bounded region, while in the mixed theory

$$\delta\int (\mathfrak{R}+\kappa\mathfrak{L})dx$$

is to vanish for similar variations of ψ, $e^p(\alpha)$ and $o_p(\alpha\beta)$. The latter variations result in a new set of field laws replacing the definitions (5). This set states that the differences

$$\Delta(\gamma;\,\alpha\beta)=[e(\alpha),\,e(\beta)](\gamma)-\{o(\alpha;\,\beta\gamma)+o(\beta;\,\gamma\alpha)\}$$

instead of being all zero, satisfy the relations

$$\Delta(\gamma;\,\alpha\beta)=\frac{\kappa}{2i}\cdot sp(\delta) \tag{6}$$

for any even permutation $\alpha\beta\gamma\delta$ of 0 1 2 3, and the relation

$$\Delta(\gamma;\,\alpha\beta)=0 \tag{6'}$$

whenever the three indices $\alpha\beta\gamma$ are not all distinct. *Thus by the influence of matter a slight discordance between affine connection and metric is created.*

Although the list of observables which describe the field in the mixed theory includes $o_p(\alpha\beta)$ beside ψ and $e^p(\alpha)$, it is possible to eliminate the $o_p(\alpha\beta)$ by means of the equations

$$o_p(\alpha\beta)=e_p(\gamma)\cdot o(\gamma;\,\alpha\beta),$$
$$2o(\gamma;\,\alpha\beta)=2o(\gamma;\,\alpha\beta)\{e\}-\Delta(\gamma;\,\alpha\beta)$$

and the laws (6), (6'). I have carried out the somewhat laborious calculations. Since the result applies to Lagrangians of a more general type than envisaged above I shall enounce it in this generality. In the expression (2) of L one has now to replace $\partial\psi/\partial x^p$, $\partial\bar{\psi}/\partial x^p$ by $\partial\psi/\partial x^p+if_p\psi$ and $\partial\bar{\psi}/\partial x^p-if_p\bar{\psi}$, respectively. Besides the terms R and L of the Einstein-Dirac theory which involve the $o_p(\alpha\beta)$ there exist four invariants which do not, to wit

$$l=(\bar{\psi}_3\psi_1+\bar{\psi}_4\psi_2)+(\bar{\psi}_1\psi_3+\bar{\psi}_2\psi_4),$$
$$l_2=(\bar{\psi}_3\psi_1+\bar{\psi}_4\psi_2)(\bar{\psi}_1\psi_3+\bar{\psi}_2\psi_4)\geq 0,$$
$$M=f_{pq}f^{pq} \quad\text{and}\quad Q=e^p(\alpha)e^q(\beta)F(\alpha\beta)f_{pq}.$$

The skew-symmetric tensor $F(\alpha\beta)$ which enters into Q is constructed out of the 16 products $\bar{\psi}\psi$ as follows:

$$F(01),\ F(23)=(\bar{\psi}_3\psi_1-\bar{\psi}_4\psi_2)\mp(\bar{\psi}_1\psi_3-\bar{\psi}_2\psi_4),$$
$$F(02),\ F(31)=(\bar{\psi}_3\psi_2+\bar{\psi}_4\psi_1)\mp(\bar{\psi}_2\psi_3+\bar{\psi}_1\psi_4),$$
$$F(03),\ F(12)=i(\bar{\psi}_3\psi_2-\bar{\psi}_4\psi_1)\pm i(\bar{\psi}_2\psi_3-\bar{\psi}_1\psi_4)$$

the upper signs to be used for $F(01)$, $F(02)$, $F(03)$, the lower for $F(23)$, $F(31)$, $F(12)$. Let H be of the general form

$$H=R+\kappa\{L+w(l,\,l_2,\,M,\,Q)\},$$

w being any function of four real variables.

THEOREM. *The mixed theory with the Lagrangian ϵH is identical with the metric theory based on the Lagrangian*

$$\epsilon H',\quad H'=R^*+\kappa\{L^*+w'(l,\,l_2,\,M,\,Q)\}$$

where

$$w'(l,\,l_2,\,M,\,Q)=w(l,\,l_2,\,M,\,Q)-\frac{3\kappa}{2}l_2.$$

To this extent then there is complete equivalence between the mixed metric-affine and the purely metric conceptions of gravitation.

Special relativity results if one lets the constant κ tend to zero. Then one obtains $e^p(\alpha)=\text{const.}$, $o_p(\alpha\beta)=0$, and one may choose

$$e^0(0)=i,\quad e^1(1)=e^2(2)=e^3(3)=1$$

and all other $e^p(\alpha)=0$. The Dirac equations and the energy-momentum tensor for the metric and the mixed theory coincide in the limit $\kappa\to 0$.

142.

Wissenschaft als symbolische Konstruktion des Menschen

Eranos-Jahrbuch 1948, 375—431 (1949)

I.

Es wäre nicht eine so üble Wahl, konstruktive Naturwissenschaft und kritische Philosophie von dem Tag zu datieren, an dem Demokritos verkündete: „Süß und bitter, kalt und warm sowie die Farben, all das existiert nur in der Meinung, nicht in Wirklichkeit (νόμῳ, οὐ φύσει); was wirklich existiert, sind unveränderliche Substanzteilchen und deren Bewegung im leeren Raum.“ In der Tat, ohne die Infragestellung des Standpunktes des naiven Realismus gibt es keine Philosphie, und eine theoretisch-konstruktive Wissenschaft von der Natur ist unmöglich, solange man die Erscheinungen, wie sie mir in der Wahrnehmung gegeben sind, für bare Münze nimmt. Die Welt, in der ich mich finde, wenn ich zum Bewußtsein meiner selbst erwache und in der ich lebe, trägt die Spannung von Subjekt und Objekt in sich. Der erste Teil des Demokritischen Satzes drückt eine kaum bestreitbare Tatsache aus, sofern er dahin interpretiert wird, daß die Qualitäten, mit denen sich für mich die Dinge der Außenwelt bekleiden, außer von ihnen selber ganz wesentlich abhängen, erstens von den näheren physischen Umständen, im Falle der Farbe z. B. von Beleuchtung und der Beschaffenheit des zwischen dem Gegenstand und meinem Auge befindlichen Mediums, und zweitens von mir, von meiner psycho-physischen Organisation. Auch die Perspektive, die Abhängigkeit dessen, was ich sehe, von meinem Standort, gehört hierher. Es handelt sich da um Fakten, die der Realist so gut wie der Idealist zugeben muß. Darüber hinaus behauptet der Idealismus, und dies ist nun freilich keine Erfahrungstatsache, sondern eine Sache der Philosophie, d. h. der Besinnung, daß Sinnesqualitäten ihrem Wesen nach nur im Bewußtsein, in der Empfindung gegeben sein können; es geht nicht an, sie losgelöst vom Bewußtsein einem Ding an sich als Eigenschaften an sich beizulegen. Demokrit macht sich anheischig, aus der im bunten Schmuck der Sinnesqualitäten prangenden phänomenalen Welt den harten Kern des wahrhaft Wirklichen herauszuschälen. Während aber die Erscheinungen mir gegeben sind, ist dieses Reale etwas zu Konstruierendes, es ist nicht gegeben, sondern aufgegeben. Er schließt aus der eben paraphrasierten Einsicht, daß die Sinnesqualitäten als Konstruktionsmaterial für den Bau der objektiven Welt unbrauchbar sind. Im zweiten Teile des oben zitierten Satzes gibt er an, woraus die letztere besteht: aus Atomen, unveränderlichen Teilchen, die sich im leeren Raum bewegen.

Der Begriff der Bewegung setzt den von Zeit und Raum voraus; daher gehören Zeit und Raum für Demokrit zu den objektiven Grundbegriffen der Weltkonstruktion. Der Raum ist Gegenstand einer mathematischen Disziplin, der Geometrie, die sich früh bei den Griechen zu einer exakten deduktiven Wissenschaft entwickelte. Von ihr sagt Kepler: „Die Raumwissenschaft ist einzig und ewig und strahlt wider aus dem Geiste Gottes. Daß die Menschen an ihr teilhaben dürfen, ist mit einer der Gründe, weshalb der Mensch das Ebenbild Gottes heißt.“ Eine ebenso ausgebildete Wissenschaft von der Zeit existiert nicht, aus dem einfachen Grunde, weil eine eindimensionale Mannigfaltigkeit wie die Zeit mathematisch viel simpler ist als eine dreidimensionale wie der Raum. Indem er Raum und Zeit als objektive Gegebenheiten a priori zugrunde legt, reduziert Demokritos allen Wechsel der Phänomene auf den absoluten Unterschied des Leeren und des Vollen (μὴ ὄν und παμπλῆρες ὄν). Das παμπλῆρες ὄν, die Materie, ist keiner qualitativen Unterschiede fähig. Bewegung besteht darin, daß die erfüllten Raumteile sich kontinuierlich ändern. Wäre der ganze Raum homogen erfüllt, so wäre keine Bewegung feststellbar; aber wenn die erfüllten Raumteile, die Atome, sich selber kongruent bleibende, voneinander durch das Leere getrennte Stücke sind, so ist es, Stetigkeit vorausgesetzt, möglich, die einzelnen Teilchen im Laufe ihrer Geschichte zu verfolgen. Die Aufgabe ist, aus den Erscheinungen die ihnen zugrunde liegende Bewegung der Atome zu erschließen. Umgekehrt müssen sich auf Grund der objektiv vor sich gehenden Bewegungen der Atome alle sich uns durch Wahrnehmung kundtuenden Erscheinungen erklären lassen; unter der Bedingung, daß das wahrnehmende Subjekt selber als objektives Naturgebilde, letzten Endes also als ein physikalischer Komplex von Atomen, mit in die Konstruktion hineingenommen wird. Und vom Erklären darf man natürlich nicht verlangen, daß z. B. aus der Wellenlänge des Lichtes der Natrium D-Linie und dem physiologischen Bau des Auges das anschaulich-spezifische Gelb meiner Farbwahrnehmung sich ableiten lasse; sondern die Forderung der restlosen Erklärung bedeutet lediglich, daß, wenn zwei verschiedene Farbempfindungen vorliegen, etwa die eines Gelb und eines Grün, dann die relevanten objektiven Bedingungen, bestehend aus der physikalischen Natur der Lichtquelle, des sich zwischen ihr und dem Auge ausbreitenden Mediums und der physikalischen Struktur des wahrnehmenden Auges, im einen und andern Fall nicht die gleichen gewesen sein können. Helmholtz formuliert dieses grundlegende Prinzip, dem unsere Konstruktion der objektiven Welt genügen muß, wenn sie die Bewußtseins-Gegebenheiten zureichend erklären soll, mit den Worten: „Wir sind, wenn verschiedene Wahrnehmungen sich uns aufdrängen, berechtigt, daraus auf Verschiedenheit der reellen Bedingungen zu schließen." Schon viel früher stellte der schweizerische Physiker und Mathematiker Lambert einen speziellen Fall davon als Axiom auf: „Eine Erscheinung ist dieselbe, sooft dasselbe Auge auf dieselbe Weise affiziert wird."

Ich glaube, daß, in solcher Weise Demokrit auslegend, ich zugleich die erkenntnistheoretische Grundeinstellung geschildert habe, die den Aufbau der klassischen Physik seit Galilei beherrschte. Gassendi erneuert Demokrits Atomtheorie zu Beginn des 17. Jahrhunderts. Nach Galilei sind weiß oder rot, bitter oder süß, tönend oder stumm, wohl- oder übel-riechend, Namen für Wirkungen auf die Sinnesorgane; sie dürfen den Dingen ebensowenig zugeschrieben werden wie der Kitzel oder Schmerz, welchen die Berührung von Gegenständen hervorrufen kann. Auch er behauptet, daß die Verschiedenheit, welche ein Körper in seinen Erscheinungen darbietet, auf bloßer Umlagerung der Teile ohne irgendwelche Neu-Entstehung oder Vernichtung beruht. Denn die Materie ist unveränderlich und immer dieselbe, da sie eine ewige und notwendige Art des Seins vorstellt. In Platos Dialog Theaitetos vertritt Sokrates die Lehre, daß die

Sinnesempfindungen zum Leben erweckt werden durch das Zusammentreffen einer vom Objekt und einer vom Subjekt ausgehenden Bewegung; indem die „Sicht" meinem Auge, die „Weiße" dem Objekt entströmt, entsteht die Gesichtswahrnehmung des weißen Gegenstandes.

Mit besonderem Nachdruck formulieren Hobbes und Locke die der Demokritischen Konstruktion zugrunde liegende Auffassung von der Subjektivität der Sinnesqualitäten und der Objektivität der geometrischen Ideen; Locke unterscheidet sie als sekundäre und primäre Qualitäten. Er sagt: „The ideas of primary qualities of bodies are resemblances of them, and their patterns do really exist in the bodies themselves; but the ideas produced in us by the secondary qualities have no resemblance of them at all... What is sweet, blue or warm in idea, is but the certain bulk, figure, and motion of the insensible parts in the bodies themselves, which we call so." Oder ein anderes Zitat, das in der naiven Entschiedenheit seiner Äußerung kaum übertroffen werden kann: „A piece of manna of a sensible bulk, is able to produce in us the idea of a round or square figure; and by being removed from one place to another, the idea of motion. This idea of motion represents it, as it really is in the manna moving; a circle or square are the same, whether in idea or existence, in the mind or in the manna. And this, both motion and figure, are really in the manna, whether we take notice of them or not. This everybody is ready to agree to. Besides manna by the bulk, figure, texture and motion of its parts has a power to produce the sensation of sickness, or sometimes of acute pains or gripings in us. That these ideas of sickness and pain are not in the manna, but effects of its operations on us, and are nowhere when we feel them not: this also everyone readily agrees to.

And yet men are hardly be brought to think, that sweetness and whiteness are not really in manna; which are but the effects of the operations of manna, by the motion, size and figure of its particles on the eyes and palate; as the pain and sickness caused by manna are confessedly nothing but the effects of its operations on the stomach and guts, by the size, motion and figure of its insensible parts." Hobbes spricht von der „great deception of sense, which is also to be by sense corrected", und er faßt seine These in die folgenden Worte: „1. that the subject wherein colour and image are inherent is not the object or thing seen; 2. that there is nothing without us (really) which we call an image or colour; 3. that the said image or colour is but an apparition unto us of the motion, agitation or alteration which the object worketh in the brain or spirit; 4. whatsoever accidents or qualities our senses make us think there be in the world, they are not there, but are seeming and apparitions only; the things that really are in the world without us, are those motions by which these seemings are caused." Auch nach Descartes darf man zwischen realem Vorgang und Wahrnehmung (etwa den Schallwellen und dem gehörten Ton) so wenig eine Ähnlichkeit fordern wie zwischen einer Sache und dem sie bezeichnenden Namen; auch er hält daran fest, daß die auf den Raum bezüglichen Ideen objektive Geltung haben. In seiner theologisch gefärbten Sprache sagt Newton am Ende seiner Optik: „All these things being consider'd, it seems probable to me, that God in the Beginning form'd Matter in solid, massy, hard, impenetrable, movable Particles, of such Sizes and Figures, and with such other Properties, and in such Proportion to Space, as most conduced to the End for which he form'd them; and that these primitive Particles are Solids, and incomparably harder than any porous Bodies compounded of them; even so hard, as never to wear or break in pieces; no ordinary Power being able to divide what God himself made one in the first Creation.. And therefore, that Nature may be lasting, the Changes of corporeal things are to be placed only in the various Separations and Motions of these permanent Particles."

Es lag mir daran, den consensus laut werden zu lassen, der hinsichtlich des grundlegenden erkenntnistheoretischen Standpunktes und der damit eng verbundenen Prinzipien der mechanistischen Weltkonstruktion zwischen allen diesen Männern herrscht, die der Physik und dann unserer ganzen modernen Naturwissenschaft die Wege gewiesen haben. Meine erste Aufgabe soll nun sein, in einem knappen historischen Abriß die Seite der Objektivität zu verfolgen und aufzuzeigen, wie sich durch die Entwicklung der Physik selbst die Ansicht über die letzten Elemente, die Bausteine der objektiven Weltkonstruktion modifiziert hat. Ich will auch sogleich verraten, zu welchem Resultat wir dabei gelangen werden; anstatt eines realen räumlich-zeitlich-materiellen Seins behalten wir nur eine Konstruktion in reinen Symbolen übrig. Das ist meine erste These. Ich werde mich dann zur Mathematik wenden, um Auskunft über Sinn und Ursprung der Symbole zu erhalten, und wir werden da den Menschen, sofern er schöpferischer Geist ist, als den Baumeister der Symbolwelt entdecken. Der Geist ist Freiheit in der Gebundenheit seines Daseins; sein Feld ist das Mögliche, das geöffnet ist zum Unendlichen hin, im Gegensatz zum geschlossenen Sein. Nur indem die Freiheit des Geistes sich selber bindet an das Gesetz, begreift der Geist nachkonstruierend die Gebundenheit der Welt und seines eigenen Daseins in der Welt. Dies ist meine zweite

These. Und in ihr liegt die Rechtfertigung dafür, daß ich im Rahmen von Vorträgen, deren allgemeines Thema der Mensch ist, über die Struktur der menschlichen Wissenschaft rede. Ich glaube, mit meinen beiden Thesen auf einem ziemlich festen Boden zu stehen und sie ziemlich einleuchtend begründen zu können. Aber im Anschluß daran werde ich am Ende des zweiten Vortrages auf Dinge zu sprechen kommen, deren Dunkel sich mir nicht im gleichen Maße gelichtet hat; da werde ich mehr als ein Fragender denn als ein Wissender zu Ihnen sprechen. Worum es sich da handeln wird, mag auch vorweg mit einigen Worten angedeutet werden.

In der gegenwärtigen Physik treten die Grenzen der Objektivität besonders prägnant hervor im allgemeinen Relativitätsprinzip und in dem aus der Quantenphysik entsprungenen Gedanken der Komplementarität. Eine Theorie der wissenschaftlichen Erkenntnis kann nicht zulänglich sein, wenn sie nicht über diese beiden so charakteristischen Züge vom Standpunkte des Menschen, des konstruierenden Geistes aus, Rechenschaft geben kann. Haben wir eine Philosophie, die in dieser Hinsicht zureichend ist? In Kopenhagen spricht man von einer Philosophie der Komplementarität, in Zürich wird einer dialektischen Philosophie der Wissenschaft das Wort geredet. Im Existentialismus kündigt sich eine philosophische Haltung an, die vielleicht besser mit der Struktur der modernen Wissenschaft verträglich ist als der Kantische Idealismus, in dem die erkenntnistheoretischen Positionen von Demokrit, Descartes, Galilei, Newton ihre volle philosophische Ausprägung gefunden zu haben schienen. Die Frage nach dem, was eine mitteilbare nachprüfbare Tatsache konstituiert, und nach dem Verhältnis zwischen Tatsache und Sprache, in der

der Mensch sein Dasein in der Welt sich artikuliert, spielt hier entscheidend hinein. Vielleicht kann es zur vorläufigen Verständigung dienen, wenn ich anstatt an den Prinzipien der Physik an der Urschöpfung der Mathematik, der Reihe der ganzen Zahlen, exemplifiziere. Vielleicht darf man sagen, daß wir kaum eine bündigere und tiefere Auskunft über die Natur des Geistes besitzen, als sie uns die vom Geist frei geschaffene Reihe der natürlichen Zahlen 1, 2, 3... gibt – obschon es eine Auskunft vom Erzeugten, nicht vom Ursprung her ist. Die Grundlagenforschung der Mathematik hat bisher immer versucht, den sachlichen Sinn der Mathematik zu erhellen; ist es ganz aussichtslos, sich dem Problem der Mathematik dadurch zu nähern, daß man das Mathematisieren als eine in Symbolen schaffende Grundtätigkeit des Geistes zu charakterisieren versucht, indem man sie abhebt gegen andere gleich ursprüngliche schöpferische Tätigkeiten wie Sprache und Musik?

Vorläufig ist dies nicht viel mehr als Geraune. Lassen Sie mich, ehe ich gegen Ende meiner Vorträge darauf zurückkomme, nun mit der ersten These beginnen und schildern, wie sich die reale Welt Demokrits und Newtons in ein rein symbolisches construct aufgelöst hat. Dort wo man die solide materielle Basis alles Geschehens zu finden glaubte, stieß man in Wahrheit auf etwas rein Spirituelles.

Zunächst freilich führte der Gang der Geschichte dazu, Demokrits Grundkonzeption, nach der die ungeheure Mannigfaltigkeit des Weltgeschehens lediglich eine Manifestation der latenten Verschiedenheit der Raumpunkte mittels des absoluten Gegensatzes des Vollen und Leeren ist, weiter zu präzisieren. Zwei Fragen erheben sich: die nach der Gestalt der Atome und die nach den Gesetzen ihrer Bewegung. Die einfachste Hypothese betreffend der Gestalt schreibt den Atomen Kugelform zu; ungleichartige Atome können sich dann lediglich durch ihre Radien voneinander unterscheiden. (Doch gibt es auch phantasiebegabte Autoren, die den Atomen Haken geben, mit denen sie sich zusammenhaken, wenn sie den schwer zu lösenden Verband eines festen Körpers bilden; es ist nicht ganz leicht sich vorzustellen, wie beim Zerbrechen eines solchen Körpers diese Ösen und Häkchen, die ja selber unzerbrechlich sind, auseinander genestelt werden.) Was die Bewegung anlangt, so stellten sich Demokrit und Epikur vor, daß die Atome, solange sie nicht kollidieren, im Raume frei von oben nach unten fallen. Seit Galilei übernimmt die Bewegung in gerader Linie mit gleichförmiger Geschwindigkeit diese Rolle der ungestörten Bewegung. Die Gesetze, nach denen zwei Atome bei der Kollision abrupt ihre Bewegung verändern, wurden zuerst (nach ganz unzureichenden Versuchen Descartes', der überhaupt als Physiker gründlich versagt) von Huygens richtig formuliert. Es sind das die für die ganze Physik grundlegenden Gesetze der Erhaltung von Energie und der Erhaltung von Impuls. Im Moment des Zusammenstoßes haben die kollidierenden Körper eine gemeinsame Tangentenebene im Berührungspunkt; zusammen mit der Annahme, daß nur in der zu ihr senkrechten Richtung ein Austausch von Impuls stattfindet, gestatten jene Erhaltungsgesetze die Bewegung nach der Kollision eindeutig aus der Bewegung vor der Kollision vorauszusagen. Es tritt in jene Gesetze ein gewisser, jedem Teilchen eigentümlicher Koeffizient ein, die Teilchenmasse. Aber für den konsequenten Mechanisten, der keine qualitativen Unterschiede der Materie anerkennt, ist die Masse des Atoms nichts Anderes als das Volumen des von ihm erfüllten Raumstücks.

In der Physik von Galilei, Newton und Huygens werden alle Geschehnisse konstruiert als Bewegungen in einem Raum, der beides, anschaulich und objektiv ist. Huygens, der bekanntlich die Wellentheorie des Lichtes entwickelte, kann mit dem besten Gewissen von der Welt sagen, daß farbige Lichtstrahlen in Wahrheit Schwingungen sind, Wellen in einem ätherischen, d. i. aus besonders feinen Partikeln bestehenden Medium. Man sollte die Einfachheit, Präzision und Überzeugungskraft der mechanistischen Weltvorstellung nicht unterschätzen! Aber wie war man zu ihr gelangt? Aus den Erscheinungen, wie wir sie unmittelbar erfahren, hatte man ein paar Züge, nämlich die geometrisch-kinematischen, ausgewählt. Aus Gründen, die wir heute ein wenig schwer finden nachzufühlen, wurden sie für besonders vertrauenswürdig gehalten. Mittels dieser Elemente baute man ein objektives Sein auf, angesichts dessen die Welt unserer täglichen Erfahrung zu bloßem Schein herabsank. Wenn Huygens und Newton recht haben, so sollten wir uns eigentlich schämen, daß wir in der Führung unseres täglichen Lebens uns immer noch an diese bunte Schein- und Schattenwelt halten. An den Prinzipien der mechanistischen Weltdeutung haben nicht nur die Philosophen, sondern auch die Physiker, wenn sie von den Grundlagen ihrer Wissenschaft Rechenschaft zu geben versuchten, hartnäckig festgehalten, als bereits die Physik selber in ihrer tatsächlichen Struktur längst darüber hinausgewachsen war. Sie haben dieselbe Entschuldigung wie der Landmensch, der zum erstenmal das offene Meer befährt: er wird sich an den Anblick der entschwindenden Küste klammern, solange keine neue Küste in Sicht ist, welcher er zusteuert. Ich werde jetzt die Reise beschreiben, auf welcher die alte Küste längst unter dem Horizont entschwunden ist. Vielleicht ist die neue Küste vor uns schon schwach erkennbar, vielleicht aber täuscht uns auch nur eine Nebelbank.

Demokrits altes Schema wurde an so vielen Stellen durchlöchert, daß ich in Verlegenheit bin, wo zu beginnen. Fangen wir von dem erkenntnistheoretischen Ende an! Die Objektivität von Raum und Zeit wurde bald kaum weniger verdächtig als die der Sinnesqualitäten. Schon Descartes fühlte die Notwendigkeit einer Rechtfertigung. Er schreibt den räumlichen Ideen Objektivität zu, weil wir sie im Gegensatz zu den Sinnesqualitäten klar und deutlich erkennen. Was wir aber so erkennen, ist nach dem Grundsatz seiner Erkenntnislehre wahr. Zur Stütze dieses Prinzips beruft er sich auf die Wahrhaftigkeit Gottes, der uns nicht täuschen wolle. Der Grundgedanke des Idealismus, daß mir nichts anderes gegeben ist als die unmittelbaren Daten meines Bewußtseins, war Descartes aufgeleuchtet. Das Problem ist, wie man von hier zu einer transzendenten realen Welt kommen kann, die allen Menschen gemeinsam ist und in welcher mein Bewußtsein die Rolle eines realen Individuums neben ungezählten anderen übernimmt. Will man diese reale Welt aufbauen aus Elementen, die im Bewußtsein liegen, z. B. in Form der räumlichen Anschauung, aber als real hingenommen werden, weil sie aus gewissen Gründen für besonders vertrauenswürdig gelten, so kommt man nur schwer um einen solchen die Realität verbürgenden Gott herum. „Er ist die Brücke", so spottet zwei Jahrhunderte später Georg Büchner, „zwischen dem einsamen, irren, nur einem, dem Selbstbewußtsein gewissen Denken und der Außenwelt. Der Versuch ist etwas naiv ausgefallen, aber man sieht doch, wie scharf schon Cartesius das Grab der Philosophie ausmaß; sonderbar ist freilich, wie er den lieben Gott als Leiter gebrauchte, um herauszukriechen.

Doch schon seine Zeitgenossen ließen ihn nicht über den Rand..."

Der Enzyklopädist D'Alembert rechtfertigt die Benutzung der Raum-Zeit-Begriffe zur Konstruktion der objektiven Welt nicht mehr, wie Descartes, durch ihre Klarheit und Deutlichkeit, sondern allein durch den Erfolg der Methode. Aber schon vorher hatte Leibniz mit größter Bestimmtheit erklärt: „Betreffs der Körper kann ich beweisen, daß nicht nur Licht, Wärme, Farbe und dergleichen, sondern auch Bewegung, Figuren und Ausdehnung nur erscheinende Qualitäten sind." Die Lehre von der Idealität von Raum und Zeit wird dann zum Eckstein von Kants transzendentalem Idealismus. Kant unterscheidet in den Erscheinungen das, was der Empfindung korrespondiert, als die Materie derselben und stellt ihr die Form gegenüber, „welche macht, daß das Mannigfaltige der Erscheinungen in gewissen Verhältnissen geordnet, angeschaut wird" und fährt fort: „Da das, worin sich die Empfindungen allein ordnen und in gewisse Formen gestellt werden können, nicht selbst wiederum Empfindung sein kann, so ist uns zwar die Materie aller Erscheinungen nur a posteriori gegeben, die Formen derselben aber müssen zu ihnen insgesamt im Gemüte a priori bereit liegen und daher abgesondert von aller Empfindung können betrachtet werden." Aus der apriorischen Natur der Raumanschauung erschließt er die apriorische Gewißheit und Notwendigkeit der Aussagen über Raum und räumliche Figuren, wie sie historisch zuerst in grandioser Weise in der Euklidischen Geometrie zusammengefaßt sind. Kant ist überzeugt, daß nur seine Auffassung von der Natur des Raumes die Möglichkeit der Geometrie als einer synthetischen Erkenntnis a priori begreiflich macht. Es ist wohl überflüssig, zu erwähnen, daß durch die ganze Entwicklung der Mathematik seit Kant diese seine Ansicht vom Wesen der Geometrie völlig untergraben ist. Aber darum mag seine Kennzeichnung des Raumes als Anschauungsform doch das Rechte treffen. Fichte findet dafür die kräftigen Worte: „Der erleuchtete, durchsichtige, durchgreifbare und durchdringliche Raum, das reinste Bild meines Wissens, wird nicht gesehen, sondern a n g e s c h a u t, und in ihm wird mein Sehen selbst angeschaut. Das Licht ist nicht außer mir, sondern in mir, und ich selbst bin das Licht."

War es nötig, beim Entwurf einer objektiven Welt auf die Sinnesqualitäten um ihrer subjektiven Natur willen zu verzichten, so ist es nun aus dem gleichen Grunde nötig, Raum und Zeit zu eliminieren. Das Mittel dazu ist von Descartes erfunden worden; es ist die analytische Geometrie. Halten wir uns um der Einfachheit willen an die ebene Geometrie. Nachdem man ein bestimmtes Koordinatensystem in der Ebene gewählt hat, kann jeder Punkt bestimmt und repräsentiert werden durch seine beiden Koordinaten x, y, somit durch ein reines Zahlsymbol (x, y). Der Umstand, daß verschiedene Punkte (x, y) auf einer geraden Linie liegen, übersetzt sich in die Aussage, daß alle diese Wertepaare (x, y) einer und derselben linearen Gleichung genügen. Die Kongruenz zweier Strecken AB, $A'B'$ wird ausgedrückt durch eine gewisse einfache arithmetische Beziehung zwischen den Koordinaten der vier Punkte A, B, A', B'. Etc. In dieser Weise gewinnen wir sozusagen ein arithmetisches Modell der Geometrie. Statt zu behaupten, daß die objektive Welt tatsächlich aus soliden, im Raum sich bewegenden Teilchen besteht, wäre es bescheidener gewesen, nur den Anspruch zu erheben, daß auf diese Art ein M o d e l l für die reale Welt konstruiert werden kann, aus dem sich ablesen läßt, was in Wirklichkeit geschieht, indem man die entsprechenden Konstruktionen am Modell ausführt. Das ist ja eigentlich genau das, was ein Ingenieur oder Architekt tut, wenn er seine Werkzeichnungen und Pläne anfertigt. Aber in der theoretischen Physik ist dieses Verfahren schon längst und so gut wie vollständig durch das der *Berechnung* ersetzt worden, und die Gesetze der Physik werden als arithmetische Gesetze zwischen den Zahlwerten variabler Größen gefaßt, in denen Raumpunkte und Zeitmomente durch deren Zahlkoordinaten vertreten werden. Größen etwa wie die Temperatur eines Körpers oder die Feldstärke eines elektrischen Feldes, die an jeder Raum-Zeit-Stelle einen bestimmten Wert haben, erscheinen als Funktionen von vier Variablen, den Raum-Zeit-Koordinaten x, y, z, t.

Während so das Faß der Welt den Boden von Raum und Zeit verlor, meldete sich eine neue Sorte von Begriffen mit dem Anspruch, wesentlich am Aufbau der objektiven Welt beteiligt zu sein. Es begann mit Galileis Einführung der *trägen Masse*. Wenn man das Gerüst abbricht, mittels dessen er zu dieser Idee aufstieg, behält man die folgende Erklärung in Händen. Solange ein Körper keine Einflüsse von außen erleidet und daher in gerader Linie mit gleichförmiger Geschwindigkeit $\mathfrak{v}$ sich bewegt, besitzt er einen gewissen Impuls $\mathfrak{J}$; dies ist ein Vektor, der dieselbe Richtung hat wie die Geschwindigkeit $\mathfrak{v}$. Der Faktor, mit welchem man $\mathfrak{v}$ multiplizieren muß, um den Impuls $\mathfrak{J}$ zu erhalten, ist die *Masse*. Darum weiß man, was Masse ist, sobald man weiß, wie der Impuls zu messen ist. Auf diese Frage aber antwortet Galilei nicht mit einer expliziten Definition, sondern mit einem Naturgesetz, dem Gesetz von der Erhaltung des Impulses: Wenn verschiedene Körper in eine Reaktion eintreten, dann ist die vektorielle Summe ihrer Impulse nach der Reaktion die gleiche wie vorher. Indem man die beobachteten Bewegungen miteinander reagierender Körper diesem Gesetz unterwirft, wobei man nötigenfalls Hilfskörper einführt, erhält man die Mittel, die relativen Werte ihrer Massen zu bestimmen. Um die Masse eines Körpers zu messen, muß man ihn daher mit andern Körpern reagieren lassen. Träge Masse ist ein verborgenes Merkmal, das wir einem Körper nicht so direkt ansehen können wie etwa seine Farbe. Außerdem ist die Bestimmung der Masse nur möglich auf Grund eines Naturgesetzes, an welches der Begriff selber gebunden ist. Das Naturgesetz erscheint so halb als Ausdruck einer Tatsache, halb als ein den Geschehnissen auferlegtes Postulat. Schließlich schreiben wir einem jeden Körper eine Masse zu, ob wir nun die zu ihrer Messung nötigen Reaktionen wirklich ausführen oder nicht; wir verlassen uns auf die Möglichkeit sie auszuführen. Diesen Punkt hebe ich besonders hervor, weil hier ein wesentlicher Ansatz für die von der Quantentheorie an der klassischen Physik geübte Kritik liegt. Die träge Masse erscheint in dieser im wesentlichen von Galilei herrürenden Analyse als ein dynamischer Koeffizient. Es hält sich daneben aber noch lange mit Hartnäckigkeit die rein verbale und eigentlich nichtssagende Erklärung der Masse als Quantität der Materie; so z. B. bei Newton und Kant.

Mit der eben geschilderten Einführung der Masse war ein Schritt von den weitgehendsten Konsequenzen vollzogen. Nachdem die Materie aller ihrer Qualitäten beraubt war, schien die Physik zunächst lediglich auf den Gebrauch rein räumlicher Attribute angewiesen – eine solche Physik suchte Descartes aufzubauen –, ja die objektive Geltung sogar dieser Attribute war durch Leibniz und Kant völlig in Frage gestellt. Aber nun entdecken wir, daß wir an einem Körper numerische Charakteristiken von ganz anderer Art, z. B. die Masse, ermitteln können, indem wir ihn mit andern Körpern reagieren lassen. Die Beobachtung der Reaktion muß zu diesem Behuf mit einer bestimmten Theorie, hier bestehend aus dem Erhaltungsgesetz des Impulses, gekoppelt werden. Die implizite Definition und die experimentelle Messung des idealen oder konstruktiven Charakteristikums sind an diese Theorie gebunden. Es ist dieses Verfahren, das die Sphäre der eigentlich mechanischen und physikalischen Begriffe jenseits der rein geometrisch-kinematischen aufschloß.

Es konnte bewiesen werden – freilich nur indem durch die Wahrscheinlichkeitsrechnung ein neues theoretisches Element in die Analyse der Naturerscheinungen eingeführt wurde –, daß ein Schwarm von kugelförmigen Atomen, welche bei der Kollision sich den Huyghens'schen Gesetzen gemäß benehmen, das an einem Gas beobachtete Verhalten zeigen. Die Huyghens'sche Theorie war daher unfähig, den flüssigen und festen Zustand der Materie zu erklären. Aber selber dem gasförmigen gegenüber versagte sie in einem entscheidenden Punkte. Indem man die Beobachtungen mit der Theorie vergleicht, erhält man einigermaßen zuverlässige Werte für den Radius und die Masse der das Gas konstituierenden Atome, die alle als gleich angenommen werden. Es stellt sich dabei aber heraus, daß für die verschiedenen chemischen Elemente die Atommassen den Atomvolumina nicht einfach proportional sind. Dies Resultat widerspricht der Vorstellung eines gleichförmigen Substanzteiges, aus dem der Schöpfer zu Beginn der Zeiten die kleinen Atomkuchen ausschnitt und diese, nachdem er sie zu absoluter Härte gebacken hatte, mit verschiedenen

Geschwindigkeiten in die Welt losließ. Mit der Galileischen Konzeption der Masse ist das Resultat eher verträglich. Jedenfalls ist es ein Faktum, daß die aus der kinetischen Theorie der Gase abgeleiteten relativen Atomgewichte gut übereinstimmen mit denjenigen, zu welchen die Chemie durch die Interpretation der chemischen Reaktionen im Lichte des Dalton'schen Gesetzes der multiplen Proportionen führt. Das ist ein außerordentlich bemerkenswerter Erfolg der kinetischen Theorie. Für lange Zeit war es nur die Chemie, die dem Atomismus eine kräftige empirische Stütze gab.

Wir sagten, daß ein Körper K, solange er unbeeinflußt ist, einen zeitlich konstanten Impuls besitzt. Steht er aber unter dem Einfluß anderer Körper $\kappa_1, \kappa_2, \ldots$, wie z. B. die Erde unter dem Einfluß der Sonne und der übrigen Planeten, so mißt die Änderung des Impulses pro Zeiteinheit die Kraft, welche jene andern Körper auf K ausüben. In der Tat fand Newton, daß die so gemessene Kraft die Summe von Teilkräften ist, die man einzeln dem Einfluß je eines der Körper $\kappa_1, \kappa_2, \ldots$ zuschreiben kann; in solcher Weise, daß z. B. die Teilkraft, welche κ_1 auf K ausübt, nur abhängt von dem Zustand der beiden Körper κ_1 und K, nicht aber von $\kappa_2, \kappa_3, \ldots$. Genauer wird jene Teilkraft abhängen von der Lage, der Geschwindigkeit und dem inneren Zustand der beiden Körper κ_1 und K. Der innere Zustand wird in das Kraftgesetz eingehen vermittels gewisser numerischer Charakteristiken, wie z. B. die elektrische Ladung eingeht in das Coulomb'sche Gesetz der elektrostatischen Attraktion und Repulsion. Auf Grund der geschilderten Tatsachen scheint es unnatürlich, die Aussage „Kraft = zeitliche Änderung des Impulses" als die Definition der Kraft anzusehen. Die Kraft ist vielmehr eine der Wechselwirkung zweier Körper zugehörige Potenz, und die auf einen Körper von allen übrigen Körpern ausgeübten Kräfte verursachen eine Änderung des Impulses, welche der Vektorsumme der einzelnen Kräfte gleich ist. Den physikalischen Tatsachen suchen wir uns durch eine kausal-metaphysische Deutung des Sachverhalts und durch Ausbildung einer passenden suggestiven Sprache anzupassen.

Hand in Hand mit dieser Entwicklung nimmt die Idee der Materie einen mehr dynamischen Zug an, während die Substanzhaftigkeit in den Hintergrund tritt. Die Atome werden zu „Kraftzentren". Boscovich, Cauchy und Ampère nehmen an, daß diese Zentren streng ausdehnungslose Punkte sind. Auch Kant sucht die Materie zu konstruieren vermittels einer attraktiven und einer ihr das Gleichgewicht haltenden repulsiven Kraft. Das alte geometrisch-substantielle Bild weicht allmählich der Physik der Zentralkräfte.

Der nächste entscheidende Schritt, wir stehen damit bereits im 19. Jahrhundert, ist die Ersetzung der elektrostatischen Fernkraft durch die kontinuierliche, mit endlicher Geschwindigkeit sich vollziehende Ausbreitung des elektromagnetischen Feldes. Maxwell machte es so gut wie gewiß, daß Licht aus hochfrequenten Schwingungen dieses Feldes besteht. Die Frage, was das elektromagnetische Feld ist, kann so wenig beantwortet werden wie die Frage, was Masse ist. Aber die Fakten, die durch diesen Begriff adäquat beschrieben werden, sind klar. Wenn man den Raum zwischen gegebenen, – nehmen wir an: ruhenden, – geladenen Konduktoren mit Hilfe eines kleinen geladenen Probekügelchen untersucht, so findet man, daß dieses an jeder Stelle P des Raumes eine bestimmte Kraft erleidet; immer die gleiche, wenn man das Probekügelchen an die gleiche Stelle P zurückbringt. An verschiedenen Stellen P ist diese Kraft im allgemeinen verschieden. Indem man das Probekügelchen variiert, findet man, daß die Kraft sich in zwei Faktoren zerlegen läßt: $e \cdot F$; der erste Faktor e ist eine richtungslose Maßzahl, die nur von dem Zustand des Probekörpers abhängt, aber weder vom Punkte P noch von Zustand und Anordnung der das Feld erzeugenden Konduktoren; der zweite Faktor F hingegen ist ein mit Richtung begabter Vektor, der nicht vom benutzten Probekörper abhängt, wohl aber von P und den Konduktoren. Auf Grund dieser Tatsache sprechen wir e als die Ladung des Probekörpers an und bezeichnen F als die elektrische Feldstärke an der Stelle P. [Sie ist nichts anderes als die Kraft, die an einem an der Stelle P befindlichen Probekörper mit der Normalladung $e = 1$ ausgeübt wird.] Wiederum scheint es natürlich, dem Feld an jeder Stelle P eine elektrische Feldstärke zuzuschreiben, ob wir nun dieser durch einen an die Stelle P gebrachten geladenen Probekörper Gelegenheit geben, sich durch eine proportionale Kraftwirkung zu verraten, oder nicht. Der frei transportierbare Probekörper dient dazu, die an sich vorhandene Feldstärke zu messen. Die Feldstärke ist an sich da, der Probekörper kann nach dem freien Entschluß des Experimentators variiert werden.

Aber dieser Kontrast ist doch wohl nur vorläufig. In Wahrheit gehört der Probekörper so gut zur vorhandenen Wirklichkeit wie die das Feld erzeugenden Konduktoren; er ist einfach ein weiterer kleiner Konduktor. Weder er noch die andern Konduktoren brauchen übrigens zu ruhen. Statt die Feldstärke durch die Kraftwirkung an dem Probekörper zu definieren, braucht man in Wahrheit ein Gesetz, nach welchem sich die ponderomotorischen Wirkungen des

Feldes auf die geladenen Konduktoren bestimmen, die das Feld erzeugen. Ob das Gesetz zu der ursprünglichen Definition zurückführt, wenn man die Ladung eines der Körper als unendlich schwach, seine Ausdehnung als unendlich klein annimmt und ihn als den vom Experimentator frei zu handhabenden Probekörper den andern Konduktoren gegenüberstellt, muß zunächst dahingestellt bleiben und bedarf einer nachträglichen Untersuchung. Man fängt daher besser überhaupt nicht mit einer Definition der elektromagnetischen Feldstärke an, sondern stellt Gleichungen auf, in denen unerklärte Symbole, wie die Raum-Zeit-Koordinaten, die Komponenten der Feldstärke, Ladungsdichte usw. auftreten. Neben den immanenten Feldgesetzen müssen die Gleichungen angegeben werden, nach denen sich die auf die erzeugenden Konduktoren wirkenden Kräfte aus den primitiven das Feld kennzeichnenden Größen bestimmen. Erst am Schluß der ganzen Konstruktion wird der Übergang von der physikalischen Welt zur Wahrnehmung vollzogen, indem man zu beschreiben versucht, in welcher Weise die Werte gewisser abgeleiteter Größen in den Wahrnehmungen sich kundgeben. Die Theorie wird so zu einem zusammenhängenden System, das nur als Ganzes mit der Erfahrung konfrontiert werden kann, während die einzelnen Aussagen und Gesetze keinen für sich realisierbaren Inhalt haben.

Ich exemplifiziere weiter an der Theorie der elektromagnetischen Erscheinungen, führe aber zwei schwer zu verteidigende Vereinfachungen ein. Erstens nehme ich instantane Ausbreitung statt Ausbreitung mit endlicher Geschwindigkeit an [obschon die endliche Ausbreitungsgeschwindigkeit gerade das entscheidende Merkmal der Maxwellschen Theorie ist, durch welche sie der Physik der

Zentralkräfte ebenso überlegen ist, wie Newtons dynamische Theorie der Gravitation der in Keplers drei Gesetzen niedergelegten kinematischen Beschreibung der Planetenbewegung überlegen war]. Zweitens nehme ich an, daß wir nur mit einer Sorte von Teilchen zu tun haben, Elektronen, die alle die gleiche Ladung und Masse besitzen, und daß ihre Bewegung das ist, was wir direkt beobachten können [eine vom modernen Standpunkt geradezu haarsträubende Idealisation!]. Wir wünschen aus Lage und Geschwindigkeit der Teilchen in einem Augenblick t ihre Lage und Gschwindigkeit in dem unmittelbar folgenden Moment t' zu bestimmen. Falls uns dies gelingt, können wir durch Integration die ganze Bewegung der Teilchen berechnen und das theoretische Resultat mit der Beobachtung vergleichen. Nach gewissen Gesetzen bestimmen Lage und Geschwindigkeit der Teilchen zur Zeit t eindeutig das elektromagnetische Feld im gleichen Augenblick. Nach weiteren Gesetzen bestimmt das Feld die kontinuierliche Verteilung von Energie und Impuls im Felde; und der Fluß des Impulses bestimmt schließlich die auf das einzelne Elektron ausgeübte Kraft, indem der in das Teilchen eintretende Feldimpuls seine Kompensation findet in einer gleich großen Änderung des Teilchenimpulses. Nach Newtons Grundgesetz der Mechanik äußert sich dies in einer bestimmten Beschleunigung des Teilchens. Aber Geschwindigkeit und Beschleunigung geben an, wie rasch sich Ort und Geschwindigkeit des Teilchens ändern, darum erlauben sie uns, Ort und Geschwindigkeit des Teilchens im nächstfolgenden Augenblick t' vorauszusagen. Nur diese ganze zusammenhängende Theorie, in die auch die Raumgeometrie als ein integrierender Bestandteil hineinverwoben ist, kann durch Beobachtung nachgeprüft werden. Die

Situation wird noch ganz wesentlich komplizierter, wenn wir unsere ungerechtfertigten Idealisationen aufgeben. Das isolierte physikalische Gesetz, wenn es aus diesem Zusammenhang losgelöst wird, hängt in der Luft. Letzten Endes verwachsen alle Teile der Physik einschließlich der Geometrie zu einer unlöslichen Einheit, von der kein Stück für sich eine selbständige und an der Beobachtung verifizierbare Bedeutung hat.

Bei jedem Schritt dieser Konstruktion, die zu immer tieferen Schichten fortschreitet – wie man z. B. von den anschaulich gegebenen perspektivischen Ansichten eines Körpers, die sich bei wechselnder Stellung des Beobachters ergeben, zu der Idee der unveränderlichen Körpergestalt selbst vordringt – versucht der Physiker eine intuitive Sprache zu erfinden, die die neuen Wesenheiten so anspricht, als wären sie uns vertraut wie Tisch und Bett. Aber in der systematischen Theorie sollte man, alle Zwischenstufen überspringend, ein Formelgerüst aus bloßen Symbolen hinstellen, ohne zu erklären, was die Symbole für Masse, Ladung, Feldstärke, usw. bedeuten, und nur am Ende beschreiben, wie diese ganze symbolische Struktur mit unserer unmittelbaren Erfahrung verknüpft ist. Es ist sicher, daß auf der Seite der Symbolik nicht Raum und Zeit, sondern vier unabhängige Variable x, y, z, t auftreten werden; von Raum ist wie von Tönen und Farben nur auf der Seite der Bewußtseinsgegebenheiten die Rede. Ein monochromatischer Lichtstrahl, der nach Huygens in Wirklichkeit aus einer Oszillation des Äthers besteht, ist nun eine mathematische Formel geworden, die ein gewisses Symbol F, genannt elektromagnetische Feldstärke, als eine rein arithmetisch konstruierte Funktion von vier andern Symbolen x, y, z, t, genannt Raum-Zeit-Koordinaten, ausdrückt. Es ist klar, daß wir nun in Huyghens' Behauptung: Licht ist „in Wirklichkeit"..., die Worte „in Wirklichkeit" zwischen Anführungsstriche setzen müssen. Wir haben eine symbolische Konstruktion, aber nichts, was wir im Ernst als die den Erscheinungen zugrunde liegende Wirklichkeit in Anspruch nehmen können. „Wir machen uns innere Scheinbilder oder Symbole der äußeren Gegenstände", so formuliert Heinrich Hertz in seinen Prinzipen der Mechanik diesen Standpunkt, „und zwar machen wir sie von solcher Art, daß die denknotwendigen Folgen der Bilder stets wieder die Bilder seien von den naturnotwendigen Folgen der abgebildeten Gegenstände". Um die Beziehung der Symbolwelt zum unmittelbar Wahrgenommenen noch etwas genauer zu bezeichnen, halte ich mich an das Beispiel des monochromatischen Lichtstrahls. Ich habe anzudeuten versucht, was für ein symbolisches construct ihm entspricht. Ich wage es nicht, auch nur andeutungsweise zu schildern, was für ein symbolisches construct eine zu Ende geführte Physiologie an die Stelle des den Lichtstrahl beobachtenden Auges zu setzen haben wird. Die Verbindung zwischen Objekt (Symbol) und Subjekt (Wahrnehmung) wird jedenfalls durch einen Satz wie diesen ausgedrückt werden müssen: Liegen diese den Lichtstrahl und das lebende sehende Auge repräsentierenden, in Symbolen ausgedrückten Verhältnisse vor, und bin Ich dies sehende Auge, so erscheint Gelb. Die objektiv gar nicht faßbare Bedingung, daß ich das sehende Auge bin, ist wesentlich; denn Bewußtsein ist, wie Heidegger sagt, jemeiniges. Gelb steht hier für die unmittelbar von mir wahrgenommene Farbqualität.

Als die Korpuskeln von Gassendi, Huyghens und Newton nicht mehr zureichten, hat man versucht, die objektive

Wirklichkeit auf andere Elemente zu basieren. So haben, durch die Relativitätstheorie veranlaßt, noch neuerdings Whitehead und Russell eine Weltkonstruktion versucht, in der das, was sie event nennen, als der harte Kern der Wirklichkeit auftritt. Ich glaube, das alles ist vergebliche Liebesmüh. Wir haben wirklich nichts in Händen behalten als die symbolische Konstruktion; und das genügt auch. Hätte es noch einer Bestätigung bedurft, so hat sie die Entwicklung der neueren Physik durch Relativitätstheorie und Quantentheorie geliefert.

Maxwell selber versuchte zunächst mechanische Modelle zu ersinnen, die seine Gesetze des elektrischen Feldes zu deuten gestatten würden. Solche Versuche wurden eine Zeitlang fortgesetzt, aber allmählich, nachdem sie immer wieder gescheitert, während der zweiten Hälfte des 19. Jahrhunderts aufgegeben. Wenn man präzis genug definiert, was ein mechanisches Modell ist, kann man wohl sogar die Unlösbarkeit der Aufgabe beweisen. Aber ein solcher showdown ist kaum nötig. Die Entscheidungen der Geschichte in Fällen wie dieser sind infallibel und unwiderrufbar. Das mechanische Weltbild ist tot, da wird es keine Restauration geben. Für eine kurze Weile waren manche geneigt, zum andern Extrem zu gehen und das kontinuierliche Feld als die einzige physikalische Realität anzunehmen, indem man die Elektronen und anderen Partikel als Gebiete geringer Ausdehnung ansprach, in denen die Feldstärke enorm hohe Werte annimmt; diese Feldknoten würden sich im Felde etwa so bewegen wie eine Welle über eine Wasserfläche fortschreitet. Aber dann kam die Quantentheorie und erkannte Feld oder Wellen einerseits, Partikel anderseits, als zwei komplementäre Aspekte des gleichen Sachverhalts. –

Entschuldigen Sie, daß ich in diesem ersten Vortrag, von der Einleitung abgesehen, kaum vom Menschen gesprochen habe. Aber es schien mir nötig, den Charakter, den die exakte Wissenschaft von der Natur heute, nach einer mehr als 2000jährigen Geschichte, angenommen hat, zunächst deutlich herauszustellen, ehe wir die Frage nach dem Baumeister dieser Wissenschaft aufwerfen. Es ist kein Zweifel, daß unsere Wissenschaft eine menschliche Wissenschaft ist; doch kommt der Mensch hier nur nach einer entscheidenden Seite seines Wesens zur Geltung, nämlich als freier mathematisch konstruierender Geist. Davon soll nun im nächsten Vortrag die Rede sein.

II.

Wir sahen in der ersten Vorlesung, daß der Aufbau der objektiven Welt gezwungen war, Raum und Zeit durch ein rein arithmetisches construct zu ersetzen, indem statt der Raum-Zeit-Punkte ihre Koordinaten in bezug auf ein gegeben vorliegendes Koordinatensystem benutzt werden. Betrachten wir eine Metallscheibe, die einen Teil der unendlichen Ebene E ausmacht. Örter auf der Scheibe kann man in concreto markieren, etwa indem man Kreuzchen in die Scheibe einritzt. Aber nachdem man zwei zueinander senkrechte Koordinatenachsen und eine Standard-Länge in die Scheibe eingeritzt hat, kann man auch ideelle Marken in die Ebene außerhalb der Scheibe setzen, indem man die Werte der beiden Koordinaten x, y einer solchen Marke angibt. Diese Zahlen beziehen sich auf eine als möglich gedachte Vermessung mittels starrer Maßstäbe. Statt einzelner in concreto markierter Örter bekommen wir so die Mannigfaltigkeit aller möglichen Örter in der Ebene in den

Griff unserer Konstruktion. Jede der beiden Koordinaten variiert nämlich über den a priori zu überblickenden Bereich der reellen Zahlen. In dieser Weise benutzt Astronomie unsere feste Erde als Basis, um die Sternräume auszuloten. Als Anaxagoras zuerst die Schatten von Erde und Mond in dem von der Sonne durchleuchteten leeren Raum konstruierte und so die Mond- und Sonnen-Finsternisse erklärte, hatte er diese Vorstellung des Raumes als der Mannigfaltigkeit aller möglichen Örter in einem imaginativen Akt von höchster Originalität vollzogen und wußte sie sogleich für die Naturwissenschaft fruchtbar zu machen. Das Kontinuum der reellen Zahlen, auf welches hier die Mannigfaltigkeit der möglichen Raumstellen reduziert wird, ist eine freie Schöpfung des Geistes. Eben weil wir die Zahlen durch freie Kombination weniger Grundzeichen erschaffen, können wir die im Prozeß ihrer Konstruktion liegenden Möglichkeiten a priori überblicken. Wir brauchen sozusagen nicht auf die uns in der Erfahrung entgegenkommenden einzelnen konkreten Raumpunkte zu warten, sondern wir stellen durch unsere arithmetische Konstruktion ein Netz bereit, in dem wir jeden möglichen Raumpunkt einfangen können. Das Seiende wird so auf den Hintergrund des Möglichen projiziert. In unserer Analysis der Natur reduzieren wir die Phänomene auf einfache Elemente, von denen jedes innerhalb eines bestimmten Bereichs möglicher Werte variiert, den wir a priori überblicken können, weil wir selber dieses Feld von Möglichkeiten auf rein kombinatorischem Wege konstruieren. So löst die Physik das Licht in ebene polarisierte monochromatische Lichtstrahlen auf; und dieses einen nicht mehr zu überbietenden Grad der Homogenität erreichende Licht kennzeichnet sie durch ein paar Charakteristiken (das wichtigste davon die Wellenlänge), die variabel sind in dem frei vom Geist geschaffenen Kontinuum der reellen Zahlen.

Bevor ich auf das Konstruktionsprinzip der reellen Zahlen etwas näher eingehe, möchte ich den hier so wesentlichen Gegensatz von Sein und Möglichkeit am räumlichen Kontinuum erläutern. Anaxagoras ist hier nochmals zu erwähnen. Denn er gab der Idee des Unendlichen zuerst eine Fassung, durch welche er in die Wissenschaft einzugreifen vermag. Ein uns überliefertes Fragment von ihm sagt: „Im Kleinen gibt es kein Kleinstes, sondern es gibt immer noch ein Kleineres. Denn was ist, kann durch keine noch so weit getriebene Teilung je aufhören zu sein.“ Das Kontinuum, sagt er, kann nicht aus diskreten Elementen zusammengesetzt sein, die „voneinander abgetrennt, wie mit dem Beile voneinander abgehauen sind“. Der Raum ist nicht nur in dem Sinne unendlich, daß man in ihm nirgendwo an ein Ende kommt; sondern an jeder Stelle ist er sozusagen nach innen hinein unendlich; ein Punkt läßt sich nur durch einen ins Unendliche fortschreitenden Teilungsprozeß von Stufe zu Stufe genauer und genauer fixieren. Das steht in Kontrast zu dem für die Anschauung ruhenden fertigen Dasein des Raumes. Gegen Anaxagoras erhebt sich die streng atomistische Theorie des Demokrit. Ich will seine Argumentation hier nicht wiederholen; aber mögen auch die wirklichen Körper aus Atomen bestehen, für den Raum selber läßt sich diese Ansicht sehr schwer durchführen (in der Tat, alle solche Versuche sind bisher gescheitert). Die Unmöglichkeit, das Kontinuum als ein starres Sein zu fassen, kann kaum prägnanter illustriert werden als durch das bekannte Paradoxon des Zenon von dem Wettlauf zwischen Achilleus und der Schildkröte. Sie wissen, worin es besteht: Die Schildkröte habe vor Achilleus einen Vorsprung von der Länge 1. Läuft Achilleus doppelt so schnell wie die Schildkröte, so wird die Schildkröte in dem Augenblick, wo Achilleus an ihrem Start ankommt, ihm um die Strecke $1/2$ voraus sein; nachdem Achilleus auch diese zurückgelegt, hat die Schildkröte einen Weg von der Länge $1/4$ durchmessen; und so fort in infinitum; woraus zu schließen ist, daß der schnellfüßige Achilleus das Reptil niemals einholt. Der Hinweis darauf, daß die sukzessiven Partialsummen der Reihe $1 + \frac{1}{2} + \frac{1}{2^2} + \frac{1}{2^3} + \ldots$ nicht über alle Grenzen wachsen, sondern gegen 2 konvergieren, durch den man heute das Paradoxon zu erledigen meint, ist gewiß eine wichtige zur Sache gehörige und aufklärende Bemerkung. Wenn aber die Strecke von der Länge 2 wirklich aus unendlich vielen Teilstrecken der Länge 1, $\frac{1}{2}$, $\frac{1}{4}$, $\frac{1}{8}$, ... als „abgehackten" Ganzen besteht, so widerstreitet es dem Wesen des Unendlichen, des „Unvollendbaren", daß Achilleus sie alle schließlich durchlaufen hat. Deshalb bemerkt auch Aristoteles zur Auflösung des Zenonischen Paradoxons, daß „das Bewegte sich nicht zählend bewege", oder genauer: „Wenn man die stetige Linie in zwei Hälften teilt, so nimmt man den einen Punkt für zwei; man macht ihn sowohl zum Anfang wie zum Ende. Indem man aber so teilt, ist nicht mehr stetig weder die Linie noch die Bewegung... In dem Stetigen sind zwar unbegrenzt viele Hälften, aber nicht der Wirklichkeit, sondern der Möglichkeit nach." Man weiß, wie diese Antinomien, von der Entfaltung der Mathematik kaum berührt und an Präzision der Fassung eher ab- als zunehmend, im philosophischen Denken fortgewirkt haben. So bezeugt Leibniz, daß das Verlangen, einen Ausweg aus dem „Labyrinth des Kontinuum" zu finden, es war, was ihn zuerst zu der Auffassung von Raum und Zeit als Ordnungen der Phänomene ohne selbständige Realität hingeführt hat.

Wenn wir jetzt daran gehen, diese alten Ideen ein wenig präziser zu fassen, so entdecken wir das Unendliche statt im Kontinuum zunächst in einer primitiveren Form, in der Reihe der natürlichen Zahlen 0, 1, 2, 3, . . . ; diese liegt aller unserer Mathematik, all unseren symbolischen Konstruktionen zugrunde. Erst mit ihrer Hilfe können wir auch an die mathematische Erfassung des Kontinuums herantreten. Die Symbole, welche wir hier verwenden, sind hintereinander gestellte Striche. Wir zählen etwa, während wir einer die Stunde schlagenden Turmuhr lauschen, die Schläge, indem wir sie durch Striche auf einem Blatt Papier vermerken. (Die Zahlwörter unserer Sprache und die üblichen auf dem Dezimalsystem beruhenden Zahlzeichen sind raffinierte Fortbildungen dieses ursprünglichen Symbolismus.) Die Objekte, wie z. B. die Schläge der Turmuhr, mögen sich auflösen, „melt, thaw and resolve themselves into a dew", aber ihre Anzahl kann in einem solchen Zahlzeichen niedergelegt und aufbewahrt werden. Was ist mehr, wir können durch einen konstruktiven Prozeß von zwei Zahlen, die durch hintereinander gesetzte Striche gekennzeichnet sind, eindeutig entscheiden, welche die größere ist; indem wir nämlich die Striche einen nach dem andern auskreuzen und dabei jedesmal, wenn wir einen Strich der ersten Reihe auskreuzen, auch einen Strich der zweiten Reihe auskreuzen. Dieser Prozeß gestattet mit Sicherheit Unterschiede zwischen Zahlen festzustellen, die sich ihnen nicht mehr direkt ansehen lassen; denn dem bloßen Anblick nach hätten wir es schon schwer, zwischen so niedrigen Zahlen wie 21 und 22 zu unterscheiden. Wir sind so vertraut mit diesen Wundern, welche die Zahlsymbole vollführen, daß wir kaum mehr darüber erstaunt sind. Aber dies ist bloß das Präludium zu dem eigentlich mathematischen Schritt.

Vier Stufen in der Entwicklung der Arithmetik heben sich voneinander ab. Zu der ersten Stufe gehört ein solches einzelnes konkretes Urteil wie $2 < 3$, d. i. das Zahlzeichen || ist als Teilstück in dem Zahlzeichen ||| enthalten. Der zweiten Stufe gehört die allgemeine Idee des $<$ an, des Enthaltenseins von einem Zahlzeichen in einem andern, und Urteile von hypothetischer Allgemeinheit wie das folgende: Wenn immer dir in concreto zwei Zahlzeichen $\mathfrak{a}$ und $\mathfrak{b}$ gegeben sind, so kannst du sicher sein, daß entweder $\mathfrak{a} < \mathfrak{b}$ oder $\mathfrak{a} = \mathfrak{b}$ oder $\mathfrak{b} < \mathfrak{a}$ ist. Der Bereich des wirklich Gegebenen wird hier nicht überschritten, insofern dieses Urteil nur etwas aussagen will, wenn bestimmte Zahlen $\mathfrak{a}$, $\mathfrak{b}$ aktuell gegeben sind. Etwas ganz Neues aber geschieht, und hier beginnt das, was ich eben die eigentliche Mathematik nannte, wenn ich nun auf der dritten Stufe die in unvoraussagbarer Weise mir aktuell begegnenden Zahlzeichen einbette in die *Reihe aller möglichen Zahlen*, welche durch einen Erzeugungsprozeß entsteht gemäß dem Prinzip, daß aus einer vorliegenden Zahl n stets eine neue, die nächstfolgende n', durch Hinzufügung eines weiteren Striches erzeugt werden kann. Hier haben Sie in deutlichster Form das vor Augen, was ich oben so beschrieb: Das Seiende wird projiziert auf den Hintergrund des Möglichen, einer nach festem Verfahren herstellbaren geordneten, ins Unendliche offenen Mannigfaltigkeit von Möglichkeiten. Die Intuition des „immer noch eins“ erscheint hier als eine letzte Grundlage der theoretischen Konstruktion. An der ins Unendliche offenen Reihe der Zahlen $0, 1, 2 \ldots$ haben wir das primitivste Beispiel eines a priori überblickbaren Variabilitätsbereiches vor uns. Die allgemeinen Aussagen der Zahlenlehre handeln auf dieser Stufe von der Freiheit, die sich entwickelnde Zahlenreihe an einer beliebigen Stelle zum Stehen zu bringen.

Während mehrerer Jahre wurden die Pausen der Philharmonic Symphony Konzerte in New York für broadcasts benutzt, in denen amerikanische Gelehrte kurz und allgemeinverständlich vor einem Auditorium von, wie man mir sagte, 6 bis 10 Millionen Hörern über ihr Forschungsgebiet berichteten. In diesem Rahmen hatte ich vor nicht langer Zeit über Mathematik zu sprechen, und ich sagte da das Folgende über die Zahlen: „Mit der Geometrie waren die Mathematiker nicht schlecht gefahren; aber wahrhaft in ihrem Element sind sie erst, wenn es zu den Zahlen kommt. Denn die Folge der natürlichen Zahlen 0, 1, 2... ist eine freie Schöpfung des menschlichen Geistes. Sie beginnt mit der 0, und auf jede Zahl folgt die nächste; das ist alles. Nach dieser einfachen Anweisung wandern die Zahlen fort ins Unendliche. 1 ist die Zahl, die auf 0 folgt, 2 die Zahl, die auf 1 folgt, usw.; nichts sonst. Sie wissen sehr wenig über Heinrich VIII., wenn Sie wissen, daß er Heinrich VII. auf dem englischen Throne folgte. Aber Sie wissen alles über die 8, wenn Sie wissen, daß sie auf 7 folgt. Der Mensch hat substantielle Existenz; die Worte der Sprache haben Bedeutungen, deren Nuancen ineinander verschwimmen, die Töne musikalischer Kompositionen haben sinnliche Qualität. Aber Zahlen haben weder Substanz noch Bedeutung noch Qualität. Sie sind nichts als Marken, und alles, was in ihnen liegt, haben wir hineingelegt, vermöge der einfachen Regel des Aufeinanderfolgens.“

Methodisch findet dieser Standpunkt seinen Ausdruck in der zuerst von Pascal und Jacob Bernoulli klar formulierten Definition und dem Schluß durch sogenannte vollständige Induktion. Beispiel einer Definition durch vollständige Induktion ist das in allen Armeen der Welt übliche Verfahren des Abzählens zu zweien, durch welches die Begriffe gerade und ungerade für Zahlen festgelegt werden. Es läßt sich in zwei Sätzen niederlegen: Die erste Zahl 0 ist gerade. Die auf eine Zahl n folgende Zahl n' ist gerade oder ungerade, je nachdem n selber ungerade oder gerade ist. Der Schluß durch vollständige Induktion beweist den Satz, daß jede mögliche Zahl eine gewisse genau umschriebene Eigenschaft E hat, dadurch daß man zeigt: 0 besitzt diese Eigenschaft; und wenn n irgend eine Zahl ist, welche die Eigenschaft E hat, so hat auch die nächstfolgende n' die Eigenschaft E. Es ist praktisch unmöglich und wäre außerdem vollkommen nutzlos, eine Zahl wie 10^{12}, welche die Europäer eine Billion, die Amerikaner tausend Billionen nennen, durch hintereinander gesetzte Striche auszuschreiben. Aber doch tragen wir kein Bedenken, von einer Ausgabe von 10^{12} cents für das amerikanische Rüstungsprogramm oder den Marshall Plan zu sprechen. Nur mittels des Durchgangs durchs Unendliche kann solchen Zahlen eine Bedeutung beigelegt werden. Man muß erst $10 \cdot n$ durch vollständige Induktion für eine beliebige Zahl n definieren und kann darauf, dem Zahlzeichen von 12 folgend, 12 Faktoren 10 sukzessive miteinander multiplizieren, indem man sich auf die allgemeine Definition der Multiplikation mit 10 stützt:

$$\begin{array}{ccccccccccccl} | & | & | & | & | & | & | & | & | & | & | & | & (=12) \\ 10 \cdot & 10 \cdot & 10 \cdot & 10 \cdot & 10 \cdot & 10 \cdot & 10 \cdot & 10 \cdot & 10 \cdot & 10 \cdot & 10 \cdot & 10 \cdot & 1 \; (=10^{12}). \end{array}$$

Archimedes ringt mit diesem Problem der rationalen Bewältigung des Uferlosen in seiner Schrift an Gelon „Über die Sandrechnung“.

Schon bei der natürlichen Zahl treten uns also die folgenden Grundzüge des konstruktiven Erkennens entgegen: 1. Das Resultat gewisser geistiger Operationen am Gegebenen (wie des Zählens), die für allgemein ausführbar gelten, wird, sofern es durch das Gegebene eindeutig bestimmt ist, als ein dem Gegebenen an sich zukommendes Merkmal aufgestellt (selbst, wenn jene Operationen, die seinen Sinn begründen, nicht wirklich ausgeführt werden). In der Physik werden in der gleichen Weise auch physische Operationen von dieser Art benutzt, so die Kollisionen von Körpern als Mittel, ihre relativen Massen zu bestimmen. 2. Durch Einführung von Zeichen wird eine Aufspaltung der Urteile vollzogen und ein Teil der Operationen durch Verschiebung auf die Zeichen vom Gegebenen und seinem Fortbestand unabhängig gemacht. Dadurch tritt das freie Schalten mit Begriffen ihrer Anwendung, die Ideen treten relativ selbständig der Wirklichkeit gegenüber. 3. Die Zeichen werden nicht einzeln für das jeweils aktuell Gegebene hergestellt, sondern sie werden dem potentiellen Vorrat einer nach festem Verfahren herstellbaren geordneten, ins Unendliche offenen Mannigfaltigkeit von Zeichen entnommen.

Ich habe bisher drei Stufen der Entwicklung der Arithmetik geschildert. Auf der vierten Stufe wird der Sprung ins Jenseits vollzogen, indem die ins Unendliche offene gesetzmäßig entstehende Reihe von Zahlen zu einem geschlossenen Inbegriff an sich seiender Elemente gemacht wird. Darüber soll erst später gesprochen werden. Wir bleiben vorerst auf der dritten Stufe stehen, die durch Worte wie Freiheit, Möglichkeit, Offenheit, gekennzeichnet ist.

Denn nun muß ich doch erst versuchen, die konstruktive Erfassung des eindimensionalen Kontinuums zu schildern. Es macht dabei wenig aus, ob wir uns von vornherein nur an die Symbole halten oder zunächst von dem räumlichen Bilde der geraden Linie ausgehen, deren Punkte ja vermöge des Koordinatenbegriffs in völlig eindeutiger Beziehung zu den reellen Zahlen stehen. Um der Anschaulichkeit willen ziehe ich das zweite Verfahren vor, will aber dabei die offene gerade Linie durch die geschlossene Kreislinie ersetzen. Es liegt im Wesen des Kontinuums, (1) daß es teilbar ist; (2) daß es nicht exakt teilbar ist (keine „wie mit dem Beil voneinander abgehauene" Stücke besitzt); wenn auch (3) die Schärfe und Feinheit der Teilung bei keiner Grenze Halt zu machen braucht. Wir fangen etwa damit an, daß wir die Kreislinie in zwei Bögen zerlegen, die wir mit + und — bezeichnen und die Elementarstücke oder Stücke 0-ter Stufe nennen. Aus den Stücken n-ter Stufe entstehen die Stücke der nächsthöheren, der $(n+1)$-ten Stufe, indem man jeden Bogen der n-ten Stufe in zwei Teile teilt, die in einem bestimmten Umlaufssinn des Kreises genommen durch 0 und 1 voneinander unterschieden werden. (Dies ist eine Definition durch vollständige Induktion. Die gleiche Methode der Diäresis diente anscheinend Plato dazu, seine idealen Zahlen aufzubauen.) Auf der 1. Stufe haben wir 4 Teilbögen, auf der 2. Stufe 8, . . ., auf der n-ten Stufe 2^{n+1}. Der einzelne Bogen der n-ten Stufe ist durch ein Symbol gekennzeichnet, das mit + oder — beginnt und von n Ziffern 0 oder 1 gefolgt ist. In dieser Weise werden die Bögen katalogisiert. Es ist im Sinne der allgemeinen Relativitätstheorie, welche die metrischen Begriffe erst nachträglich in ein schon konstruiertes Kontinuum einführt, wenn wir nicht vorschreiben, daß die einzelne Teilung in einer exakten Halbierung besteht. Die aktuelle Ausführung des Prozesses involviert also eine beträchtliche Willkür. Um aber alle Punkte zuverlässig voneinander trennen zu können, müssen wir annehmen, daß das Teilungsnetz überall schließlich unendlich fein wird. In Wirklichkeit können die sukzessiven Teilungen je nur mit einer gewissen Unschärfe ausgeführt werden; während aber die Teilung von Stufe zu Stufe fortschreitet, müssen die durch die ersten Teilungen vage gesetzten Grenzen schärfer und schärfer fixiert werden. Es ist unsinnig sich vorzustellen, daß an einem wirklichen Kontinuum dieser virtuell ins Unendliche laufende Prozeß zu Ende gekommen sei, sondern er ist immer nur bis zu einer gewissen Stufe gediehen. Aber wir müssen die Möglichkeit offen halten, daß er von jeder Stufe aus noch um einen Schritt weiter getrieben werden kann. Darum scheiden wir die arithmetische Leerform des Teilungsprozesses von seiner Verwirklichung an einem gegebenen Kontinuum; das kombinatorische Schema, nach welchem die auf jeder Stufe erhaltenen Stücke aneinander grenzen und nach welcher die Teilung fortschreitet, ist eindeutig und vollkommen fixiert. Die Leerform ist, im Gegensatz zur konkreten Ausführung, a priori ins Unendliche bestimmt; die mathematische Theorie des Kontinuums, insbesondere des Kontinuums der reellen Zahlen, handelt von dieser Leerform. Die gegebene Beschreibung läßt sich leicht auf allgemeinere Fälle ausdehnen, insbesondere auf mehrdimensionale Kontinua, wo die Grundteilung in Elementarstücke von verwickelterer Art ist als die Zweiteilung in + und —, von der wir hier ausgingen, wo aber der Fortgang von Stufe zu Stufe gleichfalls durch Wiederholung eines kombinatorisch völlig fixierten Prozesses der Unterteilung vor sich geht. Der Mathematiker hat es nur mit dem aus Symbolen bestehenden, der fortgesetzten Verfeinerung fähigen Katalog der Stücke zu tun; was diese Stücke selbst sind, braucht ihn nicht zu kümmern.

Es ist mir klar, daß ich Ihre Geduld auf eine harte Probe stelle. Aber jetzt bleibt noch zu sagen, wie ein solches Teilungsnetz dazu benutzt wird, irgend einen Punkt im Kontinuum mit größerer und größerer Genauigkeit abzufangen. Ich halte mich dabei an das Teilungsgerüst der Kreislinie, das durch immer wiederholte Bisektion entsteht. Wir vereinigen irgend zwei zusammenstoßende Teilbögen der n-ten Stufe zu einem Bogen, den wir Spannbogen n-ter Stufe nennen. Diese Spannbögen greifen so übereinander, daß, wenn immer ein Punkt mit hinreichender Genauigkeit festgelegt ist, wir mit Sicherheit einen Spannbogen n-ter Stufe angeben können, in welchem er drin liegt. Darum kann ein jeder Punkt abgefangen werden durch eine unendliche Folge von Spannbögen b_1, b_2, ... wachsender Stufe, wobei jedesmal ein Bogen b_n ganz im Innern des vorhergehenden liegt. Dies Verfahren ist prinzipiell demjenigen gleichwertig, durch welches schon im Altertum Eudoxos gelehrt hat, die Stellen im Kontinuum abzufangen und voneinander zu unterscheiden. Was die neue Zeit hinzugefügt hat, ist die Einsicht, daß man die Folge nicht nur als Mittel zu betrachten hat, um einen vorweg gegebenen und in seiner Existenz gesicherten Punkt begrifflich zu beschreiben, sondern durch sie der Punkt im Kontinuum konstruktiv erst erzeugt wird. Jede solche mögliche Folge liefert einen Punkt; im arithmetischen Leerschema werden eben hierdurch die Punkte erschaffen. Nur vermöge dieser konstruktiven Wendung ist eine mathematische Beherrschung, ist eine Analyse der Stetigkeit möglich. Der entscheidende Begriff ist hier der der Folge. Was einen Spannbogen beliebiger Stufe kennzeichnet, ist nicht so verschieden von einer natürlichen Zahl, als daß wir nicht für die weitere Diskussion die Folge von ineinander eingeschachtelten Spannbögen durch eine Folge natürlicher Zahlen a_1, a_2, ... ersetzen könnten. Objekt der Zahlentheorie, so können wir daher resumieren, sind die einzelnen möglichen natürlichen Zahlen, Objekt der Kontinuumslehre die möglichen Folgen natürlicher Zahlen.

Ein Problem liegt aber noch in der Idee der unendlichen

Folge. Eine Folge natürlicher Zahlen kommt als eine werdende zustande dadurch, daß ich willkürlich eine erste Zahl wähle, darauf eine zweite, darauf eine dritte, und so fort: freie Wahlfolge. Wenn von einer beliebigen Folge, einem beliebigen Punkt auf gegebener Geraden die Rede ist, von einer aller reellen Werte fähigen Variablen, so handelt es sich um eine solche freie Wahlfolge. Sinnvoll sind aber nur solche Eigenschaften von ihr auszusagen, über welche die Entscheidung bei einer endlichen Stelle der Entwicklung fällt. (Man kann z. B. sinnvoll fragen: kommt 5 unter den Zahlen der Folge bis zur 731-sten Stelle vor?, aber nicht: kommt 5 überhaupt vor?) Denn die Folge ist ja niemals eine vollendete! Die einzelne bestimmte, und zwar ins Unendliche hinaus bestimmte, Folge kann hieraus nicht durch bestimmte Wahl hervorgehen, sondern dazu ist ein Gesetz nötig, welches allgemein aus einer beliebigen natürlichen Zahl n die an n-ter Stelle stehende Zahl a_n der Folge zu berechnen gestattet; z. B. $a_n = n^2 + 3n + 1$.

Die freie Erschaffung der Zahlenreihe, die Möglichkeit, ihr Werden an einer beliebigen Stelle zum Stillstand zu bringen, vor allem aber die durch freie Wahl werdende Folge von natürlichen Zahlen, zeigt den konstruktiven Geist in seiner Freiheit, schöpfend aus sich selber, geöffnet für Möglichkeit, schwebend im Werden. Aus der freien Wahlfolge, der kontinuierlichen Variablen, wird der einzelne Wert, der einzelne Punkt, durch Bindung an ein Gesetz. Aber auch das Gesetz als solches muß frei konstruiert werden. Der Mensch findet sich in der Welt, auf sich selbst gestellt und doch nicht allein, zugleich frei und gebunden. Die mathematische Konstruktion ist vielleicht die markanteste Manifestation der Freiheit seines Geistes. Die Gebundenheit seines Daseins in der Welt bleibt natürlich essentiell für den

Geist etwas Undurchdringliches. Aber der die Welt in symbolischer Konstruktion nachschaffende Geist hat die Möglichkeit, selber das freie Werden durch das Gesetz zu binden. Nicht anders als im frei konstruierten mathematischen Gesetz findet das bindende Naturgesetz seinen theoretischen Ausdruck. Darum sagte ich einst: „Die Mathematik ist nicht das starre und Erstarrung bringende Schema, als das der Laie sie so gerne ansieht; sondern wir stehen mit ihr genau in jenem Schnittpunkt von Gebundenheit und Freiheit, welcher das Wesen des Menschen selbst ist.“ Und nun, denke ich, habe ich meine zweite These in einiger Deutlichkeit begründet, daß es der freie, in Symbolen schaffende Geist ist, der sich in der Physik ein objektives Gerüst baut, auf das er die Mannigfaltigkeit der Phänomene ordnend bezieht. Er bedarf dazu keiner solchen von außen gelieferten Mittel wie Raum, Zeit und Substanzpartikel: er nimmt alles aus sich selbst.

In meinen bisherigen mathematischen Ausführungen bin ich weitgehend dem holländischen Mathematiker Brouwer gefolgt, der in unsern Tagen den durch die Spannung zwischen Sein und Möglichkeit gekennzeichneten intuitionistischen Standpunkt in der Mathematik streng durchgeführt hat. Die Metaphysik hat von je her versucht, den Dualismus von Objekt und Subjekt, von Sein und Sinn, von Gebundenheit und Freiheit zu überwinden. Im Realismus erhob sie das Objekt, im Idealismus, wie ihn am kräftigsten der junge Fichte ausgesprochen hat, das Subjekt einseitig zur Würde des absoluten Seins. Sie schreibt eine auf das Transzendente verweisende Chiffre, wenn sie Gott als absolutes Sein setzt, aus dem herfließend das Licht des Bewußtseins, dem der Ursprung selber verdeckt ist, in seiner Selbstdurchdringung sich ergreift als das zwischen Sinn und Sein gespaltene und gespannte Dasein. Auch die Mathematik hat, ohne sich der Gefahr bewußt zu sein, in naiver Sachlichkeit, möchte ich sagen, den Sprung zum Absoluten vollzogen. Wer die an die unendliche Allheit der natürlichen Zahlen appellierende Alternative „n ist eine gerade oder ungerade Zahl, je nachdem es eine Zahl x gibt, für welche $n = 2x$ ist oder nicht", als sinnvoll hinnimmt, steht bereits am jenseitigen Ufer: das Zahlsystem ist ihm aus einem offenen, nur im Werden zu erfassenden Bereich von Möglichkeiten zu einem Inbegriff absoluter Existenz geworden, das „nicht von dieser Welt ist". Die Alternative „gibt es oder gibt es nicht", an eine unendliche Allheit gerichtet, läßt im allgemeinen, menschlich gesprochen, keine Entscheidung zu; stützen wir uns hier dennoch auf das logische Prinzip des tertium non datur, so nehmen wir an, daß „vor Gott" all dies entschieden sei, während in der Mathematik von Fall zu Fall versucht wird, diese verborgene in eine dem Menschen offenkundige Entscheidung zu verwandeln. Auf dem menschlichen, dem intuitionistischen Standpunkt kann man nicht erwarten, daß die Frage, ob zwei Punkte A und B im Kontinuum zusammenfallen oder nicht, immer entschieden werden könne; denn da jeder der beiden Punkte durch eine Folge ineinander geschachtelter Spannbögen definiert ist, handelt es sich hier um die Alternative: überlappen sich der n-te Spannbogen der Folge A und der Folge B für jedes n, oder gibt es ein n, für welches dies nicht der Fall ist? Darum ist die Brouwer'sche Theorie des Kontinuums im Einklang mit einer oben angeführten Bemerkung des Aristoteles, wonach es unmöglich ist, die kontinuierliche Strecke von 0 bis 1 so in die beiden Teile von 0 bis $1/2$ und von $1/2$ bis 1 zu zerlegen, daß jeder Punkt der ganzen Strecke mit Sicherheit entweder der einen oder andern Hälfte angehört.

Anders in der transzendenten oder absoluten Mathematik, der vierten Stufe der oben geschilderten Entwicklung, wo von allen Existenzfragen angenommen wird, daß sie „vor Gott" entschieden seien. Die Termini „es gibt" und „jeder" werden hier nicht nur auf die natürlichen Zahlen unbedenklich und schrankenlos angewendet, sondern auch auf die möglichen Folgen natürlicher Zahlen, d. h. auf die möglichen Stellen im Kontinuum. Auf diesem logischen Transzendentalismus beruht die eigentliche Macht der klassischen Mathematik. Brouwer gibt den größten Teil dieser Mathematik preis. Will man das nicht, so kommt man, wie die Kritik von H. Poincaré, B. Russell, Brouwer u. a. gezeigt hat, nicht um eine radikale Umdeutung des Sinnes der Mathematik herum. Man muß nämlich darauf verzichten, ihren Aussagen einzeln einen verifizierbaren Sinn beizulegen, muß vielmehr ihre Sätze in Formeln umwandeln, die aus bedeutungslosen Zeichen aufgebaut sind und darum selber nichts bedeuten noch behaupten. Die Mathematik wird dadurch, prinzipiell gesprochen, aus einem System einsichtiger Erkenntnisse zu einem nach bestimmten Regeln sich vollziehenden Spiel mit Zeichen und Formeln. Die in der Mathematik schon immer übliche Repräsentation durch Symbole muß dabei auch auf die fundamentalen logischen Operationen wie „und", „oder", „es gibt", usw. ausgedehnt werden; auch bei ihnen haben wir von ihrer logischen Bedeutung zu abstrahieren.

In der transzendenten Mathematik hatte es zunächst so geschienen, als ob dem älteren Schritt, der die Geometrie auf Arithmetik, auf Zahlen reduziert hatte, nun der weitere Schritt gefolgt sei, durch den die Zahlen und die auf sie bezüglichen arithmetischen Begriffe, wie insbesondere das Verfahren der vollständigen Induktion, auf reine Logik zurückgeführt werden, nämlich auf solche rein logischen Begriffe wie „oder“, „nicht“, „wenn, so“ und auf die allgemeine logische Idee der Funktion oder Abbildung. Dieser Versuch der Reduktion auf Logik war in der Tat erfolgreich bis zu einem erstaunlichen Grade. Ein logisches Universum übernahm da sozusagen die Rolle, welche in der Demokritischen Konzeption die im Raum sich bewegenden Atome gespielt hatten. Aber man hatte einen hohen Preis zu zahlen, nämlich den Preis der Verletzung der offenkundigsten Evidenz bezüglich des Unendlichen, und der Mathematiker wurde bestraft für seine transzendente Tollkühnheit über den eigenen Schatten zu springen, durch Widersprüche, die sich drohend, wenn auch nur an den äußersten Grenzen der Mathematik, erhoben. Ist es das Ziel der Mathematik, das Unendliche durch endliche Mittel zu meistern, so erreichte sie das in ihrer logisch-transzendenten Form, wenn wir Brouwer glauben wollen, nur durch einen Betrug – durch einen gigantischen, freilich höchst erfolgreichen Betrug, vergleichbar dem Papiergeld auf ökonomischem Gebiet.

Wollen wir ehrlich bleiben und doch auf die transzendente Mathematik nicht verzichten, so müssen wir den schon oben angedeuteten Schritt über die Logik hinaus tun: wir müssen die bedeutungsvollen Aussagen durch bedeutungslose Formeln ersetzen. Das Spiel, das der Mathematiker mit den Formeln treibt, kann man nicht unpassend dem Schachspiel vergleichen. Den Steinen des Schachspiels entspricht in der Mathematik ein beschränkter oder potentiell unbeschränkter Vorrat an Zeichen, an selbstgeschaffenen, distinkten, reproduzierbaren Zeichen; einer beliebigen Aufstellung der Steine auf dem Brett entspricht die Zusammenstellung der Zeichen zu einer Formel. Eine oder wenige

Formeln gelten als Axiome; ihr Gegenstück ist die vorgeschriebene Aufstellung der Steine zu Beginn einer Schachpartie. Und wie hier aus einer im Spiel auftretenden Stellung die nächste hervorgeht, indem ein Zug gemacht wird, der bestimmten Zugregeln zu genügen hat, so sind dort Schlußregeln anschaulich beschrieben, nach denen aus Formeln neue Formeln gewonnen, „deduziert" werden können. Unter einer spielgerechten Stellung im Schach verstehe ich eine solche, welche aus der Anfangstellung in einer den Zugregeln gemäß verlaufenen Spielpartie entstanden ist. Das Analoge in der Mathematik ist die beweisbare (oder besser, die bewiesene) Formel, welche auf Grund der Schlußregeln aus den Axiomen hervorgeht. Gewisse Formeln von anschaulich beschriebenem Charakter werden als Widersprüche gebrandmarkt. Bis hierher ist alles Spiel, nicht Erkenntnis. Immerhin spielt schon hier Bedeutung, Verstehen und Kommunikation eine Rolle, indem wir darauf rechnen, daß die Regeln für die Handhabung der Zeichen und Formeln verstanden werden. Nun aber kann das Schachspiel zum Gegenstand der Erkenntnis gemacht werden, indem sich z. B. einsehen läßt, daß in einer spielgerechten Stellung niemals mehr als 8 Bauern der gleichen Farbe vorkommen können. In ähnlichem Sinne wird in der Metamathematik, wie Hilbert sich ausdrückt, das Mathematik-Spiel zum Gegenstand der Erkenntnis gemacht: es soll nämlich erkannt werden, daß ein Widerspruch niemals als Endformel eines Beweises auftreten kann. Was Hilbert sicherstellen will, ist nur diese Widerspruchslosigkeit, nicht die inhaltliche Wahrheit der Analysis. Nur zur Gewinnung dieser einen Erkenntnis wird von Hilbert das inhaltliche, bedeutungsvolle Denken benötigt. Es ist selbstverständlich, daß bei diesen inhaltlichen Überlegungen die von Brouwer dem inhaltlichen Denken gesteckten Grenzen durchaus respektiert werden. Nach anfänglichen Erfolgen ist die Durchführbarkeit des Hilbert'schen Programms durch eine tiefe logische Entdeckung von Kurt Gödel in Frage gestellt worden; es ist äußerst zweifelhaft, ob wir auch nur imstande sind, die Widerspruchslosigkeit der klassischen Analysis auf dem von Hilbert ins Auge gefaßten Wege sicherzustellen.

Aber sei dem, wie ihm sei. Kann sich die Mathematik im Ernst um ihrer Sicherheit willen auf diese Linie des bloßen Spiels zurückziehen? Ich glaube nicht; aber ich glaube, daß nur ihr Verhältnis zur Physik deutlich macht, was es mit dieser transzendenten, rein symbolischen Mathematik auf sich hat. Die Grundlagenkrise der Mathematik, von der ich sprach, ist gewiß für das Schicksal der Erkenntnis überhaupt von grundlegender Bedeutung; doch hier tat ich ihrer Erwähnung nur, um neues Licht auf die Konstruktionen der Physik fallen zu lassen.

In der Tat, die Situation, die wir in der theoretischen Physik vorfinden, entspricht dem Ideal Brouwers von einer Wissenschaft in keiner Weise. Dieses Ideal postuliert, daß jedes Urteil seinen eigenen, in der Anschauung vollziehbaren Sinn hat. Die Aussagen und Gesetze der Physik haben jedoch, wie wir gesehen haben, einzeln genommen keinen in der Erfahrung verifizierbaren Inhalt. Nur das theoretische System als Ganzes läßt sich mit der Erfahrung konfrontieren. Was die Physik leistet, ist nicht anschauende Einsicht in singuläre oder allgemeine Sachverhalte, sondern theoretische, letzten Endes rein symbolische Konstruktion der Welt. Man hat gesagt, die Physik habe es nur mit der Feststellung von Koïnzidenzen zu tun; insbesondere hat Mach auf dem Felde der Physik einem reinen Phänomenalismus das Wort geredet. Aber wenn man ehrlich ist, so muß man doch zugestehen, daß unser theoretisches Interesse nicht ausschließlich und nicht einmal in erster Linie an den „realen Aussagen“ hängt, an den Konstatierungen, daß dieser Zeiger mit diesem Skalenteil sich deckt, sondern vielmehr an den idealen Setzungen, die laut Theorie in solchen Koïnzidenzen sich ausweisen, deren Sinn selbst aber in keiner gebenden Anschauung sich unmittelbar erfüllt, – wie z. B. der Setzung des Elektrons als eines universellen elektrischen Elementarquantums. Es ist nicht zu leugnen, daß in uns ein vom bloß phänomenalen Standpunkt schlechterdings unverständliches theoretisches Bedürfnis lebendig ist, dessen auf symbolische Gestaltung des Transzendenten gerichteter Schaffensdrang Befriedigung verlangt und das getrieben wird von dem metaphysichen Glauben an die Realität der Außenwelt (neben den sich gleichartig der Glaube an die Realität des eigenen Ich, des fremden Du und Gottes stellt).

Es ist eine tiefe philosophische Frage, welches die „Wahrheit“ oder Objektivität ist, die dieser über das Gegebene weit hinausdrängenden theoretischen Weltgestaltung zukommt. Die Entwicklung der Wissenschaft hat deutlich gezeigt, daß verschiedene theoretische Konstruktionen der Welt dem Postulat der Einstimmigkeit genügen können. Was die Entscheidung zwischen solchen konkurrierenden Theorien herbeiführt, die praktisch für jeden, der in der Wissenschaft drin steht, zwingend erscheint, ist nicht leicht zu sagen. Hier werden wir offenbar getragen von dem an uns sich vollziehenden Lebensprozeß des Geistes, und sind außerstande, den Gewinn davon in der Form toter endgültiger „Resultate“ abzuschöpfen. Diese hier nur angedeuteten erkenntnistheoretischen Überlegungen führen mich jedenfalls zu der folgenden Position. Nimmt man die Mathematik für sich allein, so beschränke man sich mit Brouwer auf die einsichtigen Wahrheiten, in die das Unendliche nur als ein offenes Feld von Möglichkeiten eingeht; es ist kein Motiv erfindlich, das darüber hinaus drängt. In der Naturwissenschaft aber berühren wir eine Sphäre, die der schauenden Evidenz sowieso undurchdringlich ist; hier wird Erkenntnis notwendig zu symbolischer Gestaltung, und es ist darum, wenn die Mathematik durch die Physik in den Prozeß der theoretischen Weltkonstruktion mit hineingenommen wird, auch nicht mehr nötig, daß das Mathematische sich daraus als ein besonderer Bezirk des anschaulich Gewissen isolieren lasse: auf dieser höheren Warte, von der aus die ganze Wissenschaft als eine Einheit erscheint, bin ich geneigt, Hilbert recht zu geben.

Lassen Sie mich in ein paar Schlagworten die Erfahrungen zusammenfassen, welche unser Weg durch die mathematisch-physikalische Erkenntnis gezeitigt hat. (1) Die Aufgabe der Erkenntnis kann durch schauende Einsicht allein sicher dort nicht geleistet werden, wo, wie in der Naturwissenschaft, eine objektive, der Vernunft von Ursprung her undurchdringliche Sphäre berührt wird. Aber selbst in der reinen Mathematik oder Logik können wir die Gültigkeit einer Formel ihr nicht mittels eines deskriptiven Merkmals ansehen, sondern sie wird gewonnen nur durch praktisches Handeln, indem man nämlich, von den Axiomen ausgehend, in beliebig oftmaliger Wiederholung und Kombination die praktischen Regeln des Schließens anwendet. Man kann darum sprechen von einer ursprünglichen Dunkelheit der Vernunft: wir haben die Wahrheit nicht, es genügt nicht, große Augen zu machen, sondern sie will durch Handeln gewonnen sein. (2) Das Unendliche ist dem Geiste, der Anschauung zugänglich in Form des ins Unendliche offenen Feldes von Möglichkeiten; aber (3) das vollendet, das aktuell Unendliche als ein geschlossenes Reich absoluter Existenz kann ihm nicht gegeben sein. (4) Doch wird der Geist durch die Forderung der Totalität und den metaphysischen Glauben an Realität unabweisbar dazu gedrängt, das Unendliche als geschlossenes Sein durch eine symbolische Konstruktion zu repräsentieren.

So lehne ich die These von der schlechthinigen Endlichkeit des Menschen ab; Geist ist Freiheit in der Gebundenheit des Daseins, er ist offen gegen das Unendliche. Gott als das vollendet Unendliche kann ihm freilich nicht gegeben sein; weder kann Gott in den Menschen durch Offenbarung einbrechen, noch kann der Mensch in mystischer Schau zu ihm durchbrechen. Das vollendet Unendliche können wir nur repräsentieren im Symbol. Aus dieser Bezogenheit empfängt alle Gestaltung, in der sich das Schöpferische des Menschen betätigt, ihren Ernst und ihre Würde.

Dem in der Einleitung des ersten Vortrags entworfenen Programm folgend, will ich nunmehr, nach Aufstellung der beiden Thesen, welche den Baustoff und den Baumeister der theoretischen Wissenschaft betreffen, einige Fragen berühren, hinsichtlich deren ich selber im Dunkeln tappe. Die neuere Entwicklung der Physik hat zwei Grundprinzipien an den Tag gebracht, das Einstein'sche Prinzip der allgemeinen Relativität und das mit der Heisenberg'schen Ungenauigkeitsrelation verknüpfte Bohr'sche Prinzip der Komplementarität. – Raum und Zeit bilden ein vierdimensionales Kontinuum, das zwecks mathematischer Erfassung (in der an der Kreislinie illustrierten Weise) mit einem Teilungsnetz übersponnen werden muß. Während das kombinatorische Schema bestimmt ist, ist die Ausführung der

Teilung bei jedem Schritt mit großer Willkür behaftet. Aber nur relativ zu einem solchen Teilungsnetz läßt sich der Zustand der Welt durch distinkte reproduzierbare Symbole objektiv festlegen. Es läßt sich mathematisch beschreiben, worin der Übergang von einem zulässigen Teilungsnetz zu einem beliebigen andern besteht. Die Naturgesetze müssen invariant sein gegenüber solchen Transformationen des Teilungsnetzes; denn die Willkür, die unsere Konstruktion hineinbringt, berührt nicht die Natur selbst. Das ist das allgemeine Relativitätsprinzip. Ich denke, es ist einigermaßen aus der Freiheit des Geistes verständlich, daß in seine Konstruktion unvermeidlich Willkür eingeht, daß aber diese nachträglich durch ein Prinzip der Invarianz unschädlich gemacht werden kann.

Eine philosophisch wesentlich prekärere und dunklere Angelegenheit aber scheint die von der Quantenphysik enthüllte Komplementarität zu sein. Wir sahen schon im ersten Vortrag, daß die idealen Merkmale, welche die Physik einem Körper zuschreibt, basiert werden auf das Verhalten, das der Körper zeigt, wenn er mit andern Körpern in Reaktion tritt. Zum Beispiel messen wir die Temperatur eines Körpers, indem wir ihn mit einem Thermometer in Berührung bringen. Die idealen Merkmale werden dem Körper selbst dann zugeschrieben, wenn die Reaktionen nicht wirklich ausgeführt werden; es genügt, daß sie als ausführbar angesehen werden. Diesem Grundsatz hat nun aber die Quantenphysik im Gebiet des atomaren Geschehens eine prinzipielle Schranke gesetzt: die präzise Messung einer Größe, z. B. des Orts eines Elektrons, vernichtet die Möglichkeit, seinen Impuls zu messen, und vice versa. Das ist ein in der Natur der Dinge liegender Sachverhalt und keine menschliche Unvollkommenheit. – Mittels eines Gitters trennt man in der Optik die verschiedenen einfachen Farben ihrer Wellenlänge nach. Jede Beobachtung oder Messung besteht in gewissem Sinne darin, daß man das zu beobachtende physikalische Geschehen ein Gitter passieren läßt. Die neue Situation in der Quantenphysik besteht darin, daß man zwei Gitter im allgemeinen nicht überlagern kann; man kann die Elektronen eines Schwarmes nicht sowohl nach Lage wie nach Impuls separieren. Außerdem läßt sich in keiner Weise voraussehen, selbst wenn hinsichtlich des Elektronenschwarms die größtmögliche Bestimmtheit obwaltet, wie die Messung durch das Gitter am einzelnen Elektron ausfallen wird; nur für sehr viele Partikeln läßt sich das Resultat statistisch voraussehen, mit dem jeder Statistik innewohnenden Maß von Unsicherheit. Jede Beobachtung oder Messung läuft auf einen im Atomaren unkontrollierbaren Eingriff in den physikalischen Vorgang hinaus. Diese Situation hat ihre erkenntnistheoretischen Konsequenzen: in dem durch die Messung ermittelten mitteilbaren Tatbestand sind Objekt und Subjekt miteinander in einer noch viel unslöslicheren Weise verbunden, als es die klassische Physik annahm. Die Situation hat ferner ihre logischen Konsequenzen. Wären z. B. rund und rot in demselben Sinne komplementäre Eigenschaften eines Körpers, wie es Lage und Impuls eines Elektrons sind, so hätte es einen Sinn zu fragen, ob ein Körper rund ist; es hätte einen Sinn zu fragen, ob er rot ist; von rot und rund zu sprechen wäre aber sinnlos, weil die konkrete Situation, in der sich Rundheit feststellen ließe, die andere ausschlösse, in der Röte beobachtbar wäre. Die logische Konjunktion von Merkmalen durch „und“ ist im allgemeinen nicht statthaft, die gewöhnliche Logik muß durch eine eigentümliche Quanten-logik ersetzt werden. Es ist recht mißlich, über diese Dinge in unserer gewöhnlichen Sprache zu reden, wenn man nicht den in sich völlig klaren und lückenlosen Symbolismus der Quatenphysik vor Augen hat.

In der Kopenhagener Schule um Niels Bohr ist eine Art Komplementaritäts-Philosphie entstanden. Es mag sein, daß das Gefühl des Handelns aus freiem Willen völlig legitim, aber an eine Art von Situation gebunden ist, welche diejenige konkrete Situation ausschließt, in der das Verhalten eines Menschen aus Naturkausalität zureichend verstanden werden kann. Es mag sein, daß, um die Aktion eines Lebewesens mit einiger Sicherheit für die nächsten Sekunden vorauszusagen, man so in sein atomares Gefüge eindringen müßte, daß man das Lebewesen mit Notwendigkeit tötet – wodurch dann in der Tat die Voraussagbarkeit auf eine drastische Weise garantiert wäre. Im psychischen Gebiet haben wir die Komplementarität zwischen dem Haben eines Erlebnisses und seinem Gewahrwerden, in dem es zum intentionalen Gegenstand eines neuen, auf den ersten reflektierenden Aktes wird. Es ist auf einem Standpunkt richtig, daß der einzelne Mensch im Kosmos eine verschwindende Rolle spielt, als wäre es von Herzen gleichgiltig, was er tut; es ist auf einem andern Standpunkt richtig, daß es im letzten Ernst auf den Einzelnen, nämlich auf mich ankommt. Auf dem Gebiet des Verhaltens der Menschen zueinander, wie es aus der ihnen innewohnenden Ethik und Religion quillt, ist Bohr geneigt, eine Komplementarität zwischen Gerechtigkeit und Liebe zu statuieren. – Aber besteht nicht doch die Gefahr, daß hier mit der Idee der Komplementarität, die innerhalb der Quantenphysik einem mathematisch genau zu präzisierenden Tatbestand entspricht, ein ähnlicher Mißbrauch getrieben wird wie mit der Idee der Relativität? Auch diese bedeutet in der physikalischen Relativitätstheorie etwas völlig Bestimmtes und hat zu Relativität in allerhand andern vagen Bedeutungen nur vage Analogien. Ich werfe nicht mehr als eine Frage auf.

Aus Zürich erklingt der Ruf nach einer Dialektisierung der Erkenntnis. Es ist mir nicht sehr klar, was damit gemeint ist. Zunächst offenbar dies: daß man zwar nicht darauf verzichten kann, ein theoretisches Bild von der Struktur der Welt zu entwerfen und eine Sprache auszubilden, in der man von diesem durch die Theorie erschlossenen Bild des verborgenen Geschehens sprechen kann; daß aber dieses Bild und diese Sprache nicht mehr durch ein festes a priori festgelegt sein soll, sondern sich den Erfahrungen der Wissenschaft und ihren aus dem Zusammenhang der Wissenschaft selbst sich aufdrängenden Deutungen anzupassen hat. Aber diese freie, der Wechselwirkung zwischen Konstruktion und Besinnung gerechtwerdende Haltung verdient wohl noch nicht den Namen Dialektik. Was darüber hinaus getroffen werden soll, hat Herr Gonseth gelegentlich durch das Beispiel der speziellen Relativitätstheorie illustriert. Hier geht man zunächst von den naiven Begriffen der Geschwindigkeit usw. aus, die sich auf einen absoluten Raum und eine absolute Zeit stützen, und findet dann durch eine genaue Analyse aller einschlägigen Tatsachen nachträglich heraus, daß diese absolute Grundlage und damit der naive Begriff der Geschwindigkeit selbst unhaltbar sind und in ganz bestimmter Weise „relativiert“ werden müssen. In dem neuen relativistischen Bild sind die ursprünglichen Begriffe, in Hegels Doppelsinne des Worts, „aufgehoben“. Das mag historisch zutreffend sein; aber es wäre eben doch nur eine „historische“ Dialektik. Denn von der Relativitätstheorie von Raum und Zeit läßt sich eine vollkommen klare und anschauliche Darstellung geben, die, ohne Anleihen bei dem absoluten Standpunkt zu machen, genau bezeichnen kann, welche Bedeutung die Begriffe Geschwindigkeit usw. in dem neuen Rahmen haben.

Die Sache liegt vielleicht anders in der Quantentheorie. Hier muß man scharf scheiden zwischen dem verborgenen physikalischen Vorgang, der nur durch den Symbolismus der Quantenphysik erfaßbar ist, auf den aber auch mit solchen Worten wie Elektron, Proton, Wirkungsquantum usw. hingewiesen wird, und der tatsächlichen Beobachtung und Messung. Über die letztere müssen wir nach Bohr sprechen in der anschaulich verständlichen Sprache der klassischen Physik; oder sollte man besser sagen: in der Sprache des täglichen Lebens? nämlich in der Sprache des natürlichen Weltverständnisses, auf das wir uns stützen, wenn wir Nagel und Hammer gebrauchen, ein Instrument an einem Brett festschrauben, uns an einen Tisch oder in einen Stuhl setzen, zu einem Freund fahren oder horchend auf eigenes und fremdes Dasein gerichtet sind. Man kann freilich auch die Messung als einen physikalischen Vorgang mit in den ursprünglichen der Messung zu unterwerfenden Vorgang einschließen; dann hat man es mit einem umfassenderen quantenphysikalischen System zu tun. Um aber über dasselbe durch Beobachtung nachkontrollierbare Aussagen zu machen, müssen wir doch wieder in einem neuen Messungsakt durch ein Gitter in diesen Gesamtvorgang einbrechen. Man mag auch darauf hinweisen, daß den in der Sprache der klassischen Physik gemachten Aussagen über die Messungsresultate im quantenphysikalischen Bild des Vorgangs Züge entsprechen, für welche die statistische Unsicherheit nahezu verschwindet. So kann man gewiß in der

Quantenphysik nicht voraussagen, ob das einzelne Elektron den Spin +1 oder —1 hat, aber man kann doch unter experimentell herstellbaren Bedingungen versichern, daß in einem Schwarm von sehr vielen Elektronen mit großer Annäherung $^1/_3$ den Spin +1 und $^2/_3$ den Spin —1 haben und darum die betreffenden Intensitäten mit großer Genauigkeit sich wie 1:2 verhalten. Aber es mag bei alledem doch dabei zu bleiben haben, daß wir das natürliche Weltverständnis und die Sprache, in der dieses Verständnis sich ausspricht, vielleicht ein wenig gereinigt und geklärt durch die klassische Physik, nimmer entbehren können und der Symbolismus der Quantenphysik keinen Ersatz dafür zu bieten vermag. Dann handelte es sich um eine echte, durch keine historische Entwicklung aufzuhebende Dialektik: der dunkle Boden kann wohl von dem höheren Standpunkt der Quantenphysik aus beleuchtet werden, aber bleibt der Grund, der nicht in das Licht der oberen Region aufgelöst werden kann.

Hier nun wittere ich gewisse Beziehungen der Quantenphysik und der Grundlagen der Mathematik zur Existentialphilosphie. Als Bertrand Russell und andere versuchten, die Mathematik auf reine Logik zurückzuführen, blieb noch ein Rest von Bedeutung, in Form der einfachen logischen Begriffe. Aber im Hilbert'schen Formalismus entschwindet auch dieser Rest. Hingegen bedürfen wir der Zeichen, wirklicher Zeichen mit Kreide auf die Tafel oder mit der Feder aufs Papier geschrieben. Wir müssen verstehen, was es heißt, einen Strich hinter den anderen zu setzen; es wäre verkehrt, diese naiv und grob verstandene räumliche Anordnung der Zeichen auf eine solche gereinigte räumliche Anschauung und Struktur, wie sie etwa in der Euklidischen Geometrie ihren Ausdruck findet, zu reduzieren. Sondern wir müssen uns hier auf das natürliche Verständnis im Umgang mit den Dingen in unserer natürlichen Umwelt stützen. Nicht reine Ideen eines reinen Bewußtseins, sondern konkrete Zeichen liegen zugrunde, die unbeschadet kleiner Variationen in der Detailausführung für uns wiedererkennbar und reproduzierbar sind und von denen wir im wesentlichen (d. h. wiederum abgesehen von irrelevanten Détails, die der Zufall unserm Handeln in der Welt unvermeidlich anhängt) wissen, wie wir mit ihnen umzugehen haben.

Als Wissenschaftler möchten wir versucht sein, so zu argumentieren: „Wie wir wissen", besteht die Kreide auf der Wandtafel aus Molekülen, und diese sind zusammengesetzt aus geladenen und ungeladenen Elementarteilchen, Elektronen, Neutronen, etc. Aber analysierend, was die theoretische Physik mit solchen Worten meint, haben wir gesehen, daß diese physikalischen Dinge sich auflösen in einen nach Regeln zu handhabenden Symbolismus; die Symbole aber sind letzten Endes wieder konkrete, mit Kreide auf die Tafel geschriebene Zeichen. Sie bemerken den lächerlichen Zirkel. Wir entrinnen ihm nur, wenn wir die Weise, in der wir im täglichen Leben die Dinge und Menschen, mit ihnen umgehend, verstehen, als ein unreduzierbares Fundament gelten lassen. So wie wir in Hilberts formalisierter Mathematik mit den konkreten Zeichen umgehen, so gehen wir in der Physik, wenn wir Messungen anstellen, mit den dazu nötigen Veranstaltungen wie Brettern, Drähten, Schrauben, Zahnrädern, Zeiger und Skala um. Wir bewegen uns hier auf demselben Niveau des Verstehens und Handhabens wie der Tischler oder der Mechaniker in seiner Werkstatt. Die Analogie zwischen formalisierter Mathematik und Quantenphysik ließe sich leicht weiter durchführen. Wie sich in der

Mathematik auch die inhaltlichen metamathematischen Überlegungen formalisieren lassen, indem man den Formalismus nötigenfalls erweitert, so kann man in der Physik das Messen mit in den quantenphysikalischen Vorgang einbeziehen. Aber im ersten Fall kann dieses erweiterte System dann doch wieder zum Gegenstand inhaltlichen Nachdenkens gemacht werden, indem wir z. B. nach seiner Widerspruchslosigkeit fragen; im zweiten Fall muß man, um über das erweiterte System Aussagen machen zu können, dasselbe der wirklichen Beobachtung und Messung unterwerfen.

Hier nun, denke ich, leuchtet die Beziehung zur Existentialphilosophie, z. B. in der Heidegger'schen Form, ein. Betont doch Heidegger mit einer gewissen Emphase, daß aller Erkenntnis dies natürliche Verstehen, die Aufgeschlossenheit vorausliegt, in der mir die Welt und meine Mitmenschen begegnen. Den zum Bewußtsein seiner selbst erwachenden Menschen beschreibt er als Dasein, als die Form des Seins, dessen Sein das Verstehen dieses Seins einschließt. Es ist jeweils ich selbst und versteht sich als in der Welt seiend und mit anderen seiend. In dieser Aufgeschlossenheit geht es verstehend mit Dingen um. Nur aus der Fülle dieses Verstehens, in dem so offenkundig und doch so schwer ausdrückbar unser ganzes Leben ruht, kann, durch eine Art von Distanzierung, die bloße neutrale Erkenntnis entstehen, oder kann die Wahrnehmung eines der entfremdeten Natur angehörigen Dinges und, als ein abstraktes Moment daran, die Empfindung von Sinnesqualitäten abgehoben werden. Zum Problem der Außenwelt bemerkt darum Heidegger, daß man nicht zu beweisen hat, daß und wie eine Außenwelt existiert, sondern aufzuzeigen, warum Dasein als Sein-in-der-Welt eine Tendenz hat, die Außenwelt erkenntniskritisch ins Nichts zu begraben und dann sie nachträglich indirekt zu beweisen. Nachdem man das ursprüngliche Phänomen des Seins-in-der-Welt unterdrückt hat, sucht man das zurückgebliebene isolierte Subjekt mit den abgerissenen Weltfetzen wieder zusammenzuleimen; aber es bleibt ein Flickwerk. Indem ich mich selbst als seiend-mit verstehe, verstehe ich auch anderes Dasein. Das ist aber nicht durch Erkenntnis gewonnenes und erschlossenes Wissen; sondern eine primäre existentielle Art des Seins, welche die conditio sine qua non ist für Erkenntnis und Wissen. – Auf den weiteren Inhalt der Heidegger'schen Lehre, wie ihm das mit den Dingen umgehende Dasein vor allem als Sorge erscheint, wie das Dasein beides, Selbstentwurf und Geworfensein ist, wie er Zeit, Gewissen, Schuld und Tod deutet, darauf will ich hier nicht eingehen. Ich wollte nur auf die Parallele hinweisen, die zu bestehen scheint zwischen den hier angedeuteten Gedankengängen der Existentialphilosophie und dem, worauf uns die Untersuchung der Grundlagen der Mathematik und Quantenphysik hingeführt hat. Ich behaupte nicht, daß diese Parallelisierung zutreffend ist. Hinter alles, was ich darüber gesagt habe, setze ich ein dickes Fragezeichen. In der Tat, die Idee des natürlichen Weltverständnisses scheint mir keineswegs klar und eindeutig. Aber die aufgeworfenen Fragen sind wohl des weiteren Nachdenkens wert.

Es wäre vielleicht nötig, auch für die Physik selbst, sorgfältig zu analysieren, was ein Experiment, eine Messung konstituiert und in welcher Weise wir uns über ihr Resultat verständigen. Die Wissenschaft hat im Laufe ihrer Geschichte immer strengere Anforderungen an das gestellt, was sie als eine nachprüfbare und mitteilbare Tatsache anerkennt. An eine Klärung dieser Frage ist wohl nicht zu denken, ohne daß wir uns zugleich über die Art der Sprache klar werden, in der solche Tatsachen kommuniziert werden. Die Bemühung um die Isolierung der Tatsachen inmitten des trüben wirbelnden Strom unseres dahingleitenden Lebens, und die Bemühung um die adäquate Sprache, in der wir Tatsachen einander mitteilen können, sind schöpferische Akte des Menschen, die Hand in Hand gehen müssen.

Elementary algebraic treatment of the quantum mechanical symmetry problem

Canadian Journal of Mathematics 1, 57—68 (1949)

§ 1. Stating the Problem

A function $\eta(i_1, \ldots, i_f)$ of f quantities i, varying over the finite range $i = 1, 2, \ldots, n$, is usually called an n-dimensional tensor of rank f. Any permutation p: $1 \to 1', \ldots, f \to f'$ changes this tensor into a tensor $\boldsymbol{p}\eta$ according to the equation $\boldsymbol{p}\,\eta(i_1, \ldots, i_f) = \eta(i_{1'}, \ldots, i_{f'})$. Thus the permutation p appears as a linear operator $\boldsymbol{p}$ in the n-dimensional space $\Sigma = \Sigma_{n,f}$ of all n-dimensional tensors of rank f. η is symmetric if $\boldsymbol{p}\,\eta = \eta$ for all permutations p, it is anti-symmetric if $\boldsymbol{p}\,\eta = \delta_p \cdot \eta$ where $\delta_p = +1$ for the even and -1 for the odd permutations. Let a linear transformation A in Σ,

$$\eta' = A\,\eta, \quad \eta'(i_1 \ldots i_f) = \sum_k a(i_1 \ldots i_f;\, k_1 \ldots k_f) \cdot \eta(k_1 \ldots k_f), \tag{1.1}$$

be called symmetric[1]) if

$$a(i_{1'} \ldots i_{f'};\, k_{1'} \ldots k_{f'}) = a(i_1 \ldots i_f;\, k_1 \ldots k_f)$$

for all permutations p. A is symmetric if and only if it commutes with all the permutation operators $\boldsymbol{p}$. The symmetric transformations A form an algebra $\mathfrak{A}$. The general symmetry problem posed by the quantum theory of an aggregate of f equal physical entities is this:

(I) *to decompose the tensor space Σ as far as possible into subspaces Π that are invariant with respect to all symmetric transformations A.*

An epistemological principle basic for all theoretical science, that of projecting the actual upon the background of the possible, is here followed by asking what happens under any *possible* Schrödinger law of dynamics $(h/i)\, d\eta/dt = A\,\eta$, before taking up the specific law involving the *actual* energy operator $A = H$. We have here ignored the further condition which physics imposes on all energy operators A, to wit their Hermitean nature,

$$a(k_1 \ldots k_f;\, i_1 \ldots i_f) = \bar{a}(i_1 \ldots i_f;\, k_1 \ldots k_f).$$

[1]) We shall adhere to this terminology and not use the word symmetric in the sense presently to be mentioned under the name Hermitean.

Essential for the theory of eigenvalues (terms) and eigenfunctions, this condition is irrelevant for our purposes. For what is invariant under all Hermitean symmetric transformations stays so even when the Hermitean restriction is lifted. As algebraists we are glad to get rid of it. For we propose to carry our investigation through in any number field in which the equation $f!\,a = 0$ for a number a implies $a = 0$ (field of characteristic 0 or of a prime characteristic dividing none of the natural numbers 1, 2, ..., f). It is no wonder that the complete solution of the above symmetry problem depends on the theory of representations of the symmetric group of all permutations and YOUNG's symmetry operators[1]).

Let Σ^+, Σ^- denote the linear manifolds of all symmetric or antisymmetric tensors respectively. Nature has most wisely put a stop to the breaking-up of Σ into isolated compartments Π by letting but one of them, the invariant subspace Σ^-, come into existence. Such at least is the case if the f entities of which the aggregate is composed are electrons (PAULI's exclusion principle). Thereby the symmetry problem (I) loses its significance for physics. Part of it, however, is restored, if the existence of the spin of the electron is taken into account but its dynamical influence disregarded—a procedure which is at least approximately permissible. The situation is then as follows. The argument i is replaced by a pair $(i\,\varrho)$ with the range $i = 1, \ldots, n$ for the 'positional' variable i and the range $\varrho = 1, \ldots, \nu$ for the 'spin' variable ϱ. (Actually $\nu = 2$ while the positional variable varies over the continuum of all possible positions in the physical three-dimensional space.) Set $N = n\nu$. The possible wave states of the aggregate of f electrons are described by the antisymmetric N-dimensional tensors $\psi(i_1\,\varrho_1, \ldots, i_f\,\varrho_f)$ of rank f, forming the space $\Sigma^-_{N,f} = \Omega$. Moreover we envisage the space $\Sigma = \Sigma_{n,f}$ of all n-dimensional tensors $\eta(i_1 \ldots i_f)$ of rank f, and the space $P = \Sigma_{\nu,f}$ of all ν-dimensional tensors $\varphi(\varrho_1, \ldots, \varrho_f)$ of rank f. Any symmetric transformation A in Σ,

$$\eta'(i_1 \ldots i_f) = \sum_k a(i_1 \ldots i_f;\, k_1 \ldots k_f) \cdot \eta(k_1 \ldots k_f)$$

induces a transformation A^* in Ω,

$$\psi'(i_1\,\varrho_1, \ldots, i_f\,\varrho_f) = \sum_k a(i_1 \ldots i_f;\, k_1 \ldots k_f) \cdot \psi(k_1\,\varrho_1, \ldots, k_f\,\varrho_f).$$

The central problem is

(II) *to decompose Ω as far as possible into subspaces that are invariant under the transformations A^* thus induced in Ω by all symmetric transformations A in Σ.*

These A^* form an algebra $\mathfrak{A}^*$. It is also true that any symmetric transformation B in P,

$$\varphi'(\varrho_1 \ldots \varrho_f) = \sum_\sigma b(\varrho_1 \ldots \varrho_f;\, \sigma_1 \ldots \sigma_f) \cdot \varphi(\sigma_1 \ldots \sigma_f), \tag{1.3}$$

---

[1]) Cf. H. WEYL, *Gruppentheorie und Quantenmechanik* (2nd ed., Leipzig, 1931) [quoted as GQ], chap. V, §§ 1–7 and 13–14.

induces a corresponding transformation B^* in Ω,

$$\psi'(i_1\,\varrho_1, \ldots, i_f\,\varrho_f) = \sum_\sigma b(\varrho_1 \ldots \varrho_f; \sigma_1 \ldots \sigma_f) \cdot \psi(i_1\,\sigma_1, \ldots, i_f\,\sigma_f). \qquad (1.4)$$

The B^* form an algebra $\mathfrak{B}^*$. Every A^* of $\mathfrak{A}^*$ commutes with every B^* of $\mathfrak{B}^*$.

Not only the problem (I), but also this new symmetry problem (II) may be solved by means of YOUNG's symmetry operators; cf. GQ, chap. v, § 12. However, as shall be discussed here in detail, a more elementary approach is available for the physically important case $\nu = 2$. Indeed the decomposition of the spin tensor space $P = \Sigma_{2,f}$ into irreducible invariant subspaces under the algebra $\mathfrak{B}$ of all its symmetric transformations B is readily derived from the classical CLEBSCH-GORDAN expansion. From the algebra $\mathfrak{B}$ in P we may pass to its representation $\mathfrak{B}^*$ in Ω. Because of the commutability of the elements A^* and B^* of $\mathfrak{A}^*$ and $\mathfrak{B}^*$, decomposition of the generic matrix of $\mathfrak{B}^*$ entails a 'dual' decomposition for $\mathfrak{A}^*$. The deeper lying fact that vice versa any linear transformation in Ω that commutes with all $B^* \in \mathfrak{B}^*$ lies in $\mathfrak{A}^*$ is needed in order to show that the latter decomposition is also one into irreducible parts.

All linear transformations (matrices) in a g-dimensional vector space Ξ form an algebra $\mathfrak{M}_g$ of order g^2, the complete matric algebra of degree g. Throughout our investigation irreducibility for matric algebras will be sharpened to completeness. Decomposition of a matrix C into two matrices $C_1 | C_2$ is defined by the equation

$$C = \begin{Vmatrix} C_1 & 0 \\ 0 & C_2 \end{Vmatrix}$$

$1 \circ C$, $2 \circ C$, $3 \circ C$, ... are the abbreviations for C, $C|C$, $C|C|C$, ..., and S is the summation sign for the addition $|$ of matrices. Let $\mathfrak{C}$ be a matric algebra of order m in a g-dimensional vector space Ξ. Suppose that, relative to a suitably chosen coordinate system for Ξ, the generic matrix C of $\mathfrak{C}$ decomposes into $m_1 \circ C_1 | m_2 \circ C_2 | \ldots$, the matrix C_r of degree g_r occurring with the multiplicity $m_r > 0$, $g = m_1 g_1 + m_2 g_2 + \cdots$. The $g_1^2 + g_2^2 + \cdots$ coefficients of the matrices $C_1, C_2, \ldots, C_h$ are linear forms of the m parameters of $\mathfrak{C}$. We speak of *complete decomposition* if these coefficients are all linearly independent and thus $m = g_1^2 + g_2^2 + \cdots$. For $\nu = 2$ we shall prove the following

Main Theorem. *Relative to a suitably chosen coordinate system for the space Ω, the generic matrix A^* of $\mathfrak{A}^*$ suffers complete decomposition*

$$A^* = \mathsf{S}\,(v+1) \circ A_u^*; \qquad (1.5)$$

u and v are two non-negative integers related by the equation $2u + v = f$. The part A_u^* of 'valence defect' u and the corresponding 'valence' v occurs with the

multiplicity $v+1$. *Set* $d = n - f$, $\bar{u} = d + u$. *The degree* g_u^* *of the matrix* A_u^* *is given by the formula*

$$g_u^* = \binom{n}{u}\binom{n}{\bar{u}} \frac{(n+1)(n+1-u-\bar{u})}{(n+1-u)(n+1-\bar{u})}, \tag{1.6}$$

$\binom{n}{u}$ *denoting the binomial coefficient* $\frac{n!}{u!\,(n-u)!}$.

Only those u *occur in the sum* (1.5) *for which* $u \geqq 0$, $\bar{u} \geqq 0$, $v = n - (u + \bar{u}) \geqq 0$.

Spectroscopically this theorem establishes the existence of non-intercombining term systems corresponding to the various valences v. The terms of valence v are of multiplicity $v+1$. Only when the actually existing weak interactions between the spins are taken into account, each term of valence v splits into a 'multiplet' of $v+1$ slightly different terms; whereas the weak interaction between position and spin accounts for weak intercombinations between the several term systems. The significance of the valence v for chemistry is sufficiently indicated by its name.

After some preliminaries in § 2 the decomposition (1.5) is derived from the Clebsch-Gordan expansion in § 3. Its completeness will be proved in § 4 and 5.

§ 2. Auxiliary Propositions

Schur's lemma for complete instead of irreducible matric algebras is a triviality; nevertheless it may be stated as our

Lemma 1. Complete decomposition of the generic matrix C of a matric algebra $\mathfrak{C}$, $C = m_1 \circ C_1 | m_2 \circ C_2 | \ldots | m_h \circ C_h$, implies the same for its commutator algebra $\mathfrak{D}$, $D = g_1 \circ D_1 | g_2 \circ D_2 | \ldots$ But degree and multiplicity are interchanged: the degree g_r of C_r is the multiplicity with which D_r occurs in the generic matrix D of $\mathfrak{D}$, and the multiplicity m_r of C_r is the degree of D_r.

As one knows, the commutator algebra of a given matric algebra $\mathfrak{C}$ consists of those matrices D that commute with all elements C of $\mathfrak{C}$. As an abstract algebra $\mathfrak{c}$ the completely decomposed matric algebra $\mathfrak{C}$ of Lemma 1 is the direct sum of a number of complete matric algebras; indeed $\mathfrak{c}$ consists of all h-uples $(C_1, \ldots, C_h)$ of arbitrary matrices $C_1, \ldots, C_h$ of the respective degrees $g_1, \ldots, g_h$. We need the following classical proposition, for the simple proof of which I refer the reader to GQ, p. 271, Satz (6.1).

Lemma 2. Every representation of the direct sum $\mathfrak{c}$ of h complete matric algebras is of the form

$$(C_1, \ldots, C_h) \to m_1^* \circ C_1 | \ldots | m_h^* \circ C_h.$$

(Here some of the multiplicities m_r^* may be zero; this will happen if the representation is not faithful and hence the representing matric algebra $\mathfrak{C}^*$ is of lower order than $\mathfrak{C}$.)

Any *antisymmetric* n-dimensional tensor η of rank f is completely characterized by its components $\eta(\iota_1 \dots \iota_f)$ with $\iota_1 < \iota_2 < \cdots < \iota_f$, and these are independent. We have

$$\eta(i_1 \dots i_f) = \delta_i \cdot \eta(\iota_1 \dots \iota_f)$$

for any permutation $i_1 \dots i_f$ of $\iota_1 \dots \iota_f$, $\delta_i = \pm 1$ distinguishing the even from the odd permutations

$$\begin{pmatrix} \iota_1 \dots \iota_f \\ i_1 \dots i_f \end{pmatrix},$$

and $\eta(i_1 \dots i_f) = 0$ if the numbers $i_1 \dots i_f$ are not all distinct. Hence $\Sigma^- = \Sigma^-_{n,f}$ does not exist unless $n \geqq f$, and its dimensionality is

$$M_n^-(f) = \frac{n!}{f!\,(n-f)!}.$$

Lemma 3. Any linear transformation in Σ^- may be written in the form (1.1) where $a(i_1 \dots i_f;\ k_1 \dots k_f)$ is antisymmetric in the f arguments i, antisymmetric in the f arguments k [and hence symmetric in the f pairs $(i\ k)$].

Indeed a linear transformation in Σ^-,

$$\eta'(\iota_1 \dots \iota_f) = \sum_{\varkappa} \alpha(\iota_1 \dots \iota_f;\ \varkappa_1 \dots \varkappa_f) \cdot \eta(\varkappa_1 \dots \varkappa_f)$$

(with the sum extending over the possible sequences $\varkappa_1 < \cdots < \varkappa_f$ chosen from the range $1, 2, \dots, n$) may be written as (1.1) when one puts

$$a(i_1 \dots i_f;\ k_1 \dots k_f) = \frac{\delta_i\,\delta_k}{f!}\,\alpha(\iota_1 \dots \iota_f;\ \varkappa_1 \dots \varkappa_f)$$

for any permutation $i_1 \dots i_f$ of $\iota_1 \dots \iota_f$ and any permutation $k_1 \dots k_f$ of $\varkappa_1 \dots \varkappa_f$, and puts $a(i_1 \dots i_f;\ k_1 \dots k_f) = 0$ in case the numbers $i_1 \dots i_f$ or $k_1 \dots k_f$ are not all distinct.

It follows from this lemma that the algebra $\mathfrak{A}$ of symmetric transformations is a complete matric algebra *in the invariant subspace* Σ^- *of* Σ.

Any *symmetric* tensor η may be completely characterized by its components $\eta(i_1\, i_2 \dots i_f)$ with $i_1 \leqq i_2 \leqq \cdots \leqq i_f$, and these are independent. On changing the labels $i_1 \dots i_f$ into $i_1 + 0,\ i_2 + 1,\ i_3 + 2,\ \dots,\ i_f + (f-1)$ one sees at once that the dimensionality of the space $\Sigma^+ = \Sigma^+_{n,f}$ of symmetric tensors equals

$$M_n^+(f) = \frac{(n+f-1)!}{f!\,(n-1)!}.$$

Set $\eta(i_1 \dots i_f) = \eta_{f_1 \dots f_n}$ if f_1 of the f arguments $i_1 \dots i_f$ equal 1, f_2 of them equal 2, …, f_n of them equal n. These numbers $\eta_{f_1 \dots f_n}$ corresponding to the various partitions $f_1 + f_2 + \cdots + f_n$ of f can also be used as the independent components of η. A typical symmetric tensor arises from a vector $(x_1, \dots, x_n)$ by the formula

$$\eta(i_1 \dots i_f) = x_{i_1} \dots x_{i_f} \quad \text{or} \quad \eta_{f_1 \dots f_n} = x_1^{f_1} \dots x_n^{f_n}. \tag{2.1}$$

A linear form $l(\eta)$ depending on a variable symmetric tensor η is to be written as

$$l(\eta) = \sum l_{f_1 \cdots f_n} \cdot \eta_{f_1 \cdots f_n}$$

with a constant coefficient $l_{f_1 \cdots f_n}$ for each partition $f_1 + \cdots + f_n$ of f. We make the altogether trivial remark that $l(\eta)$ vanishes identically in η provided it vanishes identically in x by dint of the substitution (2.1).

The symmetric transformation $B = \| b(\varrho_1 \ldots \varrho_f; \sigma_1 \ldots \sigma_f) \|$ of the algebra $\mathfrak{B}$ may be looked upon as a symmetric ν^2-dimensional tensor $b(\omega_1, \ldots, \omega_f)$ of rank f, if each pair $(\varrho\, \sigma)$ is taken as a single argument ω capable of ν^2 values. Hence the order of the matric algebra $\mathfrak{B}$ in P is $M^+_{\nu^2}(f)$ [and the order of $\mathfrak{A}$ is $M^+_{n^2}(f)$]. The linear transformation $t = \| t_{\varrho\sigma} \|$ in the ν-dimensional vector space induces the symmetric transformation $B(t)$,

$$b(\varrho_1 \ldots \varrho_f; \sigma_1 \ldots \sigma_f) = t_{\varrho_1 \sigma_1} \ldots t_{\varrho_f \sigma_f} \tag{2.2}$$

in the tensor space P. Considering the ν^2 coefficients $t_{\varrho\sigma}$ as indeterminates, we speak of t as the generic element of the linear group ζ and of $t \to B(t)$ as the representation ζ^f of ζ. Equation (2.2) is in complete analogy to (2.1), and the 'altogether trivial remark' made above amounts to the following

Lemma 4. A linear form $l(B)$ depending on an arbitrary element B of $\mathfrak{B}$ vanishes identically if it vanishes identically in the parameters $t_{\varrho\sigma}$ for $B = B(t)$.

As a final lemma we write down a simple formula for the case $\nu = 2$, $\nu^2 = 4$:

Lemma 5.

$$M_4^+(f) = \frac{(f+1)\,(f+2)\,(f+3)}{1 \cdot 2 \cdot 3} = \sum (v+1)^2 \tag{2.3}$$

where the sum extends over the non-negative members v of the sequence f, $f-2$, $f-4$, ...

Proof. Verify (2.3) for $f = 0, 1$ and the relation

$$M_4^+(f) - M_4^+(f-2) = (f+1)^2$$

for all $f \geqq 2$.

§ 3. The Clebsch-Gordan Expansion and the Decomposition of $\mathfrak{A}^*$

In this section we assume $\nu = 2$.

The symmetric 2-dimensional tensors $\varphi(\varrho_1 \ldots \varrho_v)$ $(\varrho = 1,2)$ of rank v $(\leqq f)$ form a linear manifold $P_v^+ = \Sigma_{2,v}^+$ of $v+1$ dimensions. In agreement with a usage established above denote by φ_h the component $\varphi(\varrho_1 \ldots \varrho_v)$ in which h of the v arguments ϱ have the value 1 and $h - v$ have the value 2 ($h = 0, 1, \ldots, v$). The indeterminate transformation $t = \| t_{\varrho\sigma} \|$ $(\varrho, \sigma = 1,2)$ in the 2-dimensional vector space induces the transformation

$$\varphi'(\varrho_1 \ldots \varrho_v) = \sum_{\sigma} t_{\varrho_1 \sigma_1} \ldots t_{\varrho_v \sigma_v} \cdot \varphi(\sigma_1 \ldots \sigma_v)$$

in P_v^+, and thus P_v^+ appears as the representation space of a definite representation Z_v of ζ of degree $v+1$. By multiplying the transformed components φ'_h by a fixed power Δ^u ($u = 0, 1, 2, \ldots$) of the determinant $\Delta = t_{11}t_{22} - t_{12}t_{21}$ one obtains a representation $\Delta^u Z_v$ of ζ of the same degree $v+1$. Envisage the subgroup ζ_0 of ζ, the generic element of which is the substitution

$$\begin{Vmatrix} t_{11}, & t_{12} \\ t_{21}, & t_{22} \end{Vmatrix} = \begin{Vmatrix} \lambda, & 0 \\ 0, & 1 \end{Vmatrix} \tag{3.1}$$

with one indeterminate parameter λ. That substitution multiplies φ_h by λ^h according to the representation Z_v, by λ^{u+h} according to the representation $\Delta^u Z_v$. Hence the coordinates in the representation space Π of $\Delta^u Z_v$ are so chosen that they are distinguished by a signature ('magnetic quantum number') $w = u + h$. This signature is the exponent of the factor λ^w taken on by the coordinate with the label w under the influence of (3.1) and ranges over the values $w = u, u+1, \ldots, u+v$. [Decomposition of Π into one-dimensional parts invariant with respect to the subgroup ζ_0 of ζ.]

The 2-dimensional tensors $\varphi(\varrho_1 \cdots \varrho_a, \varrho_{a+1} \cdots \varrho_{a+b})$ of rank $a+b$ ($\leqq f$) which are symmetric in the first a and symmetric in the last b arguments form the substratum of the representation $Z_a \times Z_b$ of ζ of degree $(a+1)(b+1)$. The latter breaks up into parts in accordance with the CLEBSCH-GORDAN formula

$$Z_a \times Z_b = \mathbf{S}\, \Delta^u Z_v, \tag{3.2}$$

the sum extending over all non-negative integers u, v for which $2u+v = a+b$ and $u \leqq \min(a, b)$. This follows by induction from the equation

$$Z_a \times Z_b = Z_{a+b} \mid \Delta(Z_{a-1} \times Z_{b-1}).$$

A simple proof is to be found, for instance, on pp. 115–117 of GQ.

Repeated application of (3.2) leads to a formula of this type:

$$Z_1 \times Z_1 \times \cdots \times Z_1 \ (f \text{ factors}) = \mathbf{S}\, g_u \circ \Delta^u Z_v \quad (2u+v=f).$$

$Z_1 \times \cdots \times Z_1$ is nothing but the representation ζ^f, $t \to B(t)$, of ζ in P, and our formula states that the matrix $B(t)$ breaks up in the manner described by

$$B(t) = \mathbf{S}\, g_u \circ B_u(t) \tag{3.3}$$

into partial matrices $B_u(t)$ of degree $v+1$. Here u, v range over all non-negative integers satisfying the equation $2u+v=f$, and each component $B_u(t)$ occurs with a certain multiplicity $g_u \geqq 0$.

If we now make use of Lemma 4, which also states that two linear forms $l(B)$ are identical if they become identical by the substitution $B = B(t)$, we see at once that the generic matrix B of $\mathfrak{B}$ itself breaks up in the same fashion

$$B = \mathbf{S}\, g_u \circ B_u. \tag{3.4}$$

Lemma 5 then shows that none of the valences $v = f, f-2, f-4, \ldots$ is left out, $g_u > 0$ for $0 \leqq u \leqq f/2$, and that all the coefficients of the various matrices B_u are independent linear forms of the $M_4^+(f)$ parameters $b_{f_1 f_2 f_3 f_4}$ of B. Hence (3.4) is a *complete decomposition.*

$\mathfrak{B}^*$ is a representation of $\mathfrak{B}$, and thus Lemma 2 leads to a similar formula

$$B^* = \mathbf{S}\, g_u^* \circ B_u \quad (g_u^* \geqq 0) \tag{3.5}$$

for the generic matrix B^* of $\mathfrak{B}^*$.

It is not difficult to determine the multiplicities g_u^* explicitly. Specialize the element t of ζ by (3.1) in $B = B(t)$ and the corresponding $B^*(t)$. The effect of this specialized $B^*(t)$ upon a tensor component $\psi(i_1\varrho_1, \ldots, i_f\varrho_f)$ is multiplication by λ^w if w of the f indices $\varrho_1, \ldots, \varrho_f$ are 1 (and $f-w$ of them equal 2). A complete set of independent components of ψ of that type is obtained by choosing

$$\left.\begin{matrix} \varrho_1 = \cdots = \varrho_w = 1 \\ i_1 < \cdots < i_w \end{matrix}\right\} \quad \text{and} \quad \left\{\begin{matrix} \varrho_{w+1} = \cdots = \varrho_f = 2 \\ i_{w+1} < \cdots < i_f . \end{matrix}\right.$$

Hence their number N_w equals

$$N_w = \binom{n}{w} \cdot \binom{n}{f-w} = \binom{n}{w} \cdot \binom{n}{\bar{w}} \tag{3.6}$$

where $d = n - f$ and $\bar{w} = w + d$. According to (3.5) the space Ω breaks up into subspaces Π_u^* of dimensionality $v+1$ in each of which $B^*(t)$ induces the transformation $B_u(t)$. Every one of these g_u^* subspaces Π_u^*, therefore, contributes exactly *one* coordinate of signature w to Ω provided $u \leqq w \leqq u + v = f - u$. This simple argument yields the recursive formula

$$N_w = \sum g_u^*$$

where u ranges over all integers satisfying the inequalities $u \geqq 0$ and $u \leqq w$, $u \leqq f - w$. Consequently

$$g_u^* = N_u - N_{u-1} \quad \left(0 \leqq u \leqq \frac{1}{2} f\right). \tag{3.7}$$

Put $\bar{u} = d + u$ so that $v = n - u - \bar{u}$. Now (1.6) readily follows from (3.6) and (3.7), and one sees from this explicit expression that g_u^* is positive provided $u \geqq 0$, $\bar{u} \geqq 0$ and $u + \bar{u} \leqq n$. The range of the valences v actually occurring in the decomposition of $\mathfrak{B}^*$ is thus circumscribed by the relations

$$v \geqq 0, \quad v \leqq n \pm d, \quad v \equiv n \pm d \pmod 2\text{[1]}.$$

[1]) In passing we notice that the order of the algebra $\mathfrak{B}^*$ may now be evaluated as $\Sigma(v+1)^2$, the sum extending over the non-negative v of the sequence $f', f'-2, \ldots$ where $f' = \min(n-d, n+d) = \min(f, 2n-f)$, and hence equals $(f'+1)(f'+2)(f'+3)/1 \cdot 2 \cdot 3$. It should be easily possible to confirm this directly.

$\mathfrak{B}^*$ serves merely as a jumping board for $\mathfrak{A}^*$. But since every A^* commutes with all the transformations B^* of $\mathfrak{B}^*$ the decomposition (1.5) of the generic matrix A^* of $\mathfrak{A}^*$ is now inferred from Lemma 1. A definite decomposition according to valences is thus obtained, and for physics this is the most essential result. However, as long as we have not yet convinced ourselves that $\mathfrak{A}^*$ is not only contained in, but identical with, the commutator algebra of $\mathfrak{B}^*$, completeness for the decomposition (1.5) is not ensured. In order to settle this point (5) one first has to prove that the only operators in P that commute with the symmetric transformations B are the symmetry operators (4).

§ 4. Symmetric Transformations and Permutations

Our present object is the space $\Sigma = \Sigma_{n,f}$ of the n-dimensional tensors $\eta(i_1 \dots i_f)$ of rank f. The permutations $\boldsymbol{p}$ and any linear combinations of them, $\boldsymbol{a} = \Sigma_p\, a(p)\, \boldsymbol{p}$, are linear operators in Σ, $\eta' = \boldsymbol{a}\,\eta$, which commute with all the symmetric linear transformations $\dot{\eta} = A\,\eta$,

$$\dot{\eta}(i_1 \dots i_f) = \sum_k a(i_1 \dots i_f;\, k_1 \dots k_f) \cdot \eta(k_1 \dots k_f).$$

We introduce the symmetry quantities $\underset{\cdot}{a} = \Sigma_p\, a(p)\, p$ [with arbitrary numbers $a(p)$ as coefficients][1]) quite independently from their usage as operators in Σ. They form an abstract algebra of order $f!$, the 'group ring of the symmetric group'.

Let η be a tensor and $i_1, \dots, i_f$ a given sequence of integers from the interval $1 \leqq i \leqq n$. We consider the $f!$ numbers $\boldsymbol{p}\,\eta(i_1 \dots i_f) = x(p)$ as the coefficients of a symmetry quantity $\underset{\cdot}{x} = \sim\eta(i_1 \dots i_f)$. The tensor equation $\eta' = \boldsymbol{a}\,\eta$ is equivalent with $\sim\eta' = (\sim\eta) \cdot \hat{a}$ where $\hat{a}$ is the symmetry quantity with the coefficients $\hat{a}(p) = a(p^{-1})$. Here $\sim\eta$ may be interpreted as the symmetry quantity with the tensorial coefficients $\sim\eta(p) = \boldsymbol{p}\,\eta$, or one may replace $\sim\eta$ and $\sim\eta'$ in our equation by the ordinary symmetry quantities $\sim\eta(i_1 \dots i_f)$ and $\sim\eta'(i_1 \dots i_f)$ corresponding to any argument combination $i_1, \dots, i_f$.

The group ring is an $f!$-dimensional vector space. In it we envisage those symmetry quantities $\sim\eta(i_1 \dots i_f)$ that arise from arbitrary tensors η and arbitrary argument combinations $(i_1, \dots, i_f)$, and we determine their linear closure $\varkappa = \varkappa_n$, i.e. the smallest linear subspace that comprises them all. Let γ_s $(s = 1, 2, \dots, n^f)$ be a basis for the space Σ. Then the elements $\underset{\cdot}{x}$ of $\varkappa$ are given by the equation

$$\underset{\cdot}{x} = \sum_{s;\,i} \xi_s(i_1 \dots i_f) \cdot \sim\gamma_s(i_1 \dots i_f) \tag{4.1}$$

where the $\xi_s(i_1 \dots i_f)$ are arbitrary coefficients. Write more explicitly

$$x(p) = \sum_{s;\,i} \xi_s(i_1 \dots i_f) \cdot \boldsymbol{p}\,\gamma_s(i_1 \dots i_f) = \sum_{s;\,i} \boldsymbol{p}^{-1}\,\xi_s(i_1 \dots i_f) \cdot \gamma_s(i_1 \dots i_f),$$

[1]) The dot under a letter merely serves to indicate that it stands for a symmetry quantity.

hence

$$\hat{x}=\sum_{s;i}\gamma_s(i_1\ldots i_f)\cdot\sim\xi_s(i_1\ldots i_f). \tag{4.2}$$

Since $\gamma_s'=\mathbf{a}\,\gamma_s$ implies $\sim\gamma_s'=(\sim\gamma_s)\cdot\hat{a}$ one sees that $x\,\hat{a}$ lies in $\varkappa$ if $\underset{\cdot}{x}$ does; $\varkappa$ is therefore not only an algebra, but even a right-ideal. But in (4.2) one may consider ξ_s as a tensor and the $\gamma_s(i_1\ldots i_f)$ as coefficients; consequently $\hat{x}$ lies in $\varkappa$ if $\underset{\cdot}{x}$ does, and thus $\varkappa$ is also a left-ideal. Introduce $\xi_s'=\mathbf{a}\,\xi_s$; then (4.2) yields

$$\hat{x}\cdot\hat{a}=\sum_{s;i}\gamma_s(i_1\ldots i_f)\cdot\sim\xi_s'(i_1\ldots i_f),$$

$$\underset{\cdot}{a}\cdot\underset{\cdot}{x}=\sum_{s;i}\xi_s'(i_1\ldots i_f)\cdot\sim\gamma_s(i_1\ldots i_f). \tag{4.3}$$

As a left-ideal $\varkappa$ has a generating idempotent $\underset{\cdot}{e}$. This means that $\underset{\cdot}{z}\,\underset{\cdot}{e}$ is in $\varkappa$ whatever the symmetry quantity $\underset{\cdot}{z}$, and if $\underset{\cdot}{z}$ lies in $\varkappa$ then $\underset{\cdot}{z}=\underset{\cdot}{z}\,\underset{\cdot}{e}$. Similar statements hold for multiplication by $\hat{e}$ on the left. The ensuing equations $\hat{e}=\hat{e}\cdot e$ and $\underset{\cdot}{e}=\hat{e}\cdot\underset{\cdot}{e}$ show that $\underset{\cdot}{e}=\hat{e}$. Every tensor η satisfies the equation $\mathbf{e}\,\eta=\eta$.

One more fact about $\varkappa$ is of importance. Introduce as the trace $\mathrm{tr}(a)$ of a symmetry quantity a the coefficient $a(\mathbf{1})$ corresponding to the identical permutation $\mathbf{1}$. The scalar product $\mathrm{tr}(ab)=\Sigma_p a(p^{-1})\cdot b(p)$ is clearly a symmetric and non-degenerate bilinear form of the two arbitrary symmetry quantities a and b. This non-degeneracy is preserved under restriction to $\varkappa$; i.e. an $a\in\varkappa$ such that $\mathrm{tr}(a\,b)=0$ for every $\underset{\cdot}{b}\in\varkappa$ is necessarily zero. Indeed let $\underset{\cdot}{z}$ be an arbitrary symmetry quantity; then $\underset{\cdot}{b}=\underset{\cdot}{z}\,\underset{\cdot}{e}$ is in $\varkappa$, hence $\mathrm{tr}(a\,\underset{\cdot}{z}\cdot e)=0$. But with a also $\underset{\cdot}{a}\,\underset{\cdot}{z}$ lies in $\varkappa$, therefore $a\,\underset{\cdot}{z}\cdot e=a\,\underset{\cdot}{z}$. Thus our equation turns into $\mathrm{tr}(\underset{\cdot}{a}\,\underset{\cdot}{z})=0$ for every $\underset{\cdot}{z}$, and that implies $\underset{\cdot}{a}=0$.

Theorem I. *The symmetry quantities $\mathbf{a}$ if interpreted as operators in Σ are the only ones that commute with all symmetric transformations A. The symmetry quantity $\underset{\cdot}{a}$ expressing such an operator can be uniquely normalized by requiring $\underset{\cdot}{a}$ to lie in $\varkappa$.*

Proof (cf. GQ, pp. 266–267)[1]. Let L be a linear operator in Σ, $\eta\rightarrow L\,\eta$,

---

[1]) By using deeper algebraic resources than we care to employ in this elementary approach, Theorem I could be obtained as an immediate consequence of the following two facts: (α) Every representation $\underset{\cdot}{a}\rightarrow\mathbf{a}$ of the group ring of the symmetric group breaks up into irreducible parts (is 'fully reducible'); (β) A fully reducible matric algebra coincides with the commutator algebra of its commutator algebra (R. Brauer).–Another variant: Explicit construction by means of Young's symmetry operators shows that the inequivalent irreducible parts of the representation $a\rightarrow\mathbf{a}$ are *absolutely* irreducible and inequivalent, and consequently (α) yields a *complete* decomposition. With this additional knowledge (β) can be replaced by the trivial fact that complete decomposition of a matric algebra implies its identity with the commutator algebra of its commutator algebra.

commuting with all symmetric A. Let $L\gamma_s = \beta_s$, and with the same coefficients $\xi_s(i_1 \dots i_f)$ as in (4.1) form

$$\underset{\cdot}{y} = \sum_{s;\,i} \xi_s(i_1 \dots i_f) \cdot \sim\beta_s(i_1 \dots i_f).$$

I am going to show that the equation $x = 0$ for the arbitrary coefficients $\xi_s(i_1 \dots i_f)$ implies $\underset{\cdot}{y} = 0$. Let η be any tensor and set

$$\theta = \sum_p x(p^{-1}) \cdot \boldsymbol{p}\,\eta, \quad \tilde{\theta} = \sum_p y(p^{-1}) \cdot \boldsymbol{p}\,\eta.$$

Then

$$\theta(i_1 \dots i_f) = \sum_s \sum_k a_s(i_1 \dots i_f;\, k_1 \dots k_f) \cdot \gamma_s(k_1 \dots k_f),$$

$$\tilde{\theta}(i_1 \dots i_f) = \sum_s \sum_k a_s(i_1 \dots i_f;\, k_1 \dots k_f) \cdot \beta_s(k_1 \dots k_f),$$

where

$$a_s(i_1 \dots i_f;\, k_1 \dots k_f) = \sum_p \boldsymbol{p}\,\eta(i_1 \dots i_f) \cdot \boldsymbol{p}\,\xi_s(k_1 \dots k_f)$$

is clearly the matrix of a symmetric operator A_s in Σ. As A_s commutes with L we conclude that $\tilde{\theta} = L\,\theta$. Consequently $\theta = 0$ implies $\tilde{\theta} = 0$, and $\underset{\cdot}{x} = 0$ implies $\Sigma_p\, y(p^{-1}) \cdot \boldsymbol{p}\,\eta(i_1 \dots i_f) = 0$, or $\mathrm{tr}(\underset{\cdot}{y}\, y^*) = 0$ for every $y^* \in \varkappa$. The quantity $\underset{\cdot}{y}$ itself is in $\varkappa$, and hence the last equation forces $\underset{\cdot}{y}$ to vanish.

This settled, one concludes that the correspondence $\underset{\cdot}{x} \to \underset{\cdot}{y} = R\,\underset{\cdot}{x}$ defines a linear mapping R of $\varkappa$ into itself. Formula (4.3) and its parallel

$$\underset{\cdot}{a} \cdot \underset{\cdot}{y} = \sum_{s;\,i} \xi'_s(i_1 \dots i_\cdot) \cdot \sim\beta_s(i_1 \dots i_f)$$

prove the mapping R to be a similarity; i.e. it carries $a\,\underset{\cdot}{x}$ into $\underset{\cdot}{a}\,\underset{\cdot}{y}$ whatever $\underset{\cdot}{a}$. Replace $\underset{\cdot}{x}$ and a by $\underset{\cdot}{e}$ and x. Setting $R\,\underset{\cdot}{e} = \hat{a}$ one finds that $x = x\,\underset{\cdot}{e}$ goes into $\underset{\cdot}{x} \cdot R\,\underset{\cdot}{e} = x\,\hat{a}$. This statement is equivalent with the n^f equations $\beta_s = \boldsymbol{a}\gamma_s$, or $L\eta = \boldsymbol{a}\eta$ for every tensor η. The symmetry quantities $\hat{a}$ and $\underset{\cdot}{a}$ lie in $\varkappa$.

§ 5. The Reciprocity of $\mathfrak{A}^*$ and $\mathfrak{B}^*$

In this section ν is not assumed to have the special value 2.

Theorem II. *$\mathfrak{A}^*$ is the commutator algebra of $\mathfrak{B}^*$.*

Proof. Let

$$C = \| c(i_1\varrho_1, \dots, i_f\varrho_f;\, k_1\sigma_1, \dots, k_f\sigma_f) \|$$

be the matrix of any linear transformation in Ω in the unique normalization established by Lemma 3. Hence C is antisymmetric in the f pairs $(i\,\varrho)$, antisymmetric in the f pairs $(k\,\sigma)$, and thereby symmetric in the f quadruples

$(i\,\varrho,\ k\,\sigma)$. Let $X = \| x(\varrho_1 \ldots \varrho_f;\ \sigma_1 \ldots \sigma_f) \|$ be symmetric in the f pairs $(\varrho\,\sigma)$. Then CX with the components

$$\sum_{\tau} c(i_1\,\varrho_1,\ \ldots,\ i_f\,\varrho_f;\ k_1\,\tau_1,\ \ldots,\ k_f\,\tau_f) \cdot x(\tau_1 \ldots \tau_f;\ \sigma_1 \ldots \sigma_f)$$

is certainly antisymmetric in the pairs $(i\,\varrho)$, and since it is symmetric in the quadruples $(i\,\varrho,\ k\,\sigma)$ it is also antisymmetric in the pairs $(k\,\sigma)$. The same is true for XC. Our hypothesis demands that CX and XC coincide as operators in Ω. Hence their matrices in normalized form must be identical. For fixed $i_1 \ldots i_f;\ k_1 \ldots k_f$ the coefficients

$$c(\varrho_1 \ldots \varrho_f;\ \sigma_1 \ldots \sigma_f) = c(i_1\,\varrho_1,\ \ldots,\ i_f\,\varrho_f;\ k_1\,\sigma_1,\ \ldots,\ k_f\,\sigma_f)$$

form a matrix $\| c(\varrho_1 \ldots \varrho_f;\ \sigma_1 \ldots \sigma_f) \|$ in P which may be denoted by $C(i_1 \ldots i_f;\ k_1 \ldots k_f)$. Theorem I when applied to P rather than Σ shows that this transformation is of the form $\Sigma_p\, t_p\, \boldsymbol{p}$ where

$$t(p) = t_p = t_p(i_1 \ldots i_f;\ k_1 \ldots k_f)$$

are the coefficients of a symmetry quantity

$$\underset{\cdot}{t} = \underset{\cdot}{t}(i_1 \ldots i_f;\ k_1 \ldots k_f) \tag{5.1}$$

that lies in $\varkappa = \varkappa_\nu$. Introduce the transformation

$$T_p = \| t_p(i_1 \ldots i_f;\ k_1 \ldots k_f) \| \tag{5.2}$$

in Σ. Our result may then be written in the form

$$C = \sum_{p} (T_p \times \boldsymbol{p}),$$

the cross indicating the Kronecker product of a matrix in Σ (first factor) and a matrix in P (second factor). If we are not afraid of making use of a symmetry quantity $\underset{\cdot}{T}$ whose coefficients are the matrices T_p in Σ we can express the fact that each $\underset{\cdot}{t}$ lies in $\varkappa_\nu$ by the equations

$$\underset{\cdot}{T}\,\underset{\cdot}{e} = \underset{\cdot}{e}\,\underset{\cdot}{T} = \underset{\cdot}{T}, \tag{5.3}$$

$\underset{\cdot}{e} = \underset{\cdot}{e}_\nu$ being the generating idempotent of $\varkappa = \varkappa_\nu$.

C is antisymmetric in the pairs $(k\,\sigma)$. Hence

$$C(\boldsymbol{q} \times \boldsymbol{q}) = \delta_q \cdot C \tag{5.4}$$

for any permutation q. It is antisymmetric in the pairs $(i\,\varrho)$; hence also

$$(\boldsymbol{q} \times \boldsymbol{q})\, C = \delta_q \cdot C. \tag{5.5}$$

comp

(5.4) reads

$$\sum_p \{T_p \mathbf{q} \times \mathbf{p}\,\mathbf{q}\} = \delta_q \cdot \sum_p \{T_p \times \mathbf{p}\}$$

$$\sum_p \{T_{pq^{-1}} \mathbf{q} \times \mathbf{p}\} = \delta_q \cdot \sum_p \{T_p \times \mathbf{p}\}. \tag{5.6}$$

confusion use for the moment

$$Q = \left\| q(i_1 \ldots i_f;\, k_1 \ldots k_f) \right\|$$

as a notation for the linear transformation $\mathbf{q}$ in Σ and its matrix. Set $T_p' = T_p Q$,

$$t_p'(i_1 \ldots i_f;\, k_1 \ldots k_f)$$
$$= \sum_l t_p(i_1 \ldots i_f;\, l_1 \ldots l_f) \cdot q(l_1 \ldots l_f;\, k_1 \ldots k_f).$$

Given a combination $(i_1 \ldots i_f;\, k_1 \ldots k_f)$, the symmetry quantity $\underset{\cdot}{t}'$ with the coefficients $t'(p) = t_p'(i_1 \ldots i_f;\, k_1 \ldots k_f)$ lies in $\varkappa_\nu$ because all the quantities $\underset{\cdot}{t}(i_1 \ldots i_f;\, l_1 \ldots l_f)$ do. (This is true for any linear transformation Q in Σ. What holds for $T_p' = T_p Q$ holds likewise for $T_p'' = Q\, T_p$.) For a fixed permutation q the numbers $t^*(p) = t'(p\, q^{-1})$ are the coefficients of the symmetry quantity $\underset{\cdot}{t}^* = \underset{\cdot}{t}' q$. $\| t_p^*(i_1 \ldots i_f;\, k_1 \ldots k_f) \|$ is the matrix $T'_{pq^{-1}} = T_{pq^{-1}} Q$. Hence (5.6) states that $\underset{\cdot}{t}^*$ and $\delta_q \cdot \underset{\cdot}{t}$ coincide as symmetry operators in P. But $\underset{\cdot}{t}^* = \underset{\cdot}{t}' \underset{\cdot}{q}$ lies in $\varkappa_\nu$ because $\underset{\cdot}{t}'$ does; coincidence as operators in P, therefore, implies identity of the symmetry quantities themselves, $\underset{\cdot}{t}^* = \delta_q \cdot \underset{\cdot}{t}$ or

$$T_{pq^{-1}} \mathbf{q} = \delta_q \cdot T_p.$$

Setting $q = p$, $T_1 = A$, one finds

$$T_p = \delta_p \cdot A\, \mathbf{p}.$$

In the same manner (5.5) leads to

$$T_p = \delta_p \cdot \mathbf{p} A.$$

The transformation A in Σ thus commutes with the permutation operators $\mathbf{p}$ in the same space and is therefore symmetric. Because of the antisymmetry of $\psi(i_1 \varrho_1, \ldots, i_f \varrho_f)$ in the pairs $(i\, \varrho)$ the equation

$$\psi' = C\, \psi = \sum_p (T_p \times \mathbf{p})\, \psi$$

may be written as

$$\psi' = \sum_p \delta_p \cdot (T_p \mathbf{p}^{-1} \times I)\, \psi = f!(A \times I)\, \psi$$

where I stands for identity, and thus Theorem II is proved.

The normalizing condition (5.3) takes on the form

$$\check{\boldsymbol{e}}\, A = A\, \check{\boldsymbol{e}} = A\,, \tag{5.7}$$

$\check{e}$ being the idempotent with the coefficients $\delta_p \cdot e(p) = \delta_p \cdot e(p^{-1})$. This, however, is no surprise. As a matter of fact, A induces the same transformation A^* in Ω as $\check{\boldsymbol{e}}\, A\, \boldsymbol{e}$, and hence, whether or not A satisfies (5.7), it can always be so modified as to fulfil that relation, without change in the corresponding A^*.

Application of Theorem II to $\nu = 2$ shows that the decomposition (1.5) by valences is complete.

144.

Supplementary note (anschließend an: S. Minakshisundaram: A generalization of Epstein zeta functions)

Canadian Journal of Mathematics 1, 326—327 (1949)

Supplementary Note by Hermann Weyl

At Minakshisundaram's request I add another proof of his fundamental lemma. This alternative proof is restricted to the boundary value problem I, but for this case yields a more complete result. Take for y a fixed interior point y_0 and set $l_{y_0} = l$. The maximum of the main part of Green's function along the boundary is

$$(4\pi t)^{-k/2} \exp\,(-\,l^2/4t) = j(t),$$

and this function $j(t)$ is on the increase as long as t moves from 0 to $T_0 = l^2/2k$.

Instead of splitting the integrals $\int_0^\infty$ into $\int_0^1 + \int_1^\infty$, as Minakshisundaram does, one splits into $\int_0^{T_0} + \int_{T_0}^\infty$. I maintain that in the first part, i.e. for $0 < t < T_0$, the compensating function $u(x;t) = g(x, y_0\,; t)$ satifies the inequality

$$0 \leq u(x;t) \leq j(t).$$

Proof. (1) For any t in this interval let $m(t)$ and $M(t)$ denote minimum and maximum of $u(x;t)$ as a function of x. Suppose that, contrary to the statement, $m(t_1)$ is negative for a certain t_1 between 0 and T_0, and that $u(x_1;t_1) = m(t_1)$. Since the boundary values of $u(x;t_1)$ are positive, x_1 must be an interior point, and one must have $\Delta u \geq 0$ for $x = x_1$, $t = t_1$. Hence the equation $\frac{\partial u}{\partial t} = \Delta u$ shows that $\frac{\partial u}{\partial t} \geq 0$ for the same values. Choose a positive constant a and form

$$v = e^{-at}\,.\,u, \quad \frac{\partial u}{\partial t} - au = e^{at}\,.\,\frac{\partial v}{\partial t}\,;$$

one then sees that $\frac{\partial v}{\partial t} > 0$ for $x = x_1, t = t_1$. Hence $v(x_1;t)$ actually decreases during a short time, $t_1 \geq t \geq t'_1$, while t decreases starting with the value t_1. Set $e^{-at}.\,m(t) = m^*(t)$. As $m^*(t) \leq v(x_1;t)$ we have *a fortiori* $m^*(t) \leq m^*(t_1)$ in that interval. Thus $m^*(t)$ goes down and continues to be negative while t decreases. That makes it possible to repeat the argument and to conclude that in the whole interval $t_1 \geq t > 0$ the function $m^*(t)$ decreases along with t

and thus stays less than or equal to $m^*(t_1)$. But this is a contradiction, since $v(x; t) \to 0$ for $t \to 0$.

(2) Suppose that, contrary to the proposition, $M(t_1) > j(t_1)$ for a definite value t_1 in the interval $0 < t_1 < T_0$. Then the maximum $M(t_1)$ is taken on at an interior point $x = x'_1$. We have $\Delta u \leq 0$ and hence $\frac{\partial u}{\partial t} \leq 0$ for $x = x'_1$, $t = t_1$, and thus $\frac{\partial u^*}{\partial t} < 0$ for the same values and $u^*(x; t) = u(x; t) - j(t)$. As $M(t) \geq u(x'_1; t)$, the "span" $M^*(t) = M(t) - j(t)$ thus increases during a short time while t decreases from t_1 downward. It therefore stays positive, and repetition of the argument shows that the span is increasing with decreasing t during the whole time $t_1 \geq t > 0$, a result that contradicts the limit equations $u(x; t) \to 0$, $j(t) \to 0$ for $t \to 0$.

Institute for Advanced Study

145.

Almost periodic invariant vector sets in a metric vector space

American Journal of Mathematics 71, 178—205 (1949)

1. The problem: basic notions and axioms. The situation underlying the theory of spherical harmonics deals with (complex-valued) functions $f(P)$ on the sphere Π under the influence of its rotations s. The rotations form a transitive group σ. The spherical harmonics of order l form a linear set of such functions (of dimensionality $2l+1$) that is invariant under all rotations. Here evidently the fact is used that functions $f(P)$ can be added and multiplied by numbers, in other words, that they form a vector space (of infinitely many dimensions). If the rotation s carries the point P into sP the transform $f' = sf$ of the function f is defined by $f'(sP) = f(P)$, or $sf(P) = f(s^{-1}P)$. It is obvious how to generalize this situation:

I. Given a *group* σ and a *vector space* Σ. Any two vectors can be added and a vector can be multiplied by a number; for these operations the well-known axioms of vector geometry hold. Moreover there is associated with each element s of σ a linear transformation $f \to f' = sf$ of the vector space,

$$s(f_1 + f_2) = (sf_1) + (sf_2), \qquad s(\alpha f) = \alpha \cdot sf,$$

so that

$$1 f = f, \qquad t(sf) = (ts)f$$

(α being any number, f, f_1, f_2 any vectors, s, t any group elements and 1 denoting the unit element of σ).

Definition. h linearly independent vectors $g_1, \cdots, g_h$, or their linear combinations $\xi_1 \cdot g_1 + \cdots + \xi_h \cdot g_h$ by arbitrary numerical coefficients ξ_μ, form an invariant set Γ if $g \in \Gamma$ implies $sg \in \Gamma$ for every s in the group:

$$sg_\mu = \sum_\nu \omega_{\nu\mu}(s) \cdot g_\nu \qquad (\mu, \nu = 1, \cdots, h).$$

$g = \xi_1 g_1 + \cdots + \xi_h g_h$ is changed by s into $sg = \xi_1{}^s g_1 + \cdots + \xi_h{}^s g_h$ where

$$\xi_\mu{}^s = \sum_\nu \omega_{\mu\nu}(s)\xi_\nu.$$

* Received May 11, 1948.

$s \to \Omega(s) = \| \omega_{\mu\nu}(s) \|$ is a matric representation Ω of the group σ of degree h. Change of the basis of Γ changes Ω into an equivalent representation. We speak of an invariant set of degree h and order Ω. The set is called primitive if Ω is an irreducible representation.

The salient point in the theory of spherical harmonics is the fact that they form a complete orthogonal system. Clearly this fact depends on the possibility to form a scalar product (g, f) of two functions f, g (namely by integrating $\bar{g}(P) \cdot f(P)$ over the sphere by means of an area element $d\omega_P$ that is invariant under rotations). Hence we add to I the following assumptions:

II. Σ is a metric vector space; i. e. with any two vectors f, g there is associated a number (g, f) as their scalar product, that is linear in the factor f, co-linear in the factor g:

$$(g, f_1 + f_2) = (g, f_1) + (g, f_2) \qquad (g_1 + g_2, f) = (g_1, f) + (g_2, f)$$
$$(g, \alpha f) = \alpha \cdot (g, f) \qquad (\alpha g, f) = \bar{\alpha} \cdot (g, f).$$

It is of Hermitian symmetry, $(f, g) = \overline{(g, f)}$, and $\| f \|^2 = (f, f)$ is positive except for $f = 0$. The linear transformations (s) in Σ, $f \to sf$, associated with the elements s of σ, are isometric, $(sg, sf) = (g, f)$.

The equation $\| \alpha f \| = | \alpha | \cdot \| f \|$ and the inequalities

$$|(f_2, f_1)| \leqq \| f_1 \| \cdot \| f_2 \|, \quad \| f_1 + f_2 \| \leq \| f_1 \| + \| f_2 \|$$

are an immediate consequence of these assumptions characteristic for a "unitary metric." More generally, for any numbers ξ_i and vectors f_i, g_i $(i = 1, \cdots, n)$ one has the relations

$$\| \sum_i \xi_i f_i \|^2 \leq \sum_i | \xi_i |^2 \cdot \sum_i \| f_i \|^2, \tag{1.1}$$

$$| \sum_i (g_i, f_i) |^2 \leq \sum_i \| g_i \|^2 \cdot \sum_i \| f_i \|^2, \tag{1.2}$$

which are proved in the same manner as the analogous Cauchy-Schwarz inequality for numbers.

We can now choose the basis g_μ $(\mu = 1, \cdots, h)$ of a given invariant set Γ as a unitary one, $(g_\mu, g_\nu) = \delta_{\mu\nu}$. Then $\Omega(s)$ is a unitary matrix and Ω a unitary representation. Every vector f may be split into a vector $\sum_\mu \alpha_\mu g_\mu$ $(\mu = 1, \cdots, h)$ lying in Γ and one f' that is perpendicular to Γ, by choosing α_μ as the Fourier coefficients of f, $\alpha_\mu = (g_\mu, f)$. Then

$$\| f \|^2 = \sum_\mu | \alpha_\mu |^2 + \| f' \|^2.$$

In the Parseval equation, the proof of which is our goal, only invariant sets of such functions will occur as can be prepared by linear combination from the transforms sf of f. But, of course, finite combinations alone will not do; we need some property of closure by which to pass from finite sums to their limits (integrals or mean values). But how shall we measure the degree to which a vector difference g approaches zero? If $\| g \|$ is used as this measure, then the continuous functions on the sphere will not form a closed vector space, one would have to include all Lebesgue-square-integrable functions. But it seems foolish to operate in such a wide field when all functions arising in the course of our construction are continuous. Therefore we introduce axiomatically the length $| g |$ of an arbitrary vector g as a quantity that does not necessarily coincide with the modulus $\| g \|$ and define closure in terms of this new notion.

III. *Length and Closure.* Let there be associated with every vector f a non-negative number $|f|$, the length of f, which is zero for $f=0$ and satisfies the conditions

$$|f_1+f_2| \leqq |f_1|+|f_2|, \qquad |\alpha f| = |\alpha| \cdot |f|,$$

$$\|f\| \leqq |f|. \tag{1.3}$$

The transform sf is supposed to have the same length as f, $|sf|=|f|$. Any sequence $f_1, f_2, \cdots$ of vectors is said to converge strongly if $|f_n - f_m| \to 0$ for $n, m \to \infty$. We require that any strongly convergent sequence f_n converges strongly toward a definite vector f, $|f_n - f| \to 0$ for $n \to \infty$.

It ought to be observed that, whether the space Σ is closed in this sense or not, it can always be extended into a closed space without violating any of the previous axioms. This is done by Cantor's classical procedure as follows: Any strongly convergent sequence f_n is said to define an ideal vector f^*; and two such sequences f_n, g_n are said to define the same f^* if $|f_n - g_n| \to 0$. Addition, multiplication by numbers, the definitions of (g, f), $|f|$ and of the transformation $f \to sf$ readily extend to these ideal vectors. In this way the axiom of closure can be *enforced* if it is not assumed.

We call a sequence f_n weakly convergent if $\| f_n - f_m \| \to 0$ for $n, m \to \infty$. The axioms about length are certainly fulfilled if we define $|f|$ as $\| f \|$. Strong-convergence-closure then becomes weak-convergence-closure.

Instead of Σ we may use as our field of operation any closed linear subspace Σ_f of Σ that is invariant with respect to all the transformations (s), $g \to sg$, and contains the given vector f. The "smallest" of these subspaces, $\Sigma_f{}^0$, consists of the strong limits of strongly convergent sequences, the members of which are finite linear combinations $\sum_i \xi_i \cdot s_i f$ of transforms sf of f. Our aim is to construct in $\Sigma_f{}^0$ a (finite or infinite) unitary sequence of vectors

$$g_n \quad (n = 1, 2, \cdots), \qquad (g_k, g_l) = \delta_{kl},$$

consisting of a string of invariant sets

$$g_1, \cdots, g_h \mid g_{h+1}, \cdots, g_{h+h'} \mid \cdots \quad \cdots$$

of such completeness that the Parseval equation

$$|\alpha_1|^2 + |\alpha_2|^2 + \cdots = \| f \|^2$$

holds for the Fourier coefficients $\alpha_n = (g_n, f)$ of f. This means that the partial sums of the Fourier series $\sum_n \alpha_n g_n$ converge *weakly* to f; for

$$\| f - \sum_{k=1}^{n} \alpha_k g_k \|^2 = \| f \|^2 - \sum_{k=1}^{n} |\alpha_k|^2.$$

A further axiom (Axiom IV in **4**) will be needed in order to ensure that f can also *strongly* be approximated with arbitrary accuracy by finite sums $\sum_k \beta_k g_k$.

No restriction shall be imposed on the group σ. However the proof of the fundamental theorem depends on an essential assumption concerning the vector f, an assumption to which for historical reasons the name "almost periodic" (abbreviated: a. p.) has been given. Using $|sf - tf|$ as a sort of distance between two elements s, t of the group, let us say that s lies in the f-circle of radius ϵ around a if $|sf - af| \leq \epsilon$. Any two elements s, t in this circle satisfy the relation $|sf - tf| \leq 2\epsilon$. The vector f is called *almost periodic* if the group is f-compact, i. e. if it may be covered by a finite number of f-circles of arbitrary small radius ϵ.

This definition may be replaced by the following criterion: f is a. p. if the group can be covered by a finite number of sets σ_i ("roof-tiles") such that $|sf - tf| \leq \epsilon$ for any two elements s, t in the same set σ_i. The σ_i are then said to form an (f, ϵ)-tiling. Indeed a finite number of f-circles of radius ϵ covering the whole group form an $(f, 2\epsilon)$-tiling. Vice versa, if the σ_i form an (f, ϵ)-tiling, we choose a "representative" a_i in each of the *non-empty* σ_i. The f-circles of radius ϵ around these representatives cover the group. Let f', f'' be two a. p. vectors, σ'_i $(i = 1, \cdots, n)$ an (f', ϵ)-tiling and σ''_k $(k = 1, \cdots, m)$ an (f'', ϵ)-tiling. Form the $n \cdot m$

intersections $\sigma_{ik} = \sigma'_i \cap \sigma''_k$. For any two elements s, t in the same σ_{ik} the inequalities

$$| sf' - tf' | \leq \epsilon, \qquad | sf'' - tf'' | \leq \epsilon$$

hold simultaneously. In this sense the pair f', f'' is a. p. if each of the two members is. This fact carries over to any finite number of a. p. vectors $f_1, f_2, \cdots, f_l$. Hence not only the product of an a. p. vector by a number α, but also the sum of two a. p. vectors, and thus any linear combination $\sum_i \xi_i f_i$ of a. p. vectors is a. p. Moreover the strong limit of a strongly convergent sequence of a. p. vectors is a. p. Finally the transform af of an a. p. vector f by an element a of σ is a. p. Indeed let $a_1, \cdots, a_n$ be the centers of f-circles of radius ϵ covering the whole group, and set $a^*_i = a_i a^{-1}$. Then, for any given s, at least one of the inequalities

$$| saf - a^*_1 af | \leq \epsilon, \cdots, | saf - a^*_n af | \leq \epsilon$$

holds. We thus see that with f all the vectors of the space $\Sigma_f{}^0$ are a. p.

A numerical function $\phi(s)$ is f-continuous if for every $\delta > 0$ there exists a positive ϵ such that $| \phi(s) - \phi(t) | \leq \delta$ whenever $| sf - tf | \leq \epsilon$. For a vector function $g(s)$, whose values are vectors, the length $| g(s) - g(t) |$ has to replace the absolute value $| \phi(s) - \phi(t) |$ in this definition. The chief tool of the theory is the existence of a uniquely determined *mean value* $J = \int \phi(s)$ for every f-continuous function $\phi(s)$. A number of elements $a_1, \cdots, a_n$ and corresponding weights $\alpha_i \geq 0$ $(i = 1, \cdots, n)$ of sum 1 determine a (weighted) average

$$A = A\langle\phi\rangle = \sum_i^n \alpha_i \phi(a_i).$$

Let us say that this average oscillates by less than ϵ if

$$| \sum_i \alpha_i \cdot \phi(aa_i) - A | \leq \epsilon$$

for all elements a. It can be proved [Maak **6**; see also the Appendix of this paper] that there are such averages for any given positive ϵ, and that an average A_ϵ that oscillates by less than ϵ and an average B_δ that oscillates by less than δ differ by not more than $\epsilon + \delta$,

$$| A_\epsilon - B_\delta | \leq \epsilon + \delta. \tag{1.4}$$

Hence there exists a definite number J, the mean $\int \phi(s)$ of ϕ, such that $| A_\epsilon - J | \leq \epsilon$ for every average A_ϵ that oscillates by less than ϵ and every

$\epsilon > 0$. By its very definition the function $\phi(as)$ has the same mean J as $\phi(s)$ for any fixed elements a of σ. It can be shown that the same is true for $\phi(sa)$. All this carries over to an f-continuous vector function $g(s)$ of s.

The axioms I-III remain valid if σ is replaced by any subgroup σ^0 of σ. A vector f that is almost periodic with respect to σ is also a. p. with respect to σ^0. Indeed if a finite number of subsets σ_i of σ form an (f, ϵ)-tiling on σ then the sets $\sigma^0 \cap \sigma_i$ do so on σ^0.

Next we describe in precise terms that special interpretation of our axioms, to the simplest case of which the opening paragraph alluded.

> I_0. Given a group σ and a point field Π; moreover a realization of σ by transformations of Π. In other words, with every element s of σ there is associated a mapping $\{s\}$: $P \to P' = sP$ of Π upon itself such that $1P = P$, $t(sP) = (ts)P$. The group of these transformations is supposed to be transitive, i. e. given any two points P and Q there exists an element a of σ such that $Q = aP$. The transform sf of a numerical function $f = f(P)$ is defined by $sf(P) = f(s^{-1}P)$.

An element s is said to lie in the f-circle of radius ϵ around a if $|sf(P) - af(P)| \leq \epsilon$ identically in P. The function $f(P)$ is called almost periodic if the group can be covered by a finite number of f-circles of arbitrarily small radius ϵ. Choose a point P_0. This requirement obviously implies that $sf(P_0) = f(s^{-1}P_0)$ is a bounded numerical function of s, and hence, because of the transitivity of the group of transformations, $f(P)$ is a bounded function on Π. Looking upon the function $f(P)$ as a vector f in the functional space Σ of all a. p. functions on Π, we define $|f|$ as the lowest upper bound (l. u. b.) of $|f(P)|$, with the effect that our definition of almost-periodicity for the function $f(P)$ now coincides with that for the vector f as given above. Axioms I and III are satisfied in Σ with the (possible) exclusion of the relation (1. 3).

It remains to define the scalar product. For any given point P the numerical function $\psi(s) = sf(P)$ is evidently f-continuous, and we may form the average $\int_s \psi(s) = \int_s sf(P) = \int_s f(s^{-1}P)$. Since $\int \psi(as) = \int \psi(s)$, i. e. $\int_s f(s^{-1}Q) = \int_s f(s^{-1}P)$ for $Q = a^{-1}P$, this mean has the same value for P and for $Q = a^{-1}P$ and is thus a constant J. Write $J = \int_P f(Q)$ and call J the mean value of the a. p. function $f(P)$. The relation $\int \psi(sa) = \int \psi(s)$, on the other hand, shows that the transform af has the same mean as f. (It is only at this one place where the relation $\int \psi(sa) = \int \psi(s)$ comes into play.) The pair $f(P)$, $g(P)$ of two a. p. functions is a. p., and hence also the product $\bar{g}(P)\cdot f(P)$ of the conjugate of g by f. We now define the scalar product (g,f) as the integral $\int\limits_P \bar{g}(P)\cdot f(P)$ and then find that the axioms II and the relation (1. 3) are true in Σ. (In addition all the vectors of this space are almost periodic. But this is no serious restriction. For all our axioms carry over from a space Σ to the subspace of its almost periodic vectors.) We refer to this special interpretation as the "*interpretation by scalar functions on* Π."

It can be further specialized by identifying Π with σ and associating with the element a of σ the left translation $\{a\}$: $s\rightarrow as$ of $\Pi=\sigma$. The left translations constitute a transitive group of transformations on $\Pi=\sigma$ which is isomorphic to σ. This "*interpretation by scalar functions on* σ" is used in the construction of a complete set of inequivalent irreducible unitary representations of σ.

For compact Lie groups the theory of continuous representations was developed by F. Peter and the author [**8**] in 1926, and it was at once realized [**10**] that the method, I shall call it the integral equation method, carries over to the group of translations of a straight line and thus affords a natural approach to H. Bohr's theory of almost periodic functions [**2**]. It was J. von Neumann [**7**] who discovered that mean values can be defined for almost periodic functions on any group, and thus the construction of a complete set of a. p. representations was extended to arbitrary groups. In defining the mean value and deriving its essential properties we follow here a simplified procedure due to W. Maak [**6**]. The interpretation by scalar functions on Π (for compact Lie groups) was given by E. Cartan [**3**] and the author [**11**]. Results concerning unitary group representations in Hilbert space were obtained by S. Bochner and J. von Neumann [**1**] and by Anna Hurewitsch [**5**]. But the straightforward application of the integral equation method to the general situation staked out by our axioms seems to yield more precise and complete information.—No new methodological ideas will be developed in this paper; its purpose is to circumscribe the conditions under which a known method works.

2 gives the construction of the complete string of invariant sets in $\Sigma_f{}^0$ and thus derives the Parseval equation (weak approximation of f). The result is more fully evaluated in **3**, while **4** proves the central theorem concerning strong approximations, and adds a further interpretation in terms of "vector functions on Π." The Appendix (**5**) contains some remarks about Maak's procedure and the combinatorial lemma on which it is based.

2. The construction. From now on everything is relative to a given a. p. vector f so normalized that $\| f \|^2 \leq 1$.

The Cauchy-Schwarz inequality will be used by us in three different forms:

$$(C_1) \qquad \left| \int_s \bar{\eta}(s)\xi(s) \right|^2 \leq \int_s |\eta(s)|^2 \cdot \int_s |\xi(s)|^2,$$

$$(C_2) \qquad \left\| \int_s \xi(s) \cdot g(s) \right\|^2 \leq \int_s |\xi(s)|^2 \cdot \int_s \| g(s) \|^2,$$

$$(C_3) \qquad \left| \int_s (g'(s), g(s)) \right|^2 \leq \int_s \| g'(s) \|^2 \cdot \int_s \| g(s) \|^2.$$

$\xi(s)$, $\eta(s)$ are f-continuous numerical functions, $g(s)$, $g'(s)$ are f-continuous vector functions. (C_1) arises from Cauchy's inequality for numbers by passing from finite sums to the limit of integrals, (C_2) and (C_3) stand in the same relation to the inequalities (1. 1), (1. 2).

The given vector f defines the following linear mapping $\mathfrak{f}$ of the space Ξ of f-continuous functions $\xi(s)$ upon the vector space Σ,

$$\mathfrak{f}: \quad \xi(s) \rightarrow g = \int \xi(s) \cdot sf.$$

The image $g = \mathfrak{f}\xi$ lies in $\Sigma_f{}^0$. On the other hand, consider the linear mapping $\mathfrak{f}^*$ of the vector space Σ upon the space Ξ defined by

$$\mathfrak{f}^*: \quad g \rightarrow \xi(s) = (sf, g).$$

[$\xi(s)$ is not only f-continuous, but even satisfies a strong Lipschitz condition

$$(2.\,1) \qquad |\xi(s) - \xi(t)| \leq \text{const.} \, \| sf - tf \| .]$$

If $\int_s \bar{\eta}(s) \cdot \xi(s)$ is taken as the scalar product (η, ξ) in Ξ, the operator $\mathfrak{f}^*$ is the Hermitian conjugate of $\mathfrak{f}$. Indeed $\xi' = \mathfrak{f}^* g'$ gives $\bar{\xi}'(s) = (g', sf)$ and

$$(\xi', \xi) = \int_s \bar{\xi}'(s)\xi(s) = \int_s (g', sf) \cdot \xi(s) = (g', g)$$

where $g = \int_s \xi(s) \cdot sf$, hence $(\mathfrak{f}^* g', \xi) = (g', \mathfrak{f}\xi)$. Form the operator $\mathfrak{f}^*\mathfrak{f}$ in Ξ,

$$\xi(s) \rightarrow \int_t (sf, tf) \cdot \xi(t) = \int_t H(s, t) \cdot \xi(t),$$

with the Hermitian f-continuous kernel $H(s, t) = (sf, tf)$. This kernel is positive definite in the sense that

$$\int_s \int_t \bar{\xi}(s) \cdot H(s, t) \cdot \xi(t) \geq 0$$

for every function $\xi(s)$ in Ξ. Indeed the left side has the value

$$\| \int_s \xi(s) \cdot sf \|^2 = \| \mathfrak{f}\xi \|^2.$$

Because of its invariance (under left translations),

$$H(as, at) = H(s, t), \tag{2.2}$$

the kernel $H(s, t)$ actually depends on the one argument $t^{-1}s$ only. Note that

$$|H(s, t)| \leq \| sf \| \cdot \| tf \| = \| f \|^2 \leq 1.$$

The integral equation method consists in determining the (reciprocal) positive eigenvalues $\gamma, \gamma_1, \cdots$ of this kernel by E. Schmidt's procedure and thereby proving the equation

$$\operatorname{tr}(H) = \int_s H(s, s) = \gamma + \gamma_1 + \cdots.$$

The iterations of the kernel H are formed according to

$$H_1(s, t) = H(s, t), \qquad H_{n+1}(s, t) = \int_r H_n(s, r) H(r, t) \qquad (n = 1, 2, \cdots).$$

Set $\Gamma_n = \operatorname{tr}(H_n)$. The largest positive eigenvalue γ is constructed as the limit of the quotient Γ_{n+1}/Γ_n for $n \to \infty$, and the corresponding eigenkernel

$$E(s, t) = \phi_1(s) \cdot \bar{\phi}_1(t) + \cdots + \phi_h(s) \cdot \bar{\phi}_h(t) \tag{2.3}$$

as the limit of $H_n(s, t)/\gamma^n$ for $n \to \infty$. $\phi_1(s), \cdots, \phi_h(s)$ form a unitary basis for the eigenfunctions of $H(s, t)$ that belong to the eigenvalue γ. In this construction the invariance property (2.2) plays no rôle. However the latter is decisive for the fact that the h functions $\phi_\mu(s)$ and the corresponding vectors $g_\mu = \mathfrak{f}\phi_\mu$ constitute invariant sets.

In carrying out this program one has first to establish the inequalities

$$\Gamma_{m+n} \leq \Gamma_m \cdot \Gamma_n, \tag{2.4}$$

$$\Gamma_n^2 \leq \Gamma_{n-l} \cdot \Gamma_{n+l} \tag{2.5}$$

$(m, n \geq 1,\ 0 \leq l < n)$, and this is the only point where the original exposition requires a slight new touch. Set

$$f_0(s) = sf, \qquad f_n(s) = \int_t H_n(s, t) \cdot tf \qquad (n = 1, 2, \cdots).$$

Then

$$\Gamma_{2n} = \int_s \int_t H_n(s, t) H_n(t, s) = \int_s \int_t |H_n(s, t)|^2 \geq 0 \ (n = 1, 2, \cdots) \tag{2.6}$$

and

$$(2.7) \qquad \Gamma_{2n+1} = \int_s \| f_n(s) \|^2 \geq 0 \qquad (n = 0, 1, \cdots).$$

The second relation follows from the general equation

$$(2.8) \qquad (f_m(s) . f_n(t)) = H_{m+n+1}(s, t) \qquad (n, m \geq 0).$$

Hence all $\Gamma_n \geqq 0$ $(n = 1, 2, \cdots)$. In proving (2.4) we distinguish the three cases (i) m, n even; (ii) m even, n odd; (iii) m, n odd.

(i). $H_{m+n}(s, t) = \int_r H_m(s, r) H_n(r, t)$ yields by means of the inequality (C_1):

$$| H_{m+n}(s, t) |^2 \leq \int_r | H_m(s, r) |^2 \cdot \int_r | H_n(r, t) |^2$$

and hence by integrating over (s, t) and using the definition (2.6):

$$\Gamma_{2(m+n)} \leq \Gamma_{2m} \cdot \Gamma_{2n}.$$

(ii) $f_{m+n}(s) = \int_t H_m(s, t) f_n(t)$ gives by means of (C_2):

$$\| f_{m+n}(s) \|^2 \leq \int_t | H_m(s, t) |^2 \cdot \int_t \| f_n(t) \|^2,$$

and thus in view of (2.6), (2.7), after integration over s,

$$\Gamma_{2(m+n)+1} \leq \Gamma_{2m} \cdot \Gamma_{2n+1}.$$

(iii) From (2.8) there follows

$$| H_{m+n+1}(s, t) |^2 \leq \| f_m(s) \|^2 \cdot \| f_n(t) \|^2$$

and then by integration over s and t,

$$\Gamma_{2(m+n+1)} \leq \Gamma_{2m+1} \cdot \Gamma_{2n+1}.$$

In (2.5) we distinguish the two cases (i) $n \pm l$ even; (ii) $n \pm l$ odd. Set in the first case $n - l = 2u$, $n + l = 2v$, in the second $n - l = 2u + 1$, $n + l = 2v + 1$.

$$\text{(i)} \qquad \Gamma_n = \int_s H_{u+v}(s, s) = \int_s \int_t H_u(s, t) H_v(t, s),$$

hence application of (C_1) to integrals over the pair (s, t) gives the desired result

$$| \Gamma_n |^2 \leq \int_s \int_t | H_u(s, t) |^2 \cdot \int_s \int_t | H_v(t, s) |^2 = \Gamma_{2u} \cdot \Gamma_{2v}.$$

(ii) Apply (C_3) to

$$\Gamma_n = \int_s H_{u+v+1}(s, s) = \int_s (f_u(s), f_v(s))$$

with the result

$$|\Gamma_n|^2 \leq \int_s \|f_u(s)\|^2 \cdot \int_s \|f_v(s)\|^2 = \Gamma_{2u+1} \cdot \Gamma_{2v+1}.$$

Only the case $l = 1$ of (2. 5) will be used,

$$\Gamma_n^2 \leq \Gamma_{n-1} \cdot \Gamma_{n+1} \qquad (n = 2, 3, \cdots). \tag{2. 9}$$

If $f \neq 0$ then $\Gamma_1 = \|f\|^2 > 0$ and also $\Gamma_2 = \int_s \int_t |(sf, tf)|^2 > 0$ [for otherwise $(sf, tf) = 0$ for all s and t, in particular $(f, f) = 0$]. The inequality (2. 9) then shows that one after the other of the traces $\Gamma_3, \Gamma_4, \cdots$ are likewise positive and not zero. More precisely we find that $q_n = \Gamma_{n+1}/\Gamma_n$ $(n = 1, 2, \cdots)$ is an increasing sequence, while (2. 4) shows that

$$q_n^m \leq q_n q_{n+1} \cdots q_{n+m-1} = \Gamma_{n+m}/\Gamma_n \leq \Gamma_m, \tag{2. 10}$$

or that the sequence q_n never grows beyond $\Gamma_m^{1/m}$, in particular not beyond $\Gamma_1 \leq 1$. Hence it converges to a positive number γ. Note in particular the inequality $q_1 \leq \gamma$ or

$$\Gamma_2 \leq \gamma \Gamma_1. \tag{2. 11}$$

It is even easy to make an explicit estimate of the rapidity of convergence. Considering that q_n does not grow beyond $\Gamma_n^{1/m}$ and $\Gamma_m = \Gamma_1 q_1 \cdots q_{m-1}$ $\leq q_m^{m-1}$ one finds that q_n from $n = m$ on can not grow by more than

$$q_m^{(m-1)/m}(1 - q_m^{1/m}) \leq 1 - q_m^{1/m} \leq (q_m^{-1} - 1)/m \leq (q_1^{-1} - 1)/m.$$

Γ_n/γ^n decreases with increasing n since $q_n = \Gamma_{n+1}/\Gamma_n \leq \gamma$, hence tends to a limit h. But as (2. 10) in the limit for $n \to \infty$ gives $\gamma^m \leq \Gamma_m$, the limit h of Γ_m/γ^m for $m \to \infty$ is by necessity greater than or equal to 1. (As h will turn out to be the multiplicity of the eigenvalue γ, an explicit estimate can not be expected for this second step!) Application of (C_1) to

$$\frac{H_{m+2}(s,t)}{\gamma^{m+2}} - \frac{H_{n+2}(s,t)}{\gamma^{n+2}} = \frac{1}{\gamma^2} \int_r \int_{r'} H(s,r) \left\{ \frac{H_m(r,r')}{\gamma^m} - \frac{H_n(r,r')}{\gamma^n} \right\} H(r',t)$$

yields

$$\left| \frac{H_{m+2}(s,t)}{\gamma^{m+2}} - \frac{H_{n+2}(s,t)}{\gamma^{n+2}} \right|^2 \leq \frac{1}{\gamma^4} \left\{ \frac{\Gamma_{2m}}{\gamma^{2m}} - \frac{\Gamma_{m+n}}{\gamma^{m+n}} + \frac{\Gamma_{2n}}{\gamma^{2n}} \right\}$$

and thus the uniform convergence of $H_n(s,t)/\gamma^n$ to an f-continuous limit $E(s,t)$. Invariance, (2.2), carries over from H to E, $E(as,at)=E(s,t)$. From the equations

$$\int_r H(s,r)E(r,t)=\int_r E(s,r)H(r,t)=\gamma\cdot E(s,t);$$
$$\int_r E(s,r)E(r,t), \qquad \bar{E}(s,t)=E(t,s); \qquad \operatorname{tr}(E)=h$$

there follows by a well-known elementary argument such a relation as (2.3) in which the $\phi_\mu(s)$ form a unitary-orthogonal system of f-continuous eigenfunctions of H,

$$(2.12)\qquad \int_t H(s,t)\phi_\mu(t)=\gamma\cdot\phi_\mu(s),$$
$$\int_s \bar{\phi}_\mu(s)\cdot\phi_\nu(s)=\delta_{\mu\nu} \qquad (\mu,\nu=1,\cdots,h).$$

h turns out to be a positive integer. The one equation (2.12) may be split into two,

$$(2.13)\qquad \sqrt{\gamma}\cdot g_\mu=\int\phi_\mu(s)\cdot sf, \qquad \sqrt{\gamma}\cdot\phi_\mu(s)=(sf,g_\mu),$$

by using the first as definition of the vector g_μ. The g_μ are unitary-orthogonal,

$$(g_\mu,g_\nu)=\delta_{\mu\nu} \qquad (\mu,\nu=1,\cdots,h).$$

$\mathfrak{f}\mathfrak{f}^*=\mathfrak{F}$ is a linear operator which carries the arbitrary vector g into

$$\mathfrak{F}g=\int_s (sf,g)\cdot sf.$$

This operator is invariant in the sense that

$$(2.14)\qquad \mathfrak{F}(ag)=a(\mathfrak{F}g) \qquad \text{for any group element } a.$$

Indeed

$$a(\mathfrak{F}g)=\int_s(sf,g)\cdot asf=\int_s(asf,ag)\cdot asf=\int_s(sf,ag)\cdot sf=\mathfrak{F}(ag).$$

The corresponding Hermitian form depending on two arbitrary vectors g, g' is

$$(g',\mathfrak{F}g)=\int_s(g',sf)\cdot(sf,g).$$

$(g',\mathfrak{F}g)$ is conjugate to $(g,\mathfrak{F}g')$. By combining the two equations (2.13) in inverse order one sees that the vectors g_μ obtained by our construction are eigenvectors of the operator $\mathfrak{F}$ for the eigenvalue γ,

$$(2.15)\qquad \mathfrak{F}g_\mu=\gamma\cdot g_\mu.$$

By starting with the equation

$$\phi_\mu(s) = \int_t E(s, t) \cdot \phi_\mu(t)$$

and utilizing the invariance property of E one finds that

$$\phi_\mu(a^{-1}s) = \int_t E(a^{-1}s, t) \cdot \phi_\mu(t) = \int_t E(a^{-1}s, a^{-1}t) \cdot \phi_\mu(a^{-1}t)$$
$$= \int_t E(s, t) \cdot \phi_\mu(a^{-1}t) = \sum_\nu \phi_\nu(s) \cdot \int_t \phi_\nu(t)\phi_\mu(a^{-1}t)$$

is a linear combination $\sum_\nu \omega_{\nu\mu}(a) \cdot \phi_\nu(s)$ of the $\phi_\nu(s)$, and thus

$$\int_s \phi_\mu(s) \cdot asf = \int_s \phi_\mu(a^{-1}s) \cdot sf = \sum_\nu \omega_{\nu\mu}(a) \cdot \int_s \phi_\nu(s) \cdot sf$$

or $ag_\mu = \sum_\nu \omega_{\nu\mu}(a) \cdot g_\nu$. This equation

(2.16) $$sg_\mu = \sum_\nu \omega_{\nu\mu}(s) \cdot g_\nu$$

shows that the g_μ form an invariant set $\{g_\mu\}$ belonging to the (necessarily unitary) representation Ω: $s \to \Omega(s) = \| \omega_{\mu\nu}(s) \|$. Let us say that h vectors g_μ $(\mu = 1, \cdots, h)$ transform according to Ω and call $(g_1, \cdots, g_h)$ briefly an Ω-row, if the equations (2.16) hold for every group element s. For any such row we infer from (2.16):

(2.17) $$(sf, g_\mu) = (f, s^{-1}g_\mu) = \sum_\nu \omega_{\nu\mu}(s^{-1}) \cdot (f, g_\nu).$$

(f, g_ν) is the conjugate of the Fourier coefficient $\alpha_\nu = (g_\nu, f)$. Moreover the matrix $\| \omega_{\mu\nu}(s^{-1}) \|$ is reciprocal to $\| \omega_{\mu\nu}(s) \|$ and the latter is unitary, hence $\omega_{\nu\mu}(s^{-1}) = \bar{\omega}_{\mu\nu}(s)$, whereby (2.17) turns into

(2.18) $$(sf, g_\mu) = \sum_\nu \bar{\omega}_{\mu\nu}(s) \cdot \alpha_\nu.$$

For two Ω-rows g, g' one gets

$$\sum_\mu (g'_\mu, sf) \cdot (sf, g_\mu) = \sum_\mu \alpha'_\mu \bar{\alpha}_\mu,$$

and hence by integration over s:

(2.19) $$\sum_\mu (g'_\mu, \mathfrak{F} g_\mu) = \sum_\mu \alpha'_\mu \bar{\alpha}_\mu.$$

The special set $\{g_\mu\}$ constructed above satisfies the relation (2.15), thus $(g_\mu, \mathfrak{F} g_\mu) = \gamma$, and (2.19) yields the important equation

(2.20) $$h\gamma = \sum_\mu \alpha_\mu \bar{\alpha}_\mu.$$

Let us say that the invariant set $\{g_\mu\}$ occurs in f with the (non-vanishing) *weight* $\gamma = h^{-1} \sum_\mu | \alpha_\mu |^2$.

When we subtract

$$e = \sum_{\mu=1}^{h} \alpha_\mu g_\mu$$

from f, the remainder $f' = f - e$ is orthogonal to the vectors g_μ. If it is still different from zero, we repeat for f' the process carried out before on f. One gets a new eigenvalue γ' and again a corresponding set of eigenfunctions $\phi'_{\mu'}(s)$ $(\mu' = 1, \cdots, h')$ and vectors $g'_{\mu'}$. The $\phi'_{\mu'}(s)$ turn out to be orthogonal to the $\phi_\mu(s)$, the $g'_{\mu'}$ to the g_μ. The equation $\Gamma_n = h\gamma^n + \Gamma'_n$ shows that $\Gamma'_n/\gamma^n \to 0$, hence $\gamma' < \gamma$.

It is obvious how to continue the process. It yields a string of orthogonal functions $\phi_n(s)$ and orthogonal vectors g_n $(n = 1, 2, \cdots)$ subdivided into sections. The section p, $n_{p-1} < n \leq n_p = n_{p-1} + h_p$, is characterized by a positive eigenvalue γ_p of multiplicity h_p; the vectors g_n of this section form an invariant set and are connected with their partners $\phi_n(s)$ by the relations

$$\sqrt{\gamma_p} \cdot g_n = \int_s \phi_n(s) \cdot sf, \qquad \sqrt{\gamma_p} \cdot \phi_n(s) = (sf, g_n) \qquad [n_{p-1} < n \leq n_p].$$
$$(p = 0, 1, 2, \cdots; \; h_0 = h, \; n_{-1} = 0.)$$

Form the Fourier coefficients $\alpha_n = (g_n, f)$ of f and the remainder

$$f^{(p)} = f - (e + e_1 + \cdots + e_{p-1}) = f - \sum_{k=1}^{n_{p-1}} \alpha_k g_k.$$

Should it ever happen that one of the successive remainders $f, f', \cdots$ vanishes, then the process comes to a stop. But since our final result is trivial in that case, we suppose in our argument that this never comes to pass. γ_p tends to zero, because

$$\sum_{k=1}^{n_p} | \alpha_k |^2 = h\gamma + h_1\gamma_1 + \cdots + h_p\gamma_p \leq \| f \|^2 \leq 1$$

implies

$$(h + h_1 + \cdots + h_p) \cdot \gamma_p = n_p\gamma_p \leq 1.$$

The kernel

$$H^{(p)}(s, t) = (sf^{(p)}, tf^{(p)}) = (sf, tf^{(p)})$$

depends on $t^{-1}s$ only,

$$H^{(p)}(s, t) = H^{(p)}(t^{-1}s), \qquad H^{(p)}(s) = (sf, f^{(p)}).$$

Application of the relation (2.11) to $f^{(p)}$ instead of f gives the inequality $\Gamma_2{}^{(p)} \leq \gamma_p \cdot \Gamma_1{}^{(p)}$ for

$$\Gamma_1^{(p)} = \mathrm{tr}(H^{(p)}) = H^{(p)}(1) = \| f^{(p)} \|^2 \text{ and } \Gamma_2^{(p)} = \int_s | H^{(p)}(s) |^2,$$

and thus

$$\Gamma_2^{(p)} \leq \Gamma_1^{(p)}/n_p. \tag{2.21}$$

By means of the equi-continuity of $H^{(p)}(s) = (sf, f^{(p)})$ for all p we deduce from this estimate for $\Gamma_2^{(p)}$ an estimate for $\Gamma_1^{(p)} = \| f^{(p)} \|^2 = \delta_p^2$ as follows. Choose a positive number ϵ and ascertain a finite number N_ϵ of elements a_i $(i = 1, \cdots, N_\epsilon)$ such that the f-circles C_i of radius ϵ around these elements a_i cover the whole group. We propose to show that $\delta_p \leq 2\epsilon$ as soon as $n_p \geq N_\epsilon/\epsilon^2$. Indeed the contrary hypothesis $2\epsilon < \delta_p$ will lead to $n_p < N_\epsilon/\epsilon^2$. For an element s in the f-circle of radius ϵ around 1 the following inequalities prevail:

$$\| sf - f \| \leq | sf - f | \leq \epsilon,$$

hence

$$| H^{(p)}(s) - H^{(p)}(1) |^2 = | (sf - f, f^{(p)}) |^2 \leq \epsilon^2 \cdot \| f^{(p)} \|^2 = \epsilon^2 \delta_p^2,$$

and finally, since $H^{(p)}(1) = \delta_p^2$ and by hypothesis $\delta_p > 2\epsilon$,

$$| H^{(p)}(s) | \geq \delta_p(\delta_p - \epsilon) > \epsilon\delta_p.$$

Thus $| H^{(p)}(a_i^{-1}s) |^2 > \epsilon^2\delta_p^2$ for $s \in C_i$, and consequently

$$\sum_{i=1}^{N_\epsilon} | H^{(p)}(a_i^{-1}s) |^2 > \epsilon^2 \delta_p^2$$

everywhere on σ. By integration over s one finds

$$N_\epsilon \cdot \int_s | H^{(p)}(s) |^2 > \epsilon^2 \delta_p^2, \quad \text{or} \quad N_\epsilon \cdot \Gamma_2^{(p)} > \epsilon^2 \delta_p^2.$$

Combination with (2.21), $n_p \Gamma_2^{(p)} \leq \delta_p^2$, leads to the promised conclusion $n_p < N_\epsilon/\epsilon^2$. But

$$\delta_p^2 = \| f^{(p)} \|^2 = \| f \|^2 - \sum_{k=1}^{n_p - 1} | \alpha_k |^2.$$

Hence the result can be stated as follows:

Given any $\epsilon > 0$, the inequality

$$(0 \leq) \; \| f \|^2 - \sum_{k=1}^{n-1} | \alpha_k |^2 \leq 4\epsilon^2 \tag{2.22}$$

will hold as soon as $n \geq N_\epsilon/\epsilon^2$.

This not only proves the Parseval equation

$$| \alpha_1 |^2 + | \alpha_2 |^2 + \cdots = \| f \|^2, \tag{2.23}$$

or the fact that the Fourier series $\alpha_1 g_1 + \alpha_2 g_2 + \cdots$ converges weakly to f, but also provides us with an explicit estimate for the remainder under the assumption that the invariant sets are arranged according to descending weights, $\gamma > \gamma_1 > \cdots$.

From f one easily passes to an arbitrary finite linear combination $\sum_i \lambda_i \cdot s_i f$ of transforms of f and any vector g that is the weak limit of a sequence of such combinations. It results that we can approximate every such vector g by a finite sum $v_n = \sum_{k=1}^{n-1} \beta'_k g_k$ in the weak sense, $\| g - v_n \| \leq \epsilon$, with any preassigned accuracy ϵ. But the best approximation in the weak sense by a sum v_n of a given number $n-1$ of terms is obtained by choosing β'_k as the Fourier coefficients $\beta_k = (g_k, g)$ of g. Hence the Fourier series $\sum \beta_n g_n$ of any g of the type just described, in particular of any vector g in $\Sigma_f{}^0$, converges weakly towards g. In other words, the Parseval equation extends from f to any vector g in $\Sigma_f{}^0$ without a change in the unitary sequence $g_1, g_2, \cdots$ constructed from f.

3. Evaluation of the result. Of the arbitrariness involved in the choice of a unitary basis g_μ ($\mu = 1, \cdots, h$) in a given invariant set we can make use in such a way that the set breaks up into a number of mutually orthogonal *primitive* sets. Let us do that with each of the sets obtained by our construction! Suppose, for instance, that the first set consists of $h = 8$ vectors and breaks up into two invariant sets of 5 and 3 members respectively. Since the equation (2.15) holds for each vector g_μ of the total set, the relation (2.20) will persist for the two partial sets,

$$\sum_{\mu=1}^{5} |\alpha_\mu|^2 = 5\gamma, \qquad \sum_{\mu=6}^{8} |\alpha_\mu|^2 = 3\gamma.$$

The primitive sets are therefore still arranged according to falling weights, $\gamma \geq \gamma_1 \geq \cdots$, however the equality sign can not now be excluded. In the case just mentioned γ_1 would equal γ and 8γ appear as $5\gamma + 3\gamma_1$.

A stage has now been reached where passage from the constructive to the existential standpoint becomes feasible. Let

$$\Omega: \quad s \rightarrow \Omega(s) = \| \omega_{\mu\nu}(s) \| \qquad (\mu, \nu = 1, \cdots, h)$$

be a given irreducible unitary representation of the group σ of degree h. Any representation Ω' that is equivalent to Ω, $\Omega' \sim \Omega$, is also unitary-equivalent to it. [Indeed if the non-singular square matrix A satisfies the condition $\Omega(s)A = A\Omega'(s)$ one sees that AA^* commutes with $\Omega(s)$, and hence by Schur's lemma $AA^* = \rho E$, $\rho > 0$. Here A^* is the conjugate of the transpose of A, and E denotes the unit matrix. $A/\sqrt{\rho}$ is unitary.] Thus we may see to it that, whenever a primitive invariant set obtained in the course of our construction belongs to a representation $\sim \Omega$, its orthogonal basis $g_1, \cdots, g_h$ transforms according *to* Ω *itself*. Schur's lemma further teaches the following two things: (1) If the g_μ and the g'_ν transform according to two *inequivalent* irreducible unitary representations, then they are mutually orthogonal,

$$(g_\mu, g'_\nu) = 0 \qquad (\mu = 1, \cdots, h;\ \nu = 1, \cdots, h').$$

(2) If they transform according to the *same* irreducible unitary Ω, then

$$(g_\mu, g'_\nu) = \beta \cdot \delta_{\mu\nu} \qquad (\mu, \nu = 1, \cdots, h) \tag{3.1}$$

with a factor β independent of μ and ν. Any numerical multiple $\lambda \mathbf{g} = (\lambda g_1, \cdots, \lambda g_h)$ of an Ω-row $\mathbf{g}$ is an Ω-row, and so is the sum of two such rows. Thus the Ω-rows form a linear manifold. It is natural to introduce the factor β in the equation (3.1) as the scalar product $(\mathbf{g}, \mathbf{g}')$ of the two Ω-rows $\mathbf{g}$ and $\mathbf{g}'$;

$$(\mathbf{g}, \mathbf{g}') = h^{-1} \cdot \sum_\mu (g_\mu, g'_\mu).$$

Indeed it has all the formal properties of a scalar product, in particular $(\mathbf{g}, \mathbf{g}) > 0$ unless $\mathbf{g} = 0$. Call the row $\mathfrak{a} = (\bar{\alpha}_1, \cdots, \bar{\alpha}_h)$ of the conjugates $\alpha_\mu = (f, g_\mu)$ of the Fourier coefficients the *f-component* of the Ω-row $\mathbf{g}$ and define

$$(\mathfrak{a}', \mathfrak{a}) = \sum_\mu \alpha'_\mu \bar{\alpha}_\mu, \qquad \| \mathfrak{a} \|^2 = (\mathfrak{a}, \mathfrak{a}).$$

$\lambda \mathfrak{a}$ and $\mathfrak{a} + \mathfrak{a}'$ are the f-components of λg and $g + g'$. The equation (2.19) now reads

$$(\mathbf{g}', \mathfrak{F}\mathbf{g}) = h^{-1} \cdot (\mathfrak{a}', \mathfrak{a}). \tag{3.2}$$

An Ω-row will be said to be *hidden* or flat if its f-component $\mathfrak{a}$ is zero, and *upright* if it is perpendicular to all hidden Ω-rows. Clearly there can not be more than h linearly independent upright Ω-rows. For let $\mathbf{g}^{(1)}, \cdots, \mathbf{g}^{(m)}$ be any m such rows and form the linear combination $\mathbf{g} = \lambda_1 \mathbf{g}^{(1)} + \cdots + \lambda_m \mathbf{g}^{(m)}$. Whenever its f-component $\lambda_1 \mathfrak{a}^{(1)} + \cdots + \lambda_m \mathfrak{a}^{(m)}$ vanishes, $\mathbf{g}$ itself must be zero, because such a $\mathbf{g}$ is at the same time upright and flat, therefore $(\mathbf{g}, \mathbf{g}) = 0$, $\mathbf{g} = 0$.

(2.14) shows that the operator $\mathfrak{F}$ changes an Ω-row $\mathbf{g}$ into an Ω-row $\mathfrak{F}\mathbf{g}$;

according to (3.2) this $\mathfrak{F}\boldsymbol{g}$ is necessarily upright. Any Ω-row $\boldsymbol{g}=\{g_\mu\}$ obtained by our construction satisfies an equation (2.15) with a positive factor γ, $g_\mu=\gamma^{-1}\cdot\mathfrak{F}g_\mu$, and consequently these Ω-rows themselves are upright: *among all possible Ω-rows the construction automatically selects the upright ones* as those that actually occur in the Parseval equation (2.23). Such a principle of selection is needed, since the linear manifold of *all* Ω-rows may not be of finite nor even of denumerable dimensionality.

Let $\boldsymbol{g}^{(1)},\cdots,\boldsymbol{g}^{(m)}$ $(m\leq h)$ be an orthogonal basis, $(\boldsymbol{g}^{(i)},\boldsymbol{g}^{(k)})=\delta_{ik}$, for the upright Ω-rows. Then their f-components $\mathfrak{a}^{(1)},\cdots,\mathfrak{a}^{(m)}$ are also linearly independent. Moreover we have

$$(g_\mu{}^{(i)},g_\nu{}^{(k)})=\delta_{ik}\delta_{\mu\nu}\qquad (i,k=1,\cdots,m;\ \mu,\nu=1,\cdots,h)$$

and

$$\mathfrak{F}\boldsymbol{g}^{(i)}=\sum_k\gamma_{ki}\boldsymbol{g}^{(k)}$$

where the coefficients

$$\gamma_{ki}=(\boldsymbol{g}^{(k)},\mathfrak{F}\boldsymbol{g}^{(i)})=h^{-1}(\mathfrak{a}^{(k)},\mathfrak{a}^{(i)})$$

form a Hermitian matrix. $(\boldsymbol{g},\mathfrak{F}\boldsymbol{g})$ for the arbitrary upright Ω-row $\boldsymbol{g}=\sum_i\xi_i\boldsymbol{g}^{(i)}$ is the positive-definite Hermitian form

$$G[\xi]=\sum_{i,k}\xi_k\gamma_{ki}\xi_i=h^{-1}\left\|\sum_i\xi_i\mathfrak{a}^{(i)}\right\|^2 \tag{3.3}$$

of the m variables ξ_i. One can therefore ascertain m Ω-rows $\check{\boldsymbol{g}}^{(i)}$ such that $\boldsymbol{g}^{(i)}=\sum_k\gamma_{ki}\check{\boldsymbol{g}}^{(k)}$. Then

$$g_\mu{}^{(i)}=\int_s\check{\phi}_\mu{}^{(i)}(s)\cdot sf\quad\text{where}\quad\check{\phi}_\mu{}^{(i)}(s)=(sf,\check{g}_\mu{}^{(i)}).$$

[In passing it may be observed that if m has the highest possible value h, then the functions $\omega_{\mu\nu}(s)$ will be f-continuous; for then one can express them as linear combinations of the f-continuous functions $(g_\mu{}^{(i)},sf)$ by inverting the relations (2.18),

$$(g_\mu{}^{(i)},sf)=\sum_\nu\omega_{\mu\nu}(s)\cdot\alpha_\nu{}^{(i)}\qquad(\mu,\nu,i=1,\cdots,h).]$$

The orthogonal basis $\boldsymbol{g}^{(i)}$ could be so chosen that $G[\xi]$ is on principal axes,

$$\gamma_{ki}=0\text{ for }k\neq i,\qquad\gamma_{ii}=\gamma_i>0.$$

Then $\mathfrak{F}g_\mu{}^{(i)}=\gamma_i\cdot g_\mu{}^{(i)}$. It is this normalized form which results from the construction of **2**. The greatest of the m numbers $\gamma_1,\cdots,\gamma_m$ can be defined as the maximum $\gamma(\Omega)$ of the Hermitian form (3.3) for $\sum_i|\xi_i|^2\leq1$.

The contributions

$$(3.4) \qquad e(\Omega) = \sum \alpha_\mu{}^{(i)} g_\mu{}^{(i)} \text{ and } \| e(\Omega) \|^2 = \sum | \alpha_\mu{}^{(i)} |^2 = \sum_i \| \mathfrak{a}^{(i)} \|^2$$

of Ω to f and $\| f \|^2$ are clearly independent of the choice of the orthogonal basis $\mathbf{g}^{(i)}$ for the upright Ω-rows. Hence in the final statement we shall ignore this normalization of $G[\xi]$. Nor do the contributions (3.4) change if Ω is replaced by a unitary-equivalent representation, i.e. if each of the m rows $\mathbf{g}^{(1)}, \cdots, \mathbf{g}^{(m)}$ undergoes the same unitary transformation A. The sum $\sum \| e(\Omega) \|^2$ extending over any set of inequivalent irreducible unitary Ω can not exceed $\| f \|^2$ (Bessel's inequality). It is thus further clear that the construction of $\mathbf{2}$ can not help yielding a *complete* basis $\mathbf{g}^{(i)}$ $(i = 1, \cdots, m)$ of the upright Ω-rows for each Ω. Otherwise the sum (2.23) would fall short of $\| f \|^2$ at least by the contributions of the missing $\mathbf{g}^{(i)}$. We summarize our findings in a preliminary Statement and a Theorem.

STATEMENT. *Starting from a given a.p. vector f, our construction accomplishes the following. For every irreducible unitary representation Ω of σ, degree h, it determines a complete orthogonal basis $\mathbf{g}^{(1)}, \cdots, \mathbf{g}^{(m)}$ of those Ω-rows $\mathbf{g} = (g_1, \cdots, g_h)$ which are perpendicular to all hidden Ω-rows, and it picks out a complete set of inequivalent Ω's that actually occur, i.e. for which $m > 0$. One has $m \leq h$, and the vectors*

$$g_\mu{}^{(i)} \qquad (i = 1, \cdots, m;\ \mu = 1, \cdots, h)$$

lie in $\Sigma_f{}^0$. [They are even of the special form $\int \phi(s) \cdot sf$ where $\phi(s)$ is not only f-continuous but satisfies a strong Lipschitz condition with respect to f, (2.1).] *The $g_\mu{}^{(i)}$ for two inequivalent irreducible unitary representations Ω are mutually orthongonal. The contribution of Ω to f is the sum $e(\Omega) = \sum \alpha_\mu{}^{(i)} g_\mu{}^{(i)}$ formed by means of the Fourier coefficients $\alpha_\mu{}^{(i)} = (g_\mu{}^{(i)}, f)$, the contribution of Ω to $\| f \|^2$ equals $\| e(\Omega) \|^2 = \sum | \alpha_\mu{}^{(i)} |^2$. The maximum of $h^{-1} \sum_\mu | \sum_i \alpha_\mu{}^{(i)} \xi_i |^2$ for $\sum_i | \xi_i |^2 \leq 1$ is introduced as the weight $\gamma(\Omega)$ with which Ω occurs in f.*

THEOREM. *The sum of the contributions $\sum \| e(\Omega) \|^2$ extending over the denumerable sequence of inequivalent irreducible unitary Ω actually occurring in f equals $\| f \|^2$. An explicit estimate of the convergence can be given if the terms are arranged by descending weights $\gamma(\Omega)$.*

It is perhaps not justified to speak of a completeness relation, since a host of "hidden" invariant sets are left in the dark, for the good reason that they do not contribute to f and $\| f \|^2$.

For the "interpretation by scalar functions on Π" the situation is a little simpler. Here, according to an observation made by E. Cartan [**3**, cf. also **11**], the number of linearly independent Ω-rows is at most h for any irreducible unitary representation Ω of degree h. We choose an orthogonal basis $\mathbf{g}^{(1)}, \cdots, \mathbf{g}^{(m)}$ for them in order to determine the contribution $e(\Omega)$ to f, *without rejecting the hidden ones.* But even then f must be used for picking out the denumerable sequence of Ω's that actually contribute to f.

A subgroup σ^0 of σ may be treated directly by observing that f is also almost periodic with respect to σ^0. This is a better procedure than by breaking up the σ-invariant sets obtained by our construction into primitive σ^0-invariant sets. For then one would still face the task of getting rid of all but the upright ones.

In the Main Theorem one can readily pass from a single given a. p. vector f to a finite number, or even a denumerable sequence, of such vectors, $f_1, f_2, \cdots$. An Ω-row $\mathbf{g}'$ is f-hidden when $(g'_\mu, f) = 0$ $(\mu = 1, \cdots, h)$. For any $v = 1, 2, \cdots$ we construct an orthogonal basis $\mathbf{g}^{(v,1)}, \cdots, \mathbf{g}^{(v,m_v)}$ for those Ω-rows $\mathbf{g}$ which are perpendicular to all f_v-hidden Ω-rows $\mathbf{g}'$, $(\mathbf{g}', \mathbf{g}) = 0$, and moreover satisfy the relations

$$(f_1, g_\mu) = 0, \cdots, (f_{v-1}, g_\mu) = 0 \qquad (\mu = 1, \cdots, h).$$

Of course $m_v \leq h$. If $m_1, m_2, m_3, \cdots$ all vanish then Ω "does not contribute." Otherwise the whole (finite or infinite) sequence of Ω-rows

$$\mathbf{g}^{(1,1)}, \cdots, \mathbf{g}^{(1,m_1)};\ \mathbf{g}^{(2,1)}, \cdots, \mathbf{g}^{(2,m_2)};\ \cdots \quad \cdots$$

is orthogonal. For any vector f we may determine the Fourier coefficients $\alpha_\mu{}^{(v,i)} = (g_\mu{}^{(v,i)}, f)$ and form

$$\| e_v(f;\Omega) \|^2 = \sum | \alpha_\mu{}^{(v,i)} |^2 \qquad (\mu = 1, \cdots, h;\ i = 1, \cdots, m_v).$$

The contribution of Ω to $\| f \|^2$ is defined by

$$\| e(f;\Omega) \|^2 = \| e_1(f;\Omega) \|^2 + \| e_2(f;\Omega) \|^2 + \cdots, \tag{3.5}$$

and

$$\sum_{\Omega} \| e(f;\Omega) \|^2 \leq \| f \|^2, \tag{3.6}$$

the sum extending over a complete set of inequivalent contributing irreducible unitary Ω. For $f = f_v$ the sum (3.5) is finite, $\| e_u(f_v;\Omega) \|^2$ equaling zero for $u = v + 1, v + 2, \cdots$, and *the Bessel inequality* (3.6) *changes into the corresponding equation.* As a result, an orthogonal sequence $g_1, g_2, \cdots$ of vectors, subdivided into sections that form primitive invariant sets, has been found such that the Fourier series $\alpha_1 g_1 + \alpha_2 g_2 + \cdots$ with the coefficients

$\alpha_n = (g_n, f)$ converges weakly to f *for each of the given a. p. vectors* $f = f_1, f_2, \cdots$

4. Strong approximation. Interpretation by vector functions on Π. A further axiom connecting "length" $|g|$ with "modulus" $\| g \|$ is needed for the transition from weak to strong convergence. Let us study the integral $g_\xi = \int \xi(s) \cdot sg$ in which $\xi(s)$ is any f-continuous numerical function and the vector g of such nature that sg is an f-continuous vector function of s. In the interpretation by scalar functions on Π this relation reads

$$g_\xi(P) = \int_s \xi(s) \cdot g(s^{-1}P),$$

and thus

$$| g_\xi(P) |^2 \leq \int_s | \xi(s) |^2 \cdot \int_s | g(s^{-1}P) |^2.$$

The second factor on the right is independent of P and has been denoted by $\int_P | g(P) |^2 = \| g \|^2$. If $\| \xi \|^2$ stands for $\int_s | \xi(s) |^2$, we therefore arrive at the inequality

$$| g_\xi | = \underset{P}{\text{l. u. b.}} \; | g_\xi(P) | \leq \| \xi \| \cdot \| g \| .$$

The special case suggests the following general axiom:

Axiom IV. Under the conditions specified above the integral $g_\xi = \int \xi(s) \cdot sg$ satisfies the inequality

$$| g_\xi | \leq \| \xi \| \cdot \| g \| . \tag{4.1}$$

I do not deny its arbitrary character; one would wish to reduce it to some simpler assumptions. But two things can be said for it: (1) It holds in any metric vector space Σ provided one identifies $|g|$ with $\| g \|$. Indeed for $g(s) = sg$ the inequality (C_2) gives

$$\| \int_s \xi(s) \cdot sg \|^2 \leq \| \xi \|^2 \cdot \int_s \| sg \|^2 = \| \xi \|^2 \cdot \| g \|^2.$$

(2) It holds for the interpretation by scalar functions on Π.

The following theorem is an immediate consequence of the new axiom.

$\xi(s)$ being an f-continuous function, the Fourier series of the vector $y = \int_s \xi(s) \cdot sf,$

$$\sum_n \eta_n g_n, \qquad \eta_n = (g_n, y),$$

converges strongly towards y. In the sum $\sum_n$ *members of the same set may not be separated.*

Let us indicate the contribution of the first p invariant sets, $e + e_1 + \cdots + e_{p-1}$, by $\{e + \cdots\}_p$ and carry out our calculations for one of the sets, to which the old notations

$$h, \qquad g_\mu, \qquad \alpha_\mu, \qquad e = \sum_\mu \alpha_\mu g_\mu, \qquad \omega_{\mu\nu}(s) \qquad (\mu, \nu = 1, \cdots, h)$$

may refer. We have

$$se = \sum_\mu \alpha_\mu \cdot sg_\mu = \sum_{\mu,\nu} \alpha_\mu \, \omega_{\mu\nu}(s) g_\nu = \sum_\nu (g_\nu, sf) \cdot g_\nu, \tag{4.2}$$

therefore

$$\int_s \xi(s) \cdot se = \sum_\nu (g_\nu, y) g_\nu = \sum_\nu \eta_\nu g_\nu$$

and

$$y = \{\sum_\nu \eta_\nu g_\nu + \cdots\}_p + \int_s \xi(s) \cdot sf^{(p)} \tag{4.3}$$

where $f^{(p)}$ denotes the remainder $f - \{e + \cdots\}_p$. The formula (4. 2) for se shows that

$$| se - te | \leq \sum_\nu | (g_\nu, sf - tf) | \cdot | g_\nu | \leq \| sf - tf \| \cdot \sum_\nu | g_\nu |$$

and thus

$$| sf^{(p)} - tf^{(p)} | \leq A \cdot | sf - tf |$$

with $A = 1 + \{\sum_\nu | g_\nu | + \cdots\}_p$. Thus $sf^{(p)}$ is an f-continuous vector function of s, and our axiom yields for the remainder in (4. 3) the estimate:

$$| \int_s \xi(s) \cdot sf^{(p)} | \leq \| \xi \| \cdot \| f^{(p)} \| .$$

But one knows that $\| f^{(p)} \| = \delta_p$ tends to zero with $p \to \infty$.

Theorem of strong approximation. *f can be strongly approximated with arbitrary accuracy by finite sums of the form* $\sum_{k=1}^n \alpha'_k g_k$.

The numerical function $| sf - f | = \rho(s)$ is f-continuous since $| \rho(s) - \rho(t) | \leq | sf - tf |$. Choose any $\epsilon > 0$ and let $l(\rho)$ be a non-negative uniformly continuous function of the variable ρ (≥ 0) which vanishes for $\rho \geq \epsilon$ and for which the integral of the f-continuous function $l(\rho(s)) = \lambda(s)$ is 1. Form the difference

$$\int_s \lambda(s) \cdot sf - f = \int_s \lambda(s) (sf - f).$$

Considering that $l(\rho)\cdot\rho \leq \epsilon\cdot l(\rho)$ for $\rho\geq 0$, namely both for $0\leq\rho<\epsilon$ and for $\rho\geq\epsilon$, one gets

$$|\int_s \lambda(s)\cdot sf - f| \leq \int_s \lambda(s)\cdot|sf-f| \leq \epsilon\int_s \lambda(s) = \epsilon.$$

But the Fourier series of $y=\int_s \lambda(s)\cdot sf$ converges strongly to y, and hence we obtain coefficients $\eta_k=(g_k,y)$ and an $n=n_{p-1}$ such that

$$|y-\sum_{k=1}^{n}{}' \eta_k g_k| \leq \epsilon.$$

The result is a strong approximation of f with the accuracy 2ϵ.

It follows readily that not only f, but every vector in $\Sigma_f{}^0$ can be approximated in the same manner.

It should be noted that the Axiom IV in general does not carry over from the group σ to a subgroup σ^0.

In the interpretation by functions $f(P)$ on Π it seems unnatural to limit oneself to the case where the value of the function is a number; it may itself be a vector with several components. So it is in physics, where the electromagnetic field strength has 6 and the electronic ψ has 2 or 4 components. We may even admit the value $f(P)$ to be a vector in a metric vector space of infinitely many dimensions. Let us therefore now assume that we are given a metric vector space Σ (Axioms I and II) closed in the Hilbert sense, so that any weakly convergent sequence $f_1, f_2, \cdots, \| f_n - f_m \| \to 0$, converges weakly toward a vector f, $\| f_n - f \| \to 0$. (Should the metric space not be closed, we make it so by Cantor's construction.) Suppose, moreover, we are given a point field Π and a realization of the group σ by a transitive group of transformations $P\to sP$ of this point field (Axiom I_0). We study functions $\boldsymbol{f}$ which associate with every point P of Π a vector $f(P)$ in Σ. The transform $s\boldsymbol{f}$ is then to be defined as the vector function associating with P the vector $sf(s^{-1}P)$. But it seems convenient to define a transform $(s,t)\boldsymbol{f}$ for any *pair* of elements (s,t) of σ by ascribing to $(s,t)\boldsymbol{f}$ the value $sf(t^{-1}P)$ at the point P. The pairs (s,t) form the group $\sigma\times\sigma$ in which σ itself is contained as the subgroup of the diagonal pairs (s,s); $s\boldsymbol{f}=(s,s)\boldsymbol{f}$. The restriction of almost-periodicity to be imposed upon $\boldsymbol{f}$ is twofold:

1. Let us say that the element t of σ lies in the f-disk $D_f(b;\epsilon)$ of radius ϵ around b if $\| f(t^{-1}P) - f(b^{-1}P) \| \leq \epsilon$ for all points P. We assume that for every $\epsilon>0$ the group σ may be covered by a finite number of such f-disks $D_f(b_k;\epsilon)$ $(k=1,\cdots,n)$ of radius ϵ. Choose a definite point P_0. For an $\boldsymbol{f}$ satisfying this condition the numerical function $\| f(t^{-1}P_0)\|$ of t

and hence the function $\| f(P) \|$ of P is bounded. We define the length $| f |$ of f by

$$| f | = \text{l. u. b.} \| f(P) \| .$$

This length is invariant with respect to all transformations (s, t) of $\sigma \times \sigma$, $| (s, t) f | = | f |$, in particular to all transformations (s, s) of σ. For any two vector functions $f(P)$, $g(P)$ satisfying our condition the scalar product $\psi(P) = (g(P), f(P))$ is an almost periodic numerical function on Π, and we may therefore form the constant $\int_t \psi(t^{-1}P) = \int_P \psi(P)$ and introduce it as the scalar product (g, f). Again invariance prevails: $((s, t) g, (s, t) f) = (g, f)$.

2. Let us say that the element s lies in the f-circle $C_f(a; \epsilon)$ of radius ϵ around a if $\| sf(P) - af(P) \| \leq \epsilon$ for all points P, and now assume that σ can be covered not only by a finite number of f-disks $D_f(b_k; \epsilon)$ $(k = 1, \cdots, n)$, but also by a finite number of f-circles $C_f(a_i; \epsilon)$ $(i = 1, \cdots, m)$ of arbitrarily small radius ϵ. If s lies in $C_f(a_i; \epsilon)$ and t in $D_f(b_k; \epsilon)$ we have

$$\begin{aligned} &\| sf(t^{-1}P) - a_i f(b_k^{-1}P) \| \\ \leq &\| sf(t^{-1}P) - a_i f(t^{-1}P) \| + \| a_i f(t^{-1}P) - a_i f(b_k^{-1}P) \| \\ \leq &\epsilon + \| f(t^{-1}P) - f(b_k^{-1}P) \| \leq 2\epsilon, \end{aligned}$$

or $| (s, t) f - (a_i, b_k) f | \leq 2\epsilon$. Hence f is almost periodic with respect to the group $\sigma \times \sigma$ of the pairs (s, t) and a fortiori with respect to the subgroup σ of the diagonal pairs (s, s).

All the axioms I — III are satisfied for the space Σ of the almost periodic vector functions f, both with respect to $\sigma \times \sigma$ and to σ. Also the Axiom IV introduced in this section holds. Indeed consider

$$g_\xi(P) = \int_s \int_t \xi(s, t) \cdot sg(t^{-1}P).$$

One finds by applying (C_2) to the (s, t)-means,

$$\| g_\xi(P) \|^2 \leq \int_s \int_t | \xi(s, t) |^2 \cdot \int_s \int_t \| sg(t^{-1}P) \|^2.$$

The second factor on the right equals

$$\int_t \| g(t^{-1}P) \|^2 = \int_P \| g(P) \|^2 = \| g \|^2.$$

Hence $| g_\xi | \leq \| \xi \| \cdot \| g \|$. The same argument applies to a simple mean value of the type

$$g'_\eta(P) = \int_s \eta(s) \cdot sg(s^{-1}P).$$

5. Appendix. On Maak's approach to the theory of almost periodic functions.

1. The marriage problem. Given n distinct objects $\{a\} = a_1, \cdots, a_n$ (boys) and n distinct objects $\{b\} = b_1, \cdots, b_n$ (girls); moreover a scheme of linkage Q_n according to which an a_i and a b_k are either linked (friends) or not linked. I call the number of elements in a set its rank. A set B of girls is said to be associate to a given set A of boys if no boy in A has friends outside the set B. (Then the complement $\{a\} - A$ is in the same sense an associate of $\{b\} - B$.) Question: What is the necessary and sufficient condition that the boys can be paired with the girls in such a fashion that in each of the n pairs the partners are friends? The following basic condition is obviously necessary: A set A of boys has never an associate set B of girls of lesser rank than A. The fundamental combinatorial lemma asserts that this condition is also sufficient [**9**, **4**, **6**].

Proof. I let the girls $b_1, \cdots, b_n$ choose their partners one after the other, and therefore ask b_1 first to make her choice. Suppose she chooses a_3. The choice should be fair to herself, i. e. a_3 and b_1 should be friends. But it should also be fair to the other girls by not making it impossible for them to find partners among their friends, i. e. the linkage scheme Q_{n-1} arising from Q_n by removing a_3 and b_1 should still satisfy the basic condition. Call a set A of boys of rank r (≥ 1) *distinguished* if it has an associate set $B = (b_1, b_{k_2}, \cdots, b_{k_r})$ of girls that is of the same rank r and contains b_1. Then the second postulate requires that there should be no distinguished set A which does not contain a_3, or a_3 must be in the intersection of all distinguished sets.

We make the basic observation that the intersection of two distinguished sets is again distinguished and is not empty. Indeed let A_i be a distinguished set of rank r_i and B_i an associate of the same rank $(i = 1, 2)$. Let $A = A_1 \cap A_2$ be of rank r and $B = B_1 \cap B_2$ of rank s. The rank s is at least 1, since B contains b_1. B is an associate of A, hence $r \leq s$. Moreover $B_1 \cup B_2$ is an associate of $A_1 \cup A_2$; thus $r_1 + r_2 - r \leq r_1 + r_2 - s$, or $r \geq s$. The resulting equation $r = s$ together with $s \geq 1$ proves the point.

Let $A_0 = (a_{i_1}, \cdots, a_{i_m})$ be a distinguished set *of lowest rank* $m \geq 1$ and $(b_1, b_{k_2}, \cdots, b_{k_m})$ its associate of the same rank. A_0 is contained in every distinguished set A (and therefore unique) since $A \cap A_0$ can not be of smaller rank than A_0. At least one of the boys $a_{i_1}, \cdots, a_{i_m}$ in the set A_0 is a friend of b_1; for otherwise A_0 would have an associate set $(b_{k_2}, \cdots, b_{k_m})$ of rank $m - 1$.

Let therefore b_1 pick one of her friends $a_{1'}$ in A_0, and rearrange the sequence $a_1, a_2, \cdots, a_n$ so that her mate $a_{1'}$ assumes the first place. Then the linkage scheme Q_{n-1} of $a_2, \cdots, a_n$ with $b_2, \cdots, b_n$ satisfies the basic condition. By induction the girls thus solve their marriage problem.

2. The objects to which the combinatorial lemma will be applied are pieces σ_i of the group σ, and linkage of a piece σ' and a piece τ' will mean that they have points in common. [Contrary to Maak, we do not forbid the roof-tiles σ_i to overlap.] Let f be an a. p. vector and $\phi(s)$ an f-continuous numerical function, so that for every $\delta > 0$ there is an $\epsilon > 0$ such that $|sf - tf| \leq \epsilon$ implies $|\phi(s) - \phi(t)| \leq \delta$. Then it implies also $|\phi(as) - \phi(at)| \leq \delta$ for every element a. Therefore the function $\phi(s)$ is almost periodic in the sense that for every δ one can cover σ by a finite number of tiles σ_i such that $|\phi(as) - \phi(at)| \leq \delta$ for any s, t in the same piece σ_i and any a. By a simple argument, which I shall not repeat here, Maak infers from this one-sided the two-sided almost-periodicity.

Let us say that s lies in the circlet of diameter $\rho > 0$ around s_0 if

$$|\phi(asb) - \phi(as_0b)| \leq \tfrac{1}{2}\rho \text{ for all } a \text{ and } b.$$

Our statement means that a finite number of elements c_K $(K = 1, \cdots, N)$ may be ascertained such that the circlets σ_K of diameter ρ around the c_K cover the entire group. *Choose the smallest number N of elements c_K satisfying this condition.* We speak of them as a ρ-lattice and of

$$L_\rho = N^{-1} \cdot \sum_k \phi(c_K)$$

as the lattice average of ϕ. (We differ from Maak by using no other domains but circlets. It is a task of fundamentally simpler nature to determine the minimum number of points which have a given property than the minimum number of domains.) Submit the lattice circlets σ_K to an arbitrary left-translation a, $s \to as$, $\sigma_K \to \tau_K = a\sigma_K$. The τ_K are circlets of the same diameter ρ. For any r lattice circlets $\sigma_{K_1}, \cdots, \sigma_{K_r}$ it will never happen that those τ_K that are linked with σ_{K_1} or $\sigma_{K_2} \cdots$ or σ_{K_r} are less numerous. For then we could replace the circlets $\sigma_{K_1}, \cdots, \sigma_{K_r}$ by these τ_K that are linked with them, and thereby lower the total number N. Our combinatorial lemma shows that the σ_K and the τ_K may be paired $(\tau_K, \sigma_{K'})$ in such a way that τ_K and $\sigma_{K'}$ overlap.

The centers $d_K = ac_K$ and $c_{K'}$ of two circlets of diameter ρ which overlap satisfy the inequality

$$|\phi(d_Kb) - \phi(c_{K'}b)| \leq \rho$$

for all b. Consequently

$$|N^{-1}\sum_K \phi(ac_K b) - N^{-1}\sum_K \phi(c_K b)| \leq \rho \tag{5.1}$$

for arbitrary a and b, in particular

$$|N^{-1}\sum_K \phi(ac_K) - N^{-1}\sum_K \phi(c_K)| \leq \rho. \tag{5.2}$$

By the same token the right-translation $s \to sb$ yields the inequality

$$|N^{-1}\sum_K \phi(c_K b) - N^{-1}\sum_K \phi(c_K)| \leq \rho. \tag{5.3}$$

Let a_i $(i = 1, \cdots, n)$ be any number of elements and $\alpha_i \geq 0$ corresponding weights of sum 1. The average $A = \sum_i \alpha_i \phi(a_i)$ was said to oscillate by less than ϵ if

$$|\sum_i \alpha_i \phi(aa_i) - A| \leq \epsilon$$

identically in a. Let A, B be two such averages of ϕ oscillating by less than ϵ and δ respectively. We then have

$$|\sum_i \alpha_i \phi(c_K a_i) - A| \leq \epsilon \tag{5.4}$$

for each c_K of our ρ-lattice. On the other hand (5.3) gives for $b = a_i$:

$$|N^{-1}\sum_K \phi(c_K a_i) - L_\rho| \leq \rho. \tag{5.5}$$

(5.4) and (5.5) lead to

$$|N^{-1}\sum_{K,i} \alpha_i \phi(c_K a_i) - A| \leq \epsilon, \qquad |N^{-1}\sum_{K,i} \alpha_i \phi(c_K a_i) - L_\rho| \leq \rho,$$

$$|A - L_\rho| \leq \epsilon + \rho.$$

For the same reason

$$|B - L_\rho| \leq \delta + \rho,$$

consequently $|B - A| \leq \epsilon + \delta + 2\rho$. This must be true for every $\rho > 0$, which is impossible unless

$$|B - A| \leq \epsilon + \delta,$$

cf. formula (1.4).

It is this fact on which the existence of the mean value $J = \int \phi(s)$ depends; any average A oscillating by less than ϵ differs from it by not more than ϵ. According to (5.2) and (5.1) the lattice averages

$$L_\rho\langle\phi\rangle = N^{-1}\sum_K \phi(c_K) \text{ and } L_\rho\langle\phi_b\rangle = N^{-1}\sum_K \phi(c_K b)$$

of $\phi(s)$ and $\phi_b(s) = \phi(sb)$ oscillate by less than ρ. Among themselves they differ by not more than ρ, (5.3). This proves that the absolute difference of $\int \phi(sb)$ and $\int \phi(s)$ can not exceed 3ρ, and as ρ is arbitrary, these two mean values must coincide.

INSTITUTE FOR ADVANCED STUDY.

REFERENCES.

[1] S. Bochner and J. von Neumann, "Almost periodic functions in groups," *Transactions of the American Mathematical Society*, vol. 37 (1935), pp. 21-50.

[2] H. Bohr, "Zur Theorie der fastperiodischen Funktionen I," *Acta Mathematica*, vol. 45 (1925), pp. 29-127.

[3] E. Cartan, "Sur la détermination d'un systeme orthogonal complet dans un espace de Riemann symétrique clos," *Rendiconti del Circolo Matematico di Palermo*, vol. 53 (1929), pp. 217-252.

[4] P. Hall, "On representatives of subsets," *The Journal of the London Mathematical Society*, vol. 10 (1935), pp. 26-29.

[5] Anna Hurevitsch, "Unitary representations in Hilbert space of a compact topological group," *Recueil Mathématique* (*Matematicheskii Sbornik*), vol. 13 (1943), pp. 79-86.

[6] W. Maak, "Eine neue Definition der fastperiodischen Funktionen" and "Abstrakte fastperiodische Funktionen," *Abhandlungen aus dem Mathematischen Seminar der Hamburgischen Universität*, vol. 11 (1935), pp. 240-244, and vol. 11 (1936), pp. 367-380.

[7] J. von Neumann, "Almost periodic functions in a group I," *Transactions of the American Mathematical Society*, vol. 36 (1934), pp. 445-492.

[8] F. Peter and H. Weyl, "Die Vollständigkeit der primitiven Darstellungen einer geschlossenen kontinuierlichen Gruppe," *Mathematische Annalen*, vol. 97 (1927), pp. 737-755.

[9] R. Rado, "Bemerkungen zur Kombinatorik im Anschluss an Untersuchungen von Herrn D. König," *Sitzungsberichte der Preussischen Akademie der Wissenschaft*, vol. 32 (1933), p. 68.

[10] H. Weyl, "Integralgleichungen und fastperiodische Funktionen," *Mathematische Annalen*, vol. 97 (1927), pp. 338-356.

[11] H. Weyl, "Harmonics on homogeneous manifolds," *Annals of Mathematics*, vol. 35 (1934), pp. 486-499.

146.

Inequalities between the two kinds of eigenvalues of a linear transformation

Proceedings of the National Academy of Sciences of the United States of America
35, 408—411 (1949)

With a linear transformation A in an n-dimensional vector space (matrix consisting of $n \times n$ complex numbers $a_{ii'}$) there are connected two kinds of eigenvalues: the roots $z = \alpha_1, \ldots, \alpha_n$ of the characteristic polynomial $|zE - A|$ of A (E = unit matrix) and the roots $z = \kappa_1, \ldots, \kappa_n$ of $|zE - K|$ where K is the Hermitian matrix A^*A composed of A and its Hermitian conjugate A^*. The κ_i are non-negative, and one would naturally compare the $\lambda_i = |\alpha_i|^2$ with the κ_i. If A is normal, $A^*A = AA^*$, they coincide; in general, however, they do not. Arrange the κ as well as the λ in descending order,

$$\lambda_1 \geq \lambda_2 \geq \ldots \geq \lambda_n, \qquad \kappa_1 \geq \kappa_2 \geq \ldots \geq \kappa_n.$$

I shall prove the following

THEOREM. *Let $\varphi(\lambda)$ be an increasing function of the positive argument λ, $\varphi(\lambda) \geq \varphi(\lambda')$ for $\lambda \geq \lambda' > 0$, such that $\varphi(e^\xi)$ is a convex function of ξ and $\varphi(0) = \lim_{\lambda \to 0} \varphi(\lambda) = 0$. Then the eigenvalues λ_i and κ_i in descending order satisfy the inequalities*

$$\varphi(\lambda_1) + \ldots + \varphi(\lambda_m) \leq \varphi(\kappa_1) + \ldots + \varphi(\kappa_m) \qquad (m = 1, 2, \ldots, n), \quad (1)$$

in particular

$$\lambda_1^s + \ldots + \lambda_m^s \leq \kappa_1^s + \ldots + \kappa_m^s \qquad (m = 1, 2, \ldots, n) \quad (2)$$

for any real exponent $s > 0$.

According to a familiar argument[1]

$$\lambda_1 \leq \kappa_1. \quad (3)$$

Indeed the equation $Ax = \alpha_1 x$ has a vector solution $x = a \neq 0$: $Aa = \alpha_1 a$, $a^*A^* = \bar{\alpha}_1 a^*$, hence $a^*A^*Aa = \bar{\alpha}_1\alpha_1(a^*a)$ or

$$a^*Ka = \lambda_1(a^*a), \qquad a^*a > 0.$$

Since every vector satisfies the inequality $x^*Kx \leq \kappa_1(x^*x)$, (3) follows.

The linear vector transformation A induces certain linear transformations $A^{[1]}, A^{[2]}, A^{[3]}, \ldots, A^{[n]}$ for the space elements (skew-symmetric tensors) of rank 1, 2, 3, ..., n. For instance $A^{[3]} = ||a^{[3]}_{JJ'}||$ is given by

$$a^{[3]}_{JJ'} = \begin{vmatrix} a_{ii'}, & a_{ik'}, & a_{il'} \\ a_{ki'}, & a_{kk'}, & a_{kl'} \\ a_{li'}, & a_{lk'}, & a_{ll'} \end{vmatrix}$$

where J and J' range over the triples (i, k, l) and (i', k', l') with the restrictions $i < k < l$, $i' < k' < l'$, respectively. Application of the inequality (3) to these matrices $A^{[1]}$, $A^{[2]}$, ... yields the relations

$$\lambda_1 \leq \kappa_1, \quad \lambda_1\lambda_2 \leq \kappa_1\kappa_2, \quad \ldots, \quad \lambda_1 \ldots \lambda_n \leq \kappa_1 \ldots \kappa_n \tag{4}$$

(with the equality sign prevailing in the last of them). Everything will be settled as soon as I prove the following

LEMMA: *Let κ_i, λ_i ($i = 1, \ldots, m$) be non-negative numbers such that*

$$\lambda_1 \geq \lambda_2 \geq \ldots \geq \lambda_m \tag{5}$$

and

$$\lambda_1 \leq \kappa_1, \quad \lambda_1\lambda_2 \leq \kappa_1\kappa_2, \quad \ldots, \quad \lambda_1 \ldots \lambda_m \leq \kappa_1 \ldots \kappa_m; \tag{6}$$

then

$$\sum_i \varphi(\lambda_i) \leq \sum_i \varphi(\kappa_i) \qquad (i = 1, \ldots, m) \tag{7}$$

for any function φ of the nature described in the Theorem.

Of two real numbers α, β let max.(α, β) denote α if $\alpha \geq \beta$ and β if $\beta \geq \alpha$. With a variable argument $z \geq 0$ form the functions

$$f(z) = \prod_{i=1}^{m} \max.(1, \kappa_i z) \quad \text{and} \quad g(z) = \prod_{i=1}^{m} \max.(1, \lambda_i z).$$

Then

$$g(z) \leq f(z) \text{ for } z \geq 0. \tag{8}$$

Indeed set

$$g_i(z) = 1 \text{ for } i = 0 \text{ and } g_i(z) = \lambda_1 \ldots \lambda_i z^i \quad \text{for} \quad i = 1, \ldots, m$$

and distinguish the intervals $\{0\}$, $\{1\}$, ..., $\{m-1\}$, $\{m\}$ as defined by

$$\lambda_1 z \leq 1, \ \lambda_1 z \geq 1 \geq \lambda_2 z, \ \ldots, \ \lambda_{m-1} z \geq 1 \geq \lambda_m z, \ \lambda_m z \geq 1.$$

Then $g(z) = g_i(z)$ for z in $\{i\}$. But, because of (6), $g_i(z) \leq f_i(z) \leq f(z)$, hence (8) holds in each of the $m + 1$ intervals.

With an increasing function $\psi(z)$ one can form the Stieltjes integral

$$\int_0^\infty \log g(z) \cdot d\psi(z) = \sum_i \varphi(\lambda_i), \tag{9}$$

provided $\int^\infty \log z . d\psi(z)$ converges. Here

$$\varphi(\lambda) = \int_0^\infty \log \max. (1, \lambda z) \cdot d\psi(z) = \int_{\lambda z \geq 1} \log (\lambda z) \cdot d\psi(z). \tag{10}$$

It is clear how (8) by means of (9) and the corresponding formula for (fz) leads to (7).

Set $\lambda = e^{\xi}$. If $\varphi(\lambda) = G(\xi)$ is a given function satisfying the conditions of the Theorem, it can be expressed by means of a non-decreasing function $G'(t)$ in the form

$$G(\xi) = \int_{-\infty}^{\xi} G'(t)\cdot dt = -\int_{-\infty}^{\xi} (t-\xi)\cdot dG'(t). \tag{11}$$

(The integration per partes is justified since

$$-t . G'(t) \leq 2 . \int_{t}^{t/2} G'(\tau)\cdot d\tau$$

converges to zero for $t \to -\infty$.) (10) goes over into (11) by the substitution $z = e^{-t}$, $\psi(z) = -G'(t)$.

Of the inequalities (2) thus proved, the most important is the last $m = n$, which is independent of any arrangement of the κ_i and λ_i,

$$\lambda_1^s + \ldots + \lambda_n^s \leq \kappa_1^s + \ldots + \kappa_n^s. \tag{2'}$$

Its application to $A^{[2]}, A^{[3]}, \ldots$ yields the further relations

$$\sum_{i_1 < i_2} \lambda_{i_1}^s \lambda_{i_2}^s \leq \sum_{i_1 < i_2} \kappa_{i_1}^s \kappa_{i_2}^s, \tag{2''}$$

$$\sum_{i_1 < i_2 < i_3} \lambda_{i_1}^s \lambda_{i_2}^s \lambda_{i_3}^s \leq \sum_{i_1 < i_2 < i_3} \kappa_{i_1}^s \kappa_{i_2}^s \kappa_{i_3}^s, \tag{2'''}$$

where all the indices $i_1, i_2, i_3, \ldots$ range from 1 to n. Together they state that the polynomial $Q_s(z) = \prod_{i=1}^{n} (1 + \lambda_i^s z)$ is majorized, coefficient for coefficient, by the polynomial $P_s(z) = \prod_{i=1}^{n} (1 + \kappa_i^s z)$. In the limit for $s \to \infty$ they lead back to the relations (4).

If A is non-singular, A^{-1} has the eigenvalues α_i^{-1}, and the eigenvalues of $A^{*-1}A^{-1}$ coincide with those of $A^*(A^{*-1}A^{-1})A^{*-1} = A^{-1}A^{*-1} = (A^*A)^{-1}$, i.e., with the κ_i^{-1}. Hence by application of (1) to A^{-1} corresponding inequalities

$$\sum_{i=m}^{n} \varphi(\lambda_i) \leq \sum_{i=m}^{n} \varphi(\kappa_i) \qquad (m = n, \ldots, 1)$$

will result for any *decreasing* function $\varphi(\lambda)$ for which $\varphi(\lambda) \to 0$ with $\lambda \to \infty$ and $\varphi(e^{\xi})$ is convex; in particular for $\varphi(\lambda) = \lambda^s$ with a negative exponent s. This shows that for a non-singular A the inequalities (2′) and also (2″), (2‴), ... are valid even for $s \leq 0$.

Facts and proofs, except the last remarks which depend on the consideration of A^{-1}, carry over to completely continuous linear operators A in Hilbert space, especially to continuous kernels of integral equations.

Long ago I. Schur proved (2′) for $s = 1$.[2] Recently S. H. Chang showed in his thesis[3] that, in the case of integral equations, convergence of $\sum_i \kappa_i^s$ implies convergence of $\sum_i \lambda_i^s$. These two facts led me to conjecture the relation (2′), at least for $s \leq 1$. After having conceived the simple idea for the proof, I discussed the matter with C. L. Siegel and J. von Neumann; their remarks have contributed to the final form and generality in which the results are presented here.[4]

[1] For a generalization of this inequality see A. Loewy and R. Brauer, "Ueber einen Satz für unitäre Matrizen," *Tôhoku Math. Jour.*, **32**, 44–49 (1930), formula (13) on p. 48.

[2] Schur, I., "Ueber die charakteristischen Wurzeln einer linearen Substitution, mit einer Anwendung auf die Theorie der Integralgleichungen," *Math. Ann.*, **66**, 488–510 (1909).

[3] Chang, S. H., "Theory of Characteristic Values and Singular Values of Linear Integral Equations," Thesis, Cambridge, England, 1948; also, "On the Distribution of Characteristic Values and Singular Values of L^2 Kernels," *Trans. Am. Math. Soc.* (1949).

[4] While this note was in print a result due to J. Karamata, "Sur une inégalité relative aux fonctions convex," *Publ. Math. Univ. Belgrade*, **1**, 145–148 (1932), that comes very near to our lemma, was pointed out to me.

147.

Relativity theory as a stimulus in mathematical research

Proceedings of the American Philosophical Society 93, 535—541 (1949)

ABOUT the influence which the physical theory of relativity had upon purely mathematical research different mathematicians will be of different opinions. But it is unlikely that anybody today would agree with E. Study, Felix Klein's contemporary and life-long enemy, a man of considerable merit in his field and of violent temper, who, in a book published 1923, accused the writers on relativity theory and tensor calculus of having laid waste a rich cultural domain (ein reiches Kulturgebiet der Verwahrlosung anheimgegeben zu haben), that rich cultural field being the algebraic theory of invariants. What got Study's goat was the fact that the symbolic method and the classical notations of that theory had been more or less ignored by the relativists. I shall come back to his problems later on. Anyhow I would not stand here and try to say some words about the topic which the program announces, were I not convinced that relativity theory during the last decades has been an invigorating rather than a devastating influence in the development of several branches of mathematics, including the theory of invariants. Nor is it difficult to prove my point; the facts speak a too unmistakable language.

The relativity problem is one of central significance throughout geometry and algebra and has been recognized as such by the mathematicians at an early time. It played a great role in Leibniz's philosophical-mathematical ideas. In the nineteenth century the concept of a group of transformations was devised and developed as the adequate tool for dealing with it. Suppose a realm of objects, which may be called points, is given to you. Those transformations, those one-to-one mappings of the point field into itself which leave all relations of objective significance between points undisturbed, form a group, the group of automorphisms. If in some way coordinates, self-created reproducible symbols like numbers, are assigned to the points then any automorphism carries this assignment over into a new one from which it is objectively indiscernible (to use Leibniz's word). Hence any coordinate assignment requires an act of choice by which one picks out one from a class of equally admissible coordinate systems. The class is objectively characterized, but not the individual coordinate assignment. In Galois' theory the "points" are the n roots $\alpha_1, \cdots, \alpha_n$ of an algebraic equation of degree n with rational coefficients. The objective relations are those expressible in terms of the fundamental operations of addition, multiplication, subtraction, and division, i.e., all relations of the form $F(\alpha_1, \cdots, \alpha_n) = 0$ where $F(x_1, \cdots, x_n)$ is a polynomial of the n variables $x_1, \cdots, x_n$ with rational coefficients. Labeling the roots by $1, 2, \cdots, n$ amounts to a coordinate assignment. A transformation is a permutation of the n roots, an automorphism a permutation which leaves all algebraic relations $F(\alpha_1, \cdots, \alpha_n) = 0$ with rational coefficients undisturbed; the automorphisms form the Galois group. Examples from geometry are probably more familiar to this audience. In the three-dimensional Euclidean vector space a frame of reference consists of 3 mutually perpendicular vectors of length 1. Relative to such a frame any vector can be described by the triple of its coordinates $(x_1x_2x_3)$. Transition from one such frame to an arbitrary other is effected by an orthogonal transformation of the coordinates x_1, x_2, x_3, and these transformations constitute the group of automorphisms. In affine vector geometry this group is replaced by the group of all homogeneous linear transformations with non-vanishing determinant (or, if one follows Euler, with determinant 1). In the first half of the nineteenth century projective geometry had arisen, whose group of automorphisms consists of all collineations, i.e., of all point transformations that carry straight lines into straight lines. Möbius had added his spherical geometry, the automorphisms of which carry spheres into spheres. In a space of three or more dimensions the group of Möbius transformations coincides with that of all conformal transformations. Ideas which seem to have guided Möbius implicitly in his investigations, but could not be formulated without the group concept, were made explicit by Felix Klein in his famous Erlanger program, 1872. Transitions between equally admissible coordinate assignments or frames of reference in a Klein space find their expression in a group Γ of coordinate transformations. Klein defines the geometry by this group, which the mathematician feels free to choose as he likes: point relations are then said to be of objective significance if they are invariant with respect to the group Γ, two configurations of points are considered objectively alike if one is carried into the other by an operation of the group. For instance, if the group is transitive, as we shall assume in the future, all points are alike, the space is homogeneous.

According to Einstein's special relativity theory the four-dimensional world of the space-time-points is a Klein space characterized by a definite group Γ; and that group is the one most familiar to the geometers, namely the group of Euclidean similarities—with one very important difference however. The orthogonal transformations, i.e., the homogeneous linear transformations which leave

$$(+) \qquad x_1^2 + x_2^2 + x_3^2 + x_4^2$$

unchanged have to be replaced by the Lorentz transformations leaving

$$(-) \qquad x_1^2 + x_2^2 + x_3^2 - x_4^2$$

invariant. This was certainly a surprise to the mathematicians. But it did not disturb them very greatly; Minkowski made the necessary adjustments at once. Indeed in algebraic geometry they had become used to considering their variables as capable of arbitrary complex values; for that made their theories so much simpler, owing to the fact that the field of all complex numbers is algebraically closed, i.e., that an arbitrary algebraic equation of degree n with coefficients in the field always has n roots in the field. Now in the domain of complex numbers there is no difference between (+) and (−); indeed (+) goes over into (−) by substituting ix_4 for x_4. But in the last forty odd years algebra has reversed its position: not only has it recognized the right of the field of real numbers besides that of the complex numbers, but it carries on its investigations, whenever possible, in an arbitrarily given field of numbers, no longer assuming that this underlying field, though closed with respect to the operations of addition, subtraction, multiplication, and division, be also algebraically closed. Two non-degenerate quadratic forms with coefficients in a given field $\mathfrak{F}$ belong to the same genus if one may be carried into the other by a linear transformation with coefficients in $\mathfrak{F}$. Special relativity could have taught the algebraists this lesson: do not ignore other genera of quadratic forms besides the principal genus represented by the unit form $x_1^2 + x_2^2 + \cdots$. As a matter of fact, in their arithmetical theory of quadratic forms, a classic subject from Gauss to Minkowski, they had never ignored them. I am afraid, the geometers had. Yet one can hardly say that here the mathematicians received a new stimulus from relativity theory; rather Klein's Erlanger program and the distinction of genera of quadratic forms was in happy concordance with relativity theory, and Einstein's discovery gave support to these geometric and algebraic conceptions by exhibiting one very important and quite unexpected application in physics.

We are used today to look at mathematical questions from the standpoints of abstract algebra and of topology. Between the orthogonal and the Lorentz group there is a topological difference much more incisive than the algebraic difference of genus: the one is a compact manifold, the other is not. The most systematic part of group theory deals with the representations of groups by linear transformations. Representations in a Hilbert space of finite or infinitely many dimensions are of supreme interest to quantum mechanics. If the group is finite every such representation breaks up into irreducible parts of finite dimensionality; the entire theory, one of the proudest buildings of mathematics, is dominated by the orthogonality and completeness relations. They carry over from finite to compact groups. The theory of Fourier series is nothing but the representation theory of the group of rotations of a circle. Pleased by the beauty and harmony of this theory of representations of compact groups, the mathematicians shyed away for some time from the more complicated and less harmonious situation that had to be expected for non-compact groups. But the Lorentz group and the interest which quantum mechanics has in its representations in Hilbert space finally forced the issue: V. Bargmann in this country and Gelfand and Neumark in Russia mustered enough courage to tackle the representations of the Lorentz group in Hilbert space, and the Russians went on to develop the theory for any groups that are locally but not globally compact.

There is hardly any doubt that for physics special relativity theory is of much greater consequence than the general theory. The reverse situation prevails with respect to mathematics: there special relativity theory had comparatively little, general relativity theory very considerable, influence, above all upon the development of a general scheme for differential geometry. The kind of world geometry Einstein needed to put into mathematical form his central idea of an inertial field that not only acts upon matter but is also acted upon by matter, he found ready-made in the mathematical literature: from Gauss' theory of surfaces in Euclidean space Riemann had abstracted his conception of an n-dimensional Riemannian space. Here the coordinate assignment remains quite arbitrary, subject to arbitrary (differentiable) transformations. A coordinate system usually covers only a part of the manifold; a finite or even an infinite number of partially overlapping patches are needed to cover it completely. But this is of little concern to us, as long as we are still far from overlooking the four-dimensional world in its entire extension. The fact that in an infinitesimal neighborhood of a point Pythagoras' theorem is supposed to prevail finds its expression in Riemann's formula

$$(*) \qquad ds^2 = \sum_{i,j} g_{ij} dx_i dx_j \qquad (g_{ji} = g_{ij})$$

for the square of the length ds of a line element that leads from the point $P = (x_1, \cdots, x_n)$ to an arbitrary infinitely near point $P' = (x_1 + dx_1, \cdots, x_n + dx_n)$. The coefficients g_{ij} do not depend on the line element but will in general vary from point to point. Riemann had gone some distance in developing this kind of geometry, which clearly follows a trend entirely different from Klein's Erlanger program; he had, in particular, derived what is now called the Riemann curvature tensor. An elaborate mathematical machinery for Riemannian geometry had been set up under the name of Absolute Differential Calculus by G. Ricci and T. Levi-Civita. These things, which Einstein learnt from his friend, the mathematician Marcel Grossman in Zürich, enabled him to write down the equation of motion for a planet and the differential equations for the gravitational field, without an appeal to experience, in a purely speculative and yet astonishingly compelling manner. And Nature graciously confirmed his laws with as clear an O.K. as one can ever get from her.

The great importance which Riemannian geometry acquired for Einstein's theory of gravitation gave the impetus to develop this geometry further, to study more carefully its foundations and, as a consequence of such analysis, to generalize it in various directions. The first and decisive step was Levi-Civita's discovery of the notion of infinitesimal parallel vector displacement. Let us begin with Gauss' representation of a surface in Euclidean space. The points P of the surface are referred to arbitrary coordinates x_1, x_2, the location of the point P in the embedding space is given by $\mathfrak{r} = \mathfrak{r}(x_1 x_2)$ where $\mathfrak{r} = \overrightarrow{OP}$ is the vector leading in that space from the origin O to the space point P. The increment

$$d\mathfrak{r} = \frac{\partial \mathfrak{r}}{\partial x_1} dx_1 + \frac{\partial \mathfrak{r}}{\partial x_2} dx_2$$

is the line element joining $P = (x_1, x_2)$ with the nearby surface point $P' = (x_1 + dx_1, x_2 + dx_2)$. In this way it comes about that the two-dimensional linear manifold of the tangent vectors $\mathfrak{x}$ in P is referred to the affine vector basis consisting of the two vectors

$$[\mathfrak{e}_1 = \partial \mathfrak{r}/\partial x_1,\ \mathfrak{e}_2 = \partial \mathfrak{r}/\partial x_2 : \mathfrak{x} = \xi_1 \mathfrak{e}_1 + \xi_2 \mathfrak{e}_2.$$

The metric structure of this vector compass is taken into account, only afterwards, as it were, by expressing the square of the length of $\mathfrak{x}$ as a quadratic form $\Sigma g_{ij} \xi_i \xi_j$ of its components ξ_i with certain coefficients g_{ij}. Thus the two-dimensional Euclidean vector space, the compass at P, is here treated as an affine space to which a quadratic form is joined as the "absolute." In a similar fashion affine space is sometimes treated as projective space in which one plane has been absolutely distinguished as the "plane at infinity." There is an obvious artificiality in this procedure as it falsifies the group, so to speak; but it is also obvious how Gauss' approach led to this treatment. But the natural frame for the compass would be a Cartesian one consisting of two perpendicular vectors $\vec{e}_1$, $\vec{e}_2$ of length 1 at the point P; choose such a frame without tying it to the coordinates $x_1 x_2$ to which the neighborhood P on the surface is referred. The line element $\overrightarrow{PP'}$ will then be given by an expression $\overrightarrow{PP'} = \Sigma \omega_i \vec{e}_i$ where the ω_i are linear differential forms $o_{i1} dx_1 + o_{i2} dx_2$. Invariance must prevail (1) with respect to arbitrary transformations of the coordinates x_i, (2) with respect to an arbitrary rotation of the frame $\vec{e}_i$ of the vector compass at P, arbitrary in the sense that it may depend in an arbitrary fashion on P. This scheme which E. Cartan always used is better suited to the nature of the Pythagorean metric. When one tries to fit Dirac's theory of the electron into general relativity, it becomes imperative to adopt the Cartan method. For Dirac's four ψ-components are relative to a Cartesian (or rather a Lorentz) frame. One knows how they transform under transition from one Lorentz frame to another (spin representation of the Lorentz group); but this law of transformation is of such a nature that it cannot be extended to arbitrary linear transformations mediating between affine frames.

Let us return to the surface in Euclidean space. A tangent vector $\mathfrak{x}$ at P can be transferred to the nearby surface point P' by parallel displacement in the embedding Euclidean space. The $\mathfrak{x}$ thus obtained is no longer tangential at P'; we therefore split it into its tangential and its (infinitesimal) normal component at P', $\mathfrak{x} = \mathfrak{x}' + \mathfrak{x}_n$, and throw away the latter: $\mathfrak{x} \rightarrow \mathfrak{x}'$ is Levi-Civita's process of infinitesimal parallel displacement on the surface. It first looks as if it depended on the embedment of the surface into the surrounding Euclidean space; but when one carries out the computation, it turns out to be completely determined by the intrinsic metric of the surface. Hence it must be possible to give an intrinsic definition of the process. The first obvious property of Levi-Civita's displacement is that it leaves the length of the vector $\mathfrak{x}$ unchanged. In combination with another intrinsic condition, which I shall not formulate but only allude to by the word "condition of closure," this leads indeed to a unique determination of the "affine connection" by virtue of which each vector at P goes over into a definite vector at the infinitely near point P'. After the notion is thus made independent of an embedding Euclidean space, it becomes applicable to any two-dimensional, nay to any n-dimensional Riemannian manifold.

It is natural to introduce the concept of an *affinely connected manifold* as one in which the process of infinitesimal parallel vector displacement satisfying the condition of closure is defined. It is a fact that the whole tensor analysis with its "covariant derivatives" makes use of the affine connection only, not of the metric. Also Riemann's curvature finds its place here. Indeed if one carries the compass at P by successive steps of infinitesimal parallel displacement around a circuit returning to the start P, then the compass will in general not return to its initial position, but in one that arises from the initial position by a certain rotation around P. This rotation is essentially what Riemann called curvature and what should perhaps more appropriately be called vector vortex. The affine notions of covariant differentiation and of curvature become applicable to Riemannian geometry owing to the fact that the Riemann metric uniquely determines the affine connection.

Thus an affine infinitesimal geometry has sprung up beside Riemann's metric one. One may say that the causal structure of the universe in the immediate neighborhood of the world point P is described by the equation $\Sigma g_{ij}dx_i dx_j = 0$. Hence a Riemannian metric g_{ij} and a metric g^*_{ij} of the same space lead to the same causal or conformal structure if $g^*_{ij} = \lambda g_{ij}$ with a positive factor λ depending on P in an arbitrary fashion. Such features of a Riemannian space are conformal as are not affected by the change $g_{ij} \rightarrow \lambda g_{ij}$. One can also easily describe under what conditions two affine connections are equivalent in the sense that they lead to the same geodesics and thus to the same inertial or projective structure. What happens, one may further ask, when one replaces the geodesics by any families of curves of such nature that through every point in every direction there goes one of these lines? (General geometry of paths.)

Now here is clearly rich food for mathematical research and ample opportunity for generalizations. Thus schools of differential geometers sprang up in the wake of general relativity. Here in Princeton Eisenhart and Veblen took the lead, Schouten in Holland. In France, E. Cartan's fertile geometric imagination disclosed many new aspects of the subject. Some of their outstanding pupils are Tracy Thomas and J. M. Thomas in Princeton, van Dantzig in Holland and Shiing Shen Chern of the Paris school. A lone wolf in Zurich, Hermann Weyl, also busied himself in this field; unfortunately he was all too prone to mix up his mathematics with physical and philosophical speculations. In several ways these authors soon arrived at the conclusion that it is better to establish projective differential geometry not by abstraction from the affine brand, as described above, but independently, namely by associating with each point P of the manifold a projective space Σ_P in the sense of Poncelet and Plücker, this homogeneous space taking the place of the affine vector compass in the affinely connected manifold. In the same manner a general conformal geometry may be developed by associating with each point P a Möbius space Σ_P. The generalization is evident. Let a manifold M and a homogeneous Klein space Σ, defined by a transitive group Γ of transformations, be given. Assume that with each point P there is associated a copy Σ_P of the Klein space, and that Σ_P is carried over to the space $\Sigma_{P'}$ associated with an infinitely near point P' of M by an infinitesimal operation of the group Γ that depends linearly on the relative coordinates dx_i of P' with respect to P. The manifold M, or at least part of it, is referred to coordinates x_i. In each Σ_P we must choose an admissible frame of reference, with respect to which the points of Σ_P are represented by coordinates ξ. Since the Klein space is supposed to generalize the affine tangent vector space, it is natural to assume that a definite center O in Σ_P is marked which "covers" the point P on M. The frame for Σ_P may be so chosen that O becomes the origin $\xi_1 = \xi_2 = \cdots = 0$. It is further natural to assume that the infinitesimal vectors issuing from O in Σ_P on the one hand, and those issuing from P in M on the other hand, are "in coverage" by dint of a one-to-one linear mapping. This assumption brings it about that Σ has the same number of dimensions as M. It is further natural to assume that, if the infinitesimal vector $\overrightarrow{OO_1}$ in Σ_P covers $\overrightarrow{PP'}$ on M, then the displacement $\Sigma_P \rightarrow \Sigma_{P'}$ will carry O_1 into the center O' of $\Sigma_{P'}$. But no other restrictions should be imposed. Carrying Σ_P by successive steps of infinitesimal displacement around a circuit we shall, when we return to P, have arrived at a definite mapping of Σ_P into itself, an operation of the Klein group that is independent of the choice of coordinates x_i and also, if this is understood in the proper sense, of the choice of admissible frames in all the Klein spaces associated with the various points of M. It will, of course, depend on the circuit described. This automorphism is the generalization of Riemann's curvature. Hence we have here before us the natural general basis on which that notion rests. The infinitesimal trend in geometry initiated by Gauss' theory of curved surfaces now merges with that other line of thought that culminated in Klein's Erlanger program.

It is not advisable to bind the frame of reference in Σ_P to the coordinates x_i covering the neighborhood of P in M. In this respect the old treatment of affinely connected manifolds is misleading. What I said about Cartan's method in dealing with Riemannian geometry was intended as an illustration of this lesson. In studying curves in three-dimensional Euclidean space one does not use a stationary Cartesian frame but associates with the point P traveling along the curve a mobile frame that is adapted to the curve in the most intimate manner, namely the Cartesian frame consisting of tangent, principal normal, and binormal. Freedom means adaptability. Cartan coined the phrase *méthode de répère mobile* for this procedure. Also in the modern development of infinitesimal geometry in the large, where it combines with topology and the associated Klein spaces appear under the name of fibres, it has been found best to keep the *répères*, the frames of the fibre spaces, independent of the coordinates of the underlying manifold.

The temptation is great to mention here some of the endeavors that have been made to utilize these more general geometries for setting up unified field theories encompassing the electromagnetic field beside the gravitational one or even including not only the photons but also the electrons, nucleons, mesons, and whatnot. I shall not succumb to that temptation.

Nobody can predict what sort of geometric structures may be thought up, and hence it would be foolish to claim that our pattern of associated Klein spaces and their displacement is universal. Whatever the structure, it must be described in some arithmetical way relative to a frame of reference $\mathfrak{f}$, whether that frame consists of a coordinate system for the manifold M or of a coordinate system for M plus admissible frames of references for each associated Klein space, or is something even more complicated. Always the problem of equivalence arises, i.e., the question under what conditions two such structures can be carried into each other by a change of the universal frame $\mathfrak{f}$. It was in the attempt of solving this problem for Riemannian geometry that Christoffel first introtuced his 3-indices symbols which later were interpreted by Levi-Civita as describing infinitesimal vector displacement, and that Riemann constructed his curvature tensor. This example is typical. In a number of important cases the attempt of solving the equivalence problem led to associating Klein spaces with the points of the manifold and to defining their infinitesimal displacement. Auxiliary variables that had to be introduced could be interpreted as coordinates in the fibre space Σ_P. Hence in practice the scheme has proved of fairly universal applicability.

I wish to say a word about Cartan's treatment of Riemannian geometry. At each point P of the manifold the vector compass bears a Cartesian frame $\vec{e}_i$. The infinitesimal vector $\overrightarrow{PP'}$ at P is expressed in terms of it,

$$\overrightarrow{PP'} = dP = \Sigma\omega_i\vec{e}_i \tag{1}$$

by means of coefficients ω_i that are linear differential forms of the dx_i. Passing from P to the near-by point P' we obtain two Cartesian frames at P': the one that is associated with P' and the one into which the frame associated with P goes over by parallel displacement from P to P'. These two frames in P' are linked by an infinitesimal rotation

$$d\vec{e}_i = \Sigma_j\omega_{ij}\cdot\vec{e}_j \tag{2}$$

the coefficients ω_{ij} of which are again linear differential forms. The displacement of the vector compass is now described by (1), (2). The condition of closure, which I never formulated, makes the ω_{ij} expressible in terms of the ω_i. The adequate instrument for carrying out the calculations to which such a set-up inevitably leads is a calculus fully developed by Cartan; it deals with multilinear differential forms and their multiplication and "external" derivation. This is a subject of considerable importance in several branches of mathematics including topology.

A scalar $f(x_1, \cdots, x_n)$ has a differential $df = \Sigma_i \frac{\partial f}{\partial x_i} dx_i$. From an arbitrary linear differential $\varphi = \Sigma_k\varphi_k dx_k$ we can formally derive $d\varphi = \Sigma \frac{\partial \varphi_k}{\partial x_i} dx_i dx_k$. The essential point now is that in interpreting this formal expression one assumes the antisymmetric rule $dx_k dx_i = - dx_i dx_k$ for the multiplication of the differentials dx_i. It is then perhaps more sincere to introduce two independent line elements dx and δx and let $d\varphi$ stand for the skew-symmetric bilinear differential form

$$\Sigma\left(\frac{\partial \varphi_k}{\partial x_i} - \frac{\partial \varphi_i}{\partial x_k}\right) dx_i\delta x_k. \tag{3}$$

The marvelous thing about this kind of derivative is its invariance with respect to arbitrary coordinate transformations. If φ itself is the differential df of a scalar f, then $d\varphi = 0$: the derivative of a derivative is always zero. Maxwell's theory of the electromagnetic field gives a perfect illustration for this calculus. The potentials φ_i form the coefficients of an invariant differential form $\varphi = \Sigma\varphi_i dx_i$ of rank 1; its derivative $F = d\varphi$, (3), of rank 2 gives the electromagnetic field strength $F_{ik} = \frac{\partial \varphi_k}{\partial x_i} - \frac{\partial \varphi_i}{\partial x_k}$. Since F itself is a derivative, the form dF of rank 3 must vanish, i.e.,

$$\frac{\partial F_{kl}}{\partial x_i} + \frac{\partial F_{li}}{\partial x_k} + \frac{\partial F_{ik}}{\partial x_l} = 0.$$

The symbolism here employed is not as strange as it may strike you. Indeed look at the customary form of writing a double integral, $\int\cdots dx_1dx_2$. Here also the proper meaning of dx_1dx_2 as the area of a parallelogram spanned by two line elements dx, δx would be more fully exhibited by writing it as the determinant $dx_1\delta x_2 - dx_2\delta x_1$; only this form indicates without explanation what happens to the integral under a transformation of coordinates. Multiplication of linear forms

$$\sum f_i dx_i \cdot \sum g_k dx_k = \sum f_i g_k dx_i dx_k$$

must of course also be submitted to the antisymmetric rule $dx_k dx_i = - dx_i dx_k$. It is clear how to pass on to higher ranks. All the integral theorems of vector and tensor analysis are special cases of the general Stokes formula which deals with the integral of a differential form of rank r over an r-dimensional (orientable) manifold imbedded in the space with the coordinates x_i: the integral of the derivative df over a manifold C of dimensionality $r + 1$ equals the integral of f over the r-dimensional boundary ∂C of C. The law that the derivative of a derivative vanishes is thus the dual counterpart of the topological law that the boundary ∂C of something is always closed, i.e., has the boundary zero. The question whether a form f, the derivative of which vanishes, is itself a derivative leads straight to the topological theory of homologies and cohomologies; de Rham's work is fundamental in that respect.

An impressive example demonstrating the power of this technique in which the *méthode de répère mobile* combines with the calculus of linear differential forms is a brief paper by Chern, *Annals of Mathematics*, vol. 45, in which he gives an intrinsic proof for the analogue of the Gauss-Bonnet formula in a Riemannian space of arbitrary even dimension. The classical Gauss-Bonnet formula states for a closed surface in three-dimensional Euclidean space that the integral of its Gaussian curvature equals $2\pi \cdot q$ where the even integer q is the most important topological invariant, the Euler characteristic. Allendoerfer had derived the formula for a Riemannian space of even dimension imbedded in Euclidean space, Chern freed it from the imbedding space.[1] It is this perhaps the simplest instance of a relation between the differential and the topological properties of a space, and it seems that there are still many deep problems to solve in this field.

If one tries to understand what is behind the formal apparatus of tensor calculus that is used in general relativity, one arrives with necessity at the general notion of a *covariant quantity*. Let us take those transformations of the Klein space Σ mediating between admissible frames of reference which leave the center O of Σ fixed. They form a group Γ. A covariant quantity of definite type $\mathfrak{C}$ is described relative to an admissible frame $\mathfrak{f}$ by a number of components $X_1, \cdots, X_h$ which vary independently over all real values while the quantity ranges over the manifold of all its possible values. The components X'_i of the same quantity relative to another admissible frame $\mathfrak{f}'$ are connected with the X_i by a linear transformation $X'_i = \Sigma_j t_{ij} X_j$ the coefficients t_{ij} of which are determined by the operation S of the group Γ that carries $\mathfrak{f}$ into $\mathfrak{f}'$. Composition of the group elements S must be reflected in the composition of the corresponding linear transformations $\| t_{ij}(S) \|$. We then speak of a representation of the group Γ by linear subsitutions, and that representation defines the type $\mathfrak{C}$ of the covariant quantity. In the last decades a quite elaborate theory of representations of continuous Lie groups has developed, in which algebraic, differential, and integral methods are blended with each other in a fascinating manner. Here those problems which according to Study's complaint the relativists had let go by the board are attacked on a much deeper level than the formalistically minded Study had ever dreamt of. For the representations of the linear group the symmetry characters studied in A. Young's Quantitative Analysis and in a different manner by G. Frobenius, proved of great importance. Their bearing upon the quantum mechanics of systems consisting of equal particles (e.g., electrons) has been disclosed by Wigner. For the orthogonal group Cartan found a host of double-valued irreducible representations not less numerous than the single-valued ones. Their appearance is due to the topological fact that the orthogonal group is not simply connected but has a simply connected covering manifold of two sheets extending over it without boundaries and ramifications. The most elementary of these double-valued representations is the spin representation which Dirac used in his Lorentz-invariant quantum theory of the electron.

Of course it would be foolish to maintain that all these investigations have their origin in relativity theory. Indeed Frobenius and Issai Schur's spadework on finite and compact groups and Cartan's early work on semi-simple Lie groups and their representations had nothing to do with it. But for myself I can say that the wish to understand what really is the mathematical substance behind the formal apparatus of relativity theory led me to the study of representations and invariants of groups; and my experience in this regard is probably not unique.

What is the upshot of it all? Relativity theory is intimately intertwined with a number of important branches of mathematics. Its influence in mathematics has been far less revolutionary than in physics and the epistemology of natural science; for its pattern fitted perfectly into the pattern of ideas already current in mathematics. But just because it could be absorbed so readily by mathematics it has stimulated the development and elaboration of those mathematical ideas to which it had a natural affinity.

[1] As a matter of fact, before him Allendoerfer and A. Weil had given a proof by embedding each of the cells into which the space is cut up into a Euclidean space. Chern got rid of this embedding device.

148.

Shock waves in arbitrary fluids

Communications on Pure and Applied Mathematics 2, 103—122 (1949)

1. *Introduction*

The following investigation tries to clear up the general hydrodynamical and thermodynamical foundations of the shock phenomenon.[1] The *first part*, Sections 2-5, answers the question: What are the conditions for the equation of state of a fluid under which shocks with their distinctive qualitative features may be produced. These conditions, enumerated in Section 3, are partly of differential, partly of global nature. The *second part*, Sections 6-7, investigates the physical structure of the shock layer whose "infinitesimal" width is of the order of magnitude ϵ provided heat conductivity and viscosity are small of the same order. Initial state and final state are singular points for the differential equations of the shock layer, and it is shown that they are of such a nature as to make one expect the problem to have a unique solution.

Part I. Thermodynamics and the Shock Phenomenon

2. *Shocks*

At any place and time a fluid is in a definite *thermodynamical state* Z and in a definite *state of motion*, the latter being characterized by the components u, v, w of the velocity $\vec{u}$ in a Cartesian system of coordinates x, y, z. The possible thermodynamical states form a two-dimensional manifold $\mathfrak{Z}$; examples of quantities having definite values in a definite thermodynamical state are density ρ, pressure p, temperature T; volume $\tau = \rho^{-1}$, potential energy e, heat content or enthalpy $i = e + p\tau$, and entropy S per unit mass. On $\mathfrak{Z}$ the fundamental thermodynamical equation

$$de = T\,dS - p\,d\tau \tag{1}$$

holds. It is once and for all assumed that we are dealing with a stationary state $(Z, \vec{u})$; i.e., that neither Z nor $\vec{u}$ vary in time, but depend on x, y, z only.

[1]This paper was originally submitted to the Applied Mathematics Panel in 1944 as a report. The same problem is treated in, *The theory of shock waves for an arbitrary equation of state*, H. A. Bethe, Division B, NDRC, OSRD, Report No. 545, to which the author had no access.

In the limit for vanishing viscosity and heat conductivity, a *discontinuity* is possible such that $(Z, \vec{u})$ has a constant value $(Z_0, \vec{u}_0)$ for $x < 0$ and another constant value $(Z_1, \vec{u}_1)$ for $x > 0$. The symbols [0], [1] refer to the two regions $x < 0$, $x > 0$ which are divided by the shock front $x = 0$. Because of conservation of mass, we have $\rho_0 u_0 = \rho_1 u_1$. Denoting by M this common flow of mass, we may write

$$u_0 = M\tau_0, \qquad u_1 = M\tau_1.$$

Then the laws of conservation of momentum and energy give the relations

$$Mu_0 + p_0 = Mu_1 + p_1,$$

$$Mv_0 = Mv_1, \qquad Mw_0 = Mw_1,$$

$$M\{i_0 + \tfrac{1}{2}(u_0^2 + v_0^2 + w_0^2)\} = M\{i_1 + \tfrac{1}{2}(u_1^2 + v_1^2 + w_1^2)\}$$

while the law of increasing entropy ("More entropy flows in than out") requires

$$MS_1 \geq MS_0.$$

The phenomenon that results for $M \neq 0$ is called a *shock*. The problem then splits into two parts, one involving only the normal velocity component u, the other referring to the tangential components and stating that they cross the shock front unchanged:

$$u_0 = M\tau_0, \qquad u_1 = M\tau_1, \tag{2.1}$$

$$Mu_0 + p_0 = Mu_1 + p_1, \tag{2.2}$$

$$i_0 + \tfrac{1}{2}u_0^2 = i_1 + \tfrac{1}{2}u_1^2, \tag{2.3}$$

$$v_0 = v_1, \qquad w_0 = w_1. \tag{3}$$

If, however, M vanishes, we obtain a *vortex sheet* or *slip stream* characterized by the relations

$$u_0 = u_1 = 0, \qquad p_0 = p_1.$$

The fluid in [1] glides tangentially past that in [0] and the pressure is the same on both sides of the discontinuity. Our object will largely be the study of shocks in an arbitrary fluid.

After introducing the values (2.1) for the velocities, (2.2) gives

$$M^2 = \frac{p_1 - p_0}{\tau_0 - \tau_1}, \tag{4}$$

and thereupon (2.3) yields a relation between the two thermodynamical states Z_0 and Z_1; namely,

$$i_1 - i_0 = \tfrac{1}{2}M^2(\tau_0 - \tau_1)(\tau_0 + \tau_1) = \tfrac{1}{2}(p_1 - p_0)(\tau_0 + \tau_1) \qquad \text{or}$$

$$e_1 - e_0 = \tfrac{1}{2}(p_1 + p_0)(\tau_0 - \tau_1).$$

Hence the problem of shocks is reduced to a study of this relation between two states Z_0 , Z_1 :

$$H \equiv H(Z_1 , Z_0) \equiv (e_1 - e_0) - \tfrac{1}{2}(p_1 + p_0)(\tau_0 - \tau_1) = 0 \tag{5}$$

(Hugoniot equation).

We are only interested in shocks in one and two dimensions. If in the latter case the normal unit-vector of the shock front in the direction [0] → [1] is (α, β) instead of (1, 0) the equations (2.1) and (3) for the velocities read as follows:

$$\alpha u_0 + \beta v_0 = M\tau_0 , \qquad \alpha u_1 + \beta v_1 = M\tau_1 ,$$

$$-\beta u_0 + \alpha v_0 = -\beta u_1 + \alpha v_1 .$$

A simple analysis (which will presently occupy us in much greater detail) reveals that the value of M resulting from (4) is of the order ρc where c is the acoustic velocity. Thus we are dealing with a situation like this: A certain quantity might be zero, but if it is not, it is even ≥ 1. Clearly quite different circumstances must be responsible for the two phenomena of shock and slip stream, and $M = 0$ is in no way to be considered here the limiting case of a non-vanishing M. The problem of the shock layer will be compared to that of the slip layer in Section 6 by taking viscosity and heat conductivity into account and then letting them tend to zero. In particular, this passage to the limit will explain why a shock is not conservative with respect to entropy although it conserves energy.

3. *Thermodynamical Assumptions*

Next we specify our assumptions concerning the thermodynamical behavior of our fluid. They will be enumerated, I-IV.

I. *Infinitesimal adiabatic increase of pressure produces compression*:

$$\left(\frac{d\tau}{dp}\right)_{ad} < 0.$$

II. *The rate of compression* $-\, d\tau/dp$ *diminishes in this process*:[2]

$$\left(\frac{d^2\tau}{dp^2}\right)_{ad} > 0.$$

These local hypotheses will be supplemented by two assumptions of "global" character.

[2]It can be made plausible that condition II is essential for the formation of a shock wave.

III. *In the continuous process of adiabatic compression one can raise pressure arbitrarily high.*

IV. *The state Z is uniquely specified by pressure and specific volume, and the points* (p, τ) *representing the possible states Z in a* (p, τ)*—diagram form a convex region.*

[It would be more natural but less elementary to divide IV into a local and a global part, the local postulate asserting that in the neighborhood of a given state Z_0 the variables p, τ can be used as parameters for the specification of states; in other words that the projection $Z \to (p, \tau)$ of the manifold $\mathfrak{Z}$ of states Z upon a (p,τ)-plane is locally one-to-one. The global part would assert that the projection of $\mathfrak{Z}$ covers a convex region $\bar{\mathfrak{Z}}$ in the (p,τ)-plane and that Z_0 being any given state and (p_0,τ_0) its projection, one never runs against an obstacle when, on starting at Z_0, one lets Z vary so that its projection (p, τ) follows a given path in $\bar{\mathfrak{Z}}$ beginning at (p_0,τ_0) ("continuation"). All this means that $\mathfrak{Z}$ is a covering surface over the convex $\bar{\mathfrak{Z}}$ without singularities and relative boundaries. But the convex $\bar{\mathfrak{Z}}$ being simply connected, this covering surface necessarily consists of a single sheet, or (p, τ) determines Z uniquely. We thus fall back on the more elementary formulation of IV. The assumption of convexity for the region $\bar{\mathfrak{Z}}$ will prove to be quite essential for our investigation. Henceforth we simply denote by Z the projection of Z in the (p,τ)-plane.]

Because of IV, entropy S is a (single-valued) function of p and τ, $S = S(p, \tau)$. The adiabatic process of compression is thus defined by

$$\frac{\partial S}{\partial p}\,dp + \frac{\partial S}{\partial \tau}\,d\tau = 0.$$

From the fact that (1) is a total differential we have

$$\frac{\partial}{\partial p}\left(T\,\frac{\partial S}{\partial \tau} - p\right) = \frac{\partial}{\partial \tau}\left(T\,\frac{\partial S}{\partial p}\right)$$

or

$$\frac{\partial T}{\partial p}\,\frac{\partial S}{\partial \tau} - \frac{\partial T}{\partial \tau}\,\frac{\partial S}{\partial p} = 1. \tag{6}$$

Hence $\partial S/\partial p$, $\partial S/\partial \tau$ cannot vanish simultaneously, and thus hypothesis I requires that $\partial S/\partial p$, $\partial S/\partial \tau$ are of the same sign:

$$\frac{\partial S}{\partial p} > 0, \quad \frac{\partial S}{\partial \tau} > 0 \quad \text{or} \quad \frac{\partial S}{\partial p} < 0, \quad \frac{\partial S}{\partial \tau} < 0.$$

As the region $\mathfrak{Z}$ is connected, the one set or the other will hold *everywhere*. We assume the first:

$$\frac{\partial S}{\partial p} > 0, \quad \frac{\partial S}{\partial \tau} > 0. \tag{I$'$}$$

(The other alternative would make little difference.) Infinitesimal adiabatic compression is now described by

$$dp = a^* \, dt, \qquad d\tau = b^* \, dt; \qquad a^* = \frac{\partial S}{\partial \tau}, \qquad b^* = -\frac{\partial S}{\partial p}$$

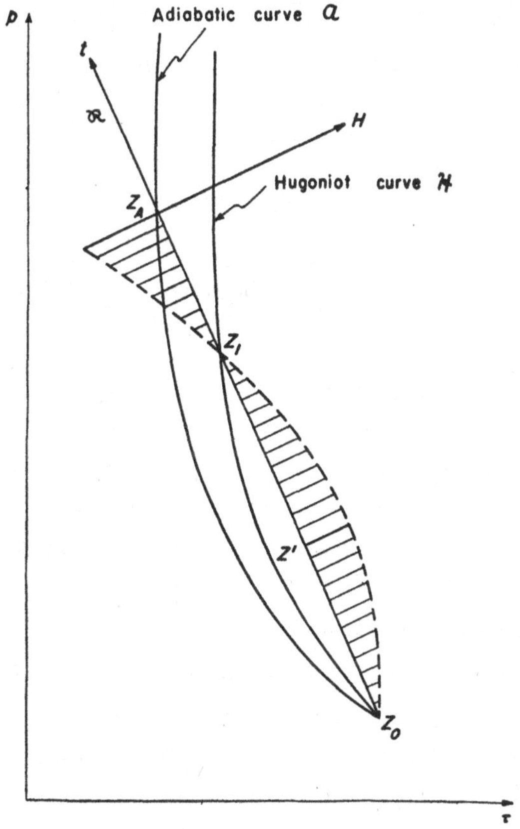

FIG. 1. This diagram shows the adiabatic through Z_0 and the Hugoniot line in a p,τ-diagram; moreover, in the form of a shaded profile, H as function of the distance t along a ray: H reaches its maximum at Z', passes through zero at Z_1, and is negative at Z_A.

(a *positive* factor of proportionality dt corresponding to an *increase* of pressure). Hypothesis II yields

$$\frac{d}{dt}\left(\frac{b^*}{a^*}\right) > 0 \qquad \text{or} \qquad b^* \frac{da^*}{dt} - a^* \frac{db^*}{dt} < 0,$$

i.e. $$b^*\left(\frac{\partial a^*}{\partial p}\,a^* + \frac{\partial a^*}{\partial \tau}\,b^*\right) - a^*\left(\frac{\partial b^*}{\partial p}\,a^* + \frac{\partial b^*}{\partial \tau}\,b^*\right)$$

$$= -a^{*2}\frac{\partial b^*}{\partial p} + a^*b^*\left(\frac{\partial a^*}{\partial p} - \frac{\partial b^*}{\partial \tau}\right) + b^{*2}\frac{\partial a^*}{\partial \tau} < 0$$

or

$$\text{(II}') \qquad \frac{\partial^2 S}{\partial p^2}\left(\frac{\partial S}{\partial \tau}\right)^2 - 2\,\frac{\partial^2 S}{\partial p\,\partial \tau}\,\frac{\partial S}{\partial \tau}\,\frac{\partial S}{\partial p} + \frac{\partial^2 S}{\partial \tau^2}\left(\frac{\partial S}{\partial p}\right)^2 < 0.$$

In the future we shall use hypotheses I and II in their analytic forms (I′) and (II′).

Our next move consists in developing a number of consequences from the assumptions I-III. Let $(a, b) \neq (0, 0)$ be two given numbers which determine the direction of a straight line through $Z_0 = (p_0\,,\,\tau_0)$,

$$p = p_0 + at, \qquad \tau = \tau_0 + bt,$$

the half-line or ray being obtained by the restriction $t \geq 0$ on the parameter t. We follow the straight line or the ray as long as it stays within $\mathcal{Z}$, and form

$$\frac{dS}{dt} = \frac{\partial S}{\partial p}\,a + \frac{\partial S}{\partial \tau}\,b = S'$$

$$\frac{dS'}{dt} = \frac{\partial S'}{\partial p}\,a + \frac{\partial S'}{\partial \tau}\,b = \frac{\partial^2 S}{\partial p^2}\,a^2 + 2\,\frac{\partial^2 S}{\partial p\,\partial \tau}\,ab + \frac{\partial^2 S}{\partial \tau^2}\,b^2.$$

Lemma 1. *S' is positive (negative) and hence S increasing (decreasing) along the whole straight line, provided $a \geq 0$, $b \geq 0$ ($a \leq 0$, $b \leq 0$).*

This follows from assumption (I′).

Lemma 2. *If $S' = 0$ for a certain value of t, then $dS'/dt < 0$ for the same value.*

Proof. Substitute the values of a and b derived from $S' = 0$ into dS'/dt and use the inequality (II′).

Lemma 3. *If $S' \leq 0$ for $t = 0$, then $S' < 0$* for *$t > 0$.*

Proof. First assume $S' < 0$ for $t = 0$. Should S change sign as Z travels along the ray, it would have to vanish somewhere. Suppose this occurs for the first time at $t = t_1$. As $S' < 0$ before t reaches this value, S' passes the zero level at t_1 ascending; hence $dS'/dt \geq 0$ for $t = t_1$. But these two relations $S' = 0$, $dS'/dt \geq 0$ for $t = t_1$, contradict Lemma 2.

If $S' = 0$ for $t = 0$ we have $dS'/dt < 0$ for $t = 0$; consequently S' becomes negative immediately after the point Z has started on its way from Z_0, and from then on, as we have seen, S' must remain negative.

We turn to III. Starting at a given point $Z_0 = (p_0\,,\,\tau_0)$ and always raising p, we follow the adiabatic through Z_0 : we thus obtain a continuous monotonely descending function $\tau(p)$, and according to III we can make p travel over the entire infinite interval $p \geq p_0$. While this happens, the directional coefficient

(7) $$s = \frac{p - p_0}{\tau_0 - \tau}$$

of the straight line joining (p_0, τ_0) with the point (p, τ) on the adiabatic $S = S_0$ increases monotonely from a certain positive value m_0 to $+\infty$. m denotes the adiabatic derivative $-dp/d\tau$ and m_0 its value at the point Z_0. Consequently, there is exactly one point on the upper branch $p > p_0$ of the adiabatic that lies on the ray from (p_0, τ_0) in a given direction s, provided s lies between m_0 and ∞.

The analytic proof for these intuitively evident statements runs as follows. Form the (non-total) differential

$$dr = (\tau_0 - \tau)\, dp + (p - p_0)\, d\tau,$$

which vanishes along any ray through Z_0. We want to show that along the adiabatic dr is positive for a positive increment dp:

(8) $$\frac{dr}{dp} = R = (\tau_0 - \tau) + (p - p_0)\frac{d\tau}{dp} > 0 \qquad \text{for} \qquad p > p_0 .$$

By hypothesis II

(9) $$\frac{dR}{dp} = (p - p_0)\frac{d^2\tau}{dp^2} > 0 \qquad \text{for} \qquad p > p_0 ,$$

and since R vanishes for $p = p_0$ it must indeed be positive for $p > p_0$:

(10) $$R = \int_{p_0}^{p} (p - p_0)\frac{d^2\tau}{dp^2}\, dp > 0.$$

Using (7) we may write

(11) $$dr = (\tau_0 - \tau)^2\, ds$$

and thus the monotone behavior of s is exhibited by the equation

$$\frac{ds}{dp} = R/(\tau_0 - \tau)^2.$$

It is clear that s tends to m_0 for $p \to p_0$, and the combination of hypothesis III with the trivial inequality $s > (p - p_0)/\tau_0$ leads to $s \to \infty$ for $p \to \infty$.

4. *The Fundamental Inequality for the Direction* $\overrightarrow{Z_0 Z_1}$

For the differential of

(12) $$H(Z, Z_0) = (e - e_0) - \tfrac{1}{2}(p + p_0)(\tau_0 - \tau)$$

we find by (1)

$$dH = de - \tfrac{1}{2}(\tau_0 - \tau)\, dp + \tfrac{1}{2}(p_0 + p)\, d\tau$$
$$= T\, dS - \tfrac{1}{2}\{(\tau_0 - \tau)\, dp + (p - p_0)\, d\tau\},$$

or

(13) $$dH = T\, dS - \tfrac{1}{2}\, dr,$$

and hence *along any ray*

(14) $$dH = T\, dS.$$

Let Z_0, Z_1 be two distinct points satisfying the Hugoniot equation $H(Z_1, Z_0) = 0$. Because of the convexity of the region Z (hypothesis IV) we may join these two points by a straight segment lying in Z. Integrating (14) along it we obtain

(15) $$H(Z_1, Z_0) = \int_{Z_0}^{Z_1} T\, dS = 0.$$

From this simple relation we are going to deduce

Theorem 1. *For any two distinct states* Z_0, Z_1 *linked by Hugoniot's equation* $H = 0$ *the following inequalities hold*

(16) $$W = (p_1 - p_0)(\tau_0 - \tau_1) > 0,$$

(17) $$(p_1 - p_0) + m_0(\tau_1 - \tau_0) > 0, \qquad (p_1 - p_0) + m_1(\tau_1 - \tau_0) < 0.$$

Before proving the theorem let us discuss its physical significance. We have seen that elimination of the velocities from the conditions for a shock lead to the *Hugoniot equation* and the inequality $W \geq 0$, cf. (4). We now realize that we may omit the supplementary relation $W \geq 0$ because it follows, in fact in the sharper form $W > 0$, from the Hugoniot equation.

The adiabatic derivative

$$\frac{dp}{d\rho} = -\tau^2 \frac{dp}{d\tau} = \tau^2 m$$

is the square of the "acoustic velocity" c. Assume $\tau_0 > \tau_1$. Then the two relations (17) and relation (4) give

(18) $$m_0 < M^2 = \frac{p_1 - p_0}{\tau_0 - \tau_1} < m_1$$

or $\rho_0 c_0 < |M| < \rho_1 c_1$, thus confirming a statement made in Section 2 on the magnitude of M. In terms of the velocities u_0, u_1 given by (2.1) or

$$u_0^2 = \frac{p_1 - p_0}{\tau_0 - \tau_1}\, \tau_0^2, \qquad u_1^2 = \frac{p_1 - p_0}{\tau_0 - \tau_1}\, \tau_1^2,$$

(18) may be written as

(19) $$|u_0| > c_0, \qquad |u_1| < c_1:$$

Relative to the shock front the flow in [0] *is supersonic, in* [1] *subsonic.*

Proceeding to the proof of our theorem, let us travel from Z_0 along that ray which passes through Z_1 and hence set in Lemma 1,

$$a = p_1 - p_0, \qquad b = \tau_1 - \tau_0.$$

Were $a \geq 0$, $b \geq 0$, then S' would be positive and hence S monotone increasing along the segment Z_0Z_1, which contradicts the relation (15). Consequently this combination (and in the same way the other, $a \leq 0$, $b \leq 0$) is ruled out, and therefore $ab < 0$, or (16) must hold.

Next consider

$$S' = a\frac{\partial S}{\partial p} + b\frac{\partial S}{\partial \tau}.$$

Were $S' \leq 0$ for $t = 0$, then S' would always be negative by Lemma 3 and hence S would monotonely decrease while Z travels along the straight segment from Z_0 to Z_1, in contradiction to equation (15). Therefore

$$a\left(\frac{\partial S}{\partial p}\right)_0 + b\left(\frac{\partial S}{\partial \tau}\right)_0 > 0,$$

and this is equivalent to the first of the inequalities (17). The second follows from the first by interchanging Z_0 and Z_1.

But our argument shows much more. While Z travels along the ray from Z_0 passing through Z_1, S' starts with a positive value; because of (15) it must change sign before Z reaches Z_1. But S' remains negative from the moment it vanishes for the first time. Any rise and fall in $H = H(Z, Z_0)$ along the ray is coupled with the same in S by the relation $dH = T\,dS$. Hence H first rises monotonely to a positive maximum and then decreases, passing on the descent through 0 for $Z = Z_1$. This description shows that H is positive from Z_0 to Z_1 and negative thereafter. In particular:

On the ray from Z_0 through Z_1 the point $Z = Z_1$ is the *only one* aside from Z_0 itself which satisfies the Hugoniot equation $H(Z, Z_0) = 0$.

According to (16) either the inequalities $p_1 > p_0$, $\tau_0 > \tau_1$ or $p_0 > p_1$, $\tau_1 > \tau_0$ hold for two distinct states Z_0, Z_1, satisfying the Hugoniot equation $H(Z_1, Z_0) = 0$. We shall indicate these alternatives by $Z_1 > Z_0$, $Z_1 < Z_0$ respectively. By the *Hugoniot contour* $\mathcal{H}$ (for a given Z_0) we understand the locus of all points $Z_1 = (p_1, \tau_1)$ for which

$$H(Z_1, Z_0) = 0 \qquad \text{and} \qquad Z_1 > Z_0.$$

It is quite essential for our argument to pick this *upper* branch $Z_1 > Z_0$. Our above result may then be stated thus: On the Hugoniot contour $s = (p_1 - p_0)/$

$(\tau_0 - \tau_1)$ lies between m_0 and ∞; s is a uniformizing parameter inasmuch as the value of s specifies uniquely the point Z_1. But it must be borne in mind that so far we have not yet proved that to every preassigned value of $s > m_0$ there actually corresponds a point on the Hugoniot contour; we only know that there cannot be more than one.

5. *Entropy and Parametrization of the Hugoniot Line*

For a moment let us return to the upper branch $\mathfrak{A}$ of the adiabatic through Z_0, on which p increases monotonely from p_0 to $+\infty$ and $s = (p - p_0)/(\tau_0 - \tau)$ from m_0 to ∞. Moving along $\mathfrak{A}$ we have by (13)

$$dH = -\tfrac{1}{2}\, dr = -\tfrac{1}{2}R\, dp \tag{20}$$

and hence (10) implies

Lemma 4. $H = H(Z, Z_0)$ *is negative along* $\mathfrak{A}$.

As a matter of fact, we even know that H is falling, $R > 0$, and by condition (9) falling with increasing rapidity, $dR/dp > 0$. The explicit formula giving $H(Z_A, Z_0)$ for any point Z_A on the adiabatic $\mathfrak{A}$ is by (20), (8) and (9),

$$H(Z_A, Z_0) = -\frac{1}{2}\int_{\mathfrak{A}^{p_0}}^{p_A} R\, dp = -\frac{1}{2}R(p - p_A)\Big|_{p_0}^{p_A} - \frac{1}{2}\int_{\mathfrak{A}^{p_0}}^{p_A} (p_A - p)\frac{dR}{dp}\, dp$$

$$= -\frac{1}{2}\int_{\mathfrak{A}^{p_0}}^{p_A} (p - p_0)(p_A - p)\frac{d^2\tau}{dp^2}\, dp < 0.$$

This lemma is instrumental in establishing the following propositions:

Theorem 2. *For any two distinct states* Z_0, Z_1 *satisfying the Hugoniot equation one has* $S_1 > S_0$ *or* $S_1 < S_0$ *according to whether* $Z_1 > Z_0$ *or* $Z_1 < Z_0$.

Theorem 3. *The Hugoniot contour is a simple line starting from* Z_0 *on which* s *and* S *are monotone increasing* (s traveling from m_0 to $+\infty$, S from S_0 to an unknown destination).

Proof. Assume the arrangement $Z_1 > Z_0$ for the two points Z_0, Z_1 linked by the Hugoniot equation. Draw the ray $\mathfrak{R}$ from Z_0 passing through Z_1; its directional coefficient $s = (p_1 - p_0)/(\tau_0 - \tau_1)$ lies between m_0 and $+\infty$, and hence $\mathfrak{R}$ meets the upper branch of the adiabatic line $\mathfrak{A}$ in a point $Z_A = (p_A, \tau_A)$ where, according to Lemma 4,

$$H = \int_{\mathfrak{R}^{Z_0}}^{Z_A} T\, dS = -\frac{1}{2}\int_{\mathfrak{A}^{p_0}}^{p_A} R\, dp < 0.$$

Hence Z on its road along the ray must have passed the point Z' where it reaches its maximum, and running downhill have passed a point Z_1 for which $H =$

$\int_{Z_0}^{Z_1} T\,dS = 0$ before coming to Z_A. Its value S_1 at the point Z_1 of the Hugoniot contour is therefore higher than its value S_0 at Z_A (Theorem 2).

Again, let s be any value between m_0 and ∞. The ray going out from Z_0 in the direction s meets $\mathfrak{A}$ in a point Z_A where H is negative. Hence H must vanish at a certain point Z_1 between Z' and Z_A (where S is already on the down grade), and thus a point on the Hugoniot contour is obtained corresponding to this preassigned value of $s = (p_1 - p_0)/(\tau_0 - \tau_1)$. We know it is the only one. The question raised at the end of Section 4 is answered in the affirmative, and we may now speak of the Hugoniot contour as a simple line $Z_1 = \Phi(s)$ with the uniformizing parameter s.

Because the point $Z_1 = \Phi(s)$ is thought of as varying along the Hugoniot line, let us drop the index 1. Along $\mathfrak{H}$ we may use the differential relation $dH = 0$ or, according to (13) and (11),

$$T\,dS - \tfrac{1}{2}\,dr = T\,dS - \tfrac{1}{2}\,(\tau_0 - \tau)^2\,ds = 0.$$

Because

$$\frac{dS}{ds} = \frac{(\tau_0 - \tau)^2}{2T}$$

thus turns out to be positive, S increases all the way along the Hugoniot line with s. This gives a new proof for the inequality $S_1 > S_0$ by integration along that line,

$$S_1 - S_0 = \int_{\mathfrak{H}\,m_0}^{s_1} \frac{(\tau_0 - \tau)^2}{2T}\,ds \qquad (s_1 > m_0)$$

Thus Theorem 3 is proved, and Theorem 2 by even two different methods.

Part II. The Problem of the Shock Layer

6. *Formulation of the Problem*

For the sake of brevity we now denote partial derivatives with respect to (p, τ) by subscripts as in

$$\frac{\partial S}{\partial p} = S_p\,, \qquad \frac{\partial S}{\partial \tau} = S_\tau\,.$$

The rest of our argument will be based on the thermodynamical assumptions I-IV, Section 3, but in its course it will become clear that one more condition of highly plausible nature is to be added to our list:

Ia. *Heating a quantity of fluid at constant volume raises its pressure and temperature.*

It must be remembered that the analytic formulation (I′), (II′) of the hypotheses I and II depended on the fixation of a sign. For the previous investigation this did not matter greatly, but it is essential for the behavior of the shock layer. We therefore require explicitly by the statement concerning pressure in Ia, that $(dp/TdS)_{T=\text{const.}} > 0$ or $S_p > 0$. The other part of Ia then adds the inequality $T_p/S_p > 0$ or $T_p > 0$.

In a stationary field the state $(Z, \vec{u})$ of a fluid of given thermodynamical nature depends on the spatial coordinates x, y, z only, and not on time. The five equations which, in the absence of an external force, express conservation of mass, momentum and energy, take heat conductivity and viscosity into account by the flow of heat $\vec{j} = -\lambda \operatorname{grad} T$ and a stress tensor S of the form

$$S_{ik} = p\delta_{ik} - S^*_{ik}, \qquad S^*_{ik} = \mu' \operatorname{div} \vec{u}\cdot\delta_{ik} + \mu\left(\frac{\partial u_i}{\partial x_k} + \frac{\partial u_k}{\partial x_i}\right)$$

(in the writing down of which we have for the moment adopted the subscript notation for the coordinates x_i and velocity components u_i). $\lambda = \lambda[Z]$, μ, μ' are given functions of the thermodynamical state Z; $\lambda > 0$, $\mu > 0$, $\mu' + \frac{2}{3}\mu > 0$. As they are small we intend to let them pass to zero. In two ways such a passage to the limit $\lambda, \mu, \mu' \to 0$ can give rise to a discontinuity in the plane $x = 0$:

(A) Write ϵx for x and $\epsilon\lambda$, $\epsilon\mu$, $\epsilon\mu'$ for λ, μ, μ', leaving all other quantities untouched, and then let the positive constant factor ϵ tend to zero (*shock layer*).

(B) Let ϵ tend to zero after writing ϵx, ϵu for x, u and $\epsilon^2\lambda$, $\epsilon^2\mu$, $\epsilon^2\mu'$ for λ, μ, μ' (*slip layer*).

The different effect is clearly exemplified by the continuity equation

$$\frac{\partial}{\partial x}(\rho u) + \frac{\partial}{\partial y}(\rho v) + \frac{\partial}{\partial z}(\rho w) = 0. \tag{21}$$

Substitution (A) changes it into

$$\frac{\partial}{\partial x}(\rho u) + \epsilon\frac{\partial}{\partial y}(\rho v) + \epsilon\frac{\partial}{\partial z}(\rho w) = 0$$

which for $\epsilon \to 0$ reduces to

$$\frac{\partial}{\partial x}(\rho u) = 0,$$

whereas process (B) leaves the three-term equation (21) unchanged. This is typical. None but the derivatives $\partial/\partial x$ survive the limiting process (A), the others $\partial/\partial y$, $\partial/\partial z$ disappear from the equations. The shock layer is thus seen to be a one-dimensional problem, no interaction taking place between the one-dimensional fibers of space y = constant, z = constant. The slip layer on the other hand constitutes a far more complicated three-dimensional problem, identical with that of Prandtl's boundary layer. We are concerned here only with the shock layer.

With the abbreviation $\mu^* = \mu' + 2\mu$, $\mu^* > 0$, the limiting process (A) leads to the following equations:

$$\rho u = \text{constant} = M, \tag{22.1}$$

$$Mu + p - \mu^* \frac{du}{dx} = \text{constant}, \tag{22.2}$$

$$Mv - \mu \frac{dv}{dx} = \text{constant}, \qquad Mw - \mu \frac{dw}{dx} = \text{constant}, \tag{22.3}$$

$$\left(p - \mu^* \frac{du}{dx} + \rho e^*\right)u - \mu\left(v \frac{dv}{dx} + w \frac{dw}{dx}\right) - \lambda \frac{dT}{dx} = \text{constant}, \tag{22.4}$$

where e^* denotes the total energy $e + \frac{1}{2}(u^2 + v^2 + w^2)$. Here the quantities λ, μ, μ' are no longer small. We assume that $(Z, \vec{u})$ approaches definite constant values $(Z_0, \vec{u}_0)$, $(Z_1, \vec{u}_1)$ for $x \to -\infty$ and $x \to +\infty$ respectively, whereas the derivatives d/dx tend to zero.

The energy equation is equivalent to the entropy equation

$$T \frac{d}{dx}\left(\rho u S - \frac{\lambda}{T}\frac{dT}{dx}\right) = \mu^*\left(\frac{du}{dx}\right)^2 + \mu\left\{\left(\frac{dv}{dx}\right)^2 + \left(\frac{dw}{dx}\right)^2\right\} + \frac{\lambda}{T}\left(\frac{dT}{dx}\right)^2, \tag{23}$$

from which one learns that the flow of entropy

$$\rho u S - \frac{\lambda}{T}\frac{dT}{dx} = MS - \frac{\lambda}{T}\frac{dT}{dx}$$

is a monotone increasing function of x and therefore $MS_1 \geq MS_0$. The equality sign could prevail only if u, v, w, T and hence by (22.2) also p were constant. Excluding this trivial case of a constant state[3] $(Z, \vec{u})$ we thus obtain the *strict inequality*

$$MS_1 > MS_0 \tag{24}$$

implying $M \neq 0$. The latter remark shows how misleading it is to link the alternative $M \neq 0$, $M = 0$ to the discrimination of shock and slip layer; rather do these two possibilities correspond to the two ways (A) and (B) of passing to the limit of vanishing viscosity and heat conductivity. Being sure now that M does not vanish, we fall back upon our old relations connecting $(Z_0, \vec{u}_0)$ and $(Z_1, \vec{u}_1)$, including the Hugoniot relation. Let us assign the labels 0, 1

[3]Indeed, in general, p = constant, T = constant, imply τ = constant. There is, however, one extremely exceptional case in which the conclusion does not stand, namely if $M = 0$ and if for some constant value of p the temperature T as function of τ stays constant in an entire τ-interval. Maybe this is a warning against the danger of a singularity in the shock layer problem which one steers clear of with certainty only by assuming $T_\tau \neq 0$ or, what is the same, if the hypotheses I–IV, Ia are preceded by another to the effect that p, T may serve as local parameters on $\mathcal{Z}$ everywhere.

to the two states Z_0, Z_1, and accordingly orient the x-axis, so that $Z_0 < Z_1$. Then Theorem 2 combined with the law of increasing entropy (24) gives $M > 0$. The meaning of this inequality is that the shock front moves in the direction $[1] \to [0]$, or that Z_0 and not Z_1 is the initial state of the fluid not yet reached by the propagating shock.

7. *Character of the Two Singular Points Z_0, Z_1: Node and Saddle*[4]

It is easy to dispose of the equations (22.3) for v and w. Single out the one for v,

$$\mu \frac{dv}{dx} = M(v - b), \qquad b = \text{constant},$$

and introduce the variable

$$\xi = M \int_0^x \frac{dx}{\mu(x)} \qquad \text{where } \mu(x) = \mu[Z(x)].$$

We find

$$v(x) - b = Ce^{\xi}, \qquad C = \text{constant}.$$

$\mu(x)$ tends to positive values $\mu_0 = \mu[Z_0]$, $\mu_1 = \mu[Z_1]$ for $x \to -\infty$, $x \to +\infty$ respectively; therefore

$$\xi = \frac{M}{\mu_0} x + o(x) \qquad \text{for } x \to -\infty,$$

$$\xi = \frac{M}{\mu_1} x + o(x) \qquad \text{for } x \to +\infty.$$

Considering that $M > 0$ one realizes that v cannot tend to a finite limit v_1 for $x \to +\infty$ unless $C = 0$. Thus v and w must be constant throughout the shock layer.

(22.1) is satisfied by $u = M\tau$. Denoting by M^2a the constant at the right of (22.2) and substituting from (22.2) the value of $p - \mu^* du/dx$ into (22.4) we are left, as is readily seen, with the following two ordinary differential equations of first order for the state $Z = Z(x)$:

$$\mu^* M \frac{d\tau}{dx} = M^2(\tau - a) + p$$

(25)

$$\frac{\lambda}{M} \frac{dT}{dx} = e - \frac{1}{2} M^2(\tau - a)^2 - c.$$

[4]For ideal gases of constant specific heat with certain special values of the adiabatic exponent γ, cf. R. Becker, Zeitschrift für Physik 8 (1922), 321–362; in particular, pp. 339–347.

The constants M^2, a, c are determined by

$$M^2\tau_0 + p_0 = M^2\tau_1 + p_1 = M^2 a,$$

$$e_0 - \tfrac{1}{2}M^2(\tau_0 - a)^2 = e_1 - \tfrac{1}{2}M^2(\tau_1 - a)^2 = c.$$

As the right members of (25) vanish for $Z = Z_0$ and $Z = Z_1$, the two points Z_0, Z_1 turn out to be *singular points* for the differential system (25). If we care only for the trajectory in the (p,τ)-diagram along which $Z(x)$ moves while x runs from $-\infty$ to $+\infty$ we are faced with one ordinary differential equation of first order in the (p,τ)-plane,

$$\frac{\lambda\, dT}{e - \frac{1}{2}M^2(\tau - a)^2 - c} = \frac{\mu^*\, d\tau}{(\tau - a) + (p/M^2)}\, .$$

But contrary to the customary viewpoint taken in the classical investigations on singular points, which were started by Briot and Bouquet almost one hundred years ago and are reproduced in all treatises on differential equations, our interest is in the *parametrized* trajectory $Z = Z(x)$.

Let us briefly recount the well-known facts about singular points relevant for our purposes.[5] Placing the singular point at the origin 0 we study two simultaneous differential equations for the "vector" $\varphi = (\varphi_1(x), \varphi_2(x))$ in a φ_1, φ_2-plane,

$$\frac{d\varphi_1}{dx} = F_1(\varphi_1, \varphi_2), \qquad \frac{d\varphi_2}{dx} = F_2(\varphi_1, \varphi_2)$$

where F_1, F_2 are given functions of φ_1, φ_2 vanishing at the origin. From the outset it is evident that any translation $x \to x + h$ carries a solution $\varphi = (\varphi_1(x), \varphi_2(x))$ into a solution ("equivalent solutions"). Suppose the Taylor expansions of F_1, F_2 begin with the terms

$$F_1(\varphi_1, \varphi_2) \sim R_{11}\varphi_1 + R_{12}\varphi_2\, , \qquad F_2(\varphi_1, \varphi_2) \sim R_{21}\varphi_1 + R_{22}\varphi_2\, .$$

The equations

$$\varphi_1' = R_{11}\varphi_1 + R_{12}\varphi_2\, , \qquad \varphi_2' = R_{21}\varphi_1 + R_{22}\varphi_2$$

define a linear mapping $\varphi \to \varphi' = R\varphi$. We determine its two eigen-values k and corresponding eigen-vectors $\varphi = \alpha$ (or rather eigen-directions) by $R\alpha = k\alpha$, explicitly

$$(k - R_{11})\alpha_1 - R_{12}\alpha_2 = 0,$$

$$-R_{21}\alpha_1 + (k - R_{22})\alpha_2 = 0,$$

[5]See for direct constructive proofs: H. Weyl, *Concerning a classical problem in the theory of singular points of ordinary differential equations.* Actas de la Academia Nacional de Ciencias de Lima, Volume 7, 1944, pp. 21–60.

and assume that the two roots k_1, k_2 of the secular equation

$$\begin{vmatrix} k - R_{11}, & -R_{12} \\ -R_{21}, & k - R_{22} \end{vmatrix} = 0$$

are real, distinct and $\neq 0$. In a suitable affine coordinate system in the φ_1, φ_2-plane our mapping is then described by $\varphi_1' = k_1\varphi_1$, $\varphi_2' = k_2\varphi_2$, so that F_1, F_2 are of the form

$$F_1(\varphi_1, \varphi_2) = k_1\varphi_1 + \cdots,$$

$$F_2(\varphi_1, \varphi_2) = k_2\varphi_2 + \cdots.$$

We distinguish three cases according to the signs of the eigen-values k_1, k_2. In the case (i) of two negative eigen-values, $k_2 < k_1 < 0$, every solution φ which comes sufficiently near to the origin plunges into it with x tending to $+\infty$; its asymptotic behavior is described by

$$\varphi_1(x) \sim a\cdot e^{k_1 x}, \qquad \varphi_2(x) \sim 0\cdot e^{k_1 x} \qquad (x \to \infty)$$

in the sense that

$$e^{-k_1 x}\varphi_1(x) = a + O(e^{-\epsilon x}),$$

$$e^{-k_1 x}\varphi_2(x) = O(e^{-\epsilon x})$$

where ϵ is a positive constant, the same for all solutions, while the constant a is specific (*node*, see Figure 2). In the case (ii) of one negative, one positive

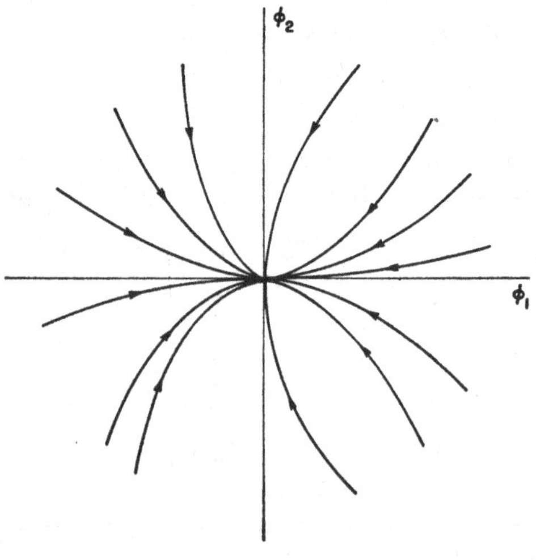

FIG. 2. Node.

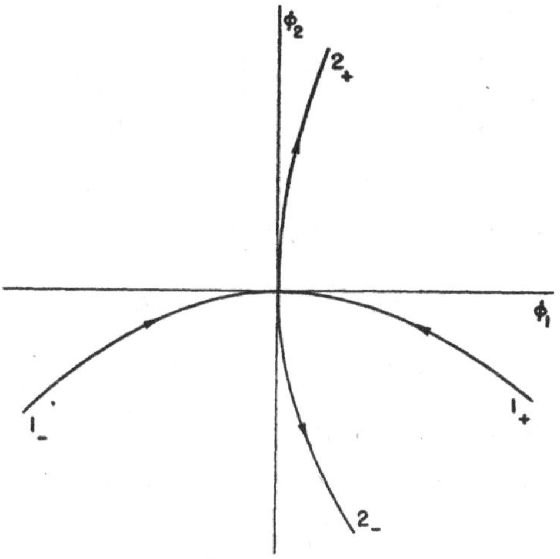

FIG. 3. Saddle.

eigen-value, $k_1 < 0$, $k_2 > 0$, there exist exactly two solutions (1_+, 1_- in Figure 3) approaching 0 with $x \to +\infty$; their asymptotic behavior is described by

$$\varphi_1(x) \sim \pm e^{k_1 x}, \qquad \varphi_2(x) \sim 0 \cdot e^{k_1 x} \qquad (x \to +\infty)$$

(*saddle*). Of course this statement is to be interpreted so that equivalent solutions are considered as one and the same. The term saddle is better understood by observing that in this case (ii) two other solutions, 2_+ and 2_-, plunge into 0 in the direction of the φ_2-axis, while x goes to $-\infty$. In case (iii) of two positive eigen-values there is no solution whatsoever which approaches the origin as x tends to $+\infty$, except the trivial one $\varphi_1 = \varphi_2 = 0$ (whereas, of course, every solution that comes sufficiently near 0 takes the plunge when x moves toward $-\infty$).

With these facts in mind we now advance our decisive

Theorem 6. *For the differential equations of the shock layer determining the transition from a state* Z_0 *to a state* $Z_1 > Z_0$, *the initial state* Z_0 *is a node, the end state* Z_1 *a saddle.*

Clearly this behavior in the neighborhood of the two singular points Z_0, Z_1 is such as to favor the existence of a unique solution of the shock layer problem. All other combinations would make one expect either no solution or an infinity of solutions

Again, the proof depends, above all, on the fundamental inequalities (18) for $M^2 = (p_1 - p_0)/(\tau_0 - \tau_1)$. Since S_τ/S_p is the adiabatic derivative $m = -dp/d\tau$ we may write them as

(26) $$M^2 S_p^{(0)} - S_\tau^{(0)} > 0, \qquad M^2 S_p^{(1)} - S_\tau^{(1)} < 0.$$

They must be combined with

(27) $$M > 0; \qquad S_p > 0, \qquad T_p > 0.$$

First investigate the neighborhood of Z_0 and therefore write

$$M^2(\tau - a) + p = M^2(\tau - \tau_0) + (p - p_0).$$

For infinitesimal $\delta p = p - p_0$, $\delta\tau = \tau - \tau_0$

$$\delta e = T_0\ \delta S - p_0\ \delta\tau, \qquad \delta\tfrac{1}{2}(\tau - a)^2 = (\tau_0 - a)\ \delta\tau, \qquad \text{hence}$$

$$\delta\{e - \tfrac{1}{2}M^2(\tau - a)^2\} = T_0\ \delta S - \{p_0 + M^2(\tau_0 - a)\}\ \delta\tau = T_0\ \delta S,$$

therefore in neglecting terms of higher order

$$e - \tfrac{1}{2}M^2(\tau - a)^2 - c \sim T_0(S - S_0).$$

The following approximate linear system for $\delta\tau$, δp results:

$$\frac{d}{dx}(\delta\tau) = g(M^2\ \delta\tau + \delta p),$$

$$\frac{d}{dx}(\delta T) = g_0'\ \delta S$$

where

$$g = 1/M\mu^* > 0, \qquad g' = TM/\lambda > 0$$

and δS, δT stand for

$$\delta S = S_p^{(0)}\ \delta p + S_\tau^{(0)}\ \delta\tau, \qquad \delta T = T_p^{(0)}\ \delta p + T_\tau^{(0)}\ \delta\tau.$$

In order to avoid unnecessary encumbrances, we drop the index 0. Then the equations determining the eigen-values k and corresponding eigen-directions $(\delta\tau, \delta p)$ at Z_0 become

$$k\cdot\delta\tau = g(M^2\delta\tau + \delta p), \qquad k\cdot\delta T = g'\delta S$$

giving rise to the following secular equation for k,

$$\begin{vmatrix} k - gM^2, & -g \\ kT_\tau - g'S_\tau, & kT_p - g'S_p \end{vmatrix} = 0 \qquad \text{or}$$

(28) $$T_p\cdot k^2 - \{g(M^2T_p - T_\tau) + g'S_p\}\cdot k + gg'(M^2S_p - S_\tau) = 0.$$

Compute the discriminant

$$\begin{aligned}\Delta &= \{g(M^2T_p - T_\tau) + g'S_p\}^2 - 4gg'T_p(M^2S_p - S_\tau) \\ &= \{g(M^2T_p - T_\tau) - g'S_p\}^2 + 4gg'(T_pS_\tau - T_\tau S_p)\end{aligned}$$

and thus verify by (6) that Δ is positive,

$$\Delta = \{g(M^2T_p - T_\tau) - g'S_p\}^2 + 4gg' \geq 4gg' = \frac{4T}{\lambda\mu^2} > 0.$$

Consequently the eigen-values are real and distinct, at the singular point Z_0 as well as at Z_1 .

Their product equals

$$gg' \cdot (M^2S_p - S_\tau)/T_p ,$$

and because of (27) and the fundamental inequalities (26) that product is positive at Z_0 , negative at Z_1 . Hence the two eigen values are of opposite sign at Z_1 , of equal sign at Z_0 . Whether the latter sign is positive or negative can be decided by the sum of the two k's for which (28) gives the value

$$\{g(M^2T_p - T_\tau) + g_0'S_p\}/T_p .$$

All constituents of this expression are positive, including

$$M^2T_p^{(0)} - T_\tau^{(0)} > (S_\tau^{(0)}/S_p^{(0)})T_p^{(0)} - T_\tau^{(0)} = 1/S_p^{(0)} > 0.$$

Thus we arrive at the result that the two roots k at Z_0 are *positive*. Our theorem has now been proved: the fundamental inequalities (17) not only show that by a shock the flow changes from supersonic to subsonic but also that one of the two points Z_0 , Z_1 is a node, the other a saddle.

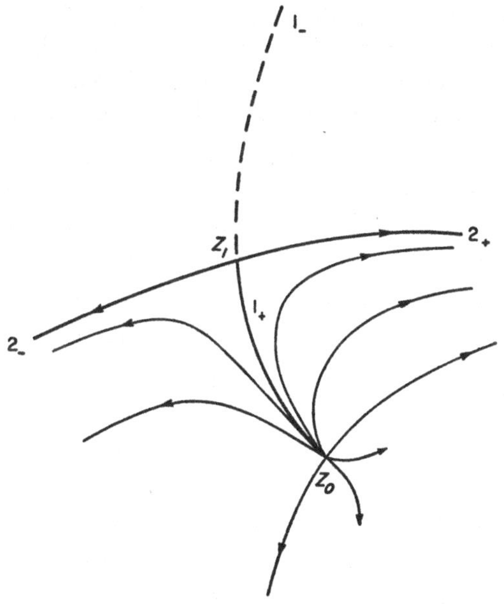

FIG. 4. The topological picture of the trajectories in the (p, τ)-plane.

In order actually to prove the existence of a unique solution of the shock layer problem, one has to bridge the gap between the neighborhoods of the two singular points Z_0, Z_1 by some sort of topological argument. In that respect one fact is of decisive importance:

Theorem 7. *The differential equations of the shock layer have no singular point besides* Z_0, Z_1.

Indeed if there existed three distinct singular points Z_0, Z_1, Z_2 they would satisfy the equations

$$M^2\tau_0 + p_0 = M^2\tau_1 + p_1 = M^2\tau_2 + p_2 = M^2a,$$

$$e_0 - \tfrac{1}{2}M^2(\tau_0 - a)^2 = e_1 - \tfrac{1}{2}M^2(\tau_1 - a)^2 = e_2 - \tfrac{1}{2}M^2(\tau_2 - a)^2.$$

The first set states that Z_0, Z_1, Z_2 lie on a straight line of negative inclination $-M^2$, and we may assume the arrangement $Z_0 < Z_1 < Z_2$. Then the second set requires that both $Z = Z_1$ and $Z = Z_2$ satisfy the Hugoniot equation $H(Z, Z_0) = 0$, in contradiction to a statement previously proved by which there cannot be more than one point Z on any ray from Z_0 satisfying that equation. The observation is urged upon us that whereas the Hugoniot relation is not transitive, the relation between Z_0, Z_1 defined by the *simultaneous* equations

$$M^2\tau_0 + p_0 = M^2\tau_1 + p_1, \qquad H(Z_1, Z_0) = 0$$

is (M^2 being a given constant).

The topological picture of the trajectories which one expects is indicated by Figure 4. If this is correct, the region covered by all trajectories running into Z_0 for $x \to -\infty$ would be bounded by the saddle line through Z_1 consisting of the two branches 2_+ and 2_- (Figure 3). The outlook is encouraging.

149.

50 Jahre Relativitätstheorie

Die Naturwissenschaften 38, 73—83 (1950)

Einleitung.

Die Historiker haben sich zuweilen der Einteilung des Geschichtsablaufs in Jahrhunderte so bedient, als wäre die Jahrhundertwende mehr als ein rein äußerlicher, durch unsere Zeitrechnung bedingter Einschnitt. So sprach und spricht man etwa vom Geist des 18. Jahrhunderts. Es ist, als wolle die Geschichte der neueren Physik diesem an sich so unwissenschaftlichen Brauch recht geben. Denn wie die Quantentheorie ziemlich genau mit dem Beginn des gegenwärtigen Jahrhunderts auf den Plan tritt, so wird um die Jahrhundertwende auch das Fundament der Relativitätstheorie in der Form gelegt, wie sie heute gilt, und die dadurch gekennzeichnet ist, daß die endliche Lichtgeschwindigkeit als die obere Grenze der Ausbreitungsgeschwindigkeit aller Wirkungen erscheint. Die historische Stufenfolge der speziellen und der allgemeinen Relativitätstheorie (SR und AR) ist auch sachlich wohl begründet. Freilich geht die SR in der AR auf, so wie ein näherungsweise in ein exakt gültiges Gesetz aufgeht. Aber jene so viel leichter zu handhabende Annäherung ist maßgebend für alle physikalischen Phänomene, in denen die Gravitation vernachlässigt werden kann, und spielt darum im aktuellen Betrieb der physikalischen Forschung, insbesondere der Atomforschung, die weitaus größere Rolle. Es ist ja eine der merkwürdigsten Tatsachen der Natur, daß die Gravitationsanziehung zweier Elektronen 10^{40}mal so klein ist wie ihre elektrostatische Abstoßung. Heute wird es wohl kaum einen Physiker geben, der daran zweifelt, daß die SR einen der wichtigsten und empirisch am sichersten gestützten Grundzüge der Natur wiedergibt. Anders steht es mit der auf einem viel schmaleren Erfahrungsfundament beruhenden AR, und wenn auch die meisten Physiker ihre Grundgedanken akzeptieren, so werden wenige anzunehmen geneigt sein, daß wir die allgemein invarianten Feldgesetze bereits in ihrer endgültigen Form besitzen. Ich brauche Ihnen nicht zu erzählen, daß in viel höherem Maße, als das in der Quantentheorie der Fall ist, Grundlagen und Ausbau der Relativitätstheorie das Werk eines Mannes sind: Albert Einstein. Wie er, ungleich den meisten andern Physikern, sich seinerzeit mit der von ihm errichteten speziellen Relativitätstheorie nicht zufrieden gab und in einer an Galilei und Newton gemahnenden Kombination von Empirie und Spekulation zur AR vorstieß, so ist er auch jetzt noch unablässig mit dem Problem einer einheitlichen, alle Naturkräfte umfassenden Feldtheorie beschäftigt [*1*].

A. Spezielle Relativitätstheorie.

1. Vorgeschichte und Begründung.

Als Max Planck 1900 die Quanten zur Erklärung der Formel für die Energieverteilung im Spektrum der schwarzen Hohlraumstrahlung einführte, trat, so darf man sagen, mit der Quantentheorie auch das Quantenproblem zum erstenmal auf den Plan. Das Relativitäts*problem* aber ist viel älter als die im 20. Jahrhundert entstandene Relativitäts*theorie*, in welcher wir seine Lösung sehen. Schon Aristoteles definiert Ort (τόπος) relativ, nämlich als Beziehung eines Körpers zu den Körpern seiner Umgebung. Locke gab in seinem Hauptwerk eine eingehende erkenntnistheoretische Analyse [*2*], Galilei illustriert die Relativität der Bewegung hübsch durch das Beispiel des Schreibers, der an Bord eines von Venedig nach Alexandrette segelnden Schiffes seine Notizen macht und dessen Feder „in Wahrheit", das ist relativ zur Erde, eine lange, glatte, nur leicht gewellte Linie beschreibt [*3*]. Offenkundig durch theologische Überzeugungen mitbestimmt, verkündet Newton am Beginn seiner Principia mit ehernen Worten die Lehre vom absoluten Raum und der absoluten Zeit. Er gibt das Beispiel zweier durch einen Faden verbundenen Kugeln, die um eine zum Faden senkrechte Mittelachse rotieren: die Spannung des Fadens zeigt Existenz und Geschwindigkeit der Rotation an. Allgemein stellt er sich die Aufgabe, aus den relativen Bewegungen der Körper und den bei der Bewegung auftretenden Kräften ihre absolute Bewegung zu bestimmen, und erklärt geradezu, daß sein *treatise* verfaßt wurde um der Lösung dieses Problems willen. „To this end it was that I composed it", heißt es am Schluß der Vorrede [*4*]. Gelingt ihm sein Vorhaben? Nicht völlig. Die von ihm aufgestellten Gesetze der Mechanik (und der Gravitation) gestatten wohl, die Bewegung eines Massenpunktes in gerader Linie mit konstanter Geschwindigkeit (gleichförmige Translation) von allen andern Bewegungen zu unterscheiden; hingegen gestatten sie nicht, unter den gleichförmigen Translationen die Ruhe auszuzeichnen. Um dies doch zu erreichen, nimmt Newton zu einer kosmologischen Hypothese und einer Begriffsunterschiebung seine Zuflucht, die sich gar fremd in dem sonst so wohlfundierten, herrlichen Aufbau der Principia ausnehmen. Seine Hypothese ist, daß das Weltall ein Zentrum habe und dieses sich in Ruhe befinde. Vom Schwerpunkt des Sonnensystems stellt er fest, daß es eine gleichförmige Translation ausführt, und fährt dann fort: „But if that center moved, the center of the world would move also against the hypothesis" [*5*]. Beides, die Hypothese vom ruhenden Zentrum des Weltalls und diese durch nichts begründete Identifizierung desselben mit dem Schwerpunkt des Sonnensystems, sind erstaunlich. Newtons Bild vom Kosmos ist offenbar noch wesentlich gebundener als das, welches schon 100 Jahre früher Giordano Brunos leidenschaftliche Seele erfüllte [*6*].

Raum und Zeit bilden ein vierdimensionales Kontinuum, das wir mit Minkowski die *Welt* nennen. Ein raum-zeitlich eng begrenztes Ereignis geschieht an einer bestimmten Raum-Zeit-Stelle, in einem bestimmten Weltpunkt, „hier-jetzt". Nur raumzeitliche Koinzidenz oder unmittelbare raum-zeitliche Nachbarschaft ist etwas in der Anschauung unzweifelhaft

---

*) Vortrag, gehalten auf der Tagung der Gesellschaft Deutscher Naturforscher und Ärzte in München am 24. Oktober 1950.

Gegebenes. Ein Massenpunkt beschreibt eine eindimensionale Linie in der Welt. Wenn zwei Personen sich treffen und sich die Hand reichen, so geschieht das an einer bestimmten Raum-Zeit-Stelle, in welcher sich ihre Weltlinien schneiden. Weil kein Raumpunkt und kein Zeitpunkt von dem andern an sich physikalisch verschieden ist, entsteht das Problem NEWTONs, wie

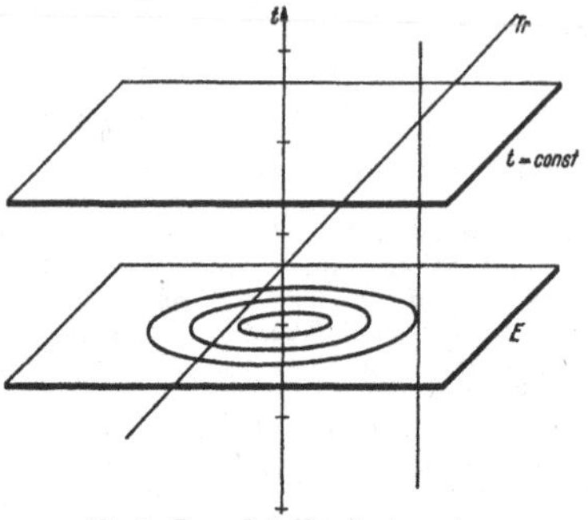

Abb. 1. Raum-Zeit-Koordinatensystem.

man von zwei Ereignissen entscheiden soll, ob sie am gleichen Ort (wenn auch zu verschiedenen Zeiten) oder zur gleichen Zeit (wenn auch an verschiedenen Orten) geschehen. Um ein graphisches Bild entwerfen zu

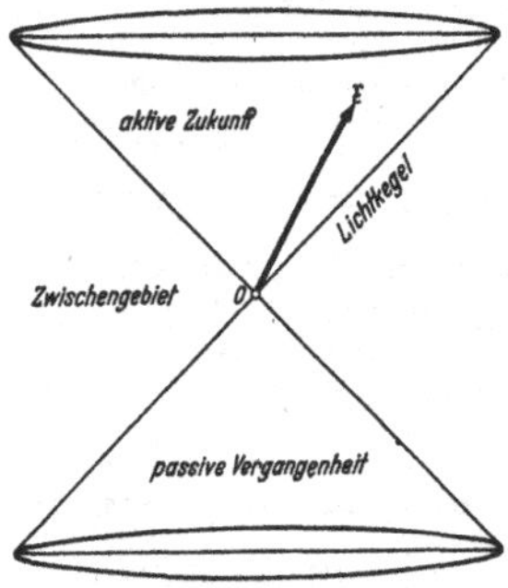

Abb. 2. Lichtkegel.

können, nehmen wir dem Raum eine Dimension, studieren also nur die Vorgänge auf einer horizontalen Ebene E und tragen im Bild die Zeit t senkrecht zu dieser Ebene E auf (Abb. 1). In unserem Bilde erscheinen gleichzeitige Weltpunkte als in einer Horizontalebene gelegen, gleichortige auf einer vertikalen Geraden. NEWTONs Glaube an einen absoluten Raum und an eine absolute Zeit bedeutet also, daß er der Welt eine Struktur zuschreibt, die sich in einer (horizontalen) Schichtung und einer quer dazu verlaufenden (vertikalen) Faserung ausdrückt. Durch jeden Weltpunkt geht eine Schicht und eine Faser. Raum und Zeit kommt ferner eine metrische Struktur zu, auf Grund deren wir von der Gleichheit von Raumstrecken und von Zeitintervallen sprechen. In unserem graphischen Bilde können wir das am anschaulichsten durch eine Schar konzentrischer Kreise in der Ebene E und eine Reihe äquidistanter Punkte auf der vertikalen Zeitachse veranschaulichen. Gibt das Bild in dieser Weise die metrischen Verhältnisse getreu wieder, so stellen (nichthorizontale) gerade Linien die möglichen gleichförmigen Translationsbewegungen dar.

Das *spezielle Relativitätsprinzip*, das sich aus der NEWTONschen Mechanik ergab und das NEWTON selber nur aus theologisch-spekulativen Gründen verleugnete, besagt, daß in einem in gleichförmiger Translation befindlichen Eisenbahnwagen alle Vorgänge genau so ablaufen, alle Tätigkeiten, wie Briefeschreiben oder Ballspielen, genau so ausgeführt werden können, wie wenn der Wagen ruhte; Störungen treten nur bei Beschleunigungen auf. Schon GALILEI hatte dieses Prinzip klar erkannt und entwickelt es in seinem *Dialogo sopra i dui massimi sistemi del mondo* [7]. Er faßt die Bewegung auf als einen Kampf zwischen *Trägheit und Kraft*; solange ein Körper durch keine äußeren Kräfte abgelenkt wird, beschreibt er eine gleichförmige Translation. In unserem graphischen Bild können wir diese Erkenntnis dahin aussprechen, daß wohl die geraden Linien unter allen andern ausgezeichnet sind, aber nicht die vertikalen Geraden.

An der objektiven Bedeutung der *Gleichzeitigkeit*, der „horizontalen Schichtung", hatte vor EINSTEIN niemand ernstlich gezweifelt. Indem es dem Menschen natürlich ist, anzunehmen, daß ein Ereignis dann geschieht, wenn er es beobachtet, dehnt er seine eigene Zeit auf die ganze Welt aus. Dieser naive Glaube war freilich durch OLAF RÖMERs Entdeckung der endlichen Lichtgeschwindigkeit erschüttert worden. Repräsentiert man in unserem graphischen Bild 1 sec auf der Zeitachse durch eine Strecke von derselben Länge wie die Raumstrecke in der Ebene E, die das Licht in 1 sec durchmißt, so wird die Ausbreitung eines in O gegebenen Lichtsignals durch einen geraden vertikalen Kreiskegel vom Öffnungswinkel 90° mit Spitze in O dargestellt (Abb. 2); er gibt den „Lichtkegel" wieder, auf welchem diejenigen Weltpunkte liegen, in denen das Signal eintrifft. Die Begriffe von Vergangenheit und Zukunft haben, wie LEIBNIZ nicht müde wurde zu betonen, kausale Bedeutung: durch „hier-jetzt", im Weltpunkt O, abgeschossene Kugeln kann ich CAIUS JULIUS CAESAR nicht mehr treffen, da seine Weltlinie ganz der vergangenen Welthälfte in bezug auf O angehört. Und man nahm an, daß es die Ebene $t=\text{const}$ durch O ist, welche diese Trennung in eine vergangene und zukünftige Welthälfte bewirkt. Die neue Erkenntnis, welche das 20. Jahrhundert brachte, war die, daß die Kausalstruktur der Welt nicht durch die ebenen Schichten $t=\text{const}$, sondern durch die von jedem Weltpunkt ausgehenden Lichtkegel beschrieben wird. Physikalisch heißt das, daß keine Wirkung sich mit größerer als Lichtgeschwindigkeit ausbreitet. Im Innern des „vorderen" von O ausgehenden Lichtkegels liegen alle Weltpunkte, auf die das, was in O geschieht, noch von Einfluß sein kann; im Innern des durch rückwärtige Verlängerung entstehenden hinteren Lichtkegels aber liegen alle Ereignisse, die auf das, was in O geschieht, möglicherweise von Einfluß waren, insbesondere solche, von denen ich jetzt hier, in O, eine auf direkte Wahrnehmung gegründete Kunde haben kann. Die beiden Teile des Kegels grenzen nicht zwischenraumlos aneinander; mit dem Zwischengebiet ist O kausal überhaupt nicht verbunden.

Wenn hier zur Beschreibung der Weltstruktur (unter Streichung einer Dimension) ein Bildraum Verwendung fand, von dem in den vertrauten geometrischen termini der euklidischen Geometrie gesprochen wurde, so geschah das nur im Interesse der leichteren Verständlichkeit. In Wahrheit brauchen wir zur begrifflichen Darstellung der Naturvorgänge eine Abbildung der Welt oder des in Frage kommenden Weltstücks auf ein Stück des vierdimensionalen *Zahlenraums*; dessen Punkte sind nichts anderes als die möglichen Zahlenquadrupel (x_1, x_2, x_3, x_4). Relativ zu einem solchen Koordinaten- oder Bezugssystem wird die Zeit eines Weltpunktes durch x_4, der Ort durch (x_1, x_2, x_3) angegeben; es legt in dieser Weise eine rein konventionelle Gleichzeitigkeit und Gleichortigkeit fest. Die Abbildung der Welt auf den Zahlenraum ist an sich ganz willkürlich; sie dient dazu, den Weltpunkten Namen zu geben. Das allgemeine Relativitätsprinzip ist so im Grunde nichts anderes als die Ablehnung des Namenzaubers, indem es behauptet, daß Ereignisse nicht davon berührt werden, in welcher Nomenklatur man sie beschreibt. Wenn eine Klasse von Koordinatensystemen objektiv herausgehoben werden soll, so kann dies nur dadurch geschehen, daß man von bestimmten physikalischen Vorgängen ausgeht und angibt, wie sie sich in einem Koordinatensystem dieser Klasse arithmetisch ausdrücken. So wird z. B. in EINSTEINs allgemein-relativistischer Gravitationstheorie das Gravitationsfeld eines Zentralkörpers gegebener Masse nach K. SCHWARZSCHILD durch bestimmte Formeln in einem Koordinatensystem dargestellt, das eben durch diese auf dasselbe bezogene Formeldarstellung weitgehend normiert ist.

Bisher haben wir uns von der NEWTONschen Ansicht über die Struktur der Welt zu einer solchen den Weg zu bahnen gesucht, die nicht in apriorischen Prinzipien, sondern auf Erfahrungen basiert ist. Nun müssen wir die so gewonnene Ansicht ohne den Umweg über NEWTON kurz systematisch darstellen. Wir benutzen zwei Grundvorgänge: die Ausbreitung eines Lichtsignals und die kräftefreie Bewegung eines Massenpunktes. Für diese gelten die folgenden beiden Grundgesetze, die, wie ich gleich bemerke, auch im Rahmen der allgemeinen Relativitätstheorie bestehen bleiben:

1. Der Lichtkegel in O, das ist der geometrische Ort der Weltpunkte, in denen ein in O gegebenes Lichtsignal eintrifft, ist durch O eindeutig bestimmt, unabhängig von dem Zustand, insbesondere dem Bewegungszustand der das Licht in O aussendenden Lichtquelle. (Dieses Gesetz wird häufig durch den etwas irreführenden Namen „Gesetz von der Konstanz der Lichtgeschwindigkeit" bezeichnet).

2. Die geodätische Weltlinie, die ein kräftefrei sich bewegender Massenpunkt beschreibt, ist durch Anfangspunkt und Anfangsrichtung in der Welt eindeutig bestimmt.

Die spezielle Relativitätstheorie ist auf die Annahme basiert, daß Bezugssysteme existieren, Abbildungen der Welt, in welcher sich diese Vorgänge in einer völlig determinierten Weise darstellen: nämlich die geodätischen Linien als gerade Linien und die Lichtkegel als Nullkegel, das ist als vertikale Kreiskegel vom Öffnungswinkel 90°. Ein solches normales Koordinatensystem ist nicht eindeutig bestimmt, sondern nur bis auf eine LORENTZ-*Transformation*; die Gruppe dieser Transformationen besteht aus allen solchen Abbildungen des Bildraums, welche Gerade in Gerade und Nullkegel in Nullkegel überführen. Es ist eine rein mathematische Aufgabe, diese Gruppe zu bestimmen. Die Theorie wird abgeschlossen durch die Behauptung, daß unter diesen gleichberechtigten Koordinatensystemen auch unter Berücksichtigung aller weiteren Naturvorgänge objektiv keine engere Auswahl getroffen werden kann: die Naturgesetze sind invariant gegenüber LORENTZ-Transformationen. Unter diesen Transformationen gibt es solche, den gegebenen Weltpunkt O festlassende, welche die vertikale Gerade durch O in eine beliebige zeitartige Gerade durch diesen Punkt überführen (d.h. in eine, die von O ins Innere des Nullkegels in O hineinführt): darin drückt sich das spezielle Relativitätsprinzip aus. MINKOWSKI erkannte, daß die Gruppe der LORENTZ-Transformationen mit der Gruppe der euklidischen Ähnlichkeiten zusammenfällt, wenn man sich nicht scheut, der einen Koordinate x_4 statt reeller rein imaginäre Werte zu erteilen. Die von der speziellen Relativitätstheorie der Welt zugeschriebene Trägheits- und Kausalstruktur ist darum von einer *metrischen Struktur* abzuleiten. Und zwar läßt sich die der Welt zukommende Geometrie, so wie die euklidische Raumgeometrie, auf den affinen Begriff des Vektors aufbauen, und wie in der euklidischen Geometrie bestimmen nach Wahl einer Maßeinheit je zwei Vektoren $\mathfrak{x}$, $\mathfrak{y}$ ein skalares Produkt $(\mathfrak{x} \cdot \mathfrak{y})$, das eine nichtausgeartete Bilinearform der beiden Argumente ist. Nur ist die zugehörige quadratische Form $(\mathfrak{x} \cdot \mathfrak{x})$, das „Quadrat der Länge" des Vektors $\mathfrak{x}$, nicht positivdefinit, $(\mathfrak{x} \cdot \mathfrak{x}) > 0$, sondern mit bezug auf eine geeignete Normalbasis $\mathfrak{e}_1, \mathfrak{e}_2, \mathfrak{e}_3, \mathfrak{e}_4$ für die Vektoren, die dem CARTESIschen Achsenkreuz in der euklidischen Geometrie entspricht, wird das Quadrat der Länge des Vektors $\mathfrak{x} = \sum x_k \mathfrak{e}_k$ durch die Formel

$$(\mathfrak{x} \cdot \mathfrak{x}) = x_1^2 + x_2^2 + x_3^2 - x_4^2$$

mit einem negativen Vorzeichen gegeben (quadratische Form der Signatur 1).

Den Zusammenhang dieser Weltgeometrie mit der physikalischen Messung von Raumstrecken durch starre Maßstäbe und von Zeitintervallen durch Uhren wollen wir erst im Zusammenhang der allgemeinen Relativitätstheorie kurz erläutern. An dieser Stelle nur zwei Bemerkungen darüber. Stößt man einen Eisenklotz gleichzeitig an verschiedenen Stellen an, so werden sich die Umgebungen dieser Stellen in Bewegung setzen; aber erst, wenn die mit Lichtgeschwindigkeit sich um die Stellen ausbreitenden Wirkungskugeln sich zu überlappen beginnen, werden diese Bewegungen sich gegenseitig beeinflussen. Diese einfache Bemerkung von Herrn VON LAUE zeigt, daß schon in der speziellen Relativitätstheorie der Begriff des starren Körpers im Prinzip hinfällig wird. — Die klassische Feldphysik läßt Körper beliebiger Ladung und Masse zu. Sie ignoriert damit, möchte man sagen, die Grundtatsache der Atomistik, daß die Materie aus Elementarteilchen fester Ladung und Masse besteht. In der Feldphysik erscheint darum auch die Wahl der Maßeinheit, in der Raum- und Zeitstrecken gemessen werden, als willkürlich; die Gruppe der LORENTZ-Transformationen umfaßt die Dilatationen. Die Unterscheidung von Ähnlichkeit und Kongruenz, auf die diese Bemerkung hinweist, ist häufig von Darstellern der Relativitätstheorie verwischt worden.

2. Anfänge und Fortbildung der Theorie.

a) Frühe Geschichte. Es ist meine Aufgabe, Ihnen ein Referat über die Entwicklung der Relativitätstheorie in der ersten Hälfte dieses Jahrhunderts zu erstatten. Es schien mir aber zweckmäßig, zunächst einen knappen systematischen Abriß der Theorie voranzustellen, in dem ich zugleich bemüht war, verkehrte Auffassungen abzuwehren. Holen wir nun in raschen Schritten die Historie nach [*8*]: Die optischen waren von MAXWELL den elektromagnetischen Erscheinungen eingegliedert worden; dementsprechend spielt die Lichtgeschwindigkeit c in den MAXWELLschen Gleichungen des elektromagnetischen Feldes eine wichtige Rolle. Solange man an einen substantiellen Lichtäther glaubte, wurde die aus diesen Gleichungen folgende Ausbreitung einer elektromagnetischen Störung mit Lichtgeschwindigkeit in konzentrischen Kugeln als eine Beschreibung des Vorganges relativ zu dem im Ganzen ruhenden Äther aufgefaßt. Dann mußte man aber erwarten, daß die Bewegung ponderabler Materie relativ zu diesem Äther sich durch bestimmte Effekte kundgeben würde. Die Erfahrung hingegen zeigte, daß das spezielle Relativitätsprinzip auch bei der Ausdehnung auf optische und elektromagnetische Vorgänge seine Richtigkeit behält. Aus der von ihm mitbegründeten Elektronentheorie heraus erschloß H. A. LORENTZ zuerst die Tatsache, daß alle Effekte 1. Ordnung in dem Verhältnis v/c zwischen Materie- und Lichtgeschwindigkeit für einen mitbewegten Beobachter herausfallen. Als weitere Versuche, wie der berühmte MICHELSONsche Interferenzversuch, auch das Fehlen von Effekten 2. Ordnung feststellten, nahm LORENTZ zu der Hypothese seine Zuflucht, daß ein Körper infolge seiner Bewegung gegen den Äther eine Längskontraktion im Verhältnis $1:\sqrt{1-(v/c)^2}$ erfährt. 1900 stellte er die heute allgemein als LORENTZ-Transformation bekannten Formeln auf. Die völlige Abklärung brachten dann drei Arbeiten von LORENTZ, H. POINCARÉ und EINSTEIN 1904/05. EINSTEIN nahm die grundsätzliche Wendung, daß er das spezielle Relativitätsprinzip als exakt gültig postulierte und nun zusah, welche Modifikationen für unsere Raum-Zeit-Vorstellungen resultieren, wenn man damit die elementaren Gesetze der Ausbreitung des Lichtes und der elektromagnetischen Wellen zusammenhält, und welche Gesetze für die Elektrodynamik bewegter Körper daraus folgen, wenn man die für ruhende Körper geltenden aus der (phänomenologischen) MAXWELLschen Theorie herübernimmt. Erst hierdurch kam es zu der radikalen Kritik des Begriffs der Gleichzeitigkeit. MINKOWSKI endlich gab die weltgeometrische Einkleidung, die insbesondere für die Weiterentwicklung zur allgemeinen Relativitätstheorie bedeutungsvoll wurde (Vortrag auf der Kölner Naturforscherversammlung 1908).

b) Elektrodynamik bewegter Körper. Leicht waren die *optischen* Konsequenzen des Relativitätsprinzips zu ziehen. Aus dem einfachen Umstand, daß die Phasendifferenz einer ebenen Lichtwelle in einem homogenen Medium an zwei Weltstellen A, B linear von dem Weltvektor $\overrightarrow{AB}$ abhängt, ergibt sich durch Vergleich zweier gleichberechtigter Spaltungen der Welt in Raum und Zeit die *Aberration* als Geschwindigkeitsperspektive, der DOPPLER-*Effekt* und der FIZEAUsche *Mitführungskoeffizient.* Betreffs Ableitung der elektromagnetischen Gleichungen für bewegte Körper sei die folgende methodische Bemerkung gemacht. Im Grunde gestattet das Prinzip, von ruhenden Körpern nur auf den Fall zu schließen, wo alle beteiligten Körper *dieselbe* gleichförmige Translation erfahren. In Wahrheit interessieren aber Aussagen über Situationen, in denen mehrere Körper mit verschiedenen Geschwindigkeiten auftreten. Nehmen wir den Fall zweier durch einen leeren Zwischenraum getrennter Körper, die sich (in unserem Bezugssystem) mit verschiedenen Geschwindigkeiten bewegen. Für den einzelnen Körper und den ihn umgebenden leeren Raum erhält man bestimmte Gleichungen, indem man die für den ruhenden Körper geltenden durch LORENTZ-Transformation in solche umwandelt, die relativ zu dem zugrunde liegenden Bezugssystem gelten. Indem man so für beide Körper verfährt, erhält man Gleichungen, die im leeren Zwischenraum miteinander übereinstimmen, da die MAXWELLschen Gleichungen im leeren Raum LORENTZ-invariant sind. Diese Gesetze kann man somit *ohne Widerspruch* als gültig annehmen, und man verwendet sie de facto als Näherungsgesetze, solange wenigstens die Distanz der beiden Körper nicht von molekularer Größenordnung ist. Die auf diese Weise von LORENTZ, EINSTEIN, MINKOWSKI, BORN u. a. gezogenen Konsequenzen haben sich durchweg in der Erfahrung bewährt.

c) Mechanik. Energie und Trägheit. In der Mechanik haben wir es mit Körpern zu tun, bei denen wir den von einem mitbewegten Beobachter zu messenden *inneren Zustand* von dem durch seine vektorielle Geschwindigkeit $\mathfrak{v}$ bestimmten *Bewegungszustand* unterscheiden können. Wir betrachten die Körper nur im Zustand gleichförmiger Translation. Das Impulsgesetz besagt, daß einem solchen Körper ein vektorieller *Impuls* $\mathfrak{J}$ zukommt, welcher der Geschwindigkeit parallel ist, $\mathfrak{J}=m\mathfrak{v}$, und daß vor und nach einer Reaktion die Summe der Impulse der beteiligten Körper die gleiche ist. Der skalare Faktor m ist die träge Masse; so war dieser Begriff im Grunde schon von GALILEI und HUYGENS eingeführt worden. Durch das Relativitätsprinzip erkennt man, daß das Impulsgesetz das weitere zur Folge hat, nach welchem auch die Massensumme $\sum m$ durch eine Reaktion sich nicht ändert. Indem man den Vorgang von zwei völlig gleich beschaffenen Körpern, die mit entgegengesetzt gleichen Geschwindigkeiten $\pm\mathfrak{v}$ gegeneinander gejagt werden und sich dabei zu einem einzigen, notwendig ruhenden Körper vereinigen, von einem beliebigen normalen Bezugssystem aus studiert, erkennt man bei Zugrundelegung der GALILEIschen Kinematik, die aus der EINSTEINschen durch den Grenzübergang zu $c\to\infty$ entspringt, daß die Masse von dem Bewegungszustand und infolgedessen auch von dem inneren Zustand eines Körpers unabhängig ist. Hingegen ergibt sich auf Grund der EINSTEINschen Kinematik die Formel

$$m=\frac{m_0}{\sqrt{1-v^2}},$$

wo m_0 die nur vom inneren Zustand abhängige Ruhmasse, v der Betrag der Geschwindigkeit ist und $c=1$ gesetzt wurde. Messungen an Kathodenstrahlen bestätigten die Formel um so genauer, je mehr sich die Messungsmethoden verbesserten. Man versteht aus dieser Formel, daß der Trägheitswiderstand eines Körpers mit seiner Geschwindigkeit so anwächst, daß diese niemals die Lichtgeschwindigkeit erreicht. Die Veränderung der Masse mit dem Bewegungszustand des Körpers hat zur Folge, daß sie auch von inneren Zustandsänderungen beeinflußt wird, z. B. bei Erwärmung eines Körpers einen Zuwachs erfährt. Natürlich ist die Massenänderung des Körpers unabhängig davon, *wie* diese Zustandsänderung zustande kommt. Zusammen mit dem Faktum, daß für ein abgeschlossenes Körpersystem die Summe der Massenänderungen verschwindet, zeigt dies an, erstens, daß träge Masse und Energie dasselbe sind, und zweitens, daß das Energiemaß einer Zustandsänderung als die Differenz von den einzelnen Zuständen zukommenden *Energieniveaus* aufzufassen ist. Dieses Gesetz von der Trägheit der Energie, das sich in den üblichen Einheiten in der heute so populär gewordenen Formel, Energie $E = mc^2$, ausdrückt, ist zweifellos die wichtigste Folgerung aus der Relativitätstheorie; sie wurde von EINSTEIN schon in seiner ersten Arbeit 1905 gezogen, und schon damals faßte er die Anwendung derselben auf Kernreaktionen ins Auge, bei denen der der freiwerdenden Energie entsprechende Massendefekt meßbare Größenordnung erreichen mag. Das ist ein Gebiet, das inzwischen ins Zentrum der Forschung gerückt ist, von dem damals aber nur die Erscheinungen des spontanen radioaktiven Zerfalls bekannt waren. Um die Größenordnung zu kennzeichnen, gebe ich als Beispiel die Bildung eines Lithium-Atoms aus drei Protonen und drei Neutronen; die Masse (das Atomgewicht) des Lithiums ist 6,01692, der Massendefekt 0,03432, entsprechend einem Energieverlust von $5{,}11 \cdot 10^{-5}$ erg je Atom. Die MAXWELLsche Theorie schreibt dem elektromagnetischen Felde Dichte und Stromdichte der Energie und des Impulses zu. Ein Stück des Feldes ist nicht ein „Körper", an dem man Bewegungs- und inneren Zustand unterscheiden kann. Hier drückt sich das Gesetz von der Trägheit der Energie darin aus, daß der Energiestrom gleich der Impulsdichte ist. An dem Beispiel eines ruhenden Körpers, der eine kugelförmige Lichtwelle vom Impuls 0 aussendet, entdeckte EINSTEIN zuerst das Gesetz von der Trägheit der Energie, indem er feststellte, daß die ausgesandte Energie durch eine entsprechende Abnahme der trägen Masse des lichtaussendenden Körpers kompensiert werden muß. Er erkannte sogleich, daß dies Gesetz dieselbe universelle Gültigkeit besitzt wie das Relativitätsprinzip, aus dem es zwingend hervorgeht.

3. *Verbindung mit der Quantentheorie.*

In eine neue Epoche tritt die SR ein durch ihre Verbindung mit der das atomare Geschehen beherrschenden *Quantentheorie*, wie sie in mathematisch präziser Form um 1925 von HEISENBERG und SCHRÖDINGER aufgestellt wurde [*10*]. Die Verbindung ist nicht ohne Schwierigkeiten zu vollziehen. Die als Funktion von Ort (x_1, x_2, x_3) und Impuls (p_1, p_2, p_3) ausgedrückte Energie p_4, welche nach den HAMILTONschen Gleichungen der klassischen Mechanik die Bewegung eines freien Elektrons bestimmt, lautet

$$p_4 = \sqrt{m^2 + (p_1^2 + p_2^2 + p_3^2)},$$

wo m die konstante Ruhmasse ist. Bei der quantentheoretischen Übersetzung dieses Ausdrucks, bei welcher p_k durch den Operator $\frac{\hbar}{i}\frac{\partial}{\partial x_k}$ zu ersetzen ist ($i = \sqrt{-1}$, $\hbar$ die durch einen Faktor 2π modifizierte PLANCKsche Wirkungskonstante), macht die Quadratwurzel Schwierigkeiten. Schreibt man aber die Gleichung in ihrer rationalen Form,

$$p_4^2 - (p_1^2 + p_2^2 + p_3^2) = m^2, \tag{1}$$

welche die relativistische Invarianz unmittelbar in Evidenz setzt, so ergibt die Übertragung nicht die von der Quantentheorie allgemein vorgeschriebene Form. DIRAC überwand dieses Hindernis durch die den Mathematikern wohlbekannte, aber in ihren Händen unfruchtbar gebliebene Bemerkung, daß die quadratische Form der Variablen p_k auf der linken Seite von (1) sich mit Hilfe gewisser hyperkomplexer Zahlen γ_k, deren Multiplikation nicht kommutativ ist, als das Quadrat einer Linearform $\sum_k \gamma_k p_k$ schreiben läßt, und er setzte darum die Gleichung an

$$\sum_k \gamma_k p_k = m,$$

die der quantenmechanischen Übersetzung keine Schwierigkeiten in den Weg legt [*11*]. Die Physiker müssen mir verzeihen, wenn ich hier um der Kürze willen diese reichlich abstrakte Fassung von DIRACs Grundidee wähle. Gemäß seinem Ansatz wird das Wellenfeld des Elektrons nicht durch eine skalare Funktion ψ der vier Weltkoordinaten x_k beschrieben, sondern durch eine Größe mit vier Komponenten ψ_q, die sich unter dem Einfluß einer LORENTZ-Transformation der Koordinaten in einer ungewöhnlichen, in der üblichen Vektor- und Tensorrechnung nicht vorgesehenen Weise transformieren. Ein gemeinsamer konstanter Phasenfaktor $e^{-i\lambda}$ vom absoluten Betrag 1 bleibt in den ψ_q unbestimmt. Aus dem HAMILTONschen Prinzip entnimmt man ferner die Regel, daß man das auf das Elektron einwirkende elektromagnetische Feld einfach dadurch berücksichtigen kann, daß man die Operatoren $\frac{\partial}{\partial x_k}$ durch $\frac{\partial}{\partial x_k} + i\varphi_k$ ersetzt, wo φ_k die mit $\varepsilon/\hbar$ multiplizierten Komponenten des elektromagnetischen Potentials sind (ε = elektrische Elementarladung). Diese Dinge werden hernach im Rahmen der AR unter dem Titel Eichinvarianz von besonderer Wichtigkeit werden. Die geschilderte LORENTZ-invariante DIRACsche Theorie des Elektrons gibt in wunderbarer Weise Rechenschaft über den aus der Analyse der Atomspektren von S. GOUDSMIT, G. E. UHLENBECK und W. PAULI erschlossenen *Elektronenspin*, über den anomalen ZEEMANN-Effekt, die Feinstruktur des Wasserstoffspektrums und viele andere Dinge. Man kann wohl sagen, daß die Tatsachen der Spektroskopie heute die zuverlässigste Stütze für das Relativitätsprinzip abgeben.

Der DIRACsche Erfolg ist um so bemerkenswerter, als bei dem Übergang von einem Teilchen, Photon oder Elektron, zu einer unbestimmten Anzahl von in Wechselwirkung miteinander stehenden Photonen und Elektronen die Verschmelzung von Quanten- und Relativitätstheorie auf Hindernisse stößt, die, trotz vielversprechender Ansätze, noch nicht aus dem Wege geräumt werden konnten. Ich glaube aber, niemand denkt daran, deswegen der Relativitätstheorie den Laufpaß zu geben; sie ist dafür zu fest in dem ganzen Gebäude unserer theoretischen Physik verankert.

B. Allgemeine Relativitätstheorie.

1. Der Grundgedanke.

Es ließ sich nicht vermeiden, daß ich hier für einen Augenblick das Gebiet der gestrigen Vorträge, die Quantentheorie[1]), berührte. Aber nun ist es höchste Zeit für mich, von der speziellen zur allgemeinen Relativitätstheorie überzugehen. Es wurde schon oben gesagt, daß wir zur begrifflichen Beschreibung der Naturvorgänge die Weltpunkte auf Koordinaten beziehen müssen, daß aber die Erscheinungen selber durch diese an sich willkürliche Namengebung natürlich nicht beeinflußt werden. Der Übergang von einem zu einem andern Koordinatensystem geschieht durch eine (stetige) Koordinatentransformation. Von was für Gesetzen auch immer die Natur beherrscht sein mag, ich kann deren Formulierung ein willkürliches Koordinatensystem zugrunde legen, und sie werden alsdann invariant sein gegenüber beliebigen Koordinatentransformationen. Freilich muß ich dabei das Trägheits- und Kausalfeld oder das metrische Feld, aus dem beide abgeleitet werden, mit unter die physikalischen Zustandsgrößen aufnehmen. Das Prinzip der Invarianz gegenüber beliebigen Koordinatentransformationen ist also an sich nichtssagend, und der physikalisch entscheidende Gedanke, der von der speziellen zur allgemeinen Relativitätstheorie führt, liegt denn auch woanders als in diesem Prinzip. Die Einsicht in die Relativität des Ortes scheint uns zu zwingen, alle Bewegungszustände und nicht nur die gleichförmigen Translationen als gleichwertig zu erachten. So haben denn schon zu NEWTONs Zeit Denker wie LEIBNIZ und HUYGENS, später EULER, sich um das Rätsel bemüht, was der offenkundigen dynamischen Ungleichwertigkeit der kinematisch gleichwertigen Bewegungszustände zugrunde liegt. In neuerer Zeit war es ERNST MACH, der mit allem Nachdruck das allgemeine Prinzip der Relativität der Bewegung verfocht, und das Studium von MACH (neben dem von HUME) war auf EINSTEIN nach seinem eigenen Geständnis von maßgebendem Einfluß [*12*].

Wir haben als Trägheitsfeld diejenige Struktur der Welt bezeichnet, die einem Körper eine durch Anfangsort und -richtung in der Welt eindeutig bestimmte Bewegung aufnötigt, in der er zu beharren bestrebt ist, solange er nicht durch äußere Kräfte abgelenkt wird. Die wirkliche Bewegung resultiert aus dem Kampf zwischen Trägheit (Beharrungstendenz) und Kraft. Lassen Sie mich an ein Beispiel anknüpfen, das, wenn ich mich recht besinne, PHILIPP LENARD auf der Naturforscherversammlung in Bad Nauheim 1920 in die Diskussion über Relativitätstheorie hineinwarf: Ein Zug stößt mit einem entgegenfahrenden Zuge zusammen, während er an dem Kirchturm eines Dorfes vorüberfährt; warum, fragte LENARD, geht der Zug in Trümmer, und nicht der Kirchturm, wo doch der Kirchturm relativ zum Zug einen ebenso starken Bewegungsruck erfährt wie der Zug relativ zum Kirchturm? Die unvoreingenommene, von keiner Relativitätstheorie angekränkelte Antwort ist wohl klar: der Zug wird zerrissen durch den Konflikt seiner eigenen Trägheit mit den auf ihn von dem zusammenstoßenden Zug ausgeübten Molekularkräften; während der Kirchturm ruhig der ihm durch das Trägheitsfeld vorgeschriebenen Bahn folgt. In dem von der SR aufgestellten, bis auf eine LORENTZ-Transformation normierten Koordinatensystem erscheint das Trägheitsfeld als eine starre, der Welt ein für allemal innewohnende geometrische Beschaffenheit: während es enorme Wirkungen auf die Materie ausübt, ist es selbst über alle Einwirkungen der Materie erhaben. Dagegen sträubt sich unser Gerechtigkeitsgefühl: was Wirkungen auf die Materie ausübt, muß auch Wirkungen von ihr erleiden. Wo aber sind in der Natur Anzeichen dafür vorhanden, daß das Trägheitsfeld nicht vorgegeben, sondern den Einwirkungen der Materie gegenüber nachgiebig ist? Hier setzt EINSTEINs fundamentaler Gedanke ein: die *Gravitation* ist dieses Anzeichen. Wenn dies stimmt, wenn in dem Dualismus von Kraft und Trägheit die Gravitation auf die Seite der Trägheit gehört, so würde auf einmal die seltsame Tatsache der *Übereinstimmung von schwerer und träger Masse* verständlich, die durch die feinsten Messungen immer wieder bestätigt wurde. Die Kraft, mit welcher ein elektrisches Feld an einem geladenen Körper angreift, ist seiner Ladung proportional; so ist die Kraft, mit welcher das Gravitationsfeld an einem Körper angreift, seinem Gewicht = schwerer Masse proportional. Aber während die elektrische Ladung ponderabler Körper in keiner Weise mit ihrer trägen Masse zusammenhängt, stimmt merkwürdigerweise ihre schwere Masse, ihre Gravitationsladung, stets mit der trägen Masse überein. Ist EINSTEINs Erklärung richtig, welche die Schwerkraft auf eine Linie mit der Zentrifugalkraft stellt, so müssen wir die an sich unbefriedigende Vorstellung eines fest vorgegebenen Trägheitsfeldes fallen lassen und müssen statt dessen nach Differentialgesetzen suchen, welche das Trägheits- = Schwere-Feld so mit den vorhandenen Massen verknüpfen wie die MAXWELLschen Gleichungen das elektromagnetische Feld mit den dasselbe erzeugenden Ladungen. Dies war das Programm, welches EINSTEIN konzipierte.

Zu seiner Durchführung war er gezwungen, mit beliebigen Koordinaten und allgemeinen invarianten Differentialgleichungen zu operieren. In mathematisch zwingender Weise ergaben sich dabei die Feldgleichungen der Gravitation, die an Stelle des NEWTONschen Gravitationsgesetzes treten. So sehr ihre mathematische Form von der NEWTONschen abweicht, führen sie doch in großer Annäherung zu den gleichen Resultaten. Nur drei kleine Abweichungen erreichen ein der Beobachtung zugängliches Maß: eine Störung des Merkur-Perihel, die sich über die von den andern Planeten der NEWTONschen Theorie gemäß verursachten Störungen überlagert [*13*], die Ablenkung eines nahe an der Sonne vorübergehenden Lichtstrahles und die Rotverschiebung der Spektrallinien im Gravitationsfeld. In allen Fällen ergaben die Messungen Übereinstimmung mit der EINSTEINschen Voraussage innerhalb der Fehlergrenzen. Die eklatanteste Bestätigung wird vielleicht von den Spektren jener Zwergsterne von enormer Dichte geliefert, von denen der lichtschwache Begleiter des Sirius ein Beispiel ist. Nach meiner Meinung hat der EINSTEINsche Grundgedanke, der mit einem Schlage das alte Rätsel der Bewegung löst, in Kombination mit diesen empirischen Resultaten, eine solche Durchschlagskraft, daß ich nicht glauben kann, daß man je zur speziellen Relativitätstheorie mit ihrer festen metrischen Trägheits- und

[1]) Vgl. die Vorträge von W. HEISENBERG, M. VON LAUE und P. HARTECK, Naturwiss. 38, 49ff. (1951).

Kausalstruktur zurückkehren wird; wozu neuerdings E. A. MILNE und G. D. BIRKHOFF uns verleiten wollten [14].

Die EINSTEINsche Lösung des Bewegungsproblems lehrt, daß es überhaupt nicht um den Gegensatz von absoluter und relativer Bewegung geht. Läßt man beliebige Koordinaten zu, so kann man nicht nur einen, sondern alle in der Welt vorhandenen Massenpunkte simultan auf Ruhe transformieren; der Begriff der relativen hat so gut wie der der absoluten Bewegung seinen Sinn verloren. Dagegen bleibt die dynamische Auszeichnung der geodätischen Weltlinien als der reinen Trägheitsbewegung bestehen, aber das sie bestimmende metrische Feld steht in Wechselwirkung mit der Materie: GALILEIs dynamische Auffassung der Bewegung erfährt dadurch eine konkretere Deutung. Solche Spekulationen wie die von MACH, wonach die Sterne des Weltalls die Ebene des FOUCAULTschen Pendels führen, die der Ausbildung der Theorie vielleicht Vorschub geleistet haben, sollte man nicht länger mit dem nüchternen physikalischen Gehalt der Theorie vermengen. Freilich ist es eine Tatsache, daß in einem geeigneten Koordinatensystem das die Gravitation mitumfassende metrische Feld wenig von dem homogenen Zustand abweicht, der durch die MINKOWSKIsche Geometrie beschrieben wurde. Gebrauchen wir mit EINSTEIN das alte Wort *Äther* für das metrische Feld, so würde dies darauf hinweisen, daß in der Wechselwirkung von Äther und Materie der Äther zwar kein die Materie bewegender und von ihr unbewegter Gott ist, aber doch ein übermächtiger Riese, und daß hierauf das nahe Zusammengehen von Trägheitskompaß und Sternenkompaß beruht.

2. *Mathematischer und physikalischer Ausbau.*

RIEMANN hatte, nach dem Muster der GAUSSschen Behandlung krummer Flächen, in der Mitte des 19. Jahrhunderts eine Infinitesimalgeometrie n-dimensionaler Mannigfaltigkeiten ausgebildet. Dieses Werkzeugs konnte sich EINSTEIN bei der Durchführung seiner Theorie bedienen. Die Linienelemente, welche einen Punkt P der Mannigfaltigkeit mit den unendlich benachbarten Punkten P' verbinden, bilden die unendlich kleinen Vektoren des dem Punkte P zugehörigen Vektorkompasses. In der Tat erleiden bei beliebiger Transformation der Koordinaten x_k die Differentiale dx_k, die relativen Koordinaten des variablen Nachbarpunktes P' mit Bezug auf den festen Punkt P, lediglich eine *lineare* Transformation. Indem RIEMANN im Unendlichkleinen die Gültigkeit der euklidischen Geometrie, das ist im wesentlichen des pythagoreischen Lehrsatzes annimmt, kann er in allgemein invarianter Weise dem Linienelement mit den Komponenten dx_k eine *Länge ds* zuschreiben, deren Quadrat eine positive quadratische Form der dx_k ist,

$$ds^2 = \sum_{i,j} g_{ij}\, dx_i\, dx_j;$$

die Koeffizienten g_{ij} hängen vom Punkte P ab. EINSTEIN konnte diesen Ansatz ohne weiteres für die vierdimensionale Welt übernehmen, mit dem Unterschied natürlich, daß hier die quadratische Form nicht definit ist, sondern die Signatur 1 besitzt. Der symmetrische Tensor mit den 10 Komponenten $g_{ij} = g_{ji}$ beschreibt das *metrische Feld* und figuriert zugleich als *Gravitationspotential*. Das metrische Feld bestimmt eindeutig die infinitesimale Parallelverschiebung eines beliebigen Vektors in P nach den unendlich benachbarten Punkten P' und damit den *affinen Zusammenhang* der Welt, welcher durch die 40 CHRISTOFFELschen Drei-Indizes-Symbole $\Gamma^i_{kl} = \Gamma^i_{lk}$ (die Komponenten des Gravitationsfeldes) beschrieben wird. Aus ihnen entspringt durch abermalige Differentiation der RIEMANNsche *Krümmungstensor* vom Range 4 (das ist mit vier Indizes). Das Wort Krümmung hat hier oft zu Mißdeutungen Anlaß gegeben, und man sollte in der Tat diesen Tensor lieber Vektorwirbel nennen. Führt man nämlich die Vektoren des Kompasses in P durch fortgesetzte infinitesimale Parallelverschiebung längs einer nach P zurückführenden Kurve herum, so kehrt der Kompaß nicht in seine Anfangslage, sondern in einer dieser gegenüber verdrehten Lage zurück; die Vektorübertragung ist, wie man sagt, nicht integrabel. Eben diese Drehung gibt der Vektorwirbel an. Wenn er verschwindet, hat der RIEMANNsche Raum oder die EINSTEINsche Welt die besondere homogene, ihr durch die euklidische bzw. MINKOWSKIsche Geometrie zugeschriebene Struktur. Nach dem EINSTEINschen Gravitationsgesetz ist aber dieser Tensor oder vielmehr ein daraus durch die mathematische Operation der Kontraktion hervorgehender Tensor R_{ik} vom Range 2 nicht Null, sondern gleich dem die Materieverteilung kennzeichnenden Energie-Impuls-Tensor, multipliziert mit einer universellen Konstanten, der *Gravitationskonstanten* $\varkappa$.

Ich hoffe, ich habe hier vom Aufbau der Relativitätstheorie ein die wesentlichen Zusammenhänge leidlich getreu wiedergebendes Bild entworfen. Natürlich mußte ich simplifizieren. So einfach, wie es hier erscheinen mag, ist die Beziehung zwischen Erfahrung und Theorie nicht; so leicht hat es die Natur dem Forscher nicht gemacht, von den beobachtbaren Größen zu den Fundamentalgrößen vorzudringen, auf welchen die Theorie aufgebaut werden muß! Lassen Sie es mit dieser allgemeinen Verwahrung sein Bewenden haben, und lassen Sie mich nun zum Ausbau und dann zu dem im Laufe der letzten Jahrzehnte vorgenommenen Erweiterungen der AR übergehen.

Beim Ausbau macht die Einführung von *Dichte und Stromdichte von Energie und Impuls der Gravitation* eine gewisse Schwierigkeit. Es liegt ja geradezu in dem von EINSTEIN erkannten Wesen der Gravitation, daß sich das Gravitationsfeld, die Komponenten Γ des affinen Zusammenhangs, lokal „wegtransformieren" lassen; damit müssen dann auch jene Energie-Impuls-Größen zum Verschwinden kommen. Dennoch ergibt sich durch Integration ein nichtverschwindender Totalbetrag von invarianter Bedeutung [15]. Wie die aktive elektrische Ladung eines Teilchens durch den Fluß definiert werden kann, den das elektrische Feld durch eine das Teilchen umschließende gedachte Hülle sendet, so kann auch die *aktive*, die gravitationsfeld-erzeugende *Masse* als Fluß des Gravitationsfeldes durch eine solche Hülle gewonnen werden. Die Berechnung des Impulsstromes ergibt im elektromagnetischen Feld, daß die aktive Ladung zugleich als passive Ladung auftritt, an der die elektrischen Kräfte angreifen; dasselbe Verfahren liefert im Gravitationsfeld die Gleichheit von aktiver und passiver oder schwerer Masse.

Hiermit hängt eine andere wesentliche Leistung der allgemeinen Relativitätstheorie zusammen: die

Herleitung der Bewegungsgleichungen eines mit Ladung und Masse begabten Teilchens aus den Feldgleichungen. Gestatten Sie mir, mich einer Ausdrucksweise zu bedienen, welche die vierdimensionale Welt mit ihrem metrischen Feld durch eine ziemlich, aber doch nicht völlig ebene zweidimensionale Fläche ersetzt. Darin beschreibt ein Teilchen wie ein Elektron eine feine, aber tiefe Furche. Wir wissen nicht, was diese Furche birgt, doch ihre Böschung ist uns zugänglich. Ohne uns also Gedanken über die innere Konstitution des Teilchens zu machen, kennzeichnen wir das Teilchen durch das dasselbe umgebende lokale Feld. Indem wir ausdrücken, daß dieses Feld sich in den Gesamtverlauf des den Feldgleichungen unterworfenen Feldes einbettet, erhalten wir die Bewegungsgleichungen [*16*].

Es ist nicht richtig, daß das Wirkungsprinzip, aus welchem die EINSTEINschen Gravitationsgesetze entspringen, durch die Forderung der Invarianz (zusammen mit der Forderung einer möglichst niedrigen Differentiationsordnung) *völlig* eindeutig bestimmt ist. Zu der von EINSTEIN ursprünglich angenommenen Wirkungsgröße kann ein zweites, besonders einfach gebautes Glied, mit einer willkürlichen Konstanten Λ multipliziert, hinzugefügt werden. EINSTEIN führte dieses „*kosmologische Glied*" zuerst ein, um den schon von der NEWTONschen Theorie her bekannten Schwierigkeiten zu entgehen, die sich aus der Annahme eines im großen ganzen gleichförmig mit Sternen erfüllten unendlichen Weltraums ergeben. Später hat er dieses sein Kind wieder verleugnet; aber man wird es wohl in der Diskussion der kosmologischen Fragen zulassen müssen, solange kein zwingender formaler oder empirischer Grund für seine Ausschließung ersichtlich ist. Die Ohnmacht der Gravitation im Haushalt der Atome wurde am Beginn durch eine reine Zahl 10^{40} ausgedrückt. Die ungewöhnliche Größenordnung dieser Zahl hat zu Spekulationen Anlaß gegeben, die sie mit dem Mißverhältnis zwischen Ausdehnung oder Masse der Elementarteilchen einerseits, des Universums andererseits, und damit letzten Endes mit der zufälligen Anzahl der in der Welt vorhandenen Teilchen zusammenbringen, oder die in der Gravitationskonstanten $\varkappa$ eine von dem Alter des Universums abhängige und mit ihm veränderliche Größe sehen. Aber dies sind Fragen, deren Diskussion ich gerne meinem Nachfolger an diesem Pult überlasse.

3. Versuche einer einheitlichen Feldtheorie.

Die MAXWELLschen Gleichungen für das elektromagnetische Feld im leeren Raum fließen aus einem sehr einfachen Wirkungsprinzip, das sich sofort von der speziellen in die allgemeine Relativitätstheorie übertragen läßt. Aber beide Felder, das metrische oder Gravitationsfeld und das elektromagnetische, stehen unverbunden nebeneinander. Es entstand natürlicherweise das Desideratum einer einheitlichen Feldtheorie, welche alle Erscheinungen umspannt. Von vornherein verband sich damit die Hoffnung, durch eine solche Theorie auch die atomare Konstitution der Materie erklären zu können. Noch vor der Entstehung der AR und mit Beschränkung auf die elektromagnetischen Erscheinungen hatte GUSTAV MIE 1912 das Programm einer reinen Feldtheorie der Materie entworfen. Das Ziel, das ihm vorschwebte, war, die MAXWELLschen Gleichungen so zu modifizieren, daß sie eine oder wenige singularitätenfreie statische kugelsymmetrische Lösungen besitzen; diese würden dann dem Elektron und den Atomkernen der in der Natur vorkommenden Elemente entsprechen. DAVID HILBERT hatte zur selben Zeit, als EINSTEIN seine Grundgleichungen des Gravitationsfeldes aufstellte, dieses MIEsche Programm auf die allgemeine Relativitätstheorie übertragen [*17*]. EINSTEIN selber war weise genug, in seiner Fassung der Gravitationsgleichungen dem Beispiel der NEWTONschen Theorie zu folgen: wie hier in der Gleichung $\Delta\Phi = k\varrho$ für das Gravitationspotential Φ auf der rechten Seite die (mit der Gravitationskonstanten k multiplizierte) Massendichte ϱ auftritt, so stellte er auf die rechte Seite seiner Gleichungen (deren linke der kontrahierte Krümmungstensor ist) einen Energie-Impuls-Tensor, der, wie er sagt, „eine formale Zusammenfassung aller Dinge war, deren Erfassung im Sinne einer Feldtheorie noch problematisch war. Natürlich", fügt er hinzu, „war ich keinen Augenblick im Zweifel, daß diese Fassung nur ein Notbehelf war" [*18*]. Viele Versuche sind seither unternommen worden, zu einer einheitlichen Feldtheorie zu gelangen, insbesondere auch von EINSTEIN selbst. Ich glaube nicht, daß das Ziel erreicht ist, oder auch nur, daß wir dem Ziel in den letzten drei Dezennien wesentlich näher gekommen sind. Jede die Gravitation mitumfassende Feldtheorie, welche die Atome nicht als Fremdkörper einführt, steht dem Rätsel der reinen Zahl 10^{40} gegenüber, des Verhältnisses von elektrischem und Gravitations Radius des Elektrons. Dennoch möchte ich mit einer kurzen Übersicht über diese Versuche mein Referat beschließen. Ich strebe keine Vollständigkeit an. Insbesondere soll die von EINSTEIN eine Zeitlang verfolgte, aber dann aufgegebene Idee des Fernparallelismus unberücksichtigt gelassen werden, weil sie fast einer Rückkehr zur SR gleichkommt. Im übrigen unterscheide ich drei Gruppen durch die Stichworte: Eichinvarianz, Affintheorie, Preisgabe der Symmetrie.

a) Eichinvarianz. Die mathematische Aufgabe, als welche MIE und HILBERT das Problem angriffen, war die Bestimmung aller Invarianten, die von den vier elektromagnetischen Potentialen φ_k und ihren ersten Ableitungen sowie von den 10 Gravitationspotentialen g_{ij} und deren ersten und zweiten Ableitungen abhängen. Unter ihnen, nahmen sie an, müsse sich die Wirkungsgröße befinden. Aber die Auswahl war groß; es galt, ein Prinzip zu finden, das darunter eine engere, womöglich eine eindeutige Wahl traf. Sprecher glaubte 1918 dies im Prinzip der Eichinvarianz gefunden zu haben [*19*]. Beim Herumfahren eines Vektors längs einer geschlossenen Kurve durch fortgesetzte infinitesimale Parallelverschiebung kehrt dieser im allgemeinen in einer andern Lage zurück; seine Richtung hat sich geändert. Warum nicht auch seine Länge? Dies war mein Einfall. Ich nahm also, an die Relativität der Länge glaubend, an, daß ein willkürliches Eichmaß zur Messung der Längen von Linienelementen lokal festgelegt werden muß, und daß wohl eine infinitesimale Übertragung desselben von Weltpunkt zu Weltpunkt statt hat, daß aber diese so wenig integrabel zu sein braucht wie die Parallelübertragung der Richtungen von Vektoren. Es zeigte sich dann, daß zur Beschreibung des metrischen Feldes neben dem Tensorfeld g_{ij} noch ein Vektorfeld φ_k nötig ist, daß aber Invarianz statt hat bei gleichzeitiger

Ersetzung der g_{ij} durch $e^{-\lambda} \cdot g_{ij}$ und der φ_k durch $\varphi_k + \frac{\partial \lambda}{\partial x_k}$, wo λ eine willkürliche Ortsfunktion in der Welt ist („Eichinvarianz"). Da man weiß, daß eine solche Willkür wie die durch die Substitution

$$\varphi_k \to \varphi_k + \frac{\partial \lambda}{\partial x_k} \qquad (2)$$

ausgedrückte in den elektromagnetischen Potentialen steckt — eine Erfahrung, welche MIE und HILBERT beim Aufbau ihrer Theorie ausdrücklich verleugnet hatten, — schien es plausibel, diese φ_k mit den (in einer unbekannten kosmischen Einheit gemessenen) elektromagnetischen Potentialen zu identifizieren. In der Tat ergab das Wirkungsprinzip, das durch die Forderung der Eichinvarianz wenigstens nahezu eindeutig festgelegt ist, daß die φ_k diese Rolle spielen. Die resultierenden Gleichungen sind den EINSTEIN-MAXWELLschen Gleichungen genügend ähnlich, um das erkennen zu lassen, weichen aber doch genügend davon ab, um der Hoffnung Raum zu geben, daß sie singularitätenfreie statische kugelsymmetrische Lösungen gestatten. Die entgegenstehenden mathematischen Schwierigkeiten haben es freilich verunmöglicht, dies zu entscheiden; aber in keiner der noch zu erwähnenden konkurrierenden Theorien steht es damit besser, und darum sind sie alle physikalisch ohne Frucht geblieben. EINSTEIN machte sogleich den Einwand, daß mein Prinzip von der Nichtintegrabilität der Längenübertragung mit der absoluten Stabilität der Frequenzen von Spektrallinien in Widerspruch stehe. Die Definition des Maßfeldes im Äther mit Hilfe von wirklichen Maßstäben und Uhren kann natürlich nur als eine vorläufige Anknüpfung an die Erfahrung gelten. Erst wenn die physikalischen Wirkungsgesetze aufgestellt sind, muß man aus ihnen ableiten, in welcher Beziehung die an jenen Körpern abgelesenen Meßresultate zu den Fundamentalgrößen der Theorie stehen. Die Erfahrungen, auf die sich EINSTEIN mir gegenüber berief, zeigen gewiß, daß die physikalisch gemessenen Längen nicht der kongruenten Verpflanzung von Strecken folgen, die zum Fundament meiner Theorie gehört. Ich habe keine Lust, diese Theorie, an die ich längst nicht mehr glaube, zu verteidigen. Aber ich konnte damals doch mit Recht auf das Faktum hinweisen, daß sie im Krümmungsradius der Welt, sozusagen nachträglich, ein absolutes lokales Eichmaß liefert, auf das sich spektrale Frequenzen und andere Längengrößen einstellen können und vielleicht gemäß dem geltenden Wirkungsprinzip wirklich einstellen.

Heute, nach Einführung der SCHRÖDINGER-DIRACschen ψ_q durch die Quantentheorie, glaube ich, können wir mit großer Bestimmtheit den Finger auf den Punkt legen, in welchem meine Theorie irrte: die Eichinvarianz verbindet die elektromagnetischen Potentiale nicht mit den g_{ij} der Gravitation, sondern mit den ψ_q des Materiefeldes. Das konnte ich freilich 1918 nicht wissen! Damals waren diese ψ noch völlig unbekannt. Im Rahmen der AR wird der willkürliche Phasenfaktor $e^{-i\lambda}$, der den ψ anhaftet, von einer Konstanten zu einer willkürlichen Ortsfunktion in der Welt. Es muß dann notwendig dem Differentialoperator $\partial/\partial x_k$, um ihm eine invariante Bedeutung zu sichern, die allgemeinere Form $\frac{\partial}{\partial x_k} + i\varphi_k$ gegeben werden, wobei die φ_k ein Vektorfeld bilden: verwandelt man ψ_q in $e^{-i\lambda} \cdot \psi_q$, so geht φ_k in $\varphi_k + \frac{\partial \lambda}{\partial x_k}$ über. Genau diese Vorschrift ist es aber, nach welcher die DIRACsche Theorie die Einwirkung des elektromagnetischen Feldes auf das Elektron wiedergibt, wenn φ_k als das mit $\varepsilon/\hbar$ multiplizierte elektromagnetische Potential gedeutet wird. Hier sind wir nicht im Gebiet der Spekulation, sondern der Erfahrung, und die Einheit, in welcher die φ_k gemessen werden, ist nicht eine unbekannte kosmische, sondern eine bekannte atomare Größe. Man sollte freilich jetzt lieber von Phasen- statt von Eichinvarianz sprechen.

Die durch (2) zum Ausdruck kommende Unbestimmtheit in den elektromagnetischen Potentialen φ_k ist jedenfalls, auch wenn man die φ mit keinen anderen Größen verknüpft, eine gesicherte Tatsache, und die Invarianz gegenüber der Substitution (2) mit der willkürlichen Ortsfunktion λ ist auf die gleiche Weise mit dem Gesetz von der Erhaltung der Ladung verknüpft wie die Invarianz gegenüber Koordinatentransformation mit dem der Erhaltung von Energieimpuls. Dem Umstand, daß nicht die Potentiale φ_k, sondern nur die daraus abgeleiteten Feldgrößen

$$f_{ik} = \frac{\partial \varphi_k}{\partial x_i} - \frac{\partial \varphi_i}{\partial x_k}$$

eine physikalische Bedeutung haben, kann man innerhalb des MIE-HILBERTschen Schemas Rechnung tragen und dadurch wenigstens eine gewisse Einschränkung in der Auswahl der zur Verfügung stehenden invarianten Wirkungsgrößen erzielen. So verfuhr BORN [*20*]. Statt der MAXWELLschen Wirkungsgröße L schlägt er insbesondere eine vor, welche unter Vernachlässigung der Gravitation so lautet:

$$\sqrt{1 + 2\beta L} - 1 \qquad (3)$$

(β ist eine kleine Konstante). Damit errang er wenigstens einen partiellen Erfolg, insofern die statischen kugelsymmetrischen Lösungen seiner Gleichungen zwar nicht singularitätenfrei sind, aber doch zu einer endlichen Energie führen.

KALUZA hatte 1921 den Gedanken, ob sich nicht die Invarianz gegenüber der Substitution (2) als Erweiterung der Invarianz gegenüber Koordinatentransformation auf eine fiktive 5. Weltkoordinate x_0 deuten ließe [*21*]. Er machte die spezielle Annahme, daß die Koordinaten x_1, x_2, x_3, x_4 sich wie bei EINSTEIN nur untereinander transformieren, während für x_0 ein beliebiges Transformationsgesetz von der besonderen Form

$$x_0 \to x_0 + \lambda(x_1, x_2, x_3, x_4) \qquad (4)$$

zugelassen wird. Setzt man dann eine quadratische Differentialform der fünf Variablen für die Beschreibung des metrischen Feldes an,

$$ds^2 = \sum_{\alpha,\beta} g_{\alpha\beta}\, dx_\alpha\, dx_\beta \quad (\alpha, \beta = 0, 1, 2, 3, 4),$$

so stellt sich heraus, daß g_{00} eine Invariante ist, die KALUZA durch $g_{00} = 1$ normiert (dies scheint zulässig, wenn man annimmt, daß nicht die Form ds^2 selber, sondern nur die Gleichung $ds^2 = 0$ eine physikalische

Bedeutung hat), während die vier Größen $\varphi_k = g_{k0}$ sich gegenüber Transformationen von x_1, x_2, x_3, x_4 so wie die Komponenten eines Vektors verhalten, bei der Transformation (4) aber die Substitution (2) erleiden. (Der Index k läuft hier immer nur von 1 bis 4.) Man macht die zusätzliche Annahme, daß alle $g_{\alpha\beta}$ von x_0 unabhängig sind. Man kommt so in der Tat auf natürliche Weise auf die MAXWELL-EINSTEINschen Feldgleichungen, und die Bewegung nicht nur von ungeladenen, sondern auch von geladenen Teilchen verläuft längs geodätischer Weltlinien. Dennoch liegt hier kaum mehr vor als eine formale Zusammenfassung der beiden Felder, die zu keinem Erkenntnisfortschritt führen kann, da sie dem von BORN korrigierten MIE-HILBERTschen Schema keinerlei Einschränkungen auferlegt. Eine ansprechende geometrische Einkleidung des Formalismus liefert die projektive Geometrie. Anstatt der vier Weltkoordinaten x_k benutzt der projektive Geometer die durch $x_k = X_k/X_0$ eingeführten fünf homogenen Koordinaten X_α. Ein Punkt bestimmt nur die Verhältnisse dieser Koordinaten; indem man diese selbst festlegt, erteilt man einem Punkt ein *Gewicht*. Setzt man $X_0 = e^{x_0}$, so bedeutet der Umstand, daß x_1, x_2, x_3, x_4 sich nur untereinander transformieren, dies, daß das Zusammenfallen von Punkten verschiedenen Gewichts eine invariante Bedeutung hat, während die Transformation (4) die Willkür des Gewichts ausdrückt. Neben KALUZA hat OSKAR KLEIN diese Ansätze verfolgt; er ist später, im Zusammenhang mit quantentheoretischen Erwägungen, dazu übergegangen, die Annahme, daß Zustandsgrößen und Transformationsfunktionen einschließlich λ von x_0 nicht abhängen, dahin zu verallgemeinern, daß sie periodische Funktionen von x_0 sind, mit einer durch die universellen Naturkonstanten vorgeschriebenen Periode [22]. Die angedeutete projektive Form wurde, von 1930 ab, ausgebaut von VAN DANTZIG und SCHOUTEN, ferner von PAULI. Sie war in einer etwas anderen Gestalt schon vorher von O. VEBLEN und B. HOFFMANN entwickelt worden. PAULI hat auch die Ausdehnung der Theorie auf die ψ-Größen verfolgt. EINSTEIN zusammen mit W. MAYER hat in den gleichen Jahren eine nahverwandte Theorie konstruiert, deren Formalismus gleichfalls fünf unabhängige Variable benutzt [23]. Aus jüngster Zeit wären Arbeiten von PASCUAL JORDAN zu erwähnen.

b) Reine und gemischte Affintheorien. EDDINGTON dehnte meine „suggestion", daß der Krümmungsradius der Welt das Eichmaß liefert, von der skalaren Krümmung auf den Krümmungstensor R_{ij} aus und wurde so dazu geführt, der Welt von Hause aus keine Metrik, sondern einen durch 40 Größen $\Gamma^i_{kl} = \Gamma^i_{lk}$ ausgedrückten affinen Zusammenhang zuzuschreiben. In der Tat entstehen die Krümmungsgrößen R_{ij} aus ihnen allein. EINSTEINs Gravitationsgleichungen im leeren Raum, $R_{ij} = 0$, hatten sich durch Hinzufügung des kosmologischen Gliedes in

$$R_{ij} = \Lambda \cdot g_{ij}$$

verwandelt; diese werden nun für EDDINGTON aus einem Naturgesetz zu einer *Definition* des metrischen Tensors g_{ij}. Diesen Gedanken aufgreifend, wies EINSTEIN alsbald darauf hin, daß dann die Gleichungen, welche die Komponenten Γ des affinen Zusammenhangs durch die g_{ij} ausdrücken, nicht länger als Definitionen aufgefaßt werden können, sondern aus einem Wirkungsprinzip abzuleitende Naturgesetze sein müssen. Und er fand in der Tat, daß sie sich ergeben, wenn man die einfachste Invariante, die im Rahmen der EDDINGTONschen Affintheorie möglich ist, als Wirkungsgröße wählt; das ist die Quadratwurzel aus der Determinante der R_{ij} [24]. Es treten dabei auch Terme auf, die sich als elektromagnetisches Potential deuten ließen, und die resultierenden Gleichungen sind einschließlich der kleinen kosmologischen Glieder mit ihren numerischen Koeffizienten genau mit den Feldgleichungen meiner metrischen Theorie identisch. Es ist mir schleierhaft, worauf diese merkwürdige Übereinstimmung beruht [25].

In dem Dilemma, ob man der Welt ursprünglich eine metrische oder eine affine Struktur zuschreiben soll, ist vielleicht der beste Standpunkt der neutrale, der sowohl die g wie die Γ als unabhängige Zustandsgrößen behandelt. Dann werden die beiden Sätze von Gleichungen, welche sie verbinden, zu Naturgesetzen, ohne daß die eine oder andere Hälfte als Definitionen eine bevorzugte Stellung bekommen. In der Tat zeigte EINSTEIN, daß ein Wirkungsprinzip von besonders simpler Bauart hier dieselben Gesetze liefert wie seine ursprüngliche rein-metrische Theorie [26]. Freilich führt dieser neutrale Standpunkt auch nicht über die rein-metrische Theorie hinaus, selbst nicht bei Einbeziehung des ψ-Feldes der Elektronen.

c) Preisgabe der Symmetrie. Ein beliebiger Tensor h_{ij} vom Range 2 spaltet in invarianter Weise in einen symmetrischen und einen schiefsymmetrischen Bestandteil:

$$h_{ij} = g_{ij} + f_{ij}; \qquad g_{ji} = g_{ij}, \qquad f_{ji} = -f_{ij}. \tag{5}$$

Diese kann man bzw. als Gravitationspotential und elektromagnetisches Feld deuten, die so zu einem einzigen Tensor h_{ij} zusammengefaßt erscheinen. Bei der Aufstellung seiner besonderen Wirkungsgröße des elektromagnetischen Feldes (3) war schon BORN von diesem Gedanken ausgegangen [27]. Systematisch haben dann EINSTEIN und SCHRÖDINGER untersucht, was geschieht, wenn man für die Γ^i_{kl} (wie auch für die g_{ij}) die Symmetrieannahme $\Gamma^i_{kl} = \Gamma^i_{lk}$ fallen läßt. SCHRÖDINGER stellt sich dabei auf den rein-affinen, EINSTEIN auf den neutralen metrisch-affinen Standpunkt [28]. SCHRÖDINGER glaubt durch seine Theorie zum mindesten eine Art von Mesonen mitzuumfassen, erwartet aber wohl, daß das ganze System von Feldgleichungen erst dem Quantisierungsprozeß unterworfen werden muß, ehe es die atomaren Erscheinungen zu erklären fähig ist. EINSTEIN nährt in seinem Busen noch immer die kühne Hoffnung, daß die Feldgleichungen selber ohne quantentheoretische Umdeutung dies leisten.

Ich gestehe, daß ich als Mathematiker mir von einer so formalen Verallgemeinerung wie dem Fallenlassen der Symmetriebedingungen nichts versprechen kann. Die Symmetrie der g_{ij} und der Γ^i_{kl} hat eine über das Formale weit hinausgehende Bedeutung, nämlich die, daß die *Natur* der Metrik und des affinen Zusammenhangs *eine* und allerorten die gleiche ist. Statt an der Symmetrie zu rütteln, sollte man nach einer andersartigen reicheren Struktur fahnden, deren Natur aber wiederum überall die gleiche sein müßte.

Wenn man der Mathematik eine für die Aufstellung physikalischer Theorien wichtige Lehre entnehmen kann, so ist es die, daß nur Größen, die unter ihrem spezifischen Transformationsgesetz *unzerlegbar* sind, eine einheitliche physikalische Entität darstellen; eine solche Größe ist der symmetrische und der schiefsymmetrische Tensor, g und f, aber nicht ihre Zusammenfassung (5). PAULI formuliert dieses Prinzip so: Was Gott getrennt hat, soll der Mensch nicht zusammenfügen. Durch die Preisgabe der Symmetrie ist die Mannigfaltigkeit der als Wirkungsgröße zur Verfügung stehenden Invarianten gewaltig gewachsen, während doch das Bestreben sein sollte, die Möglichkeiten einzuschränken. Offenbar sind wir doch nicht klug genug, um durch reines Denken — „aus dem hohlen Bauch", glaube ich, war früher der Ausdruck der Münchener Physiker dafür — die universelle Struktur der Welt und die sie beherrschenden Feldgesetze zu finden. Und ich glaube auch nicht, daß unser gegenwärtiges Wissen über die Wellenfelder der Elementarteilchen dafür irgend zureichend ist. Hier wie anderswo ist dafür gesorgt, daß unsere Bäume nicht in den Himmel (oder in die Hölle) wachsen.

Um aber nicht mit bloßer Kritik zu enden, will ich zum Schluß noch mein Scherflein zur Spekulation beitragen [*29*]. Die MAXWELLschen Gleichungen, in denen die Potentiale φ_k als die unabhängigen Zustandsgrößen figurieren, sind linear, und es besteht Invarianz gegenüber der Substitution (2). EINSTEINs Gravitationstheorie ergibt für die 10 Potentiale $\gamma_{ij}=\gamma_{ji}$ eines unendlich schwachen Gravitationsfeldes ebenfalls lineare Gleichungen, und diese sind invariant gegenüber der zu (2) analogen Substitution

$$\gamma_{ij} \rightarrow \gamma_{ij} + \left(\frac{\partial \xi_i}{\partial x_j} + \frac{\partial \xi_j}{\partial x_i}\right)$$

mit vier willkürlichen Funktionen ξ_i; eine Invarianzeigenschaft, die die Koordinateninvarianz der strengen Gleichungen widerspiegelt. Vielleicht sollte man zunächst einmal auf diesem *linearen* Niveau nach einer Vereinigung von γ_{ij} und φ_k fahnden. Die Quantentheorie läßt die elektromagnetischen Potentiale φ_k als ein vom Prinzip der Phaseninvarianz gefordertes Anhängsel an die das Elektron darstellenden Feldgrößen ψ erscheinen. Frage: Sind die γ_{ij} in analogem Sinne ein Appendix an das Wellenfeld X eines unbekannten Elementarteilchens „Graviton"? Erst nachdem diese Frage beantwortet ist, sollte man jenen Übergang zur nichtlinearen Theorie versuchen, bei welchem die γ_{ij} sich in die wirklichen Zustandsgrößen g_{ij} des metrischen Feldes (zurück-)verwandeln; ein Prozeß, der dann notwendig auch die φ_k (samt den ψ und dem unbekannten X) mitergreifen würde und so in organischer Weise eine nichtlineare Theorie des MAXWELLschen Feldes ergäbe. Ich bin weit davon entfernt, dieses Programm durchführen zu können.

Literatur.

[*1*] Über die Entwicklung von EINSTEINS Ideen vgl. seine „Autobiographical Notes" in Albert Einstein Philosopher-Scientist, Bd. VII der Library of Living Philosophers, herausgeg. von PAUL A. SCHILPP. Evanston, Ill., 1949. (In der Folge zitiert als AE.) — [*2*] Enquiry concerning Human Understanding, Book II, Chap. 13, Sections 7—10. — [*3*] „Dialogo sopra i dui massimi sistemi del mondo", in Bd. VII, S. 198, der Opere, Edizione nazionale. Florenz 1890 bis 1909. Neudruck 1929—. — [*4*] Siehe z. B. S. 10—12 der englischen Ausgabe der Philosophiae naturalis principia mathematica von F. Cajori, Berkeley, Calif., 1934, zweiter Druck 1946; auch das den Definitionen I, II und IV folgende Scholium auf S. 6—7; betreffs der Beziehung zu NEWTONS Theologie: COTES' Vorrede zur 2. Aufl. der Principia, S. XXXII, (Cajori) und Principia, S. 546, (Cajori); ferner NEWTONS Optics, S. 370, (Ausgabe von E. T. Whittaker, London 1931) und der Briefwechsel zwischen Leibniz und S. Clarke in G. W. LEIBNIZ, Philosophische Schriften, Bd. VII, S. 352—440, (Ausgabe von Gerhardt). — [*5*] Principia, S. 419, Ausgabe von Cajori. — [*6*] BRUNOS Del infinito universo e mondi erschien 1584, die erste Aufl. von NEWTONS Principia 1687. — [*7*] Opere, ed. naz., Bd. VII, S. 212—214. — [*8*] Über die Entwicklung bis 1920 orientiert am besten der Artikel V 19: Relativitätstheorie von W. PAULI, in der Enzyklopädie der mathematischen Wissenschaften, Bd. V, Teil 2, S. 539—775, 1904—1922 (zitiert als „Pauli, Enc."). — [*9*] Ich entnehme diese Zahlangaben dem schönen Artikel von MAX VON LAUE über Inertia and Energy, AE, S. 503—533. Andere frühe Hinweise auf kernphysikalische Anwendungen siehe bei PAULI, Enz., S. 681. — [*10*] Die grundlegenden Arbeiten von HEISENBERG und von ihm, BORN und JORDAN erschienen in der Z. Physik **33—35** (1925/26); die von E. SCHRÖDINGER sind zusammengefaßt in den Abhandlungen zur Wellenmechanik. Leipzig 1927. — [*11*] Proc. Roy. Soc. Lond. A **117**, 610; **118**, 351 (1928). — [*12*] Siehe AE, S. 52. — [*13*] Siehe die genaue Diskussion in dem Vortrag über Relativity Effects in Planetary Motion, den G. M. CLEMENCE im Rahmen des Princetoner Symposium aus Anlaß von EINSTEINS siebzigstem Geburtstag, hielt: Proc. Amer. Philos. Soc. **93**, 532—534 (1949). — [*14*] Siehe für MILNE: AE, S. 415—435; für BIRKHOFF: Proc. Nat. Acad. Sci. USA **29**, 231 (1943) und die Bemerkungen von H. WEYL über mögliche lineare Theorien der Gravitation in Amer. J. Math. **66**, 591—604 (1944). — [*15*] EINSTEIN: Sitzgsber. preuss. Akad. Wiss., Math.-naturwiss. Kl. **1918**, 448. — [*16*] WEYL, H.: Raum, Zeit, Materie, 5. Aufl. Berlin 1923, § 38. Mit viel größerer Sorgfalt, auch zur Bestimmung der Wechselwirkung mehrerer Teilchen, wurde dieser Weg dann in mehreren Arbeiten von EINSTEIN und INFELD beschritten; vgl. die letzte abschließende Arbeit, die im Canadian J. Math. **1**, 209—241 (1949) erschien. — [*17*] MIE, G.: Ann. Physik **37**, **39**, **40** (1912/13). — HILBERT, D.: Die Grundlagen der Physik. Nachr. Ges. Wiss. Göttingen, Math.-physik. Kl. **1915** u. **1917**. — [*18*] AE, S. 74. — [*19*] Vgl. die Darstellung in Raum, Zeit, Materie, 5. Aufl., S. 298—308. — [*20*] BORN, M.: Proc. Roy. Soc., Lond. A **143**, 410 (1934). — SCHRÖDINGER, E.: Proc. Roy. Soc., Lond. A **150**, 465 (1935). — [*21*] KALUZA: Sitzgsber. preuss. Akad. Wiss., Physik.-math. Kl. **1921**, 966. — [*22*] KLEIN, O.: Z. Physik **37**, 895 (1926); **46**, 188 (1927). — Vgl. ferner: O. KLEIN, Ark. Mat., Astronom. Fysik, Ser. A **34**, Nr 1 (1946). Weitere Publikationen stehen in Aussicht. — [*23*] SCHOUTEN u. VAN DANTZIG: Proc. Amsterdam **34**, 1398 (1931). — Z. Physik **78**, 639 (1932). — DANTZIG, VAN: Math. Ann. **106**, 400 (1932). — PAULI, W.: Ann. Physik (5) **18**, 305—372 (1933). — VEBLEN, O., u. B. HOFFMANN: Projective relativity. Physic. Rev. **36**, 810—822 (1930). — VEBLEN, O.: Projektive Relativitätstheorie, Ergebnisse der Mathematik, Bd. II/1. Berlin 1933. — EINSTEIN, A., u. W. MAYER: Sitzgsber. preuss. Akad. Wiss., Math.-naturwiss. Kl. **1931**, 541—557; **1932**, 130—137. — [*24*] EINSTEIN: Sitzgsber. preuss. Akad. Wiss., Math.-naturwiss. Kl. **1923**, 32, 76, 137. — EDDINGTON, A. S.: Mathematical Theory of Relativity, 2. Aufl., Note 14. — [*25*] Vgl. zu allen diesen Ausführungen H. WEYL, Geometrie und Physik. Naturwiss. **19**, 49—58 (1931). — [*26*] EINSTEIN, A.: Sitzgsber. preuss. Akad. Wiss., Math.-naturwiss. Kl. **1925**, 414. — WEYL, H.: Physic. Rev. **77**, 699—701 (1950). — [*27*] BORN, M.: Proc. Roy. Soc., Lond. A **144**, 425—451 (1934). — [*28*] Von SCHRÖDINGER zitiere ich die drei Arbeiten betitelt „The final affine Field Laws" in Proc. Roy. Irish Acad. A **51**, 163—171, 205—216; **52**, 1—9 (1947/48); voraufgehende Arbeiten sind dort angeführt. Besonders nützlich ist die Übersicht der verschiedenen Theorien in der 2. Abh. EINSTEINS letzte Version seiner Theorie findet man im Appendix II von The Meaning of Relativity, 3. Aufl., S. 109—147. Princeton, N. J. 1949. Vorbereitet war sie durch die Arbeiten von EINSTEIN und STRAUS in ~~Amer. J.~~ Ann. Math. **46**, 578 (1945) und **47**, 731 (1946). Nach Abschluß dieses Artikels erschien E. SCHRÖDINGERS Buch Space-Time Structure. Cambridge 1950. — [*29*] WEYL, H.: Amer. J. Math. **66**, 602 (1944).

150.

Ramifications, old and new, of the eigenvalue problem

Bulletin of the American Mathematical Society 56, 115—139 (1950)

Since this is a lecture dedicated to the memory of Josiah Willard Gibbs let me start with that purely mathematical discovery which Gibbs contributed to the theory of Fourier series. Fourier series have to do with the eigenvalues and eigenfunctions of the oldest, simplest, and most important of all spectrum problems, that of the vibrating string. In preparing this lecture, the speaker has assumed that he is expected to talk on a subject in which he had some first-hand experience through his own work. And glancing back over the years he found that the one topic to which he has returned again and again is the problem of eigenvalues and eigenfunctions in its various ramifications. It so happens that right at the beginning of my mathematical career I wrote two papers on what we now call the Gibbs phenomenon.

1. **Gibbs phenomenon.** Take a simple periodic function with a discontinuity, for example, the function $1^0(x)$ of period 2π which equals 0 for $-\pi<x<0$ and 1 for $0<x<\pi$. In a letter to the editor of Nature published on April 27, 1899, Gibbs, correcting a statement in a previous letter, pointed out that the limit of the graphs of the partial sums $y=1_n^0(x)$ of the Fourier series of $1^0(x)$ includes not only the vertical ascent from the level 0 to the level 1 at $x=0$, but extends vertically beyond it by a specific amount. A. A. Michelson had started the discussion in Nature by criticizing the way in which the mathematicians are wont to describe the limit of the sequence of those partial sums; he had pleaded for adding to the two horizontal levels the vertical precipice. Today we find in the notion of uniform convergence the most adequate analysis of the phenomenon. Introduce the sinus integral

$$\mathrm{Si}\,(x) = \frac{1}{\pi}\int_{-\infty}^{x} \frac{\sin \xi}{\xi}\, d\xi$$

and consider a closed interval I, say $-\pi/2 \leqq x \leqq \pi/2$, containing only the one discontinuity at $x=0$. It is, of course, not true that the difference between $1^0(x)$ and the nth partial sum $1_n^0(x)$ converges

The twenty-second Josiah Willard Gibbs lecture delivered at Columbus, Ohio, December 28, 1948, under the auspices of the American Mathematical Society; received by the editors January 17, 1949.

uniformly to zero in I, but it is true that $1_n^0(x) - \mathrm{Si}\,(nx)$ does so. Thus the graph of $1_n^0(x)$ for large n is essentially that of the undulating function Si (x) compressed at the ratio $1:n$ in the x-direction.

Instead of direct summation one can apply to infinite series, in particular to Fourier series, other methods of summation. Let me mention only one here: in the Fourier series $\sum_{n=-\infty}^{\infty} a_n e^{inx}$ add the factor $e^{-n^2 t}$ to the nth term and then let the positive parameter t = time in the resulting sum

$$\sum_n a_n e^{-n^2 t} e^{inx}$$

converge to zero. Since that sum is a solution of the equation of heat conduction

$$\partial^2 u/\partial x^2 - \partial u/\partial t = 0,$$

I call this the heat conduction summation. The Gibbs phenomenon for this summation is ruled by the function

$$\mathrm{Er}\,(x) = \frac{1}{\pi^{1/2}} \int_{-\infty}^{x} e^{-\xi^2} d\xi$$

in the same sense as it is ruled by Si (x) for the direct summation. This is an immediate consequence of the fact that Er $(x/2t^{1/2})$ is actually a solution—though not a periodic one—of the heat equation.

In the two papers just mentioned **[1]**[1] I considered the Gibbs phenomenon for a certain general type of summation methods. My chief concern was with the simplest two-dimensional case of eigenfunctions, namely Laplace's expansion of functions on a sphere in terms of spherical harmonics. Nothing new occurs if the function has a discontinuity along a smooth line with a continuous tangent. But quite an impressive mountain landscape develops in the neighborhood of a point where this line makes an angle. There was one specific one-dimensional problem which attracted my attention. It deals with a circular metal ring consisting of two halves of different conductivities α and β. Assume the normalization $\alpha + \beta = 1$. If one of the halves has a temperature of 100° C. at the time $t = 0$, the other of 0°, how will the temperature level off in the progress of time? The solution can be easily expressed by means of the function Er, and this gives the Gibbs phenomenon for the heat conduction summation of the corresponding eigenfunction expansion. The Gibbs phenomenon for direct sum-

[1] The letters (A), (B), · · · refer to a number of notes printed at the end of the paper (and not included in the actual lecture as delivered at the Columbus meeting). The bold face numerals **[1]**, · · · refer to the bibliography.

mation is a more subtle question because the distribution of the eigenvalues λ depends on the arithmetical character of the numbers α and β. Only if they are rational, the problem may be settled by fairly direct computation (**A**). The result would carry over to an arbitrary irrational α if it were true that α can be approximated by a sequence of fractions a_n/c_n $(n=1, 2, \cdots)$ such that

$$n(\alpha - a_n/c_n) \to 0, \qquad c_n/n \to 0 \qquad \text{for } n \to \infty.$$

It is simple enough to show that this can be done, and thus to determine the Gibbs phenomenon for arbitrary conductivities α, β. Our lemma on Diophantine approximations proves at the same time the equidistribution mod. 1 of the multiples $n\alpha$ of an irrational number α.

When, not so long after, I learned through Felix Bernstein about the problem of mean motion in Lagrange's linear theory of perturbation for the planetary system, a problem P. Bohl had connected with that of equidistribution mod. 1, I remembered this investigation and tried to settle the question of equidistribution in a more general form [2]. This is an example of how experience in one field of mathematics may give one the lead in an entirely different field. It is chiefly for this lesson that I have mentioned here this early work of mine, by which, in a very modest way, I paid homage to the genius of Gibbs about forty years ago.—Incidentally, Fourier series provided the basis for the analytic method which I brought to bear on the general problem of deciding whether a given sequence of real numbers $\xi_1, \xi_2, \cdots$ is equidistributed mod. 1. Such equidistribution can be formulated as a statement concerning the mean value of any Riemann integrable function $f(x)$ of period 1 for the argument values $x=\xi_1, \xi_2, \cdots$; and the gist of the method lies in the observation that verification of the statement for the special periodic functions $e^{2\pi nix}$ $(n=0, \pm 1, \pm 2, \cdots)$ is sufficient.

2. **Limit circle and limit point.** As one knows, Fourier series were generalized by Sturm and Liouville so as to cover the eigenvalues λ and eigenfunctions $\phi(s)$ of the self-adjoint differential equation

$$\text{(1)} \qquad L_\lambda(\phi) \equiv \left\{\frac{d}{ds}\left(p(s)\frac{d\phi}{ds}\right) - q(s)\phi(s)\right\} + \lambda\phi(s) = 0,$$

subject to a real linear boundary condition at either end of the interval $0 \leqq s \leqq l$. The coefficients $p(s) > 0$ and $q(s)$ are given real continuous functions in this interval. Let the abbreviation ϕ' be used for $p(s)d\phi/ds$. A real linear boundary condition for $s=l$ is of the form

(2) $$\phi'(l) - h\cdot\phi(l) = 0$$

with a real constant h (not excluding $h=\infty$). For a solution ϕ of $L_\lambda(\phi)=0$ and a solution ϕ_* of $L_{\lambda_*}(\phi_*)=0$ we have the simple Green's formula

(3) $$\left[\phi\phi_*' - \phi_*\phi'\right]_0^l = (\lambda - \lambda_*)\int_0^l \phi\phi_* ds.$$

It shows that eigenfunctions ϕ, ϕ_* belonging to two distinct eigenvalues λ, λ_* are necessarily orthogonal, $\int_0^l\phi\phi_*\cdot ds=0$. On taking $\lambda_*=\bar\lambda$, $\phi_*=\bar\phi$ one finds that for a non-real λ the function $\phi(s)$ cannot satisfy a real linear boundary condition at both ends without vanishing identically; for under these circumstances our equation would give $\int_0^l\phi\bar\phi ds=0$. The positive-definite character of the integrand $\phi\bar\phi$ is decisive here.

The spectrum of the eigenvalues is discrete provided the differential equation is regular at both ends, that is, provided $p(s)$, $q(s)$ are continuous and $p(s)$ actually positive throughout the closed interval $0\leqq s\leqq l$. If we make this assumption only for the right-open interval $0\leqq s<l$, as we shall now do, then the end $s=0$ stays regular, but the end $s=l$ is (possibly) singular. Let us throw the singular end into $s=+\infty$. Under these circumstances one must expect that a continuous spectrum will appear side by side with the point spectrum. Moreover it seems that sometimes a boundary condition is required at the singular end, just as it would be for a regular one, but sometimes not.

The very first result by which I added my mite to our stock of mathematical knowledge had to do with the clarification of this issue [3]. Since one cannot vouch that the spectrum will not cover the entire real λ-axis, I had the simple idea (not as trivial at that time as it has now become) to determine Green's function $G(s, t)$ neither for $\lambda=0$ nor for any real λ, but for a λ in the upper half-plane, $\Im\lambda>0$, for example for $\lambda=i$. Let $\eta(s)$, $\theta(s)$ designate the two solutions $\phi(s)=\phi(s;\lambda)$ of (1) determined by the initial conditions

$$\eta(0) = 1,\ \eta'(0) = 0 \quad\text{and}\quad \theta(0) = 0,\ \theta'(0) = 1$$

respectively, and then consider the solution

(4) $$\phi(s) = w\cdot\eta(s) - \theta(s)$$

which combines them by means of an arbitrary constant w. The question naturally arises for which values of w this $\phi(s)$ satisfies a real linear boundary condition (2) at $s=l$. The answer is: for those w

that lie on a certain circle C_l in the complex w-plane. Indeed the condition (2) gives

$$(w\eta' - \theta') - h(w\eta - \theta) = 0 \qquad \text{for } s = l$$

or

$$w = \frac{\theta'(l) - h\theta(l)}{\eta'(l) - h\eta(l)},$$

and this fractional linear or Möbius transformation $h \rightarrow w$ maps the real h-axis upon a circle in the w-plane. Here λ is a given value in the upper half-plane. We now compare two such values λ, λ_0. Whereas $\phi(s) = \phi(s; \lambda)$, η, θ, and so on refer to λ, let ϕ_0, η_0, θ_0, and so on refer to λ_0. By picking a point w^0 on $C_l^0 = C_l(\lambda_0)$ one fixes the coefficient h of the real boundary condition (2). Clearly the point $w = w(\lambda)$ on $C_l = C_l(\lambda)$ for which (4) satisfies the same boundary condition as $\phi_0 = w^0 \cdot \eta_0 - \theta_0$ at $s = l$ proceeds from w^0 by a certain Möbius transformation $w^0 \rightarrow w$. Points w^0 and w on C_l^0 and C_l thus related may be called homologous points.

We now face the task of transferring these obvious answers to the limit $l \rightarrow \infty$. For that purpose the definitions of the circle C_l and of the homology between the two circles $C_l(\lambda_0)$ and $C_l(\lambda)$ must first be given a new form, one that looks more complicated but is in fact more instructive. Put $\lambda_* = \bar{\lambda}$, $\phi_* = \bar{\phi}$ in (3):

$$[\phi\bar{\phi}' - \bar{\phi}\phi']_0^l = (\lambda - \bar{\lambda}) \cdot \int_0^l \phi\bar{\phi} ds.$$

On account of this identity the requirement that ϕ satisfies a real linear boundary condition at $s = l$, $\phi\bar{\phi}' - \bar{\phi}\phi' = 0$ for $s = l$, is equivalent to the relation

$$\Im\lambda \int_0^l \phi\bar{\phi} ds = \Im w \qquad \text{for } \phi = w\eta - \theta.$$

This is indeed the equation of a circle C_l in the upper half w-plane. The points of the circular disk (C_l) bounded by C_l are characterized by the inequality

$$\int_0^l \phi\bar{\phi} ds \leqq \Im w / \Im\lambda.$$

This shows at once that $C_{l'}$ lies inside C_l if $l' > l$. Hence with l tending to infinity, C_l shrinks either to a limit circle or a limit point $C = C(\lambda)$. This alternative, limit circle or limit point, is clearly the correct formulation of the question whether or not to impose a boundary condition at $s=\infty$.

Next we have to put the description of the homology mapping $w^0 \rightarrow w$ of the circle C_l^0 upon C_l into a form suitable for passage to the limit $l \rightarrow \infty$. Pick a point w^0 on $C_l^0 = C_l(\lambda_0)$ and form

$$\phi_0(s) = \phi(s;\lambda_0) = w^0 \cdot \eta(s;\lambda_0) - \theta(s;\lambda_0). \tag{5}$$

The solution

$$\phi(s) = \phi(s;\lambda) = w \cdot \eta(s;\lambda) - \theta(s;\lambda), \tag{6}$$

satisfying at the end $s=l$ the same real linear boundary condition as $\phi_0(s)$, is obtained from $\phi_0(s)$ by solving the linear integral equation

$$\phi_0(s) = \phi(s) - (\lambda - \lambda_0)\int_0^l G^0(s,t)\cdot\phi(t)dt$$

the kernel $G^0(s,t)$ of which is Green's function for λ_0:

$$G^0(s,t) = \begin{cases} \phi_0(s)\cdot\eta_0(t) & (t \leqq s), \\ \eta_0(s)\cdot\phi_0(t) & (s \leqq t). \end{cases}$$

From the solution $\phi(s)$, (4), one gets $w=w(\lambda)$ as its initial value $\phi(0)$. This prescription at once carries over to the limit $l \rightarrow \infty$; one has simply to replace integration from 0 to l in our integral equation by one extending from 0 to ∞. Of course in (5) the factor w^0 is now supposed to be a point on the circle C^0 (whether that is a real circle or degenerates into a point). Indeed, on trying to solve in the simplest way the integral equation thus resulting, namely by the Neumann series, one finds by direct estimates that the series converges within the circle around λ_0 in the λ-plane that touches the real axis. Hence analytic continuation encounters no obstacle, and $\phi(0;\lambda)=w(\lambda)$ is a regular analytic function in the entire upper half λ-plane. The method works in both the limit-circle and the limit-point cases. In the latter w^0 is *the* limit point for λ_0, and the construction gives *the* limit point $w=w(\lambda)$ for λ. Whether the singular end $s=\infty$ is of the limit circle or limit point type does not depend on the value of λ as long as λ is restricted to the upper half-plane. In the limit circle case it follows easily that the homology mapping $w^0 \rightarrow w = w(\lambda)$ of C^0 onto C is a Möbius transformation with coefficients depending analytically on λ (for $\Im\lambda > 0$).

If one replaces the differential equation (1) by the corresponding difference equation, one arrives at a neat formulation for the theory of Stieltjes' continued fractions and his moment problem. It was treated by E. Hellinger in a manner analogous to the one outlined here for the differential equation [4]. The moment problem is a very special limiting case of an interpolation problem in the theory of analytic functions $w = w(\lambda)$ of a complex variable λ first studied by G. Pick and R. Nevanlinna [5]. It concerns analytic functions $w(\lambda)$ defined in the upper half λ-plane, $\Im\lambda > 0$, the values of which have themselves positive imaginary parts. For the moment let us call them positive functions. How far is such a function determined if its values $w(\alpha_n)$ are prescribed for a sequence of points $\lambda = \alpha_1, \alpha_2, \cdots$ in the upper half-plane? The differential problem corresponding to this interpolation or difference problem can be put in the form of a system of two linear differential equations of the first order for two unknowns ϕ, ϕ' containing the spectral parameter λ in broken linear fashion,

$$(\mathrm{S}_\lambda) \qquad \begin{cases} \dfrac{d\phi}{ds} = \dfrac{\lambda a_1(s) - b_1(s)}{\lambda a(s) - b(s)} \cdot \phi'(s), \\[2ex] \dfrac{d\phi'}{ds} = \dfrac{\lambda a_2(s) - b_2(s)}{\lambda a(s) - b(s)} \cdot \phi(s) \end{cases}$$

with real coefficients which satisfy the inequalities

$$k'(s) = a_1(s)b(s) - b_1(s)a(s) > 0, \quad k(s) = a(s)b_2(s) - b(s)a_2(s) > 0.$$

This general system now replaces our former system (1) or

$$(\mathrm{L}_\lambda) \qquad \frac{d\phi}{ds} = \frac{1}{p(s)} \cdot \phi'(s), \qquad \frac{d\phi'}{ds} = (q(s) - \lambda) \cdot \phi(s).$$

The theory of limit circles and of the homology mapping of the limit circles for different values of λ carries over practically without alteration to this more general problem [6]. The decisive point is the positive definite character of the integrand

$$\frac{k\phi\bar{\phi} + k'\phi'\bar{\phi}'}{(\lambda a - b)(\bar{\lambda} a - b)}$$

that appears in Green's formula and takes over the role played by $\phi\bar{\phi}$ in the problem (L_λ). By direct constructive solutions of integral equations one thus proves the fundamental facts about the Nevanlinna interpolation problem, which Nevanlinna himself had derived with the aid of some of the strong "existential" methods characteristic for the theory of analytic functions, such as the Vitali theorem (**B**).

3. **Expansion theorem for ordinary self-adjoint linear differential**

equations of second order with singular end. For the classical problem (L_λ) the investigation of the singular end $s=\infty$ is merely a preliminary to the study of expansions by eigenfunctions. What one has to expect can be predicted when one first replaces the singular end ∞ by the regular end l. But one has to obliterate the feature of a discrete spectrum by writing the sum over the eigenvalues in the expansion formula as a Stieltjes integral involving a non-decreasing step function. Rather than attempt to carry out the passage to the limit $l\to\infty$ one seeks to verify the formula thus obtained directly for the interval $0\leqq s<\infty$ with the singular end ∞. For simplicity's sake let us prescribe the boundary condition $\phi'(0)=0$ at the regular end $s=0$. Then we know a priori that for any eigenvalue $\lambda=\lambda_n$ the function $\eta(s;\lambda_n)$ must be the eigenfunction. The eigenvalues are those real values of λ for which $\eta(s;\lambda)$ satisfies a given real linear boundary condition $\eta'(l;\lambda)-h\cdot\eta(l;\lambda)=0$ at the end $s=l$.

Choose a definite λ_0 with positive imaginary part, for example $\lambda_0=i$. Fixing the coefficient h amounts to fixing a definite point w^0 on the circle $C_l(\lambda_0)=C^0$. The expansion of the arbitrary function $f(s)$ is then given by

$$f(s)\sim\sum_n a_n\cdot\eta(s;\lambda_n)\quad\text{where}\quad a_n=\int_0^l\eta(s;\lambda_n)r_l(\lambda_n)f(s)ds$$

and

$$r_l(\lambda_n)=1\Big/\int_0^l(\eta(s;\lambda_n))^2ds.$$

Let $\Delta=(\lambda_1,\lambda_2)$ be any interval on the real λ-axis, and, $x(\lambda)$ being any function of the real λ, let Δx stand for the difference $x(\lambda_2)-x(\lambda_1)$. We may now define a nondecreasing step function $\rho_l(\lambda)$ by the equation

$$\Delta\rho_l=\sum_{\lambda_n\in\Delta}r_l(\lambda_n). \tag{7}$$

(If one of the ends λ_1,λ_2 of the interval is an eigenvalue, the summand $r_l(\lambda_1)$ or $r_l(\lambda_2)$ in (7) should be counted with the weight 1/2 only.) After forming

$$\Delta P(s)=\int_\Delta\eta(s;\lambda)d\rho(\lambda),\qquad\Delta\alpha=\int_0^l\Delta P(s)\cdot f(s)ds \tag{8}$$

our expansion appears as the following Stieltjes integral extending over the real λ-axis:

$$f(s) \sim \int_{-\infty}^{+\infty} \eta(s;\lambda)\,d\alpha(\lambda). \tag{9}$$

One has good reason to hope that in this form the expansion theorem will carry over to the interval $(0, \infty)$ with the singular end ∞, and in anticipation of this result we have dropped the subscript l in (8). The whole problem boils down to determining the nondecreasing function $\rho(\lambda)$.

For a finite interval $0 \leqq s \leqq l$ and a non-real λ one easily proves (C) the following expansion, which is uniformly convergent with respect to s:

$$\mathfrak{J}\phi(s;\lambda) = \int_{-\infty}^{+\infty} \mathfrak{J}\,\frac{1}{\mu-\lambda}\cdot\eta(s;\mu)\,d\rho(\mu) \tag{10}$$

in particular $(s=0)$

$$\mathfrak{J}w(\lambda) = \int_{-\infty}^{+\infty} \mathfrak{J}\,\frac{1}{\mu-\lambda}\cdot d\rho(\mu). \tag{11}$$

We expect these equations to hold even for the infinite interval—although $\rho(\lambda)$ may then cease to be a step function—if $w(\lambda)$ and $\phi(s;\lambda)$ are constructed from a (or the) point w^0 on the limit circle C^0 according to the prescription given before. Denote by Δ_ϵ the segment Δ after it has been raised by the positive amount ϵ in the direction of the imaginary axis. One computes $\Delta\rho$ and $\Delta P(s) = \int_\Delta \eta(s;\ \lambda)d\rho(\lambda)$ from (11) and (10) as the limits

$$\Delta\rho = \lim_{\epsilon\to 0}\frac{1}{\pi}\int_{\Delta_\epsilon} \mathfrak{J}w(\lambda)\cdot d\lambda, \tag{12}$$

$$\Delta P(s) = \lim_{\epsilon\to 0}\frac{1}{\pi}\int_{\Delta_\epsilon} \mathfrak{J}\phi(s;\lambda)\cdot d\lambda.$$

The first formula determines ρ.

When turning these heuristic arguments into an actual proof one should first endeavor to prove the existence of the limit (12) and then to establish the expansion (9), (8), with the density differential $d\rho$ thus constructed. Mean convergence of the expansion is to be expected for any square integrable $f(s)$; certain slight restrictions imposed upon $f(s)$ will insure ordinary uniform convergence.

Many authors have written on our subject. My own first approach was based on Hilbert's general theory of spectral decomposition of a bounded symmetric linear operator and specialized it by taking advantage of the particular circumstances prevailing for the differential problem in question, above all, of the fact that for any eigenvalue λ the eigenfunction, $\eta(s;\lambda)$, is known a priori [**7**]. M. H. Stone's procedure in his book on *Linear transformations in Hilbert space* [**8**] is of the same character, but he is able to utilize the machinery of general concepts developed in the twenty intervening years for axiomatized Hilbert space. Earlier, the Stieltjes method, on which Hellinger had founded Hilbert's general theory [**9**], had been directly applied to the special differential problem by E. Hilb [**10**]; but he did not carry it so far as to obtain the explicit construction of the differential $d\rho$. Recently E. C. Titchmarsh in several papers and in his book on *Eigenfunction expansions* [**11**] resumed this direct approach. The basic equation (12) is due to him. Yet his construction of $w(\lambda)$ and of $d\rho$ is not as direct as I should wish them. Also a number of contributions made by A. Wintner and P. Hartman during the last two years ought to be mentioned [**12**]. The formula (12) was rediscovered by Kunihiko Kodaira (who of course had been cut off from our Western mathematical literature since the end of 1941); his construction of ρ and his proofs for (12) and the expansion formula (9), still unpublished, seem to clinch the issue. It is remarkable that forty years had to pass before such a thoroughly satisfactory direct treatment emerged; the fact is a reflection on the degree to which mathematicians during this period got absorbed in abstract generalizations and lost sight of their task of finishing up some of the more concrete problems of undeniable importance.

4. **Inequalities and asymptotic laws for eigenvalues.** But let us drop this matter now and turn to another subject, that of the asymptotic distribution of the eigen-frequencies for the two- or more-dimensional membrane and for other oscillating continua. H. A. Lorentz had impressed upon the mathematicians the urgency for physics of a settlement of this question. For a pupil of Hilbert around 1910 it was natural to visualize the question as one concerning integral equations. By means of a real symmetric kernel $K(s, t)$ one introduces the linear operator $u \rightarrow Ku$, more explicitly $u(s) \rightarrow \int_0^1 K(s, t)u(t)dt$, in the vector space of all real-valued continuous functions $u=u(s)$ defined over the interval $0 \leqq s \leqq 1$. If the integral $\int_0^1 u(s)v(s)ds$ is taken as the scalar product (u, v) of any two vectors u, v in this space, then the quadratic integral form

$$K\langle u\rangle = \int_0^1 \int_0^1 K(s, t)u(s)u(t) \cdot dsdt$$

is the scalar product of u and Ku. The reciprocal eigenvalues κ and corresponding eigenvectors ϕ are the solutions of the equation $K\phi = \kappa\ \phi$. Let the positive κ's be arranged in descending order, $\kappa_1 \geqq \kappa_2 \geqq \cdots$, and the corresponding eigenfunctions ϕ_n so chosen as to form an orthonormal system, $(\phi_m, \phi_n) = \delta_{mn}$. Then κ_n is the maximum of the form K under the auxiliary conditions

$$\|u\|^2 = (u, u) \leqq 1, \qquad (u, \phi_i) = 0 \qquad (i = 1, \cdots, n-1).$$

Let $\omega_1(s), \cdots, \omega_{n-1}(s)$ be any $n-1$ functions. The fundamental lemma which made my investigation possible states that there exists a vector u of length $\|u\| = 1$ which is orthogonal to $\omega_1, \cdots, \omega_{n-1}$,

$$(u, \omega_1) = 0, \cdots, (u, \omega_{n-1}) = 0,$$

such that $K\langle u \rangle \geqq \kappa_n$. This characterizes κ_n independently of the preceding eigenvalues and eigenfunctions as the "minimum of a maximum." The construction of such a u as the lemma requires is easy enough: a suitable linear combination $c_1\phi_1(s) + \cdots + c_n\phi_n(s)$ of the first n eigenvectors will do the trick. I used this lemma (in a slightly different form) for the purpose of carrying over to all κ_n statements that are evident for the first reciprocal eigenvalue κ_1 [**13**].

Here is an example. Suppose you add to a kernel K a positive-definite one k, that is, one for which $k\langle u \rangle \geqq 0$. It is clear that the first reciprocal positive eigenvalue κ_1^* of $K^* = K + k$ is greater than or equal to κ_1; for κ_1 is the maximum of $K\langle u \rangle$ and κ_1^* the maximum of $K\langle u \rangle + k\langle u \rangle$ under the condition $\|u\|^2 \leqq 1$. Our lemma carries the inequality over to all κ's arranged in descending order: $\kappa_n \leqq \kappa_n^*$.

This result is of immediate application to the two-dimensional membrane problem. Let the membrane cover a region S of (Jordan) area V. With the argument P ranging over the points of S the eigenvalues λ and eigenfunctions $\phi(P)$ satisfy the differential equation $\Delta\phi + \lambda\phi = 0$ in S and the condition $\phi = 0$ along the boundary S' of S. The differential equation together with the boundary condition is equivalent to the integral equation

$$\phi(P) - \lambda \int_S G(P, Q) \cdot \phi(Q) \cdot dQ = 0,$$

the symmetric kernel of which is the Green's function $G(P, Q)$. Divide S by a line l into two parts S_1, S_2 and let G_1, G_2 be their Green's functions (setting $G_i(P, Q) = 0$ if one of the argument points P, Q or both are outside S_i). It can easily be shown that the kernel $G - (G_1 + G_2)$ is positive-definite. Hence the nth eigenvalue λ_n of G is less than or equal to the nth eigenvalue λ_n' of $G_1 + G_2$. The eigenvalues of $G_1 + G_2$ are the eigenvalues of a membrane covering S which is kept fixed not only along the boundary S', but also along the line l. By combining this result with the known asymptotic distribution of the membrane eigenvalues for a square it could readily be deduced that the fixing of the membrane along l does not alter the asymptotic distribution,

$$\lambda_n/\lambda_n' \to 1 \qquad \text{for} \quad n \to \infty,$$

and moreover to establish the law of asymptotic distribution, according to which the number $N(\lambda)$ of eigenvalues less than λ equals asymptotically $(V/4\pi)\cdot\lambda$.

After the first world war Courant resumed this sort of problems [14]. If I see correctly, his essential contribution is not the minimum-maximum principle formulated in our fundamental lemma, but its application to a fairly general typical situation. The maximum of a quadratic form $K\langle u\rangle$ depending on a vector u of length not greater than 1 in a vector space $\mathfrak{H}$ is lowered if additional restrictions are imposed upon u, for example, if u is restricted to a linear subspace $\mathfrak{H}'$ of $\mathfrak{H}$. If the quadratic form is what Hilbert calls completely continuous, this obvious statement may be put into the inequality $\kappa_1' \leqq \kappa_1$. The quadratic form $K\langle u\rangle$ in $\mathfrak{H}'$ is $(u, K'u)$ where K' is the operator K followed by perpendicular projection upon $\mathfrak{H}'$. While κ_1 is the first reciprocal eigenvalue of the operator K in $\mathfrak{H}$, κ_1' has the same significance for the operator K' in $\mathfrak{H}'$. The lemma carries the inequality $\kappa_1' \leqq \kappa_1$ over to all reciprocal (positive) eigenvalues, $\kappa_n' \leqq \kappa_n$. It is clear that the fixing of a membrane along the line l introduces a new restriction, and hence Courant's observation at once gives rise to the inequality $\lambda_n \leqq \lambda_n'$, derived before in another way, for the eigenvalues λ_n, λ_n' of the undivided and divided membrane. Anyone familiar with the abstract concept of Hilbert space who ponders a little more closely upon the situation to which the principle is applied here, will describe it as follows. All continuous functions $u(P)$ in the closed region S with continuous first derivatives in the interior, such that $u=0$ at the boundary and the Dirichlet integral $D\langle u\rangle = \int_S (\text{grad } u)^2 dP$ is finite, form a functional space $\mathfrak{H}$. Define the square $\|u\|^2$ of the length of a vector u in this space by $D\langle u\rangle$, and not by $I\langle u\rangle = \int_S u^2 dP$. The closure of $\mathfrak{H}$ with respect to this metric is a Hilbert space $\overline{\mathfrak{H}}$ and $I\langle u\rangle$ is a completely continuous quadratic form in $\overline{\mathfrak{H}}$. Without altering the other conditions for the functions $u \in \mathfrak{H}$ add the restriction $u=0$ along the line l. The closure of the subspace $\mathfrak{H}'$ thus obtained is a closed subspace $\overline{\mathfrak{H}}'$ of $\overline{\mathfrak{H}}$, and the relation $\kappa_n' \leqq \kappa_n$ holds for the reciprocal eigenvalues of the form $I\langle u\rangle$ in $\overline{\mathfrak{H}}$ and in $\overline{\mathfrak{H}}'$ respectively. (Forming the closure is essential since the eigenfunctions of the divided membrane are in $\bar{\mathfrak{H}}'$ but in general not in $\mathfrak{H}'$, because of the jump of their normal derivative along l. To be sure, by going back to the proof of the fundamental lemma, one can avoid this whole abstract set-up. But for the moment we are interested in the general formulation.)

Let then $K\langle u\rangle$ again be a completely continuous quadratic form in a Hilbert space $\mathfrak{G}$, and $\mathfrak{H}$ be a closed subspace of $\mathfrak{G}$. For simplicity's sake we assume $K\langle u\rangle$ to be positive-definite and denote the eigenvalues of $K\langle u\rangle$ in $\mathfrak{G}$ and in $\mathfrak{H}$ by λ_n and μ_n respectively. We have seen that

$$(13) \qquad \lambda_n \leqq \mu_n.$$

Splitting $\mathfrak{G}$ into $\mathfrak{H}$ and its perpendicular subspace and choosing a basis $p_1, p_2, \cdots$ for the latter (preferably an orthonormal basis), we can pass from $\mathfrak{H}$ to $\mathfrak{G}$ by a sequence of intermediary subspaces $\mathfrak{H}\subset\mathfrak{H}'\subset\mathfrak{H}''\subset\cdots$, $\mathfrak{H}^{(\nu)}\rightarrow\mathfrak{G}$ with $\nu\rightarrow\infty$, by adding one vector of this basis after the other, $\mathfrak{H}^{(\nu)}=\mathfrak{H}+\{p_1, \cdots, p_\nu\}$. Or we can submit the vectors u of $\mathfrak{G}$ to one after the other of the conditions $(u, p_1)=0$, $(u, p_2)=0, \cdots$, and thus obtain a descending sequence $\mathfrak{G}\supset\mathfrak{G}'\supset\mathfrak{G}''\supset\cdots$ with $\mathfrak{G}^{(\nu)}\rightarrow\mathfrak{H}$ for $\nu\rightarrow\infty$. If λ_n is known, the inequality (13) gives a lower bound for μ_n, and the second of our sequences gives rise to an increasing sequence of such lower bounds,

$$\lambda_n \leqq \lambda_n' \leqq \lambda_n'' \leqq \cdots, \qquad \lim_{\nu\rightarrow\infty} \lambda_n^{(\nu)} = \mu_n.$$

If μ_n is known, the same inequality gives an upper bound for λ_n, which by the first sequence of subspaces may be extended into a whole sequence of decreasing upper bounds, $\mu_n\geqq\mu_n'\geqq\mu_n''\geqq\cdots$, $\lim_{\nu\rightarrow\infty}\mu_n^{(\nu)}=\lambda_n$. With some right we may call the first procedure the Rayleigh-Ritz method and with more right ascribe the second to A. Weinstein [**15**].

The basic situation encountered here is that of a Hilbert space $\mathfrak{H}$ that splits into a subspace $\mathfrak{H}'$ and a perpendicular ν-dimensional space $\{p_1, \cdots, p_\nu\}$ spanned by the vector basis $p_1, \cdots, p_\nu$. N. Aronszajn recently developed two neat formulas concerning this situation [**16**]. I shall here mention but the one that corresponds to the Weinstein process, the descent from $\mathfrak{H}$ to $\mathfrak{H}'$, and omit the other which refers to the inverse Rayleigh-Ritz process. Let $R(\zeta)$ be the resolvent of the operator K in $\mathfrak{H}$, so that $u=R(\zeta)v$ is the solution of the equation $u-\zeta\cdot Ku=v$. Moreover let λ', u' be an eigenvalue and corresponding eigenvector of K' in $\mathfrak{H}'$. This fact is expressed by the relations

$$u' \in \mathfrak{H}', \qquad u' - \lambda' \cdot Ku' = \beta_1 p_1 + \cdots + \beta_\nu p_\nu.$$

If we suppose that λ' coincides with no eigenvalue λ_n of K then the constants β_i cannot all be zero. By means of the resolvent $R(\zeta)$ the second relation takes on the form $u' = \sum_{j=1}^{\nu} \beta_j \cdot R(\lambda') p_j$. The first relation requires that the vector u' is perpendicular to all the p_i,

$$\sum_j \beta_j (p_i, R(\lambda') p_j) = 0 \qquad (i, j = 1, \cdots, \nu).$$

These ν equations for the ν unknowns β_j have a nontrivial solution only if the determinant

$$\det (p_i, R(\zeta) p_j) = W_p(\zeta)$$

vanishes for $\zeta = \lambda'$. Divide $W_p(\zeta)$ by Gram's determinant $G_p = \det (p_i, p_j)$. The quotient

$$W(\zeta) = W_p(\zeta)/G_p$$

is clearly independent of the choice of the basis $p_1, \cdots, p_\nu$. From the theory of the resolvent one knows that $W(\zeta)$ is a meromorphic function with simple poles at the eigenvalues λ_n of K. On the other hand, we have seen that $W(\zeta)$ vanishes for a value $\zeta = \lambda'$ different from all the λ_n if and only if λ' is an eigenvalue of K' in $\mathfrak{H}'$. Hence we shall not find the following equation of Aronszajn too surprising:

$$W(\zeta) = \left(\frac{-1}{\zeta}\right)^\nu \cdot \prod_n \frac{\zeta - \lambda_n'}{\zeta - \lambda_n}.$$

I regret that shortness of time prevents me from illustrating these general developments by their applications to the classical problems of elastic and electromagnetic oscillations (**D**).

5. **Zeta-function of the membrane and asymptotic laws for its eigenfunctions.** The physicist will not be satisfied with a knowledge of the asymptotic behavior of the eigenvalues alone; also that of the eigenfunctions should be investigated. Carleman was the first to attack this more difficult problem by a new powerful method [**17**]. Last year we had the good fortune to have with us at the Institute Dr. Åke Pleijel, who had extended Carleman's investigations [**18**], and Dr. Minakshisundaram, who independently of Carleman had just found a modification of Carleman's method shedding new light on the whole problem. Let us again envisage the two-dimensional membrane. Carleman made use of the Green's function of the "meson" equation $\Delta u - k^2 u = 0$ with the positive parameter k^2 (for

the boundary condition $u=0$). Minakshisundaram [19] used instead Green's function $G(P, Q; t)$ of the heat equation

$$\Delta u - \partial u/\partial t = 0 \tag{14}$$

with the positive time parameter t.

$$u(P; t) = \int_S G(P, Q; t)\cdot f(Q)\cdot dQ$$

is the temperature of the disk S at the point P and at the moment t if $f(P)$ describes the initial distribution of temperature and the boundary of S is kept constantly on the temperature zero. For the infinite plane $G(P, Q; t)$ is

$$G_0(P, Q; t) = \frac{1}{4\pi t}\cdot \exp\left(-\frac{r_{PQ}^2}{4t}\right)$$

where r_{PQ} denotes the distance of the two points P, Q. For an arbitrary domain S we write

$$G(P, Q; t) = G_0(P, Q; t) - g(P, Q; t).$$

For a fixed P in the interior of S, the compensating term $g(P, Q; t)$ is a solution of (14) which vanishes for $t\to 0$ and on the surface S' of S has the same boundary values as the principal part G_0.

The maximum which the principal term G_0 assumes when P and t are given, but Q varies over the boundary S', is

$$H_0(P; t) = \frac{1}{4\pi t} \exp\left(-\frac{l_P^2}{4t}\right),$$

l_P denoting the shortest distance of P from the boundary. As a function of t this H_0 is on the increase from $t=0$ to $t=T=4/l_P^2$ and then decreases. A simple argument shows that, as long as $0<t\leqq T$, the compensating function $g(P, Q; t)$ of Q is positive and reaches its maximum at the boundary, or

$$0 < g(P, Q; t) \leqq H_0(P; t) \tag{15}$$

for all Q in S and $0<t\leqq T$.

In terms of the orthonormal system of eigenfunctions $\phi_n(P)$ and their eigenvalues λ_n Green's function is expressible as the sum

$$\sum_n e^{-\lambda_n t}\phi_n(P)\phi_n(Q). \tag{16}$$

The series

$$\zeta(P, Q; s) = \sum_n \frac{\phi_n(P)\phi_n(Q)}{\lambda_n^s} \tag{17}$$

may be called the ζ-function and accordingly (16) the θ-function of the membrane problem. We know that (17) converges uniformly in P and Q for all complex s the real part σ of which is greater than or equal to 2. For $s=1, 2, 3, \cdots$ the ζ-function gives Green's function $G(P, Q)$ of the membrane and its successive iterations.

One of Riemann's methods for deriving the properties of the ordinary ζ-function was based on a connection between the θ-function and the ζ-function, which at once carries over to the general functions here considered as follows:

$$\Gamma(s)\cdot\zeta(P, Q; s) = \int_0^\infty G(P, Q; t)\cdot t^{s-1}\cdot dt. \tag{18}$$

On the basis of this relation Minakshisundaram proves that $\Gamma(s)\cdot\zeta(P, Q; s)$ is a regular function of s in the entire s-plane if P, Q are two distinct inner points of S, but that it is regular except for a simple pole at $s=1$ with residue $1/4\pi$ if Q and P coincide. Still following Riemann, Minakshisundaram splits the integral $\int_0^\infty$ on the right of (18) into $\int_0^T+\int_T^\infty$. Because of the uniform convergence of $\zeta(P, Q; 2)$, the θ-function (16) falls off exponentially with $t\to\infty$, and hence the integral

$$\int_T^\infty G(P, Q; t)\cdot t^{s-1}dt$$

is a regular-analytic function of s in the whole s-plane. If $P\neq Q$ the formula

$$G(P, Q; t) = \frac{1}{4\pi t}\exp\left(-\frac{r_{PQ}^2}{4t}\right) - g(P, Q; t)$$

together with the estimate (15) proves that $G(P, Q; t)$ also goes down exponentially to zero with $1/t\to\infty$, and hence the integral $\int_0^T$ is likewise regular-analytic in s. This proves the result for $P\neq Q$. However if $Q=P$ then

$$G(P, P; t) = \frac{1}{4\pi t} - g(P, P; t).$$

The second part is positive and does not exceed $H_0(P; t)$; therefore

$\int_0^T g(P, P; t) \cdot t^{s-1} dt$ is regular in s, however

$$\int_0^T \frac{1}{4\pi t} t^{s-1} dt = \frac{1}{4\pi} \cdot \frac{T^{s-1}}{s-1}$$

has a pole at $s=1$ with the residue $1/4\pi$. That completes the proof. $\zeta(P, Q; s)$ itself has zeros for the same values for which the regular function $1/\Gamma(s)$ has zeros, namely for $s=0, -1, -2, \cdots$.

Integration over P, one might think, would give for

$$\zeta(s) = \int_S \zeta(P, P; s) \cdot dP = \sum_n \lambda_n^{-s}$$

the following result

$$\Gamma(s) \cdot \zeta(s) = \frac{V}{4\pi} \cdot \frac{1}{s-1} + \text{a regular function } R(s). \tag{19}$$

Unfortunately this conclusion is too hasty: if P is near the boundary, $H_0(P; t)$ is not so small. By carrying out the integration over P in the relation

$$G(P, P; t) = \frac{1}{4\pi t} - g(P, P; t), \qquad 0 \leqq g(P, P; t) \leqq H_0(P; t),$$

one finds that the remainder $R(s)$ in (19) is regular at least for $\sigma > 1/2$; it seems difficult to go beyond the vertical $\sigma = 1/2$.

Standard devices familiar from the theory of Riemann's ζ-function permit one to deduce from this behavior of the ζ-function of the membrane Carleman's asymptotic formulas

$$\sum_{\lambda_n \leqq \lambda} (\phi_n(P))^2 \sim \lambda/4\pi, \qquad N(\lambda) = \sum_{\lambda_n \leqq \lambda} 1 \sim V\lambda/4\pi$$

and also the "incoherence relation"

$$\sum_{\lambda_n \leqq \lambda} \phi_n(P)\phi_n(Q) = o(\lambda) \qquad \text{for } P \neq Q.$$

I feel that these informations about the proper oscillations of a membrane, valuable as they are, are still very incomplete. I have certain conjectures on what a complete analysis of their asymptotic behavior should aim at; but since for more than 35 years I have made no serious attempt to prove them, I think I had better keep them to myself.

In general, it can not be expected that our ζ-function satisfies a functional equation of the Riemann type; one may guess that this feature depends on the homogeneity of the domain of integration. Such a domain is the circumference of the unit circle. Functions on it are functions $f(x)$ of period 2π. The periodic eigenfunctions ϕ and corresponding eigenvalues λ of $d^2\phi/dx^2+\lambda\phi=0$ are

$$\phi(x) = e^{inx}, \qquad \lambda_n = n^2 \qquad (n = 0, \pm 1, \pm 2, \cdots).$$

This leads straight to the Riemann ζ-function $\sum_{n=1}^{\infty} n^{-2s}$ usually denoted not by $\zeta(s)$ but by $\zeta(2s)$. It was therefore natural that Minakshisundaram should investigate the spherical harmonics on a k-dimensional sphere (in $k+1$-dimensional space). Here he found indeed a sort of Riemann functional equation, the structure of which is, however, essentially more complicated than in the classical Riemann case $k=1$.

The two-sphere is homogeneous because it permits a compact transitive Lie-group σ of transformations s into itself, namely the group of rotations. The spherical harmonics of order l form a $(2l+1)$-dimensional linear manifold that is invariant with respect to the group of rotations and has the property of irreducibility in this regard. Consider arbitrary (complex-valued) continuous functions on the sphere and define the scalar product of two such functions f and g by the integral $\int \bar{g}(P) f(P) \cdot d\omega_P$ formed by means of the invariant area element $d\omega_P$. It is obvious how to generalize this situation to any homogeneous manifold S of points P, that is, any manifold that permits a compact transitive Lie-group σ of transformations s, $P \to sP$. The existence of an invariant volume element on such a manifold (which itself is of necessity compact) follows easily from the fact that a compact Lie-group has an invariant volume element ds. We normalize the unit for measuring volumes on the group so that the total volume $\int ds$ of the group becomes 1. The integrals with respect to s are then in truth mean values. The transform sf of a function $f=f(P)$ on S is defined by $sf(sP)=f(P)$ or $sf(P)=f(s^{-1}P)$. A set $\phi_1(P), \cdots, \phi_h(P)$ of functions on S, or the manifold of their linear combinations $\phi(P)=x_1 \cdot \phi_1(P)+ \cdots +x_h \cdot \phi_h(P)$, is invariant if each $s\phi_i(P)$ is a linear combination $\sum_j \omega_{ji}(s) \cdot \phi_i(P)$ of the ϕ_i themselves. Then $s \to \|\omega_{ij}(s)\|$ is a representation of degree h of the group σ. Inequivalent irreducible invariant sets are orthogonal to each other. Besides orthogonality there is the completeness relation, to which we shall presently return. Thus we are in possession of the "eigenfunctions" of the homogeneous manifold, the sequence of which is subdivided into irreducible invariant sets of finite length. Theorems about summability of expansions in terms of these eigenfunctions have been proved by S. Bochner [20]. But so far they are eigenfunctions without eigenvalues. It was the young Dutch physicist H. B. G. Casimir who, prompted by the applications of group theory to quantum mechanics, found the eigenvalues. Indeed he constructed an invariant self-adjoint differential operator Δ working on arbitrary functions $f(P)$ which is the analogue of the Laplace operator on a sphere, and he was able to show that the functions of a given irreducible invariant set satisfy an equation $\Delta\phi+\lambda\phi=0$ with a constant λ characteristic for the entire set [21]. Having the eigenfunctions, one can, following a suggestion by Bochner, form the ζ-function

$$\zeta(P, Q; z) = \sum_n \frac{\phi_n(P)\phi_n(Q)}{\lambda_n^z}$$

and might expect that this function, in addition to having the properties quite generally established by Minakshisundaram, will satisfy a functional equation of Riemann's type. But this is a question that remains to be investigated.

6. **Integral equations and the group-theoretic completeness relation.** The proof of the completeness relation for invariant sets on a homogeneous manifold S is one of the most surprising applications of the eigenvalue theory of integral equations. If the manifold S is the compact Lie group itself under the influence of its left translations, then this theorem states the completeness of the totality of all irreducible representations of the group. But the method for its proof, developed in 1927 by F. Peter and the speaker [22], not only carries over to the homogeneous manifolds, but applies to a far more general situation, that is best described in axiomatic terms [23]. We replace the functions on the homogeneous manifold by vectors f in a vector space Σ and suppose that Σ bears a Hermitian metric defined by a scalar product (g, f) with the usual properties including the positive character of $(f, f) = \|f\|^2$. An abstract compact Lie group σ is given and a representation of its elements s by linear transformations $f \rightarrow sf$ in our vector space. The invariance of the metric is assumed, $(sg, sf) = (g, f)$. Let f be a given vector. All vectors that will occur in our construction are prepared from f by forming linear combinations of its transforms sf. Besides Σ we envisage the "vector space" Ξ of all continuous functions $\xi = \xi(s)$ on the group manifold σ and define a linear mapping $\xi \rightarrow g$ of Ξ onto Σ by

$$g = \mathfrak{f}\xi = \int \xi(s) \cdot sf \cdot ds,$$

and its Hermitian conjugate, a mapping $g \rightarrow \xi$ of Σ onto Ξ, by

$$\xi = \mathfrak{f}^* g = (sf, g).$$

The mapping $\mathfrak{f}^*\mathfrak{f} = \mathfrak{F}$ of Ξ into itself has the positive-definite Hermitian kernel $H(s, t) = (sf, tf)$. By Erhard Schmidt's method we construct its largest reciprocal eigenvalue γ and an orthonormal set of eigenfunctions $\phi_1(s), \cdots, \phi_h(s)$ for it. Repetition of the construction gives the reciprocal eigenvalues in descending order, $\gamma > \gamma' > \cdots$, and the sought-for completeness relation results from the well known fact that the trace of H equals the sum of the reciprocal eigenvalues,

$$\|f\|^2 = (f, f) = h\gamma + h'\gamma' + \cdots .$$

Indeed $\int H(s, t)\cdot\phi_i(t)\cdot dt = \gamma\cdot\phi_i$ may be written in the form

$$\mathfrak{f}\phi_i = \gamma^{1/2}\cdot g_i, \qquad \mathfrak{f}^* g_i = \gamma^{1/2}\cdot\phi_i$$

where the first equation is to be taken as the definition of the vector g_i. The g_i then form an orthonormal invariant set, and if the Fourier coefficients $(g_i, f) = \alpha_i$ are introduced one finds that

$$h\gamma = |\alpha_1|^2 + \cdots + |\alpha_h|^2.$$

Thus the completeness relation follows, stating that the orthonormal sequence $g_1, g_2, \cdots, (g_m, g_n) = \delta_{mn}$, resulting from our construction and consisting of sections of finite length, each of which is an invariant set, makes the absolute square sum $|\alpha_1|^2 + |\alpha_2|^2 + \cdots$ of the Fourier coefficients $\alpha_i = (g_i, f)$ not only $\leqq \|f\|^2$, as is trivial (Bessel's inequality), but actually $= \|f\|^2$. The construction picks out those invariant sets that contribute to f and $\|f\|^2$.

This feature is quite essential when, with Harald Bohr and J. von Neumann [24], all restrictions concerning the group σ are abandoned. The construction still works, provided one supposes f to be "almost periodic." But in general there are under these circumstances more than denumerably many inequivalent irreducible invariant sets of vectors; but f itself picks out those among them that matter for f.— The assumption of almost periodicity is highly restrictive. One may instead impose some slight restriction on the group, e.g., local compactness, and at the same time admit a far wider class of vectors f. In that case nothing resembling completeness is to be expected unless one includes also representations of infinite degree. This step has recently been taken by D. Rykov and I. Gelfand in Russia, by V. Bargmann and I. Segal in this country [25].

I think it is time for me to stop here. I have not even touched on the extension of Hilbert's theory of bounded to non-bounded linear operators, which came about under the pressure of quantum mechanics, nor to the connection between spectral decomposition and ergodic theory. Other mathematicians at other times have spoken or will speak on these subjects with more competence than I could. I hope you have taken this lecture for what it was meant to be: a *Plauderei*, the chat of a man who has reached the age where it is more pleasant to remember the past than to look forward into the future. Even so, it gives him a little satisfaction to see that the issues to which the efforts of his youth were dedicated have kept alive over the years and are still in the process of unfolding their implications.

NOTES

(A) There are two classes of eigenfunctions. But since one of them does not contribute to the expansion of the discontinuous function $1^0(x)$, only the eigenvalues λ^2 belonging to the other class have to be taken into account; they are determined by the transcendental equation

$$\beta\cdot\tan\frac{\alpha\lambda\pi}{2}+\alpha\cdot\tan\frac{\beta\lambda\pi}{2}=0.$$

It is easily seen that for every integer n this equation has exactly one root λ_n of the form $\lambda_n=2n+\theta_n$, $-1<\theta_n<1$. If α and β are rational,

$$\alpha=a/c,\quad \beta=b/c,\quad a+b=c;\quad a,\, b,\, c \text{ integers},\quad c>0,$$

then θ_n has the period c, $\theta_{n+c}=\theta_n$, and this circumstance makes a fairly explicit evaluation of the nth partial sum $1_n^0(x)$ possible.

(B) The main fact is as follows: The given values $\beta_1=w(\alpha_1)$, $\beta_2=w(\alpha_2)$, $\cdots$ have to satisfy a sequence of inequalities of which the first, $\Im\beta_1>0$, involves only β_1, the second β_1 and β_2, and so on. If these inequalities are fulfilled then there are two possibilities, which are distinguishable by a convergence criterion. In the first, the limit point case, the problem has a unique solution; in the second, the limit circle case, the manifold of all solutions $w(\lambda)$ is obtained from that of all positive functions $z(\lambda)$ by a certain Möbius transformation

$$w(\lambda)=\frac{A(\lambda)\cdot z(\lambda)+B(\lambda)}{C(\lambda)\cdot z(\lambda)+D(\lambda)}$$

with coefficients A, B, C, D that are regular analytic functions of λ in the upper half λ-plane.

(C) Indeed replacing λ_0, λ by λ, $\bar\lambda$ in (6) one finds

$$\Im\phi(s)=\Im\lambda\cdot\int_0^l G(s,t)\cdot\bar\phi(t)\cdot dt,$$

and therefore $\mathfrak{J}\phi(s)$ has a uniformly convergent expansion in terms of the eigenfunctions $\eta(s; \lambda_n)$. For the integral $\int_0^l \phi(s)\eta(s; \lambda_n)\cdot ds$ one obtains from (3) the value $1/(\lambda_n - \lambda)$.

(D) The eigenvalues $\lambda = \mu_n$ of $\Delta u + \lambda u = 0$ corresponding to the boundary condition $\partial u/\partial n = 0$ (normal derivative of u equal to zero; "acoustic eigenvalues") are the reciprocals of the successive maxima of $I\langle u\rangle$ under the restriction $D\langle u\rangle = 1$, or the successive minima of $D\langle u\rangle$ under the restriction $I\langle u\rangle = 1$. The boundary condition $\partial u/\partial n = 0$ gets lost, as it were, in the process of closure under the metric defined by $D\langle u\rangle$. Hence $\mu_n \leqq \lambda_n$, where λ_n, as before, are the membrane eigenvalues corresponding to the boundary condition $u = 0$.

The equation for the oscillations of a plate,

$$\Delta\Delta u - \lambda^2 \cdot u = 0 \qquad \text{in } S,$$

arises from minimizing

(20)
$$\int_S (\Delta u)^2 \cdot dP$$

under the auxiliary condition $I\langle u\rangle = 1$. Let μ_n^2 be the eigenvalues of the clamped plate,

$$\text{boundary conditions } u = 0, \quad \frac{\partial u}{\partial n} = 0,$$

and λ_n^2 those for the "half-free" plate,

(21)
$$\text{boundary conditions } u = 0, \quad \Delta u = 0.$$

Again the boundary condition $\Delta u = 0$ gets lost in the process of closure under the metric defined by (20). Hence $\lambda_n^2 \leqq \mu_n^2$; one can further expect that λ_n and μ_n follow the same asymptotic law. Weinstein observed that the eigenvalues of the half-free plate (as their notation indicates) are simply the squares of the membrane eigenvalues. Indeed if $\Delta\Delta u = \lambda^2 u$ $(\lambda > 0)$ set $\Delta u = -\lambda v$, so that $\Delta v + \lambda u = 0$. The boundary conditions (21) give $u = 0$, $v = 0$ along S', hence $(u+v)/2$ is a membrane eigenfunction with the eigenvalue λ and $(u-v)/2$ for $-\lambda$. But the membrane has no negative eigenvalues; consequently $u - v = 0$ and $(u+v)/2 = u$.

It was by a somewhat similar remark that I had previously reduced the elastic oscillations of a three-dimensional body asymptotically to the three-dimensional membrane problem [26]. The vector field $\mathfrak{v}(P)$ describing a proper oscillation of the elastic body satisfies an equation

$$\text{(E)} \qquad a \cdot \text{grad div}\, \mathfrak{v} - b \cdot \text{rot rot}\, \mathfrak{v} + \lambda \mathfrak{v} = 0 \qquad \text{in } S$$

with two elastic constants a, b. Impose the boundary conditions

$$\text{(E')} \qquad \mathfrak{v} \text{ normal}, \qquad \text{div}\, \mathfrak{v} = 0 \quad \text{on the surface } S' \text{ of } S.$$

For $\phi = \text{div}\, \mathfrak{v}$ one finds $a \cdot \Delta\phi + \lambda\phi = 0$ in S and the boundary condition $\phi = 0$; hence if ϕ is an eigenfunction of the membrane problem with the eigenvalue λ/a then $\mathfrak{v} = \text{grad}\, \phi$ satisfies (E), (E′). If, however, div $\mathfrak{v} = 0$ throughout S we have

$$b \cdot \Delta\mathfrak{v} + \lambda\mathfrak{v} = 0, \qquad \text{div}\, \mathfrak{v} = 0 \text{ in } S; \qquad \mathfrak{v} \text{ normal on } S'$$

(since $\Delta\mathfrak{v} = \text{grad div}\, \mathfrak{v} - \text{rot rot}\, \mathfrak{v}$), or λ/b is an eigenvalue of the problem of radiation in a Hohlraum S whose wall S' is a perfect mirror. Denoting the numbers of eigenvalues $\leqq \lambda$ of the membrane, the radiation, and the elastic problem (E) & (E′) by $N_m(\lambda)$, $N_r(\lambda)$ and $N_e(a,b;\lambda)$ respectively, we thus find the relation

$$N_e(a, b; \lambda) = N_m(\lambda/a) + N_r(\lambda/b).$$

For $a = b = 1$ the left side of (E) turns into $\Delta\mathfrak{v} + \lambda\mathfrak{v}$. Since asymptotically the boundary conditions are of no influence we must have the asymptotic relation

$$N_e(1, 1; \lambda) \sim 3N_m(\lambda), \quad \text{that is,} \quad N_m(\lambda) + N_r(\lambda) \sim 3N_m(\lambda)$$

or $N_r(\lambda) \sim 2N_m(\lambda)$ and thus

$$N_e(a, b; \lambda) \sim N_m(\lambda/a) + 2N_m(\lambda/b).$$

Bibliography

1. H. Weyl, *Die Gibbs'sche Erscheinung in der Theorie der Kugelfunktionen*, Rend. Circ. Mat. Palermo vol. 29 (1910) pp. 308–328; *Ueber die Gibbs'sche Erscheinung und verwandte Konvergenzphänomene*, Rend. Circ. Mat. Palermo vol. 30 (1910) pp. 1–31.

2. ———, *Ueber die Gleichverteilung von Zahlen mod. Eins*, Math. Ann. vol. 77 (1916) pp. 313–352.

3. ———, *Ueber gewöhnliche Differentialgleichungen mit Singularitäten und die zugehörigen Entwicklungen willkürlicher Funktionen*, Math. Ann. vol. 68 (1910) pp. 220–269; *Ueber gewöhnliche Differentialgleichungen mit singulären Stellen und ihre Eigenfunktionen*, Nachr. Ges. Wiss. Göttingen (1910) pp. 442–467.

4. E. Hellinger, *Zur Stieltjes'schen Kettenbruchtheorie*, Math. Ann. vol. 86 (1922) pp. 18–29.

5. G. Pick, *Ueber die Beschränkungen analytischer Funktionen, welche durch vorgegebene Funktionswerte bewirkt werden*, Math. Ann. vol. 77 (1916) pp. 7–23. R. Nevanlinna, *Ueber beschränkte Funktionen, die in gegebenen Punkten vorgeschriebene Werte*

annehmen, Annales Academiae Scientiarium Fennicae vol. 13, No. 1, 1919; *Ueber beschränkte analytische Funktionen*, ibid. vol. 32, No. 7, 1929.

6. H. Weyl, *Ueber das Pick-Nevanlinna'sche Interpolationsproblem und sein infinitesimales Analogon*, Ann. of Math. vol. 36 (1935) pp. 230–254.

7. ———, Chapter 3 of the first, and §1 of the second, paper quoted under [3].

8. M. H. Stone, *Linear transformations in Hilbert space*, Amer. Math. Soc. Colloquium Publications, vol. 15, 1932.

9. E. Hellinger, *Neue Begründung der Theorie quadratischer Formen von unendlichvielen Veränderlichen*, J. Reine Angew. Math. vol. 136 (1909) pp. 265–326.

10. E. Hilb, *Ueber gewöhnliche Differentialgleichungen mit Singularitäten und die dazugehörigen Entwicklungen willkürlicher Funktionen*, Math. Ann. vol. 76 (1915) pp. 333–339.

11. E. C. Titchmarsh, *Eigenfunction expansions associated with second-order differential equations*, Oxford, Clarendon Press, 1946.

12. A. Wintner, *Stability and spectrum in the wave mechanics of lattices*, Physical Review vol. 72 (1947) pp. 81–82; *On the normalization of characteristic differentials in continuous spectra*, ibid. vol. 72 (1947) pp. 516–517; *On the location of continuous spectra*, Amer. J. Math. vol. 70 (1948) pp. 22–30; *Asymptotic integrations of the adiabatic oscillator*, Amer. J. Math. vol. 69 (1947) pp. 251–272, and Duke Math. J. vol. 15 (1948) pp. 53–67. P. Hartmann and A. Wintner, *An oscillation theorem for continuous spectra*, Proc. Nat. Acad. Sci. U.S.A. vol. 33 (1947) pp. 376–379.

13. H. Weyl, *Das asymptotische Verteilungsgesetz der Eigenwerte linearer partieller Differentialgleichungen*, Math. Ann. vol. 71 (1911) pp. 441–469; *Ueber die Abhängigkeit der Eigenschwingungen einer Membran von deren Begrenzung*, J. Reine Angew. Math. vol. 141 (1912) pp. 1–11.

14. See his account in: R. Courant and D. Hilbert, *Methoden der mathematischen Physik*, vol. 1, 2d ed., Berlin, 1931, in particular Chap. 6, §4.

15. A. Weinstein, *Étude des spectres des équations aux dérivées partielles de la théorie des plaques élastiques*, Mémorial des Sciences Mathématiques, vol. 88, 1937.

16. N. Aronszajn, *Rayleigh-Ritz and A. Weinstein methods for approximation of eigenvalues*, I: *Operators in a Hilbert space*, Proc. Nat. Acad. Sci. U.S.A. vol. 34 (1948) pp. 474–480; II: *Differential operators*, ibid. pp. 594–601.

17. T. Carleman, *Propriétés asymptotiques des fonctions fondamentales des membranes vibrantes*, Förhandlingar Skandinaviska Matematikerkongressen Stockholm, 1934, pp. 34–44; *Ueber die asymptotische Verteilung der Eigenwerte partieller Differentialgleichungen*, Berichte über die Verhandlungen der Sächsischen Akademie der Wissenschaften zu Leipzig vol. 88 (1936) pp. 119–132.

18. Å. Pleijel, *Propriétés asymptotiques des fonctions et valeurs propres de certains problèmes de vibration*, Arkiv för Matematik, Astronomi och Fysik vol. 27 A, No. 13, 1940; *Sur la distribution des valeurs propres de problèmes régis par l'équation $\Delta u + \lambda k(x, y)u = 0$*, ibid. vol. 29 B, No. 7, 1943; *On Hilbert-Schmidt's theorem in the theory of partial differential equations*, Fysiografiska Sälskapets i Lund Förhandlingar vol. 17, No. 2, 1946; *Asymptotic relations for the eigenfunctions of certain boundary problems of polar type*, Amer. J. Math. vol. 70 (1948) pp. 892–907.

19. S. Minakshisundaram, *A generalization of Epstein zeta functions*, Canadian Journal of Mathematics vol. 1 (1949) pp. 320–327. S. Minakshisundaram and Å Pleijel, *Some properties of the eigenfunctions of the Laplace-operator on Riemannian manifolds*, Canadian Journal of Mathematics vol. 1 (1949) pp. 242–256.

20. S. Bochner, *Summation of derived Fourier series (An application to Fourier expansions on compact Lie groups)*, Ann. of Math. vol. 37 (1936) pp. 345–356.

21. H. B. G. Casimir, *Rotation of a rigid body in quantum mechanics*, Leiden thesis, 1931.

22. F. Peter and H. Weyl, *Die Vollständigkeit der primitiven Darstellungen einer geschlossenen kontinuierlichen Gruppe*, Math. Ann. vol. 97 (1927) pp. 737–755.

23. H. Weyl, *Almost periodic invariant vector sets in a metric vector space*, Amer. J. Math. vol. 71 (1949) pp. 178–205.

24. I quote but the two most important papers: H. Bohr, *Zur Theorie der fastperiodischen Funktionen* I, Acta Math. vol. 45 (1925) pp. 29–127. J. von Neumann, *Almost periodic functions in a group* I, Trans. Amer. Math. Soc. vol. 36 (1934) pp. 445–492. For further references see [**23**].

25. I. Gelfand and D. Rykov, *Irreducible unitary representations of locally compact groups*, Rec. Math. (Mat. Sbornik) N. S. vol. 3 (1943) pp. 301–316; also: Academy of Sciences of the USRR. Journal of Physics vol. 10 (1946) pp. 93–94. V. Bargmann, *Irreducible unitary representations of the Lorentz group*, Ann. of Math. (2) vol. 48 (1947) pp. 568–640. I. E. Segal, *Irreducible representations of operator algebras*, Bull. Amer. Math. Soc. vol. 53 (1947) pp. 73–88.

26. H. Weyl, *Das asymptotische Verteilungsgesetz der Eigenschwingungen eines beliebig gestalteten elastischen Körpers*, Rend. Circ. Mat. Palermo vol. 39 (1915) pp. 1–49

151.

Elementary proof of a minimax theorem due to von Neumann

Contributions to the theory of games I, Annals of Mathematics Studies, Princeton University Press 24, 19—25 (1950)

J. von Neumann's minimax problem in the theory of games belongs to the theory of linear inequalities and can be approached in the same elementary way in which I proved the fundamental facts about convex pyramids. As elementary are considered such operations in an ordered field K of numbers as require nothing but addition, subtraction, multiplication and division, and the decision whether a given number is > 0 or $= 0$ or < 0. Decisions about a set of numbers are elementary only if they concern a finite set, the members of which are exhibited one by one. In such a sequence of numbers $\alpha_1, \ldots, \alpha_n$ we can find the smallest, $\min \alpha_k$, and the biggest, $\max \alpha_k$. As for the field K no continuity axioms, not even the axiom of Archimedes, are assumed.

Let there be given a matrix of numbers

$$a_{ik}(i = 1, \ldots, m;\ k = 1, \ldots, n) .$$

For any point $\eta = (\eta_1, \ldots, \eta_n)$ in n-space set

$$m(\eta) = \min_i (\Sigma_k a_{ik}\eta_k) ,$$

and for any point $\xi = (\xi_1, \ldots, \xi_m)$ in m-space:

$$M(\xi) = \max_k (\Sigma_i a_{ik}\xi_i) .$$

After suitably constructing a number λ_o, a point in n-space $\eta = \eta^o$ satisfying the conditions

$$\eta_k^o \geq 0 (k = 1, \ldots, n), \Sigma_k \eta_k^o = 1$$

and a point in m-space $\xi = \xi^o$ satisfying the corresponding conditions

$$\xi_i^o \geq 0 (i = 1, \ldots, m), \Sigma_i \xi_i^o = 1 ,$$

we are going to establish the following

Fundamental facts:

$$m(\eta) \leq \lambda_o$$

---

*Accepted as a direct contribution to ANNALS OF MATHEMATICS STUDY No. 24.

whenever

(1) $$\eta_k \geq 0 (k = 1, \ldots, n), \Sigma_k \eta_k = 1 ,$$

the upper bound λ_0 being obtained for $\eta = \eta^0$, and

$$M(\xi) \geq \lambda_0$$

whenever

(2) $$\xi_i \geq 0 (i = 1, \ldots, m), \Sigma_i \xi_i = 1 ,$$

the lower bound being obtained for $\xi = \xi^0$.

One part of the main theorem about convex pyramids which I proved in elementary fashion (and independently of the rest) and which I am now going to repeat as Lemma 1 is concerned with a configuration Σ of points

$$b_j = (b_{j1}, \ldots, b_{jn}) \quad (j = 1, \ldots, r)$$

which span n-space, and the extreme supports of this configuration. Given a linear form $(\gamma x) = \gamma_1 x_1 + \ldots + \gamma_n x_n$, let us say that the point x lies in the plane γ or in the half-space γ if $(\gamma x) = 0$ or $(\gamma x) \geq 0$ respectively. A form $\gamma \neq 0$ such that all points b_j of the set Σ lie in the half-space γ is called a support of the set, and this support is extreme if $n - 1$ linearly independent b_j, for instance

$$b_{j_1}, \ldots, b_{j_{n-1}} \quad (j_1 < j_2 < \ldots < j_{n-1}) ,$$

lie in the plane γ. In this case (γx) is proportional to

$$|b_{j_1}, \ldots, b_{j_{n-1}}, x| .$$

LEMMA 1. Either the configuration Σ has an extreme support, or every point p is representable in the form

$$p = \sum_{j=1}^{r} \mu_j b_j \quad (\mu_j \geq 0) .$$

One may cancel here the adjective extreme and then add the corollary: If Σ has a support it has an extreme support.

We are interested in the case where the configuration Σ consists of the n unit points

$$e_1 = (1, 0, \ldots, 0), e_2 = (0, 1, \ldots, 0), \ldots, e_n = (0, 0, \ldots, 1)$$

and some further points

$$a_i = (a_{i1}, \ldots, a_{in}) \quad (i = 1, \ldots, m):$$

$$b_k = e_k (k = 1, \ldots, n); \; b_{n+i} = a_i (i = 1, \ldots, m) .$$

Set $e = e_1 + \ldots + e_n = (1, \ldots, 1)$. Every support γ $(\neq 0)$ satisfies the inequalities

$$(\gamma e_k) = \gamma_k \geq 0 \quad (k = 1, \ldots, n)$$

and hence we may normalize by

$$(\gamma e) = \gamma_1 + \ldots + \gamma_n = 1 \ .$$

If the second alternative of Lemma 1 takes place, represent $-e$ and add

$$e = 1 \cdot e_1 + \ldots + 1 \cdot e_n + 0 \cdot a_1 + \ldots + 0 \cdot a_m \ .$$

One thus finds a representation of zero,

$$0 = \sum \rho_k e_k + \sum \xi_i a_i \quad \text{with} \quad \rho_k \geq 1, \ \xi_i \geq 0 \ ,$$

hence

$$\sum_i \xi_i a_{ik} = -\rho_k < 0 \quad (k = 1, \ldots, n) \ .$$

Thus we may state the following two facts:

LEMMA 2. If the system Σ consisting of the points $e_k (k = 1, \ldots, n)$ and $a_i (i = 1, \ldots, m)$ has a support then it has an extreme support.

LEMMA 3. If it has no support, i.e. if the inequalities

$$\eta_k \geq 0 \quad (k = 1, \ldots, n), \sum_k a_{ik} \eta_k \geq 0 \quad (i = 1, \ldots, m)$$

have none but the trivial solution $\eta = 0$, then the n inequalities

$$\sum_i \xi_i a_{ik} < 0 \quad (k = 1, \ldots, n)$$

are solvable by non-negative ξ_i.

We find it convenient to apply Lemma 3 to the matrix $-||a_{ki}||$ rather than to $||a_{ik}||$ and to change the notation i, m, ξ into k, n, η. Then Lemma 3 takes on the (nearly identical) form:

LEMMA 4. Either the system

$$\sum_k a_{ik} \eta_k > 0 \quad (i = 1, \ldots, m), \quad \eta_k \geq 0 \quad (k = 1, \ldots, n)$$

or the system

$$\sum_i \xi_i a_{ik} \leq 0 \ (k = 1, \ldots, n), \ \xi_i \geq 0 \ (i = 1, \ldots, m), \sum \xi_i = 1$$

is solvable.

One needs but form $\sum_{i,k}\xi_i a_{ik}\eta_k$ in order to see that this is a disjunctive either-or.

Let us now introduce a parameter λ called time and try to find out for which values of λ the system Σ_λ consisting of the n points e_k and the m points $a_i - \lambda e$ has or has not a support. In the first case we say that λ makes good. If λ makes good so does every $\lambda' \leq \lambda$, and the same support γ will do for all of them. Indeed a support γ of Σ_λ satisfies the inequalities

$$\gamma_1 \geq 0, \ldots, \gamma_n \geq 0, (\gamma e) > 0$$

and thus

$$(\gamma, a_i - \lambda' e) = (\gamma, a_i - \lambda e) + (\lambda - \lambda')(\gamma e) \geq 0$$

holds as a consequence of $(\gamma, a_i - \lambda e) \geq 0$ and $\lambda - \lambda' \geq 0$.

Take any point η satisfying (1). For

$$\lambda = \min_i (a_i\eta) = \min_i (\Sigma_k a_{ik}\eta_k) = m(\eta)$$

one has

$$(\eta, a_i - \lambda e) \geq 0 (i = 1, \ldots, m), (\eta e_k) \geq 0 (k = 1, \ldots, n) ;$$

hence this λ makes good. On the other hand, given λ and a $\gamma \neq 0$ satisfying the relations

$$(\gamma, a_i - \lambda e) \geq 0 (i = 1, \ldots, m), \gamma_k = (\gamma e_k) \geq 0 (k = 1, \ldots, n)$$

one finds for every point of the form

$$p = \Sigma_i \xi_i a_i \quad (\xi_i \geq 0, \xi_1 + \ldots + \xi_m = 1)$$

the inequality

$$(\gamma, p - \lambda e) \geq 0 \quad \text{or} \quad \Sigma_k \gamma_k (p_k - \lambda) \geq 0 .$$

Hence at least <u>one</u> of the $p_k - \lambda \geq 0$, or

$$\lambda \leq \max_k p_k = \max_k (\Sigma_i \xi_i a_{ik}) = M(\xi) .$$

Any $M(\xi)$ corresponding to a point ξ fulfilling the relations (2) is therefore an upper bound for those λ that make good.

Use the unified notation

$$\delta_k = 0 \quad \text{for} \quad k = 1, \ldots, n; \quad \delta_{n+i} = 1 \quad \text{for} \quad i = 1, \ldots, m;$$

$$b_j(\lambda) = b_j - \lambda\delta_j e \quad (j = 1, \ldots, r = n + m) .$$

In view of Lemma 2 pick any (n - 1)-combination $J = \{j_1 < j_2 < \ldots < j_{n-1}\}$ out of the indices $j = 1, \ldots, r$. If $b_{j_1}(\lambda), \ldots, b_{j_{n-1}}(\lambda)$ are linearly independent and the plane going through them is a supporting plane for the entire set Σ_λ of the $b_j(\lambda)$ then we shall say that the combination J is alive at the moment λ. Let us express that the combination[1] $J = \{1, 2, \ldots, n-1\}$ is alive at the moment λ', or that we have a $\gamma \neq 0$ such that

$$(\gamma, b_j(\lambda')) \geq 0 \quad \text{for all} \quad j$$
$$\text{and} = 0 \quad \text{for} \quad j = 1, \ldots, n-1 .$$

This implies $\gamma_k \geq 0$, hence $(\gamma e) > 0$, and we may normalize by $(\gamma e) = 1$. Therefore e is independent of the $n-1$ linearly independent points $b_1(\lambda'), \ldots, b_{n-1}(\lambda')$, or

$$D = |b_1, \ldots, b_{n-1}, e| = |b_1(\lambda'), \ldots, b_{n-1}(\lambda'), e| \neq 0 .$$

The normalized expression of (γx) is

$$(\gamma x) = \frac{1}{D} |b_1(\lambda'), \ldots, b_{n-1}(\lambda'), x| .$$

Consequently one has to form the r functions

$$\varphi_j(\lambda) = \frac{1}{D} |b_1(\lambda), \ldots, b_{n-1}(\lambda), b_j(\lambda)| .$$

[$\varphi_1(\lambda), \ldots, \varphi_{n-1}(\lambda)$ vanish identically.] If J is alive at the moment λ' the simultaneous inequalities

$$\varphi_j(\lambda') \geq 0 \quad (j = 1, \ldots, r)$$

hold (and vice versa). Notice that

$$|b_1(\lambda), \ldots, b_{n-1}(\lambda), b_n(\lambda)| = \begin{vmatrix} b_{11}, & \ldots, & b_{1n}, & \lambda\delta_1 \\ \ldots & \ldots & \ldots & \ldots \\ b_{n1}, & \ldots, & b_{nn}, & \lambda\delta_n \\ 1, & \ldots, & 1, & 1 \end{vmatrix}$$

is a linear function of λ. Write therefore

$$\varphi_j(\lambda) = \alpha_j^0 - \alpha_j \lambda$$

and distinguish between indices j of the first and second class by $\alpha_j \leq 0$ and $\alpha_j > 0$ respectively.

For an index j of the first class

I) $$\varphi_j(\lambda) \geq \varphi_j(\lambda') \geq 0 \quad \text{for} \quad \lambda \geq \lambda' .$$

---

[1]This combination serves merely as an example for an arbitrary J. Hence the reader should ignore the special value 0 which δ_k takes on for $k = 1, \ldots, n$.

For an index j of the second class the inequality $\varphi_j(\lambda') \geq 0$ states that

$$\lambda' \leq \alpha_j^0/\alpha_j .$$

The second class cannot be vacuous; for then J would stay alive at all times $\lambda \geq \lambda'$, and every λ would make good, which we know to be impossible. Determine the minimum λ_J of the quotients α_j^0/α_j that correspond to the indices j of the second class. Then

II) $$\lambda' \leq \lambda_J$$

and

III) $$\varphi_j(\lambda_J) \geq 0, \ \varphi_j(\lambda) \geq 0 \quad \text{for} \quad \lambda \leq \lambda_J$$

and every j of the second class. The relations I), II), III) show: if J is alive at the moment λ' then $\lambda' \leq \lambda_J$ and J keeps alive during the whole period $\lambda' \leq \lambda \leq \lambda_J$; in other words, the uninterrupted life of J terminates at $\lambda = \lambda_J$.

Rearranging the argument, I describe once more how the "admissible" combinations J are separated from the others which never come to life. The combination $J = \{1, \ldots, n-1\}$ is ruled out (1) if the determinant D vanishes. Otherwise form the functions $\varphi_j(\lambda) = \alpha_j^0 - \alpha_j\lambda$ and rule out J if (2) all $\alpha_j \leq 0$. But if some α_i are positive take the least of the quotients α_j^0/α_j corresponding to those j for which $\alpha_j > 0$ (j's of the second class) and call it λ_J. The combination J is thrown away unless (3) $\varphi_j(\lambda_J) \geq 0$ for all j, even for those of the first class. A J which passes the three consecutive tests is admissible. Then we may state our result as follows:

> LEMMA 5. If λ makes good then there is an admissible combination J whose death occurs at a time $\lambda_J \geq \lambda$.

Indeed if the system Σ_λ has a support it also has an extreme support determined by $n-1$ linearly independent points $b_{j_1}(\lambda), \ldots, b_{j_{n-1}}(\lambda)$. The combination $J = \{j_1, \ldots, j_{n-1}\}$ is then alive at the moment λ.

Lemma 5 shows that there are admissible combinations J. Choose now for λ_0 the largest of the λ_J that correspond to the admissible combinations J and let $J = J_0$ be that admissible combination for which $\lambda_{J_0} = \lambda_0$. This $J_0 = \{j_1, \ldots, j_{n-1}\}$ yields an extreme support

$$(\eta^o x) = \frac{|b_{j_1}(\lambda_o), \ldots, b_{j_{n-1}}(\lambda_o), x|}{|b_{j_1}(\lambda_o), \ldots, b_{j_{n-1}}(\lambda_o), x|}$$

for Σ_{λ_o} (and hence a support η^o for Σ_λ, $\lambda \leq \lambda_o$). Thus λ_o makes good, but according to Lemma 5 no $\lambda > \lambda_o$ does.

We have seen that $m(\eta)$ is a λ that makes good whenever the point η satisfies the conditions (1), and therefore $m(\eta) \leq \lambda_o$. The upper bound is actually reached for $\eta = \eta^o$; indeed

$$(\eta^o, a_i - \lambda_o e) \geq 0 \quad \text{or} \quad \lambda_o \leq (\eta^o a_i) \quad (i = 1, \ldots, m), \lambda_o \leq m(\eta^o) .$$

This is the first of our Fundamental facts.

On the other hand there can be no γ such that

$$(\gamma, a_i - \lambda_o e) > 0 \ (i = 1, \ldots, m), \ (\gamma e_k) \geq 0 \ (k = 1, \ldots, n), \ (\gamma e) = 1 .$$

For then $\lambda_1 = \min_i (\gamma a_i) > \lambda_o$ would make good, against our better knowledge. Hence by applying Lemma 4 to the matrix $a_{ik} - \lambda_o$ we find a solution $\xi = \xi^o$ of

$$\Sigma_i \xi_i (a_{ik} - \lambda_o) \leq 0, \ \xi_i \geq 0, \Sigma \xi_i = 1 ,$$

so that we have

$$M(\xi^o) \leq \lambda_o, \ \xi_i^o \geq 0 \quad (i = 1, \ldots, m), \ \Sigma \xi_i^o = 1 . \tag{3}$$

But for every point ξ satisfying (2) and every λ that makes good, $\lambda \leq M(\xi)$, in particular $\lambda_o \leq M(\xi)$. According to (3) the lower bound λ_o is actually reached for $\xi = \xi^o$. This completes the proof.

152.

A half-century of mathematics

The American Mathematical Monthly 58, 523—553 (1951)

1. Introduction. Axiomatics. Mathematics, beside astronomy, is the oldest of all sciences. Without the concepts, methods and results found and developed by previous generations right down to Greek antiquity, one cannot understand either the aims or the achievements of mathematics in the last fifty years. Mathematics has been called the science of the infinite; indeed, the mathematician invents finite constructions by which questions are decided that by their very nature refer to the infinite. That is his glory. Kierkegaard once said religion deals with what concerns man unconditionally. In contrast (but with equal exaggeration) one may say that mathematics talks about the things which are of no concern at all to man. Mathematics has the inhuman quality of starlight, brilliant and sharp, but cold. But it seems an irony of creation that man's mind knows how to handle things the better the farther removed they are from the center of his existence. Thus we are cleverest where knowledge matters least: in mathematics, especially in number theory. There is nothing in any other science that, in subtlety and complexity, could compare even remotely with such mathematical theories as for instance that of algebraic class fields. Whereas physics in its development since the turn of the century resembles a mighty stream rushing on in one direction, mathematics is more like the Nile delta, its waters fanning out in all directions. In view of all this: dependence on a long past, other-worldliness, intricacy, and diversity, it seems an almost hopeless task to give a non-esoteric account of what mathematicians have done during the last fifty years. What I shall try to do here is, first to describe in somewhat vague terms general trends of development, and then in more precise language explain the most outstanding mathematical notions devised, and list some of the more important problems solved, in this period.

One very conspicuous aspect of twentieth century mathematics is the enormously increased role which the axiomatic approach plays. Whereas the axiomatic method was formerly used merely for the purpose of elucidating the foundations on which we build, it has now become a tool for concrete mathematical research. It is perhaps in algebra that it has scored its greatest successes. Take for instance the system of real numbers. It is like a Janus head facing in two directions: on the one side it is the field of the algebraic operations of addition and multiplication; on the other hand it is a continuous manifold, the parts of which are so connected as to defy exact isolation from each other. The one is the algebraic, the other the topological face of numbers. Modern axiomatics, simple-minded as it is (in contrast to modern politics), does not like such ambiguous mixtures of peace and war, and therefore cleanly separated both aspects from each other.

In order to understand a complex mathematical situation it is often convenient to separate in a natural manner the various sides of the subject in question, make each side accessible by a relatively narrow and easily surveyable group of notions and of facts formulated in terms of these notions, and finally return to the whole by uniting the partial results in their proper specialization. The last synthetic act is purely mechanical. The art lies in the first, the analytic act of suitable separation and generalization. Our mathematics of the last decades has wallowed in generalizations and formalizations. But one misunderstands this tendency if one thinks that generality was sought merely for generality's sake. The real aim is simplicity: every natural generalization simplifies since it reduces the assumptions that have to be taken into account. It is not easy to say what constitutes a natural separation and generalization. For this there is ultimately no other criterion but fruitfulness: the success decides. In following this procedure the individual investigator is guided by more or less obvious analogies and by an instinctive discernment of the essential acquired through accumulated previous research experience. When systematized the procedure leads straight to axiomatics. Then the basic notions and facts of which we spoke are changed into undefined terms and into axioms involving them. The body of statements deduced from these hypothetical axioms is at our disposal now, not only for the instance from which the notions and axioms were abstracted, but wherever we come across an interpretation of the basic terms which turns the axioms into true statements. It is a common occurrence that there are several such interpretations with widely different subject matter.

The axiomatic approach has often revealed inner relations between, and has made for unification of methods within, domains that apparently lie far apart. This tendency of several branches of mathematics to coalesce is another conspicuous feature in the modern development of our science, and one that goes side by side with the apparently opposite tendency of axiomatization. It is as if you took a man out of a milieu in which he had lived not because it fitted him but from ingrained habits and prejudices, and then allowed him, after thus setting him free, to form associations in better accordance with his true inner nature.

In stressing the importance of the axiomatic method I do not wish to exaggerate. Without inventing new constructive processes no mathematician will get very far. It is perhaps proper to say that the strength of modern mathematics lies in the interaction between axiomatics and construction. Take algebra as a representative example. It is only in this century that algebra has come into its own by breaking away from the one universal system Ω of numbers which used to form the basis of all mathematical operations as well as all physical measurements. In its newly-acquired freedom algebra envisages an infinite variety of "number fields" each of which may serve as an operational basis; no attempt is made to embed them into the one system Ω. Axioms limit the possibilities for the number concept; constructive processes yield number fields that satisfy the axioms.

In this way algebra has made itself independent of its former master analysis and in some branches has even assumed the dominant role. This development in mathematics is paralleled in physics to a certain degree by the transition from classical to quantum physics, inasmuch as the latter ascribes to each physical structure its own system of observables or quantities. These quantities are subject to the algebraic operations of addition and multiplication; but as their multiplication is non-commutative, they are certainly not reducible to ordinary numbers.

At the International Mathematical Congress in Paris in 1900 David Hilbert, convinced that problems are the life-blood of science, formulated twenty-three unsolved problems which he expected to play an important role in the development of mathematics during the next era. How much better he predicted the future of mathematics than any politician foresaw the gifts of war and terror that the new century was about to lavish upon mankind! We mathematicians have often measured our progress by checking which of Hilbert's questions had been settled in the meantime. It would be tempting to use his list as a guide for a survey like the one attempted here. I have not done so because it would necessitate explanation of too many details. I shall have to tax the reader's patience enough anyhow.

Part I. Algebra. Number Theory. Groups.

2. Rings, Fields, Ideals. Indeed, at this point it seems impossible for me to go on without illustrating the axiomatic approach by some of the simplest algebraic notions. Some of them are as old as Methuselah. For what is older than the sequence of *natural numbers* 1, 2, 3, · · · , by which we count? Two such numbers a, b may be added and multiplied ($a + b$ and $a \cdot b$). The next step in the genesis of numbers adds to these positive *integers* the negative ones and zero; in the wider system thus created the operation of addition permits of a unique inversion, subtraction. One does not stop here: the integers in their turn get absorbed into the still wider range of *rational numbers* (fractions). Thereby division, the operation inverse to multiplication, also becomes possible, with one notable exception however: division by zero. (Since $b \cdot 0 = 0$ for every rational number b, there is no inverse b of 0 such that $b \cdot 0 = 1$.) I now formulate the fundamental facts about the operations "plus" and "times" in the form of a table of axioms:

Table T

(1) The commutative and associative laws for addition,

$$a + b = b + a. \qquad a + (b + c) = (a + b) + c.$$

(2) The corresponding laws for multiplication.

(3) The distributive law connecting addition with multiplication

$$c \cdot (a + b) = (c \cdot a) + (c \cdot b).$$

(4) The axioms of subtraction: (4_1) There is an element o (0, "zero") such that $a + o = o + a = a$ for every a. (4_2) To every a there is a number $-a$ such that $a + (-a) = (-a) + a = o$.

(5) The axioms of division: (5_1) There is an element e (1, "unity") such that $a \cdot e = e \cdot a = a$ for every a. (5_2) To every $a \neq o$ there is an a^{-1} such that $a \cdot a^{-1} = a^{-1} \cdot a = e$.

By means of (4_2) and (5_2) one may introduce the difference $b - a$ and the quotient b/a as $b + (-a)$ and $b \cdot a^{-1}$, respectively.

When the Greeks discovered that the ratio ($\sqrt{2}$) between diagonal and side of a square is not measurable by a rational number, a further extension of the number concept was called for. However, all measurements of continuous quantities are possible only approximately, and always have a certain range of inaccuracy. Hence rational numbers, or even finite decimal fractions, can and do serve the ends of mensuration provided they are interpreted as approximations, and a calculus with approximate numbers seems the adequate numerical instrument for all measuring sciences. But mathematics ought to be prepared for any subsequent refinement of measurements. Hence dealing, say, with electric phenomena, one would be glad if one could consider the approximate values of the charge e of the electron which the experimentalist determines with ever greater accuracy as approximations of one definite *exact* value e. And thus, during more than two millenniums from Plato's time until the end of the nineteenth century, the mathematicians worked out an exact number concept, that of *real numbers*, that underlies all our theories in natural science. Not even to this day are the logical issues involved in that concept completely clarified and settled. The rational numbers are but a small part of the real numbers. The latter satisfy our axioms no less than the rational ones, but their system possesses a certain completeness not enjoyed by the rational numbers, and it is this, their "topological" feature, on which the operations with infinite sums and the like, as well as all continuity arguments, rest. We shall come back to this later.

Finally, during the Renaissance *complex numbers* were introduced. They are essentially pairs $z = (x, y)$ of real numbers x, y, pairs for which addition and multiplication are defined in such a way that all axioms hold. On the ground of these definitions $e = (1, 0)$ turns out to be the unity, while $i = (0, 1)$ satisfies the equation $i \cdot i = -e$. The two members x, y of the pair z are called its real and imaginary parts, and z is usually written in the form $xe + yi$, or simply $x + yi$. The usefulness of the complex numbers rests on the fact that every algebraic equation (with real or even complex coefficients) is solvable in the field of complex numbers. The analytic functions of a complex variable are the subject of a particularly rich and harmonious theory, which is the show-piece of classical nineteenth century analysis.

A set of elements for which the operations $a + b$ and $a \cdot b$ are so defined as to satisfy the axioms (1)–(4) is called a *ring*; it is called a *field* if also the axioms (5) hold. Thus the common integers form a ring I, the rational numbers form a field ω; so do the real numbers (field Ω) and the complex numbers (field Ω^*). But these are by no means the only rings or fields. The polynomials of all possible degrees h,

(1) $$f = f(x) = a_0 + a_1x + a_2x^2 + \cdots + a_hx^h,$$

with coefficients a_i taken from a given ring R (*e.g.* the ring I of integers, or the field ω), called "polynomials over R," form a ring $R[x]$. Here the variable or indeterminate x is to be looked upon as an empty symbol; the polynomial is really nothing but the sequence of its coefficients $a_0, a_1, a_2, \cdots$. But writing it in the customary form (1) suggests the rules for the addition and multiplication of polynomials which I will not repeat here. By substituting for the variable x a definite element ("number") γ of R, or of a ring P containing R as a subring, one projects the elements f of $R[x]$ into elements α of P, $f \rightarrow \alpha$: the polynomial $f = f(x)$ goes over into the number $\alpha = f(\gamma)$. This mapping $f \rightarrow \alpha$ is ***homomorphic***, *i.e.*, it preserves addition and multiplication. Indeed, if the substitution of γ for x carries the polynomial f into α and the polynomial g into β then it carries $f + g, f \cdot g$ into $\alpha + \beta, \alpha \cdot \beta$, respectively.

If the product of two elements of a ring is never zero unless one of the factors is, one says that the ring is without null-divisor. This is the case for the rings discussed so far. A field is always a ring without null-divisor. The construction by which one rises from the integers to the fractions can be used to show that any ring R with unity and without null-divisor may be imbedded in a field k, the quotient field, such that every element of k is the quotient a/b of two elements a and b of R, the second of which (the denominator) is not zero.

Writing $1a, 2a, 3a, \cdots$, for $a, a + a, a + a + a$, *etc.*, we use the natural numbers $n = 1, 2, 3, \cdots$, as multipliers for the elements a of a ring or a field. Suppose the ring contains the unity e. It may happen that a certain multiple ne of e equals zero; then one readily sees that $na = 0$ for every element a of the ring. If the ring is without null divisors, in particular if it is a field and p is the least natural number for which $pe = 0$, then p is necessarily a prime number like 2 or 3 or 5 or 7 or 11 $\cdots$. One thus distinguishes fields of prime characteristic p from those of characteristic 0 in which no multiple of e is zero.

Plot the integers $\cdots, -2, -1, 0, 1, 2, \cdots$ as equidistant marks on a line. Let n be a natural number $\geqq 2$ and roll this line upon a wheel of circumference n. Then any two marks a, a' coincide, the difference $a - a'$ of which is divisible by n. (The mathematicians write $a \equiv a' \pmod{n}$; they say: a congruent to a' modulo n.) By this identification the ring of integers I goes over into a ring I_n consisting of n elements only (the marks on the wheel), as which one may take the "residues" $0, 1, \cdots, n - 1$. Indeed, congruent numbers give congruent results under both addition and multiplication: $a \equiv a', b \equiv b' \pmod{n}$ imply $a + b \equiv a' + b', a \cdot b \equiv a' \cdot b' \pmod{n}$. For instance, modulo 12 we have $7 + 8 = 3, 5 \cdot 8 = 4$ because 15 leaves the residue 3 and 40 the residue 4 if divided by 12. The ring I_{12} is not without null divisors since $3 \cdot 4$ is divisible by 12, but neither 3 nor 4 is. However, if p is a natural prime number, then I_p has no null divisor and is even a field; for as the ancient Greeks proved by an ingenious procedure (Euclid's algorism), for every integer a not divisible by

p there is one, a', such that $a \cdot a' \equiv 1 \pmod{p}$. This Euclidean theorem is at the basis of the whole of number theory. The example shows that there are fields of any given prime characteristic p.

In any ring R one may introduce the notions of unit and prime element as follows. The ring element a is a unit if it has a reciprocal a' in the ring, such that $a' \cdot a = e$. The element a is composite if it may be decomposed into two factors $a_1 \cdot a_2$, neither of which is a unit. A prime number is one that is neither a unit nor composite. The units of I are the numbers $+1$ and -1. The units of the ring $k[x]$ of polynomials over a field k are the non-vanishing elements of k (polynomials of degree 0). According to the Greek discovery of the irrationality of $\sqrt{2}$ the polynomial $x^2 - 2$ is prime in the ring $\omega[x]$; but, of course, not in $\Omega[x]$, for there it splits into the two linear factors $(x - \sqrt{2})(x + \sqrt{2})$. Euclid's algorism is also applicable to polynomials $f(x)$ of one variable x over any field k. Hence they satisfy Euclid's theorem: Given a prime element $P = P(x)$ in this ring $k[x]$ and an element $f(x)$ of $k[x]$ not divisible by $P(x)$, there exists another polynomial $f'(x)$ over k such that $\{f(x) \cdot f'(x)\} - 1$ is divisible by $P(x)$. Identification of any elements f and g of $k[x]$, the difference of which is divisible by P, therefore changes the ring $k[x]$ into a field, the "residue field κ of $k[x]$ modulo P." Example: $\omega[x]$ mod $x^2 - 2$. (Incidentally the complex numbers may be described as the elements of the residue field of $\Omega[x]$ mod $x^2 + 1$.) Strangely enough, the fundamental Euclidean theorem does not hold for polynomials of two variables x, y. For instance, $P(x, y) = x - y$ is a prime element of $\omega[x, y]$, and $f(x, y) = x$ an element not divisible by $P(x, y)$. But a congruence

$$x \cdot f'(x, y) \equiv 1 \pmod{x - y}$$

is impossible. Indeed, it would imply $-1 + x \cdot f'(x, x) = 0$, contrary to the fact that the polynomial of one indeterminate x,

$$-1 + x \cdot f'(x, x) = -1 + c_1 x + c_2 x^2 + \cdots,$$

is not zero. Thus the ring $\omega[x, y]$ does not obey the simple laws prevailing in I and in $\omega[x]$.

Consider κ, the residue field of $\omega[x]$ mod $x^2 - 2$. Since for any two polynomials $f(x)$, $f'(x)$ which are congruent mod $x^2 - 2$ the numbers $f(\sqrt{2})$, $f'(\sqrt{2})$ coincide, the transition $f(x) \to f(\sqrt{2})$ maps κ into a sub-field $\omega[\sqrt{2}]$ of Ω consisting of the numbers $a + b\sqrt{2}$ with rational a, b. Another such projection would be $f(x) \to f(-\sqrt{2})$. In former times one looked upon κ as the part $\omega[\sqrt{2}]$ of the continuum Ω or Ω^* of all real or all complex numbers; one wished to embed everything into this universe Ω or Ω^* in which analysis and physics operate. But as we have introduced it here, κ is an algebraic entity the elements of which are not numbers in the ordinary sense. It requires for its construction no other numbers but the rational ones. It has nothing to do with Ω, and ought not to be confused with the one or the other of its two projections into Ω. More generally, if $P = P(x)$ is any prime element in $\omega[x]$ we can form the residue field κ_P of $\omega[x]$ modulo P. To be sure, if δ is any of the real or complex roots of the equation $P(x)=0$ in Ω^* then $f(x) \rightarrow f(\delta)$ defines a homomorphic projection of κ_P into Ω^*. But the projection is not κ_P itself.

Let us return to the ordinary integers $\cdots, -2, -1, 0, 1, 2, \cdots$, which form the ring I. The multiples of 5, *i.e.*, the integers divisible by 5, clearly form a ring. It is a ring without unity, but it has another important peculiarity: not only does the product of any two of its elements lie in it, but all the integral multiples of an element do. The queer term *ideal* has been introduced for such a set: Given a ring R, an R-ideal $(\mathfrak{a})$ is a set of elements of R such that (1) sum and difference of any two elements of $(\mathfrak{a})$ are in $(\mathfrak{a})$, (2) the product of an element in $(\mathfrak{a})$ by any element of R is in $(\mathfrak{a})$. We may try to describe a divisor $\mathfrak{a}$ by the set of all elements divisible by $\mathfrak{a}$. One would certainly expect this set to be an ideal $(\mathfrak{a})$ in the sense just defined. Given an ideal $(\mathfrak{a})$, there may not exist an actual element a of R such that $(\mathfrak{a})$ consists of all multiples $j = m \cdot a$ of a (m any element in R). But then we would say that $(\mathfrak{a})$ stands for an "*ideal* divisor" $\mathfrak{a}$: the words "the element j of R is divisible by $\mathfrak{a}$" would simply mean: "j belongs to $(\mathfrak{a})$." In the ring I of common integers all divisors are actual.

But this is not so in every ring. An algebraic surface in the three-dimensional Euclidean space with the Cartesian coordinates x, y, z is defined by an equation $F(x, y, z) = 0$ where F is an element of ${}^3\Omega = \Omega[x, y, z]$, *i.e.*, a polynomial of the variables x, y, z with real coefficients. F is zero in all the points of the surface; but the same is true for every multiple $L \cdot F$ of F (L being any element of ${}^3\Omega$), in other words, for every polynomial of the ideal (F) in ${}^3\Omega$. Two simultaneous polynomial equations

$$F_1(x, y, z) = 0, \qquad F_2(x, y, z) = 0$$

will in general define a curve, the intersection of the surface $F_1 = 0$ and the surface $F_2 = 0$. The polynomials $(L_1 \cdot F_1) + (L_2 \cdot F_2)$ formed by arbitrary elements L_1, L_2 of ${}^3\Omega$ form an ideal (F_1, F_2), and all these polynomials vanish on the curve. This ideal will in general not correspond to an actual divisor F, for a curve is not a surface. Examples like this should convince the reader that the study of algebraic manifolds (curves, surfaces, *etc.*, in 2, 3, or any number of dimensions) amounts essentially to a study of polynomial ideals. The field of coefficients is not necessarily Ω or Ω^*, but may be a field of a more general nature.

3. Some achievements of algebra and number theory. I have finally reached a point where I can hint, I hope, with something less than complete obscurity, at some of the accomplishments of algebra and number theory in our century. The most important is probably the freedom with which we have learned to manage these abstract axiomatic concepts, like field, ring, ideal, *etc.* The atmosphere in a book like van der Waerden's *Moderne Algebra*, published about 1930, is completely different from that prevailing, *e.g.*, in the articles on algebra written for the *Mathematical Encyclopaedia* around 1900. More specifically, a general theory of ideals, and in particular of polynomial ideals, was developed. (However, it should be said that the great pioneer of abstract algebra, Richard Dedekind, who first introduced the ideals into number theory, still belonged to the nineteenth century.) Algebraic geometry, before and around 1900 flourishing chiefly in Italy, was at that time a discipline of a type uncommon in the sisterhood of mathematical disciplines: it had powerful methods, plenty of general results, but they were of somewhat doubtful validity. By the abstract algebraic methods of the twentieth century all this was put on a safe basis, and the whole subject received a new impetus. Admission of fields other than Ω^*, as the field of coefficients, opened up a new horizon.

A new technique, the "primadic numbers," was introduced into algebra and number theory by K. Hensel shortly after the turn of the century, and since then has become of ever increasing importance. Hensel shaped this instrument in analogy to the power series which played such an important part in Riemann's and Weierstrass's theory of algebraic functions of one variable and their integrals (Abelian integrals). In this theory, one of the most impressive accomplishments of the previous century, the coefficients were supposed to vary over the field Ω^* of all complex numbers. Without pursuing the analogy, I may illustrate the idea of p-adic numbers by one typical example, that of quadratic norms. Let p be a prime number, and let us first agree that a congruence $a \equiv b$ modulo a power p^h of p for rational numbers a, b has this meaning that $(a - b)/p^h$ equals a fraction whose denominator is not divisible by p;

$$e.g., \frac{39}{4} - \frac{12}{5} \equiv 0 \text{ (mod. } 7^2) \text{ because } \frac{39}{4} - \frac{12}{5} = 7^2 \cdot \frac{3}{20}.$$

Let now a, b be rational numbers, $a \neq 0$, and b not the square of a rational number. In the quadratic field $\omega[\sqrt{b}]$ the number a is a *norm* if there are rational numbers x, y such that

$$a = (x + y\sqrt{b})(x - y\sqrt{b}), \qquad \text{or} \qquad a = x^2 - by^2.$$

Necessary for the solvability of this equation is (1) that for every prime p and every power p^h of p the congruence $a \equiv x^2 - by^2 \pmod{p^h}$ has a solution. This is what we mean by saying the equation has a p-adic solution. Moreover there must exist rational numbers x and y such that $x^2 - by^2$ differs as little as one wants from a. This is what we mean by saying that the equation has an ∞-adic solution. The latter condition is clearly satisfied for every a provided b is positive; however, if b is negative it is satisfied only for positive a. In the first case every a is ∞-adic norm, in the second case only half of the a's are, namely, the positive ones. A similar situation prevails with respect to p-adic norms. One proves that these necessary conditions are also sufficient: if a is a norm locally everywhere, *i.e.*, if $a = x^2 - by^2$ has a p-adic solution for every "finite prime spot p" and also for the "infinite prime spot ∞," then it has a "global" solution, namely an exact solution in rational numbers x, y.

This example, the simplest I could think of, is closely connected with the theory of genera of quadratic forms, a subject that goes back to Gauss' *Disquisitiones arithmeticae*, but in which the twentieth century has made some decisive progress by means of the p-adic technique, and it is also typical for that most fascinating branch of mathematics mentioned in the introduction: class field theory. Around 1900 David Hilbert had formulated a number of interlaced theorems concerning class fields, proved some of them at least in special cases, and left the rest to his twentieth century successors, among whom I name Takagi, Artin and Chevalley. His norm residue symbol paved the way for Artin's general reciprocity law. Hilbert had used the analogy with the Riemann-Weierstrass theory of algebraic functions over Ω^* for his orientation, but the ingenious, partly transcendental methods which he applied had nothing to do with the much simpler ones that had proved effective for the functions. By the primadic technique a rapprochement of methods has occurred, although there is still a considerable gap separating the theory of algebraic functions and the much subtler algebraic numbers.

Hensel and his successors have expressed the p-adic technique in terms of the non-algebraic "topological" notion of ("valuation" or) *convergence*. An infinite sequence of rational numbers $a_1, a_2, \cdots$ is convergent if the difference $a_i - a_j$ tends to zero, $a_i - a_j \to 0$, provided i and j independently of each other tend to infinity; more explicitly, if for every positive rational number ϵ there exists a positive integral N such that $-\epsilon < a_i - a_j < \epsilon$ for all i and $j > N$. The completeness of the real number system is expressed by Cauchy's convergence theorem: To every convergent sequence $a_1, a_2, \cdots$ of rational numbers there exists a *real* number α to which it converges: $a_i - \alpha \to 0$ for $i \to \infty$. With the ∞-adic concept of convergence we have now confronted the p-adic one induced by a prime number p. Here the sequence is considered convergent if for every exponent $h = 1, 2, 3, \cdots$, there is a positive integer N such that $a_i - a_j$ is divisible by p^h as soon as i and $j > N$. By introduction of p-adic numbers one can make the system of rational numbers complete in the p-adic sense as the introduction of real numbers makes them complete in the ∞-adic sense. The rational numbers are embedded in the continuum of all real numbers, but they may be embedded as well in that of all p-adic numbers. Each of these embedments corresponding to a finite or the infinite prime spot p is equally interesting from the arithmetical viewpoint. Now it is more evident than ever how wrong it was to identify an algebraic number field with one of its homomorphic projections into the field Ω of real numbers; along with the (real) infinite prime spots one must pay attention to the finite prime spots which correspond to the various prime ideals of the field. This is a golden rule abstracted from earlier, and then made fruitful for later, arithmetical research; and here is one bridge (others will be pointed out later) joining the two most fascinating branches of modern mathematics: abstract algebra and topology.

Besides the introduction of the primadic treatment and the progress made in the theory of class fields, the most important advances of number theory during the last fifty years seem to lie in those regions where the powerful tool of analytic functions can be brought to bear upon its problems. I mention two such fields of investigation: I. distribution of primes and the zeta function, II. additive number theory.

I. The notion of prime number is of course as old and as primitive as that of the multiplication of natural numbers. Hence it is most surprising to find the distribution of primes among all natural numbers is of such a highly irregular and almost mysterious character. While on the whole the prime numbers thin out the further one gets in the sequence of numbers, wide gaps are always followed again by clusters. An old conjecture of Goldbach's maintains that there even come along again and again pairs of primes of the smallest possible difference 2, like 57 and 59. However, the distribution of primes obeys at least a fairly simple *asymptotic* law: the number $\pi(n)$ of primes among all numbers from 1 to n is asymptotically equal to $n/\log n$. [Here $\log n$ is not the Briggs logarithm which our logarithmic tables give, but the natural logarithm as defined by the integral $\int_1^n dx/x$.] By asymptotic is meant that the quotient between $\pi(n)$ and the approximating function $n/\log n$ tends to 1 as n tends to infinity. In antiquity Eratosthenes had devised a method to sift out the prime numbers. By this sieve method the Russian mathematician Tchebycheff had obtained, during the nineteenth century, the first non-trivial results about the distribution of primes. Riemann used a different approach: his tool is the so-called zeta-function defined by the infinite series

$$\zeta(s) = 1^{-s} + 2^{-s} + 3^{-s} + \cdots. \tag{2}$$

Here s is a complex variable, and the series converges for all values of s, the real part of which is greater than 1, $\Re s > 1$. Already in the eighteenth century the fact that every positive integer can be uniquely factorized into primes had been translated by Euler into the equation

$$1/\zeta(s) = (1 - 2^{-s})(1 - 3^{-s})(1 - 5^{-s}) \cdots$$

where the (infinite) product extends over all primes $2, 3, 5, \cdots$. Riemann showed that the zeta-function has a unique "analytic continuation" to all values of s and that it satisfies a certain functional equation connecting its values for s and $1 - s$. Decisive for the prime number problem are the zeros of the zeta-function, *i.e.*, the values s for which $\zeta(s) = 0$. Riemann's equation showed that, except for the "trivial" zeros at $s = -2, -4, -6, \cdots$, all zeros have real parts between 0 and 1. Riemann conjectured that their real parts actually equal $\frac{1}{2}$. His conjecture has remained a challenge to mathematics now for almost a hundred years. However, enough had been learned about these zeros at the close of the nineteenth century to enable mathematicians, by means of some profound and newly-discovered theorems concerning analytic functions, to prove the above-mentioned asymptotic law. This was generally considered a great triumph of mathematics. Since the turn of the century Riemann's functional equation with the attending consequences has been carried over from the "classical" zeta-function (ii) of the field of rational numbers to that of an arbitrary algebraic number field (E. Hecke). For certain fields of prime characteristic one succeeded in confirming Riemann's conjecture, but this provides hardly a clue for the classical case. About the classical zeta-function we know now that it has infinitely many zeros on the critical line $\Re s = \frac{1}{2}$, and even that at least a fixed percentage, say 10 per cent, of them lie on it. (What this means is the following: Some percentage of those zeros whose imaginary part lies between arbitrary fixed limits $-T$ and $+T$ will have a real part equal to $\frac{1}{2}$, and this percentage will not sink below a certain positive limit, like 10 per cent, when T tends to infinity.) Finally about two years ago Atle Selberg succeeded, to the astonishment of the mathematical world, in giving an "elementary" proof of the prime number law by an ingenious refinement of old Eratosthenes' sieve method.

II. It has been known for a long time that every natural number n may be written as the sum of at most four square numbers, *e.g.*,

$$7 = 2^2 + 1^2 + 1^2 + 1^2, \qquad 87 = 9^2 + 2^2 + 1^2 + 1^2 = 7^2 + 5^2 + 3^2 + 2^2.$$

The same question arises for cubes, and generally for any k^{th} powers ($k=2, 3, 4, 5, \cdots$). In the eighteenth century Waring had conjectured that every non-negative integer n may be expressed as the sum of a limited number M of kth powers,

$$n = n_1^k + n_2^k + \cdots + n_M^k, \tag{3}$$

where the n_i are also non-negative integers and M is independent of n. The first decade of the twentieth century brought two events: first one found that every n is expressible as the sum of at most 9 cubes (and that, excepting a few comparatively small n, even 8 cubes will do); and shortly afterwards Hilbert proved Waring's general theorem. His method was soon replaced by a different approach, the Hardy-Littlewood circle method, which rests on the use of a certain analytic function of a complex variable and yields asymptotic formulas for the number of different representations of n in the form (3). With some precautions demanded by the nature of the problem, and by overcoming some quite serious obstacles, the result was later carried over to arbitrary algebraic number fields; and by a further refinement of the circle method in a different direction Vinogradoff proved that every sufficiently large n is the sum of at most 3 primes. Is it even true that every even n is the sum of 2 primes? To show this seems to transcend our present mathematical powers as much as Goldbach's conjecture. The prime numbers remain very elusive fellows.

III. Finally, a word ought to be said about investigations concerning the arithmetical nature of numbers originating in analysis. One of the most elementary such constants is π, the area of the circle of radius 1. By proving that π is a transcendental number (not satisfying an algebraic equation with rational coefficients) the age-old problem of "squaring the circle" was settled in 1882 in the negative sense; that is, one cannot square the circle by constructions with ruler and compass. In general it is much harder to establish the transcendency of numbers than of functions. Whereas it is easy to see that the exponential function

$$e^x = 1 + \frac{x}{1} + \frac{x^2}{1 \cdot 2} + \frac{x^3}{1 \cdot 2 \cdot 3} + \cdots$$

is not algebraic, it is quite difficult to prove that its basis e is a transcendental number. C. L. Siegel was the first who succeeded, around 1930, in developing a sort of general method for testing the transcendency of numbers. But the results in this field remain sporadic.

4. Groups, vector spaces and algebras. This ends our report on number theory, but not on algebra. For now we have to introduce the *group* concept, which, since the young genius Evariste Galois blazed the trail in 1830, has penetrated the entire body of mathematics. Without it an understanding of modern mathematics is impossible. Groups first occurred as *groups of transformations*. Transformations may operate in any set of elements, whether it is finite like the integers from 1 to 10, or infinite like the points in space. *Set* is a premathematical concept: whenever we deal with a realm of objects, a set is defined by giving a criterion which decides for any object of the realm whether it belongs to the set or not. Thus we speak of the set of prime numbers, or of the set of all points on a circle, or of all points with rational coordinates in a given coordinate system, or of all people living at this moment in the State of New Jersey. Two sets are considered equal if every element of the one belongs to the other and vice versa. A *mapping* S of a set σ into a set σ' is defined if with every element a of σ there is associated an element a' of σ', $a \rightarrow a'$. Here a rule is required which allows one to find the "image" a' for any given element a of σ. This general notion of mapping we may also call of a premathematical nature. Examples: a real-valued function of a real variable is a mapping of the continuum Ω into itself. Perpendicular projection of the space points upon a given plane is a mapping of the space into the plane. Representing every space-point by its three coordinates x, y, z with respect to a given coordinate system is a mapping of space into the continuum of real number triples (x, y, z). If a mapping S, $a \rightarrow a'$ of σ into σ', is followed by a mapping S', $a' \rightarrow a''$ of σ' into a third set σ'', the result is a mapping SS': $a \rightarrow a''$ of σ into σ''. A *one-to-one mapping* between two sets σ, σ' is a pair of mappings, S: $a \rightarrow a'$ of σ into σ', and S': $a' \rightarrow a$ of σ' into σ, which are inverse to each other. This means that the mapping SS' of σ into σ is the identical mapping E of σ which sends every element a of σ into itself, and that $S'S$ is the identical mapping of σ'. In particular, one is interested in one-to-one mappings of a set σ into itself. For them we shall use the word *transformation*. Permutations are nothing but transformations of a finite set.

The inverse S' of a transformation S, $a \to a'$ of a given set σ, is again a transformation and is usually denoted by S^{-1}. The result ST of any two transformations S and T of σ is again a transformation, and its inverse is $T^{-1}S^{-1}$ (according to the rule of dressing and undressing: if in dressing one begins with the shirt and ends with the jacket, one must in undressing begin with the jacket and end with the shirt. The order of the two "factors" S, T is essential.) *A group of transformations* is a set of transformations of a given manifold which (1) contains the identity E, (2) contains with every transformation S its inverse S^{-1}, and (3) with any two transformations S, T their "product" ST. Example: One could define congruent configurations in space as point sets of which one goes into the other by a congruent transformation of space. The congruent transformations, or "motions," of space form a group; a statement which, according to the above definition of group, is equivalent to the threefold statement that (1) every figure is congruent to itself, (2) if a figure F is congruent to F', then F' is congruent to F, and (3) if F is congruent to F' and F' congruent to F'', then F is congruent to F''. This example at once illuminates the inner significance of the group concept. *Symmetry* of a configuration F in space is described by the group of motions that carry F into itself.

Often manifolds have a structure. For instance, the elements of a field are connected by the two operations of plus and times; or in Euclidean space we have the relationship of congruence between figures. Hence we have the idea of structure-preserving mappings; they are called *homomorphisms*. Thus a homomorphic mapping of a field k into a field k' is a mapping $a \to a'$ of the "numbers" a of k into the numbers a' of k' such that $(a + b)' = a' + b'$ and $(a \cdot b)' = a' \cdot b'$. A homomorphic mapping of space into itself would be one that carries any two congruent figures into two mutually congruent figures. The following terminology (suggestive to him who knows a little Greek) has been agreed upon: homomorphisms which are one-to-one mappings are called isomorphisms; when a homomorphism maps a manifold σ into itself, it is called an endomorphism, and an automorphism when it is both: a one-to-one mapping of σ into itself. Isomorphic systems, *i.e.*, any two systems mapped isomorphically upon each other, have the same structure; indeed nothing can be said about the structure of the one system that is not equally true for the other.

The *automorphisms* of a manifold with a well-defined structure form a *group*. Two sub-sets of the manifold that go over into each other by an automorphism deserve the name of *equivalent*. This is the precise idea at which Leibniz hints when he says that two such sub-sets are "indiscernible when each is considered in itself"; he recognized this general idea as lying behind the specific geometric notion of similitude. The general problem of relativity consists in nothing else but to find the group of automorphisms. Here then is an important lesson the mathematicians learned in the twentieth century: whenever you are concerned with a structured manifold, study its group of automorphisms. Also the inverse problem, which Felix Klein stressed in his famous Erlangen program (1872), deserves attention: Given a group of transformations of a manifold σ, determine such relations or operations as are invariant with respect to the group.

If in studying a group of transformations we ignore the fact that it consists of transformations and look merely at the way in which any two of its transformations S, T give rise to a composite ST, we obtain the abstract composition schema of the group. Hence an *abstract group* is a set of elements (of unknown or irrelevant nature) for which an operation of composition is defined generating an element st from any two elements s, t such that the following axioms hold:

(1) There is a unit element e such that $es = se = s$ for every s.

(2) Every element s has an inverse s^{-1} such that $ss^{-1} = s^{-1}s = e$.

(3) The associative law $(st)u = s(tu)$ holds.

It is perhaps the most astonishing experience of modern mathematics how rich in consequences these three simple axioms are. A realization of an abstract group by transformations of a given manifold σ is obtained by associating with every element s of the group a transformation S of σ, $s \rightarrow S$, such that $s \rightarrow S$, $t \rightarrow T$ imply $st \rightarrow ST$. In general, the commutative law $st = ts$ will not hold. If it does, the group is called commutative or Abelian (after the Norwegian mathematician Niels Henrik Abel). Because composition of group elements in general does not satisfy the commutative law, it has proved convenient to use the term "ring" in the wider sense in which it does not imply the commutative law for multiplication. (However, in speaking of a field one usually assumes this law.)

The simplest mappings are the linear ones. They operate in a vector space. The vectors in our ordinary three-dimensional space are directed segments AB leading from a point A to a point B. The vector AB is considered equal to $A'B'$ if a parallel displacement (translation) carries AB into $A'B'$. In consequence of this convention one can add vectors and one can also multiply a vector by a number (integral, rational or even real). Addition satisfies the same axioms as enumerated for numbers in the table **T**, and it is also easy to formulate the axioms for the second operation. These axioms constitute the general axiomatic notion of vector space, which is therefore an algebraic and not a geometric concept. The numbers which serve as multipliers of the vectors may be the elements of any ring; this generality is actually required in the application of the axiomatic vector concept to topology. However, here we shall assume that they form a field. Then one sees at once that one can ascribe to the vector space a natural number n as its dimensionality in this sense: there exist n vectors $e_1, \cdots, e_n$ such that every vector may be expressed in one and only one way as a linear combination $x_1e_1 + \cdots + x_ne_n$, where the "coordinates" x_i are definite numbers of the field. In our three-dimensional space n equals 3, but mechanics and physics give ample occasion to use the general algebraic notion of an n-dimensional vector space for higher n.

The endomorphisms of a vector space are called its *linear mappings*; as such they allow composition ST (perform first the mapping S, then T), but they also allow addition and multiplication by numbers γ: if S sends the arbitrary vector x into xS, T into xT, then $S + T$, γS are those linear mappings which send x into $(xS) + (xT)$ and $\gamma \cdot xS$, respectively. We must forego to describe how in terms of a vector basis $e_1, \cdots, e_n$ a linear mapping is represented by a square matrix of numbers.

Often rings occur—they are then called *algebras*—which are at the same time vector spaces, *i.e.*, for which three operations, addition of two elements, multiplication of two elements and multiplication of an element by a number, are defined in such manner as to satisfy the characteristic axioms. The linear mappings of an n-dimensional vector space themselves form such an algebra, called the complete matric algebra (in n dimensions). According to quantum mechanics the observables of a physical system form an algebra of special type with a non-commutative multiplication. In the hands of the physicists abstract algebra has thus become a key that unlocked to them the secrets of the atom. A realization of an abstract group by linear transformations of a vector space is called *representation*. One may also speak of representations of a ring or an algebra: in each case the representation can be described as a homomorphic mapping of the group or ring or algebra into the complete matric algebra (which indeed is a group and a ring and an algebra, all in one).

5. Finale. After spending so much time on the explanation of the notions I can be brief in my enumeration of some of the essential achievements for which they provided the tools. If g is a subgroup of the group G, one may identify elements s, t of G that are congruent mod g, *i.e.*, for which st^{-1} is in g; g is a "self-conjugate" subgroup if this process of identification carries G again into a group, the "factor group" G/g. The group-theoretic core of Galois' theory is a theorem due to C. Jordan and O. Hölder which deals with the several ways in which one may break down a given finite group G in steps $G = G_0, G_1, G_2, \cdots$, each G_i being a self-conjugate subgroup of the preceding group G_{i-1}. Under the assumption that this is done in as small steps as possible, the theorem states, the steps (factor groups) G_{i-1}/G_i $(i = 1, 2, \cdots)$ in one such "composition series" are isomorphic to the steps, suitably rearranged, in a second such series. The theorem is very remarkable in itself, but perhaps the more so as its proof rests on the same argument by which one proves what I consider the most fundamental proposition in all mathematics, namely the fact that if you count a finite set of elements in two ways, you end up with the same number n both times. The Jordan-Hölder theorem in recent times received a much more natural and general formulation by (1) abandoning the assumption that the breaking down is done in the smallest possible steps, and (2) by admitting only such subgroups as are invariant with respect to a given set of endomorphic mappings of G. It thus has become applicable to infinite as well as finite groups, and provided a common denominator for quite a number of important algebraic facts.

The theory of representations of finite groups, the most systematic and substantial part of group theory developed shortly before the turn of the century by G. Frobenius, taught us that there are only a few irreducible representations, of which all others are composed. This theory was greatly simplified after 1900 and later carried over, first to continuous groups that have the topological property of compactness, but then also to all infinite groups, with a restrictive imposition (called almost-periodicity) on the representations. With these generalizations one trespasses the limits of algebra, and a few more words will have to be said about it under the title analysis. New phenomena occur if representations of finite groups in fields of prime characteristic are taken into account, and from their investigation profound number-theoretic consequences have been derived. It is easy to embed a finite group into an algebra, and hence facts about representations of a group are best deduced from those of the embedding algebra. At the beginning of the century algebras seemed to be ferocious beasts of unpredictable behavior, but after fifty years of investigation they, or at least the variety called semi-simple, have become remarkably tame; indeed the wild things do not happen in these superstructures, but in the underlying commutative "number" fields. In the nineteenth century geometry seemed to have been reduced to a study of invariants of groups; Felix Klein formulated this standpoint explicitly in his Erlangen program. But the full linear group was practically the only group whose invariants were studied. We have now outgrown this limitation and no longer ignore all the other continuous groups one encounters in algebra, analysis, geometry and physics. Above all we have come to realize that the theory of invariants has to be subsumed under that of representations. Certain infinite discontinuous groups, like the unimodular and the modular groups, which are of special importance to number theory, witness Gauss' class theory of quadratic forms, have been studied with remarkable success and profound results. The macroscopic and microscopic symmetries of crystals are described by discontinuous groups of motions, and it has been proved for n dimensions, what had long been known for 3 dimensions, that in a certain sense there is but a finite number of possibilities for these crystallographic groups. In the nineteenth century Sophus Lie had reduced a continuous group to its "germ" of infinitesimal elements. These elements form a sort of algebra in which the associative law is replaced by a different type of law. A Lie algebra is a purely algebraic structure, especially if the numbers which act as multipliers are taken from an algebraically defined field rather than from the continuum of real numbers Ω. These Lie groups have provided a new playground for our algebraists.

The constructions of the mathematical mind are at the same time free and necessary. The individual mathematician feels free to define his notions and to set up his axioms as he pleases. But the question is, will he get his fellow-mathematicians interested in the constructs of his imagination. We can not help feeling that certain mathematical structures which have evolved through the combined efforts of the mathematical community bear the stamp of a necessity not affected by the accidents of their historical birth. Everybody who looks at the spectacle of modern algebra will be struck by this complementarity of freedom and necessity.

PART II. ANALYSIS. TOPOLOGY. GEOMETRY. FOUNDATIONS.

6. Linear operators and their spectral decomposition. Hilbert space. A mechanical system of n degrees of freedom in stable equilibrium is capable of oscillations deviating "infinitely little" from the state of equilibrium. It is a fact of fundamental significance not only for physics but also for music that all these oscillations are superpositions of n "harmonic" oscillations with definite frequencies. Mathematically the problem of determining the harmonic oscillations amounts to constructing the principal axes of an ellipsoid in an n-dimensional Euclidean space. Representing the vectors x in this space by their coordinates $(x_1, x_2, \cdots, x_n)$ one has to solve an equation

$$x - \lambda \cdot Kx = 0,$$

where K denotes a given linear operator (= linear mapping); λ is the square of the unknown frequency ν of the harmonic oscillation, whereas the "eigenvector" x characterizes its amplitude. Define the scalar product (x, y) of two vectors x and y by the sum $x_1y_1 + \cdots + x_ny_n$. Our "affine" vector space is made into a metric one by assigning to any vector x the length $\|x\|$ given by $\|x\|^2 = (x, x)$, and this metric is the Euclidean one so familiar to us from the 3-dimensional case and epitomized by the "Pythagoras." The linear operator K is symmetric in the sense that $(x, Ky) = (Kx, y)$. The field of numbers in which we operate here is, of course, the continuum of all real numbers. Determination of the n frequencies ν or rather of the corresponding eigen-values $\lambda = \nu^2$ requires the solution of an algebraic equation of degree n (often known as the secular equation, because it first appeared in the theory of the secular perturbations of the planetary system).

More important in physics than the oscillations of a mechanical system of a finite number of degrees of freedom are the oscillations of continuous media, as the mechanical-acoustical oscillations of a string, a membrane or a 3-dimensional elastic body, and the electromagnetic-optical oscillations of the "ether." Here the vectors with which one has to operate are continuous functions $x(s)$ of a point s with one or several coordinates that vary over a given domain, and consequently K is a linear *integral* operator. Take for instance a straight string of length 1, the points of which are distinguished by a parameter s varying from 0 to 1. Here (x, x) is the integral $\int_0^1 x^2(s) \cdot ds$, and the problem of harmonic oscillations (which first suggested to the early Greeks the idea of a universe ruled by harmonious mathematical laws) takes the form of the integral equation

[1] $$x(s) - \lambda \int_0^1 K(s, t)x(t)dt = 0, \qquad (0 \leqq s \leqq 1),$$

where

$$[1'] \qquad K(s, t) = \left(\frac{a}{\pi}\right)^2 \cdot \begin{cases} s(1 - t) & \text{for } s \leqq t \\ (1 - s)t & \text{for } s \geqq t \end{cases},$$

and a is a constant determined by the physical conditions of the string. The solutions are

$$\lambda = (na)^2, \qquad x(s) = \sin n\pi s,$$

where n is capable of all positive integral values 1, 2, 3, $\cdots$. This fact that the frequencies of a string are integral multiples na of a ground frequency a is the basic law of musical harmony. If one prefers an optical to an acoustic language one speaks of the *spectrum* of eigen-values λ.

After Fredholm at the very close of the 19th century had developed the theory of linear integral equations it was Hilbert who in the next decade established the general *spectral theory of symmetric linear operators* K. Only twenty years earlier it had required the greatest mathematical efforts to prove the existence of the ground frequency for a membrane, and now constructive proofs for the existence of the whole series of harmonic oscillations and their characteristic frequencies were given under very general assumptions concerning the oscillating medium. This was an event of great consequence both in mathematics and theoretical physics. Soon afterwards Hilbert's approach made it possible to establish those asymptotic laws for the distribution of eigen-values the physicists had postulated in their statistical treatment of the thermodynamics of radiation and elastic bodies.

Hilbert observed that an arbitrary continuous function $x(s)$ defined in the interval $0 \leqq s \leqq 1$ may be replaced by the sequence

$$x_n = \sqrt{2} \int_0^1 x(s) \cdot \sin n\pi s \cdot ds, \qquad n = 1, 2, 3, \cdots,$$

of its Fourier coefficients. Thus there is no inner difference between a vector space whose elements are functions $x(s)$ of a continuous variable and one whose elements are infinite sequences of numbers $(x_1, x_2, x_3, \cdots)$. The square of the "length," $\int_0^1 x^2(s) \cdot ds$ equals $x_1^2 + x_2^2 + x_3^2 + \cdots$. Between the two forms in which one may pass from a finite sum to a limit, the infinite sum $a_1 + a_2 + a_3 + \cdots$ and the integral $\int_0^1 a(s) \cdot ds$, there is therefore here no essential difference. Thus an axiomatic formulation is called for. To the axioms for an (affine) vector space one adds the postulate of the existence of a scalar product (x, y) of any two vectors (x, y) with the properties characteristic for Euclidean metric: (x, y) is a number depending linearly on either of the two argument vectors x and y; it is symmetric, $(x, y) = (y, x)$; and $(x, x) = \|x\|^2$ is positive except for $x = 0$. The axiom of finite dimensionality is replaced by a denumerability axiom of more general character. All operations in such a space are greatly facilitated if it is assumed to be complete in the same sense that the system of real numbers is complete; *i.e.*, if the following is true: Given a "convergent" sequence $x', x'', \cdots$ of vectors, namely, one for which $\|x^{(m)} - x^{(n)}\|$ tends to zero with m and n tending to infinity, there exists a vector a toward which this sequence converges, $\|x^{(n)} - a\| \to 0$ for $n \to \infty$. A non-complete vector space can be made complete by the same construction by which the system of rational numbers is completed to form that of real numbers. Later authors have coined the name "Hilbert space" for a vector space satisfying these axioms.

Hilbert himself first tackled only integral operators in the strict sense as exemplified by [1]. But soon he extended his spectral theory to a far wider class, that of bounded (symmetric) linear operators in Hilbert space. Boundedness of the linear operator requires the existence of a cónstant M such that $\|Kx\|^2 \leqq M \cdot \|x\|^2$ for all vectors x of finite length $\|x\|$. Indeed the restriction to integral operators would be unnatural since the simplest operator, the identity $x \to x$, is not of this type. And now one of those events happened, unforeseeable by the wildest imagination, the like of which could tempt one to believe in a pre-established harmony between physical nature and mathematical mind: Twenty years after Hilbert's investigations *quantum mechanics* found that the observables of a physical system are represented by the linear symmetric operators in a Hilbert space and that the eigen-values and eigen-vectors of that operator which represents *energy* are the energy levels and corresponding stationary quantum states of the system. Of course this quantum-physical interpretation added greatly to the interest in the theory and led to a more scrupulous investigation of it, resulting in various simplifications and extensions.

Oscillations of continua, the boundary value problems of classical physics and the problem of energy levels in quantum physics, are not the only titles for applications of the theory of integral equations and their spectra. One other somewhat isolated application is the solution of *Riemann's monodromy problem* concerning analytic functions of a complex variable z. It concerns the determination of n analytic functions of z which remain regular under analytic continuation along arbitrary paths in the z-plane provided these avoid a finite number of singular points, whereas the functions undergo a given linear transformation with constant coefficients when the path circles one of these points.

Another surprising application is to the establishment of the fundamental facts, in particular of the completeness relation, in the theory of *representations of continuous compact groups*. The simplest such group consists of the rotations of a circle, and in that case the theory of representations is nothing but the theory of the so-called Fourier series, which expresses an arbitrary periodic function $f(s)$ of period 2π in terms of the harmonic oscillations

$$\cos ns, \qquad \sin ns, \qquad\qquad n = 0, 1, 2, \cdots.$$

In Nature functions often occur with hidden non-commensurable periodicities. The mathematician Harald Bohr, the brother of the physicist Niels Bohr, prompted by certain of his investigations concerning the Riemann zeta function, developed the general mathematical theory of such *almost periodic func-*

tions. One may describe his theory as that of almost periodic representations of the simplest continuous group one can imagine, namely, the group of all translations of a straight line. His main results could be carried over to arbitrary groups. No restriction is imposed on the group, but the representations one studies are supposed to be almost-periodic. For a function $x(s)$, the argument s of which runs over the group elements, while its values are real or complex numbers, almost-periodicity amounts to the requirement that the group be compact in a certain topology induced by the function. This relative compactness instead of absolute compactness is sufficient. Even so the restriction is severe. Indeed the most important representations of the classical continuous groups are not almost-periodic. Hence the theory is in need of further extension, which has busied a number of American and Russian mathematicians during the last decade.

7. Lebesgue's integral. Measure theory. Ergodic hypothesis. Before turning to other applications of operators in Hilbert space I must mention the, in all probability final, form given to the idea of integration by Lebesgue at the beginning of our century. Instead of speaking of the area of a piece of the 2-dimensional plane referred to coordinates x, y, or the volume of a piece of the 3-dimensional Euclidean space, we use the neutral term *measure* for all dimensions. The notions of measure and *integral* are interconnected. Any piece of space, any set of space points can be described by its characteristic function $\chi(P)$, which equals 1 or 0 according to whether the point P belongs or does not belong to the set. The measure of the point set is the integral of this characteristic function. Before Lebesgue one first defined the integral for continuous functions; the notion of measure was secondary; it required transition from continuous to such discontinuous functions as $\chi(P)$. Lebesgue goes the opposite and perhaps more natural way: for him measure comes first and the integral second. The one-dimensional space is sufficient for an illustration. Consider a real-valued function $y = f(x)$ of a real variable x which maps the interval $0 \leqq x \leqq 1$ into a finite interval $a \leqq y \leqq b$. Instead of subdividing the interval of the argument x Lebesgue subdivides the interval (a, b) of the dependent variable y into a finite number of small subintervals $a_i \leqq y < a_{i+1}$, say of lengths $< \epsilon$, and then determines the measure m_i of the set S_i on the x-axis, the points of which satisfy the inequality $a_i \leqq f(x) < a_{i+1}$. The integral lies between the two sums $\Sigma_i\, a_i m_i$ and $\Sigma_i\, a_{i+1} m_i$ which differ by less than ϵ, and thus can be computed with any degree of accuracy. In determining the measure of a point set—and this is the more essential modification—Lebesgue covers the set with infinite sequences, rather than finite ensembles, of intervals. Thus, to the set of rational x in the interval $0 \leqq x \leqq 1$ no measure could be ascribed before Lebesgue. But these rational numbers can be arranged in a denumerable sequence $a_1, a_2, a_3, \cdots$, and, after choosing a positive number ϵ as small as one likes, one can surround the point a_n by an interval of length $\epsilon/2^n$ with the center a_n. Thus the whole set of rational points is enclosed in a sequence of intervals of total length

$$\epsilon(1/2 + 1/2^2 + 1/2^3 + \cdots) = \epsilon;$$

and according to Lebesgue's definition its measure is therefore less than (the arbitrary positive) ϵ and hence zero. The notion of *probability* is tied to that of measure, and for this reason mathematical statisticians are deeply interested in measure theory. Lebesgue's idea has been generalized in several directions. The two fundamental operations one can perform with sets are: forming the intersection and the union of given sets, and thus sets may be considered as elements of a "*Boolean algebra*" with these two operations, the properties of which may be laid down in a number of axioms resembling the arithmetical axioms for addition and multiplication. Hence one of the questions which has occupied the more axiomatically minded among the mathematicians and statisticians is concerned with the introduction of measure in abstract Boolean algebras.

Lebesgue's integral is important in our present context, because those real-valued functions $f(x)$ of a real variable x ranging over the interval $0 \leqq x \leqq 1$, the squares of which are Lebesgue-integrable, form a complete Hilbert space —provided two functions $f(x)$, $g(x)$ are considered equal if those values x for which $f(x) \neq g(x)$ form a set of measure zero (Riesz-Fischer theorem).

The mechanical equations for a system of n degrees of freedom in Hamilton's form uniquely determine the state tP at the moment t if the state P at the moment $t = 0$ is given. Such is the precise formulation of the law of causality in mechanics. The possible states P form the points of a $(2n)$-dimensional phase space, and for a fixed t and an arbitrary P the transition $P \rightarrow tP$ is a measure-preserving mapping (t). These transformations form a group: $(t_1)(t_2) = (t_1 + t_2)$. For a given P and a variable t the point tP describes the consecutive states which this system assumes if at the moment $t = 0$ it is in the state P. Considering P as a particle of a $(2n)$-dimensional fluid which fills the phase-space and ascribing to the particle P the position tP at the time t, one obtains the picture of an incompressible fluid in stationary flow. The statistical derivation of the laws of thermodynamics makes use of the so-called *ergodic hypothesis* according to which the path of an arbitrary individual particle P (excepting initial states P which form a set of measure zero) covers the phase-space (or at least that $(2n - 1)$-dimensional sub-space of it where the energy has a given value) everywhere dense, so that in the course of its history the probability of finding it in this or that part of the space is the same for any parts of equal measure. Nineteenth century mathematics seemed to be a long way off from proving this hypothesis with any degree of generality. Strangely enough it was proved shortly after the transition from classical to quantum mechanics had rendered the hypothesis almost valueless, and it was proved by making use of the mathematical apparatus of quantum physics. Under the influence of the mapping (t), $P \rightarrow tP$, any function $f(P)$ in phase-space is transformed into the function $f' = U_t f$, defined by the equation $f'(tP) = f(P)$. The U_t form a group of operators in the Hilbert space of arbitrary functions $f(P)$, $U_{t_1}U_{t_2} = U_{t_1+t_2}$, and application of spectral decomposition to this group enabled J. von Neumann to deduce the ergodic hypothesis with two provisos: (1) Convergence of a sequence of functions $f_n(P)$ toward a function $f(P)$, $f_n \to f$, is understood (as it would in quantum mechanics, namely) as convergence in Hilbert space where it means that the total integral of $(f_n - f)^2$ tends to zero with $n \to \infty$; (2) one assumes that there are no subspaces of the phase-space which are invariant under the group of transformations (t) except those spaces that are in Lebesgue's sense equal either to the empty or the total space. Shortly afterwards proofs were also given for other interpretations of the notion of convergence.

The laws of nature can either be formulated as differential equations or as "principles of variation" according to which certain quantities assume extremal values under given conditions. For instance, in an optically homogeneous or non-homogeneous medium the light travels along that road from a given point A to a given point B for which the time of travel assumes minimal value. In potential theory the quantity which assumes a minimum is the so-called Dirichlet integral. Attempts to establish directly the existence of a minimum had been discouraged by Weierstrass' criticism in the 19th century. Our century, however, restored the direct methods of the Calculus of Variation to a position of honor after Hilbert in 1900 gave a direct proof of the Dirichlet principle and later showed how it can be applied not only in establishing the fundamental facts about functions and integrals ("algebraic" functions and "abelian" integrals) on a compact Riemann surface (as Riemann had suggested 50 years earlier) but also for deriving the basic propositions of the *theory of uniformization*. That theory occupies a central position in the theory of functions of one complex variable, and the first decade of the 20th century witnessed the first proofs by P. Koebe and H. Poincaré of these propositions conjectured about 25 years before by Poincaré himself and by Felix Klein. As in an Euclidean vector space of finite dimensionality, so in the Hilbert space of infinitely many dimensions, this fact is true: Given a linear (complete) subspace E, any vector may be split in a uniquely determined manner into a component lying in E (orthogonal projection) and one perpendicular to E. Dirichlet's principle is nothing but a special case of this fact. But since the function-theoretic applications of orthogonal projection in Hilbert space which we alluded to are closely tied up with topology we had better turn first to a discussion of this important branch of modern mathematics: topology.

8. Topology and harmonic integrals. Essential features of the modern approach to *topology* can be brought to light in its connection with the, only recently developed, theory of *harmonic integrals*. Consider a stationary magnetic field h in a domain G which is free from electric currents. At every point of G it satisfies two differential conditions which in the usual notations of vector analysis are written in the form div $h = 0$, rot $h = 0$. A field of this type is called harmonic. The second of these conditions states that the line integral of

h along a closed curve (cycle) C, $\int_C h$, vanishes provided C lies in a sufficiently small neighborhood of an arbitrary point of G. This implies $\int_C h = 0$ for any cycle C in G that is the boundary of a surface in G. However, for an arbitrary cycle C in G the integral is equal to the electric current surrounded by C.

Let the phrase "C homologous to zero," $C \sim 0$, indicate that the cycle C in G bounds a surface in G. One can travel over a cycle C in the opposite sense, thus obtaining $-C$, or travel over it $2, 3, \cdots$ times, thus obtaining $2C, 3C, \cdots$; and cycles may be added and subtracted from each other (if one does not insist that cycles are of one piece). Two cycles C, C' are called homologous, $C \sim C'$, if $C - C' \sim 0$. Note that $C \sim 0$, $C' \sim 0$ imply $-C \sim 0$, $C + C' \sim 0$. Hence the cycles form a commutative group under addition, the "Betti group," if homologous cycles are considered as one and the same group element. These notions of *cycles and their homologies* may be carried over from a three-dimensional domain in Euclidean space to any n-dimensional manifold, in particular to closed (compact) manifolds like the two-dimensional surfaces of the sphere or the torus; and on an n-dimensional manifold we can speak not only of 1-dimensional, but also of 2-, 3-, $\cdots$, n-dimensional cycles. The notion of a harmonic vector field permits a similar generalization, harmonic tensor field (harmonic form) of rank r ($r = 1, 2, \cdots, n$), provided the manifold bears a Riemannian metric, an assumption the meaning of which will be discussed later in the section on geometry. Any tensor field (linear differential form) of rank r may be integrated over an r-dimensional cycle.

The fundamental problem of *homology theory* consists in determining the structure of the Betti group, not only for 1-, but also for 2-, $\cdots$, n-dimensional cycles, in particular in determining the number of linearly independent cycles (Betti number). [ν cycles $C_1, \cdots, C_\nu$ are linearly independent if there exists no homology $k_1C_1 + \cdots + k_\nu C_\nu \sim 0$ with integral coefficients k except $k_1 = \cdots = k_\nu = 0$.] The fundamental theorem for harmonic forms on compact manifolds states that, given ν linearly independent cycles $C_1, \cdots, C_\nu$, there exists a harmonic form h with pre-assigned periods

$$\int_{C_1} h = \pi_1, \cdots, \quad \int_{C_\nu} h = \pi_\nu.$$

H. Poincaré developed the algebraic apparatus necessary to formulate exactly the notions of cycle and homology. In the course of the twentieth century it turned out that in most problems co-homologies are easier to handle than homologies. I illustrate this for 1-dimensional cycles. A line C_1 leading from a point p_1 to p_2, when followed by a line C_2 leading from p_2 to a third point p_3, gives rise to a line $C_1 + C_2$ leading from p_1 to p_3. The line integral $\int_C h$ of a given vector field h along an arbitrary (closed or open) line C is an additive function $\phi(C)$ of C, $\phi(C_1 + C_2) = \phi(C_1) + \phi(C_2)$. If moreover rot h vanishes everywhere, then $\phi(C) = 0$ for any closed line C that lies in a sufficiently small neighborhood of a point, whatever this point may be. Any real-valued function ϕ satisfying these two conditions may be called an abstract integral. The co-homology $\phi \sim 0$ means that $\phi(C) = 0$ for any closed line C, and thus it is clear what the co-homology $k_1\phi_1 + \cdots + k_r\phi_r \sim 0$ with arbitrary real coefficients $k_1, \cdots, k_r$ means. The homology $C \sim 0$ could now be defined, not by the condition that the cycle C bounds, but by the requirement that $\phi(C) = 0$ for every abstract integral ϕ. With the convention that any two abstract integrals ϕ, ϕ' are identified if $\phi - \phi' \sim 0$, these integrals form a vector space, and the dimensionality of this vector space is now introduced as the Betti number. And the fundamental theorem for harmonic integrals on a compact manifold now asserts that for any given abstract integral ϕ there exists one and only one harmonic vector field h whose integral is co-homologous to ϕ, $\int_C h = \phi(C)$, for every cycle C (realization of the abstract integral in concreto by a harmonic integral).

J. W. Alexander discovered an important result connecting the Betti numbers of a manifold M that is embedded in the n-dimensional Euclidean space R_n with the Betti numbers of the complement $R_n - M$ (*Alexander's duality theorem*).

The difficulties of topology spring from the double aspect under which one can consider continuous manifolds. Euclid looked upon a figure as an assemblage of a finite number of geometric elements, like points, straight lines, circles, planes, spheres. But after replacing each line or surface by the set of points lying on it one may also adopt the set-theoretic view that there is only one sort of elements, points, and that any (in general infinite) set of points can serve as a figure. This modern standpoint obviously gives geometry far greater generality and freedom. In topology, however, it is not necessary to descend to the points as the ultimate atoms, but one can construct the manifold like a building from "blocks" or cells, and a finite number of such cells serving as units will do, provided the manifold is compact. Thus it is possible here to revert to a treatment in Euclid's "finitistic" style (combinatorial topology).

On the first standpoint, manifold as a point set, the task is to formulate that *continuity* by which a point p approaching a given point p_0 becomes gradually indistinguishable from p_0. This is done by associating with p_0 the *neighborhoods* of p_0, an infinite shrinking sequence of sub-sets $U_1 \supset U_2 \supset U_3 \supset \cdots$, all containing p_0. [$U \supset V$ means: the set U contains V.] For example, in a plane referred to Cartesian coordinates x, y we may choose as the nth neighborhood U_n of a point $p_0 = (x_0, y_0)$ the interior of the circle of radius $1/2^n$ around p_0. The notion of convergence, basic for all continuity considerations, is defined in terms of the sequence of neighborhoods as follows: A sequence of points $p_1, p_2, \cdots$ converges to p_0 if for every natural number n there is an N so that all points p_ν with $\nu > N$ lie in the nth neighborhood U_n of p_0. Of course, the choice of the neighborhoods U_n is arbitrary to a certain extent. For instance, one could also have chosen as the nth neighborhood V_n of (x_0, y_0) the square of side $2/n$ around (x_0, y_0), to which a point (x, y) belongs, if

$$- 1/n < x - x_0 < 1/n, \qquad - 1/n < y - y_0 < 1/n.$$

However the sequence V_n is equivalent to the sequence U_n in the sense that for every n there is an n' such that $U_{n'} \subset V_n$ (and thus $U_\nu \subset V_n$ for $\nu \geqq n'$), and also for every m an m' such that $V_{m'} \subset U_m$; and consequently the notion of convergence for points is the same, whether based on the one or the other sequence of neighborhoods. It is clear how to define continuity of a mapping of one manifold into another. A one-to-one mapping of two manifolds upon each other is called topological if continuous in both directions, and two manifolds that can be mapped topologically upon each other are topologically equivalent. Topology investigates such properties of manifolds as are invariant with respect to topological mappings (in particular with respect to continuous deformations).

A continuous function $y=f(x)$ may be approximated by piecewise linear functions. The corresponding device in higher dimensions, the method of *simplicial approximations* of a given continuous mapping of one manifold into another, is of great importance in set-theoretic topology. It has served to develop a general *theory of dimensions*, to prove the topological invariance of the Betti groups, to define the decisive notion of the degree of mapping ("Abbildungsgrad," L. E. J. Brouwer) and to prove a number of interesting fixed point theorems. For instance, a continuous mapping of a square into itself has necessarily a fixed point, *i.e.*, a point carried by the mapping into itself. Given two continuous mappings of a (compact) manifold M into another M', one can ask more generally for which points p on M both images on M' coincide. A famous formula by S. Lefschetz relates the "total index" of such points with the homology theory of cycles on M and M'.

Application of fixed point theorems to functional spaces of infinitely many dimensions has proved a powerful method to establish the existence of solutions for non-linear differential equations. This is particularly valuable, because the hydrodynamical and aerodynamical problems are almost all of this type.

Poincaré found that a satisfactory formulation of the homology theory of cycles was possible only from the second standpoint where the n-dimensional manifold is considered as a conglomerate of n-dimensional cells. The boundary of an n-dimensional cell (n-cell) consists of a finite number of $(n - 1)$-cells, the boundary of an $(n - 1)$-cell consists of a finite number of $(n - 2)$-cells, *etc.* The *combinatorial skeleton* of the manifold is obtained by assigning symbols to these cells and then stating in terms of their symbols which $(i - 1)$-cells belong to the boundary of any of the occurring i-cells ($i = 1, 2, \cdots, n$). From the cells one descends to the points of the manifold by a repeated process of sub-division which catches the points in an ever finer net. Since this subdivision proceeds according to a fixed combinatorial scheme, the manifold is in topological regard completely fixed by its combinatorial skeleton. And at once the question arises under what circumstances two given combinatorial skeletons represent the same manifold, *i.e.*, lead by iterated sub-division to topologically equivalent manifolds. We are far from being able to solve this fundamental problem. Algebraic topology, which operates with the combinatorial skeletons, is in itself a rich and beautiful theory, linked in various ways with the basic notions and theorems of algebra and group theory.

The connection between algebraic and set-theoretic topology is fraught with serious difficulties which are not yet overcome in a quite satisfactory manner. So much, however, seems clear that one had better start, not with a division into cells, but with a covering by patches which are allowed to overlap. From such a pattern the fundamental topologically invariant concepts are to be developed. The above notion of an abstract integral, which relates homology and co-homology, is an indication; it can indeed be used for a direct proof of the invariance of the first Betti number without the tool of simplicial approximation.

9. Conformal mapping, meromorphic functions, Calculus of Variation in the large. Homology theory, in combination with the Dirichlet principle or the method of orthogonal projection in Hilbert space, leads to the theory of harmonic integrals, in particular for the lowest dimension $n=2$ to the theory of abelian integrals on Riemann surfaces. But for Riemann surfaces the Dirichlet principle also yields the fundamental facts concerning uniformization of analytic functions of one variable if one combines it with the *homotopy* (not homology) *theory* of closed curves. Whereas a cycle is homologous to zero if it bounds, it is homotopic to zero if it can be contracted into a point by continuous deformation. The homotopy theory of 1- and more-dimensional cycles has recently come to the fore as an important branch of topology, and the group-theoretic aspect of homotopy has led to some surprising discoveries in abstract group theory. Homotopy of 1-dimensional cycles is closely related with the idea of the *universal covering manifold* of a given manifold. Given a continuous mapping $p \rightarrow p'$ of one manifold M into another M', the point p' may be considered as the trace or projection in M' of the arbitrary point p on M, and thus M becomes a manifold covering M'. There may be no point or several points p on M which lie over a given point p' of M' (which are mapped into p'). The mapping is without ramifications if for any point p_0 of M it is one-to-one (and continuous both ways) in a sufficiently small neighborhood of p_0. Let p_0 be a point on M, p_0' its trace on M', and C' a curve on M' beginning at p_0'. If M covers M' without ramifications we can follow this curve on M by starting at p_0, at least up to a certain point where we would run against a "boundary of M relative to M'." Of chief interest are those covering manifolds M over a given M' for which this never happens and which therefore cover M' without ramifications and relative boundaries. The best way of defining the central topological notion "simply connected" is by describing a simply connected manifold as one having no other unramified unbounded covering but itself. There is a strongest of all unramified unbounded covering manifolds, the universal covering manifold, which can be described by the statement that on it a curve C is closed only if its trace C' is (closed and) homotopic to zero. The proof of the fundamental theorem on uniformization consists of two parts: (1) constructing the universal covering manifold of the given Riemann surface, (2) constructing by means of the Dirichlet principle a one-to-one conformal mapping of the covering manifold upon the interior of a circle of finite or infinite radius.

All we have discussed so far in our account of analysis, is in some way tied up with operators and projections in Hilbert space, the analogue in infinitely many dimensions of Euclidean space. In H. Minkowski's *Geometry of Numbers* distances $|AB|$, which are different from the Euclidean distance but satisfy the axioms that $|BA| = |AB|$ and that in a triangle ABC the inequality $|AC| \leq |AB| + |BC|$ holds, were used to great advantage for obtaining numerous results concerning the solvability of inequalities by integers. We do not find time here to report on the progress of this attractive branch of number theory during the last fifty years. In infinitely many dimensions spaces endowed with a metric of this sort, of a more general nature than the Euclid-Hilbert metric, have been introduced by Banach, not however for number-theoretic but for purely analytic purposes. Whether the importance of the subject justifies the large number of papers written on *Banach spaces* is perhaps questionable.

The Dirichlet principle is but the simplest example of the direct methods of the Calculus of Variation as they came into use with the turn of the century. It was by these methods that the theory of *minimal surfaces*, so closely related to that of analytic functions, was put on a new footing. What we know about non-linear differential equations has been obtained either by the topological fixed point method (see above) or by the so-called continuity method or by constructing their solutions as extremals of a suitable functional.

A continuous function on an n-dimensional compact manifold assumes somewhere a minimum and somewhere else a maximum value. Interpret the function as altitude. Besides summit (local maximum) and bottom (local minimum) one has the further possibility of a saddle point (pass) as a point of "stationary" altitude. In n dimensions the several possibilities are indicated by an inertial index k which is capable of the values $k = 0, 1, 2, \cdots, n$, the value $k = 0$ a minimum and $k = n$ characterizing a maximum. Marston Morse discovered the inequality $M_k \geqq B_k$ between the number M_k of stationary points of index k and the Betti number B_k of linearly independent homology classes of k-dimensional cycles. In their generalization to functional spaces these relations have opened a line of study adequately described as *Calculus of Variation in the large.*

Development of the theory of uniformization for analytic functions led to a closer investigation of *conformal mapping* of 2-dimensional manifolds in the large, which resulted in a number of theorems of surprising simplicity and beauty. In the same field there is to register an enormous extension of our knowledge of the behavior of *meromorphic functions*, *i.e.*, single-valued analytic functions of the complex variable z which are regular everywhere with the exception of isolated "poles" (points of infinity). Towards the end of the previous century Riemann's zeta function had provided the stimulus for a deeper study of "entire functions" (functions without poles). The greatest stride forward, both in methods and results, was marked by a paper on meromorphic functions published in 1925 by the Finnish mathematician Rolf Nevanlinna. Besides meromorphic functions in the z-plane one can study such functions on a given Riemann surface; and in the way in which the theory of algebraic functions (equal to meromorphic functions on a compact Riemann surface) as a theory of algebraic curves in two complex dimensions may be generalized to any number of dimensions, so one can pass from meromorphic functions to meromorphic curves.

The theory of *analytic functions of several complex variables*, in spite of a number of deep results, is still in its infancy.

10. Geometry. After having dealt at some length with the problems of analysis and topology I must be brief about geometry. Of subjects mentioned before, minimal surfaces, conformal mapping, algebraic manifolds and the whole of topology could be subsumed under the title of geometry. In the domain of *elementary axiomatic geometry* one strange discovery, that of von Neumann's pointless "continuous geometries" stands out, because it is intimately interrelated with quantum mechanics, logic and the general algebraic theory of "lattices." The 1-, 2-, $\cdots$, n-dimensional linear manifolds of an n-dimensional vector space form the 0-, 1-, $\cdots$, $(n - 1)$-dimensional linear manifolds in an $(n - 1)$-dimensional projective point space. The usual axiomatic foundation of projective geometry uses the points as the primitive elements or atoms of which the higher-than-zero-dimensional manifolds are composed. However, there is possible a treatment where the linear manifolds of all dimensions figure as elements, and the axioms deal with the relation "B contains A" ($A \subset B$) between these elements and the operation of intersection, $A \cap B$, and of union, $A \cup B$, performed on them; the union $A \cup B$ consists of all sums $x + y$ of a vector x in A and a vector y in B. In quantum logic this relation and these operations correspond to the relation of implication ("The statement A implies B") and the operations 'and,' 'or' in classical logic. But whereas in classical logic the distributive law

$$A \cap (B \cup C) = (A \cap B) \cup (A \cap C)$$

holds, this is not so in quantum logic; it must be replaced by the weaker axiom:

$$\text{If } C \subset A \text{ then } A \cap (B \cup C) = (A \cap B) \cup C.$$

On formulating the axioms without the implication of finite dimensionality one will come across several possibilities; one leads to the Hilbert space in which quantum mechanics operates, another to von Neumann's continuous geometry with its continuous scale of dimensions, in which elements of arbitrarily low dimensions exist but none of dimension zero.

The most important development of geometry in the twentieth century took place in differential geometry and was stimulated by general relativity, which showed that the world is a 4-dimensional manifold endowed with a Riemannian metric. A piece of an n-dimensional manifold can be mapped in one-to-one continuous fashion upon a piece of the n-dimensional "arithmetical space" which consists of all n-uples $(x_1, x_2, \cdots, x_n)$ of real numbers x_i. A *Riemann metric* assigns to a line element which leads from the point $P = (x_1, \cdots, x_n)$ to the infinitely near point $P' = (x_1 + dx_1, \cdots, x_n + dx_n)$ a distance ds the square of which is a quadratic form of the relative coordinates dx_i,

$$ds^2 = \sum g_{ij} dx_i dx_j, \qquad (i, j = 1, \cdots, n)$$

with coefficients g_{ij} depending on the point P but not on the line element. This means that, in the infinitely small, Pythagoras' theorem and hence Euclidian geometry are valid, but in general not in a region of finite extension. The line elements at a point may be considered as the infinitesimal vectors of an n-dimensional vector space in P, the tangent space or the compass at P; indeed an arbitrary (differentiable) transformation of the coordinates x_i induces a linear transformation of the components dx_i of any line element at a given point P. As Levi-Civita found in 1915 the development of Riemannian geometry hinges on the fact that a Riemannian metric uniquely determines an infinitesimal parallel displacement of the vector compass at P to any infinitely near point P'. From this a general scheme for differential geometry arose in which each point P of the manifold is associated with a homogeneous space Σ_P described by a definite group of "authomorphisms," this space now taking over the role of the tangent space (whose group of authomorphisms consists of all non-singular linear transformations). One assumes that one knows how this associated space Σ_P is transferred by infinitesimal displacement to the space $\Sigma_{P'}$ associated with any infinitely near point P'. The most fundamental notion of Riemannian geometry, that of curvature, which figures so prominently in Einstein's equations of the gravitational field, can be carried over to this general scheme. Thus one has erected general differential affine, projective, conformal, geometries, *etc.* One has also tried by their structures to account for the other physical fields existing in nature beside the gravitational one, namely the electromagnetic field, the electronic wave-field and further fields corresponding to the several kinds of elementary particles. But it seems to the author that so far all such speculative attempts of building up a unified field theory have failed. There are very good reasons for interpreting gravitation in terms of the basic concepts of differential geometry. But it is probably unsound to try to "geometrize" all physical entities.

Differential geometry in the large is an interesting field of investigation which relates the differential properties of a manifold with its topological structure. The schema of differential geometry explained above with its associated spaces

Σ_P and their displacements has a purely topological kernel which has recently developed under the name of *fibre spaces* into an important topological technique.

Our account of progress made during the last fifty years in analysis, geometry and topology had to touch on many special subjects. It would have failed completely had it not imparted to the reader some feeling of the close relationship connecting all these mathematical endeavors. As the last example of fibre spaces (beside many others) shows, this unity in diversity even makes a clear-cut division into analysis, geometry, topology (and algebra) practically impossible.

11. Foundations. Finally a few words about the *foundations of mathematics*. The nineteenth century had witnessed the critical analysis of all mathematical notions including that of natural numbers to the point where they got reduced to pure logic and the ideas "set" and "mapping." At the end of the century it became clear that the unrestricted formation of sets, sub-sets of sets, sets of sets *etc.*, together with an unimpeded application to them as to the original elements of the logical quantifiers "there exists" and "all" [cf. the sentences: the (natural) number n is even if there exists a number x such that $n = 2x$; it is odd if n is different from $2x$ for all x] inexorably leads to antinomies. The three most characteristic contributions of the twentieth century to the solution of this Gordian knot are connected with the names of L. E. J. Brouwer, David Hilbert and Kurt Gödel. Brouwer's critique of "mathematical existentialism" not only dissolved the antinomies completely but also destroyed a good part of classical mathematics that had heretofore been universally accepted.

If only the historical event that somebody has succeeded in constructing a (natural) number n with the given property P can give a right to the assertion that "there exists a number with that property" then the alternative that there either exists such a number or that all numbers have the opposite property non-P is without foundation. The principle of excluded middle for such sentences may be valid for God who surveys the infinite sequence of all natural numbers, as it were, with one glance, but not for human logic. Since the quantifiers "there is" and "all" are piled upon each other in the most manifold way in the formation of mathematical propositions, Brouwer's critique makes almost all of them meaningless, and therefore Brouwer set out to build up a new mathematics which makes no use of that logical principle. I think that everybody has to accept Brouwer's critique who wants to hold on to the belief that mathematical propositions tell the sheer truth, truth based on evidence. At least Brouwers' opponent, Hilbert, accepted it tacitly. He tried to save classical mathematics by converting it from a system of meaningful propositions into a game of meaningless formulas, and by showing that this game never leads to two formulas, F and non-F, which are inconsistent. Consistency, not truth, is his aim. His attempts at proving consistency revealed the astonishingly complex logical structure of mathematics. The first steps were promising indeed. But then Gödel's discovery cast a deep shadow over Hilbert's enterprise. Consistency itself may be expressed by a formula. What Gödel showed was this: If the game of mathematics is actually consistent then the formula of consistency cannot be proved within this game. How can we then hope to prove it at all?

This is where we stand now. It is pretty clear that our theory of the physical world is not a description of the phenomena as we perceive them, but is a bold symbolic construction. However, one may be surprised to learn that even mathematics shares this character. The success of the anti-phenomenological constructive method is undeniable. And yet the ultimate foundations on which it rests remain a mystery, even in mathematics.

153.

Radiation capacity

Proceedings of the National Academy of Sciences of the United States of America
37, 832—836 (1951)

Capacity is a classical notion in the theory of electrostatic fields. This note deals with its analog for radiating fields (unquantized!). The static theory will be reproduced here briefly in a form evincing the points where it carries over to radiating fields or where essential modifications are needed.

1. The static theory in Euclidean 3-space is governed by Newton's potential

$$Q(pp') = 1/2\pi r \tag{1}$$

where $r = r(pp')$ is the distance of any two points p, p'. Suppose the space is divided into two complementary parts, the connected "exterior" W and the bounded "interior" V. The latter may exist of a finite number h of connected components V_i, the conductors. Their surfaces Ω_i constitute the common boundary Ω of V and W, which we assume to be smooth, i.e., the unit vector of the external normal at the variable point s of Ω is not only a continuous function of s, but also satisfies a Hölder condition. The external boundary problem asks for a harmonic function $u(p)$ in W, $\Delta u = 0$, tending to zero uniformly in all directions if p approaches infinity, and assuming given boundary values $\gamma(s)$ on Ω. Since uniqueness holds, i.e., since $\gamma(s) = 0$ on Ω implies $u(p) = 0$ in W, one expects that the problem has a unique solution whatever the given function $\gamma(s)$ on Ω. One tries to obtain it in the form of the potential of a dipole layer,

$$u(p) = \int_\Omega (\partial Q(ps')/\partial n_{s'}) \cdot \mu(s')ds'. \tag{D}$$

Here $\mu(s)$ is an unknown continuous function on Ω, and $\partial/\partial n$ stands for differentiation along the external normal. Integration with respect to the variable point s or s' of Ω always extends over the whole of Ω; ds or ds' is the area element. The condition that $u(p)$ on the external side of Ω assumes the values $\gamma(s)$ leads to the integral equation

$$\mu(s) + \int K(ss')\mu(s')ds' = \gamma(s), \quad \text{briefly } (E + K)\mu = \gamma \tag{2}$$

for the layer density $\mu(s)$. The Hölder condition keeps the singularity of its kernel

$$K(s, s') = \partial Q(ss')/\partial n_{s'} \tag{3}$$

for $s = s'$ so low that Fredholm's theory is applicable. The difficulty stems from the fact that the corresponding homogeneous equation

$$(E + K)\varphi = 0 \tag{4}$$

and hence also the homogeneous equation with the transposed kernel,

$$\eta(E + K) = 0 \qquad \text{or} \qquad \eta(s') + \int \eta(s)K(ss')\, ds = 0, \quad (4')$$

each have an h-dimensional manifold Φ and H of solutions. (Our form of writing emulates the matrix calculus as if K were a square matrix and the values of φ arranged in a column and those of η in a row.) The solutions of (4) are the functions $\varphi(s)$ which assume arbitrary constant values q_i on each of the conductor surfaces Ω_i. A solution of (4′) gives rise to a field of simple layer

$$v(p) = \int Q(ps')\cdot\eta(s')ds', \qquad (5)$$

harmonic both in V and W, continuous even on the boundary Ω while the normal derivative $\partial v/\partial n$ vanishes on the inner side of Ω, and assumes the value $-2\eta(s)$ at the point s of the outer side. The vanishing of the inner derivative $\partial v/\partial n$ implies that v has a constant value q_i in each conductor V_i including its surface: (5) is the general solution of the conductor problem. Thus we have as our

First fact: The formula

$$\varphi(s) = \int Q(ss')\cdot\eta(s')\, ds'$$

stablishes a linear mapping M, $\eta \to \varphi = M\eta$, of H into Φ.

From the uniqueness theorem there follows the

Second fact: This mapping is non-singular, i.e., $M\eta = 0$ implies $\eta = 0$.

Hence M is a one-to-one mapping of H upon Φ. Two elements η of H and φ of Φ have a product

$$(\eta\varphi) = \int \eta(s)\varphi(s)\, ds.$$

Our

Third fact is the law of symmetry $C(\eta, \eta^*) = C(\eta^*, \eta)$ for the bilinear form

$$C(\eta, \eta^*) = (\eta, M\eta^*)$$

of two arbitrary elements η, η^* of H.

Indeed

$$(\eta, M\eta^*) = \int\int \eta(s)Q(ss')\eta^*(s')\, ds'\, ds.$$

Fourth fact: The symmetric bilinear form $C(\eta, \eta^*)$ is non-singular; i.e., if an element η^* of H satisfies the condition $C(\eta, \eta^*) = 0$ for every $\eta \in$ H then $\eta^* = 0$.

This follows since $C(\eta, \eta)$ is nothing but the Dirichlet integral of (5) extended over the whole space and therefore positive unless $v = 0$, $\eta = 0$.

One may choose the basis φ_i for Φ such that $\varphi_i(s) = 1$ on Ω_i, $= 0$ on all other Ω_j $(j \neq i)$. Form the linear combination $\eta = q_1\eta_1 + \ldots + q_h\eta_h$ of the

$\eta_i = M^{-1}\varphi_i$ with constant coefficients q_i: the solution (5) of the conductor problem with this η corresponds to the values q_i for the potentials of the several conductors V_i. The total charge e_i of V_i is given by $2(\eta\varphi_i)$, and hence potentials and charges are connected by the linear relations

$$e_i = 2\cdot\sum_j c_{ji}q_j \qquad \{c_{ji} = (\eta_j\varphi_i)\}.$$

Thus c_{ji}, the symmetric coefficients of the quadratic form $C(\eta\eta) = \sum c_{ji}q_jq_i$, are the (halved) capacity coefficients.

From the fourth fact follows the unique existence of an element $\eta^* \in$ H such that $\varphi^* = M\eta^*$ satisfies the condition $(\eta\varphi^*) = (\eta\gamma)$ for every $\eta \in$ H. For the boundary values $\gamma^* = \gamma - \varphi^*$ one can therefore construct a potential $u^*(p)$ by means of a suitable double layer $\mu^*(s)$. Adding to it the potential $v^*(p)$ of the simple layer η^*, chosen according to the above construction from the special manifold H, one obtains the desired field $u(p)$.

2. Let us now pass from static potential to scalar radiation. Given a positive constant k, we are concerned with scalar fields $u(p)$ in W satisfying the equation $\Delta u + k^2u = 0$ in W and the condition of outgoing radiation:

$$R(\partial u/\partial n + iku)_{\text{on } \Sigma_R} \to 0 \qquad \text{with } R \to \infty$$

(admissible fields). i is the imaginary unit, Σ_R the sphere of radius R around the origin. The main question is: can one assign arbitrary boundary values on Ω for such a field?

The place of (1) is now taken by

$$Q(pp') = e^{-ikr}/2\pi r. \tag{6}$$

The uniqueness theorem has been established by W. Magnus and F. Rellich in 1943.[1]

The formula (2) (with the new Q) gives an admissible field in W with the boundary values

$$(E + K)\mu = \gamma \tag{2}$$

where $K(s, s')$ is the kernel (3). What stands in the way of solving the external boundary problem in this fashion is the circumstance that the homogeneous equations

$$(E + K)\varphi = 0 \qquad \text{and} \qquad \eta(E + K) = 0$$

may have non-vanishing solutions. They form linear manifolds Φ and H of the same number h of dimensions, and the equation (2) is solvable for a given γ only if γ is orthogonal to all the elements η of H, $(\eta\gamma) = 0$. (The number h has now nothing to do with the number of connected components of V.)

The external boundary problem for scalar and also for electromagnetic radiation has recently been treated by W. K. Saunders.[2] He overcame the difficulty just mentioned by an ingenious but highly artificial device which he ascribes to a suggestion by H. Lewy. I asked myself whether the construction of capacity coefficients for radiating fields would not provide a better and more natural way out. Checking the above four facts I found that the first, second and third hold good, but since the argument by means of the definite Dirichlet integral fails there is no reason to expect the universal validity of the fourth fact. Nor is this supported by the general theory of integral equations—at least not unless one extends the manifold Φ of the eigen-functions to that of the "principal functions" which the theory of elementary divisors deals with.

Let $K(ss')$ be an arbitrary kernel regular enough for the applicability of Fredholm's theory, and let A be the operator $E + K$. We form the finite-dimensional manifolds $\Phi_0, \Phi_1, \Phi_2, \ldots$ of the solutions φ of the successive equations

$$\varphi = 0, \qquad A\varphi = 0, \qquad A^2\varphi = 0, \ldots .$$

The sequence $\Phi_0 \subset \Phi_1 \subset \Phi_2 \subset \ldots$ becomes stable after a number l of steps: $\Phi_l = \Phi_{l+1} = \Phi_{l+2} = \ldots$ while Φ_{l-1} is still a proper part of Φ_l. The manifold H_m of the solutions η of $\eta A^m = 0$ is of the same dimensionality h_m as Φ_m. Set $h_1 = h$, $\Phi_1 = \Phi$, $\mathrm{H}_1 = \mathrm{H}$ and $h_l = n$. According to the general theory an η of the n-dimensional manifold H_l vanishes if it satisfies $(\eta\varphi) = 0$ for all $\varphi \in \Phi_l$: the bilinear form $(\eta\varphi)$ for $\eta \in \mathrm{H}_l$, $\varphi \in \Phi_l$ is non-singular.

Use the abbreviations $A\varphi = \varphi'$, $\eta A = \eta'$, and observe the simple rule

$$(\eta'\varphi) = (\eta\varphi').$$

The elements of the form φ' ($\varphi \in \Phi_l$) constitute a manifold Φ_l' of dimensionality $n - h$. Hence let $\varphi_1^*, \ldots, \varphi_h^*$ be h elements of Φ_l that are linearly independent mod. Φ'_l and $\eta_1, \ldots, \eta_h$ a basis of H. I maintain that the square matrix

$$d_{ij} = (\eta_i\varphi_j^*) \qquad (i, j = 1, \ldots, h) \tag{7}$$

is non-singular, det d_{ij}, $\neq 0$. Indeed every element $\eta \in \mathrm{H}$ is orthogonal to the elements φ' of Φ'_l because $(\eta\varphi') = (\eta'\varphi) = 0$; hence if it is also orthogonal to $\varphi_1^*, \ldots, \varphi_h^*$ it is orthogonal to all $\varphi \in \Phi_l$ and therefore zero.

With this in mind we return to the kernel (3) derived from (6) and again examine the four facts, now for Φ_l and H_l instead of Φ_1 and H_1. Does M map H_l into Φ_l? The answer is yes. For any $\eta \in \mathrm{H}_l$ form $\varphi = M\eta$ and again set $\eta' = \eta(E + K) = \eta A$. One infers from Green's formula that for points p in W the equation holds

$$\int (\partial Q(ps')/\partial n_{s'}) \cdot \varphi(s')\, ds' = \int Q(ps')\eta'(s')\, ds'. \tag{8}$$

Letting p tend to a point s on Ω from the exterior one gets

$$(E + K)\varphi = M\eta';$$

in other words: if M carries η into φ it carries $\eta' = \eta A$ into $\varphi' = A\varphi$. This proves even considerably more than the "first fact."

Fact 2 holds good for the mapping M of H_l upon Φ_l because of Rellich's uniqueness theorem. Fact 3 is trivial, and fact 4 is identical with the universally valid statement that $(\eta\varphi)$ is non-singular for $\eta \in H_l$, $\varphi \in \Phi_l$. Thus the external problem can be solved in the manner outlined for the static case if only one uses the manifolds H_l and Φ_l instead of H and Φ.

Still there remains something to be desired. For we know that the "potential" $u(p)$, equation [D], of a suitable double layer $\mu(s)$ solves the problem for given boundary values $\gamma = \gamma(s)$ if γ is orthogonal to the elements η of $H = H_1$; it need not be orthogonal to all the elements of H_l! But for every element η' of H_l' the formula (8) shows that the potential of the simple layer η' is identical with that of the double layer $\varphi = M\eta$. This remark clarifies the situation completely and suggests the following procedure. One chooses $\eta_1^*(s), \ldots, \eta_h^*(s)$ as h elements of H_l that are linearly independent modulo H_l'. They are mapped by M into h elements $\varphi_1^*, \ldots, \varphi_h^*$ of Φ_l that are linearly independent mod. Φ_l'. If $\eta_1, \ldots, \eta_h$ is a basis of H the determinant of the coefficients (7) does not vanish. We can therefore choose h constants α_j such that

$$\gamma^*(s) = \gamma(s) - \sum_{j=1}^{h} \alpha_j \varphi_j^*(s)$$

is orthogonal to $\eta_1, \ldots, \eta_h$. For the boundary values $\gamma^*(s)$ we have a solution in form of the potential of a double layer; to it we add the linear combination $\sum_j \alpha_j v_j^*(p)$ of the potentials $v_j^*(p)$ of the simple layers $\eta_j^*(s)$. These are exactly those potentials of simple layers chosen from the manifold H_l which cannot be carried over into double layer potentials.

The analysis may be pursued to finer details. It is not without interest that here we have a case in mathematical physics where proof of non-degeneracy of a quadratic form is not based on its definite character, and where one can cope successfully with the complications of elementary divisors of higher degree than 1.

3. The best choice of surface layers for the treatment of electromagnetic radiation is set forth in §3 of an old paper of mine[3] on the eigen-frequencies of elastic bodies. It superseded an earlier paper[4]. Since, unfortunately, Mr. Saunders followed the bad example set by me in the latter I propose to come back to the external boundary problem of outgoing electromagnetic waves in a more systematic paper on radiation capacity.

[1] Rellich, F., *Jahresber. d. Deutschen Mathematiker-Vereinigung*, **53**, 57–65 (1943).

[2] Saunders, W. K., Reports No. 175 and 176 of series 7 (1950–1951) issued by the Univ. of California, Dept. of Engineering, Antenna Laboratory.

[3] Weyl, H., *Rend. Circolo Mat. di Palermo*, **39**, 1–49 (1914).

[4] Weyl, H., *J. f. d. reine u. angew. Math.*, **143**, 177–202 (1913).

154.

Kapazität von Strahlungsfeldern

Mathematische Zeitschrift 55, 187—198 (1952)

Diese Arbeit behandelt in breiterer Ausführung, welche die Beweise nicht unterschlägt, denselben Gegenstand wie eine jüngst von mir in den Proceedings der National Academy of Sciences der USA. veröffentlichte Note über skalare Strahlungsfelder (vol. **37**, Dezember 1951). Den Anstoß dazu gaben zwei bisher nur in vorläufiger Form vom Antenna Laboratory der University of California bekannt gegebene Mitteilungen (Nr. 175 und 176 der Serie 7, 1950/51) von Herrn W. K. SAUNDERS, die sich vor allem mit dem elektromagnetischen Strahlungsfeld befassen. Darauf ließe sich meine Theorie, bei richtigem Ansatz der „Belegungs-Potentiale", leicht ausdehnen[1]), wie denn überhaupt im n-dimensionalen EUKLIDischen Raum das skalare Feld durch eine lineare Differentialform beliebigen Ranges ersetzt werden könnte. Mir ist es genug, das Prinzip am skalaren Fall (Rang 0) zu erläutern.

Bezeichnungen. Ein dreidimensionaler EUKLIDischer Raum mit den CARTESischen Koordinaten x_1, x_2, x_3 liegt zugrunde. Der vom beliebigen Punkte p' zum Punkte p führende Vektor mit den Komponenten $x_i - x'_i$ werde mit $\mathfrak{r}(pp')$, seine Länge, die Distanz der beiden Punkte, mit $r = r(pp')$ bezeichnet. Σ_R ist die Kugel vom Radius R um den Ursprung $O = (0, 0, 0)$. Runde Klammern finden für das skalare Produkt zweier Vektoren Verwendung. Die Bezeichnungen grad und div werden aus der Vektoranalysis übernommen. $\Delta = \operatorname{div} \operatorname{grad}$ ist der LAPLACEsche Operator $\frac{\partial^2}{\partial x_1^2} + \frac{\partial^2}{\partial x_2^2} + \frac{\partial^2}{\partial x_3^2}$. Da wir im Gebiet komplexer Zahlen operieren, tritt die imaginäre Einheit i auf.

Es wird angenommen, daß der Raum in zwei komplementäre Teile, das „Innere" V und daß „Äußere" W, geteilt ist. V wird als beschränkt,

---

[1]) Erst nach dem Druck der Proceedings-Note (und nach Abschluß des Manuskripts dieser Arbeit) ist mir die jüngst veröffentlichte Arbeit von CLAUS MÜLLER „Über die Beugung elektromagnetischer Schwingungen an endlichen homogenen Körpern" (Math. Annalen **123**, 1951, S. 345—378) zugänglich geworden. Er verwendet, für den Fall der vollkommenen Reflexion, einen ähnlichen Kunstgriff wie Herr SAUNDERS: Modifikation der Fläche Ω. Ich glaube, daß der hier entwickelte Kapazitäts-Begriff die natürliche Lösung der Schwierigkeit liefert, welche aus der Existenz von Eigenfunktionen des Schwingungsproblems für den Innenraum entsteht.

W als zusammenhängend vorausgesetzt. Ihre gemeinsame Grenze Ω möge aus einer oder mehreren getrennt verlaufenden geschlossenen glatten Flächen bestehen. Glatt bedeutet, daß der Einheitsvektor $\mathfrak{n} = \mathfrak{n}(s)$ der äußeren Normalen im variablen Punkt s von Ω nicht nur eine stetige Funktion von s ist, sondern auch einer HÖLDER-Bedingung genügt. Die normale Ableitung eines skalaren Feldes $u(p)$ wird mit $\partial u/\partial n = (\operatorname{grad} u, \mathfrak{n})$, ihr Wert im Punkte s von Ω zuweilen der Deutlichkeit halber mit $\partial u/\partial n_s$ oder gar (vielleicht nicht sehr glücklich) mit $\partial u(s)/\partial n_s$ bezeichnet. Bei Integration über den Raum mit Bezug auf den variablen Raumpunkt p bedeutet dp das Volumelement, bei Integration über einen variablen Flächenpunkt s ist ds das Oberflächenelement.

Das Randwertproblem und der Eindeutigkeitssatz. k sei eine gegebene positive Konstante. Wir betrachten skalare (komplex-wertige) Felder $u(p)$, welche der Schwingungsgleichung

$$\Delta u + k^2 u = 0 \tag{1}$$

genügen. Für ein derartiges Feld, das außerhalb einer genügend großen Kugel Σ_{R_0} definiert ist, gilt das folgende

Lemma von RELLICH. *u verschwindet identisch, vorausgesetzt, daß*

$$\int_{\Sigma_R} |u|^2 \, ds \to 0 \qquad \textit{für } R \to \infty$$

gleichmäßig in allen Richtungen.

Nur solche Lösungen von (1) gelten als zulässig, welche der Ausstrahlungsbedingung

$$R(\partial u/\partial n + iku)_{\text{auf } \Sigma_R} \to 0 \qquad \text{für } R \to \infty$$

gleichmäßig in allen Richtungen genügen. Die auf einen festen „Aufpunkt“ p' sich beziehende Funktion

$$Q(pp') = Q(r) = e^{-ikr}/2\pi r \qquad [r = r(pp')]$$

von p ist eine solche Lösung (Grundlösung), die einer punktförmigen Quelle in p' entspricht. Die Aufgabe ist, eine zulässige Lösung in W zu bestimmen, die am Rande von W, d. i. auf Ω, vorgegebene Randwerte $\gamma(s)$ annimmt. Vor der Existenz- stellt sich die Eindeutigkeitsfrage. Sie wurde von W. MAGNUS und F. RELLICH auf Grund des oben erwähnten Lemmas im Jahre 1943 bejahend beantwortet, indem sie zeigten, daß $\gamma(s) = 0$ auf Ω die Gleichung $u(p) = 0$ in W impliziert[2]).

[2]) Jahresber. d. Deutschen Math.-Vereinigung **53**, S. 57—65. — Der RELLICHsche Beweis liefert, genau genommen, die Gleichung $u = 0$ nur im Außenraum einer ganz V einschließenden Kugel Σ_{P_0} um O, nicht schon in ganz W. Um sie auf ganz W auszudehnen, muß man sich des Prozesses der analytischen Fortsetzung bedienen, und hier kommt die Voraussetzung zur Geltung, daß W zusammenhängend ist. Sei σ eine samt ihrer Oberfläche ganz innerhalb W liegende

Zurückführung auf eine Fredholmsche Integralgleichung. Man kann den Gradienten von $Q(pp')$ sowohl mit Bezug auf p als p' bilden, und es ist

$$\operatorname{grad}_p Q(pp') = \frac{dQ}{dr}\frac{\mathfrak{r}(pp')}{r(pp')} = -\operatorname{grad}_{p'} Q(pp').$$

So ist klar, was mit

$$\partial Q(ps')/\partial n_{s'} = (\operatorname{grad}_{s'} Q(ps'), \mathfrak{n}(s'))$$

für irgend einen Raumpunkt p und einen Punkt s' auf Ω gemeint ist.

In Analogie zur eigentlichen Potentialtheorie, die mit dem Fall $k = 0$ zu tun hat, bilden wir mit Hilfe einer beliebigen stetigen Funktion $\mu(s)$ auf Ω das „Potential" der Doppelbelegung $\mu(s)$:

(2)
$$u(p) = \int_\Omega (\partial Q(ps')/\partial n_{s'})\,\mu(s')\,ds'.$$

Es genügt sowohl in W wie in V der Gleichung (1). Es erfüllt ferner die Ausstrahlungsbedingung. Seine Randwerte sind

$$\pm\,\mu(s) + \int_\Omega K(s, s')\,\mu(s')\,ds',$$

wobei das Zeichen + für Annäherung an Ω vom Äußeren W, das Zeichen – für Annäherung vom Inneren V gilt. Auf den Kern

$$K(s, s') = (\operatorname{grad}_{s'} Q(ss'), \mathfrak{n}(s'))$$

ist die Fredholmsche Theorie anwendbar, da wegen der Glattheit von Ω das Produkt $(\mathfrak{r}(ss'), \mathfrak{n}(s'))$ für $s = s'$ von höherer als erster Ordnung verschwindet. Integration mit Bezug auf s oder s' erstrecke sich stets auf die ganze Fläche Ω. Das Feld (2) im Äußern wird somit die vorgegebenen stetigen Randwerte $\gamma(s)$ annehmen, falls $\mu(s)$ eine Lösung der Integralgleichung

(3)
$$\mu(s) + \int K(ss')\,\mu(s')\,ds' = \gamma(s)$$

ist. Diese Gleichung besitzt für beliebig vorgegebenes $\gamma(s)$ stets eine (eindeutige) Lösung $\mu(s)$, vorausgesetzt, daß die homogene Gleichung

---

Kugel vom Radius a um den Punkt p_0. Für einen Punkt p' innerhalb σ erhält man in bekannter Weise die Gleichung

$$2\,u(p') = \int_\sigma \{(\partial Q(sp')/\partial n_s)\,u(s) - Q(sp')\,\partial u/\partial n_s\}\,ds.$$

Daraus ergibt sich eine Entwicklung von $u(p)$ nach Potenzen der relativen Koordinaten $x_i - x_i^0$ von p in bezug auf p_0, die zum mindesten innerhalb der konzentrischen Kugel σ' um p_0 vom Radius $a' = a(\sqrt{2} - 1)$ Gültigkeit besitzt. Darauf fußt die kettenartige analytische Fortsetzung der Gleichung $u(p) = 0$ in alle „Buchten" des „Meeres" W hinein.

mit dem transponierten Kern

(4) $$\eta(s') + \int_{\Omega} \eta(s)\, K(s\,s')\, d\,s = 0$$

keine andere Lösung gestattet als die triviale $\eta = 0$. Indem wir die Gleichungen (3) und (4) in der abgekürzten Form

$$(E + K)\,\mu = \gamma, \quad \eta(E + K) = 0$$

schreiben, ahmen wir den Symbolismus des Matrixkalküls so nach, als ob K eine quadratische Matrix wäre und die Werte von μ in einer Spalte, die von η in einer Reihe arrangiert sind. E ist der Operator „Identität".

Aber die Möglichkeit, daß die Gleichung (4) nnd dann auch die Gleichung

(5) $$(E + K)\,\varphi = 0 \quad \text{oder} \quad \varphi(s) + \int K(s\,s')\,\varphi(s')\,d\,s' = 0$$

nicht-triviale Lösungen η oder φ besitzt, kann natürlich nicht ausgeschlossen werden. Die inhomogene Gleichung (3) ist dann und nur dann lösbar, wenn ihre rechte Seite $\gamma(s)$ zu allen Lösungen η von (4) orthogonal ist,

$$(\eta\,\gamma) = 0, \quad \text{d. i.} \quad \int \eta(s)\,\gamma(s)\,d\,s = 0;$$

und die Lösung $\mu(s)$ ist in dem Sinne vieldeutig, daß zu ihr eine beliebige Lösung $\varphi(s)$ von (5) addiert werden kann. Nach dem Eindeutigkeitssatz hat das freilich auf das zugehörige Doppelbelegungs-Potential (2) keinen Einfluß. Die Lösungen η von (4) bilden eine lineare Mannigfaltigkeit H von endlichvielen, sagen wir h, Dimensionen; die lineare Mannigfaltigkeit $\boldsymbol{\Phi}$ der Lösungen φ von (5) besitzt die gleiche Dimensionszahl. Es ist nötig, in Analogie zu dem elektrostatischen Problem der Elektrizitätsverteilung auf vorgegebenen Konduktoren zunächst diese homogenen Gleichungen zu studieren. Man gelangt aber zum Ziel nur, wenn man dabei den Kreis der Eigenfunktionen in der aus der Elementarteilertheorie geläufigen Weise überschreitet. Wir stellen zunächst die einschlägigen Tatsachen der allgemeinen Theorie kurz zusammen.

Die charakteristischen Funktionen eines Kerns. Sei also $K(s\,s')$ ein beliebiger „hinreichend regulärer" Kern. [Es genügt z. B. anzunehmen, daß $K(s\,s')$ gleichmäßig stetig ist, falls man eine beliebig kleine Umgebung $r(s, s') < \varepsilon$ der Diagonale $s = s'$ auf der Mannigfaltigkeit der Punkte (s, s') von $\Omega \times \Omega$ ausnimmt, und daß die (uneigentlichen) Integrale

$$\int_{\Omega} |K(s\,s')|\,d\,s' \quad \text{und} \quad \int_{\Omega} |K(s\,s')|\,d\,s$$

gleichmäßig in s bzw. s' konvergieren. Die ERHARD SCHMIDTsche Methode der Approximation von $K(s\,s')$ durch eine endliche Produktsumme von der Gestalt $\sum f_i(s)\,g_i(s')$ macht es möglich, den insbesondere für singuläre Kerne so schwerfälligen FREDHOLMschen Determinanten-Formalismus ganz zu umgehen.] Für den Operator $E+K$ werde die Abkürzung A verwendet. Man bilde die endlichdimensionalen Mannigfaltigkeiten $\mathsf{H}_0, \mathsf{H}_1, \mathsf{H}_2, \ldots$ der Lösungen η der sukzessiven Gleichungen

$$\eta = 0 | \quad \eta A = 0 | \quad \eta A^2 = 0 | \ldots .$$

Offenbar ist $\mathsf{H}_0 \subset \mathsf{H}_1 \subset \mathsf{H}_2 \subset \cdots$. Diese Sequenz wird stabil nach einer gewissen Anzahl l von Schritten; d. h. es ist $\mathsf{H}_l = \mathsf{H}_{l+1} = \cdots$, während H_{l-1} noch ein echter Teil von H_l ist. Die Mannigfaltigkeit Φ_m der Lösungen φ von $A^m \varphi = 0$ hat dieselbe Dimensionszahl h_m wie H_m $(m = 0, 1, 2, \ldots)$. Die aus einem willkürlichen Element η von H_l und φ von Φ_l gebildete Bilinearform

$$(\eta \varphi) = \int \eta(s)\, \varphi(s)\, d s$$

ist nicht-ausgeartet; d. h. $\varphi = 0$ ist das einzige Element φ von Φ_l, das der Gleichung $(\eta \varphi) = 0$ für alle $\eta \in \mathsf{H}_l$ genügt[3]). Wir verwenden die Bezeichnungen H, Φ, h für H_1, Φ_1, h_1 und setzen $h_l = n$. Die Elemente φ von Φ_1 sind die Eigenfunktionen des Kernes K (sc. zum Eigenwert -1); die Elemente von Φ_l tragen in der Literatur verschiedene Namen, sie mögen hier charakteristische Funktionen heißen.

Unter Verwendung der Abkürzungen $\varphi' = A\varphi$, $\eta' = \eta A$ gilt die einfache Regel

$$(\eta' \varphi) = (\eta \varphi'); \tag{6}$$

denn beide Ausdrücke sind gleich $\eta A \varphi$. Die Elemente von der Form φ' $(\varphi \in \Phi_l)$ bilden eine Mannigfaltigkeit Φ_l' von $n-h$ Dimensionen. In der Tat, ergänzt man die Basis $\varphi_1, \ldots, \varphi_h$ von Φ durch Hinzufügung weiterer Elemente $\varphi_{h+1}, \ldots, \varphi_n$ zu einer Basis von Φ_l, so ist offenbar $\varphi_{h+1}', \ldots, \varphi_n'$ eine Basis von Φ_l'. Durch h Elemente $\varphi_1^*, \ldots, \varphi_h^*$, welche linear unabhängig mod Φ_l' sind, kann man diese ihrerseits zu einer vollen Basis von Φ_l ergänzen. Es sei ferner $\eta_1, \ldots, \eta_h$ eine Basis von H. Ich behaupte, daß die Matrix

$$d_{ij} = (\eta_i \varphi_j^*) \qquad (i, j = 1, \ldots, h)$$

nicht-ausgeartet ist, d. h. det $d_{ij} \neq 0$. Dies folgert man entweder aus der Tatsache, daß die Bilinearform $(\eta\varphi)$ für $\eta \in \mathsf{H}_l$, $\varphi \in \Phi_l$ nicht-ausgeartet ist, oder man zeigt direkt, daß eine lineare Kombination $\varphi = \sum_{j=1}^{h} a_j \varphi_j^*$, welche den Gleichungen $(\eta_i \varphi) = 0$ für $i = 1, \ldots, h$ genügt, Null sein muß. In der Tat ergibt sich aus diesen Bedingungen nach dem Fundamentalsatz über die Lösung von Integralgleichungen die Existenz einer Funktion ψ, für die $A\psi = \varphi$ gilt. Da somit $A^{l+1}\psi = 0$

---

[3]) Hier ist ein rascher Beweis: Nach dem Hauptsatz über Integralgleichungen hat die Gleichung $A^l \psi = \varphi$ eine Lösung ψ, falls φ orthogonal zu allen $\eta \in \mathsf{H}_l$ ist. Liegt φ in Φ_l, so ψ in Φ_{2l}. Es ist aber $\Phi_{2l} = \Phi_l$, somit $\psi \in \Phi_l$, $A^l \psi = \varphi = 0$.

ist, liegt ψ in $\Phi_{l+1} = \Phi_l$ und darum φ in Φ'_l. Das hat aber, da $\varphi_1^*, \ldots, \varphi_h^*$ linear unabhängig mod Φ'_l sind, die Gleichungen $a_1 = \cdots = a_h = 0$ zur Folge.

Die Abbildung *M*. Wir kehren nunmehr zu unserm speziellen Kern K zurück. Neben (2) betrachten wir das Potential einer einfachen Belegung $\nu(s)$,

$$v(p) = \int Q(ps)\,\nu(s)\,ds. \tag{7}$$

Hier ist $\nu(s)$ eine beliebige stetige Funktion auf Ω; und wie immer erstreckt sich die Integration über ganz Ω. Die Funktion $v(p)$ genügt sowohl in V als in W der Gleichung $\Delta v + k^2 v = 0$, sie erfüllt die Ausstrahlungsbedingung, und sie geht stetig über Ω hinüber. Hingegen erleidet die normale Ableitung $\partial v/\partial n$ auf Ω den Sprung $2\nu(s)$, indem dieselbe auf der Außen- und Innen-Seite im Punkte s' von Ω bzw. die Werte

$$\mp \nu(s') + \int \nu(s)\,K(ss')\,ds, \qquad \text{kürzer} \quad \nu(\mp E + K)$$

annimmt.

Man bilde das Potential (7) insbesondere für den Fall, daß $\nu(s)$ eine charakteristische Funktion η des transponierten Kernes ist:

$$v(p) = \int Q(ps')\,\eta(s')\,ds', \tag{8}$$

und bezeichne seine Randwerte mit

$$\varphi(s) = \int Q(ss')\,\eta(s')\,ds'. \tag{9}$$

Ich behaupte:

Satz 1. *Die Gleichung* (9) *definiert eine lineare Abbildung* $\eta \to \varphi = M\eta$ *von* H_l *in* Φ_l.

An Stelle dieses werde sogleich ein schärferer Satz bewiesen, in dem der Angelpunkt unserer ganzen Methode liegt:

Satz 2. *Führt die Abbildung M das Element $\eta \in \mathsf{H}_l$ in φ über, so geht durch dieselbe Abbildung $\eta' = \eta(E+K)$ in $\varphi' = (E+K)\varphi$ über.*

M verwandelt also auch ηA^m in $A^m\varphi$; und wenn $\eta A^m = 0$ ist, muß $A^m\varphi = 0$ sein. So folgt Satz 1 in der Form, daß M nicht nur H_l in Φ_l verwandelt, sondern auch jede der Teilmannigfaltigkeiten H_m $(m = 1, \ldots, l)$ in das entsprechende Φ_m.

Beweis von Satz 2. Sei p' ein Punkt im Äußern W. Man wende Greens Formel

$$\int_\Omega \left(u\,\frac{\partial v}{\partial n} - v\,\frac{\partial u}{\partial n}\right) ds = \int_V \{u(\Delta v + k^2 v) - v(\Delta u + k^2 u)\}\,dp$$

an auf das Innere V und die beiden folgenden daselbst regulären Funktionen von p: $Q(pp')$ und (8). Man findet

$$\int (\partial Q(sp')/\partial n_s)\,v(s)\,ds = \int Q(sp')\,(\partial v/\partial n)(s)\cdot ds.$$

Hier ist mit $\partial v/\partial n$ die normale Ableitung auf der *inneren* Seite von Ω gemeint, welche den Wert $\eta(E+K) = \eta A = \eta'$ besitzt. Ändert man die Bezeichnungen p', s in p, s' um, so erhält man folgende für alle Punkte $p \in W$ gültige Gleichung

$$(10)\qquad \int_\Omega (\partial Q(p s')/\partial n_{s'})\, \varphi(s')\, d s' = \int_\Omega Q(p s')\, \eta'(s')\, d s'.$$

Läßt man jetzt den Punkt p von außen in einen Randpunkt s rücken, so geht die linke Seite in $\varphi'(s)$ über, wo $\varphi' = (E+K)\varphi$, während man für die rechte Seite

$$\int Q(s s')\, \eta'(s')\, d s',$$

d. i. das Bild $M\eta'$ von η', erhält. Darum ist, zugleich mit $\varphi = M\eta$, in der Tat

$$\varphi' = M\eta'.$$

Satz 3. *Die Abbildung M ist nicht-ausgeartet, d. h. $\eta = 0$ ist das einzige Element von H_l, das durch M in $\varphi = 0$ übergeht.*

Daraus folgt sogleich, daß die Abbildung M eineindeutig ist.

Beweis. Für das Potential (8) bedeutet die Gleichung $M\eta = 0$, daß seine (inneren und äußeren) Randwerte verschwinden. Nach dem Rellichschen Eindeutigkeitssatz verschwindet darum $v(p)$ im ganzen Außenraum, und darum ist auch die normale Ableitung $(\partial v/\partial n)_a$ auf der Außenseite von Ω gleich Null, $\eta(-E+K) = 0$. Infolgedessen gilt

$$\eta(E+K) = 2\eta \quad \text{oder} \quad \eta = \tfrac{1}{2}\eta A.$$

Daraus folgt aber

$$\eta A = \tfrac{1}{2}\cdot\eta A^2, \quad \eta = \tfrac{1}{2}\eta A = \tfrac{1}{4}\cdot\eta A^2,$$

und durch Wiederholung dieses Verfahrens schließlich

$$\eta = 2^{-l}\cdot\eta A^l = 0,$$

q. e. d.

Wir bilden nun aus irgend zwei Elementen η, η^* von H_l die Bilinearform

$$C(\eta^*, \eta) = (\eta^*, M\eta)$$

und behaupten:

Satz 4. *Die Kapazitätsform $C(\eta^*, \eta)$ ist symmetrisch und nicht-ausgeartet.*

Beweis. Die Symmetrie ergibt sich ohne weiteres aus dem expliziten Ausdruck

$$C(\eta^*, \eta) = \int\int \eta^*(s)\, Q(s s')\, \eta(s')\, d s'\, d s.$$

Die zweite Behauptung, wonach $\eta = 0$ das einzige Element von H_l ist, das für alle Elemente η^* von H_l die Gleichung $C(\eta^*, \eta) = 0$ erfüllt, entspringt aus der Kombination der beiden Tatsachen, daß 1) für ein festes $\eta \in \mathsf{H}_l$ das Element $\varphi = M\eta$ von Φ_l der Gleichung $(\eta^* \varphi) = 0$ für alle $\eta^* \in \mathsf{H}_l$ nur dann genügt, wenn $\varphi = 0$ ist, und daß 2) die Gleichung $M\eta = 0$ die andere $\eta = 0$ impliziert (Satz 3).

Wegen der Symmetrie genügt es, die quadratische Form $C(\eta\eta)$ statt der Bilinearform $C(\eta^*, \eta)$ zu betrachten. Mit Bezug auf eine Basis $\eta_1, \ldots, \eta_n$ von H_l läßt sich jedes $\eta \in \mathsf{H}_l$ als eine lineare Kombination

$$\eta(s) = x_1\, \eta_1(s) + \cdots + x_n\, \eta_n(s)$$

der η_i schreiben, und $C(\eta\eta)$ geht dann in die nicht-ausgeartete quadratische Form $\sum c_{ij}\, x_i\, x_j$ der Variablen x_i über mit den symmetrischen Koeffizienten

$$c_{ij} = C(\eta_i, \eta_j) \qquad (i, j, = 1, \ldots, n). \tag{11}$$

Diese nennen wir die Kapazitätskoeffizienten. [Im Fall der eigentlichen Potentialtheorie, $k = 0$, ist die Form $C(\eta\eta)$ sogar nicht-ausgeartet, wenn wir uns auf die Elemente η von $\mathsf{H} = \mathsf{H}_1$ beschränken, da sie gleich dem über den ganzen Raum erstreckten Dirichletschen Integral von (8) und darum positiv-definit ist. Dies hat übrigens, wie man sofort sieht, zur Folge, daß $l = 1$ ist[4]). Im Falle der Strahlungstheorie versagt dieser Schluß, da hier an Stelle des Dirichlet-Integrals das Integral über den indefiniten Ausdruck $(\operatorname{grad} v)^2 - k^2 v^2$ tritt; zudem ist v komplex und das Integral von zweifelhafter Konvergenz.]

Sei $\eta_1, \ldots, \eta_h$ eine Basis von $\mathsf{H} = \mathsf{H}_1$. Wir wählen h Elemente $\eta_1^*, \ldots, \eta_h^*$ von H_l, die linear unabhängig sind mod H_l'; ihre Bilder $\varphi_i^* = M\eta_i^*$ $(i = 1, \ldots, h)$ in Φ_l sind linear unabhängig mod Φ_l', und darum ist die quadratische Matrix der

$$d_{ij} = (\eta_i, \varphi_j^*) = C(\eta_i, \eta_j^*) \qquad (i, j = 1, \ldots, h) \tag{12}$$

nicht-ausgeartet. (Symmetrie kann man für sie natürlich nicht erwarten.) Wir nennen (12) die Kapazitätskoeffizienten im engeren Sinne. Wir werden sogleich noch näher analysieren, in welcher Weise diese „kleine“ h-zeilige Kapazitätsmatrix $D = \|d_{ij}\|$ in der „großen“ n-zeiligen symmetrischen Matrix C der c_{ij} enthalten ist.

Lösung der Randwertaufgabe. Nunmehr sind wir in der Lage, eine zulässige Lösung u der Gleichung (1) im Außenraum W für beliebig vorgegebene stetige Randwerte $\gamma(s)$ auf Ω zu konstruieren. Wir können uns dabei entweder der „großen“ oder der „kleinen“ Matrix der Kapazitätskoeffizienten bedienen. Im ersten Falle verfährt man so: Aus einer Basis $\eta_1, \ldots, \eta_n$ von H_l bilde man die Basis $\varphi_i = M\eta_i$ von Φ_l. Die Funktion $\varphi_i(s)$ besteht aus den (inneren und äußeren) Randwerten des Potentials

$$v_i(p) = \int G(p s')\, \eta_i(s')\, ds'.$$

[4]) In der Tat zeigt der erwähnte Umstand, daß ein $\eta \in \mathsf{H}$, das zu allen φ in Φ orthogonal ist, notwendig verschwindet. Für ein η in H_2 ist aber η' ein solches Element, darum ist $\eta' = 0$, d. h. η liegt in H_1, oder $\mathsf{H}_2 = \mathsf{H}_1$.

Von der vorgegebenen Randfunktion $\gamma(s)$ subtrahiere man eine solche lineare Kombination der $\varphi_j(s)$, daß die Differenz

$$\gamma^*(s) = \gamma(s) - \sum_{j=1}^{n} a_j \varphi_j(s)$$

zu allen η_i orthogonal wird. Das ergibt die n Gleichungen

$$(\eta_i \gamma) = \sum_j c_{ij} a_j \qquad (i, j = 1, \ldots, n).$$

Diese besitzen eine eindeutige Lösung a_j, da die symmetrische Matrix der Kapazitätskoeffizienten

$$c_{ij} = (\eta_i, \varphi_j) = (\eta_i, M\eta_j) = C(\eta_i, \eta_j)$$

nicht-ausgeartet ist. Für $\gamma^*(s)$ hat die Integralgleichung (3), $(E+K)\mu^* = \gamma^*$ eine Lösung $\mu^*(s)$, und diese führt mittels (2) zu einem Doppelbelegungs-Potential $u^*(p)$ in W mit den Randwerten $\gamma^*(s)$. Indem wir der Mannigfaltigkeit H_l das Element $\eta = \sum_j a_j \eta_j$ entnehmen und zu u^* das der einfachen Belegung $\eta(s)$ entspringende Potential addieren, erhalten wir in

$$u(p) = u^*(p) + \sum_j a_j v_j(p)$$

eine zulässige Lösung in W mit den vorgegebenen Randwerten $\gamma(s)$.

Dies Verfahren ist darum unbefriedigend, weil ja zur Lösbarkeit der inhomogenen Gleichung (3) gar nicht die Orthogonalität von $\gamma(s)$ zu allen Elementen η von H_l, sondern nur zu allen Elementen η von $\mathsf{H} = \mathsf{H}_1$ notwendig ist. Darum ziehe ich vor, h Elemente $\eta_1^*, \ldots, \eta_h^*$ von H_l zu wählen, die mod H_l' linear unabhängig sind, und außerdem eine Basis $\eta_1, \ldots, \eta_h$ von $\mathsf{H} = \mathsf{H}_1$. Dann ist die Matrix (12) nicht-ausgeartet. Ich subtrahiere also von $\gamma(s)$ eine solche lineare Kombination $\sum_j a_j \varphi_j^*$ der h Funktionen $\varphi_j^* = M\eta_j^*$, daß die Differenz γ^* zu den Basiselementen $\eta_1, \ldots, \eta_h$ von H orthogonal wird; das ergibt die in der Tat lösbaren Gleichungen

$$(\eta_i \gamma) = \sum_{j=1}^{h} d_{ij} a_j \qquad (i = 1, \ldots, h).$$

Für γ^* bekommen wir wie vorhin ein μ^* und ein zugehöriges Doppelbelegungs-Potential $u^*(p)$ in W mit den Randwerten $\gamma^*(s)$. Wenn wir

$$\sum_{j=1}^{h} a_j v_j^*(p), \quad v_j^*(p) = \int G(ps') \eta_j^*(s')\, ds',$$

dazu addieren, resultiert eine zulässige Lösung $u(p)$ in W mit den Randwerten $\gamma(s)$.

Die Aufklärung über das Verhältnis der beiden Verfahrensweisen bringt die Gleichung (10). Sie zeigt nämlich, daß das aus einer zu H_l' gehörigen Belegung η' entspringende einfache Potential dem von der

Doppelbelegung $\varphi = M\eta$ erzeugten Potential gleich ist. Das zweite Verfahren sondert aus H_l eine lineare Schar $\sum_j a_j \eta_j^*(s)$ von einfachen Belegungen aus, deren Potentiale sich durchaus nicht in Doppelbelegungs-Potentiale umwandeln lassen und die darum das Minimum dessen darstellen, was man zum allgemeinen Doppelbelegungs-Potential hinzufügen muß, um die Lösung der Randwertaufgabe für beliebiges $\gamma(s)$ zu erzwingen.

Genauere Analyse des Enthaltenseins der kleinen in der großen Kapazitäts-Matrix. Eine Basis („Normalbasis") von H_l kann in folgender Weise konstruiert werden. Man fängt an mit einer Basis $\sigma_0 = \{\eta_{01}, \ldots, \eta_{0r}\}$ von H_l mod H_{l-1}. Die Unabhängigkeit mod H_{l-1} hat zur Folge, daß $\eta_{01}A^{l-1}, \ldots, \eta_{0r}A^{l-1}$ linear unabhängig sind, und darum sind $\sigma_0' = \{\eta_{01}', \ldots, \eta_{0r}'\}$ Elemente in H_{l-1}, die linear unabhängig sind mod H_{l-2}, $\sigma_0'' = \{\eta_{01}'', \ldots, \eta_{0r}''\}$ besteht aus Elementen von H_{l-2}, die linear unabhängig sind mod H_{l-3}, und so fort bis $\sigma_0^{(l-1)}$. So erhalten wir $r \cdot l$ linear unabhängige Elemente, welche in die Abschnitte $\sigma_0, \sigma_0', \ldots, \sigma_0^{(l-1)}$ je von der Länge $r = r_0$ zerfallen. Darauf ergänzen wir $\eta_{01}', \ldots, \eta_{0r}'$ durch Hinzufügung von r_1 weiteren Elementen $\sigma_1 = \{\eta_{11}, \ldots, \eta_{1r_1}\}$ zu einer vollen Basis von H_{l-1} mod H_{l-2} und bilden wiederum $\sigma_1, \sigma_1', \ldots, \sigma_1^{(l-2)}$. In dieser Weise fortfahrend lassen wir eine Normalbasis von H_l entstehen, die aus den Abschnitten

$$\sigma_f^{(i)} \qquad (i = 0, 1, \ldots, l-1-f;\ f = 0, \ldots, l-1)$$

je von der Länge r_f besteht. Die Elemente η von σ_f gehören zu H_{l-f} und genügen darum der Gleichung $\eta^{(l-f)} = 0$. Die Reihe der Indizes, durch die unsere Basiselemente voneinander unterschieden sind, zerfällt in $L = \frac{l(l+1)}{2}$ Sektionen (f, i). Die Basis selbst besteht aus Ketten

$$\eta, \eta', \ldots, \eta^{(l-1-f)}$$

verschiedener Länge $l-f$ $(f = 0, 1, \ldots, l-1)$; das Anfangselement η einer Kette von der Länge $l-f$ gehört zu σ_f.

Bei Wahl dieser Normalbasis steht in der aus den Kapazitätskoeffizienten (11) gebildeten „großen Matrix" C vom Grade n dort, wo sich die Sektion (f, i) des Zeilenindex mit der Sektion (g, j) des Kolonnenindex kreuzt, die rechteckige Matrix $C_{f,g}^{i,j}$ der Elemente

$$C(\eta_f^{(i)}, \eta_g^{*(j)}), \quad \text{wo} \quad \eta_f \in \sigma_f,\ \eta_g^* \in \sigma_g.$$

Nach der Regel (6) und Satz 2 hängt diese Matrix nur von $i + j = \varrho$ ab und werde darum mit $C_{f,g}^{\varrho}$ bezeichnet. Da $\eta_f^{(\varrho)} = 0$ ist, sobald $\varrho \geqq l-f$, und $\eta_g^{*(\varrho)} = 0$, sobald $\varrho \geqq l-g$, verschwindet diese Matrix, wenn $\varrho \geqq l - \max(f, g)$. Sie verschwindet insbesondere, wenn entweder

$$i + j \geqq l - f \quad \text{oder} \quad i + j = l - 1 - f,\ g > f$$

ist. Nehmen wir für einen Augenblick in der Matrix C diejenige Umstellung der Zeilen vor, welche zustande kommt, wenn man die Nummern $i = 0, 1, \ldots, l-1-f$ des Index der Zeilensektionen $(f\,i)$ rückwärts durchläuft,

$$i' = l-1-f-i, \quad \tilde{C}^{i,j}_{f,g} = C^{i',j}_{f,g},$$

so kann man die für das Verschwinden von $C^{i',j}_{f,g}$ hinreichenden Bedingungen

$$i'+j > l-1-f \quad \text{oder} \quad i'+j = l-1-f,\ g > f,$$

in der Form schreiben:

$$j > i \quad \text{oder} \quad j = i,\ g > f. \tag{13}$$

Bei alphabetischer Anordnung der Sektionen $(f\,i)$ nach der Regel, daß $(g\,j)$ auf $(f\,i)$ folgt, wenn (13) besteht, tritt also für die Matrix $\tilde{C}$ Reduktion ein, indem alle Rechtecke von $\tilde{C}$, die auf einer Seite der durch $j = i$, $g = f$ gekennzeichneten Hauptdiagonale stehen, mit Nullen besetzt sind. Darum zerfällt die Determinante von $\tilde{C}$ in die Determinanten der in der Hauptdiagonale von $\tilde{C}$ stehenden Matrizen $\tilde{C}^{i,i}_{f,f} = C^{i',i}_{f,f} = C^{l-1-f}_{f,f}$. Setzen wir $C^{l-1-f}_{f,f} = C_f$, so besteht demnach die Gleichung

$$|C| = \pm\, |C_0|^l \cdots |C_{l-1}|^1. \tag{14}$$

Sie beruht auf dem Umstand, daß in den L „Hauptfeldern" von C, wo sich (f, i') mit (f, i) kreuzt, die Matrizen C_f stehen, während in einer Hälfte der $L(L-1)$ Nebenfelder Nullen stehen, in solcher Verteilung, daß Reduktion eintritt. Die Regel

$$C^{i,j}_{f,g} = 0 \qquad \text{für } i+j \geqq l - \max(f, g)$$

ergibt, wie man leicht ausrechnet, als Anzahl der leerstehenden Felder in dem Raster von C den Wert

$$N_l^* = \frac{(l-1)\cdot l\cdot (l+1)^2}{6},$$

während

$$N_l = \frac{L(L-1)}{2} = \frac{(l-1)\cdot l\cdot(l+1)\cdot(l+2)}{8}$$

ist. Also tragen noch

$$\frac{(l-2)\cdot(l-1)\cdot l\cdot(l+1)}{1\cdot 2\cdot 3\cdot 4}$$

mehr Rechtecke $(f\,i) \times (g\,j)$ die Matrix 0 als für die Reduktion nötig wäre.

Die Endelemente unserer Ketten, d. i.

$$\sigma_f^{(l-1-f)} \qquad (f = 0, 1, \ldots, l-1),$$

bilden zusammen eine Basis $\eta_1, \ldots, \eta_h$ für $\mathsf{H} = \mathsf{H}_1$, während ihre Anfangselemente, d. i. die σ_f, zusammen eine Basis $\eta_1^*, \ldots, \eta_h^*$ von H_l mod H_l' ergeben. So zerfällt die Reihe der hier einen Augenblick benutzten Indizes $1, 2, \ldots, h$ in natürlicher Weise in l durch den Index $f = 0$,

$1, \ldots, l-1$ unterschiedene Sektionen von der Länge r_f. Für die kleine Matrix D der

$$d_{ij} = C(\eta_i, \eta_j^*) \qquad (i, j = 1, \ldots, h)$$

ergibt sich ein Raster, daß im Durchschnitt $f \times g$ der Sektion f mit g die Matrix $C_{f,g}^{l-1-f}$ trägt. Diese verschwindet für $g > f$, während längs der Hauptdiagonale die gleichen Matrizen C_f auftreten, die in den Hauptfeldern von C stehen. Somit

$$(15) \qquad |D| = |C_0| \cdot |C_1| \cdots |C_{l-1}|.$$

Wenn man will, mag man die Formeln (14), (15) dazu benutzen, um aus $|D| \neq 0$ die Ungleichung $|C| \neq 0$ zu erschließen. Die Analyse zeigt, daß D bereits alle in C wesentlichen Bestandteile C_f enthält; während aber die Matrix C_f in den Hauptfeldern von C $(l-f)$mal vorkommt, tritt sie in D nur einfach auf.

155.

Die natürlichen Randwertaufgaben im Außenraum für Strahlungsfelder beliebiger Dimension und beliebigen Ranges

Mathematische Zeitschrift 56, 105—119 (1952)

Die von mir in einer voraufgehenden Note[1]) gegebene Behandlung der „Kapazität" von Strahlungsfeldern hat inzwischen durch Herrn C. MÜLLER eine wesentliche Vereinfachung erfahren, indem es ihm gelang zu zeigen, daß hier so wenig wie in der Potentialtheorie Elementarteiler höherer Ordnung auftreten[2]). Im n-dimensionalen Euklidischen Raum $\mathfrak{R}_n (n \geqq 2)$ existieren zwei zueinander duale natürliche Randwertprobleme, das „elektrische" und „magnetische", für ein Schwingungsfeld vom Range q; wobei q der Werte $0, 1, \ldots, n$ fähig ist. Das magnetische Problem vom Range $n-q$ ist mit dem elektrischen vom Range q identisch. Nachdem die Herren W. K. SAUNDERS und C. MÜLLER dem für $n = 3$, $q = 1$ sich ergebenden Randwertproblem des elektromagnetischen Feldes von neuem ihre Aufmerksamkeit geschenkt haben[3]), möchte ich auch die von mir für den allgemeinen Fall eines beliebigen n und q vorgesehene Behandlung bekannt geben, mit der Modifikation jedoch, welche die schöne MÜLLERsche Entdeckung möglich gemacht hat. Dafür ist der heute vielen Mathematikern geläufige CARTANsche Kalkül der linearen Differentialformen, wie ich ihn kurz in § 2 auseinandersetze, das angemessene Werkzeug. Im Unterschied von neueren Untersuchungen über harmonische Integrale ist das Operationsgebiet hier euklidisch, aber nicht-kompakt.

§ 1.

Die Grundlösung.

Im $\mathfrak{R}_n$ seien $x_1, \ldots, x_n$ die Cartesischen Koordinaten eines variablen Punktes p, k eine positive Konstante. Für irgend zwei Punkte

---

[1]) Kapazität von Strahlungsfeldern, diese Zeitschrift **55** (1952), 187—198.

[2]) C. MÜLLER, Zur Methode der Strahlungskapazität von H. WEYL, diese Zeitschrift **56** (1952), 80—83.

[3]) W. K. SAUNDERS, On solutions of MAXWELL's equations in an exterior region, Proc. Nat. Ac. of Sciences, USA., **38** (1952), 342—348. C. MÜLLER, Randwertprobleme der Theorie elektro-magnetischer Schwingungen (erscheint demnächst in dieser Zeitschrift).

p, p' bezeichne $\mathfrak{r}(pp') = \overrightarrow{p'p}$ den von p' nach p führenden Vektor, r seine Länge. Σ_R sei die Kugel um den Ursprung o vom Radius R. Indem wir zunächst p' mit o zusammenfallen lassen, bestimmen wir eine kugelsymmetrische, d. i. nur von r abhängige Lösung (Grundlösung) $G(r)$ der Schwingungsgleichung $\Delta u + k^2 u = 0$ aus der HANKELschen Zylinderfunktion $H(r)$ der Ordnung $\frac{n}{2} - 1$, die sich im Unendlichen asymptotisch wie $e^{-ir}/\sqrt{r}$ verhält, indem wir $G(r) = H(kr)/r^{\frac{n}{2}-1}$ setzen. Im Nullpunkt wird $G(r)$ singulär wie const. $r^{-(n-2)}$ (man weiß, wie diese Aussage für $n = 2$ zu modifizieren ist). Ein zunächst in $G(r)$ willkürlich bleibender konstanter Faktor werde so normiert, daß der Fluß von G, nämlich das Integral von $-dG/dr$ über die Kugel vom Radius r, mit $r \to 0$ gegen 2 konvergiert. Die Wahl der besonderen HANKELschen Zylinderfunktion hat zur Folge, daß

$$r^{\frac{n-1}{2}} \cdot \left(\operatorname{grad} G + ikG\frac{\mathfrak{r}}{r}\right)$$

mit $r \to \infty$ gegen 0 strebt (Ausstrahlungsbedingung).

Ist $r = r(pp')$ die Entfernung des variablen Punktes p von einem beliebigen festen Aufpunkt p', so werde die Grundlösung $G(r)$ mit $G(pp')$ bezeichnet, und $\Gamma = \Gamma(pp')$ sei der Vektor $\operatorname{grad}_p G(pp')$ mit den Komponenten $\partial G/\partial x_i$. Offenbar ist $\Gamma' = \operatorname{grad}_{p'} G(pp') = -\Gamma$.

§ 2.

Kalkül der schiefsymmetrischen Tensorfelder im Euklidischen Raum.

Ein schiefsymmetrischer Tensor mit den Komponenten $f_{i_1 \ldots i_q}$ vom Range q werde kurz als q-Tensor, ein derartiges Tensorfeld als q-Feld bezeichnet. Das innere Produkt $(f \cdot g)$ zweier q-Tensoren f und g ist die Summe

$$\sum f_{i_1 \ldots i_q} \cdot g_{i_1 \ldots i_q},$$

erstreckt über die $\binom{n}{q}$ Kombinationen verschiedener Ziffern $i_1, \ldots, i_q$ aus der Reihe $1, 2, \ldots, n$. Ist α ein Vektor mit den Komponenten α_i, so ist $(f\alpha)$ ein Tensor vom Range $q-1$, $[f\alpha]$ einer vom Range $q+1$, mit den Komponenten

$$(1) \quad (f\alpha)_{i_1 \ldots i_{q-1}} = \sum_{i=1}^{n} f_{i_1 \ldots i_{q-1} i} \cdot \alpha_i, \quad [f\alpha]_{i_1 \ldots i_{q+1}} = \mathbf{S} \pm f_{i_1 \ldots i_q} \cdot \alpha_{i_{q+1}}.$$

$\mathbf{S} \pm$ zeigt eine alternierende Summe von $q+1$ Gliedern an, die dadurch zustande kommt, daß der letzte i_{q+1} in der Reihe der Indizes schrittweise von hinten nach vorn durchgezogen wird. Man beachte die Regeln

$$(2) \qquad ((f\alpha)\alpha) = 0 \quad \text{und} \quad [[f\alpha]\alpha] = 0$$

Eine triviale Rechnung ergibt für zwei Vektoren α, β

(3) $$([f\beta]\alpha) = f(\beta\alpha) - [(f\alpha)\beta],$$

insbesondere für $\beta = \alpha$:

(4) $$f(\alpha\alpha) = ([f\alpha]\alpha) + [(f\alpha)\alpha].$$

Wird der Vektor α vom Betrage 1, $(\alpha\alpha) = 1$, als „Normale" angesprochen, so liefert diese Gleichung die Zerlegung von f in einen „tangentiellen" Tensor $f_t = ([f\alpha]\alpha)$ und einen „normalen" $f_\nu = [(f\alpha)\alpha]$; denn wir werden f normal oder tangentiell nennen, je nachdem $[f\alpha] = 0$ oder $(f\alpha) = 0$ ist. Wird α als $(0, \ldots, 0, 1)$ normiert, so faßt f_t die $\binom{n-1}{q}$ Komponenten $f_{i_1 \ldots i_q}$ zusammen, für die keiner der q Indizes i gleich n ist, f_ν die $\binom{n-1}{q-1}$ Komponenten $f_{i_1 \ldots i_{q-1} n}$. Daher das Pythagoreische Gesetz

$$(f \cdot g) = (f_t \cdot g_t) + (f_\nu \cdot g_\nu).$$

Die folgende Regel, in der f vom Range q, w vom Range $q + 1$ ist, findet vielfache Verwendung:

(5) $$(f \cdot (w\alpha)) = ([f\alpha] \cdot w).$$

Nimmt man $\alpha_i = \partial/\partial x_i$, so erhält man statt (1) die beiden auf ein q-Feld f wirkenden Differentialoperatoren div und rot,

$$(\operatorname{div} f)_{i_1 \ldots i_{q-1}} = \sum_i \frac{\partial f_{i_1 \ldots i_{q-1} i}}{\partial x_i}, \quad (\operatorname{rot} f)_{i_1 \ldots i_{q+1}} = \mathbf{S} \pm \frac{\partial f_{i_1 \ldots i_q}}{\partial x_{i_{q+1}}}.$$

Den Formeln (2), (4) entsprechen die Gleichungen

(6) $$\operatorname{div} \operatorname{div} f = 0, \quad \operatorname{rot} \operatorname{rot} f = 0;$$

(7) $$\Delta f = \operatorname{div} \operatorname{rot} f + \operatorname{rot} \operatorname{div} f.$$

Aus einem q-Tensor f entsteht in einer gegenüber *eigentlichen* orthogonalen Koordinaten-Transformationen invarianten Weise der duale $(n - q)$-Tensor f^* nach der Gleichung

$$f_{i_1 \ldots i_q} = f^*_{i_{q+1}, \ldots, i_n},$$

in der $i_1, \ldots, i_q, i_{q+1}, \ldots, i_n$ irgend eine gerade Permutation von $1, \ldots, n$ ist. Ersetzt man konsequent alle Tensoren durch ihre dualen, so vertauschen sich die Operationen $(f\alpha)$ und $[f\alpha]$ sowie $\operatorname{div} f$ und $\operatorname{rot} f$. .

§ 3.

Greensche Formel.

Seien u, v zwei Felder vom Range q. Man bilde die folgenden beiden Vektorfelder F und $\tilde{F}$:

$$F_i = \sum_{i_1, \ldots, i_{q-1}} u_{i_1 \ldots i_{q-1} i} (\operatorname{div} v)_{i_1 \ldots i_{q-1}},$$

$$\tilde{F}_i = \sum_{i_1, \ldots, i_q} u_{i_1 \ldots i_q} \cdot (\operatorname{rot} v)_{i_1 \ldots i_q i},$$

(die Summen erstrecken sich über alle Kombinationen $(i_1, \ldots)$ zu $q-1$ bzw. q). Es sei V ein beschränktes Gebiet, das von einer oder mehreren glatten Oberflächen Ω begrenzt wird. Auf Ω bezeichne $\boldsymbol{\nu}$ den Vektor der äußeren Einheitsnormalen. Ein variabler Punkt auf Ω werde mit s (oder s') bezeichnet; bei Integration über den Raumpunkt p oder den Flächenpunkt s bedeute dp das Volumen-, ds das Flächen-Element. Anwendung der Gaussschen Integralformel auf F und $\tilde{F}$ ergibt die Gleichungen

$$\int_V \{(\operatorname{div} u \cdot \operatorname{div} v) + (u \cdot \operatorname{rot} \operatorname{div} v)\} \cdot dp = \int_\Omega ((u\boldsymbol{\nu}) \cdot \operatorname{div} v) \cdot ds,$$

$$\int_V \{(\operatorname{rot} u \cdot \operatorname{rot} v) + (u \cdot \operatorname{div} \operatorname{rot} v)\} \cdot dp = \int_\Omega (u \cdot (\operatorname{rot} v, \boldsymbol{\nu})) \cdot ds,$$

aus denen durch Addition nach (7) die fundamentale Greensche Formel folgt:

$$\int_V \{(\operatorname{div} u \cdot \operatorname{div} v) + (\operatorname{rot} u \cdot \operatorname{rot} v) + (u \cdot \Delta v)\} \cdot dp$$
$$= \int_\Omega \{((u\boldsymbol{\nu}) \cdot \operatorname{div} v) + (u \cdot (\operatorname{rot} v, \boldsymbol{\nu}))\}\, ds.$$

Auf der rechten Seite kann man, da die Felder $(u\boldsymbol{\nu})$, $(\operatorname{rot} v, \boldsymbol{\nu})$ tangentiell sind, $\operatorname{div} v$ und u auch durch ihre tangentiellen Teile $(\operatorname{div} v)_t$, u_t ersetzen. Vertauscht man in dieser Formel u mit v und subtrahiert, so ergibt sich die zweite Greensche Gleichung

$$\int_\Omega \mathfrak{L}(u, v)\, ds = \int_V \{u(\Delta v + k^2 v) - v(\Delta u + k^2 u)\}\, dp.$$

Hier ist

$$\mathfrak{L}(u, v) = \begin{matrix} ((u\boldsymbol{\nu}) \cdot (\operatorname{div} v)_t) + (u_t \cdot (\operatorname{rot} v, \boldsymbol{\nu})) \\ -((v\boldsymbol{\nu}) \cdot (\operatorname{div} u)_t) - (v_t \cdot (\operatorname{rot} u, \boldsymbol{\nu})). \end{matrix}$$

Insbesondere gilt

(8) $$\int_\Omega \mathfrak{L}(u, v)\, ds = 0,$$

wenn u und v Lösungen der Schwingungsgleichung $\Delta u + k^2 u = 0$ in V sind.

§ 4.

Eindeutigkeit.

Der Raum $\mathfrak{R}_n$ sei durch eine oder mehrere glatte Flächen Ω in ein beschränktes Inneres V und ein zusammenhängendes Äußeres W geteilt. Zur Untersuchung stehen (komplex-wertige) q-Felder u im Außenraum W, die daselbst der Schwingungsgleichung $\Delta u + k^2 u = 0$ genügen. Eine derartige Lösung gilt als zulässig, wenn sie der Ausstrahlungsbedingung genügt. Diese besagt, daß die auf der Kugel Σ_R gebildeten Ausdrücke

$$R^{\frac{n-1}{2}} \cdot \{(\operatorname{div} u)_t + ik(uv)\} \quad \text{und} \quad R^{\frac{n-1}{2}} \cdot \{(\operatorname{rot} u, v) + iku_t\}$$

gleichmäßig in allen Richtungen gegen 0 streben, wenn der Radius R über alle Grenzen wächst. Wir suchen eine zulässige Lösung u, für welche $(\operatorname{div} u)_t$ und u_t auf Ω mit einem vorgegebenen tangentiellen $(q-1)$-Feld γ bzw. q-Feld g zusammenfallen (elektrisches Randwertproblem). Der Eindeutigkeitssatz behauptet, daß eine zulässige Lösung u in ganz W verschwindet, falls $(\operatorname{div} u)_t$ und u_t auf Ω verschwinden. Der Beweis folgt dem RELLICHschen Schema. Neben u benutzt man das konjugierte Feld $\bar{u}$. Gemäß der Ausstrahlungsbedingung konvergiert das über Σ_R erstreckte Integral von

$$\left\{\begin{aligned} &((\operatorname{div} u)_t + ik(uv),\ (\operatorname{div} \bar{u})_t - ik(\bar{u}v)) \\ &+ ((\operatorname{rot} u, v) + iku_t,\ (\operatorname{rot} \bar{u}, v) - ik\bar{u}_t) \end{aligned}\right. \tag{9}$$

mit $R \to \infty$ gegen Null. Sobald Σ_R ganz V einschließt, kann man die GREENsche Formel (8) auf die beiden Felder $u, \bar{u}$ in dem von Σ_R begrenzten Teil von W anwenden. Wegen der auf Ω geltenden Randbedingungen folgt alsdann, daß das über Σ_R erstreckte Integral von $\mathfrak{L}(u, \bar{u})$ exakt gleich 0 ist, und aus unserer Aussage ergibt sich demnach, daß das Integral von

$$(|(\operatorname{div} u)_t|^2 + |(\operatorname{rot} u, v)|^2) + k^2(|u_v|^2 + |u_t|^2)$$

mit $R \to \infty$ gegen 0 strebt und demnach a fortiori das Integral von

$$|u_v|^2 + |u_t|^2 = (u \cdot \bar{u}) = \Sigma |u_{i_1 \ldots i_q}|^2.$$

Anwendung des RELLICHschen Lemmas auf jede der Komponenten $u_{i_1 \ldots i_q}$ führt zu dem gewünschten Resultat.

§ 5.

Integralgleichung des elektrischen Randwertproblems.

Die Größen

$$G \cdot \begin{vmatrix} \delta_{i_1 k_1}, & \ldots, & \delta_{i_1 k_q} \\ \cdot & \cdot\ \cdot\ \cdot & \cdot \\ \delta_{i_q k_1}, & \ldots, & \delta_{i_q k_q} \end{vmatrix} = G_{i_1 \ldots i_q,\, k_1 \ldots k_q} \tag{10}$$

können als die Komponenten einer Matrix betrachtet werden, für welche $(i_1, \ldots, i_q)$ als Zeilen-, $(k_1, \ldots, k_q)$ als Kolonnen-Index figuriert. Für variables p und festes p' ist die einer festen Kombination $(i_1, \ldots, i_q)$ entsprechende Zeile $G_{i_1 \ldots i_q}(pp')$ ein q-Feld w mit den Komponenten $w_{k_1, \ldots, k_q} = G_{i_1 \ldots i_q,\, k_1 \ldots k_q}$.

Die Greensche Formel (8) weist den Weg zur Konstruktion der gesuchten Lösung $u(p)$ des elektrischen Randwertproblems mittels zweckmäßig gewählter Oberflächenbelegungen. Sei $\varphi = \varphi(s)$ ein stetiges $(q-1)$-Feld und $f = f(s)$ ein ebensolches q-Feld auf Ω; beide seien tangentiell, so daß $\varphi = ([\varphi \nu]\nu)$ und $f = ([f\nu]\nu)$, sonst aber beliebig. Man bilde aus dem q-Feld $w(p) = G_{i_1 \ldots i_q}(pp')$ und dem Paar $\boldsymbol{\varphi} = \{\varphi(s), f(s)\}$ dieser beiden Felder den Ausdruck[4])

(11) $$-((w\nu)\cdot\varphi) + ((\operatorname{rot} w, \nu)\cdot f),$$

(wobei in w und $\operatorname{rot} w$ der variable Flächenpunkt s für p eintritt) und integriere alsdann nach s über ganz Ω: die entstehenden Zahlen $u_{i_1 \ldots i_q}(p')$ sind die Komponenten eines q-Feldes, das sowohl im Innen- wie im Außenraum als Funktion von p' der Gleichung $\Delta u + k^2 u = 0$ genügt und im Außenraum der Ausstrahlungsbedingung genügt. Es ist bequem, die Bezeichnungen s, p' gegen s', p zu vertauschen. Wir gebrauchen die Abkürzungen φ, φ' für $\varphi(s)$ und $\varphi(s')$, und analog für alle Funktionen auf Ω. $\Gamma' = \Gamma'(s'p)$ bedeutet nunmehr den Vektor $\operatorname{grad}_{p'} G(p'p)$ für $p' = s'$. Integration über Ω nach s oder s' werde durch $\int$ bzw. $\int'$ angedeutet. Das beschriebene Verfahren ergibt den folgenden Ansatz:

(12) $$u(p) = u(p; \boldsymbol{\varphi}) = -\int' [\varphi', \nu'] G + \int' ([f', \nu'] \Gamma').$$

Es gilt, die äußeren Randwerte der Tangentialteile von u und $\operatorname{div} u$ zu bestimmen. Da $\Gamma' = -\Gamma$ und somit

$$\int' ([f', \nu'] \Gamma') = -\operatorname{div} \int' [f', \nu'] G$$

ist, kommt wegen (6):

$$\operatorname{div} u = -\int' ([\varphi', \nu'] \Gamma) = \int' ([\varphi', \nu'] \Gamma').$$

Um die gesuchten Randwerte an der Stelle s zu finden, errichte man die Normale in s, wähle für p den Punkt auf dieser Normalen, der von s den Abstand $+\varepsilon$ oder $-\varepsilon$ hat, und lasse dann ε gegen 0 konvergieren. Indem wir für einen Augenblick ψ für $[\varphi\nu]$ schreiben, handelt es sich also zunächst darum, den Grenzwert von

$$[\operatorname{div} u, \nu] = \int' [(\psi' \Gamma')\nu]$$

zu ermitteln. Nach (3) ist

(13) $$[(\psi'\Gamma')\nu] = \psi'(\Gamma'\nu) - ([\psi'\nu], \Gamma').$$

Es ist wohl bekannt, was aus dem Integral des ersten Teiles, $\int' \psi'(\Gamma'\nu)$, wird, wenn ε gegen 0 strebt, nämlich

$$\pm\psi(s) + \int' \psi'(\Gamma'\nu).$$

---

[4]) So verfuhr ich schon in einer frühen Arbeit über die Eigenschwingungen elastischer Körper, Rend. Circ. Mat. di Palermo **39** (1915), 1—49, insbesondere S. 16.

Hier gilt das obere Vorzeichen bei Annäherung von außen, das untere bei Annäherung von innen, und in Γ' ist $p = s$ zu setzen. Der Kern $(\Gamma'(s's), \nu(s))$ ist regulär genug, da wegen der Glattheit der Oberfläche $\left(\frac{\mathfrak{r}}{r}(s's), \nu(s)\right)$ von positiver Ordnung für $s' = s$ verschwindet. Da die Komponenten von $[\psi'\nu] = [[\varphi'\nu']\nu]$ die Größen ν'_i, ν_i nur in der „schiefen" Kombination $\nu_i(s')\nu_k(s) - \nu_k(s')\nu_i(s)$ enthalten, kann im zweiten Teil der Grenzübergang von p zu s unmittelbar ausgeführt werden. Wenn man im Endresultat die Verwandlung (13) wieder rückgängig macht, so erhält man

$$[\operatorname{div} u, \nu](s) = \pm\, \psi(s) + \int' [(\psi(s'), \Gamma'(s's)), \nu(s)]$$

und damit für den Tangentialteil $(\operatorname{div} u)_t$ von $\operatorname{div} u$ an der Stelle s den Wert

$$\pm\, \varphi + \int' ([([\varphi'\nu']\Gamma')\nu]\nu).$$

Auf genau die gleiche Weise bekommt man für den Tangentialteil u_t von u selber den Wert

$$\pm\, f + \int' ([([f'\nu']\Gamma')\nu]\nu) - \int' G([[\varphi'\nu']\nu]\nu).$$

Sei $\boldsymbol{\gamma} = \{\gamma, g\}$ ein gegebenes Paar bestehend aus einem tangentiellen $(q-1)$-Feld γ und einem tangentiellen q-Feld g auf Ω. Das elektrische Randwertproblem, ein zulässiges Außenfeld $u(p)$ zu finden, für welches auf dem Rande Ω die Tangentialteile von $\operatorname{div} u$ und u bzw. die vorgegebenen Werte γ und g annehmen, ergibt dann das System von FREDHOLMschen Integralgleichungen

$$(14)\qquad \begin{cases} \gamma = \varphi + \int' ([([\varphi'\nu']\Gamma')\nu]\nu), \\ g = f + \int' ([([f'\nu']\Gamma')\nu]\nu) - \int' G([[\varphi'\nu']\nu]\nu), \end{cases}$$

die wir symbolisch zu

$$\boldsymbol{\gamma} = (\boldsymbol{E} + \boldsymbol{K})\boldsymbol{\varphi}$$

zusammenfassen mit dem Einheitsoperator $\boldsymbol{E}$. Wenn γ und g tangentiell sind, sorgt diese Integralgleichung von selbst dafür, daß die Lösung $\{\varphi, f\}$ aus zwei tangentiellen Feldern besteht.

Die Möglichkeit der Existenz von nicht-trivialen Lösungen $\boldsymbol{\eta}$ der homogenen transponierten Gleichung $\boldsymbol{\eta}(\boldsymbol{E} + \boldsymbol{K}) = 0$ erfordert das Eingehen auf

§ 6.

Das duale Problem.

Dazu verfolgen wir den zu (12) dualen „magnetischen" Ansatz. Wiederum wählen wir eine stetige Belegung $\boldsymbol{\eta} = \{\eta, e\}$, die aus zwei tangentiellen Feldern η und e auf Ω vom Range $q-1$ bzw. q besteht, und integrieren diesmal an Stelle von (11) den Ausdruck

$$(\operatorname{div} w \cdot \eta) + (w \cdot e)$$

nach s über Ω, unter w wiederum eine der Zeilen $G_{i_1 \ldots i_q}(s p')$ von (10) verstehend. Wir unterlassen diesmal die Abänderung der Bezeichnung von $s p'$ in $s' p$. Wir erhalten so die Komponenten $v_{i_1 \ldots i_q}(p')$ des folgenden q-Feldes

$$v(p') = \int [\eta \Gamma] + \int e G, \tag{15}$$

wo Γ und G für $\Gamma(s p')$ und $G(s p')$ stehen. Es genügt als Funktion von p' im Innen- und Außenraum der Schwingungsgleichung, und der Ausstrahlungsbedingung im Außenraum. Jetzt lassen wir den Punkt p' von außen oder innen gegen den Punkt s' von Ω auf der in s' errichteten Normalen $\nu' = \nu(s')$ rücken. Es ist zu bestimmen, was dabei aus $-(v(p'), \nu')$ und $(\operatorname{rot} v(p'), \nu')$ wird. Was $-(v(p'), \nu')$ anlangt, so ist klar, daß der zweite Summand des Ausdrucks (15) von $v(p')$ zum Grenzwert den Beitrag $-\int (e \nu') G$ leistet, wo G jetzt für $G(s s')$ steht. Im ersten Teil $([\eta \Gamma] \nu')$ nehme man zunächst wieder die Umwandlung

$$-([\eta \Gamma] \nu') = -\eta (\Gamma \nu') + [(\eta \nu') \Gamma]$$

vor und nutze dann die tangentielle Natur von η aus, indem man η durch $([\eta \nu] \nu)$ ersetzt; dann hat man mit dem Integral $\int$ von

$$-\eta (\Gamma \nu') + [(([\eta \nu] \nu) \nu') \Gamma]$$

zu tun. Aus dem gleichen Grunde wie früher ergibt sich für dieses Integral im Limes $\mp \eta(s')$ plus dem Integral über den gleichen Ausdruck, in dem nun p' durch den Punkt s' auf Ω ersetzt ist, und aus dem gleichen Grunde ist auch hier der Kern hinreichend regulär. Indem man den Ausdruck in seine ursprüngliche Form zurückverwandelt, erhält man darum

$$-(v \nu)' = \mp \eta' + \int ([([\eta \nu] \nu) \Gamma'] \nu') - \int (([e \nu] \nu) \nu') G.$$

Dabei ist noch Γ durch $-\Gamma'$ ersetzt worden.

Für ein Feld $v(p')$ wird man unter rot' natürlicherweise den Operator $\operatorname{rot}_{p'}$ verstehen. Wegen

$$\int [\eta \Gamma] = -\int [\eta \Gamma'] = -\operatorname{rot}' \int \eta G$$

ist

$$\operatorname{rot}' v = \int [e \Gamma'] = -\int [e \Gamma],$$

und durch dieselbe kleine Rechnung, die oben für $[\eta \Gamma]$ ausgeführt wurde, kommt

$$(\operatorname{rot}' v, \nu') = \mp e' + \int ([([e \nu] \nu) \Gamma'] \nu').$$

Ist darum $\boldsymbol{\lambda} = (\lambda, l)$ das Paar, das aus den inneren Randwerten von $-(v \nu)$ und $(\operatorname{rot} v, \nu)$ besteht, so resultiert die Gleichung

$$\boldsymbol{\lambda} = \boldsymbol{\eta} (\boldsymbol{E} + \boldsymbol{K}).$$

In der Tat ist der Operator, der $\boldsymbol{\eta}$ in $\boldsymbol{\lambda}$ überführt, transponiert zu demjenigen, der nach den Gleichungen (14) $\boldsymbol{\gamma}$ aus $\boldsymbol{\varphi}$ hervorgehen läßt. Denn gemäß (5) ist ja z. B.

$$(f' \bullet ([([e\boldsymbol{\nu}]\,\boldsymbol{\nu})\,\Gamma']\,\boldsymbol{\nu}')) = (([([f'\boldsymbol{\nu}']\,\Gamma')\,\boldsymbol{\nu}]\,\boldsymbol{\nu}) \bullet e),$$

da nach jener Gleichung die Vektor Faktoren von e in umgekehrter Reihenfolge an f' angehängt werden können, indem man jede runde in eine eckige, jede eckige in eine runde Klammer verwandelt. Die Gleichung

$$\boldsymbol{\eta}\,(\boldsymbol{E}+\boldsymbol{K}) = 0 \tag{16}$$

bedeutet also, daß für $v(p) = v(p; \boldsymbol{\eta})$ die inneren Randwerte von $-(v\boldsymbol{\nu})$ und (rot v, $\boldsymbol{\nu}$) verschwinden. Auf jeden Fall erleiden diese Größen den Sprung -2η bzw. $-2e$ beim Durchgang durch die Fläche Ω von innen nach außen.

§ 7.

Der Operator M und die Kapazitätsmatrix.

Die Lösungen $\boldsymbol{\varphi}$, $\boldsymbol{\eta}$ von $(\boldsymbol{E}+\boldsymbol{K})\boldsymbol{\varphi} = 0$ bzw. $\boldsymbol{\eta}(\boldsymbol{E}+\boldsymbol{K}) = 0$ bilden lineare Mannigfaltigkeiten Φ und H von derselben Dimension h. Es sei p' ein Punkt im Äußern W. Dann ist $G(pp')$ als Funktion von p eine reguläre Lösung der Schwingungsgleichung für $p \in V$. Wir wenden die Gleichung (8) an auf das q-Feld $w(p) = G_{i_1 \ldots i_q}(pp')$, das an Stelle von u tritt, und das q-Feld $v(p) = v(p; \boldsymbol{\eta})$, das mittels der H entnommenen Belegung $\boldsymbol{\eta}$ durch (15) definiert ist. Wir werden im nächsten Paragraphen zeigen, daß die Randwerte f und φ der Tangentialteile von v und div v existieren und auf der Innen- und Außenseite von Ω die gleichen Werte haben. M bedeutet den linearen Operator, durch den $\boldsymbol{\eta}$ in $\boldsymbol{\varphi} = \{\varphi, f\}$ übergeht, $\boldsymbol{\varphi} = M\boldsymbol{\eta}$. Berücksichtigt man, daß die inneren Randwerte von $-(v\boldsymbol{\nu})$ und (rot v, $\boldsymbol{\nu}$) verschwinden, so ergibt unsere Gleichung

$$\int \{-((w\boldsymbol{\nu})\cdot\varphi) + ((\text{rot } w, \boldsymbol{\nu})\cdot f)\} = 0$$

oder

$$u(p; \boldsymbol{\varphi}) = 0 \quad \text{für } \boldsymbol{\varphi} = M\boldsymbol{\eta},\ \boldsymbol{\eta} \in H \text{ und } p \in W.$$

Läßt man jetzt p von außen in einen Flächenpunkt s hineinrücken, so kommt

$$(\boldsymbol{E}+\boldsymbol{K})\boldsymbol{\varphi} = 0.$$

Verfährt man in gleicher Weise mit $\bar{v}(p)$, so zeigt sich, wie Herr Müller zuerst bemerkte, daß nicht nur $\boldsymbol{\varphi} = M\boldsymbol{\eta}$, sondern auch das konjugierte $\bar{\boldsymbol{\varphi}}$ derselben Gleichung genügt.

Sind $\boldsymbol{\eta} = \{\eta, e\}$, $\boldsymbol{\varphi} = \{\varphi, f\}$ irgend zwei Belegungen von der hier durchgehend benutzten Art, die erste als Zeile, die zweite als Spalte aufgefaßt, so kann man das innere Produkt

$$(\eta, \varphi) = \int (\eta \cdot \varphi) + \int (e \cdot f)$$

bilden. Aus zwei Belegungen η, η^* von der Art wie η entsteht die Bilinearform

$$C(\eta, \eta^*) = (\eta, M\eta^*).$$

Ihre *Symmetrie* beweist man am besten, indem man die GREENsche Formel (8) auf $v = v(p; \eta)$ und $v^* = v(p; \eta^*)$ anwendet, nicht jedoch für V, sondern für das Innere der Kugel Σ_R. Sobald der Radius R so groß ist, daß Σ_R ganz V umschließt, erhält man dann, bedenkend, daß $-(v\nu)$ und $(\mathrm{rot}\; v, \nu)$ den Sprung -2η bzw. $-2e$ über Ω hinüber erleiden, während $\varphi = (\mathrm{div}\; v)_t$ und $f = v_t$ stetig hindurchgehen, eine Gleichung, durch welche $2\{(\eta, M\eta^*) - (\eta^*, M\eta)\}$ in das über Σ_R sich erstreckende Integral von $\mathfrak{L}(v, v^*)$ verwandelt wird. Schreibt man hier $\mathfrak{L}(v, v^*)$ in der Form

$$\begin{aligned} &((v\nu), (\mathrm{div}\; v^*)_t + ik(v^*\nu)) + (v_t, (\mathrm{rot}\; v^*\nu) + ikv_t^*) \\ &-((v^*\nu), (\mathrm{div}\; v)_t + ik(v\nu)) - (v_t^*, (\mathrm{rot}\; v, \nu) + ikv_t), \end{aligned} \tag{17}$$

so ergibt sich aus den Ausstrahlungsbedingungen, daß $\int\limits_{\Sigma_R} \mathfrak{L}(vv^*)\,ds$ mit $R \to \infty$ gegen 0 konvergiert, folglich

$$(\eta, M\eta^*) - (\eta^*, M\eta) = 0. \tag{18}$$

Das entsprechende Verfahren für v und $\bar{v}$ liefert zunächst

$$2(\eta, \overline{M\eta}) - 2(\bar{\eta}, M\eta) = \int\limits_{\Sigma_R} \mathfrak{L}(v, \bar{v})\,ds.$$

Parallel zu (17) bildet man [5])

$$\begin{cases} ((v\nu), (\mathrm{div}\; \bar{v})_t - ik(\bar{v}\nu)) + (v_t, (\mathrm{rot}\; \bar{v}, \nu) - ik\bar{v}_t) \\ -((\bar{v}\nu), (\mathrm{div}\; v)_t + ik(v\nu)) - (\bar{v}_t, (\mathrm{rot}\; v, \nu) + ikv_t). \end{cases} \tag{19}$$

Gemäß der Ausstrahlungsbedingung konvergiert das Integral dieses Ausdrucks über Σ_R mit $R \to \infty$ gegen Null, somit

$$k \cdot \int\limits_{\Sigma_R} |v|^2\,ds \to 2 \cdot \mathfrak{J}(\eta, \overline{M\eta}).$$

Daraus folgt die Ungleichung

$$\mathfrak{J}(\eta, \overline{M\eta}) \geqq 0, \tag{20}$$

und nach RELLICHS Lemma kann hier das Gleichheitszeichen nur gelten, wenn $v(p)$ in ganz W verschwindet; diese Bedingung hat das Ver-

[5]) Dieser Gedanke rührt von Herrn MÜLLER her, nur benutzt er statt (19) den RELLICHschen Ausdruck (9) für v.

schwinden der *äußeren* Randwerte von $-(v\nu)$ und (rot v, ν) oder die Gleichung

(21) $$\eta(-\boldsymbol{E}+\boldsymbol{K}) = 0$$

zur Folge.

Die Beziehungen (18), (20) samt diesem Zusatz gelten unabhängig davon, ob η zu H gehört. Ist aber $\eta(\boldsymbol{E}+\boldsymbol{K}) = 0$, so führt (21) auf $\eta = 0$. Für die Elemente η von H gilt somit

(20*) $$\mathfrak{J}(\eta, \overline{M\eta}) > 0 \text{ außer für } \eta = 0.$$

Diese Ungleichung lehrt, daß $M\eta$ für $\eta \in H$ nur dann Null sein kann, wenn $\eta = 0$ ist, und darum ist die Abbildung $\eta \to \varphi = M\eta$ von H auf Φ ein-eindeutig. Aus einer Basis η_i $(i = 1, \ldots, h)$ von H entspringt darum eine Basis $\varphi_i = M\eta_i$ für Φ. Es gilt der fundamentale

Satz: *Die Kapazitätsmatrix der* $c_{ij} = (\eta_i \varphi_j)$ *ist nicht-ausgeartet.*

Zum Beweise muß man zeigen, daß $\eta = 0$ die einzige Linearkombination der η_i ist, die zu allen φ_j orthogonal ist. Da $\overline{M\eta}$ zu Φ gehört, folgt aus $(\eta\varphi_j) = 0$ $(j = 1, \ldots, h)$ in der Tat $(\eta, \overline{M\eta}) = 0$, was der Ungleichung (20*) widerspricht, es sei denn $\eta = 0$.

Danach führt das klassische Vorgehen zur Lösung der elektrischen Randwertaufgabe: Von dem gegebenen γ subtrahiert man ein solches Element $\Sigma\alpha_j\varphi_j$ von Φ, daß die Differenz γ^* zu allen η_i orthogonal ist.

$$(\eta_i\gamma) = \Sigma_j c_{ij}\alpha_j \qquad (i, j = 1, \ldots, h).$$

Für γ^* hat die inhomogene Integralgleichung $(\boldsymbol{E}+\boldsymbol{K})\varphi^* = \gamma^*$ eine Lösung φ^*. Zum „elektrischen Schwingungspotential" $u^*(p) = u(p; \varphi^*)$ der so gefundenen Belegung φ^* addiert man das magnetische Potential $v(p; \eta)$ der H entnommenen Belegung $\eta = \Sigma_j\alpha_j\eta_j$. Die Summe $u(p)$ ergibt die gewünschten äußeren Randwerte $\{\gamma, g\}$ für $(\operatorname{div} u)_t$ und u_t.

§ 8.

Die Randwerte von v_t und $(\operatorname{div} v)_t$.

Ein heikler Punkt, der im Falle $q = 0$ noch nicht auftaucht, ist der Nachweis der Existenz und des stetigen Durchgangs der tangentiellen Teile v_t und $(\operatorname{div} v)_t$ des im Innen- und Außenraum durch (15) bestimmten Potentials v. In der Tat, für $q = 0$ hat man einfach

$$v(p') = \int e(s)\, G(s p'),$$

und es ist sofort klar, daß die Randwerte von v auf der Innen- und Außenseite durch $v(s') = \int e(s)\, G(s s')$ gegeben sind. Für $q \geqq 1$ erfordert die Bestimmung des Oberflächenwertes des tangentialen Teils von $\int'[\eta'\Gamma]$ eine zusätzliche Überlegung, die aber aus der klassischen Potentialtheorie wohlbekannt ist. Wenn p längs der Normalen in s

in den Oberflächenpunkt s hineinrückt, ergibt sich von innen und außen der gleiche Limes, nämlich der *Hauptwert* des uneigentlichen Integrals

$$\int' ([[\eta(s'),\ \Gamma(s' s)],\ \nu(s)]\, \nu(s)).$$

Dabei wird es dann freilich nötig, von $\eta(s)$ vorauszusetzen, daß es nicht nur stetig ist, sondern auch einer HÖLDER-Bedingung genügt. Der Hauptwert des Integrals über Ω wird bestimmt, indem man aus Ω um den Punkt s zunächst ein kleines *kreisförmiges* Loch ausstanzt, dessen Radius ε man dann gegen 0 geben läßt. Genauer gesagt: in der Tangentenebene im Punkte s von Ω beschreibt man um s einen $(n-1)$-dimensionalen Kreis vom Radius ε und errichtet über ihm in Richtung der Normalen ν einen geraden Kreiszylinder; dieser wird zum Ausstanzen des Loches benutzt.

Die Rechnung gestaltet sich am einfachsten, wenn man s als den Ursprung des Koordinatensystems annimmt und der x_n-Achse die Richtung der Normalen ν gibt, $\nu = (0, \ldots, 0, 1)$. Dann besteht $[\eta' \Gamma]_t$ aus denjenigen Komponenten

$$[\eta' \Gamma]_{i_1 \ldots i_q} = \frac{1}{r} \frac{dG}{dr} \mathbf{S} \pm \eta'_{i_1 \ldots i_{q-1}} x'_{i_q}$$

von $[\eta' \Gamma]$, in denen alle Indizes $i_1, \ldots, i_q \neq n$ sind. Daraus ergibt sich sofort das gewünschte Resultat. Ohne Normierung des Koordinatensystems kann man das Argument auch darauf stützen, daß jede Komponente $i_1, \ldots, i_{q-1}, i, k$ von $[[\eta' \Gamma]\nu]$ die Vektoren Γ und ν nur in einer der schiefen Kombinationen $\Gamma_i \nu_k - \Gamma_k \nu_i$ enthält.

Für $q \geqq 1$ ist ferner der Oberflächenwert von $(\operatorname{div} v)_t$ zu bestimmen. Hier ziehen wir zunächst vor, die Bezeichnung $s p'$ beizubehalten und nicht mit $s' p$ zu vertauschen. Wir operieren also mit $v(p')$ und schreiben z. B. $\operatorname{div}_{p'} v(p')$ als $\operatorname{div}' v$. Zu dieser Divergenz liefert der Teil $\int e G$ von $v(p')$ den Beitrag $\int (e \Gamma') = -\int (e \Gamma)$. Führt man $[e \nu] = c$ ein, so ist $(e \Gamma) = ((c \nu) \dot{\Gamma})$, und dies ist wiederum ein Ausdruck, der ν und Γ nur in der schiefen Kombination $\nu_i \Gamma_k - \nu_k \Gamma_i$ enthält. Setzt man also voraus, daß e und darum auch c einer HÖLDER-Bedingung genügt, so ergibt sich als Grenzwert von $\int (e \Gamma')$ der Hauptwert des Integrals

$$\int ((c(s)\, \nu(s))\, \Gamma'(s s')),$$

wenn p' auf der Normalen ν' in s' dem Oberflächenpunkt s' zustrebt, ob dies nun von der Innen- oder Außenseite her geschieht.

Bleibt noch der Beitrag zu diskutieren, den der Teil $\int [\eta \Gamma]$ von $v(p')$ zu $\operatorname{div}' v$ liefert. Hier hat man es mit der gleichen Schwierigkeit zu tun, auf die man in der klassischen dreidimensionalen Potentialtheorie bei der normalen Ableitung des Potentials einer Doppelbelegung stößt. Ich verfahre so, wie ich es seit langem in Vorlesungen über Potentialtheorie zu tun pflegte. Es ist

$$\int [\eta \Gamma] = -\int [\eta(s), \Gamma'(s p')] = -\operatorname{rot}' \int (\eta(s) \cdot G(s p')).$$

Darum ist $\operatorname{div}' \int [\eta \Gamma] = -\operatorname{div}' \operatorname{rot}' \int (\eta G)$; da aber $\operatorname{div} \operatorname{rot} = \Delta - \operatorname{rot} \operatorname{div}$ und $\Delta' G = -k^2 G$ ist, so kommt

$$\operatorname{div}' \int [\eta \Gamma] = k^2 \cdot \int (\eta G) + \operatorname{rot}' \int (\eta \Gamma'). \tag{22}$$

Was mit dem ersten Summanden rechts geschieht, wenn p' in den Oberflächenpunkt s' hineinrückt, ist klar: er geht in

$$k^2 \cdot \int \eta(s)\, G(s s')\, d s$$

über.

Wir untersuchen jetzt das nur für $q \geqq 2$ auftretende zweite Glied in (22), $\int (\eta \Gamma') = -\int (\eta \Gamma) = -\int ((\vartheta \nu) \Gamma)$, wo $\vartheta = [\eta \nu]$. Um das über eine geschlossene Oberfläche Ω sich erstreckende Integral zu berechnen, bedeckt man die Fläche „dachziegel-artig" mit sich überlappenden Gebieten, die je auf ein Koordinatensystem $\xi_1, \ldots, \xi_{n-1}$ bezogen sind. Man kann dann ϑ additiv in Bestandteile zerlegen (Dieudonné), deren jedes nur in einem der Stücke, genauer in einem kompakten „Kern" eines solchen Stückes nicht verschwindet. Sei $x_i = x_i(\xi_1, \ldots, \xi_{n-1})$ die Parameterdarstellung der Fläche in dem von dem Parametersystem ξ bedeckten Stück. Wir haben angenommen, daß diese Darstellungsfunktionen $x_i(\xi)$ stetig differentiierbar sind, daß die Ableitungen einer Hölder-Bedingung genügen und daß die Jacobische Matrix

$$\left\| \frac{\partial x_1}{\partial \xi}, \ldots, \frac{\partial x_n}{\partial \xi} \right\|,$$

— die aus der hingeschriebenen Zeile entsteht, indem man $n-1$ Zeilen untereinander schreibt, in welchen ξ durch $\xi_1, \ldots, \xi_{n-1}$ ersetzt ist —, vom Maximalrang $n-1$ sei. Das ist, was wir eine glatte Fläche nannten. Wir fügen jetzt die Voraussetzung hinzu, daß auch die zweiten Ableitungen von $x_i(\xi)$ existieren und einer Hölder-Bedingung genügen. Wir sprechen dann von einer *glatt-gekrümmten* Fläche. Es gilt z. B. (richtige Orientierung des Systems $\xi_1, \ldots, \xi_{n-1}$ vorausgesetzt)

$$\nu_1\, d s = \left| \frac{\partial x_2}{\partial \xi}, \ldots, \frac{\partial x_n}{\partial \xi} \right| \cdot d\xi_1 \ldots d\xi_{n-1}.$$

So ergibt sich für $(\nu_1 \Gamma_2 - \nu_2 \Gamma_1) \cdot d s$ der Ausdruck

$$\left| \frac{\partial G}{\partial x_1} \frac{\partial x_1}{\partial \xi} + \frac{\partial G}{\partial x_2} \frac{\partial x_2}{\partial \xi}, \frac{\partial x_3}{\partial \xi}, \ldots, \frac{\partial x_n}{\partial \xi} \right| \cdot d\xi_1 \ldots d\xi_{n-1}.$$

Indem man zu der ersten Spalte dieser Determinante die bzw. mit $\frac{\partial G}{\partial x_3}, \ldots, \frac{\partial G}{\partial x_n}$ multiplizierten folgenden Spalten hinzuaddiert, verwandelt sie sich in

$$\left| \frac{\partial G}{\partial \xi}, \frac{\partial x_3}{\partial \xi}, \ldots, \frac{\partial x_n}{\partial \xi} \right|.$$

Ich nehme jetzt an, daß η und darum auch ϑ als Funktion der ξ differentiierbar ist und die Ableitungen einer Hölder-Bedingung genügen. Dann kann man auf das Integral

$$\text{(23)} \qquad \int \vartheta_{i_1 \ldots i_{q-2} 12} \cdot \left| \frac{\partial G}{\partial \xi}, \frac{\partial x_3}{\partial \xi}, \ldots, \frac{\partial x_n}{\partial \xi} \right| \cdot d\xi_1 \ldots d\xi_{n-1}$$

partielle Integration so anwenden, daß G von der Differentiation befreit wird. Auf diese Weise verwandelt sich (23) in

$$-\int G \cdot \left| \frac{\partial \vartheta_{i_1 \ldots i_{q-2} 12}}{\partial \xi}, \frac{\partial x_3}{\partial \xi}, \ldots, \frac{\partial x_n}{\partial \xi} \right| \cdot d\xi_1 \ldots d\xi_{n-1}.$$

Schreibt man

$$\text{(24)} \quad \mu_{i_1 \ldots i_{q-2}}(s)\, ds = \mathbf{S} \pm \left| \frac{\partial \vartheta_{i_1 \ldots i_{q-2} 12}}{\partial \xi}, \frac{\partial x_3}{\partial \xi}, \ldots, \frac{\partial x_n}{\partial \xi} \right| \cdot d\xi_1 \ldots d\xi_{n-1},$$

wo die Summe sich alternierend über die $n(n-1)/2$ Mischungen der Ziffern 1, 2 mit $3, \ldots, n$ erstreckt, so erhält man

$$\text{(25)} \qquad \int (\eta\, \Gamma') = \int \mu(s) \cdot G(s p') \cdot ds = \int \mu\, G.$$

In Wahrheit bezieht sich die Zwischenrechnung auf ein durch die Koordinaten ξ bedecktes Stück von Ω und den zugehörigen Bestandteil von ϑ. Aber wegen der Invarianz des durch (24) eingeführten μ gegenüber Transformationen der Koordinaten ξ ist im Endresultat (25) die Zerstückelung von Ω und ϑ wieder überwunden.

In diesem Stadium finden wir es zweckmäßig, die Bezeichnung $s p'$ wieder gegen $s' p$ zu vertauschen. Es handelt sich dann darum, während p auf der Normalen ν im Flächenpunkt s gegen s rückt, den Grenzwert des tangentiellen Teils von rot $\int'(\mu' G) = \int'[\mu' \Gamma]$ zu ermitteln $[G = G(s' p),\ \Gamma = \Gamma(s' p)]$. Das geschieht genau so, wie vorhin für $\int'[\eta' \Gamma]$ verfahren wurde. Das Resultat ist der Hauptwert des Integrals $\int'[\mu(s'), \Gamma(s' s)]_t$.

Genügt η der Gleichung $\eta(\boldsymbol{E} + \boldsymbol{K}) = 0$, so ist $\eta = -\eta \boldsymbol{K} = \eta \boldsymbol{K}^2 = \ldots$, und daraus schließt man leicht, daß η unsern Forderungen genügt, unter der oben gemachten Voraussetzung, daß Ω glatt-gekrümmt ist.

§ 9.

Schlußbemerkungen.

Wird in der elektrischen Randwertaufgabe $\gamma(s) = 0$ angenommen, so erfüllt die Divergenz $w = \operatorname{div} u$ der im Außenraum konstruierten zulässigen Lösung $u(p)$ die Randbedingung $w_t = 0$ auf Ω. Außerdem ist $\operatorname{div} w = 0$ in ganz W. Da man aus dem Ausdruck von u entnimmt, daß auch w der Ausstrahlungsbedingung

$$R^{\frac{n-1}{2}} \{(\operatorname{rot} w, \nu) + i k w_t\}_{\text{auf } \Sigma_R} \to 0 \text{ mit } R \to \infty$$

genügt, so ergibt sich nach dem auf w anzuwendenden Eindeutigkeits-Argument, daß w im Außenraum identisch verschwindet, $\operatorname{div} u = 0$. Damit reduziert sich die Schwingungsgleichung auf $\operatorname{div} \operatorname{rot} u + k^2 u = 0$. Ohne diese Bemerkung wäre die Behandlung des elektromagnetischen Feldes unvollständig.

156.

Über den Symbolismus der Mathematik und mathematischen Physik

Studium generale 6, 219—228 (1953)

Die vertrauteste und wohl auch grundlegendste Form, in der in unserem geistigen Leben die symbolische Funktion, die Repräsentation durch Zeichen auftritt, ist die *Sprache*. Darum meint *H. Noack* (*11*, p. 97)[1]: „Das mit der Ausbildung der Sprache einhergehende Symbolverständnis kann als der entscheidende Schritt des Menschen über das animalische Leben hinaus bezeichnet werden." Wie in einem Brennpunkt treffen sich hier die großen philosophischen Probleme: das des Verhältnisses von Sachverhalt — Gedanke — Aussage, an dem noch jeder Versuch einer Beruhigung im Realismus bloßen Seins gescheitert ist; aber auch des Verhältnisses vom Ich des immanenten, im Strom des nur-eigenen Erlebens stehenden Bewußtseins zu dem individuellen, der Welt und dem Tode verhafteten, mit Wesen seinesgleichen kommunizierenden Menschen. „Viele andere sinnliche Gebilde, die zum Zwecke der Bezeichnung, Vergegenwärtigung, Mitteilung usw. verwendet werden, verdanken ihre ‚semantische' oder ‚symbolische' Bedeutung erst der Sprache, insofern sie nach deren Analogie geformt oder mit deren Hilfe verabredet werden." (*11*, p. 19). Häufig wird die Ansicht vertreten, daß das begriffliche Denken an die Sprache gebunden sei; das Tier (das ja sehr wohl in seiner Welt sich zu orientieren vermag) entbehre darum mit der Sprache auch des Begriffs. So sagt *Wilhelm v. Humboldt* in der Einleitung zu seinem Kawi-Werk (Bd. I, Berlin 1836, pp. 68/69): „Indem in ihr (der Sprache) das geistige Streben sich Bahn durch die Lippen bricht, kehrt das Erzeugnis desselben zum eignen Ohr zurück. Die Vorstellung wird also in wirkliche Objektivität hinüberversetzt, ohne darum der Subjektivität entzogen zu werden. Dies vermag nur die Sprache; und ohne diese, wo Sprache mitwirkt auch stillschweigend immer vorgehende, Versetzung in zum Subjekt zurückkehrende Objektivität ist die Bildung des Begriffs, mithin alles wahre Denken, unmöglich." Aus einer kälteren Zone klingt das Gleiche wieder in *Ludwig Wittgensteins Tractatus logico-philosophicus* (*15*, 5.6 und 5.62): „Die Grenzen meiner Sprache bedeuten die Grenzen meiner Welt ... daß die Welt *meine* ist, das zeigt sich darin, daß die Grenzen der Sprache (der Sprache, die ich allein verstehe) die Grenzen meiner Welt bedeuten ...". Merkwürdig, wie solipsistisch hier die Sprache gesehen wird; deren existentieller Ursprung und Aufgabe doch wohl in erster Linie in der *Kommunikation* liegt[2]. Hierin stimmen Denker so weiter geistiger Distanz wie *Jaspers* und *Hilbert* überein. (Ich denke an die Rolle der Sprache zur Mitteilung über die Weise, wie mit den selber bedeutungslosen

[1] Die Nummern in Schrägschrift verweisen auf das am Ende stehende Literaturverzeichnis.

[2] Ich weiß, daß dies auf das Ganze der *Humboldt*schen Sprachphilosophie nicht zutrifft. An der angeführten Stelle fährt er, nachdem noch einmal eingeschärft ist, daß „das Sprechen eine notwendige Bedingung des Denkens des Einzelnen in abgeschlossener Einsamkeit" ist, so fort: „In der Erscheinung entwickelt sich jedoch die Sprache nur *gesellschaftlich*, und der Mensch versteht sich selbst nur, indem er die Verstehbarkeit seiner Worte an Andren versuchend geprüft hat." Man lese den ganzen Absatz bis zu dem mächtigen Finale: „Alles Sprechen, von dem einfachsten an, ist ein Anknüpfen des einzeln Empfundenen an die gemeinsame Natur der Menschheit."

Zeichen und Formeln in *Hilberts* formalisierter Mathematik zu verfahren ist.) Über die wissenschaftliche Kunstsprache im Gegensatz zur Sprache des täglichen Lebens, in der sich der Mensch in seinem Umgang mit Welt und Mitmenschen ausspricht, heißt es bei *Karl Voßler* (*12*, p. 225): „Vor den mathematischen und naturwissenschaftlichen Begriffen gelten alle Sprachen als etwas Äußeres gleich; diese Begriffe sind fähig, sich in jeder Sprache anzusiedeln, da sie nur in der äußeren Sprachform Wohnung nehmen, die innere aber aufzehren und entleeren."

Nicht alle *Symbole* sind sprachlicher Natur. *E. Cassirer* handelt in seiner Theorie der symbolischen Formen der Reihe nach von Sprache, Mythos und den Konstruktionen der wissenschaftlichen Erkenntnis. Zählen wir einige besondere Formen auf, wie Wort, Bilder- und Buchstabenschrift, Rangabzeichen, Fahnen, allegorische Attribute, Traumsymbole, Kunstwerke, magische und religiös-kirchliche Symbole sowie die Zahlen und andere Begriffssymbole der exakten Wissenschaften, so wird uns klar, wie mannigfaltig der Sinn ist, in dem ein Zeichen auf das Bezeichnete hinweisen kann. Immer bleibt die *Deutung* ein Problem; diese mag sogar mehrschichtig sein. Eine steinzeitliche Höhlenzeichnung etwa, auf der Büffel und Boot und Baum dargestellt sind, wird vielleicht durch diese gegenständlichen Dinge hindurch auf Geister und Dämonen zielen. *Leibniz* erklärt die Zeichen für „gewisse Dinge, durch welche die gegenseitigen Beziehungen anderer Dinge ausgedrückt werden und deren Behandlung leichter ist als die der letzteren". Das macht sie zugleich zu einem Werkzeug für die Entdeckung neuer, in den dargestellten Beziehungen gründender Zusammenhänge. Das Operieren mit Zahlzeichen ist das eindruckvollste Beispiel. In der reichen Skala von modi der Bedeutung, an die wir eben erinnerten, sind die extremen Fälle: auf der einen Seite das Zeichen, das eine getreue Reproduktion des Bezeichneten ist (oder anstrebt), auf der andern das rein konventionelle oder gar das „leere" Zeichen, von dem *Hilbert* versichert, daß es überhaupt nicht über sich hinausweist. *Helmholtz* sagt von der Qualität unserer Empfindung, „insofern sie uns von der Eigentümlichkeit der äußeren Einwirkung, durch welche sie erregt ist, eine Nachricht gibt, könne sie als ein *Zeichen* derselben gelten, aber nicht als ein *Abbild*", da sie ja zugleich ganz wesentlich von der Natur des Sinnesapparats abhängt, auf den gewirkt wird. *E. Cassirer* (*6*, pp. 34—41) sieht in der Weise, wie sich für unsere Anschauung der *Raum* aus systematisch verknüpften und aufeinander bezogenen Perzeptionen aufbaut, die „Urfunktion der Repräsentation" tätig; nicht ganz verständlich ist es mir, wenn er dann fortfährt, daß man auf diese ‚natürliche Symbolik' zurückgehen müsse, wenn man „die künstliche Symbolik begreifen wolle, die sich das Bewußtsein in Sprache, Kunst und Mythos schafft".

In der Deutung der uns gegebenen Empfindungsdata und Wahrnehmungen, die (nach *Helmholtz*) „Zeichen" für die Wirklichkeit sind, mußte die Wissenschaft bestrebt sein, die innewohnende Subjektivität zu überwinden. Die von ihr dazu entwickelten Grundbegriffe enthüllen sich letzten Endes als vom Geist in freier Schöpfung dem Gegebenen gegenübergestellte *Symbole*; und es ist gerade diese zur symbolischen Konstruktion gewordene theoretische Erkenntnis, nicht eine reine Phänomenologie der Natur, die es uns erlaubt, Ereignisse *vorauszusagen*. Um ein Beispiel zu nennen: „Die abstrakte chemische Formel, die als Bezeichnung eines bestimmten Stoffes gebraucht wird, enthält nichts mehr von dem, was die direkte Beobachtung und die sinnliche Wahrnehmung uns an diesem Stoff kennen lehrt; — aber statt dessen stellt sie den besonderen Körper in einen außerordentlich reichen und fein gegliederten Beziehungskomplex ein, von dem die Wahrnehmung als solche überhaupt noch nichts weiß." „Was hier die eigentliche Kraft des Zeichens ausmacht, ist eben dies: daß in dem Maße, als die unmittelbaren Inhaltsbestimmungen zurücktreten, die allgemeinen Form- und Relationsmomente zu um so schärferer und reinerer Ausprägung gelangen." (*6*, pp. 44—45.)

Die Symbole, deren sich der Mathematiker am häufigsten bedient, sind *Schriftzeichen*; z. B. die aus hintereinander gestellten Strichen bestehenden „natürlichen" Zeichen für die natürlichen Zahlen, wie ||| für drei und ||||| für fünf. So tritt für ihn die fixierende Schrift neben die der Kommunikation dienende Sprache. Sie ist wichtig, wenn ich recht sehe, vor allem vermöge der durch sie möglichen *Dokumentierung*. Die Sprachlaute verhallen, die Schriftzeichen beharren. Ich zähle etwa, während ich einer die Stunde schlagenden Turmuhr lausche, die Schläge, indem ich sie durch Bleistiftstriche auf einem Blatt Papier vermerke. Die Objekte, wie hier die Schläge der Turmuhr, mögen sich auflösen, „melt, thaw and resolve themselves into a dew", aber ihre Anzahl kann in einem solchen Zahlzeichen niedergelegt und aufbewahrt werden. Tue ich dies für zwei Reihen von Tönen, so kann ich mich der Zeichen zu der nachträglichen Feststellung bedienen, daß es „das zweite Mal mehr Töne waren als das erste Mal" — was mir beim direkten Anhören vielleicht zweifelhaft geblieben wäre. Als Zeichen dienen sichtbare Gebilde von einer gewissen Beständigkeit (nicht etwa Schälle und Rauchwolken; zum mindesten so lange standhaltend, als zur Ausführung der an ihnen vorzunehmenden Operationen benötigt wird). Sie müssen leicht und immer wieder herstellbar sein; ihre Gestalt muß sich, wie *Hilbert* sagt, „unabhängig von Ort und Zeit und von den besonderen Bedingungen der Herstellung des Zeichens sowie von geringfügigen Unterschieden in der Ausführung, von uns allgemein und sicher wiedererkennen lassen" (*9*, p. 163). Eine solche Beschreibung ist nicht gleichgültig, wenn einem (wie *Hilbert*) „die Zeichen selbst die Gegenstände der Zahlentheorie" sind. Der idealistischen Einstellung, für welche die Zahlen ideale Objekte sind oder aus einem Akt des reinen Bewußtseins entspringende Möglichkeiten, tritt hier eine „anthropistische" Einstellung gegenüber, die das konkrete Tun des Menschen ins Auge faßt. Da geht der Mathematiker nicht viel anders mit seinen aus Zeichen gebauten Formeln um wie der Tischler in seiner Werkstatt mit Holz und Hobel, Säge und Leim. Mit einem Unterton von Ironie sagt *Brouwer*, der Intuitionist, darüber: „Op de vraag, waar de wiskundige exactheid dan wel bestaat, antwoorden beide partijen verschillend; de intuitionist zegt: in het menschelijk intellect, de formalist: op het papier." (*4*, p. 7.)

Wenn *Newton* die wahrnehmungsmäßig erlebte Welt erklären will durch die Bewegung fester Partikeln im Raum, so benutzt er den Raum, der ihm zugleich anschaulich gegeben und objektiv ist, zur Konstruktion der hinter den Erscheinungen verborgenen wirklichen Welt. Er verwirft, wie schon *Demokrit*, die Sinnesqualitäten ob ihrer Subjektivität als ungeeignet zu ihrem Aufbau, aber behält den Raum bei. Als *Leibniz* die Phänomenalität von Raum und Zeit erkannte, wurde man gezwungen, auch diese zu eliminieren. Glücklicherweise stand das Mittel dazu in *Descartes*' analytischer Geometrie bereit; denn diese lehrt, wie man (mit Bezug auf ein gegebenes Koordinatensystem) einen jeden Raumpunkt durch seine drei Koordinaten x, y, z, ein Tripel reeller Zahlen, repräsentieren kann. Nicht etwas in der Natur Gegebenes wie der Raum (in dem *Newton* das *sensorium Dei* erblickte), sondern etwas frei Erschaffenes wie die Zahl, ist jetzt das Material zur Konstruktion der objektiven Welt. Es liegt mir an diesem Gegensatz zwischen wirklichem Raum und frei geschaffener Zahl. Oben betonte ich die konkrete sinnliche Gestalt des Zahl-Zeichens, jetzt betone ich die Freiheit des Geistes, die sich in der Schaffung dieser Symbole („auf deren sinnlichen Gehalt es nicht eigentlich ankommt") und in der Weltinterpretation durch sie kundtut. Nichts mehr, nicht einmal Raum und Zeit, entnimmt der Geist für seine symbolische Repräsentation der Welt dem Gegebenen. Es ist wesentlich, daß das Symbol *als Symbol* und nicht als Bestandstück der zu repräsentierenden Wirklichkeit verstanden wird. *Huygens* konnte noch mit gutem Gewissen sagen, daß ein monchromatischer Lichtstrahl *in Wirklichkeit* aus einer Oszillation des aus besonders feinen Partikeln bestehenden Lichtäthers besteht. *Wir* repräsentieren den Strahl durch eine Formel, in der ein gewisses Symbol F, elektromagnetische Feldstärke genannt, als eine rein arithmetisch konstruierte Funktion von vier anderen Symbolen x, y, z, t, genannt Raum-Zeit-Koordinaten, ausgedrückt wird. Das symbolische *construct*, das uns so in Händen bleibt, kann niemand mehr im Ernst als eine den Erscheinungen zugrunde liegende Wirklichkeit in Anspruch nehmen. Natürlich braucht darum das Band zwischen Symbol und wahrnehmungsmäßig Gegebenem nicht durchschnitten zu werden; der Physiker versteht, wie der Symbolismus „gemeint" ist, wenn er die in ihm niedergelegten physikalischen Gesetze an der Erfahrung prüft. (Vergl. hierzu *13*.) Die hier kurz angedeutete Entwicklung der Physik zu einer rein symbolischen Konstruktion gipfelt in unserem Jahrhundert in der Relativitäts- und Quantentheorie. Die Weise, wie die Quantenphysik die beobachtbaren Größen durch Hermitesche Formen in einem unendlich-dimensionalen *Hilbert*schen Raum darstellt, ist ein besonders markantes Beispiel symbolischer Repräsentation.

Nach dieser allgemeinen Orientierung ist es nun an der Zeit, etwas konkreter auf die Konstruktionen der Mathematik und ihre Symbolik einzugehen (vergl. *1*). Da finden wir zunächst *Zeichen für einzelne Zahlen*, wie 2 oder π, und Zeichen für einzelne bestimmte *Operationen* oder *Relationen*, wie + (plus), = (gleich), < (kleiner als). Damit lassen sich bereits bestimmte Aussagen in Formeln fassen, wie $2 < 3$, $2 + 2 = 4$. Auf dieser Stufe stellt die mathematische Symbolik offenbar noch kein eignes Problem über das allgemeine von Sprache und Schrift hinaus. Wissen wir, was wir generell unter Zahl verstehen und erfassen die im Symbol < ausgedrückte Idee des „kleiner als", so können wir zu solchen *Urteilen von hypothetischer Allgemeinheit* fortschreiten wie das folgende: Wenn dir in concreto Zahlen a, b, c gegeben sind und findest du, daß $a < b$ und $b < c$ ist, so kannst du sicher sein, daß auch $a < c$ ist. Hier treten die Buchstaben a, b, c auf als Zeichen für „*irgendeine Zahl*". Auch darin erblicke ich noch nichts spezifisch Mathematisches. Ein Gesetz über eine Rechtshandlung, an der verschiedene Personen beteiligt sind, bezieht sich gleichfalls auf „beliebige“ Personen, und es wäre zweckmäßig, diese durch Buchstaben zu unterscheiden, wenn in der Formulierung des Gesetzes die eine oder andere Person öfter genannt werden muß; die im englischen Recht übliche Benennung 'the first party, the second party', etc., ist nur durch ihre Schwerfälligkeit davon verschieden. Sie wird unerträglich, wenn, wie in einem mathematischen Beweis, dieselbe willkürliche Zahl vielleicht zweihundertmal statt bloß viermal vorkommt; da ist dann wirklich ein Buchstabe viel bequemer. Die Verwendung von Buchstaben für Punkte einer Figur — in der aber das etwa vorkommende Dreieck A B Γ als *irgendein* Dreieck gemeint ist (variable Punkte) — ist den griechischen Mathematikern geläufig. Daß sie durchweg nur Geometrisches, nicht aber unbestimmte Zahlen durch Buchstaben bezeichnen, hängt mit der ihnen durch die Entdeckung inkommensurabler Streckenverhältnisse aufgezwungenen Geometrisierung der Algebra zusammen. Doch bedient sich *Diophant* regelmäßig des Buchstabens ς zur Bezeichnung der Unbekannten. Die freie Buchstabenbezeichnung für Zahlen tritt ziemlich unvermittelt im abendländischen Mittelalter um 1200 auf. Aber eine konsequente Buchstaben-Algebra, die das Hinschreiben wortfreier Formeln erlaubt, wie $a + b = b + a$, ist erst von *Vieta* (1591) entwickelt worden. Wenn es dadurch auch möglich geworden ist, Formelrechnungen ohne die Dazwischenkunft von Worten auszuführen, so ist es doch irreführend, zu behaupten, daß auf der hier erreichten Stufe die Formulierung mathematischer Sätze der Worte nicht mehr bedarf. Eine Formel wie die oben angeführte $a + b = b + a$ besagt vollständig ausgesprochen dies: Sind a, b irgend zwei Zahlen, so besteht die Beziehung $a + b = b + a$. Wir erleben es immer wieder an den jungen Studenten der Mathematik, daß wir ihnen einschärfen müssen, die Worte zu den Formeln, die ihnen erst Sinn verleihen, nicht zu vergessen. Z. B. in der folgenden Definition der Stetigkeit sind die schwerwiegenden Worte „es gibt“ und „alle“ mindestens so wichtig wie die Formeln: „Die Funktion $y = f(x)$, die jeder reellen Zahl x eine reelle Zahl y zuordnet, ist stetig für den Argumentwert $x = a$, wenn es zu jeder positiven Zahl ε eine positive Zahl Zahl δ gibt, so daß $|f(x) - f(a)| < \varepsilon$ ist für alle Zahlen x, für welche $|x - a| < \delta$ ist.“ In diesem Beispiel steht der Buchstabe f für eine willkürliche Funktion (warum sollte man nicht, wenn Punkte und Zahlen durch Buchstaben bezeichnet werden, das gleiche mit einer Funktion tun dürfen?); während die beiden Klammern in $f(x)$ ein Individualzeichen bilden und die universelle Operation andeuten, die aus einer Funktion f und einem Argumentwert x den zugehörigen Funktionswert y entspringen läßt.

Kehren wir zu den gewöhnlichen Zahlen zurück, deren wir uns beim Zählen bedienen. Wir machten halt auf der Stufe, wo wir in hypothetischer Allgemeinheit von irgendwelchen gegebenen Zahlen sprechen können. Etwas radikal Neues geschieht, und dies ist *die Geburt der Mathematik*, wenn man die Zahlen nicht hinnimmt, wie sie einem zufällig in der Wirklichkeit begegnen, sondern einbettet in die *Reihe* **|**, **||**, **|||**, ... *aller möglichen Zahlen*. Diese entsteht durch einen *Erzeugungsprozeß*, in welchem die gleiche Operation, der Übergang von einer Zahl n zur nächstfolgenden n', immer wiederholt wird. Am Zeichen wird diese Operation ausgeführt durch Anhängen eines weiteren Striches. Es ist nicht so, daß die unendliche Zahlenreihe wirklich hergestellt werden kann, aber man glaubt an die *Möglichkeit*, den Prozeß über jeden schon erreichten Punkt hinaus fortsetzen zu können. Hier wird *das Wirkliche projiziert auf den Hintergrund des Möglichen, einer nach festem Verfahren frei vom Geiste erschaffenen, ins Unendliche offenen Mannigfaltigkeit.* Den *Raum* als das Kontinuum möglicher Örter konstruieren wir mittels des Koordinatenbegriffs aus der nicht minder frei von uns erschaffenen *Mannigfaltigkeit aller möglichen reellen Zahlen.* Nur so gelingt es, wie es insbesondere in der Astronomie nötig ist, „Raummarken“ auch in den die Erde umgebenden leeren Raum hinauszusetzen. Ich sehe eben hierin, in dieser Projektion des zufällig begegnenden *Wirklichen* auf das durch einen Konstruktionsprozeß gewonnene a priori *Mögliche*, das entscheidende Kennzeichen der theoretischen Wissenschaft.

Die Reihe der natürlichen Zahlen ist das primitivste Beispiel eines solchen zur symbolischen Konstruktion dienenden, a priori überblickbaren, weil von uns selbst geschaffenen, Variabilitätsgebiets; alle höheren Fälle gehen im Grund auf diesen einfachsten zurück. Es scheint mir darum vollständig berechtigt, die Idee des *„immer noch eins“*, aus der die Zahlenreihe entspringt, mit *Brouwer* als die *mathematische Urintuition* anzusprechen. Sie ist die Grundlage, auf der *Brouwer* seine aus intuitiv einsichtigen Sätzen bestehende Mathematik aufbaut *(4, 5)*; sie ist es auch, auf die sich *Hilbert* stützt, wenn er beschreibt, was eine Formel seiner „formalisierten Mathematik“ ist, oder durch „metamathematische“ Überlegungen die Widerspruchslosigkeit seines Systems sicherzustellen sucht *(9, 10)*. Aus ihr heraus verstehen wir auch erst, was wir meinen, wenn wir von einer *beliebigen* Zahl sprechen — ob nun die Zahlen ideale Wesen *sui generis* sind oder nur in Form der Zahlzeichen existieren. Kein Wunder, daß auch in der Symbolik hier zuerst die für die Mathematik charakteristische *systematische Bezeichnungsweise* auftritt, vermöge deren eine jede vorkommende Zahl durch hintereinander gesetzte Striche (oder durch eine Ziffernreihe unseres dezimalen Positionssystems) gekennzeichnet werden kann. Halten wir uns an die natürliche Bezeichnung durch eine Folge von Strichen, so sehen wir, daß diese uns in Stand setzt, von irgend zwei in Zeichen vorliegenden Zahlen zu entscheiden, welche die größere ist; indem wir nämlich die Striche einen nach dem andern auskreuzen und dabei jedesmal, wenn wir einen Strich der ersten Reihe auskreuzen, auch einen Strich der zweiten Reihe auskreuzen. Dies ist eine erstaunliche Leistung; denn dem bloßen Anblick nach haben wir es schon schwer, zwischen so niedrigen Anzahlen wie 20 und 21 zu unterscheiden. Es ist also nicht bloß so, daß das Operieren mit Zeichen eine viel größere Sicherheit gewährt als das anschaulich-inhaltliche Denken, sondern in völlig legitimer Weise über dessen Bereich weit hinausdringt. Die Verifizierung von Aussagen über Zahlen ist an die Zahlzeichen gebunden. Das Zeichen repräsentiert die Zahl *vollständig;* denn nichts läßt sich über Zahlen sagen, das nicht an den Zahlzeichen verifiziert werden könnte.

Methodisch findet die Erzeugung der Zahlen durch den immer wiederholten Übergang von einer Zahl n zur nächstfolgenden n' ihren Ausdruck in der *Definition und dem Schluß durch vollständige Induktion,* die zuerst von *Pascal* und *Jacob Bernoulli* ins ausdrückliche Bewußtsein der Mathematiker gehoben wurden. Hier liegt also der eigentliche Lebensnerv des mathematischen Beweisens. Beispiel einer Definition durch vollständige Induktion ist die Unterscheidung von gerade und ungerade durch „Abzählen zu zweien"; sie läßt sich in zwei Sätzen niederlegen: i) die erste Zahl 1 ist ungerade; ii) die auf eine Zahl n folgende n' ist gerade oder ungerade, je nachdem n ungerade oder gerade ist. Nachdem die Bedeutung von $2 \cdot n$ durch vollständige Induktion festgelegt ist, kommt man mittels eines Schlusses durch vollständige Induktion zu dem Ergebnis, daß $2 \cdot n$ für jedes n gerade ist. Und da, wie sich durch Abzählen herausstellt, 13 ungerade ist, kann man also sicher sein, daß, wie weit man in der Reihe der natürlichen Zahlen auch geht, man niemals eine, n, antreffen wird, deren Doppeltes $2 \cdot n = 13$ ist. So kommt man zu Aussagen, die für „alle Zahlen" gültig sind, obschon es deren unendlich viele gibt und man sie darum nicht einzeln durchprüfen kann.

Durch das Beispiel der natürlichen Zahlen belehrt, kann man die Grundzüge des konstruktiv-symbolischen Erkennens, das die ganze Wissenschaft beherrscht, zusammenfassend etwa so beschreiben (cf. *13*): α) Das Resultat gewisser Operationen am Gegebenen (z. B. des Durchzählens gegebener Mengen), die für allgemein ausführbar gelten, wird, sofern es durch das Gegebene eindeutig bestimmt ist (am Beispiel: die Anzahl ist unabhängig davon, in welcher Reihenfolge man beim Durchzählen einer Menge ihre Elemente vornimmt), als ein dem Gegebenen an sich zukommendes Merkmal aufgestellt (selbst wenn jene Operationen, die seinen Sinn begründen, nicht wirklich ausgeführt werden). In der Physik werden als solche Operationen auch Reaktionen der zu untersuchenden Körper untereinander sowie mit wirklichen oder fingierten Hilfskörpern benutzt (z. B. die Kollision von Körpern als Mittel, ihre träge Masse zu bestimmen). β) Durch Einführung von Zeichen wird eine Aufspaltung der Urteile vollzogen (z. B. spaltet man die Aussage: Die eben gehörte Tonfolge bestand aus mehr Tönen als die vorige, in die Aussagen: Das erste Mal waren es 7 Töne; jetzt waren es 12; $12 > 7$) und ein Teil der Operationen durch Verschiebung auf die Zeichen vom Gegebenen und seinem Fortbestand unabhängig gemacht (am Beispiel: $12 > 7$). γ) Die Zeichen werden nicht einzeln für das jeweils aktuell Gegebene hergestellt, sondern sie werden dem potentiellen Vorrat einer nach festem Verfahren herstellbaren, geordneten, ins Unendliche offnen Mannigfaltigkeit von Zeichen entnommen.

In seinem Entwurf einer *Mathesis universalis,* die er auch als allgemeine Charakteristik oder *ars combinatoria* beschreibt, hat Leibniz den Plan verfolgt, den Symbolismus der Algebra auf alle Gebiete des Erkennens auszudehnen. Was er aber in dieser Richtung ausgeführt hat, bleibt bei recht primitiven Ansätzen stehen. Doch die Mathematik selbst, indem sie den alten Rahmen einer „Lehre von Raum und Zahl" sprengte und solche Disziplinen wie *Mengenlehre, kombinatorische Topologie, Gruppentheorie, abstrakte Algebra* aus sich hervortrieb, verwirklichte in ihrer historischen Entwicklung, mittels eines allmählich sich anreichernden und immer größere Geschmeidigkeit erwerbenden Symbolismus, ein gut Teil des *Leibniz*schen Programms. In Anpassung an die in diesem weiteren Rahmen ausgebildeten Bräuche wird es nun vernünftig, die Operation des Übergangs von einer Zahl zur nächstfolgenden durch einen dem Zahlzeichen *vor*gestellten Buchstaben, etwa σ, zu bezeichnen (und außerdem die Reihe der natürlichen Zahlen mit der Null beginnen zu lassen). Das Zahlsymbol der *drei* sieht dann also so aus: $\sigma\sigma\sigma 0$. Der prinzipiell wichtigste Schritt, der in dieser Richtung geschieht, ist die Einbeziehung der Logik in den Symbolismus, indem Buchstaben dazu dienen, auch (willkürliche) *Aussagen* zu benennen, auf die sich die logischen Grundoperationen der Verbindung durch „und" (∩), „oder" (∪), der Negation (~) und der Implikation (→) erstrecken (so daß z. B. $a \rightarrow b$ für die Aussage steht, daß die Aussage a die Aussage b impliziert). Es stellt sich nämlich heraus, daß sich solche logische Formeln manipulieren lassen, ohne daß man auf den Sinn der logischen Grundbegriffe zu rekurrieren braucht, ja daß auch das Beweisen sich auf ein Umgehen mit Formeln reduziert, dessen Regeln leicht und vollständig an der äußeren Gestalt der Formeln beschrieben werden können, ohne daß ihr Sinn in Frage kommt. Hier ist dann eine Stufe erreicht, wo wirklich der Inhalt der Mathematik durch einen *systematischen* Symbolismus gänzlich in wortfreien Formeln niedergelegt werden kann. Das mathematische Denken gewinnt damit die größte Sicherheit und Spannweite. Es sei beiläufig angemerkt, daß nun nicht mehr Operationen und Relationen nebeneinander verwendet zu werden brauchen. Das Relationszeichen < in der Formel $a < b$ (die Zahl a ist kleiner als b) kann jetzt als die Operation interpretiert werden, die aus zwei beliebigen *Zahlen* a, b die *Aussage* „a ist kleiner als b" erzeugt. Eine besondere Rolle in der Logik spielen die Quantifikatoren „es gibt" und „alle". Ist z. B. $A(x, y)$ eine Aussage über zwei beliebige Zahlen x, y, etwa $x = y^2$, so kann ich zu einer Aussage über x allein nicht nur dadurch gelangen, daß ich für y eine bestimmte Zahl einsetze, z. B. $A(x, 2)$, x ist gleich 2^2, sondern auch indem ich bilde $\varepsilon_y A(x, y)$: *es gibt* eine Zahl y, so daß $A(x, y)$ gilt, „x ist eine Quadratzahl". Der Index y an dem Individualsymbol ε_y des Quantifikators zeigt an, daß es die Variable y ist, welche durch ihn aus $A(x, y)$ eliminiert wird.

Man kann in diesem Symbolismus schildern, wie Formeln hergestellt werden, in welcher Weise in ihnen Variable durch Formeln substituiert werden können, wie man sich „Axiome" beschafft, und schließlich die Regel des Syllogismus angeben, durch deren fortgesetzte Anwendung man, mit Axiomen beginnend, aus schon gewonnenen „richtigen" Formeln neue „richtige" Formeln gewinnt. Dabei hat die Beschreibung des Aufbaus einer Formel notwendig den Charakter der vollständigen Induktion: eine neue Formel entsteht, indem man an das Symbol einer Operation (oder eines Quantifikators), je nachdem, eine oder mehrere schon gebildete Formeln anhängt. Eben darin drückt sich die *spezifische Systematik der mathematischen Symbolik* aus, die vor einer noch so großen Komplikation der Formeln nicht zurückzuschrecken braucht. Man kann es einer fertigen Kombination von Zeichen ansehen, ob sie eine Formel ist. *Unvoraussehbar aber bleibt, welche Formeln sich in dem „Beweisspiel" als richtig herausstellen werden:* wir *haben* die Wahrheit nicht, sie muß durch Handeln von Fall zu Fall gewonnen werden. Dies liegt daran, daß durch den Syllogismus aus zwei Formeln eine dritte hergeleitet wird, die *kürzer* ist als eine der Prämissen, und daß infolgedessen im Beweis Expansion und Kontraktion der Formeln miteinander abwechseln. Man mag dies als ein weiteres entscheidendes Faktum δ) der oben in α), β), γ) versuchten Kennzeichnung des konstruktiven Erkennens anfügen. Ein *Widerspruch* entstünde, wenn *ein* Beweis zu einer Endformel $\mathfrak{a}$ führte und ein anderer zu ihrer Negation $\sim \mathfrak{a}$.

Während der Ausbildung des Formalismus war die Einstellung durchaus die, daß die Formeln die Abbilder *sinnvoller*, insbesondere *wahrer* mathematischer Aussagen sind: die Formeln und ihre Manipulation waren nicht Selbstzweck, sondern dienten dazu, mathematische Tatsachen auszudrücken und zu ermitteln. Da traten, noch bevor der Formalismus eine einigermaßen endgültige Gestalt gewonnen hatte, zu Anfang des zwanzigsten Jahrhunderts zwei Ereignisse ein: erstens stellte es sich heraus, daß durch die schrankenlose Verwendung der Quantifikatoren tatsächlich Widersprüche entspringen, und zweitens machte *Brouwer* klar (schon in seiner Dissertation *Over de grondslagen der wiskunde*, Amsterdam und Leipzig 1907), daß das Prinzip des *tertium non datur* sich auf keine Evidenz berufen kann, wenn es auf Aussagen angewendet wird, in denen „es gibt" und „alle" sich nicht auf eine Menge einzeln aufgewiesener Objekte beziehen, sondern auf *unendliche* Mengen, wie die Menge aller natürlichen Zahlen oder gar die Menge aller möglichen unendlichen Sequenzen solcher Zahlen. Danach standen zwei Wege offen: *der Brouwersche Intuitionismus*, der sich auf die anschaulich evidenten (in der mathematischen Urintuition gründenden) Aussagen beschränkt und die ins Unendliche offne Reihe möglicher Zahlen nicht in einen geschlossenen Bereich an sich existierender Elemente verwandelt; und *der Hilbertsche Formalismus*, in welchem die Aussagen durch sinnleere Formeln ersetzt sind und darum der Gebrauch der Quantifikatoren nur durch die Rücksicht darauf beschränkt ist, daß sich keine Widersprüche ergeben dürfen. Durch diese Umdeutung, in welcher die *Wahrheit* der einzelnen mathematischen Sätze preisgegeben ist und nur noch auf die *Widerspruchslosigkeit* ihres Systems Wert gelegt wird, macht *Hilbert* sich anheischig, die klassische Mathematik in ihrem ganzen Umfang zu retten (*9, 10*).

Es ist klar, daß die Symbole für die eine und die andere Richtung eine ganz verschiedene Rolle spielen: Für *Brouwer* gehören sie wie die Worte zur Sprache, sind lediglich Hilfsmittel zur Vergegenwärtigung und Mitteilung eines mathematischen Sachverhalts oder Gedankens. Bei *Hilbert* sind sie, obschon oder gerade weil nichts bezeichnend, die Substanz der Mathematik. Emphatisch heißt es: Am Anfang ist das Zeichen. Zwar kennt auch er Mitteilungszeichen, die im Gefüge der sprachlichen Beschreibung des Handhabens von Formeln auftreten. Sie haben aber nichts zu tun mit den Symbolen, aus denen die Formeln selbst aufgebaut sind, und man tut gut, sie durch ihre äußere Gestalt vor Verwechslungen damit zu schützen. Die Widerspruchslosigkeit ist zu erweisen durch inhaltlich-anschauliche Überlegungen, deren Gegenstand die mathematischen Formeln sind. Hier wird *gedacht* und nicht mehr *gespielt*, und es werden in diesen „metamathematischen" Überlegungen die von *Brouwer* dem inhaltlichen Denken gesetzten Grenzen durchaus respektiert.

Das Unternehmen, so die Widerspruchslosigkeit der klassischen Mathematik zu erweisen, wurde, nach anfänglichen Erfolgen, durch eine tiefsinnige Entdeckung *K. Gödels* ernstlich in Frage gestellt. Der Beweis gelang nur für beschränkte Teile der Mathematik, und selbst da schleichen sich Zweifel ein. *Evidenz* bleibt, wie wir uns auch wenden mögen, letzte Quelle von Wahrheit und Erkenntnis. *Brouwer* gründet auf sie die Mathematik, *Hilbert* die (erhoffte) Einsicht in die Widerspruchslosigkeit der Mathematik. Aber Evidenz läßt sich nie endgültig auf Regeln bringen und gegen den Irrtum schützen. Darum bleiben die Grenzen, bis zu denen sich die *Brouwer*sche Mathematik erstreckt, vage; und fraglich bleibt auch, ob die von verschiedenen Autoren in Ausführung des Hilbertschen Programms angestellten metamathematischen Überlegungen nicht manchmal den Bogen der Evidenz überspannen.

Das Faktum aber bleibt: wir besitzen einen einfachen Formalismus, der die gesamte Mathematik, soweit wir sie heute kennen, umspannt und *der bisher zu keinen Widersprüchen geführt hat*. Er verbürgt die unvergleichliche Sicherheit des Operierens in der Mathematik. Könnten wir uns nicht daran genügen lassen? Muß uns wirklich die Widerspruchslosigkeit *für alle Ewigkeit* garantiert werden, oder können wir nicht eine etwa nötige Revision des Formalismus vertagen bis zu jenem Augenblick, da tatsächlich aus ihm ein Widerspruch hervorgegangen ist? Dem Mathematiker fällt dieser Verzicht schwer. Der Physiker arbeitet mit physikalischen Gesetzen, deren Bewährung darin liegt, daß sie mit allen bekannten Erscheinungen in Einklang stehen; ihm ist es selbstverständlich, daß er stets darauf gefaßt sein muß, sie einmal durch neu entdeckte Tatsachen über den Haufen geworfen zu sehen!

Vorahnungen der ganzen hier geschilderten Entwicklung der Mathematik finden sich bei *Nicolaus Cusanus*, bestimmter bei *Leibniz*. Sie sehen im Symbolismus eine Repräsentation der vom menschlichen Denken unvollziehbaren göttlichen Ideenwelt. Das intuitive Erkennen des Menschen ist beschränkt, Gottes alldurchdringend. Natürlich sind der Cusaner wie *Leibniz* von der scharfen Fassung *Hilberts* noch weit entfernt. Es fehlt die Schichtung in die freie, rein symbolische „formale Mathematik" und die darüberliegende „intuitionistische Metamathematik", deren inhaltliche Überlegungen nur das eine Anliegen haben, die Widerspruchslosigkeit der unteren Schicht zur Evidenz zu bringen — obschon *Cusanus* der Gedanke nicht fern zu liegen scheint, am endlichen Symbolismus durch seine Konsistenz wenigstens die *Möglichkeit* der uns verschlossenen transzendenten Gotteswelt zu erweisen.

Es ist nicht ohne Interesse, vom Standpunkt der modernen Grundlagenforschung auf den alten Streit zurückzukommen, ob die Zahlen, wie *Dedekind*, *Frege* und *Brouwer* meinen, selbständige ideale Objekte sind, deren Ursprung in anschaulich vollziehbaren Bewußtseinsakten liegt, oder ob, wie *Helmholtz* und *Hilbert* behaupten, die Zahlentheorie es allein mit den konkreten Zahlzeichen zu tun hat. Das Wesen Zahl mag uns in reiner Anschauung gegeben sein; es ist aber schwer zu glauben, daß uns z. B. die Zahl 10^{11} (ungefähr mit soviel Dollars rechnet das Jahresbudget des U.S. Government) so gegeben sein sollte. Es ist auch praktisch unmöglich und wäre nutzlos, eine solche Zahl durch hintereinandergesetzte Striche auszuschreiben. Um zu wissen, was damit gemeint ist, muß man (nach *Dedekind* sowohl als *Brouwer*) durchs *Unendliche* hindurchgehen, indem man zunächst $10 \cdot n$ durch vollständige Induktion für eine beliebige Zahl n definiert und darauf, mit 1 beginnend, 11mal hintereinander die Multiplikation mit 10 ausführt. Auch hier bleibt — wenigstens für den, der nicht an die Zahlen als ein fertiges System von Seinsgegebenheiten zu glauben vermag — eine Unklarheit, da ja die vollständige Induktion auf eine ins Unbegrenzte sich erstreckende Möglichkeit verweist, die aber in dem durch das Zeichen 10^{11} geforderten Umfang praktisch nicht auszuführen ist. Die bündigste Antwort erhält man bei *Hilbert:* für ihn ist 10^{11} die Abkürzung für ein Symbol, das sich leicht im Formalismus explicite hinschreiben läßt. Wie man aber auch entscheiden mag: sobald die Sicherheit im Umgehen mit den Zeichen erreicht ist, „weiß man, woran man ist", und es verstummt der Streit um Worte und Auffassungen.

Wenn die formale Mathematik nicht mehr den Anspruch erhebt, wahre Behauptungen aufzustellen, so muß man sich fragen, was sie dann überhaupt bezweckt. Die Antwort von *Cusanus* und *Leibniz*, daß sie im endlichen Symbol die Gott in unmittelbarer Intuition des Unendlichen gegenwärtigen Ideen spiegelt, findet heute wenig Anklang und ist zumindest zu einseitig theologisch. Überzeugender ist der Hinweis auf ihren *naturwissenschaftlichen* Gebrauch, auf die Rolle, die sie beim konstruktiven Aufbau einer Theorie der wirklichen Welt durch die Physik spielt. Denn hier können wir uns auf die Bewährung der theoretischen Konstruktion durch Erfahrung und Voraussage berufen. Die Situation, die wir in der *theoretischen Physik* vorfinden, entspricht in keiner Weise dem von *Brouwer* geforderten und in seiner Mathematik erfüllten Ideal, daß nämlich jedes Urteil seinen eigenen, in der Anschauung vollziehbaren Sinn hat. Die Gesetze der Physik haben, einzeln genommen, keinen in der Erfahrung verifizierbaren Inhalt. Nur das theoretische System als Ganzes läßt sich mit der Erfahrung konfrontieren. Mag es wahr sein, daß die physikalische Messung es nur mit der Feststellung von Koinzidenzen zu tun hat, wie oft behauptet wurde: man muß doch zugestehen, daß unser Interesse nicht in erster Linie an der Konstatierung hängt, daß dieser Zeiger diesen Skalenteil deckt, sondern an den idealen Setzungen, die laut Theorie in solchen Koinzidenzen sich ausweisen, deren Sinn selbst aber in keiner gebenden Anschauung sich erfüllt, — wie z. B. der Setzung des Elektrons als eines universellen elektrischen Elementarquantums. Vom metaphysischen Glauben an die Realität der Außenwelt getrieben, versuchen wir eine symbolische Gestaltung des Transzendenten — und erleben die Genugtuung, daß sie sich in der Erfahrung bewährt. So komme ich zu folgender Position (*13*, p. 420): Nimmt man die Mathematik für sich allein, so beschränke man sich mit *Brouwer* auf die einsichtigen Wahrheiten, in die das Unendliche nur als ein offenes Feld von Möglichkeiten eingeht; es ist kein Motiv erfindlich, das darüber hinaus drängt. In der Naturwissenschaft aber berühren wir eine Sphäre, die der schauenden Evidenz sowieso undurchdringlich ist; hier wird Erkenntnis notwendig zu symbolischer Gestaltung, und es ist darum, wenn die Mathematik durch die Physik in den Prozeß der theoretischen Weltkonstruktion mit hineingenommen wird, auch nicht mehr nötig, daß das Mathematische sich daraus als ein besonderer Bezirk des anschaulich Gewissen isolieren lasse; auf dieser höheren Warte, von der aus die ganze Wissenschaft als eine Einheit erscheint, bin ich geneigt, *Hilbert* grundsätzlich recht zu geben.

Dabei möchte ich aber folgendes zu bedenken geben: Der im Symbolismus der Quantifikatoren sich ausdrückende „mathematische Existentialismus" ist recht und gut, solange es sich um die Entwicklung *allgemeiner Theorien* handelt. Sobald aber in einem *konkreten Fall* eine bestimmte numerische (wenn auch niemals exakte, sondern immer nur angenäherte) Voraussage gemacht werden soll, muß man versuchen, die symbolisch sichergestellte Existenz durch eine explizite Auswertung auszufüllen, wie das die Brouwersche Mathematik grundsätzlich verlangt. Schon an einem rein mathematischen Theorem, wie dem von der Existenz der Wurzeln einer algebraischen Gleichung, läßt sich das klarmachen. Gewisse in der formalen Mathematik durchaus akzeptable Beweise für diesen „Fundamentalsatz der Algebra" versagen, wenn es sich darum handelt, die Wurzeln aus den Koeffizienten wirklich zu berechnen; oder etwas präziser, wenn es gilt, ein Verfahren zu finden, durch welches die Wurzeln genauer und genauer bestimmt werden, wenn die Genauigkeit, mit der die Koeffizienten bekannt sind, unbegrenzt wächst. Darum finde ich es am Platze, an den heutigen Mathematiker die Mahnung zu richten: Wenn immer du eine Frage durch explizite Konstruktion erledigen kannst, begnüge dich nicht mit rein existentiellen Argumenten!

Die Zahlen und mathematischen Symbole bilden nicht nur den Baustoff, aus dem eine echte theoretische Wissenschaft von der Natur ihr Gebäude zu errichten strebt, sondern daneben läuft in der ganzen Geschichte des menschlichen Geistes die *Zahlenmagie*, welche die Zahl in ganz anderem Sinne zum Symbol der irdischen und göttlichen Wirklichkeit macht. Beides findet sich bereits in einfacher Ausprägung und seltsamer Mischung bei *Pythagoras*, der rätselhaftesten Gestalt der griechischen Geistesgeschichte. Die ungeraden und die geraden Zahlen repräsentieren ihm das männliche und das weibliche Prinzip; die Zahl 4 oder das Quadrat wird zum Symbol der Gerechtigkeit (ist noch die englische Redensart '*a square deal*' ein Überbleibsel davon?). Für jede der Zahlen von 2 bis 7 kann man mannigfache magische Bedeutungen bei den Völkern aller Zeiten und Zonen nachweisen; 3 und 7 spielen eine besonders bevorzugte Rolle, aber auch die „Zahl der Engel" 9 ist manchenorts beliebt. *Dante* sagt in seiner *Vita Nuova* (XXX, 26—27) von Beatrice, die Zahl 9 wäre ihr eigentliches Selbst. Auch bei raffinierterer Ausbildung bleiben es immer *simple* (ein Mathematiker würde sagen, allzu simple) *zahlentheoretische* Eigenschaften von Zahlen, die als Quelle ihrer magischen Kraft angesprochen werden. *Pythagoras'* „vollkommene Zahlen" (deren Auffindung, beiläufig bemerkt, ein nicht-triviales, obschon unfruchtbares mathematisches Problem ist)[8] gehören schon zum Kompliziertesten, was vorkommt. *Plato* übernimmt ein gut Stück der pythagoreischen Zahlenweisheit; aber die Zahl der Bürger der idealen Stadt, welche er zu $5040 = 7!$ ansetzt, und die sehr unklar beschriebene Hochzeitszahl im „Staat" scheinen seine eigene numerologische Erfindung zu sein. *Augustin* und *Philo* tragen manches zur „zahlentheoretischen" Exegese der Heiligen Schriften bei. Das Mittelalter schwelgt in Zahlenmagie. Im populären Aberglauben ist einiges davon bis auf den heutigen Tag lebendig geblieben, so die Scheu vor der Zahl 13. Auch die Astrologie, selbst wenn sich ihr ein so erleuchteter und tief um Wahrheit ringender Geist ergibt wie *Kepler*, rechne ich dahin. Es mag sich lohnen, dem allen in seinen historischen Zusammenhängen nachzugehen; aber man wird nicht erwarten, daß diese Seite des mathematischen Symbolismus hier eingehender besprochen wird. Ich möchte nur auf einen Zug hinweisen, der für diese Denkweise kennzeichnend zu sein scheint: was in der Zahlenmagie gilt, sind die *zahlentheoretischen* Eigenschaften der Zahlen; was in der Naturwissenschaft gilt, sind ihre *Größen*eigenschaften. Vom Größen-Standpunkt macht es wenig Unterschied, ob die Zahl der Bürger einer Stadt 5040 beträgt oder 5039, vom zahlentheoretischen Standpunkt ist da ein himmelweiter Abstand; z. B. besitzt $5040 = 2^4 \cdot 3^2 \cdot 5 \cdot 7$ viele Teiler, während 5039 eine Primzahl ist. Wenn in *Platos* idealer Stadt *ein* Bürger über Nacht stirbt und dadurch die Bürgerzahl sich auf 5039 erniedrigt, ist sie sogleich völlig korrumpiert. Es ist wohl eine der fundamentalsten Tatsachen, der *Leibniz* in seinem Kontinuitätsprinzip Ausdruck zu leihen versuchte, daß in die Naturerklärung die Zahlen durch ihren Größencharakter und nicht durch ihre zahlentheoretischen Eigenschaften eingehen; der moderne Algebraiker würde sagen, nicht irgendeine endliche, sondern die unendliche Primstelle des rationalen Zahlkörpers ist entscheidend. Es wäre vielleicht ganz hübsch, wenn es anders wäre; aber es ist nicht so.

Aus dem amüsanten Buch *The Magic of Numbers* von *E. T. Bell* (*2*) spricht die unverhohlene Angst, daß in *Eddingtons* Apriorismus und seinem Versuch, die Feinstruktur-Konstante 137 oder die Anzahl der in der Welt vorhandenen Elementarteilchen durch erkenntnistheoretische Überlegungen abzuleiten, aber auch in der modernen Quantenphysik, die Numerologie von neuem auf dem Wege ist, wie einst bei *Pythagoras*, echte quantitative Naturwissenschaft zu verdunkeln. Er gesteht dem Prinzip der numerologischen Weltdeutung zu: „It was a great dream, as simple and as childlike as it was great." „But it was only a dream", und Numerologie bleibt für ihn eine Afterwissenschaft; es beunruhigt ihn tief, daß „ihre Popularität seit 1920 rascher gewachsen ist als zu irgendeiner früheren Zeit seit dem sechzehnten Jahrhundert". Anders wertet O. *Becker*. Am Ende eines schon früher zitierten Artikels (*1*) macht er im Anschluß an Ausführungen von *Hilbert*, *v. Neumann* und *Nordheim* über die Grundlagen der Quantenmechanik die folgenden Bemerkungen: „Man springt also gewissermaßen mit dem vollständigen, ontologisch unverständlichen ‚mathematischen Apparat' in die ‚Deutung' der Natur hinein; der Apparat ist wie ein magischer Schlüssel, der das physikalische Problemgebiet erschließt, — aber nur erschließt im Sinne einer symbolischen Repräsentation, nicht im Sinne einer die Phänomene in ihrem Zusammenhang wirklich ‚entdeckenden' Interpretation." Er schließt mit den Worten: „Die Grundrichtung dieses symbolischen Weges ist uralt, archaisch, ja geradezu ‚prähistorisch': die modernste ‚exakte' Wissenschaft wird wieder zur Magie, aus der sie ursprünglich abstammt." Wer hat recht, *Becker* oder *Bell*? Wat den Eenen sin Ul, is den Annern sin Nachtigall!

[8] Eine Zahl n ist vollkommen, wenn die Summe ihrer Teiler, 1 eingeschlossen, sie selbst ausgeschlossen, gleich n ist. Die niedrigste vollkommene Zahl ist $6 = 1 + 2 + 3$.

Literatur.

(*1*) *Oskar Becker*, „Das Symbolische in der Mathematik", in einem Heft über Symbolik der *Blätter für deutsche Philosophie*, *1*, Berlin 1927/28 (pp. 329—348).

(*2*) *Eric Temple Bell*, The Magic of Numbers, McGraw-Hill, New York 1946.

(*3*) *P. Bernays*, „Die Philosophie der Mathematik und die *Hilbert*sche Beheistheorie", in einem Doppelheft über „Philosophische Grundlegung der Mathematik" der *Blätter für deutsche Philosophie*, *4*, 1930, pp. 326—367.

(*4*) *L. E. J. Brouwer*, Intuitionisme en Formalisme, Groningen 1912 (englische Übersetzung in *Bull. Am. Math. Soc.*, *20*, 1913/14).

(*5*) *L. E. J. Brouwer*, „Zur Begründung der intuitionistischen Mathematik", *Mathematische Annalen*, *93*, *95*, *96* (1924/27).

(*6*) *E. Cassirer*, Philosophie der symbolischen Formen, Teil I: Die Sprache, Berlin 1923.

(*7*) *K. Gödel*, „Über formal unentscheidbare Sätze der Principia Mathematica und verwandter Systeme", *Monatsh. Math. Phys.* *38*, 1931, pp. 173—198.

(8) *H. von Helmholtz*, „Zählen und Messen" in: *Wissenschaftliche Abhandlungen*, III, p. 356.

(9) *David Hilbert*, *Gesammelte Abhandlungen*, Bd. 3, Berlin (Julius Springer), 1935, namentlich „Neubegründung der Mathematik, Erste Mitteilung", pp. 157—177.

(10) *D. Hilbert* und *P. Bernays*, Grundlagen der Mathematik (2 Bände), Berlin, Julius Springer, 1934—1939.

(11) *Hermann Noack*, Symbol und Existenz der Wissenschaft, Halle/Saale (Max Niemeyer), 1936.

(12) *Karl Voßler*, Sprache und Wissenschaft, in der Sammlung: *Geist und Kultur in der Sprache*, Heidelberg 1925.

(13) *H. Weyl*, „Wissenschaft als symbolische Konstruktion des Menschen", *Eranos-Jahrbuch* 1948, Rhein-Verlag Zürich, 1949, pp. 375—431.

(14) *H. Weyl*, Philosophy of Mathematics and Natural Science, Princeton University Press, Princeton, N. J., second printing 1950.

(15) *L. Wittgenstein*, Tractatus logico-philosophicus, London, New York, und (deutsche Ausgabe) Berlin, 1922.

157.

Universities and Science in Germany

The Mathematics Student (Madras, India) 21, Nos. 1 und 2, 1—26 (March—June 1953)

You have heard that I am a mathematician. My interest has always been concentrated on science itself and not on its organization. If, nevertheless, the major part of my talk tonight will deal with the organization of the universities and scientific research in Germany, I beg you to realize that I am not an expert in the subject on which I am talking. But I know Germany and know German higher education through personal experience ; for I was born and educated in Germany, I taught at the University of Göttingen, first as Privatdozent for three and a half years before the first World War and then again during the years 1930-1933 as full professor, including one-half year under the Nazis. In the intervening period, from 1913 to 1930, I was Professor at the Technische Hochschule in Zürich, Switzerland.

I have tried to fortify my own knowledge by a few books on the subject which I recommend to your attention :

Friedrich Paulsen, *The German Universities :* A German classic published in German in 1902, in English in 1906.

James Morgan Hart, *German Universities, A Narrative of Personal Experience,* New York 1874. As this torn copy from the Princeton Library indicates, it must have been a favourite among American students for many years.

Abraham Flexner, the spiritual founder of our Institute for Advanced Study, published in 1930 a book entitled, *Universities, American, English, German.*

It is surprising to observe how Hart and Flexner agree almost point by point on what they find praiseworthy and what they blame.

This is the text of two lectures given in 1945 by Professor Hermann Weyl in a course on European history and civilization arranged by the Princeton School of International Affairs.

For the history of the German universities under the Nazis, I consulted and you may consult

E. Y. Hartshorne, Jr., *The German Universities and National Socialism*, 1937.

I shall first discuss the organization and history of the German universities and of scientific research in Germany up to 1933, and conclude with a very brief account of the fate that befell both under the Nazis.

I believe this to be true : *As disastrous as Germany's political history has been through the centuries, so fortunate is her history of higher education.* The German people has been politically unfree during almost its entire historical career, but the spirit of intellectual freedom has been strong in their universities ever since they assumed their present character in the 18th century, until 1933. It was an American, Stanley Hall, who said in 1890 : " The German university is today the freest spot on earth."

Were I to describe the scale of values underlying the social structure of the Germany of William II in which I spent my student days, I should say that two classes enjoyed a prestige far higher than corresponds to American standards : the military and the scholars. Germany was unquestionably militaristically minded; but that is only one side of the picture : the German nation gloried in her army *and* in her universities. It was Palmerston who spoke of Germany as the land of damned professors. The Germans have a gift and passion for intellectual and artistic endeavors. I much preferred life in Switzerland with her old democracy to that in class-ridden imperial Germany ; but there is one trait for which I have always loved my old countrymen in preference to the more sober Swiss : for their genuine and passionate interest in all things of the mind. Everyone and everything connected with the universities was held in the greatest popular esteem.

A little anecdote of German university life in the nineties may illustrate the point. Kuno Fischer, a second-rate philosopher at

Heidelberg, was one day disturbed by the noise of workers who were putting in new cobblestones in the street before his house. He had at that time been offered a professorship in Berlin. So he opened his window and shouted to the workmen, "If you don't stop that noise at once, I'll accept the call to Berlin." Whereupon the foreman ran to the mayor, he summoned the Stadtbaumeister and they decided to postpone repair of the street until after the beginning of the academic vacation.

Practically every one of the men in the Government in Germany had gone through the thorough German system of higher education, the gymnasium and the university. On the other hand it is true that here, in the political sphere, the prestige of the university professors found its limit; their influence on public affairs was almost nil. In the imperial Germany it was unthinkable that a professor of mathematics would serve as minister of war as Paul Painlevé did in France. Such learned men in Germany as could have been great leaders not only in the world of thought but also of action—I think for instance of the economist Max Weber, whose death in 1920 was a great loss to the young German Republic—found themselves barred from political activity before 1918.

Though the imperial Germany was certainly not a democratic country, admission to the institutions of higher learning was more liberal and democratic in Germany than in France or England. No tuition had to be paid by the student in the universities and the fees were moderate. The work student, however, who paid his own way in whole or part by working during the vacations, made his appearance only after the end of the first World War. In my time, a student was supposed to devote his vacations to home study, and manual labor was considered as not befitting his social rank. There were some but not many who earned money by private tutoring. These conditions changed fundamentally after 1918. On the other side, the universities were definitely meant for an intellectual élite, not for the masses. There are about twenty universities in Germany. The Nazis limited the number of university students admitted each year in the whole Reich to 15,000; before 1933 the number of yearly admissions had risen slightly over 20,000. These figures should be held against a total population of sixty to seventy millions.

I shall now start on a little more systematic description of the German university system, under three headings: (1) union of teaching and research; (2) Lehr-and Lern-freiheit; that is, freedom of learning and teaching; (3) autonomous corporation of scholars versus state institution.

Unity of teaching and research was one of the fundamental characteristics of the modern German university. The normal state in Germany was that every university professor was a scholar engaged in independent research, and vice versa, that all scholars and scientists of significance were university professors. Therefore *the great men of science were at the same time the actual day by day teachers of the academic youth.* This was not the case in England where in the 19th century teaching was mostly left in the hands of fellows and tutors, while the scientific life of the country stood but in loose relationship to the ancient seats of learning, Oxford and Cambridge. Nor was this the case, or only to a much lesser degree, in France where, among many other historical institutions, the old universities had been destroyed by the great Revolution. What took their place in the Napoleonic era was no longer rooted in the tradition of the mediaeval universities, which both England and Germany have carried on to this day. With a little exaggeration one could say: The leading minds in England and France stood outside, in Germany inside, the universities. Accordingly here the universities exerted a wider and more vital influence upon the nation's life. If union of research and teaching seems a less distinctive feature of the German universities today than a few decades ago, the reason is simply that the developments in other countries have tended in their direction. The graduate school in America is an example.

It is impossible to understand the German university apart from its history. Its tradition, as that of most European universities, is derived from the University of Paris as it came into being in the last quarter of the 12th century. The privileges of a mediaeval university were established by a papal bull; the Church and the Church alone was in charge of all education. Only later the Emperor or the sovereigns of the several states concurred as representatives of the Roman law under which the university operated. It was essentially a self-governing corporation with its own jurisdiction, a corporation of scholars, both professors and students, who lived together in colleges. They elected their rector and the university council for a limited period; in the earliest times the rector was not necessarily a professor but could also be a student. The teachers were organized in four faculties,—theology, law, medicine and the facultas artium ; the student body in nations. Teaching consisted in lectures and disputations. The facultas artium had a propaedeutic character : here the student was, during three or four years, instructed in the basic sciences, logic, physics, mathematics including astronomy, psychology, ethics and politics,—according to Aristotle, Euclid, Ptolemy. He acquired the degrees of Baccalaureus and Magister Artium. From there he passed on to one of the three more professional faculties. The teachings and degrees of these mediaeval universities were recognized throughout the Christian world without any national boundaries. They were entirely dedicated to the tradition of a *fixed body* of beliefs, knowledge and thoughts about God and the world : *truth* is considered as *given once for all* and passed on by teaching to the next generation.

The modern German university has preserved quite a number of features of its mediaeval ancestor, in particular the division into the four faculties. But the last, the facultas artium, which is now called the philosophical faculty, completely changed its role and significance in the first half of the 18th century. Reformation and that hideous principle emerging from the religious wars, *hujus regio ejus religio*, meaning the religion of the sovereign determines the religion of his subjects, deprived the universities of their universality; they became territorial and, still being ecclesiastic institutions, therefore, denominational in character. The blight of narrow denominational orthodoxy settled down on them. But the end result of the process of secularization in the 17th and 18th centuries is that the power of the confessional church is taken over by the state or sovereign. Ever since, the German universities have been state institutions. But it is not this political development with which we are concerned at the moment. A spiritual development indigenous to Germany set in at the beginning of the 18th century, and from it the typical German union of teaching and research has sprung. Under the influence of the philosophy of Leibniz and Christian Wolff and under the personal influence of Wolff himself, the philosophical faculty of the Prussian University of Halle became a center of free research in the physical sciences, in mathematics, in the humanities, history and philosophy. While the other three faculties continued to be, in the first line, professional schools for clergymen, physicians, judges and barristers, the philosophical faculty ceases to be their servant. From a subservient, it passes to a directing role, as the home and source of that basic scientific research and knowledge on which the professions depend. Carrying on of independent research and training for such research is now its main task. Soon afterwards, George II of England as the ruler of Hanover, founded the Georgia Augusta, the University of Göttingen, which from the beginning followed the same course as its rival Halle. By degrees the spirit of independent research spread from the philosophical to the other faculties. Truth is no longer given but something to be *sought*; and the university teacher, instead of teaching old wisdom from textbooks, has to educate and train his students in the art of discovering new truth. It is the glory of the German universities, especially Halle and Göttingen, that they initiated this movement and first proclaimed the principle of the *libertas philosophandi*, the freedom of research, teaching and learning. With the Wolffian revolution, textbooks began to disappear from the German universities ; after more than 200 years they are still one of the worst impediments to progress in our American institutions. Lectures remain as the main form of instruction but the disputation is replaced by the seminar, the purpose of which is the introduction of the student into, and his training for, research. In the seminar the student is an active partner and there is personal contact, give-and-take, between teacher and student.

Through the reception of the Wolffian rationalistic philosophy, the German universities gained their ruling position in the intellectual life of the nation. In later times, the Kantian philosophy, and finally the Romantics from Fichte to Hegel, played a similar role. In France and England on the other hand, the universities proved incapable of absorbing the contemporaneous philosophy and remained on the confessional standpoint. This philosophic penetration has modified German religiosity : such a clash between conventional observance of religion and crude, purely negative atheism as we have in this country was unknown there (at least among the intellectuals). Let me add at once, neither a chapel nor a stadium is part of a German university.

What Halle and Göttingen began was completed by the university in Berlin founded in 1810 under the eyes of Napoleon's occupational armies. It was a time of utter political impotence for Germany, but also a time of high achievements in the cultural sphere, in literature and philosophy; it suffices to mention the names of Kant, Goethe, Schiller. The man who drafted the blueprint of the new university was Wilhelm von Humboldt. He and his brother Alexander were great scholars in their own fields and at the same time far-sighted statesmen of wide influence, liberal-minded men of high character, men of the Jeffersonian type, who happened to belong to the ruling class. At their side in Berlin stood the philosophers, Fichte, Schleiermacher and Hegel. " On the founding of the University of Berlin ", says Dr. Flexner, " new wine was poured into old bottles...Never before or since have ancient institutions been so completely remodeled to accord with an idea ". Let me quote two sentences from Humboldt's memorandum. He wants to preserve unity of general and professional education as well as that of teaching and scientific research. "Science ", he says, " is the fundamental thing ; for when she is pure she will be adequately and sincerely pursued, notwithstanding exceptional aberrations. *Solitude and freedom* are the principles prevailing in her realm ". " Solitude and freedom",—I like that. And then when he advocates establishing scientific research firmly in the universities instead of leaving it to the academies, he remarks : " The unhampered oral lecture before an audience, in which there will always be a considerable number of independent thinkers, is certainly apt to enthuse one as much as the lonely desert of authorship or the loose connection of an academic society ". The University of Berlin set the example first for the other Prussian, then for all German universities. I cannot help hoping that far-sighted men of Humboldt's stature will again arise in Germany after her present downfall and that the Allies, as Napoleon did, will let them start their work of regeneration.

In summarizing, I may say the German university does four closely connected things : (1) it provides general scientific instruction, passing on to the young generation the cultural and intellectual inheritance in its most mature and sublime form; (2) it provides professional training for the clergy, for the judges and lawyers, the physicians, the secondary school teachers and the higher branches of the civil service (in particular, the philosophical faculty trains the secondary school teachers); (3) it conducts research and (4) provides training for independent research. The functions (3) and (4) are considered the most important, above all in the philosophical faculty. The German university professor looks upon himself in the first line as a scientific investigator. His talent for research makes his reputation. Says Paulsen : " The philologist, the historian, the mathematician, the physicist, proceeds in his class as if his audience consisted exclusively of scholars and professors; he ignores, as it were, the fact that in reality the great majority of them are destined for a practical profession, to become school teachers, or rather he does not ignore it but he is convinced that the teacher's greatest asset is a real scholarly education." So it was indeed : for all the advanced mathematics which the student of mathematics learned at the university he would have little use as a teacher in a secondary school. Hence others, as for instance Carl Heinrich Becker, who was Prussian Minister of Education during the Republic, criticized the system.

He thought that a wrong professional ideal was implanted in the students of the philosophical faculty by this kind of advanced instruction, which tended to make them unhappy in their future profession.

Hart says in the book mentioned before: "To the German mind the idea of a university implies an object and two conditions; the object is *Wissenschaft*, knowledge in the most exalted sense of that term, namely, the ardent, methodical, independent search after truth in any and all of its forms but wholly irrespective of utilitarian application. The conditions are Lehrfreiheit and Lernfreiheit. Lehrfreiheit means that the one who teaches is free to teach what he chooses as he chooses. Lernfreiheit denotes the emancipation of the student from all compulsion and compulsory drill, recitations, quizzes, tests, etc." Dr. Flexner criticized the American colleges and universities at great length for their attempt to be of service to the people by doing many incompatible things, —so that Columbia University, for instance, would offer a course in History of Philosophy along with one on Running Privately Owned Rural Drugstores. Germans like to set up institutions for clearly circumscribed purposes. They have Hochschulen for engineering, for agriculture, for commerce, for forestry, for mining, for music, and pedagogical seminaries and academies; but the university is something different. Its object is as Hart says, theoretical knowledge on the most advanced level. Athletics has no place there. Nor have teachers' colleges or departments for hotel administration (one of the most recent additions to Cornell University). I have here a catalogue of the German universities for the winter semester, 1932-1933. If you look up any university, for instance Göttingen, you will realize what I mean. A few auxiliary teachers, who however did not belong to the academic staff, gave lessons in Greek for beginners, in drawing, the practice of music, shorthand and fencing. But the numerous courses given by the members of the staff have such titles as: Theology of the New Testament, Roman civil law, general pathology, history of the Mediterranean countries until Pompey, Hellenistic philosophy, theory of electricity and magnetism, colloidal chemistry, partial differential equations. In mathematics the nucleus consisted of a number of systematic courses on an advanced level, each covering a wide field; they formed the backbone of the education of a mathematician. Such courses could not be offered in an American college because they are too advanced; and there is little room for them in the curriculum of our graduate students because they specialize too thoroughly on some subject on which they plan to do their research work. There is much to be said for the German method.

Three forms of instruction were commonly employed : lectures to large groups, practical exercises in the classroom or laboratory in which assistants cooperate, and seminars for the training in research. Lectures in the main subjects were attended by large audiences. A professor of mathematics in Berlin would have four to five hundred students in his class on advanced algebra. This great number of hearers made the system of lectures highly economical. But the professors also attached to themselves smaller groups of devoted students thus forming schools, of disciples, who, one by one, carried new ideas into old chairs. On the whole the German professor was much more accessible to his students and to young scholars than his French counterpart. Many of the leading American mathematicians of my generation passed through the school of David Hilbert in Göttingen. They all remember their Göttingen days with enthusiasm.

I will now say a few more words about the German student's *Lernfreiheit*. Once he has matriculated in the university he chooses his own course. For the following years he will be subjected to no quizzes or oral or written examinations. He selects his own teachers and the lectures, exercises, seminars which he wants to attend. Usually he changes his university one or two times. What attracts him to a special university is often the fame of a great teacher or scientist under whom he wants to study. He may take advice from his professors or neglect it at his own peril. The university exerts no control over his private life. There is no discipline whatsoever. If he does not like a particular professor he can hear another; he has often the choice between several teachers in the same field. If he does not like a particular university he can go to another. If he does not feel disposed to attend on a particular day he can stay away; nobody bothers. He does not live in dormitories on a campus but rents a room in the university town and prepares his meals at home or goes to a restaurant. There is nothing resembling the mediaeval Bursen or the English and American colleges. He pays no tuition, but a moderate fee for each course he takes; that is the only external stimulus for not cutting the course: of his own free will he chose to attend and paid for it. He has two or three or four weeks' time to make up his mind before he registers and pays his fees; during this time he can *hospitieren*, that is he drops into this or that lecture to find out for himself how he likes the lecturer—and will not return and will not register for it if he does not like it.

This seemingly boundless freedom of the student is practically abridged by the necessity of passing a *state examination* at the end of his university career, provided he wants to enter one of the academic professions or the higher branches of civil service. It is the *state* which examines the future physicians, judges, barristers, school teachers, pastors and public servants, or at least the state exerts the supreme control over these examinations. For instance, the board of examiners for secondary school teachers, appointed by the state, would consist of university professors and schoolmen; the examination would usually be held at the seat of one of the universities. Examination requirements were set down in general terms; the examination would cover certain fields, not particular courses given by the examiner at the university. One of the requirements is that the canditate should have studied for a definite period, three or four years, at a university. It is clear that this requirement implies a privilege for the universities, probably more important than all their other privileges together; but the examinations as such are not the university's business.

While the German student in the universities enjoyed this complete freedom which made him responsible only to himself, the secondary schools which led up to the universities followed a fixed plan of study rather than the elective system, and they adhered to a rather strict school discipline. Their organization was based on the conviction that a national culture, which the school is supposed to transmit to the young minds, is an integrated whole that would be destroyed if the pupil were left free to pick the raisins out of the cake according to his own fancy. But he (or his parents) could choose between three school types : the Gymnasium, with its main emphasis on the classics, Latin and Greek ; the Oberrealschule, with its main emphasis on science and modern languages ; and a middle type, the Realgymnasium. Passing of the final examination in one of these three schools gave the student a right to matriculate in a university. After what has been said, the transition from school to university was a very abrupt change, a change from strict discipline and guidance to complete freedom and self-responsibility. No doubt it carried with it great dangers for the youth.

Indeed the dangers involved in the whole system of German university education are so obvious as to need no particular emphasis. May be too many of the average students went by the board or profited too little ; may be the prevailing standard was too high and some adjustment should have been made to the widening circle of people who sought and gained admission to the universities. From 1840 to 1940 the number of students increased five times, while the population did not more than double. But the system worked extremely well for the able ; and as far as the service to science and research is concerned, it had one great advantage, namely this : *The men who actually were destined to carry on research could be chosen from a great reservoir of young men* who while they studied had been in touch with the spirit of research and thus been given an opportunity to show their mettle. I think this is one of the main reasons for the success and high quality of scientific research in Germany.

The Lernfreiheit of the student is paralleled by the *Lehrfreiheit* of the professor. To quote Dr. Flexner once more: "The German university teacher pursues his own course, unhindered. He is perfectly free in the choice of topics, in the manner of presentation, in the formation of his seminar, in his way of life. Neither the faculty nor the ministry supervises him : he has the dignity that surrounds a man who, holding an intellectual post, is under no one's order." Or Hart: "The university is a law unto itself, each professor is a law unto himself, each student revolves on his own axis and at his own rate of speed." My appointment in Göttingen merely stated that I was obliged to give one private course of lectures every semester and one public course of lectures every two years. So I could have gotten away with one or two lectures weekly. I think six to nine hours weekly of lectures and exercises was considered normal. To make sure that a reasonably coordinated program could be offered to our students of mathematics and physics, the teachers in these fields at the University of Göttingen used to hold an informal meeting where each teacher in the order of seniority (that is the number of years spent in the university's service) announced his plans for the next semester, and the desirable adjustments were made by informal discussion. Differences of rank or age played no role in this meeting, no dean presided over it or influenced the agreements thus reached.

In this context the institution of the *Privatdozent* deserves mention. You begin your university career in Germany as Privatdozent as you would as instructor in this country. On the ground of an examination by the faculty, the Privatdozent receives the Venia legendi, the right to lecture. Thus a state examination was the entrance to all academic professions in Germany *with one notable exception* : the keys to the career of a university teacher were in the possession of the university itself, subject to no supervision by the state or administrative officers outside the university's teaching staff. The Privatdozent, unlike the professor, is therefore not an appointee of the state and receives no salary. The income from his teaching consists merely of the fees of the students who attend his courses. He has the right to lecture but no obligations whatsoever. Therefore, he can devote his whole time and energy to research and to giving lectures, perhaps two or three per week, on such topics as are of special interest to him. Since these topics are often too advanced for all examination purposes, they will, as a rule, attract only a limited number of students but students who have themselves a deep and genuine interest in the special subject. In the mediaeval universities you acquired the right to teach when you became Magister; the Magister has sometimes even the duty to teach in the facultas artium without receiving compensation. These Magisters, who usually continued to be students in one of the three higher faculties, were in a different class from the professors. They are the direct ancestors of the German Privatdozenten.

You see the positive side of this institution: in his younger years a scholar has to carry only the slightest burden of teaching. He is absolutely his own master. He can grow scientifically, develop his ideas, and within a small circle of students attracted by his subject and his personality learn the art of teaching. The Privatdozent represents, as Dr. Flexner says, the "sheerest and purest form of the academic type—a highly honorable introduction to a highly honoured career." Hart calls him the "life blood of the institution". The seamy side of the Privatdozentur is also evident: economic insecurity. At the time of your habilitation you did not know whether you would remain Privatdozent for two or for fifteen years. Some solved their economic problems by marrying into wealthy families, a solution favored by the high social prestige of the university teacher. Other less objectionable correctives were provided. In Göttingen the full professors saw to it that the Privatdozent now and then gave one of the big courses, say in calculus, from which in one year he would earn an income lasting him for the next three years. Sometimes, and much more frequently in the days of the Republic when fortunes had been wiped out by inflation, the Privatdozentur was combined with an assistantship or a salaried

Lehrauftrag (teaching assignment) for some special field. But, of course, all such compromises reduced the freedom.

After discussing the union of research and teaching, then the Lehr-and Lern-freiheit, I now come to a third point : to what extent were the German universities autonomous? The university elected its own rector and council from among its professors, usually for a term of one year. I think it was not without importance that in all social respects, in its ceremonies and celebrations, in the entertainment of illustrious guests, etc. the university was represented by the rector, one of its own scholars, not by a president or an official of the state. The faculties elected the deans from their own members likewise for a period of from one to two years. The scholars themselves, organized in faculties, conducted the major part of the university's affairs without interference from administrative officers. I think I have never seen in operation a better functioning and more democratic administrative body than was the mathematisch-naturwissenschaftliche Facultät in Göttingen of which I was a member from 1930-1933. The faculty also conferred the accademic degrees, the doctorate and, more important, the Venia legendi for Privatdozent.

On the other hand, the universities were state-endowed institutions; the ministry of education administered its funds, played a decisive role in the creation of new chairs, and had the final voice in the appointment of new professors for already existing chairs. The procedure in the latter case is as follows: The faculty suggests three candidates, primo, secundo and tertio loco. The minister is not bound to these suggestions; they are sometimes disregarded, and sometimes the minister would return the nominations to the faculty with his objections and ask for new recommendations. But as a rule he appointed one of the faculty nominees and most frequently the one proposed primo loco. The only statistics which I could find about this tell that between 1882 and 1902, 125 appointments in the faculty of law were made upon recommendation and 15 without or contrary to it. Dr. Flexner has this to say about the relation of university and government in Germany: "Taking matters as they stood in 1914, the Germans with their state monopoly, in which the university is *a legal partner on equal terms,* did better than either English or American organizations in their respective countries; their universities were more highly developed, more nearly autonomous, far more highly respected and exerted wider influence." Humboldt was of the opinion that the state has no other duty towards the university than to supply the necessary means and cooperate in the selection and appointment of the right men. "The state should bear in mind," he says, "that it does not and cannot do her (the university's) work and always becomes a hindrance when it interferes." Up to the Nazis, the German states on the whole followed his advice. They had incomparably less influence on the internal affairs of the universities than the president and trustees in American institutions of higher learning.

There existed two classes of professors, the ordinary and the extraordinary, but the number of the latter was much smaller. Thus one had essentially but two categories, the Privatdozenten and the professors. Promotion within a university played a very minor role and hence a teacher did not expect such promotion. A Privatdozent in Göttingen would expect to be offered at some time a professorship say in Heidelberg, a professor in Heidelberg expected to be offered the same position in some other university of greater importance in his field. Whether their hopes were fulfilled depended almost exclusively on their scientific reputation among their colleagues in all German-speaking universities. The result is a nationwide competition: each university strives to attract the best scholars which it can hope to win in order to raise its own standard and its attraction for the students. Even the Austrian and Swiss universities, not different in organization from the German ones, participate in this universal exchange. Take my own example: Privatdozent in Göttingen for three and a half years, then Professor in Zurich, from there calls to Karlsruhe, Breslau, Göttingen, Berlin,

Amsterdam, Leipzig, all of which I declined, and finally Göttingen again.

The Privatdozent has only the fees of his students to live on. The income of the professor is two-fold : he receives a fixed salary from the state, and also the fees or a certain part of them paid by the students for such courses of his as they attend. At least this is the basic scheme of which several modifications were in existence. His appointment is for life, and even after retirement, when he is released from his teaching duties, he continues to receive his full salary. The state had no legal power to deprive him of his professorship or his salary, to transfer or dismiss him once he had been appointed. These regulations reflect the high prestige accorded to the university professors.

Comparing it with the mediaeval institutions we may now characterize the modern German university as follows : it has retained the old division into four faculties yet completely changed the character of the philosophical faculty; it has retained much of its autonomy as a privileged corporation in spite of its having become a state institution ; but in contrast to England, it has abandoned entirely the living together of students, or students and teachers, in colleges. Since 1810 it has been firmly based on the union of teaching and research. Philosophical thought had an unusally large share in this development.

The picture drawn here may impress you as pretty rosy. I admit that I have described, so to speak, the ideal German university ; the usual human shortcomings were missing here as little as anywhere else. World opinion, as long as it had not been antagonized by Germany's disastrous power politics, fairly unanimously supported the positive view here taken. I could adduce a long list of quotations from competent Americans and Englishmen in praise of the German universities, from George Bancroft who wrote to President Kirkland of Harvard in 1820 : " No government knows so well how to create universities and high schools as the Prussian ";

to Dr. Flexner's verdict in 1930: "As a well thought out institution for the doing of certain definite and difficult things the German university was a better piece of mechanism than any other nation has as yet created." Among the Germans themselves, especially those in responsible positions, one could hear more critical voices. It was widely admitted that student life in England and America was more wholesome. It is notable that, whereas Flexner favors changes of our American institutions in the direction of the German type, the Prussian Minister of Education, Becker, in a pamphlet written in 1919 suggested radical reforms of the German universities that would have made them more than half American.

In my mind there is no doubt that the German university organization as it was molded by Humboldt was one of the factors responsible for the eminence of *German scholarship and science* that once had the respect of the entire world. Vansittart in his well-known book *Lessons of my Life* devotes a chapter, "The probing of a bubble," to the thesis that the significance of German science has been vastly exaggerated. His arguments, as I could show you by more than a few instances, are often slipshod, not to say dishonest. The Frenchman Ferdinand Lot, in his book *L'enseignement supérieure en France,* says in 1892: "The scientific hegemony of Germany in all fields without exception is today recognized by all peoples. German superiority in science is the counterpart of England's superiority in trade and on the seas. Perhaps it is relatively even greater." Amusing is the Spanish philosopher Ortega y Gasset's comment in "Mission of the University" (1930, English translation 1944): "It happens, at once luckily and unluckily, that the nation which stands gloriously and indisputably in the van of science is Germany. The German, in addition to his prodigious talent and inclination for science, has a congenital weakness which it would be extremely hard to extirpate: he is, a nativitate, pedantic and impervious of mind."

These statements may go too far in the opposite direction; but Lot and Ortega probably know better than Vansittart.

Let us look at a few items of the record. The Germans are very strong in the Geisteswissenschaften as they call them, the intellectual sciences and humanities. One must go back to the Athenian Thucydides to find a historian on a level with *Ranke*, and *Mommsen* is not far behind. The critical methods of philology, above all in the classical field, were developed almost exclusively by the Germans *F. A. Wolf*, *Boekh*, *Lachmann*, the brothers *Grimm*, *Bopp* (the founder of comparative linguistics), and so on. Germany has produced a number of first rate philosophers; I mention only Meister Eckhart, Leibniz, Kant, Hegel. In the second half of the 19th century the Germans were leading in *medicine*, up to the first World War in *chemistry*.

But I had better limit myself to my own field, mathematics, and the neighboring science of physics. In mathematics the man acclaimed by universal consensus as the princeps mathematicorum is *Carl Friedrich Gauss*, who was active in Göttingen during the first half of the 19th century. But I hasten to add that the over-all picture in mathematics is different. There has never been such an outburst of mathematical activity in Germany as there was in the France of the late 18th and early 19th centuries. The prevailing opinion among us German students of mathematics around 1908 was that France and Germany were the two leading nations in our field, but that France was a little ahead of us. She fell behind after 1918, recovered later. Meanwhile, of course, America has come to the fore, and also Russia has made great strides. There is, however, one branch of mathematics, so to speak its inner sanctum, the theory of numbers, in which the Germans excelled all others. The same is true for the foundations of mathematics, which are so closely tied up with the problem of the infinite. The leading mathematicians in this development, which started about 1870, are *Dedekind*, *Cantor*, *Brouwer*, *Hilbert*, *Gödel*—all Germans with the exception of the Dutch Brouwer.

As to physics, Germany cannot equal the British galaxy of *Newton*, *Faraday*, *Maxwell*, *Lord Rutherford*. *Helmholtz* may be put on the same level as *Lord Kelvin*. But again the Germans excelled, especially in the 20th century, in one important direction, theoretical physics. There is *Einstein*. His case shows how silly it is to classify men by their nationality; but Einstein was born in Germany, and after going through the early stages of his scientific career in Switzerland was a member of the Berlin Academy for nearly twenty years. German physicists had a large share in that recent development of atomic physics that goes by the name of quantum mechanics: beginning with *Max Planck's* introduction of the universal quantum of action in 1900 and leading up to a radical modification of classical physics in the theoretical work of *Heisenberg* and *Schrödinger* (1925-26). Besides quantum mechanics and the theory of relativity, which overshadow everything else, I may mention two other outstanding events in physics of which I was a witness. The X-rays were, as you know, discovered by a German, *Röntgen*; and it was *Max von Laue* who first used them to disclose the inner atomic structure of crystals. His method, of paramount importance for crystallography, is also of great technical importance for the investigation and testing of all sorts of materials, metals, fibres, etc. *Ludwig Prandtl* in Göttingen became the father of modern fluid dynamics, when by his theory of the boundary layer he made us understand what causes the resistance of a solid immersed in a current of air or water.

But of course what the Germans achieved in mathematics and physics, they achieved by collaboration and exchange of ideas with scientists of all nationalities. Except in times of war, this give-and-take knows no national boundaries. It is silly to try to separate the several strands, and utterly absurd to talk of German mathematics or German physics, as some of the Nazi fanatics did. Nothing indeed could be more international than mathematics and the natural sciences. A mathematical concept that is clear, a proposition that is true, a theory that is coherent, for an American mathematician is so for a Chinese or a Hindu mathematician, and *vice versa*. That is a trite observation. There are individual differences of style, but no difference in substance, and none along national or racial lines.

But let us return to institutions !

The universities were not the only seats of learning and research in Germany. Besides them there had arisen during the 19th century ten *Technische Hochschulen*, not too different in character from our Institutes of Technology. It is more or less an historical accident that engineering was not added as a fifth faculty to the universities but that separate institutions were founded for it. For after all, engineering stands in very much the same relation to physics as medicine stands to biology. I think that the influence of Napoleon's Ecole Polytechnique was the essential factor in deciding the issue. The Swiss federal Technische Hochschule in Zürich was an adaptation of the Ecole Polytechnique to a country whose main universities were organized along the German lines, and thus the pattern of the Zürich institute could easily be taken over by the Germans. Thus I should describe the German Technische Hochschule as a cross breed of the Ecole Polytechnique and the German university.

Then there are the Academies. Academies, learned societies, first sprung up in Italy during the renaissance. The Academia dei Lincei in Rome, the Académie française and the Académie des Sciences in Paris as well as the London Royal Society, all of which still flourish, came into being in the 17th century. Germany is comparatively late (aftermath of the Thirty Years War !). The Berlin Academy was founded in 1700 by Leibniz. The other German Academies, in Göttingen, Leipzig, Heidelberg, Munich, as well as the academies of Turin, Stockholm and Leningrad, are also, at least indirectly, his offspring. At a time when free research had not yet entered the somewhat decrepit universities, when there existed no scientific journals, and ideas were stimulated and spread mainly by correspondence between the few leading European savants, Leibniz planned to span Europe by a net of academies, centers of research which he expected to become a strong combine for the promotion of enlightenment and peace among nations. He dreamt of a union of European nations based on a reunited Christian Church and the common concern of all people for truth. —Today the German academies grow, as it were, in the shadow of the universities and play only a subsidiary role. They still have —like their sister institutions in the other European countries—a definite, indispensable task in the quick publication of the results of research, and engage in such teamwork enterprises as for instance the Thesaurus linguae latinae or the Mathematical Encyclopedia. As there is nothing typically German about them, this brief mention may be sufficient.

A new development for research started in 1911 with the foundation by Emperor William II of the *Kaiser Wilhelm Gesellschaft* for the advancement of science, the purpose of which is the organization of independent research institutes. The Kaiser coaxed the leading men of German industry into giving and underwriting vast sums of money for the promotion of scientific research. The idea of national competition in the field of science and the aim of strengthening Germany in that competition were pretty evident. The society was founded as a private organization ; only later, during the Republic, when the resources of the society had dwindled, the state began to participate. The sound principle was followed that each institute was built around a man, an investigator of great and proven ability. The first president of the Kaiser Wilhelm Gesellschaft, who wrote its charter, was queerly enough a theologian, Harnack. He was succeeded by the theoretical physicist Planck. Both were learned men of great distinction, not administrators. In a sense these institutes marked a breach in the old principle of union of research and teaching ; in some quarters they were viewed as a symbol for the triumph of efficiency and specialization over culture. There were Kaiser Wilhelm Institutes for physics, for fluid mechanics, for chemistry, for metal, coal and fiber research, for biology, anthropology, for the study of the brain, tuberculosis, psychiatry, international law and public law of foreign countries, etc., etc. The discovery of the fission of the uranium atom on which the atomic bomb is based was made in the Kaiser Wilhelm Institute for Chemistry. In 1933 the research output of these institutes, more than thirty in number, had become comparable to that of the universities; they may even have been on the way to over-shadowing the latter. Because they were essentially beyond the control of the state, the Nazis did not succeed in coordinating them as rapidly and as thoroughly as the universities. Lise Meitner left the Kaiser Wilhelm Institute for Chemistry only in 1938, and as far as I remember was not dismissed but left of her own free will, and fled to Sweden.

When I planned this lecture I considered contrasting my sketch of German universities and science before 1933 with an account of the fate that befell them under the gangster regime of the Nazis. But who cares any longer? The basic autonomous structure developed without abrupt changes in a history of six centuries is far more important than the ephemeral disfigurement which it suffered in the last twelve catastrophical years. I shall limit myself therefore to a few remarks.

The Nazis promulgated the theory that the *ecclesiastical* university of the Middle Ages had been followed by the *humanistic* university of the Humboldt type which in its turn had now reached a state of decay and was, under their own glorious leadership, to be replaced by the *political* university. Consequently political reliability as certified by the Gauleiters of the Party, evidence of service in Nazi organizations, physical prowess, became, in addition to intellectual ability, criteria for the admission of a student to the university. Service in labour camps and compulsory athletics were imposed on him. As they phrased it, the key test is whether the candidate would later be " capable when in high or leading positions, of doing his part in molding the political, cultural and economic life of the people. " Likewise for the teacher, the Venia legendi was merely a first condition ; he had to attend community camps and special training academies where by simple fare, physical labour and lectures on the Nazi Weltanschauung he was imbued with the heroic völkische outlook on life. In the processions which the Nazis staged on old and newly invented national holidays, the university would march like a guild in the Meistersinger in one body comprising the teachers, the students, the clerks, the officials and workmen.

The autonomous administration of the universities was recast according to the leadership principle. The details are of little interest. Every decision conferring the right to teach requires the approval of the minister. The Party is represented by inner cells in the two corporations that embrace all teachers and all students in a given university. Professors can be dismissed, pensioned and transferred, and ample use was made of this new possibility. Sober estimates put the number of university teachers dismissed for political and racial reasons (between 1933 and 1938) at 15 to 20 per cent. The Lernfreiheit is greatly abridged by official study plans. For instance the study plan for economics states : " In the first two semesters the student is to become acquainted with the racial foundations of science. Lectures on race and tribe, anthropology and pre-history, on the political development of the German people, especially in the last 100 years, belong to the beginning of every study in the humanities". An American observer in 1936 summarized his impressions as follows : " The German universities, which once bestowed their best features on their sister institutions in the U. S. A., have now taken over the worst features of these. They have become obsessed with athletics, they have succumbed to the limited ideal of vocationalism, and have subjected the curriculum to a regimented uniformity surpassing the worst evils of the textbook mania. " All this, it ought to be added, in the service of a savage ideology entirely alien to America.

The enrollment sank to 60 per cent. of its former figure. One must admit, however, that part of this reduction was due to deliberate measures, the quota system and new selective tests mainly of a political nature. But it was also due to the diminished prestige of the educated classes and of scientific work under a regime which considered scientific objectivity as prejudice, a prejudice belonging to the same age as the belief in free competition between labourers, businesses and countries.

The students, many of them ardent Nazis, had hoped for a fulfillment of their old dreams that, as in the Middle Ages, they would again become partners in the conduct of the affairs of the university, would have a definite voice in appointments, etc. During the first months of the Nazi revolution the ministry gave in to the pressure of the students. But soon this was stopped. In 1934 the student body received a constitution barring it from all such interference and assigning to them as their main task "to guarantee the indissoluble bond between university and people, as well as a generation of university graduates rooted in their people and strong in body and soul." Thus the Nazis betrayed the students as they betrayed their allies in the first cabinet under Hitler's chancellorship, the conservatives, and the aspirations of German labor. The students obeyed the Führer's order and returned to their work. In a speech at the beginning of the winter semester of 1934 the leader of the Studentenschaft of the Technische Hochschule Darmstadt admitted that the students had been on the wrong road and promised that they would again accept good scientific workmanship and efficiency as their supreme maxim. Nothing in the world is entirely bad: the Nazis destroyed the colorful student fraternities, the Corps and Burschenschaften, one of the most unpleasant features of German student life. Instead "houses of comradeship" were founded in which students lived together without dividing loyalties to separate fraternities (it seems, however, that this institution proved a dismal failure).

As the net effect of the Nazification of the universities, the standard in the intellectual sciences suffered greatly; psychology, history, sociology, economics, etc., became to a large extent instruments of Nazi propaganda. Not so in mathematics, physics, chemistry and engineering. In spite of their racial and völkische ideology, the Nazis soon realized that there is simply no substitute for scientific technology. Thus the mathematical, physical, technological journals continued to publish good papers and the output was not reduced to any considerable extent. It would have taken a longer time to make the destruction wrought by the Nazis visible in these fields. Ultimately, I have no doubt, decay would have set in. For a system founded on the equilibrium between the authority of the universities as corporations of independent thinkers and the authority of government as the guardian of law and public welfare cannot survive when the first authority is denied and destroyed by a totalitarian government recognizing neither law nor responsibility to its people.

158.

A simple example for the legitimate passage from complex numbers to numbers of an arbitrary field

Scientific papers presented to Max Born, Hafner Publishing Co Inc. New York, 75—79 (1953)

In his great book on the Foundations of Algebraic Geometry (*Am. Math. Soc. Colloquium Publ.*, vol. 29, New York, 1946) André Weil, after pointing out that "there is but one geometry of characteristic 0, to which the methods of the theory of analytic functions and topology, ..., etc., may legitimately be applied", makes the remark: "It would be very convenient to have, in addition to this, a principle of 'reduction modulo p' by which one could derive theorems in the algebraic geometry of characteristic p, for $p \neq 0$, from the corresponding results for characteristic 0" (l.c., p. 243). In an article on the future of mathematics (*Am. Math. Monthly*, 1950) he seems to hold out hope that the discovery of such a universal method is not too far-off. As a matter of fact, there exists a principle of the desired kind, of such special and trivial nature, however, that one may be tempted to push it aside as valueless. It is this:

T. Let K be the field of all complex numbers, k an arbitrary field, and $f(x_1, \ldots, x_n)$ a polynomial of n indeterminates x *having ordinary integers as coefficients.* If the equation $f(x_1, \ldots, x_n) = 0$ holds for all values of the variables x_i in K, then it also holds for arbitrary numbers x_i in k.

Indeed, under the hypothesis $f(x_1, \ldots, x_n)$ vanishes identically (*i.e.* all its coefficients vanish), and hence the equation $f(\xi_1, \ldots, \xi_n) = 0$, since it makes sense, also holds, for any $\xi_i \in k$. The field K of complex numbers may here be replaced by any field containing infinitely many elements.

I will now show by an example that this principle, trivial as it is, sometimes has useful applications. The example was suggested to me by Max Deuring's note "Eine Bemerkung über die Bürmann-Lagrangesche Reihe" (*Nachr. d. Akad. d.*

Wissensch. in Gottingen, Math.-physik. Klasse, 1946, pp. 33-35). It deals with the inversion of an analytic function

$$u = \phi(z) = c_0 z + c_1 z^2 + \ldots \qquad (c_0 \neq 0)$$

of the complex variable z in the neighbourhood of the origin $z = 0$. Since one may replace u by u/c_0 it is no restriction to assume $c_0 = 1$. Let us write

$$\phi(z) = z\eta(z), \qquad \eta(z) = 1 + c_1 z + c_2 z^2 + \ldots \quad . \quad . \quad (1)$$

In every lecture course on function theory the problem is solved on the basis of Cauchy's integral theorem as follows. Let $0 < r < r_0$ be such that $\eta(z)$ converges for $|z| < r_0$ and

$$|c_1 z + c_2 z^2 + \ldots| \leqslant \tfrac{1}{2} \qquad \text{for } |z| \leqslant r,$$

and denote by C_r the circle of radius r around the origin. Since $|\eta(\zeta)| \geqslant \frac{1}{2}$, $|\phi(\zeta)| \geqslant \frac{1}{2}r$ for $|\zeta| = r$, the integral

$$\frac{1}{2\pi i}\int \frac{\phi'(\zeta)}{\phi(\zeta) - u}\, d\zeta$$

extending over C_r like all subsequent integrals, is a continuous function of u for $|u| < \frac{1}{2}r$. Being an integer, namely the number $N(u)$ of solutions ζ of the equation $\phi(\zeta) = u$ within C_r, it must be constant $= N(0) = 1$ for $|u| < \frac{1}{2}r$. In other words, if $|u| < \frac{1}{2}r$, then the equation for z, $\phi(z) = u$, has a unique solution $z = \psi(u)$ within C_r. Let $F(z)$ be a power series $\Sigma_{\nu=0}^{\infty} A_\nu z^\nu$, convergent for $|z| < r_0$. Then one has

$$\frac{1}{2\pi i}\int \frac{F(\zeta)\phi'(\zeta)}{\phi(\zeta) - u}\, d\zeta = F(z) \qquad (|u| < \tfrac{1}{2}r)$$

if z is that solution. The left side is a power series of u, $G(u) = \Sigma_{\nu=0}^{\infty} B_\nu u^\nu$, with the coefficients

$$B_\nu = \frac{1}{2\pi i}\int \frac{F(\zeta)\phi'(\zeta)}{\eta(\zeta)^{\nu+1}} \cdot \frac{d\zeta}{\zeta^{\nu+1}};$$

hence B_ν is the coefficient of z^ν in the power series of

$$F(z) \,.\, \phi'(z)/\eta(z)^{\nu+1},$$

briefly

$$B_\nu = \mathrm{co}_\nu\{F(z) \,.\, \phi'(z)/\eta(z)^{\nu+1}\}. \quad . \quad . \quad . \quad (2)$$

For $\phi'(z)$ one may write $\eta(z) + z \,.\, \eta'(z)$.

It is sufficient to take for $F(z)$ any power z^n, where $n = 0, 1, 2, \ldots$. The resulting equation then reads

$$z^n = u^n(b_0^{(n)} + b_1^{(n)}u + \ldots) \qquad . \quad . \quad (3)$$

where

$$b_\nu^{(n)} = \text{co}_\nu\{\phi'(z)/\eta(z)^{n+\nu+1}\}; \qquad . \quad . \quad (4)$$

in particular ($n = 0$ and 1)

$$1 = \beta_0 + \beta_1 u + \ldots, \qquad . \quad . \quad . \quad (5)$$

$$\psi(u) = u(b_0 + b_1 u + \ldots) \quad . \quad . \quad . \quad (6)$$

with

$$\beta_\nu(= b_\nu^{(0)}) = \text{co}_\nu\{\phi'(z)/\eta(z)^{\nu+1}\}, \qquad . \quad . \quad (7)$$

$$b_\nu(= b_\nu^{(1)}) = \text{co}_\nu\{\phi'(z)/\eta(z)^{\nu+2}\}. \qquad . \quad . \quad (8)$$

The fact that $z = \psi(u)$ is a solution z of $\phi(z) = u$ is expressed by the equation

$$\phi(\psi(u)) = u \qquad (\text{for } |u| < \tfrac{1}{2}r). \quad . \quad . \quad (9)$$

On the other hand, let z be any value of modulus less than $\frac{1}{3}r$. Put $u = \phi(z)$. Since then $|u| \leqslant \frac{3}{2}|z| < \frac{1}{2}r$, the equation $\phi(\zeta) = u = \phi(z)$ has the only solution $\zeta = z$ within the circle C_r. Hence the relations $F(z) = \sum_{\nu=0}^{\infty} B_\nu u^\nu$ and (3) hold for $|z| < \frac{1}{3}r$ after $\phi(z)$ has been substituted for u; in particular for $n = 1$:

$$\psi(\phi(z)) = z \qquad (|z| < \tfrac{1}{3}r). \quad . \quad . \quad (10)$$

Having thus repeated a familiar and "classical" performance, let us now ignore features of convergence and look upon our equations as relations for formal power series. Considering $c_1, c_2, \ldots$ as indeterminates, set

$$\eta(z)^n = \sum_{\nu=0}^{\infty} P_\nu^{(n)}(c_1, \ldots, c_\nu) z^\nu.$$

Here $P_\nu^{(n)}(c_1, \ldots, c_\nu)$ is a polynomial of $c_1, \ldots, c_\nu$ with integral coefficients. Similarly for the quantities $b_\nu^{(n)} = f_\nu^{(n)}(c_1, \ldots, c_\nu)$ defined by (4). Comparison of the coefficients of the same power $z^{n+\nu}$ on both sides of Lagrange's series (3),

$$z^n = \sum_{i=0}^{\infty} b_i^{(n)} \phi(z)^{n+i}$$

results in the relations

$$\sum_{i+j=\nu}^{\infty} f_i^{(n)}(c_1 \ldots c_i) P_j^{(n+i)}(c_1 \ldots c_j) = \begin{cases} 1 \text{ for } \nu = 0 \\ 0 \text{ for } \nu \geqslant 1. \end{cases}$$

The left side is a polynomial $g_\nu(c_1 \ldots c_\nu)$ of $c_1, \ldots, c_\nu$ with integral coefficients. The equation for $\nu = 0$, $g_0 = 1$, is trivial. Application of the proof to

$$\phi_m(z) = z(1+c_1z+\ldots+c_mz^m)$$

instead of $\phi(z)$ yields the equations

$$g_\nu(c_1 \ldots c_\nu) = 0 \qquad \text{for } 1 \leqslant \nu \leqslant m$$

and any complex numbers $c_1, c_2, \ldots$. Our principle then carries them over to any number field k. By treating (5) and (9) in the same fashion one, therefore, obtains the following

THEOREM: Let

$$\phi(z) = z \,.\, \eta(z) = z \,.\, \Sigma_{\nu=0}^{\infty} c_\nu z^\nu \qquad (c_0 = 1),$$

$$F(z) = \Sigma_{n=0}^{\infty} A_n z^n$$

be power series with coefficients in k. Then Lagrange's equation

$$F(z) = \Sigma_{\nu=0}^{\infty} B_\nu(\phi(z))^\nu \quad . \quad . \quad . \quad (11)$$

with the coefficients (2) holds, in the sense that the coefficients of the same power of z on both sides coincide. In a similar sense the power series $\psi(u)$, (6), with the coefficients (8), satisfies the equation (9), while the numbers β_ν defined by (7) have the values

$$\beta_0 = 1,\ \beta_1 = \beta_2 = \ldots = 0. \quad . \quad (12)$$

On purpose I have based this explicit, not to say pedantic, exposition of Lagrange's theorem and its transfer to arbitrary number fields on the classical function-theoretic argument. More algebraically one could have proceeded as follows. Set $\eta(z)^{-\nu} = \mathrm{H}(z)$. With Deuring one observes that $\nu\beta_\nu$ is the coefficient of z_ν in the expansion of $\nu\mathrm{H}(z) - z\,.\,\mathrm{H}'(z)$, a coefficient that vanishes whatever the power series $\mathrm{H}(z)$. By an ingenious device Deuring overcomes the difficulty of inferring $\beta_\nu = 0$ from $\nu\beta_\nu = 0$ in case of a field of prime characteristic p and exponents ν that are multiples of p. Instead I would simply point out that β_ν is a polynomial $f_\nu^{(0)}(c_1 \ldots c_\nu)$ of the indeterminates $c_1, \ldots, c_\nu$ with ordinary integers as coefficients and thus pass from $\nu\,.\,f_\nu^{(0)} = 0$ to $f_\nu^{(0)} = 0$ $(\nu > 1)$, first for indeterminate c_i and then for arbitrary numbers c_i in the field k. [The equation $\beta_0 = 1$ is trivial.]

Operating in k one can, for a given power series $F(z)$, determine by recursion the coefficients B_ν of a power series $G(u) = \Sigma B_\nu u^\nu$ such that $F(z) = G(\phi(z))$. Since $F(z)$ thus may be written as a linear combination of powers $\phi(z)^n$ it is sufficient to verify Lagrange's equation for $F(z) = \phi(z)^n$. As one easily sees, it then boils down to the relation

$$\phi(z)^n = \Sigma_{\nu=0}^{\infty} \beta_\nu \, . \, \phi(z)^{n+\nu}$$

and is thus a consequence of (12). In particular for $F(z) = z$:

$$\psi(\phi(z)) = z. \qquad \text{. . . . (10)}$$

Finally one ascertains by recursive construction the coefficients b_ν^* of a power series $\psi^*(u) = u \, . \, \Sigma_{\nu=0}^{\infty} b_\nu^* u^\nu$ such that $\phi(\psi^*(u)) = u$. This implies $\psi(\phi(\psi^*(u))) = \psi(u)$, and since (10) reduces the left side to $\psi^*(u)$, one finds $\psi^*(u) = \psi(u)$ and thus also

$$\phi(\psi(u)) = u. \qquad \text{. . . (9)}$$

The disadvantage of this procedure is that one first determines the coefficients B_ν, b_ν^* in k by recursion and only afterwards verifies that they are given by (2) and (8) respectively.

I am afraid André Weil had in mind a much profounder principle than the triviality **T**. Nevertheless it may be worth looking for other applications of it, perhaps even in algebraic geometry.

159.

Über die kombinatorische und kontinuumsmäßige Definition der Überschneidungszahl zweier geschlossener Kurven auf einer Fläche

Zeitschrift für angewandte Mathematik und Physik 4, 471—492 (1953)

Auf einer zweiseitigen Fläche bezeichnet man als Charakteristik zweier geschlossener Wege α, β die algebraische Summe der Überkreuzungen von α über β, wobei in dieser Summe eine Überkreuzung von links nach rechts mit $+1$, von rechts nach links mit -1 in Ansatz gebracht wird. Die Verkehrsregel des «Vorfahrrechts» beruht auf der Tatsache, dass eine Überkreuzung von α über β von links nach rechts zugleich eine Überkreuzung von β über α von rechts nach links ist. Dies besagt, dass die Charakteristik $\mathrm{ch}(\alpha, \beta)$ schiefsymmetrisch ist:

$$\mathrm{ch}(\beta, \alpha) = -\mathrm{ch}(\alpha, \beta)\,. \tag{1}$$

Wenn man die Möglichkeiten bedenkt, die der allgemeine Begriff der stetigen Kurve offen lässt, so ist es klar, dass die Definition in der obigen Form unbrauchbar ist. Die kombinatorische Topologie ergreift den Ausweg, dass sie sich auf eine *Triangulation* der Fläche stützt, statt beliebiger Wege zunächst nur *Kantenzüge* ins Auge fasst und von da aus durch eine Art Approximation die allgemeine Situation zu meistern sucht. Ihr darin folgend, werden wir zu einer kombinatorischen Definition der Charakteristik für beliebige geschlossene Wege gelangen. Freilich erscheint der Begriff dadurch an eine Triangulation gebunden. Um zu zeigen, dass er in Wahrheit davon unabhängig ist, werde ich dann eine andere kontinuumsmässige Erklärung der Charakteristik aufstellen und zeigen, dass sie mit der kombinatorischen übereinstimmt. Das Verfahren ist eng mit demjenigen verwandt, das ich in meinem 1913 bei Teubner erschienenen Buch *Die Idee der Riemannschen Fläche* zur Einführung des Zusammenhangsgrades mit Hilfe der Integralfunktion benutzte. So wie die Integralfunktionen aus einer «Topologisierung» der Integrale analytischer Funktionen entspringen, ist die hier gegebene Kontinuumsdefinition der Charakteristik durch Topologisierung derjenigen Konstruktion entstanden, mit Hilfe deren ich an dem angegebenen Ort die abelschen Integrale erster Gattung gewann. Zur näheren Ausführung schreitend, muss ich zunächst an die wichtigsten Grundbegriffe der Topologie erinnern.

---

[1]) Institute for Advanced Study.

Fläche

Eine *Mannigfaltigkeit* besteht aus Elementen, die Punkte genannt werden. Jedem Punkt p sind gewisse p enthaltende Teilmengen der Mannigfaltigkeit als *Umgebungen* von p zugeordnet, die den zuerst von F. HAUSDORFF vollzählig aufgestellten Axiomen genügen müssen:

1. Zu zwei Umgebungen von p gibt es immer eine Umgebung von p, die in beiden enthalten ist;
2. liegt p in der Umgebung U_0 von p_0, so gibt es eine Umgebung von p, die ganz in U_0 enthalten ist;
3. zwei verschiedene Punkte besitzen Umgebungen, die zueinander punktfremd sind.

Mit Hilfe der Umgebungen lassen sich alle Stetigkeitsbegriffe definieren, insbesondere der Begriff der topologischen Abbildung (so heisst ein Paar zueinander inverser Abbildungen $p \to p'$, $p' \to p$, die beide stetig sind). Eine Mannigfaltigkeit wird als zweidimensionale *Fläche* bezeichnet, wenn sich jede Umgebung topologisch auf das Innere E des Einheitskreises der (x, y)-Ebene abbilden lässt. Dabei ist E natürlich selber als Mannigfaltigkeit dadurch definiert, dass als Umgebung eines Punktes p_0 von E das Innere eines jeden um p_0 beschriebenen, ganz in E gelegenen Kreises gilt. Eine *Kurve* (Weg) auf der Fläche $\mathfrak{F}$ ist gegeben, wenn jedem Wert des in den Grenzen $0 \leqq \lambda \leqq 1$ variierenden Parameters λ ein Punkt $p(\lambda)$ von $\mathfrak{F}$ in stetiger Weise zugeordnet ist. Wir legen der Fläche die Bedingung auf, *zusammenhängend* zu sein; das heisst, wir nehmen an, dass jeder Punkt auf $\mathfrak{F}$ mit jedem durch eine Kurve verbunden werden kann. Einer der wichtigsten Stetigkeitsbegriffe ist der der *kompakten Menge* (eine Zeitlang, während das Wort kompakt in anderm Sinne verwendet wurde, hiessen sie bikompakt). Eine Punktmenge M auf $\mathfrak{F}$ ist kompakt, wenn folgendes der Fall ist: Ist jedem Punkt p von M irgendwie eine Umgebung $U(p)$ von p zugeordnet, so lassen sich stets endlichviele Punkte p unter den Punkten von M auswählen, deren zugeordnete Umgebungen $U(p)$ ganz M bedecken. Für Flächen, die selber kompakt sind, ist der Name *geschlossen* in Gebrauch. In diesem Sinne ist zum Beispiel die Kugeloberfläche geschlossen, die Ebene nicht.

Orientierung

Sei A ein Punkt der (x, y)-Ebene, $\mathfrak{C}$ eine nicht durch A hindurchgehende Kurve in der Ebene. Wir bezeichnen mit $2\pi\,\varphi(p)$ den Winkel, den der Strahl $\overrightarrow{Ap}$ von A nach einem variablen Punkt p auf $\mathfrak{C}$ mit einer festen, von A ausgehenden Halbgeraden einschliesst. $\varphi(p)$ ist nur modulo 1 eindeutig bestimmt. Man kann aber die *stetige* Änderung von $\varphi(p)$ verfolgen, während p die Kurve $\mathfrak{C}$ durchläuft; der Zuwachs, welchen φ dabei erfährt, wird eine ganze Zahl n sein. Wir werden sagen, dass $\mathfrak{C}$ den Punkt A im ganzen n-mal im positiven Sinne umschlingt oder dass n die *Ordnung* von A in bezug auf $\mathfrak{C}$ ist. Entscheidend für das Vorzeichen von n ist, dass ein bestimmter Drehungssinn in der Ebene als positiver zugrunde gelegt ist. Sind die Punkte A und B durch eine stetige, $\mathfrak{C}$ nicht treffende Kurve miteinander verbunden, so hat B dieselbe Ordnung in bezug auf $\mathfrak{C}$ wie A. Es gilt der folgende fundamentale Satz: Es sei ein ebenes Gebiet $\mathfrak{G}$ auf ein anderes $\mathfrak{G}'$ topologisch abgebildet. Der Punkt A in $\mathfrak{G}$ gehe durch diese Abbildung in A' über. Zu einem gegebenen Drehsinn ∂ in A gehört dann ein Bild-Drehsinn ∂' in A', der so zu kennzeichnen ist: Wenn $\mathfrak{C}$ irgendeine A nicht passierende geschlossene Kurve ist, die in dem ganz zu $\mathfrak{G}$ gehörigen Innern $\mathfrak{K}$ eines Kreises um A verläuft, so stimmt die auf Grund von ∂' ermittelte Ordnung von A' in bezug auf die Bildkurve $\mathfrak{C}'$ von $\mathfrak{C}$ stets mit der auf Grund von ∂ ermittelten Ordnung von A in bezug auf $\mathfrak{C}$ überein.

Hierauf beruht die Möglichkeit, einen Drehsinn ∂ in einem Punkte p_0 einer gegebenen *Fläche* $\mathfrak{F}$ festzulegen, indem man ihn in dem topologischen Bild irgendeiner Umgebung von p_0 festlegt. Ein den sämtlichen Punkten p von $\mathfrak{F}$ zugewiesener Drehsinn $\partial(p)$ wird *stetig* in p_0 heissen, wenn er im topologischen ebenen Bild einer hinreichend kleinen Umgebung von p_0 überall als der *gleiche* Drehsinn erscheint. Eine Fläche heisst *zweiseitig* oder orientierbar, wenn sich auf ihr ein einheitlicher Drehsinn festlegen lässt, das heisst, wenn sich jedem Punkt ein solcher Drehsinn zuweisen lässt, dass derselbe überall stetig ist. Durch diese Festlegung wird die Fläche zur *orientierten Fläche*. Indem wir eine Drehung im positiven Sinne als «Wendung linksum» bezeichnen, gelingt es auf einer orientierten Fläche (aber nur auf einer solchen), die Unterscheidung zwischen links und rechts durchzuführen. Die Kugeloberfläche ist zum Beispiel zweiseitig, die projektive Ebene, die aus ihr durch Identifizierung antipodischer Punkte entsteht, ist es nicht. Wir beschäftigen uns fortan nur mit zweiseitigen Flächen.

Integralfunktion

Eine *Kurvenfunktion* F ist gegeben, wenn jeder Kurve γ auf der Fläche $\mathfrak{F}$ eine reelle Zahl $F(\gamma)$ zugeordnet ist. Aus einer von a nach b führenden Kurve γ' und einer von b nach c führenden Kurve γ'' kann man die von a nach c führende Kurve $\gamma = \gamma' + \gamma''$ zusammensetzen. Die Kurvenfunktion ist *linear*, wenn unter diesen Umständen stets

$$F(\gamma' + \gamma'') = F(\gamma') + F(\gamma'')$$

ist. Die Kurvenfunktion F heisst *kohomolog Null*, $F \sim 0$, falls $F(\gamma) = 0$ ist für jede geschlossene Kurve γ. Alsdann existiert eine Punktfunktion $f(p)$, so dass für irgendeine von a nach b führende Kurve γ die Gleichung

$$F(\gamma) = f(b) - f(a)$$

gilt. Wir beschäftigen uns mit solchen linearen Kurvenfunktionen, die im kleinen überall kohomolog Null[1] sind; das heisst, zu jedem Punkt p_0 soll eine Umgebung existieren, derart dass $F(\gamma) = 0$ gilt für jede in dieser Umgebung verlaufende geschlossene Kurve γ. Derartige lineare Kurvenfunktionen mögen *Integralfunktionen* heissen. Mehrere Integralfunktionen $F_1, \ldots, F_l$ sind linear abhängig, wenn es nicht sämtlich verschwindende reelle Zahlen $c_1, \ldots, c_l$ gibt, so dass die Relation

$$c_1 F_1 + \cdots + c_l F_l \sim 0$$

besteht. Die Maximalzahl der linear unabhängigen Integralfunktionen auf einer Fläche heisst ihr *Zusammenhangsgrad* (derselbe kann natürlich auch unendlich sein). In dieser von mir 1913 aufgestellten, von jeder Triangulation unabhängigen Definition tauchte wohl zum erstenmal der Gedanke auf, die Homologietheorie der geschlossenen Wege auf die «Kohomologietheorie» der Integralfunktionen zu gründen. Zwischen mehreren geschlossenen Wegen $\gamma_1, \ldots, \gamma_l$ besteht die Homologie

$$c_1 \gamma_1 + \cdots + c_l \gamma_l \simeq 0 ,$$

wenn für jede Integralfunktion F die Gleichung

$$c_1 F(\gamma_1) + \cdots + c_l F(\gamma_l) = 0$$

besteht. Führt man formal lineare Kombinationen $c_1 \gamma_1 + \cdots + c_l \gamma_l$ geschlossener Wege γ_i als «*Ströme*» ein, so bilden die Ströme und die Integralfunktionen zwei zueinander duale Vektorräume. Der Zusammenhangsgrad ist zugleich die Maximalzahl der im Sinne der Homologie linear unabhängigen geschlossenen Wege.

Ich will die Integralfunktion F *beschränkt* nennen, wenn es eine kompakte Teilmenge M auf $\mathfrak{F}$ gibt, so dass $F(\gamma) = 0$ ist für jede ganz ausserhalb M verlaufende Kurve γ. Die zwischen geschlossenen Wegen $\gamma_1, \ldots, \gamma_l$ bestehende *schwache Homologie*

$$c_1 \gamma_1 + \cdots + c_l \gamma_l \sim 0$$

bedeutet, dass

$$c_1 F(\gamma_1) + \cdots + c_l F(\gamma_l) = 0$$

ist für jede beschränkte Integralfunktion F. So kann man denn auch von einem schwachen Zusammenhangsgrad sprechen; er kann niemals grösser sein als der Zusammenhangsgrad. Zum Beispiel hat ein Kreisring in der Ebene den Zusammenhangsgrad 1, aber den schwachen Zusammenhangsgrad 0, indem ein zu den Rändern konzentrischer Kreis im Ring, obschon nicht homolog Null, doch schwach homolog Null ist. – Für geschlossene Flächen besteht natürlich kein Unterschied zwischen Homologie und schwacher Homologie.

Triangulation

Ich komme jetzt zu dem für die kombinatorische Topologie grundlegenden Begriff der Triangulation und fasse diesen Begriff etwas schärfer als in dem oben zitierten Buch. Ein ebenes Dreieck lässt sich am leichtesten mit Hilfe der zu seinen Eckpunkten 1, 2, 3 gehörigen Schwerpunktskoordinaten beschreiben. Darum beginnen wir mit dieser Erklärung: Ein Dreieck Δ auf $\mathfrak{F}$ ist durch eine stetige Funktion $p = \Delta(\xi_1, \xi_2, \xi_3)$ gegeben, die jedem Tripel von reellen nichtnegativen Zahlen ξ_1, ξ_2, ξ_3 von der Summe 1 in stetiger Weise einen Punkt p auf $\mathfrak{F}$ zuordnet, in solcher Weise, dass verschiedenen Tripeln stets verschiedene Punkte p korrespondieren. Von der Gesamtheit der den angegebenen Bedingungen genügenden reellen Zahlentripel sprechen wir auch als von dem *Zahldreieck* $Z = Z_\xi$. $\xi_1 = 0$ definiert die Kante 1 des Dreiecks, (1, 0, 0) ist die Ecke 1, (0, 1/2, 1/2) die Mitte der Kante 1, (1/3, 1/3, 1/3) der Mittelpunkt des Dreiecks. Da Z kompakt ist, ist auch sein stetiges Abbild Δ kompakt, und nach einem wichtigen Satz über die stetige Abbildung kompakter Mengen ist nicht nur die Abbildung $Z \to \Delta$, sondern auch die inverse $\Delta \to Z$ stetig. Man beachte, dass ein Dreieck auf der Fläche nicht als Menge der zu ihm gehörigen Punkte definiert ist, sondern durch die Z auf Δ stetig abbildende Funktion $\Delta(\xi_1, \xi_2, \xi_3)$.

Zur *Triangulation* einer Fläche gehört, dass auf ihr gewisse Punkte e als Ecken ausgezeichnet sind, die in gewisser Weise zu Paaren (= Kanten) (ef) und zu Tripeln (= Dreiecken) (efg) zusammengefasst sind. Zu einem Paar (ef) gibt es genau zwei Ecken g, so dass ef mit g ein Tripel bilden. Zu einer Ecke e gibt es nur endlichviele von e verschiedene Ecken $f = f_1, \ldots, f_r$ derart, dass (ef) als Kante auftritt. Diese bilden einen Zykel, indem ef_i die gemeinsame Kante der beiden Dreiecke

$$\Delta_{i-1} = (ef_{i-1}f_i) \quad \text{und} \quad \Delta_i = (ef_i f_{i+1}) \quad (i = 1, \ldots, r;\ f_0 = f_r,\ f_{r+1} = f_1)$$

ist. Zu jedem vorkommenden Tripel (efg) gehört eine stetige Funktion

$$p = \Delta_{efg}(\xi_e, \xi_f, \xi_g)\ ,$$

die jedem zulässigen, das heisst den Bedingungen

$$\xi_e \geqq 0, \quad \xi_f \geqq 0, \quad \xi_g \geqq 0, \quad \xi_e + \xi_f + \xi_g = 1$$

genügenden Wertetripel (ξ_e, ξ_f, ξ_g) einen Punkt p auf der Fläche so zuordnet, dass verschiedenen Tripeln verschiedene Punkte entsprechen. Die Funktion definiert das Dreieck Δ_{efg} der Triangulation. Die Bezeichnung ist so zu verstehen, dass, wenn efg im Index des Funktionszeichens Δ und zugleich als Indizes der Variablen ξ_e, ξ_f, ξ_g der gleichen Permutation unterworfen werden, der Wert p der Funktion sich nicht ändert. Auf den beiden Dreiecken $\Delta_{efg}, \Delta_{efg'}$

mit der gemeinsamen Kante ef gelte

$$\Delta_{efg}(\xi_e, \xi_f, 0) = \Delta_{efg'}(\xi_e, \xi_f, 0) .$$

Damit ist also die die Kante (ef) definierende Funktion $p = \delta_{ef}(\xi_e, \xi_f)$ eindeutig erklärt. Für jedes vorkommende Tripel (efg) mit der Ecke e gelte $\Delta_{efg}(1, 0, 0) = e$. Jeder Punkt der Fläche erscheint als Wert mindestens einer dieser den Tripeln (efg) zugehörigen Funktionen Δ_{efg}. Die Topologie kommt in den folgenden Forderungen zum Ausdruck: Zu einem inneren Punkt eines Dreiecks

$$\Delta_{efg}: \ \xi_e > 0, \ \xi_f > 0, \ \xi_g > 0 ,$$

gibt es eine Umgebung, deren Punkte keinem andern als diesem Dreieck angehören; zu einem innern Punkt einer Kante δ_{ef}: $\xi_e > 0$, $\xi_f > 0$, gibt es eine Umgebung, in der keine andern Punkte liegen als solche, die den beiden Dreiecken Δ_{efg}, $\Delta_{efg'}$ mit der gemeinsamen Kante ef angehören (Dreieckspaar); zu einem Eckpunkt e gibt es eine Umgebung, zu der keine andern Punkte als solche gehören, die in dem zyklischen Stern der Dreiecke

$$\Delta_0 = \Delta_{ef_rf_1}, \ldots, \Delta_{r-1} = \Delta_{ef_{r-1}f_r}$$

um e liegen.

Für die Folge nehmen wir an, dass die Fläche $\mathfrak{F}$ in einer bestimmten Triangulation ζ vorliegt.

Der Umstand, dass eine Fläche *zusammenhängend* ist, gibt sich jetzt in der kombinatorischen Eigenschaft kund, dass die Ecken sich in keiner Weise so in zwei Klassen teilen lassen, dass nur Paare und Tripel solcher Ecken vorkommen, die der gleichen Klasse angehören. Eine Fläche ist dann und nur dann *geschlossen*, wenn das Schema der Triangulation nur aus *endlichvielen* Ecken (und darum auch aus nur endlichvielen Kanten und Dreiecken) besteht. Ein einheitlicher, auf der Fläche festgelegter Drehsinn gibt sich in jedem Dreieck Δ_{efg} der Triangulation dadurch kund, dass ihm eine bestimmte der beiden *Indikatrizen* (= Ecken-Reihenfolgen) $efg = fge = gef$ oder $gfe = feg = egf$ zugewiesen ist; in solcher Weise, dass die Indikatrizen zweier in einer Kante zusammenstossender Dreiecke (efg) und (efg') kohärent sind. Die Indikatrix efg eines Dreiecks induziert auf den begrenzenden Kanten je einen Durchlaufungssinn $\vec{ef}$, $\vec{fg}$, $\vec{ge}$. *Kohärente* Indikatrizen der in einer Kante ef zusammenstossenden beiden Dreiecke induzieren auf der gemeinsamen Kante *entgegengesetzte* Durchlaufungssinne. Jedesmal ist hier eine Kontinuumseigenschaft (zusammenhängend, geschlossen, orientierbar) in eine kombinatorische umgesetzt; und es ergibt sich daraus, dass die betreffende kombinatorische Eigenschaft von der Triangulation unabhängig ist.

Aus der Beschreibung geht ferner hervor, dass das Innere des um eine Ecke e sich gruppierenden Dreieckssterns $\Delta_0 + \Delta_1 + \cdots + \Delta_{r-1}$ sich topologisch auf das

Innere eines ebenen regulären Polygons so abbilden lässt, dass die von e ausgehenden, in den Dreiecken verlaufenden geradlinigen Strahlen in geradlinige Strahlen in der Ebene übergehen («Strahlabbildung»). Da das Innere eines solchen Polygons einfach zusammenhängend, nämlich dem Inneren eines Kreises topologisch äquivalent ist, gilt $F(\gamma) = 0$ für eine jede Integralfunktion F und eine jede geschlossene Kurve γ, die ganz im Innern des Dreieckssterns verläuft.

Ein *Kantenzug* $e_0 e_1 e_2 \ldots e_n$ (der Länge n) ist eine Folge von Kanten unserer Triangulation, in welcher der Endpunkt e_i der (i-ten) Kante $e_{i-1}e_i$ zugleich der Anfangspunkt der nächsten Kante $e_i e_{i+1}$ ist. Der Kantenzug ist *einfach*, wenn alle seine Ecken $e_0, e_1, \ldots, e_n$ voneinander verschieden sind. Er ist *geschlossen*, wenn e_n mit e_0 zusammenfällt. Statt der Kanten, welche die Ecken miteinander verbinden, kann man auch die *Cokanten* (sit venia verbo!) $\varkappa^*$ betrachten, deren jede die Mittelpunkte i, i' der beiden Dreiecke Δ, Δ' eines Dreieckpaars miteinander verbindet; $\varkappa^*$ besteht aus der Strecke ic in Δ, die von i zu der Mitte c der den beiden Dreiecken gemeinsamen Kante $\varkappa$ läuft, und aus der Strecke ci' in Δ'. Da wir annehmen, dass unsere Fläche zweiseitig ist, können wir jedem Dreieck der Triangulation eine positive Indikatrix so erteilen, dass die beiden Dreiecke eines Dreieckpaars stets kohärente Indikatrizen bekommen. Eine Kante $\varkappa = \overrightarrow{ee'}$ sei mit dem durch die Schreibweise angedeuteten Durchlaufungssinn versehen; von den beiden Dreiecken Δ, Δ' mit der gemeinsamen Kante ee' sei Δ dasjenige, dessen positive Indikatrix auf $\varkappa$ den Durchlaufungssinn $\overrightarrow{ee'}$ induziert. Wir sagen dann, dass die oben beschriebene Cokante $\varkappa^*$, von i nach i' durchlaufen, die Kante $\varkappa$ *von links nach rechts* (oder im positiven Sinne) überkreuzt.

Ansatz des Problems

Jetzt sind wir in der Lage, anzudeuten, wie wir das Problem der Charakteristik in Angriff nehmen wollen. Wenn β ein geschlossener Cokantenzug und α ein geschlossener Kantenzug ist, so ist der Begriff der algebraischen Summe der Überkreuzungen von β über α vollkommen klar. Wir hoffen, dass folgende Tatsachen wahr sind: Jede geschlossene Kurve ist sowohl einem geschlossenen Kantenzug wie einem geschlossenen Cokantenzug homolog. Wenn immer α, α' zwei zueinander homologe geschlossene Kantenzüge und β, β' zwei homologe geschlossene Cokantenzüge sind, besteht die Gleichung

$$\mathrm{ch}(\beta, \alpha) = \mathrm{ch}(\beta', \alpha') \,.$$

Damit ist es dann möglich, $\mathrm{ch}(\beta, \alpha)$ für irgend zwei geschlossene Wege α, β dadurch zu definieren, dass man α durch einen beliebigen zu α homologen geschlossenen Kantenzug, β durch einen beliebigen zu β homologen geschlossenen Cokantenzug ersetzt. Wir hoffen endlich, dass die so allgemein erklärte

Charakteristik das Gesetz der Antisymmetrie (1) erfüllt. Freilich haben wir uns den Nachweis dieses Faktums dadurch erschwert, dass wir die beiden Argumente α, β ungleichartig behandeln, nämlich für α Kantenzüge, für β Cokantenzüge einsetzen.

Der Poincarésche Formalismus. Man erteile willkürlich jeder Kante $\varkappa$ einen bestimmten «positiven» Durchlaufungssinn und der zugehörigen Cokante $\varkappa^*$ einen solchen, dass sie $\varkappa$ von links nach rechts überkreuzt. Ein geschlossener Kantenzug α durchläuft jede Kante $\varkappa$ in summa eine bestimmte Anzahl von Malen $x_\varkappa$, wobei eine positive Durchlaufung mit $+1$, eine negative mit -1 in Ansatz gebracht wird. Wir schreiben symbolisch $\alpha = \Sigma\, x_\varkappa\, \varkappa$ und sprechen von α als einem *Zykel* in dem Sinne, dass zwei geschlossene Kantenzüge α und α' identifiziert werden, wenn sie gleiche Durchlaufungszahl, $x_\varkappa = x'_\varkappa$, für eine jede Kante $\varkappa$ haben. Von den ganzen Zahlen $x_\varkappa$ verschwinden alle bis auf endlich-viele. Sei $\varkappa = \overrightarrow{e e'}$. Wir setzen für eine beliebige Ecke a: $\varepsilon(a\,\varkappa) = +1$, wenn $a = e'$, $\varepsilon(a\,\varkappa) = -1$, wenn $a = e$, $\varepsilon(a\,\varkappa) = 0$, wenn a verschieden von e und e' ist. Die Bedingung der Geschlossenheit bedeutet, dass unser Kantenzug α in irgendeine Ecke a ebensooft ein- wie ausläuft, und das drückt sich in der Gleichung aus:

$$\sum_\varkappa \varepsilon(a\,\varkappa)\; x_\varkappa = 0\,. \tag{2}$$

Ist Δ irgendein Dreieck, so setzen wir $\varepsilon(\varkappa\Delta) = 0$, ausser wenn Δ eines der beiden Dreiecke Δ_1, Δ_2 mit der gemeinsamen Kante $\varkappa$ ist. Hingegen setzen wir $\varepsilon(\varkappa\Delta_1) = +1$, $\varepsilon(\varkappa\Delta_2) = -1$, wenn die positive Indikatrix von Δ_1 auf $\varkappa$ den positiven, die von Δ_2 den negativen Durchlaufungssinn induziert. Es besteht für jede Ecke a und jedes Dreieck Δ der Triangulation die Gleichung

$$\sum_\varkappa \varepsilon(a\,\varkappa)\; \varepsilon(\varkappa\,\Delta) = 0\,. \tag{3}$$

Ein geschlossener Cokantenzug kann als ein *Cozykel* $\Sigma\, y_\varkappa\, \varkappa^*$ angesprochen werden, indem man angibt, wie oft er, $y_\varkappa$-mal, in summa die Cokante $\varkappa^*$ im positiven Sinne durchläuft. Die Geschlossenheit des Cozykels findet in den Gleichungen

$$\sum_\varkappa y_\varkappa\, \varepsilon(\varkappa\,\Delta) = 0 \tag{4}$$

ihren Ausdruck. Dies ist die Weise, wie H. Poincaré uns gelehrt hat, das kombinatorische Schema der Triangulation den Methoden der linearen Algebra zu unterwerfen. Schreibt man die Zahlen $\varepsilon(a\,\varkappa)$ als eine Matrix E, in welcher a der Zeilen-, $\varkappa$ der Spaltenindex ist, die Zahlen $x_\varkappa$ als eine Spalte x, die Zahlen $y_\varkappa$ als eine Zeile y, die Grössen $\varepsilon(\varkappa\Delta)$ endlich als eine Matrix E^* mit $\varkappa$ als Zeilen- und Δ als Spaltenindex, so lauten die Gleichungen (2), (3) und (4) im Matrizenkalkül beziehungsweise

$$E\,x = 0\,, \quad E\,E^* = 0, \quad y\,E^* = 0\,.$$

Der Cozykel $\beta = \Sigma\, y_\varkappa\, \varkappa^*$ überschneidet den Zykel $\alpha = \Sigma\, x_\varkappa\, \varkappa$ so oft, wie die folgende Charakteristik angibt:

$$\mathrm{ch}(\beta, \alpha) = \Sigma\, y_\varkappa x_\varkappa = y\, x\,.$$

Algebraisierung der Integralfunktionen

Eine Integralfunktion F ordnet jeder Cokante $\varkappa^*$ einen bestimmten Wert $f_\varkappa = F(\varkappa^*)$ zu. Da der geschlossene Cokantenzug, der die um eine Ecke a herumliegenden Elementardreiecke

$$\Delta_0, \Delta_1, \ldots, \Delta_{r-1} \quad (\Delta_r = \Delta_0)$$

von Mittelpunkt zu Mittelpunkt miteinander verbindet, im Innern des einfach zusammenhängenden Dreieckssterns verläuft, muss der Wert von F für diesen Zug, das ist

$$\Sigma\, \varepsilon(a\, \varkappa)\, f_\varkappa = 0 \tag{5}$$

sein. Der Wert $F(\beta)$ der Integralfunktion F für den Cozykel $\beta = \Sigma\, y_\varkappa\, \varkappa^*$ ist $\Sigma\, y_\varkappa\, f_\varkappa$. Ich behaupte zweierlei:

1. Durch die Bestimmungszahlen $f_\varkappa$ ist die Integralfunktion im Sinne der Kohomologie eindeutig festgelegt.
2. Unter der Voraussetzung, dass sie den sämtlichen, den Ecken a entsprechenden Gleichungen (5) genügen, können diese Bestimmungszahlen beliebig vorgegeben werden.

Daraus folgt, dass es zu jedem Zykel $\alpha = \Sigma x_\varkappa\, \varkappa$ eine beschränkte Integralfunktion F_α gibt (mit den Bestimmungszahlen $f_\varkappa = x_\varkappa$), so dass für beliebige geschlossene Cozykel $\beta = \Sigma y_\varkappa\, \varkappa^*$ die Gleichung

$$F_\alpha(\beta) = \mathrm{ch}(\beta, \alpha)$$

besteht. Da der Wert $F_\alpha(\beta)$ sich nicht ändert, wenn man β durch irgendeinen ihm schwach homologen geschlossenen Weg ersetzt, überträgt sie die Definition der Charakteristik ch sofort von Cozykeln auf beliebige geschlossene Wege β und liefert das Gesetz

$$\mathrm{ch}(\beta, \alpha) = \mathrm{ch}(\beta', \alpha)$$

für irgend zwei geschlossene Wege β, β', die einander schwach homolog sind; insbesondere $\mathrm{ch}(\beta, \alpha) = 0$, wenn immer $\beta \sim 0$. Jedoch bleibt das Argument α einstweilen auf Zykeln beschränkt. Da die Gleichungen (5) ganzzahlige Koeffizienten besitzen, ist jede beschränkte Integralfunktion einer linearen Kombination endlichvieler beschränkter Integralfunktionen kohomolog, deren Bestimmungszahlen $f_\varkappa$ ganze Zahlen sind. Darum gilt auch die umgekehrte Tatsache:

A. *Hat ein geschlossener Weg β die Eigenschaft, dass* $\mathrm{ch}(\beta, \alpha) = 0$ *ist für jeden Zykel α, so ist β schwach homolog Null.*

Um die beiden oben erwähnten Tatsachen betreffend die Bestimmungszahlen einer Integralfunktion zu erweisen, haben wir an erster Stelle zu zeigen, dass jeder geschlossene Weg γ einem Cozykel homolog ist. Da jeder Punkt der Fläche im Innern mindestens eines Dreieckssterns liegt, folgt aus den Grundlagen der Analysis, dass sich γ in endlichviele Teilbögen

$$\gamma_1, \gamma_2, \ldots, \gamma_n \quad (\gamma_{n+1} = \gamma_1)$$

zerlegen lässt, deren jeder ganz im Innern eines Dreieckssterns verläuft. Aufeinanderfolgende Bögen, die im gleichen Stern liegen, vereinigen wir zu einem einzigen Teilbogen. Dann hat der Dreiecksstern Σ_1, innerhalb dessen γ_1 liegt, ein Dreieckspaar mit demjenigen Stern gemein, in welchem γ_2 liegt. Der Endpunkt a_2 von γ_1, welcher zugleich der Anfangspunkt von γ_2 ist, liegt in einem, Δ, der beiden Dreiecke dieses Paars (liegt er in beiden, so nehmen wir für Δ eines davon). Wir hängen dann an γ_1 die in Δ verlaufende Strecke an, die von a_2 nach dem Mittelpunkt i von Δ führt, und bringen dieselbe im umgekehrten Sinne durchlaufene Strecke vor γ_2 an. Sei F eine Integralfunktion. Durch das geschilderte Verfahren verwandeln sich die Bögen $\gamma_1, \gamma_2, \ldots$ durch Hinzufügung je einer Strecke am Anfang und am Ende in Bögen $\gamma_1', \gamma_2', \ldots$, die je innerhalb eines Sterns vom Mittelpunkt eines Dreiecks zum Mittelpunkt eines andern führen. Dabei bleibt die Summe

$$\gamma = \gamma_1 + \gamma_2 + \cdots + \gamma_n$$

ungeändert und damit auch der Wert von $F(\gamma)$, da ja die zwischengeschalteten Strecken, zweimal in entgegengesetztem Sinne durchlaufen, sich aufheben. Der im Stern Σ_1 vom Mittelpunkt des Dreiecks Δ zum Mittelpunkt des Dreiecks Δ' führende Weg γ_1' kann durch einen dieselben beiden Punkte verbindenden, im Dreiecksstern im einen oder andern Sinne herumlaufenden Cokantenzug γ_1'' ersetzt werden, und wegen des einfachen Zusammenhangs des Sterns ist dabei $F(\gamma_1') = F(\gamma_1'')$. So erhalten wir schliesslich einen geschlossenen Cokantenzug $\gamma'' = \gamma_1'' + \gamma_2'' + \cdots$, der homolog γ ist; denn er genügt für jede Integralfunktion F der Gleichung $F(\gamma) = F(\gamma'')$. Ebenso ergibt sich, dass jeder geschlossene Weg homolog einem geschlossenen Kantenzug ist.

Die zweite Tatsache, dass die Zahlen $f_\varkappa$ unter Einhaltung der sämtlichen Bedingungen (5) beliebig vorgegeben werden können, ergibt sich etwa so. Für die r in a mündenden Kanten $\varkappa_1, \ldots, \varkappa_r$ setzen wir $f_i = \varepsilon(a\,\varkappa_i)\, f_{\varkappa_i}$. Dann ist die Summe $f_1 + \cdots + f_r = 0$, und es lassen sich also Zahlen $g_0, \ldots, g_{r-1}$ $(g_r = g_0)$ so finden, dass $f_i = g_i - g_{i-1}$. Im positiv umlaufenen Zykel der Dreiecke mit der Ecke a trennt $\varkappa_i$ das Dreieck Δ_{i-1} von Δ_i. Wir ordnen dann den inneren Punkten p von Δ_i den Funktionswert $g(p) = g_i$ zu, den inneren Punkten p der

Kante $\varkappa_i$ den Wert

$$g(p) = \frac{1}{2} (g_{i-1} + g_i) ,$$

schliesslich dem Zentrum a den Wert

$$g(a) = \frac{1}{r} (g_0 + \cdots + g_{r-1}) .$$

Für eine Kurve γ, die innerhalb dieses Sterns vom Punkte p zum Punkte p' läuft, setze man

$$F(\gamma) = g(p') - g(p) .$$

In die Definition von $g(p)$ geht eine willkürliche additive Konstante ein, die aber bei der Bestimmung von $F(\gamma)$ wieder herausfällt. Liegt der Kurvenbogen γ in zwei Dreieckssternen, die dann ein Dreieckspaar gemein haben, so ist der resultierende Wert $F(\gamma)$ auch unabhängig davon, welchen dieser beiden Sterne man der Berechnung zugrunde legt. Ist schliesslich γ irgendeine Kurve, geschlossen oder nicht, so teile man sie in endlichviele Bögen γ_i ($i = 1, \ldots, n$), deren jeder ganz in einem Dreiecksstern verläuft, und bilde dann

$$F(\gamma) = F(\gamma_1) + \cdots + F(\gamma_n) .$$

Dieser Wert ist von der benutzten Teilung unabhängig, wie man sofort sieht, wenn man zwei Einteilungen in Bögen überlagert. Auf diese Weise erhält man eine Integralfunktion F mit den Bestimmungszahlen $f_\varkappa$.

Berechnung des Zusammenhangsgrades einer geschlossenen Fläche

F ist dann und nur dann homolog Null, wenn sich jedem Elementardreieck Δ eine Zahl g_Δ so zuordnen lässt, dass allgemein für die von Δ nach Δ' führende Cokante $\varkappa^*$ der Wert $F(\varkappa^*) = f_\varkappa$ gleich $g_{\Delta'} - g_\Delta$ wird; oder

$$f_\varkappa = -\Sigma_\Delta\, \varepsilon(\varkappa\, \Delta)\, g_\Delta . \tag{6}$$

Im Falle einer geschlossenen Fläche seien die Anzahlen der Ecken, Kanten und Dreiecke $\mathfrak{e}$, $\mathfrak{k}$ und $\mathfrak{d}$. Alsdann hat man $\mathfrak{k}$ Unbekannte $f_\varkappa$, denen die $\mathfrak{e}$ Gleichungen (5) auferlegt sind. Da zwischen ihren linken Seiten nur *eine* Identität besteht, nämlich diejenige mit den Koeffizienten $\lambda_a = 1$, so ist die Anzahl der linear unabhängigen Lösungen $\mathfrak{k} - \mathfrak{e} + 1$. Auf der andern Seite ist die Anzahl der linear unabhängigen Lösungen, die aus (6) mittels beliebiger Zahlen g_Δ hervorgehen, gleich $\mathfrak{d} - 1$. Der Überschuss

$$(\mathfrak{k} - \mathfrak{e} + 1) - (\mathfrak{d} - 1) = \mathfrak{k} - \mathfrak{e} - \mathfrak{d} + 2$$

ist darum die Maximalzahl h der linear unabhängigen Integralfunktionen, und daraus folgt, dass diese «Eulersche Charakteristik» $\mathfrak{k} - \mathfrak{e} - \mathfrak{d} + 2$ von der Triangulation der Fläche unabhängig ist.

Ein Satz über primitive geschlossene Wege

Für geschlossene Flächen kann der Satz A in bemerkenswerter Weise verschärft werden. Wir nennen einen geschlossenen Weg β *primitiv*, wenn er nicht dem Vielfachen $n\gamma$ ($n = 2$ oder 3 oder 4 oder ...) eines geschlossenen Weges γ homolog ist. Ich behaupte:

A'. *Zu einem primitiven Weg* β *gibt es einen Zykel* α, *für den* $\mathrm{ch}(\beta, \alpha) = 1$ *ist* (der also von β in summa genau einmal überkreuzt wird).

Beweis. Zykeln α und Cozykeln β werden durch ihre $\mathfrak{k}$ Durchlaufungszahlen $x_\varkappa$ bzw. $y_\varkappa$ gekennzeichnet; die $x_\varkappa$ werden als Spalte x, die $y_\varkappa$ als Zeile y geschrieben und x und y als Vektoren in zueinander dualen Vektorräumen angesehen, für welche das innere Produkt

$$\Sigma\, y_\varkappa\, x_\varkappa = y\, x$$

eine invariante Bedeutung hat. Die Vektoren mit ganzzahligen Koordinaten sollen Gittervektoren heissen. Werden die Koordinaten $x_\varkappa$ einer beliebigen unimodularen Transformation unterworfen (das ist einer linearen Transformation mit ganzzahligen Koeffizienten, deren Determinante $= \pm 1$ ist), so bleibt die Kennzeichnung der Gittervektoren als der Vektoren mit ganzzahligen Koordinaten erhalten. Werden die Koordinaten $y_\varkappa$ des Vektors y der kontragredienten linearen Transformation unterworfen, so bleibt auch der Ausdruck des inneren Produkts der gleiche:

$$y\, x = y_1\, x_1 + \cdots + y_{\mathfrak{k}}\, x_{\mathfrak{k}}\,.$$

Man kann nach dem Verfahren, das von MINKOWSKI als Adaptation eines Gitters an ein enthaltenes bezeichnet wurde, eine unimodulare Transformation im x-Raum so ausführen, dass die den Gleichungen (2) genügenden «geschlossenen» Vektoren $(x_1, \ldots, x_{\mathfrak{k}})$ durch das Verschwinden der $\mathfrak{k} - l$ letzten Koordinaten gekennzeichnet sind. Mit andern Worten: Wir bekommen l Zykel $\alpha_1, \ldots, \alpha_l$ so, dass jeder Zykel sich in einer und nur einer Weise als eine Summe $\sum_{i=1}^{l} x_i\, \alpha_i$ mit ganzzahligen Koeffizienten x_i schreiben lässt. Man führe die kontragrediente Transformation im dualen y-Raum aus. Wegen der Gleichung $E\, E^* = 0$ für die Matrizen E und E^* verschwinden nach dieser Transformation in jeder Spalte von E^* die letzten $\mathfrak{k} - l$ Glieder, das heisst die letzten $k - l$ Zeilen von E^* sind mit Nullen besetzt. Die die Geschlossenheit eines y-Vektors zum Ausdruck bringenden Gleichungen (4) im dualen Raum betreffen demnach nur die Koordinaten $y_1, \ldots, y_l$; und nach einer geeigneten unimodularen Transformation dieser l Koordinaten besagen jene Gleichungen, dass die letzten $l - h$ der

Koordinaten $y_1, \ldots, y_l$ verschwinden. Nachdem $x_1, \ldots, x_l$ der kontragredienten Transformation unterworfen sind, nehmen die geschlossenen Gittervektoren im x-Raum die Form

$$(x_1, \ldots, x_h, x_{h+1}, \ldots, x_l, 0, \ldots, 0)$$

an, die im y-Raum die Form

$$(y_1, \ldots, y_h, 0, \ldots, 0, y_{l+1}, \ldots, y_{\mathfrak{l}}),$$

und die Charakteristik wird

$$\mathrm{ch}(\beta, \alpha) = y\,x = y_1\,x_1 + \cdots + y_h\,x_h\,. \tag{7}$$

Der Cozykel β wird dann und nur dann homolog Null sein, wenn dieser Ausdruck für beliebige ganzzahlige Werte von $x_1, \ldots, x_l$ verschwindet, das heisst, wenn $y_1 = \cdots = y_h = 0$ ist. Wir haben also h Cozykel $\beta_1, \ldots, \beta_h$ gefunden, so dass jeder geschlossene Weg β homolog einer eindeutig bestimmten *ganzzahligen* linearen Kombination $y_1\,\beta_1 + \cdots + y_h\,\beta_h$ derselben ist, und damit erweist sich h als der Zusammenhangsgrad. β ist primitiv, falls $y_1, \ldots, y_h$ den grössten gemeinsamen Teiler 1 besitzen. Dann kann man aber ganze Zahlen $x_1, \ldots, x_h$ und damit einen Zykel $\sum_{i=1}^{h} x_i\,\alpha_i$ finden, für welche (7) gleich 1 wird, und damit ist unser Satz bewiesen.

Das Gesetz der Antisymmetrie

Wir haben die Regel angegeben, nach welcher der Wert der zum Zykel $\alpha = \sum_{\varkappa} x_\varkappa\,\varkappa$ gehörigen Integralfunktion

$$F_\alpha(\beta) = \mathrm{ch}(\beta, \alpha)$$

für einen beliebigen Weg β zu berechnen ist. Hier spezialisieren wir nun zunächst auch β als einen *Zykel* $\Sigma\, y_\varkappa\,\varkappa$. Eine Kante $\varkappa$ mit den beiden Endpunkten e, e' zerlegen wir durch ihre Mitte c in die zwei Halbkanten $(\varkappa\,e) = \overrightarrow{ce}$ und $(\varkappa\,e') = \overrightarrow{ce'}$, so dass

$$\varkappa = \sum_e \varepsilon(e\,\varkappa)\,(\varkappa\,e)\,.$$

Die r in e einmündenden Kanten, wie die Halbkanten $\overrightarrow{c_i e}$, numerieren wir in zyklischer Reihenfolge, e im positiven Sinne umlaufend, mit $i = 1, \ldots, r$. Es ist zweckmässig, den Index i als einen solchen zu verstehen, der alle ganzen Zahlen durchläuft, aber so, dass Werte i, die einander kongruent modulo r sind, als gleich gelten. Die den in e *einlaufenden* Kanten $\varkappa_i$ zugehörigen Zahlen $x_\varkappa$, genauer also die Zahlen $\varepsilon(e\,\varkappa_i)\,x_{\varkappa_i}$ wollen wir, wie schon früher, mit $x_1, \ldots, x_r$ bezeichnen. Ihre Summe ist 0, und wir können darum Zahlen g_i so bestimmen, dass $x_i = g_i - g_{i-1}$. Der Beitrag zu $F_\alpha(\beta) = \mathrm{ch}(\beta\,\alpha)$, der von den in e mündenden Halbkanten (der «Spinne in e») herrührt, ist dann

$$w_e(\beta\,\alpha) = \sum_i y_i\, F_\alpha(\overrightarrow{c_i\, e}) .$$

Hier ist

$$F_\alpha(c_i\,\vec{e}) = -\frac{1}{2}\,(g_i + g_{i-1}) + g ,$$

wo g der Mittelwert der r Zahlen $g_0, \ldots, g_{r-1}$ ist. Da auch die Summe $\Sigma\, y_i = 0$ ist, können wir für diesen Beitrag schreiben:

$$-\frac{1}{2} \sum_i y_i\,(g_i + g_{i-1}) .$$

Eben wegen der Gleichung $\Sigma\, y_i = 0$ kann die in die g_i eingehende willkürliche additive Konstante irgendwie gewählt werden, etwa gemäss $g_0 = 0$. Dann ist

$$\frac{1}{2}\,(g_i + g_{i-1}) = \sum_{j>i} x_j + \frac{1}{2}\,x_i = -\sum_{j<i} x_j - \frac{1}{2}\,x_i .$$

Daher kommt

$$w_e(\beta\,\alpha) = \sum_{1 \leqq j < i \leqq r} y_i\,x_j + \frac{1}{2}\sum_{i=1}^{r} y_i\,x_i = -\sum_{1 \leqq i < j \leqq r} x_j\,y_i - \frac{1}{2}\sum_{1 \leqq i \leqq r} x_i\,y_i . \quad (8)$$

Der von der Spinne in e herrührende Beitrag zur Charakteristik ist also eine halbganze Zahl. Wenn wir in der zweiten Formel die Indizes i und j vertauschen, so folgt ohne weiteres das für die Verkehrsregelung so wichtige *Gesetz der Antisymmetrie*

$$w_e(\beta\,\alpha) = -w_e(\alpha\,\beta) .$$

Setze einen Augenblick

$$L_m = \Sigma\, y_i\,x_j \quad (m + 1 \leqq j < i \leqq m + r) .$$

Aus unserer Berechnung geht hervor, dass L_0 unabhängig davon ist, wo wir im Zykel der Kanten, die in e hineinlaufen, mit der Numerierung durch die Ziffern 1 bis r beginnen. Dies ist sofort zu verifizieren. Denn L_m entsteht aus L_{m-1} dadurch, dass man

$$(y_{m+1} + \cdots + y_{m+r-1})\,x_m$$

fortlässt und dafür

$$y_{m+r}\,(x_{m+1} + \cdots + x_{m+r-1})$$

hinzufügt, diese Terme sind aber beide gleich $-y_m\,x_m$. Darum ist $L_0 = L_1 = L_2 = \cdots$.

Man mache sich klar, dass unsere Formel für die Schnittpunktmultiplizität w_e der beiden Zykeln β und α in e dem entspricht, was man anschaulich erwartet. Ein der Fahrbahn α folgendes Auto fahre längs der Strasse $\varkappa_i$ in die Kreuzung e hinein und verlasse sie längs $\varkappa_{i'}$; ein zweites Auto β komme zur gleichen Zeit längs der Strasse $\varkappa_j$ auf die Kreuzung zu und verlasse sie längs $\varkappa_{j'}$. Dann wird man erwarten, dass die Schnittmultiplizität 0 ist, falls in der zyklischen Anordnung der in e einmündenden Strassen das Paar $(\varkappa_i, \varkappa_{i'})$ das Paar $(\varkappa_j, \varkappa_{j'})$ nicht trennt; man wird die Multiplizität ± 1 erwarten, falls die beiden Paare sich trennen, und zwar sollte 1 herauskommen, wenn β den Weg α in e von links nach rechts überkreuzt, -1 im entgegengesetzten Falle. Es kann aber auch geschehen, dass die beiden Autos die Kreuzung längs derselben Strasse verlassen. Dann ergibt unsere Formel einen Wert $\pm 1/2$. In der Tat bleibt in diesem Falle die Entscheidung, ob die eine Bahn die andere kreuzt, in e noch suspendiert. Erst wenn die Wege sich später in einer andern Kreuzung trennen, wird durch deren Beitrag $\pm 1/2$ entschieden werden, ob Kreuzung stattgefunden hat (falls nämlich die Summe der beiden Beiträge nicht 0 ist), und wenn ja, in welchem Sinne. Unsere Formel umfasst in präziser algebraischer Form diese und alle andern Möglichkeiten.

Natürlich muss die über die Ecken e erstreckte Summe der Beiträge $w_e(\beta\,\alpha)$ eine ganze Zahl sein. Auch dies ist sofort zu verifizieren. Der möglicherweise halbganze Anteil von w_e, $(1/2)\,\Sigma\, y_i\, x_i$, ist ja gleich

$$\frac{1}{2}\sum_{\varkappa} \varepsilon(e\,\varkappa)\, y_\varkappa\, \varepsilon(e\,\varkappa)\, x_\varkappa\,,$$

und darum ist die über alle Ecken e erstreckte Summe dieser Anteile

$$\frac{1}{2}\sum_{e,\varkappa} \varepsilon^2(e\,\varkappa)\, y_\varkappa\, x_\varkappa \quad \text{wegen} \quad \sum_{e} \varepsilon^2(e\,\varkappa) = 2 \quad \text{gleich} \quad \sum_{\varkappa} y_\varkappa\, x_\varkappa\,.$$

Das für *Zykeln* α, β nunmehr erwiesene Gesetz der Antisymmetrie (1) ermöglicht es uns, in $\mathrm{ch}(\beta, \alpha)$ die Beschränkung von α, und nicht bloss von β, auf geschlossene Kantenzüge aufzuheben. Seien nämlich α_1, α_1' zwei geschlossene Kantenzüge, welche dem geschlossenen Weg α schwach homolog sind, und β_1 ein Zykel, der schwach homolog dem geschlossenen Weg β ist. Dann gilt $\alpha_1 \sim \alpha_1'$ und darum

$$\mathrm{ch}(\beta_1, \alpha_1) = -\mathrm{ch}(\alpha_1, \beta_1) = -\mathrm{ch}(\alpha_1', \beta_1) = \mathrm{ch}(\beta_1, \alpha_1')\,,$$

folglich $\mathrm{ch}(\beta, \alpha_1) = \mathrm{ch}(\beta, \alpha_1')$. Man kann demnach $\mathrm{ch}(\beta, \alpha)$ eindeutig dadurch definieren, dass man den geschlossenen Weg α durch irgendeinen ihm schwach homologen Zykel (α_1 oder α_1') ersetzt. Sind α_1, β_1 zwei Zykel, die beziehungsweise den geschlossenen Wegen α, β schwach homolog sind, so gilt dann $\mathrm{ch}(\beta, \alpha) = \mathrm{ch}(\beta_1, \alpha_1)$. Das Gesetz der Antisymmetrie bleibt für beliebige α, β

erhalten. Das Ziel einer universellen Definition der Charakteristik ist damit erreicht. Immerhin ist zu beachten, dass diese Definition *sich auf eine feste Triangulation* ζ *der Fläche* $\mathfrak{F}$ *stützt.* Soll dies in Evidenz gesetzt werden, so schreibe man ch_ζ statt ch.

Ist die Fläche geschlossen und h ihr Zusammenhangsgrad, so können irgend h voneinander linear unabhängige geschlossene Wege $\gamma_1, \ldots, \gamma_h$ als eine Wegebasis benutzt werden, indem jeder geschlossene Weg α homolog einer linearen Kombination $x_1 \gamma_1 + \cdots + x_h \gamma_h$ ist. Für dieses α und $\beta \sim y_1 \gamma_1 + \cdots + y_h \gamma_h$ ergibt sich

$$\mathrm{ch}(\beta, \alpha) = \sum_{i,j=1}^{h} s_{ij}\, y_i\, x_j\,, \quad s_{ij} = \mathrm{ch}(\gamma_i, \gamma_j)\,.$$

Der Satz A besagt, dass diese schiefsymmetrische bilineare «*Charakteristikenform*» nicht ausgeartet ist und dass somit ihre Determinante $d = \det s_{ij}$ nicht verschwindet. Dies kann aber für eine schiefsymmetrische Form nur dann stattfinden, wenn die Dimension h gerade ist. *Darum ist der Zusammenhangsgrad h einer geschlossenen zweiseitigen Fläche stets eine gerade Zahl*; die ganze Zahl $h/2$ heisst *Geschlecht.* Die Zahlen x_i sind nicht notwendig ganz, doch folgt aus den Gleichungen

$$\mathrm{ch}(\gamma_i, \alpha) = \sum_{j=1}^{h} s_{ij}\, x_j\,,$$

dass sie ganze Zahlen dividiert durch d sind. Durch den schärferen Satz A′ und seinen Beweis hat sich gezeigt, 1. dass die geschlossenen Wege eine *Integritätsbasis* $\gamma_1, \ldots, \gamma_h$ besitzen, so dass in der Homologie

$$\alpha \sim x_1 \gamma_1 + \cdots + x_h \gamma_h$$

für jeden geschlossenen Weg α die Koeffizienten x_i *ganze Zahlen* werden, und 2. dass bei Zugrundelegung einer solchen Basis $d = 1$ wird. Aus dem letzten Umstand folgt, dass man die Integritätsbasis so wählen kann, dass die Charakteristikenform die *kanonische* Gestalt

$$(y_1 x_2 - y_2 x_1) + \cdots + (y_{h-1} x_h - y_h x_{h-1})$$

annimmt (RIEMANNS «kanonische Rückkehrschnittpaare»).

Wir haben jetzt alle wesentlichen Eigenschaften der Charakteristik beisammen und möglichst einfach kombinatorisch abgeleitet. Um aber nun festzustellen, dass die Charakteristik in Wahrheit von der Triangulation, auf die sich ihre Definition stützte, unabhängig ist, benutze ich eine ganz andere Definition derselben, die für die funktionentheoretischen Anwendungen viel geeigneter ist und mir durch diese Anwendungen nahegelegt wurde. Zur Vorbereitung stellen wir zunächst fest, dass die Charakteristik sich nicht ändert, wenn die Triangulation ζ in gewisser elementarer Weise einer Unterteilung unterworfen wird.

Unterteilungen

Sind $a_i = (\alpha_{i1}, \alpha_{i2}, \alpha_{i3})$ drei (nicht in einer Geraden liegende) Punkte ($i = 1, 2, 3$) im Zahlendreieck Z_ξ mit den Ecken 1, 2, 3, so wird durch die Substitution

$$\xi_i = \sum_j \alpha_{ji} \eta_j \quad (i, j = 1, 2, 3);$$

$$\eta_1 \geqq 0, \ \eta_2 \geqq 0, \ \eta_3 \geqq 0, \ \eta_1 + \eta_2 + \eta_3 = 1,$$

ein Zahlendreieck Z_η als das Teildreieck von Z_ξ mit den Ecken a_1, a_2, a_3 festgelegt. Eine Triangulation ζ' ist eine *Unterteilung* der Triangulation ζ, wenn die Elementardreiecke von ζ' Teile der Elementardreiecke von ζ sind. Wir wollen hier aber nur zwei ganz spezielle Typen von Unterteilungen betrachten. Ist a ein Punkt von Z_ξ, so zerlegen wir Z_ξ in die drei Dreiecke $23a$, $31a$, $12a$. Ist a nicht im Innern von Z_ξ, sondern etwa auf der Kante 23 gelegen, so fällt hier das erste Dreieck, $23a$, fort. Ist a ein innerer Punkt des Elementardreiecks Δ der Triangulation ζ, so führen wir die angegebene Teilung nur in diesem Dreieck aus. Ist aber a ein Kantenpunkt, so muss sie in den beiden an diese Kante anstossenden Dreiecken bewerkstelligt werden. Diese «*Elementarteilung erster Art*» wird dazu benutzt, um einen Punkt a, der noch nicht Eckpunkt ist, zu einem Eckpunkt der Triangulation zu machen. Die «*Elementarteilung zweiter Art*» ist die *Normalteilung*, durch welche Z_ξ in die vier Teildreiecke $(23c_1)$, $(31c_2)$, $(12c_3)$, $(c_1, c_2 c_3)$ zerlegt wird, wo

$$c_1 = \left(0, \frac{1}{2}, \frac{1}{2}\right), \quad c_2 = \left(\frac{1}{2}, 0, \frac{1}{2}\right), \quad c_3 = \left(\frac{1}{2}, \frac{1}{2}, 0\right).$$

Auf der Fläche wird diese Normalteilung in allen Elementardreiecken Δ der gegebenen Triangulation ζ gleichzeitig ausgeführt. Durch Iteration dieses Prozesses der Normalteilung erhält man Triangulationen, die schliesslich «*beliebig fein*» werden.

Geht die Triangulation ζ' aus ζ durch einen der beiden Elementarprozesse hervor, so ist ein ζ-Kantenzug zugleich ein ζ'-Kantenzug, und aus der Formel (8) für den Beitrag w_e geht ohne weiteres hervor, dass, wenn α, β irgend zwei geschlossene ζ-Kantenzüge sind, die Gleichung

$$\mathrm{ch}_{\zeta'}(\beta, \alpha) = \mathrm{ch}_\zeta(\beta, \alpha) \tag{9}$$

besteht. In der Tat, wird das Strassennetz einer Stadt durch neu angelegte Strassen erweitert, so kann man diese sowie die etwa dadurch neu entstehenden Kreuzungen ignorieren, *solange niemand die neuen Strassen benutzt.* Die Gleichung (9) überträgt sich dann aber durch Homologie auf beliebige geschlossene Wege α, β; und in derselben Allgemeinheit gilt sogar die Regel

$$\mathrm{ch}_{\zeta^*}(\beta, \alpha) = \mathrm{ch}_{\zeta}(\beta, \alpha) \tag{10}$$

für eine Triangulation ζ^*, die durch *wiederholte* elementare Unterteilungen aus ζ entsteht.

Die «Kontinuumsdefinition» der Charakteristik

zu welcher wir jetzt übergehen, benutzt keine Triangulation, sondern die Bedeckung einer Fläche mit Umgebungen und die topologische Abbildung der Umgebungen auf das Innere von Kreisen. Es sei α eine gegebene geschlossene Kurve auf $\mathfrak{F}$. Durch endlichviele Teilpunkte

$$\mathit{0}, \mathit{1}, \mathit{2}, \ldots, n-\mathit{1} \quad (n = \mathit{0})$$

werde sie so in Bögen

$$\alpha_1 = \mathit{01}, \quad \alpha_2 = \mathit{12}, \quad \ldots, \quad \alpha_n = n-\mathit{1},\mathit{0}$$

geteilt, dass der Bogen α_i in der Umgebung U_i eines Flächenpunktes liegt, und diese Umgebung sei durch die topologische Abbildung A_i auf das Innere E des Einheitskreises der (x, y)-Ebene bezogen. Wir nehmen zunächst $i = 1$. Es sei K das Innere eines zu E konzentrischen Kreises von einem Radius ϱ, der kleiner als 1, aber doch so gross ist, dass die beiden Punkte $\mathit{0}, \mathit{1}$, oder vielmehr ihre durch die Abbildung $A = A_1$ erzeugten Bilder noch in K liegen. Wir sprechen von K als der «geschrumpften Umgebung $U = U_1$». Für einen von $\mathit{0}$ und $\mathit{1}$ verschiedenen Punkt p in E sei $2\pi\,\varphi(p)$ der Winkel, um den man den Strahl $\overrightarrow{p\mathit{0}}$ im positiven Sinne drehen muss, damit er in die Lage $\overrightarrow{p\mathit{1}}$ kommt. (φ ist gleich $\varphi_1 - \varphi_0$, wenn $2\pi\,\varphi_0$, $2\pi\,\varphi_1$ die Winkel bedeuten, welche die Strahlen $\overrightarrow{p\mathit{0}}$, $\overrightarrow{p\mathit{1}}$ mit der positiven x-Achse bilden.) $\varphi(p)$ ist nur mod 1 bestimmt. Doch auf der Peripherie k von K ist $\varphi(p)$ eine eindeutige stetige Funktion (deren Werte absolut kleiner als 1/2 sind). Diese kann man leicht als eine eindeutige stetige Funktion $u(p)$ auf die ganze gelochte Fläche $\mathfrak{F} - K$ so ausdehnen, dass sie ausserhalb einer kompakten Teilmenge von $\mathfrak{F}$ verschwindet (zum Beispiel ausserhalb einer konzentrischen Kreisscheibe K_1, deren Radius $> \varrho$, aber < 1 ist). Wir haben dann auf ganz $\mathfrak{F}$ eine mehrdeutige stetige Funktion $v(p)$, die in K gleich $\varphi(p)$ ist und deren Werte für irgendeinen nicht in K gelegenen Punkt p von $\mathfrak{F}$ die Zahlen sind, die sich von $u(p)$ um eine ganze Zahl unterscheiden[1]).

---

[1]) Für die funktionentheoretischen Anwendungen ist es zweckmässig, dass zu $v(p)$ noch irgendeine ausserhalb einer kompakten Menge verschwindende eindeutige stetige Funktion hinzugefügt wird.

Ist β irgendeine nicht durch *0* und *1* hindurchgehende Kurve, so kann man die stetige Änderung von $v(p)$ längs dieses Weges verfolgen und erhält einen eindeutig durch β bestimmten Zuwachs $s_1(\beta)$. Dies liefert eine (beschränkte) Integralfunktion s_1 auf $\mathfrak{F}$ mit den ***singulären Punkten*** *0* und *1*; das heisst eine Integralfunktion auf der «punktierten Fläche», die aus $\mathfrak{F}$ entsteht, wenn man die beiden Punkte *0* und *1* heraussticht. Da $v(p)$ mod 1 eindeutig ist, ist $s_1(\beta)$ für eine geschlossene (nicht durch *0* und *1* hindurchgehende) Kurve β eine ganze Zahl ν; diese gibt an, «wie oft der Weg β in summa im positiven Sinne zwischen *0* und *1* hindurchgeht». Von der besonderen Konstruktion der Funktion $u(p)$ ist ν unabhängig, da ja die Differenz der beiden, aus zwei verschiedenen Wahlen von $u(p)$ entspringenden mehrdeutigen Funktionen $v(p)$ eindeutig ist.

Indem wir dieselbe Konstruktion für jeden der Teilbögen $\alpha_1 = 01$, $\alpha_2 = 12$, ... von α ausführen, erhalten wir eine beschränkte Integralfunktion

$$s = s_1 + s_2 + \cdots + s_n$$

mit den Singularitäten *0, 1, 2, ...*. Nach dem, was am Anfang bei Einführung des Drehsinns über die Invarianz der Ordnung gegenüber topologischen Abbildungen gesagt wurde, sind diese Singularitäten aber jetzt zu ***hebbaren*** Singularitäten geworden. Dies bedeutet, dass sich zum Beispiel um *1* eine – keine der von *1* verschiedenen Singularitäten enthaltende – Umgebung $\mathfrak{u}_1$ abgrenzen lässt, derart, dass für jede in $\mathfrak{u}_1$ verlaufende und *1* nicht passierende geschlossene Kurve γ die Integralfunktion s den Wert 0 hat[1]). Daher existiert eine in allen Punkten p von $\mathfrak{u}_1$ ausser in *1* definierte Punktfunktion $\psi(p)$, so dass für eine jede zwei Punkte p und p' verbindende, in $\mathfrak{u}_1$ verlaufende Linie γ, die nicht durch *0* und *1* hindurchgeht,

$$s(\gamma) = \psi(p') - \psi(p)$$

ist. Der Umstand, dass wir an die Integralfunktionen keine Stetigkeitsforderungen gestellt haben, ermöglicht es uns, die Singularität im Punkte *1* einfach dadurch zu beseitigen, dass wir der Punktfunktion ψ daselbst irgendeinen bestimmten Wert geben, zum Beispiel $\psi(1) = 0$. Dann ist die Integralfunktion $s(\beta)$ für beliebige Wege β definiert; ihr Wert für eine *geschlossene* Kurve β aber ist stets eine ganze Zahl. Diese ganze Zahl $s_\alpha(\beta) = s(\beta, \alpha)$ bezeichnen wir jetzt als die ***Charakteristik*** s. Sie hängt von α und β ab. In ihre Berechnung geht aber ausser der Kurve α noch eine bestimmte Einteilung der Kurve α in Teilbögen α_i ein, ferner die Einbettung jedes Teilbogens α_i in eine Umgebung U_i und deren topologische Abbildung A_i auf E. Es bleibt zu zeigen, dass das Resultat von diesen Hilfskonstruktionen unabhängig ist.

---

[1]) Denn *1* hat in beiden Abbildern A_1 und A_2 in bezug auf jede solche Kurve dieselbe Ordnung.

Gleichheit der Charakteristiken ch und s

Die Schritte, welche nötig sind, um die Übereinstimmung der Charakteristik s mit der früher auf Grund einer Triangulation ζ definierten ch_ζ zu erweisen, sind die folgenden.

1. Durch mehrere Elementarteilungen erster Art verwandelt man ζ in eine feinere Triangulation, in welcher die Teilpunkte $0, 1, 2, \ldots$ zu Eckpunkten geworden sind. Indem man darauf den Prozess der Normalteilung hinreichend oft ausführt, bekommt man eine Triangulation ζ^* von solcher Art, dass sich 0 und 1 innerhalb der geschrumpften Umgebung K durch einen einfachen ζ-Kantenzug $\alpha_1' = 0ab \ldots e1$ verbinden lassen und Entsprechendes auch für die andern Teilbögen gilt. Da $\alpha_1 - \alpha_1'$ eine in K verlaufende geschlossene Kurve ist, gilt $\alpha_1 - \alpha_1' \simeq 0$, und darum ist der geschlossene Kantenzug

$$\alpha' = \alpha_1' + \cdots + \alpha_n'$$

homolog $\alpha = \alpha_1 + \cdots + \alpha_n$. Wir operieren weiter mit dieser Triangulation ζ^*.

2. Ein einfacher Kantenzug wie α_1' zerlegt die Fläche $\mathfrak{F}$ nicht; das heisst, je zwei nicht auf α_1' gelegene Punkte können durch eine α_1' nicht treffende Kurve miteinander verbunden werden. Es folgt daraus auch, dass in der abgeschlossenen Kreisscheibe $\overline{K}$ (die aus K durch Hinzufügung seiner Peripherie k entsteht) jeder nicht auf α_1' gelegene Punkt p von der Peripherie aus auf einem α_1' nicht treffenden Wege erreicht werden kann. Durch stetige Fortsetzung längs dieses Weges erhält man von den auf k herrschenden Werten von φ aus einen bestimmten Wert $\varphi(p)$ in p. Dieser ist unabhängig von dem Weg in $\overline{K}$, der von k aus zu p führt. In der Tat, sind γ_1, γ_2 zwei Wege, die von den Punkten q_1 bzw. q_2 auf k zu p führen, so durchlaufe man γ_1 rückwärts von p nach q_1, darauf den Bogen $q_1 q_2$ auf k und schliesslich den Weg γ_2 von q_2 nach p. So entsteht eine geschlossene Kurve $\mathfrak{C}$ in E, und unsere Behauptung ist, dass $\varphi = \varphi_1 - \varphi_0$, stetig längs $\mathfrak{C}$ fortgesetzt, zu seinem Ausgangswert zurückkehrt oder dass 0 und 1 dieselbe Ordnung in bezug auf $\mathfrak{C}$ haben. Das ergibt sich aber daraus, dass 0 und 1 durch die $\mathfrak{C}$ nicht treffende Kurve α_1' verbunden sind. Man erhält also eine in der längs α_1' aufgeschnittenen Fläche $\mathfrak{F}$ *eindeutige* stetige Funktion $v(p)$.

3. Den «Sprung» von v über eine Kante $\varkappa$ der Triangulation ζ^* kann man dadurch bestimmen, dass man je einen inneren Punkt p, p' im Innern der beiden längs $\varkappa$ zusammenhängenden Dreiecke Δ, Δ' der Triangulation wählt, diese innerhalb $\Delta + \Delta'$ durch eine Linie γ verbindet, und dann

$$s_1(\gamma) - \{ v(p') - v(p) \}$$

bildet. Der Sprung über eine jede nicht zu α_1' gehörige Kante ist 0. Indem man den Dreiecksstern um 1 herum betrachtet, findet man mit Hilfe der Strahlabbildung dieses Sterns, dass der Sprung über die letzte zu α_1' gehörige Kante $e1$

gleich 1 ist; darauf, mittels des Dreieckssterns um e, dass der Sprung über die vorletzte Kante gleich dem über die letzte Kante, also auch $= 1$ ist; und so fort. Ist daher β^* ein geschlossener Cokantenzug, so ergibt sich als Wert $s_1(\beta^*)$ die Anzahl von Malen, die β^* in summa die Kanten des Kantenzugs α_1' im positiven Sinne überschreitet.

4. Durch Addition über $i = 1, 2, \ldots, n$ findet man daraus die Gleichung

$$s_\alpha(\beta^*) = \mathrm{ch}(\beta^*, \alpha')\ ,$$

wo die Berechnung der Charakteristik ch sich natürlich auf die Triangulation ζ^* stützt. Ist jetzt β eine beliebige geschlossene Kurve und β^* ein ζ^*-Cozykel, der homolog β ist, so folgt, da s_α auf der linken Seite eine Integralfunktion ist und $\alpha' \simeq \alpha$,

$$s_\alpha(\beta) = \mathrm{ch}_{\zeta^*}(\beta, \alpha)\ .$$

5. Endlich macht es die Gleichung (10) möglich, von der Triangulation ζ^* auf die ursprüngliche ζ zurückzugreifen:

$$s(\beta, \alpha) = \mathrm{ch}_\zeta(\beta, \alpha)\ .$$

Von links nach rechts gelesen, lehrt diese Gleichung, dass $s(\beta, \alpha)$ von den an die Kurve α anschliessenden Hilfskonstruktionen, die bei der Definition von $s(\beta, \alpha)$ benutzt wurden, unabhängig ist; darauf lehrt sie, von rechts nach links gelesen, dass die Charakteristik ch_ζ von der Triangulation ζ unabhängig ist. Alle Eigenschaften von ch_ζ, vor allem die schiefe Symmetrie, übertragen sich vermöge dieser Gleichung auf $s(\beta, \alpha)$.

Zur Bestimmung der Charakteristik stehen somit zwei Wege offen: der eine, «kombinatorische», benutzt eine Triangulation und approximiert die geschlossenen Kurven im Sinne der Homologie durch Zykeln (Kantenzüge) der triangulierten Fläche. Der andere, «kontinuumsmässige», fragt, wie oft eine geschlossene Kurve zwischen zwei «nahe gelegenen» Punkten *0* und *1* hindurchgeht, und bestimmt diese Anzahl, indem sie die beiden Punkte in eine Umgebung einschliesst und diese topologisch auf das Innere eines ebenen Kreises abbildet. Auf Grund der zweiten Definition erkennt man übrigens leicht, dass die Charakteristik immer dann Null ist, wenn die beiden Wege α, β sich nicht treffen. Es fragt sich, ob sich nicht aus der zweiten Definition die wesentlichen Eigenschaften der Charakteristik direkt ableiten lassen, ohne den Durchgang durch eine Triangulation. Dies ist in der Tat, wenigstens wenn man die Mannigfaltigkeit als differenzierbar annimmt, leicht möglich; durch die Ausführung dieses Gedankens bin ich zu der Überzeugung gekommen, dass für die Theorie der Riemannschen Flächen und der Funktionen und Integrale auf ihnen das Heranziehen der kombinatorischen Topologie ein Umweg ist, der sich nicht lohnt. In einer in Vorbereitung begriffenen neuen Auflage meines alten Buches über Riemannsche Flächen wird dieser Erkenntnis Rechnung getragen werden.

Es lag mir daran, an diesem einfachen Begriff der algebraischen Schnittpunktszahl zweier geschlossener Kurven auf einer zweiseitigen Fläche zu zeigen, welche Schwierigkeiten der Mathematiker bei der präzisen Fassung solcher Begriffe antrifft und durch welche Methoden er sie überwindet. Ich hoffe, diese Ausführungen sind nicht ganz fehl am Platze in einem Heft, das dem Andenken PAUL NIGGLIS gewidmet ist, des Mannes, der so viel für die saubere Fassung der kristallographischen Grundbegriffe getan hat.

Summary

The characteristic of two closed curves on an oriented surface is, roughly speaking, the algebraic sum of their intersections, counting a crossing from left to right as $+1$, a crossing in the opposite sense as -1. To make this definition precise, combinatorial topology replaces the surface by an aggregate of 0-, 1- and 2-cells and first assumes one curve to be a chain (joining the 0-cells), the other to be a co-chain (joining the 2-cells). The idea of homology (as defined by means of 'integral functions') and the law of antisymmetry then serve to pass from these specialized to arbitrary curves. This definition is here contrasted with another that arose from 'topologizing' the procedure by which the author in his *Idee der Riemannschen Fläche* (1913) constructed the Abelian integrals of the first kind: it makes no use of triangulation, but instead of the coverage of the surface by neighborhoods and their topological mapping onto circular disks. Finally the paper indicates the steps by which one proves the coincidence of both definitions.

160.

Bauer on Theory of Groups

Bulletin of the American Mathematical Society 40, 515—516 (1934)

The concept of group is one of those very few primitive and fundamental ideas which play the part of an ordering, suggestive, and guiding principle in all branches of mathematics and wherever mathematics is applied. It is therefore a great satisfaction to the mathematician that not only the concept and some obvious statements about it, but also the deepest and most interesting part of group theory, dealing with the representation of groups by linear transformations, prove to be of paramount import to quantum mechanics. This close relationship is a subject so attractive from all points of view—mathematical, physical, philosophical, and even aesthetic—that it has allured writer after writer. Group theory accounts for the kinematical or combinatorial aspect of quantum mechanics, in particular for the schemes of atomic and molecular spectra, leaving aside the manner in which dynamics vests this skeleton with flesh and blood. Two groups stand in the foreground: first, the group P of permutations (of the electrons in a given atomic structure)—because of the essential likeness of all electrons; second, the group R of rotations around the origin O in 3-dimensional euclidean space—because the kinematic and dynamic constitution of an atom with the fixed nucleus in O is spherically symmetric.

After the reviewer's treatise on group theory and quantum mechanics, E. Wigner, who first discovered and explored the whole domain, gave his competent account, and was followed by van der Waerden's comprehensive and perspicuous presentation. The book under review, addressing the French scientist, cannot claim the same originality as Wigner's or van der Waerden's treatments. Nor does it cover the whole ground; it omits the more difficult parts: the general theory of the exchange phenomenon (group of permutations) and its applications to the classification of spectral lines and to chemical binding, quantum theory of radiation and quantization of the material and electromagnetic field equations, finally all things that are linked up with relativity including Dirac's dynamics of the spinning electron. The book intends to give an introduction only, by developing the fundamental ideas and illustrating them by the easiest accessible physical instances. Within these limits it is a thoroughly readable, clear, and reliable narrative of the mating between groups and quanta.

Chapters I (Vector space unitary geometry), II (Principles of quantum mechanics) and III (Group theory), are arranged very similarly to the same chapters of the reviewer's book mentioned above. When defining the moment of momentum in Chapter II, the author grazes its group theoretical significance at once: namely, that it consists of the operators corresponding to the infinitesimal rotations of the functional space of wave functions. The theory of perturbation is given only as an answer to the question how the energy terms shift and split under the influence of the perturbing forces; quasi degeneracy (energy levels differ by quantities of the same order as the perturbing energy) is discussed along with actual degeneracy. The other problem, how the probability of a quantum state ascertaining a definite value to the unperturbed energy varies in time under the influence of perturbation, is not discussed. The examples in this part are scarce. The reader looks in vain for treatment of the electron moving in a Coulomb field, for any Stoss-problems. Heisenberg's general scheme corresponding to Hamilton's canonical equations in classical mechanics, with its commutation rules and canonical transformations, is likewise omitted. The quantum theory of interaction between matter and radiation lies beyond the scope of the present work. Zeeman and Stark effects are taken up in the last chapter of the book in connection with the group of rotations.

Chapter III makes use of the transcendental methods of group theory and the calculus of characters. In an appendix, however, all irreducible representations of the group R of rotations are derived by Cartan's infinitesimal, purely algebraic, and hence "elementary" approach. (The last point at which one still needed recourse to a transcendental consideration in the quantum applications of group theory—full reducibility of the representations of R—has recently been settled by Casimir and Pauli; more generally for all semi-simple infinitesimal groups by van der Waerden.)

Chapter IV is dedicated to "General applications to quantum mechanics." It centers around Wigner's theorem: a transformation in the configuration space leaving the Hamiltonian invariant sends every quantum state into a quantum state of the same energy level. The little that is said about permutations finds its place here.

Chapter V (Rotations in space) deals in detail with the quantum mechanical consequences of the rotational symmetry of space: inner quantum number j, rules of selection and intensity. The Clebsch-Gordan decomposition of the Kronecker product of two irreducible representations D_j and $D_{j'}$ of R:

$$D_j \times D_{j'} = D_{j+j'} + D_{j+j'-1} + \cdots + D_{|j-j'|}$$

shows how the moment of momentum behaves when two kinematically independent systems are added to form a single system. The spin phenomenon is predicted in a general way from the double-valued representations of R (half integral j's) only after this has been discussed in detail from the mathematical standpoint. Pauli's idea of two-component wave function turns up and accounts for the distinction between azimuthal and inner quantum number (orbital, spin, and total momentum). Here the narrative breaks off just before the next event of the drama, after Schroedinger's great discovery: Dirac's relativistically invariant wave equation.

The book will prove useful for both mathematical and physical readers.

161.

Cartan on Groups and Differential Geometry

Bulletin of the American Mathematical Society 44, 598—601 (1938)

This book, which originated from a course of lectures given in 1931–1932 at the Sorbonne, covers in a somewhat more explicit form essentially the same material as no. 194 (1935) of the Actualités Scientifiques et Industrielles (see the review, this Bulletin, vol. 41 (1935), p. 774). By means of the method of the repère mobile the author studies arbitrary manifolds M_λ in a Klein space R whose geometry is described by its group of automorphisms. The chief aim of this review shall be to bring out the axiomatic foundations of the theory.

Coördinatization of a space R consists in a one-to-one mapping of the points A of R upon a manifold Σ of (numerical) symbols x serving as coördinates. In a Klein space such coördinatization is possible only with respect to a frame of reference, or briefly frame, $\mathfrak{f}$. An abstract group G and a realization of it by means of one-to-one transformations of Σ are supposed to be given. We thus deal with four kinds of objects: points A, symbols x, frames $\mathfrak{f}$, and group elements s, their mutual relation being established by two axioms (a) and (b):

(a) Any pair of frames $\mathfrak{f}, \mathfrak{f}'$ determines a group element $s=(\mathfrak{f}\to\mathfrak{f}')$ called the transition from $\mathfrak{f}$ to $\mathfrak{f}'$. Vice versa, any element s of G carries a given frame $\mathfrak{f}$ into a uniquely determined frame $\mathfrak{f}'$ such that $s=(\mathfrak{f}\to\mathfrak{f}')$. Succession of transitions $\mathfrak{f}\to\mathfrak{f}'\to\mathfrak{f}''$ corresponds to the composition of group elements:

$$s = (\mathfrak{f}\to\mathfrak{f}'),\quad t = (\mathfrak{f}'\to\mathfrak{f}'')\quad \text{imply}\quad ts = (\mathfrak{f}\to\mathfrak{f}'').$$

(The identical element is $\mathfrak{f}\to\mathfrak{f}$, and $\mathfrak{f}'\to\mathfrak{f}$ is the inverse of $\mathfrak{f}\to\mathfrak{f}'$.)

(b) With respect to a given frame $\mathfrak{f}$ each point A determines a symbol $x=(A, \mathfrak{f})$ as its coördinate, thus setting up a one-to-one correspondence $A\rightleftarrows x$ between R and Σ. The coördinate $x'=(A, \mathfrak{f}')$ of A in another frame $\mathfrak{f}'$ arises from x by the transformation associated with the group element $s=(\mathfrak{f}\to\mathfrak{f}')$ in the given realization.

Consequences: $\mathfrak{f}, \mathfrak{f}^*$ being any two frames, the equation $(A^*, \mathfrak{f}^*)=(A, \mathfrak{f})$ defines a one-to-one mapping $A\to A^*$ of R upon itself, the *space automorphism* $\{\mathfrak{f}, \mathfrak{f}^*\}$. If a group element t changes $\mathfrak{f}, \mathfrak{f}^*$ into $\mathfrak{g}, \mathfrak{g}^*$, one evidently has at the same time $(A^*, \mathfrak{g}^*) =(A, \mathfrak{g})$. Hence the space automorphisms $\{\mathfrak{f}, \mathfrak{f}^*\}$ form a group isomorphic with G; but their isomorphic mapping onto the elements of G is fixed except for an arbitrary inner automorphism of the group G. Figures in R which arise from each other by space automorphisms are considered *equal*.

We are concerned with λ-dimensional parametrized manifolds

$$M_\lambda:\quad x = x(t_1, \cdots, t_\lambda),$$

where t_α are real parameters. Let us employ for a moment ordinary real coördinates in R, $x=(x_1, \cdots, x_n)$. At a given point $A=(t_1, \cdots, t_\lambda)$ of M_λ the functions $x_i(t_1, \cdots, t_\lambda)$ and their derivatives up to a given order p constitute a contact element of order p, or briefly a p-spread. We obtain a succession of such spreads of orders $p=0, 1, 2, \cdots$, each of which is contained in the following. The central problem consists in deciding when two parametrized manifolds M_λ, M_λ' are equal. In the analytic case one may ask instead under what circumstances two given p-spreads are equal, and then apply such a criterion to the aforementioned succession of spreads of orders $p=0, 1, 2, \cdots$. We start with the lowest case, $p=0$, equality of two points.

All symbols arising from a given x by the operations of our group form a *layer* $\phi=\Sigma(x)$ in Σ. The *point* A belongs to ϕ if its coördinate with respect to some, and hence to every, frame $\mathfrak{f}$ lies in ϕ; in a manner independent of the frames the stratification of Σ transfers to R. Let us assume that Σ, when considered as the manifold of its layers, is m-dimensional and that $k_1, \cdots, k_m$ are parameters whose simultaneous values characterize and distinguish the individual layers within this manifold. They can be used as point invariants (invariants of order 0) thus solving the central problem for points. Cartan demands construction in the simplest possible way of a manifold Σ_0 in Σ intersecting each layer at exactly one point x_0; $k_1, \cdots, k_m$ are introduced as parameters on Σ_0.

With every x_0 on Σ_0 there is associated the subgroup $G^0=G(x_0)=G(k_1, \cdots, k_m)$ of G whose elements are realized by transformations leaving x_0 fixed. We assume G to be an r-parameter continuous Lie group, and we denote by $\omega_1, \cdots, \omega_r$ a basis for the components of the generic infinitesimal element of G. Moreover, we ascertain a basis

$$\pi_i = c_{i1}(k)\omega_1 + \cdots + c_{ir}(k)\omega_r, \qquad i = 1, \cdots, m,$$

for those components whose vanishing characterizes the infinitesimal elements of $G(x_0)$; Cartan calls them the *principal components of order* 0. For a point A belonging to the layer $\Sigma(x_0)$ the *frames* $\mathfrak{f}$ *of order* 0 are introduced by the requirement

$$(A, \mathfrak{f}) = x_0.$$

Transition among frames of order 0 is accomplished by means of the elements of the subgroup $G(x_0)$.

From a single point A we pass on to a pair of points AA', or more specifically to a line element AA' issuing from A (A' infinitely near A), asking again when two such line elements AA', BB' are equal. Let A, A' belong to the layers $\Sigma(x_0)$, $\Sigma(x_0')$, respectively; the frames $\mathfrak{f}$, $\mathfrak{f}'$ shall be restricted by the equations

$$(1) \qquad (A, \mathfrak{f}) = x_0, \qquad (A', \mathfrak{f}') = x_0'.$$

We want to know when there exists a frame $\mathfrak{g}$ such that

$$(A, \mathfrak{f}) = (B, \mathfrak{g}), \qquad (A', \mathfrak{f}) = (B', \mathfrak{g}).$$

We introduce the frame $\mathfrak{g}'$ arising from $\mathfrak{g}$ by the transition $\omega=(\mathfrak{f}\rightarrow\mathfrak{f}')$ and then obtain

$$(A, \mathfrak{f}) = x_0, \qquad (A', \mathfrak{f}') = x_0' \mid (B, \mathfrak{g}) = x_0, \qquad (B', \mathfrak{g}') = x_0',$$
$$(\mathfrak{f}\rightarrow\mathfrak{f}') = (\mathfrak{g}\rightarrow\mathfrak{g}').$$

A full system of invariantive characteristics of AA' therefore consists of the values k_i, $k_i' = k_i+dk_i$ of the invariants at the points A and A', and of the set of transitions $\omega=(\mathfrak{f}\rightarrow\mathfrak{f}')$ among frames $\mathfrak{f}$, $\mathfrak{f}'$ satisfying (1). We choose $\mathfrak{f}'$ infinitely near to $\mathfrak{f}$ so that ω is infinitesimal. With $\mathfrak{f}$ fixed, the variation of $\mathfrak{f}'$ leaves the m components π_i untouched while assigning perfectly arbitrary values to the remaining $r-m$ components. If one passes from $\mathfrak{f}$, $\mathfrak{f}'$ to another pair by means of group elements s, $s+ds$ in $G(x_0)$, $G(x_0')$, respectively, then ω changes into

$$\bar{\omega} = s^{-1}\omega s + s^{-1}ds.$$

In other words, the π_i undergo a nonhomogeneous linear substitution

$$\bar{\pi}_i = \sum_k \sigma_{ik}\pi_k + \delta\sigma_i$$

depending on s only.

We apply our remark to all points $A'=(t_\alpha+dt_\alpha)$ in the neighborhood of a given point $A=(t_\alpha)$ on a parametrized manifold M_λ. On emphasizing dependence on the line element (dt_α) we shall get equations

$$dk_i = k_{i1}dt_1 + \cdots + k_{i\lambda}dt_\lambda, \qquad i = 1, \cdots, m \tag{2}$$

and with respect to an arbitrary frame $\mathfrak{f}$ of order 0 in A,

$$\pi_i = b_{i1}dt_1 + \cdots + b_{i\lambda}dt_\lambda. \tag{3}$$

The matrix $\|b_{i\alpha}\|$ appears here as coördinate in a certain Klein space $R^{(1)}$ with the group $G(x_0)=G^0$, each element s of G^0 being realized by a transformation of the coördinate of the special type

$$\bar{b}_{i\alpha} = \sum_k \sigma_{ik}b_{k\alpha} + \overset{0}{\sigma}_{i\alpha}.$$

With the new Klein space we may proceed as before, and we then will obtain a number of invariants $k_j^{(1)}$ of order 1 for our 1-spreads, a subgroup $G^{(1)}$ of G^0 depending on the values of the new invariants, and the "frames of order 1" forming a subclass of the frames of order 0 and turning into each other by means of the elements of $G^{(1)}$. A basis for those components of the generic infinitesimal element of G whose vanishing characterizes the elements of $G^{(1)}$ may be ascertained by adding to $\pi_1, \cdots, \pi_m$ some further components

$$\overset{(1)}{\pi_j} = \overset{(1)}{c_{j1}}\omega_1 + \cdots + \overset{(1)}{c_{jr}}\omega_r. \tag{4}$$

The full invariantive characterization of a 1-spread consists in giving the values of k_i, $k_{i\alpha}$, and $k_j^{(1)}$. The coefficients $c^{(1)}$ in (4) depend on these arguments.

In the same fashion as from 0 to 1, one passes from order $p-1$ to p. The shrinkage of the group $G\supset G^0\supset G^{(1)}\supset \cdots$ must come to a standstill after a finite number of steps, and with the next step the production of new invariants will stop. On a parametrized manifold M_λ all these invariants are functions of the parameters t_α. Their coincidence for two such manifolds M_λ, M_λ' is necessary and sufficient for the manifolds to be equal. This unicity theorem is accompanied by a corresponding existence theorem. For one-dimensional manifolds the invariants are subject to no restriction, while in the case of several parameters t_α certain integrability conditions must be satisfied.

It is possible, and becomes sometimes desirable, to investigate special classes of manifolds by imposing conditions on the invariants.

Curves in euclidean 3-space E_3 are one of the simplest examples of our theory. With respect to a cartesian frame, we have triples of numbers (x_1, x_2, x_3) as coördinates; the space is homogeneous, and Σ_0 consists of the one symbol (0, 0, 0). The frames with the point A of the curve as their origin are of order 0; those whose first axis, moreover, coincides with the tangent, are of order 1; and finally the Frenet trihedral is the one frame of order 2. Here the process comes to a standstill. There is no invariant of order 0, but there is one each of the orders 1, 2, 3, namely ds/dt (t being the parameter, ds the arc element), curvature ρ, and torsion τ, respectively. For the minimal curves in the (complex) E_3 for which the invariant ds/dt vanishes, the above normalization of the frames of order 1 becomes impossible; they therefore require a separate treatment illustrating our remark in the preceding paragraph. The same is true for the plane curves with $\rho=0$.

Slightly more complicated is the study of manifolds without a fixed parametrization. One then wants to know when two manifolds,

$$M_\lambda:\ x = \phi(t_1, \cdots, t_\lambda) \qquad \text{and} \qquad M_\lambda:\ x = \phi'(\tau_1, \cdots, \tau_\lambda),$$

may be changed into each other by a suitable automorphism combined with a suitable transformation of the parameters. Both influences, of the arbitrary frame $\mathfrak{f}$ and of the arbitrary parameters t_α, have to be taken into account. In that case one must combine the coefficients $k_{i\alpha}$ with the $b_{i\alpha}$ in (2) and (3) as coördinates in a Klein space $R^{(1)}$, since the transformation of the t_α affects the $b_{i\alpha}$ as well as the $k_{i\alpha}$ (although the latter are indifferent towards a change of frame). It is this problem with which Cartan deals in the present book, and in some way he reduces the second influence, the choice of parameters, to the choice of the frame. I did not quite understand how he does this in general, though in the examples he gives the procedure is clear. To me it seems advisable to keep both factors apart from the beginning; the process itself tends to normalize both in mutual interdependence as it advances to higher and higher orders. The same situation is met with everywhere in differential geometry. For instance, riemannian spaces could be treated by introducing coördinates and attaching to each point A a frame, that is, a cartesian set of axes. Invariance is required with respect to arbitrary transformations of the coördinates and to orthogonal transformations of the frames which may depend arbitrarily on the point A. One knows how Gauss, Riemann, and Einstein got around the frames: the parameters once chosen define uniquely at each point an affine set of axes, and one takes advantage of it by treating cartesian geometry in terms of affine frames and a fundamental metric form rather than in terms of cartesian frames. Cartan goes here to the opposite extreme by normalizing the parameters in terms of the frames. I should advocate full impartiality on both sides as long as one deals with fairly general differential geometric problems.

The book under review pursues a three-fold purpose: it contains (1) an exposition of the general theory of finite continuous Lie groups in a terminology adapted to its differential geometric applications; (2) a general description of the method of repères mobiles; and (3) its application to a number of important examples. The arrangement is didactic rather than systematic. Thus the first examples, Chapters 1–3, on curves in E_3, minimal curves in E_3, ruled surfaces in E_3 (considered as one-dimensional manifolds of lines), precede the general formulation. Chapters 4–9, 11, 13, 14, deal with Lie groups. While the topics (1) and (3) are given in full detail, the central problem (2) emphasized by this review is but briefly dealt with, at the beginning of Chapter 10 more from the standpoint of transformation groups, at the beginning of Chapter 12 more from the abstract standpoint. In both chapters there follow further applications: curves in the affine and the projective planes, and arbitrary surfaces in E_3. With the last example, the only one concerned with manifolds of more than one dimension, the integrability conditions turn up; although their rôle in the theory of Lie groups is amply discussed, their general formulation as an intrinsic part of the existence theorem in the theory of repèrage is omitted.

All of the author's books, the present one not excepted, are highly stimulating, full of original viewpoints, and profuse in interesting geometric details. Cartan is undoubtedly the greatest living master of differential geometry. This review is incomplete in so far as it has tried to lay bare the roots, rather than describe the rich foliage of the tree which his book unfolds before its reader. We should not let pass unmentioned Jean Leray's merit in molding the lecture notes he took into something which is a true book and yet catches some of the vividness of the original lectures. Nevertheless, I must admit that I found the book, like most of Cartan's papers, hard reading. Does the reason lie only in the great French geometric tradition on which Cartan draws, and the style and contents of which he takes more or less for granted as a common ground for all geometers, while we, born and educated in other countries, do not share it?

162.

Courant and Hilbert on Partial Differential Equations

Bulletin of the American Mathematical Society 44, 602—604 (1938)

Thirteen years have elapsed between the publication of the first volume of Courant-Hilbert, *Methoden der mathematischen Physik*, and this concluding second volume. The two volumes are a beautiful, lasting, and impressive monument of what Courant, inspired by the example of his great teacher Hilbert and supported by numerous talented pupils, accomplished in Göttingen, both in research and advanced instruction. Courant came to Göttingen at a time of enormous political and economic difficulties for Germany, on a difficult inheritance, with the day of the heroes, Klein, Hilbert, and Minkowski drawing to a close. But by research and teaching, by personal contacts, and by creating and administering in an exemplary manner the new Mathematical Institute, he did all that was humanly possible to propagate and develop Göttingen's old mathematical tradition. How his fatherland rewarded him is a known story. The publication of the present volume seems to the reviewer a fitting occasion for expressing the recognition his work has earned him in the rest of the mathematical world.

The first volume treated a closed, relatively unramified field: the doctrine of eigen values and eigen functions. It met with enormous success, in particular among the physicists, because shortly after its publication these matters, due to Schrödinger's wave equation, gained an unexpected importance for quantum physics. This second volume covers the theory of partial differential equations in all of its aspects which are of importance for the problems of physics. Its table of contents is therefore necessarily much more variegated, preventing the book from attaining an equally perfect esthetic unity. But it makes up for this by its wealth of material, and it shares the first volume's high didactic accomplishments. Nowadays many mathematical books do not seem to be written by living men who not only know, but doubt and ask and guess, who see details in their true perspective—light surrounded by darkness—who, endowed with a limited memory, in the twilight of questioning, discovery, and resignation, weave a connected pattern, imperfect but growing, and colored by infinite gradations of significance. The books of the type I refer to are rather like slot machines which fire at you for the price you pay a medley of axioms, definitions, lemmas, and theorems, and then remain numb and dead however you shake them. Courant imparts an insight into a situation which has manifold aspects and develops methods without disintegrating them into a discontinuous string of theorems; and nevertheless, the essential results stand out in clear relief. Numerous interesting examples help to enliven and clarify the general theories. In still another respect I found this volume comforting: when one has lost himself in the flower gardens of abstract algebra or topology, as so many of us do nowadays, one becomes aware here once more, perhaps with some surprise, of how mighty and fruitbearing an orchard is classical analysis.

Here follows a brief characterization of what the several chapters of the book contain.

Chapter 1. The usual basic concepts, preliminaries, and isolated elementary methods, namely: Discussion of the manifold of solutions for typical examples, the partial differential equation of a given family of functions, irreducibility of systems to a single equation, geometric interpretation of the equation of first order, complete and singular integrals. Special methods of integration (separation of variables, superposition), the Legendre transformation, linear and quasi-linear equations. The existence theorem for given initial values in the analytic case, treated by means of majorizing power series.

Chapter 2. Theory of characteristics, first illustrated by the quasi-linear equations. Solution of the equation of first order by means of its characteristic strips. The Hamilton-Jacobi theory, including its relationship to the calculus of variations and canonical transformations. The treatment exhibits all sides of these interrelated topics, which were favored subjects for Hilbert's lectures at Göttingen.

Chapters 3–7 deal in the main, though not exclusively, with differential equations of second order.

Chapter 3. The distinction between the hyperbolic, elliptic, and parabolic cases, even for systems of differential equations; the latter receive a more complete treatment than in any other place of which I know. Linear equations with constant coefficients. Plane waves with or without dispersion. Initial value and radiation problems, construction of more general solutions by the method of Fourier integrals, retarded potentials. Illustrative material: heat conduction, wave and telegraph equations. The appendix contains a particularly illuminating exposition of the Heaviside calculus, justifying it to a sufficiently wide extent by Laplace transformations and by a proper application of the process of dislocating paths of integration in the complex plane.

Chapter 4. Potential theory and elliptic equations. Potentials of mass distributions. Green's formula, the Poisson integral, the mean value theorem with its inversion. The boundary value problem is dealt with both by the alternating process and by integral equations. Boundary value problem and parametrix for general elliptic equations. N. Wiener's limit theorems.

The most original part of the book is probably the study of hyperbolic equations contained in Chapters 5–6. The physical ideas of a wave front, of propagation, of rays, the iconal, Huygens' principle and construction have to be rendered in a mathematically tenable and sufficiently general form. Riemann's method of integration, Picard's application of successive approximations, Hans Lewy's integration by introducing the parameters of the characteristic curves as independent variables, and finally Hadamard's profound investigations, which here find a simplified and very lucid exposition, pass muster one after the other and serve to discuss the initial value and the radiation problems, and to settle the questions of unicity and "causality," that is, the question on which part of the initial data (Abhängigkeitsgebiet) the value of the solution at a given point actually depends. Linear equations with constant coefficients, in particular the classical wave equation, receive their due special attention. Huygens' principle is proved both for the initial value and the radiation problem, with emphasis on the difference existing in this respect between even and odd dimensions. H. Lewy's method is used for proving the analyticity of the solutions of analytic elliptic equations. Very interesting are the mean value theorems, in particular one of considerable generality due to Courant's pupil, Asgeirsson, and their applications. The physical side of the problems is not neglected. The relations between physical intuition and mathematical theory are carefully exhibited and illustrated in detail by such important examples as the differential equations of hydrodynamics and of crystal optics.

The seventh chapter deals with the boundary and eigen value problems by means of the calculus of variations. This direct method is here worked out in more detail than has been done before and is applied to transversal deformations and vibrations of plates and to the general 2-dimensional problem of elasticity. The chapter properly concludes with Plateau's problem which has recently entered upon a new phase through J. Douglas' ideas, and to which Courant himself has contributed largely during his last New York years.

The present volume is sprinkled throughout with a wealth of little new illuminating observations which this review had to skip. The author apologizes that lack of time prevented him from fitting out his book with a full sized index of literature and such paraphernalia. The same reason may be responsible for quite a few misprints on which the reader will occasionally stumble. But perhaps even these minor faults deserve praise rather than blame. Although I know that a craftsman's pride should be in having his work as perfect and shipshape as possible, even in the most minute and inessential details, I sometimes wonder whether we do not lavish on the dressing-up of a book too much time that would better go into more important things.

163.

Review: The philosophy of Bertrand Russell

The American Mathematical Monthly 53, 208—214 (1946)

This is the fifth volume of *The Library of Living Philosophers*, the previous volumes dealing with John Dewey, George Santayana, Alfred North Whitehead, and G. E. Moore. The basic thought behind the enterprise is to help remove one of the greatest obstacles to fruitful discussion in philosophy, namely "the curious etiquette which apparently taboos the asking of questions about a philosopher's meaning while he is alive" (F. S. C. Schiller). There is no denying the fact that different contemporary interpreters find different ideas in the writings of the same philosopher. Could not at least some clarification be brought about by putting these varying interpretations and critiques before the philosopher while he is still in our midst, and let him act both as defendant and judge? That is the way in which the articles in this volume approach Bertrand Russell's philosophy. It is opened by one of the most charming essays Mr. Russell ever wrote, a brief autobiography entitled "My Mental Development." There follow 21 descriptive and critical articles on R's philosophy, covering the same extremely wide range as the protagonist's work: mathematics and logic, theory of knowledge, the philosophical aspects of science, of language and history; social, political and educational philosophy; psychology, metaphysics, ethics and religion. In 61 pages Russell boldly faces and responds to this not always too consonant chorus of philosophical voices, answering their questions, meeting a fair part of their many criticisms, and rectifying misinterpretations. A carefully compiled bibliography of his publications (45 pages) greatly increases the usefulness of the book for those who are anxious to study Russell's philosophy at the source. Also the detailed index deserves a word of praise.

The articles—how could it be else?—vary greatly in tone and value. Some, as, *e.g.*, Reichenbach's on Logic, are content to outline R's essential ideas in the field under discussion and ask a few relevant questions. Others attempt a detailed critical evaluation. Such, as one would expect, is Gödel's article on Mathematical Logic. An example of the best one can hope to accomplish in a philosophical dispute is the exchange of views on Philosophy of Science between E. Nagel and R. Here the disputants, critical but fair-minded, meet as it were on the same level. Einstein's criticisms are less detailed but sustained by his profound experience in the actual creation and operation of physical theories,—an experience which Russell lacks. Einstein is a great admirer of Russell's work, to the reading of which, as he confesses, he owes innumerable happy hours. Let us pass in silence other articles where the critics show themselves clearly incapable of doing justice to R's thoughts or are more anxious to ride their own hobby-horse into the arena than to face the issues raised by R's philosophy. In spite of such shortcomings, inevitable in a community where freedom of thought is treasured, the book proves that philosophical discussion can benefit greatly by concentration on the work of one philosopher, especially if that philosopher can still be asked for his own interpretation and is as versatile and frank as Russell in his answers. The book, however, does not provide, nor does it intend to provide, a substitute for first-hand contact with the original thought of the philosopher himself.

Great is the temptation for the reviewer to quote profusely from Russell's introduction which, in sketching his own life and intellectual adventures, conjures up a whole period. Speaking of his friends at Cambridge in the nineties and their activities he exclaims: "For those who have been young since 1914 it must be difficult to imagine the happiness of those days." Here is a little illuminating story: "I remember the precise moment, one day in 1894," so R tells us, "as I was walking along Trinity Lane, when I saw in a flash (or thought I saw) that the ontological argument is valid. I had gone out to buy a tin of tobacco; on my way back, I suddenly threw it up in the air and exclaimed, as I caught it: Great Scott, the ontological argument is sound." Can anyone, after this anecdote, doubt that Russell is constitutionally a philosopher and not a mathematician?

In his "Reply to Criticisms" Russell draws a sharp line between that part of philosophy which is concerned with facts, and that which depends on ethical considerations or value judgments. In his opinion ethical propositions should be expressed in the optative mood, not in the indicative. They express desires rather than facts, although he agrees with Kant that beyond expressing a desire "ethical judgments must have an element of universality. I should interpret '*A* is good' as 'Would that all men desired *A*'." For this reason, he continues, "I do not offer the same kind of defence for what I have said about values as I do for what I have said on logical or scientific questions." In his replies touching on social, ethical and educational matters Russell occasionally uses the weapon of irony, though more sparingly than one would perhaps expect. Even when he begins with a sentence like this: "Mr. Brightman's essay on my philosophy of religion is a model of truly Christian forbearance," he continues with a dispassionate and frank restatement of his religious attitude. In only one instance, Bode's paper on Educational Philosophy, he shoots back with sharp arrows.

From among the articles a few more which concern the theory of scientific knowledge ought to be cited, at least by title: G. E. Moore, to whom Russell in his youth owed his emancipation from Hegelianism, writes about R's Theory of Description, J. Feibleman on the Introduction to the Second Edition of the Principles of Mathematics, P. P. Wiener on the Method of R's (early) work on Leibnitz, A. P. Ushenko on his Critique of Empiricism, and R. M. Chisholm on the Foundations of Empirical Knowledge. In the face of such an *embarras de* *richesse* the reviewer plans to focus his attention on the two articles of the most undeniable interest to mathematicians, Reichenbach's and Gödel's essays on Logic, and on E. Nagel's discussion of R's Philosophy of Science. If ***Principia Mathematica*** can be called a critique of naïve set-theory, then Gödel's article is a critique of this critique, and what I, as the reviewer, am expected to write now, is a critique of a critique of a critique. This makes my task so difficult that I see no other way out than to base my review on a brief independent survey of the situation; it adds one more feature to the picture: the historical perspective.*

In the essay on Logic Reichenbach gives Russell his due credit for improving the formal apparatus of symbolic logic, in particular for the consistent use of propositional functions with their variables, and for his interpretation of the "If then" as material implication $\rightarrow$. Indeed, it is used in this sense throughout mathematics, and only material implication is a primitive logical operator like $\sim$, $\cap$, $\cup$. However, Reichenbach's account of the ramified theory of types misses the essential point. In all questions concerning the logical foundations of mathematics he has the tendency to oversimplify the issues. His remarks on pp. 43–44 turn Brouwer's wine into water. Neither Reichenbach nor Nagel challenges the Frege-Russell thesis that logic and pure mathematics are essentially identical and that the theory of natural numbers with its definitions and inferences by complete induction, is reducible to logic. The Survey shows why I do not believe in this thesis at all; I find, on the contrary, that since Burali-Forti's and Cantor's discovery of the antinomies, the whole drift of research has been away from it.

In his reply Russell argues for the law of excluded middle, and against the definition of truth in terms of verifiability. But he argues on a level that has little to do with Brouwer's deep insight into the fallacy of that law with regard to existential propositions. Russell wishes to keep room for such unverifiable truths as he feels are "necessary for the interpretation of beliefs which none of us, if we are sincere, are prepared to abandon." The following dictum of his should be taken to heart: "As logic improves, less and less can be proved." Reichenbach ventures beyond deductive logic by opening a discussion with Russell on the principles of inductive logic on which the inductive methods of natural science depend; but they find little common ground.

Kurt Gödel's paper on R's Mathematical Logic is the work of a pointillist: a delicate pattern of partly disconnected, partly interrelated, critical remarks and suggestions. It must have been very difficult to arrange them in the one-dimensional sequence to which the writer, unlike the painter, is forced by his medium of words. With caution and circumspection the argument touches on a great variety of possibilities (and pitfalls) of interpretation. Russell varies the formulation of his vicious circle principle as follows: "No totality can contain members *definable* only in terms of this totality, or members *involving* or *presupposing* this totality." He himself obviously thinks that it makes little differ-

* Mathematics and Logic, this MONTHLY, January 1946 [quoted as *Survey*].

ence whether one says "definable" or "involving" or "presupposing." But Gödel sees here three different principles, and the acceptance or rejection of one or the other depends above all on whether the members are *given* or *constructed*. I think, with the latter contrast he touches on a very significant point. As far as the Survey preceding this review elaborates the same point, it is nothing but a comment on Gödel's paper. The constructive definition of the lowest level in the hierarchy of levels, first given in my book *Das Kontinuum* some thirty years ago, is in full accord with this view. About Russell's axiom of reducibility Gödel says that "it is demonstrably false unless one assumes either the existence of classes or of infinitely many 'qualitates occultae'."

In addition to Russell's own work concerning the solution of the paradoxes, two other possible directions are mentioned in which, as Russell himself had pointed out, a "criterion for the existence of classes" might be sought, namely the theory of limitation of size exemplified by the Zermelo-Fraenkel axioms, and a zigzag theory toward which Gödel sees Quine's recent system moving.* After disposing of Ramsay's escape by means of propositions of infinite length, he envisages as a further possibility, though somewhat vaguely, a theory in which "every concept is significant everywhere except for certain 'singular points' . . . so that the paradoxes would appear as something analogous to dividing by zero. . . . Our logical intuitions would then remain correct up to certain minor corrections, *i.e.*, they could then be considered to give an essentially correct, only somewhat 'blurred' picture of the real state of affairs".

In order to appreciate his last sentence one must know that Gödel takes the paradoxes very seriously; they reveal to him "the amazing fact that our logical intuitions are self-contradictory." This attitude toward the paradoxes is of course at complete variance with the view of Brouwer who blames the paradoxes not on some transcendental logical intuition which deceives us, but on a gross error inadvertently committed in the passage from finite to infinite sets. I confess that in this respect I remain steadfastly on the side of Brouwer.

On the ground of all his experience Gödel makes a strong plea for the realistic standpoint where classes are conceived as real objects, namely as "pluralities of things," or as structures consisting of such pluralities, and he adds: "It seems to me that the assumption of such objects [classes or concepts] is quite as legitimate as the assumption of physical bodies and there is quite as much reason to believe in their existence. They are in the same sense necessary to obtain a satisfactory system of mathematics as physical bodies are necessary for a satisfactory theory of our sense perceptions and in both cases it is impossible to interpret the propositions one wants to assert about these entities as propositions about the 'data.' Logic and mathematics (just as physics) are built upon axioms with a real content which cannot be 'explained away'." In his opinion the two major attempts in this realistic direction are "the simple theory of types (in an appropriate interpretation) and axiomatic set theory." He adds the warning: "Many symptoms show only too clearly however that the primitive concepts need further clarification."

---

* This MONTHLY, vol. 44, 1937, pp. 70–80.

Gödel hardly mentions Brouwer and Hilbert, and the last remarks sound as if he were prepared to retire from their advanced positions to some axiomatic setup of the Zermelo type. Does this mean that he has given up all hope of settling things along the lines of Hilbert's approach, and is content to adjourn the problem of consistency *ad calendas Graecas*? It is impossible to discuss realism in logic without drawing in the empirical sciences. In defending his program against Brouwer, Hilbert pointed emphatically to the situation in theoretical physics. The individual physical statements and laws have no "meaning" verifiable in immediate observation; only the system as a whole can be confronted with experience. Here consistency is absorbed into the farther-reaching requirement of "*concordance.*" Every indirect determination of the value of a physical quantity in a concrete case (*e.g.*, the charge of an electron) is based on theories establishing functional relations between that quantity and others amenable to direct observation (as, *e.g.*, position of a pointer on a scale). Concordance means that, with due regard to the inaccuracy of measurements, all such determinations must lead to the same value. We never can be sure whether such concordance, however complete for the moment, will still survive when our observations expand and become more accurate. Why then should we wish that consistency, the mathematical part of concordance, be assured *a priori* for all future? Is it not enough that the system has met the test of all our elaborate mathematical experiments so far? Will it not be early enough to change the foundations when, at a later stage, discrepancies appear? That is a position against which I cannot find much to say. But what shall be the guiding principles of our theoretical constructions? Gödel, with his basic trust in transcendental logic, likes to think that our logical optics is only slightly out of focus and hopes that after some minor correction of it we shall see *sharp*, and then everybody will agree that we see *right*. But he who does not share this trust will be disturbed by the high degree of arbitrariness involved in a system like Z, or even in Hilbert's system. What is the supporting faith? Success alone cannot be the answer. How much more convincing and closer to facts are the heuristic arguments and the subsequent systematic constructions in Einstein's general relativity theory, or the Heisenberg-Schrödinger quantum mechanics! A truly realistic mathematics should be conceived, in line with physics, as a branch of the theoretical construction of the one real world, and should adopt the same sober and cautious attitude toward hypothetic extensions of its foundations as physics exhibits. Theoretical physics today is in a healthier state than mathematics. But neither Gödel nor I can offer concrete suggestions for a cure, although Gödel seems to hope for clues from a careful study of Leibnitz's notes on his project of a *characteristica universalis*.

By talking about physics we have already passed into the domain of Nagel's controversy with Russell on Philosophy of Science. Unfamiliar with R's pertinent systematic publications, I must limit my discussion to this controversy as it reflects itself in my own mind. The following is typical for the pointed arguments by which Russell sometimes likes to reduce an intricate epistemological problem to a simple formula: "Common sense says: 'I see a brown table.' It will agree to both the statements: 'I see a table' and 'I see something brown.' Since, according to physics, tables have no color, we must either (a) deny physics, or (b) deny that I see a table, or (c) deny that I see something brown. It is a painful choice; I have chosen (b), but (a) or (c) would lead to at least equal paradoxes." In looking upon things as "bundles of qualities" and letting physics tell us that tables have no color, Russell remains pretty close to the favored epistemological slogan of the 19th century that only the raw material of sensations is immediately given to us, and to the doctrine, as old as Democritus and Huyghens, that nothing but "the atoms and their motions exist in reality" or that light and color are "in reality" oscillations of the ether. But in the development of physics the physical concepts have revealed themselves more and more as free constructions, mere *symbols* handled according to certain rules; theoretical physics becomes a system as thoroughly formalized as Hilbert's mathematics. Concerning the other side, the "given," many philosophers are now willing to agree that the sensory data, far from being given in their naked purity, are theoretical abstractions, while the true raw material is the manifest disclosed world in which man finds himself in his everyday life. It is quite characteristic that Hilbert bases his mathematics on the practical manipulation of concrete symbols rather than on some "pure consciousness" and its data. Thus quantum mechanics sets the physical phenomena symbolized by the Schrödinger-Dirac ψ's against the actual measurements for whose description our common sense "workshop" language is sufficient and adequate.

"Occam's razor," the search for a "minimum vocabulary," are some of the favored terms in which Russell describes the gist of his method. He pronounces as his supreme maxim of scientific philosophizing: "When possible substitute constructions out of known entities for inferences to unknown entities." Russell's goal is "to find an interpretation of physics which gives a due place to perceptions." Nagel criticizes in detail the steps by which he arrives (allegedly) at the twofold conclusion that (1) all the objects of common sense and developed science are logical constructions out of *events*, and that (2) our perceptions are events, but there exist many events which are not perceptions. I am inclined to side with Nagel, in particular when he says (p. 345, footnote) with regard to R's definition of point-instances in terms of events (dubious even from a purely mathematical standpoint): "One need only compare [this definition] with such analyses as those of Mach concerning mass and temperature, or those of N. R. Campbell concerning physical measurements to appreciate the difference between an analysis which is quasi-mathematics and an analysis which is directed toward actual usage." When Russell speaks of perceptions as events he does not seem to use the term in its "immanent" sense for which it is true that an actual perception is always *my* perception.

The physicist of today sees no need to use as material of his constructions some such ultimate entities as events; the pure symbols are enough for him. But he can handle these symbols and carry out his measurements only as a man living in and accepting our common sense world. Of course the connection between theory and observation must be understood, and it is the main task of an epistemological analysis to make this understanding explicit. Nagel comes to a similar conclusion: "It seems to me, therefore, that, instead of making the elimination of symbols for constructs the goal of the logical analysis of physics, a more reasonable and fruitful objective would be the following: to render explicit the pattern of interconnections between constructs and observations, on the strength of which these latter can function as relevant evidence for theories about the former." With this sober statement I think both Gödel and Einstein would agree.

But the debate on questions of this sort will probably go on forever. The volume under review is likely to stir up a host of thoughts in the mind of every reader. Even a mathematician must at times grapple with the problem to understand "what it is all about."

Verschiedenes

164.

Address of the President of the Fields Medal Committee 1954

Proceedings of the International Congress of Mathematicians (1954)

That at each International Mathematical Congress two gold medals be presented to two young mathematicians who have won distinction in recent years by outstanding work in our science: this was the intention of the late Professor J. C. Fields, the donor of the trust fund for these medals, and such the resolution adopted at the International Congresses in Toronto 1924 and Zurich 1932. In his Toronto memorandum Professor Fields expressed the wish that scientific merit be the only guide for the award of the medals, and he expressed the hope that the prizes would be considered by the recipients not only as an acknowledgment of past, but also as an encouragement for future, work.

Owing to the turbulent conditions of the world, award of the Fields Medals has up to now taken place but twice, at the Oslo Congress in 1936 and at Harvard, in 1950. As Chairman of a Committee consisting of the following members: E. Bompiani, F. Bureau, H. Cartan, A. Ostrowski, Å. Pleijel, G. Szegö, E. C. Titchmarsh and the speaker, I have the great honor and pleasure to perform this ceremony here-now for the third time. Our Committee owes its origin to the action of Prof. J. G. van der Corput as nominee of Het Wiskundig Genootschap for the presidency of this Congress, Prof. J. A. Schouten as Chairman, Prof. H. D. Kloosterman as Vice-Chairman and Prof. J. F. Koksma as Secretary of its Organizing Committee. Not for the composition, but for the judgment of our Committee the responsibility is ours, and we assume it gladly and with a good conscience indeed. After much deliberation during which many names were discussed and which made us fully aware of the arbitrariness involved in the selection of just two from the array of mathematicians of outstanding merit, we finally agreed, and this decision was reached as unanimously as anyone could reasonably expect, to present the two gold medals, together with an honorarium of $ 1,500 each, to:

Professor *Kunihiko Kodaira,* and

Dr. *Jean-Pierre Serre.*

I hope the Congress as a whole will approve our choice. In justification of it let me say this: by study and information we became convinced that Serre and Kodaira had not only made highly original and important contributions to mathematics in recent years, but that these hold out great promises for future fruitful non-analytic (will say: non-foreseeable) continuation. Carrying out the Committee's resolution, I now call upon Professor Kodaira and Dr. Serre to receive from my hands the prizes awarded them — awarded them, to repeat the donor's words, as recognition of past, and encouragement for future, research work. In the name of the Committee I extend to you, Professor Kodaira, and to you, Dr. Serre, my heartiest congratulations.

Two precedents have established the custom to combine with the award of the Fields Medals a brief survey of the recipients' main mathematical achievements, in particular of those which most attracted the Committee's attention. In Oslo it was not its Chairman, Elie Cartan, but another of the Committee's members, Carathéodory, who discharged this duty. In 1950, at Harvard, Harald Bohr was Chairman of the Fields Medals Committee, and he did both: handed over the medals to Atle Selberg and Laurent Schwartz and delivered the *laudatio*. I am going to follow his example, though with considerable hesitation; for I realize how difficult it is for a man of my age to keep abreast of the rapid development in methods, problems and results which the young generation forces upon our old science; and without the help of friends inside and outside the Committee I could not have shouldered this burden at all. It rests more heavily on my than on my predecessors' shoulders; for while they reported on things within the circle of classical analysis, where every mathematician is at home, I must speak on achievements that have a less familiar conceptual basis. A report like this cannot help reflecting personal impressions. Hence I now speak for myself and no longer as Chairman of the Committee. Thus freed from irksome bonds, I start by confessing that I am deeply satisfied by the Committee's choice — though I hope they will admit that it was reached without undue pressure from my side.

In view of the difficulties mentioned and in view of certain common foundations of Serre's and Kodaira's work, I find it convenient to explain as briefly as I can a number of universal concepts before entering upon some of our laureates' individual achievements. Be prepared then to have to listen now to a short lecture on cohomology, linear differential forms, faisceaux or sheaves Kähler manifolds and complex line bundles [1]).

An n-dimensional manifold M may be covered by a finite or denumerable sequence of neighborhoods U^ϱ. Their non-empty intersections are denoted by $U^{\varrho\sigma} = U^\varrho \cap U^\sigma$, $U^{\varrho\sigma\tau} = U^\varrho \cap U^\sigma \cap U^\tau, \ldots$. *Co-chains* of dimension 0, 1, 2, ... are defined by associating with each $U^\varrho, U^{\varrho\sigma}, U^{\varrho\sigma\tau}, \ldots$ numbers $c^\varrho, c^{\varrho\sigma}, c^{\varrho\sigma\tau}, \ldots$ depending skew-symmetrically on the indices. Take a 1-dimensional cochain c with the coefficients $c^{\varrho\sigma}$ as example. Its 2-dimensional *co-boundary* ∂c is defined by associating with each $U^{\varrho\sigma\tau}$ the number

$$(\partial c)^{\varrho\sigma\tau} = c^{\varrho\sigma} - c^{\varrho\tau} + c^{\sigma\tau}.$$

A cochain c is *co-closed* and called a *co-cycle* if its co-boundary is zero, $\partial c = 0$.

---

[1]) Only an abbreviated version of this part was read in the oral address.

The co-boundary $\partial c'$ of a $(q-1)$-dimensional cochain c' is a q-dimensional cocycle, of which we say that it is *co-homolog* zero (~ 0) or a *bounding* cocycle. The linear manifold of the cocycles of dimension q modulo the bounding cocycles form the additive abelian cohomology group $\mathfrak{C}_q$. We write $\mathfrak{C}_q(R)$ or $\mathfrak{C}_q(I)$ according as to whether we admit to the "sheaf" of numbers c all real numbers or only the integers. (Unfortunately history forbids us to drop the prefix *co-* in all these terms.) The groups $\mathfrak{C}_q$ apparently depend on a definite coverage of the manifold by overlapping neighborhoods U^ϱ. But either by normalizing it in a suitable way, or by passing to ever more refined coverings one makes himself independent thereof.

The n-dimensional manifold M_n may be locally referred to n realvalued coordinates $x_1, x_2, \ldots, x_n$. With respect to them we can define covariant skew-symmetric tensors (or rather: tensor fields) of rank $q = 0, 1, 2, \ldots$. For instance, such a field f of rank 2 has skew-symmetric numerical components f_{ik} which are functions of the coordinates (while the indices i, k, of course, run from 1 to n). It is convenient to write f as the "linear differential form"

$$f = \frac{1}{2!} \sum_{i,k} f_{ik} (dx_i \wedge dx_k)$$

with a skew-symmetric product $dx_i \wedge dx_k$ of the differentials, because this suggests the transformation under transition to other coordinates. We therefore speak simply of *forms* of rank $q = 0, 1, 2, \ldots$. The invariant *derivative* df, called *rotation* in tensor analysis, of a form f of rank 2, is defined by

$$(df)_{ikl} = \frac{\partial f_{ik}}{\partial x_l} - \frac{\partial f_{il}}{\partial x_k} + \frac{\partial f_{kl}}{\partial x_i}$$

and is of the next higher rank 3. Similarly for all ranks. f is said to be *closed* if $df = 0$. The derivative $d\varphi$ of a form φ of rank $q-1$ is a closed form of rank q, which we call a *derived* form (or ~ 0). The closed forms of rank q modulo the derived forms constitute a linear manifold, the additive abelian group Γ_q of rank q. The parallelism to the theory of cochains is obvious. In his thesis de Rham replaced this parallelism by a true group-isomorphism. Today this connection is best established in terms of the new-fangled notion of faisceau first introduced by Leray. Princeton has decreed that 'sheaf' should be the English equivalent of the French 'faisceau'. Sheaves play an important part in Kodaira's as well as Serre's investigations.

The fact that a form f of rank q is closed may also be expressed by saying that the equation $d\varphi = f$ has a *local* solution φ everywhere; i.e. in each neighborhood U^ϱ there is a $(q-1)$-form φ^ϱ such that $f = d\varphi^\varrho$. Such a form φ^ϱ defined only in a neighborhood U^ϱ I call, slightly modifying Chern's terminology, germ of a form, or simply *germ*. The closed germs of rank q belong, so I will say, to the sheaf G_q. The local φ^ϱ is not uniquely determined by f, but one may add to it any closed germ ω^ϱ of rank $q-1$, $\omega^\varrho \in G_{q-1}$. Under what conditions will f be homolog zero? Provided this arbitrariness may be used in such a way as to make $\varphi^\varrho = \varphi^\sigma$ in each $U^{\varrho\sigma}$. One therefore forms $\Phi^{\varrho\sigma} = \varphi^\varrho - \varphi^\sigma$ in $U^{\varrho\sigma}$; being closed, this germ belongs to G_{q-1}. But it is determined only up to an additive term $(\partial\omega)^{\varrho\sigma} = \omega^\varrho - \omega^\sigma$. The several $\Phi^{\varrho\sigma}$ and $(\partial\omega)^{\varrho\sigma}$ make up a 1-dimensional cocycle Φ and bounding cocycle $\partial\omega$ respectively, not in the sheaf of numbers however, but in G_{q-1}. Thus f determines uniquely an element of the 1-dimensional cohomology group $\mathfrak{C}_1(G_{q-1})$ in the sheaf of germs G_{q-1}, in such a way that the corresponding element vanishes if and only if f is homolog zero. Following the argument in the opposite direction, one realizes that the resulting homomorphism $\Gamma_q \to \mathfrak{C}_1(G_{q-1})$ is in fact an isomorphism: $\Gamma_q \simeq \mathfrak{C}_1(G_{q-1})$. The isomorphic mapping of the one group onto the other has been constructed explicitly.

A similar argument leads to the sequence of isomorphisms

$$\mathfrak{C}_1(G_{q-1}) \simeq \mathfrak{C}_2(G_{q-2}) \simeq \ldots \simeq \mathfrak{C}_q(G_0),$$

and since G_0 is identical with the domain R of real numbers, we thus arrive at de Rham's result

$$\Gamma_q \simeq \mathfrak{C}_q(R). \tag{1}$$

The usefulness of faisceaux for algebraic geometry was recognized early in 1953 by Dr. Serre and D. C. Spencer.

I need a little more of tensor analysis in the form in which Einstein used it for his general relativity theory. On our n-dimensional manifold M we may also consider skew-symmetric contravariant tensor densities or "*co-forms*" of rank 0, 1, 2, ... I denote their components by German letters and upper indices: $\mathfrak{g}, \mathfrak{g}^i, \mathfrak{g}^{ik}, \ldots$. For them the invariant operator ∂ of derivation is the one called *divergence* in tensor analysis, leading for instance from a tensor density $\mathfrak{g}$ of rank 2 to a vector density $\partial\mathfrak{g}$ according to the equation

$$(\partial\mathfrak{g})^i = \sum_k \frac{\partial\mathfrak{g}^{ik}}{\partial x_k}.$$

If M is compact, a scalar density $\mathfrak{j}$ has an invariant integral $\int\mathfrak{j}$ over M.

Assume now that we are dealing with a Riemannian manifold M_n carrying a metric by dint of an invariant quadratic differential form

$$ds^2 = \sum_{i,k} \gamma_{ik} dx_i dx_k \quad (\gamma_{ik} = \gamma_{ki}) \tag{*}$$

with symmetric coefficients γ_{ik} depending on the variables x. Then one has the algebraic operation $g \leftrightarrows \mathfrak{g}$, which changes a form g into a co-form $\mathfrak{g}$ of the same rank, and vice versa, so that we can identify the forms with the co-forms. A relation like $\partial\mathfrak{f} = \mathfrak{s}$ for co-forms $\mathfrak{f}$, $\mathfrak{s}$ may therefore be written as $\partial f = s$ for the corresponding forms f, s. The derivative d *increases*, the co-derivative ∂ *decreases* the rank by 1. Incidentally, in this symbolism Maxwell's equations for the electromagnetic field f of rank 2 in the empty four-dimensional world may be simply written as $df = 0$, $\partial f = s$, where s is the electric current. With a form f and a co-form $\mathfrak{g}$ of rank 2 one can construct the scalar density

$$f \cdot \mathfrak{g} = \frac{1}{2!} \sum f_{ik}\, \mathfrak{g}^{ik}$$

and thus form the integral

$$\int f \cdot \mathfrak{g} = (f, g).$$

In the functional "space" of infinitely many dimensions, the elements or "vectors" of which are the forms f of rank p, we may introduce this number (f, g) as the scalar product of the two "vectors" f, g. It depends symmetrically on f and g, and if the metric ground form (*) is positive-definite (not indefinite as in the physical four-dimensional world), then (f, f) is also positive-definite, i.e. $(f, f) > 0$ except for $f = 0$. The fact that the integral over M of the divergence of a vector density is zero, entails the equation

$$(d\varphi, g) + (\varphi, \partial g) = 0 \tag{2}$$

for any $(p-1)$-form φ and p-form g.

A form h satisfying both conditions $dh = 0$ and $\partial h = 0$ is called *harmonic*. The central existence theorem in Hodge's book on the "Theory and application of harmonic integrals" published in 1941 asserts that every closed form f, $df = 0$, is homolog to a uniquely determined harmonic form h, $f - h \sim 0$. Consequently de Rham's result (1) leads to a definite *isomorphic mapping between the cohomology group $\mathfrak{C}_p(R)$ and the additive abelian group of harmonic forms of rank p*. A proof of Hodge's theorem is ready at hand if one operates in the space Σ of all closed forms f of rank p. The closed forms of rank p which are homolog zero, i.e. derivatives $d\varphi$ of forms φ of rank $p - 1$, form a linear subspace Σ_0 of Σ. As in a vector space of finite dimension, we split any vector f in Σ into one, f_0, lying in Σ_0, $f_0 \sim 0$, and one, h, perpendicular to Σ_0, i.e. satisfying the condition $(d\varphi, h) = 0$ for every form φ of rank $p - 1$. According to (2) the latter equation amounts to $(\varphi, \partial h) = 0$ for every φ, hence to $\partial h = 0$. Thus as a vector in Σ, the form h satisfies the condition $dh = 0$, as one perpendicular to Σ_0 the condition $\partial h = 0$; it is therefore harmonic, and we have

$$f = f_0 + h, \qquad f_0 \sim 0, \qquad \text{or } f \sim h.$$

Of course, this method of orthogonal projection needs justification in a functional space of infinite dimension, a justification that is essentially identical with that of Dirichlet's minimal principle. In passing you may notice that I have avoided speaking about cycles and integrals of closed forms over cycles (periods); instead I talked of co-cycles only.

Hodge applied the theory of harmonic forms to algebraic varieties. These are special *complex-analytic varieties*. Such a variety V_n of n complex and $2n$ real dimensions can be locally referred to n complex-valued coordinates $z_r = x_r + iy_r$; and two sets of such coordinates, z and z', in two overlapping neighborhoods U and U', are related to each other in $U \cap U'$ by a regular-analytic transformation of non-vanishing Jacobian. A *metric* can be introduced into the complex variety by an invariant positive-definite Hermitean form

$$\alpha = \sum_{r,s} \alpha_{rs} dz_r d\bar{z}_s \qquad (\bar{\alpha}_{rs} = \alpha_{sr}). \tag{3}$$

The coefficients α_{rs} are functions of the real variables x_r and y_r, or of z_r and $\bar{z}_r$. The conditions of symmetry $\bar{\alpha}_{rs} = \alpha_{sr}$ guarantee that the form is *real*. Such a metric is clearly a special Riemannian metric. The Hermitean symmetry of the coefficients α_{rs} evidently implies skew symmetry for the quantities $\omega_{rs} = i \,.\, \alpha_{rs}$ arising through multiplication by $i = \sqrt{-1}$. Hence with the metric ground form α there is associated a real linear differential form of rank 2,

$$\omega = \sum_{r,s} \omega_{rs}(dz_r \wedge d\bar{z}_s) \qquad (\bar{\omega}_{rs} = -\,\omega_{sr})$$

If ω satisfies the condition $d\omega = 0$, we speak of a *Kähler metric*. In introducing this notion, Kähler observed that any manifold imbedded into a Kähler manifold by substituting for $z_1, \ldots, z_n$ analytic functions of new variables is again a Kähler manifold. By the way, the equation $d\omega = 0$ implies $\partial\omega = 0$; ω is therefore harmonic. Since the complex projective space is obviously Kählerian, so is any algebraic variety imbedded in this space in a non-singular fashion. From the transcendental standpoint it is easier to deal with Kähler varieties in general than with the special algebraic ones, and experience has shown that many notions and results of algebraic geometry apply to them. In the complex domain we must distinguish not only ranks, but also *types* among the linear differential forms. For instance, the form

$$\sum \varphi_{rst}(dz_r \wedge dz_s \wedge d\bar{z}_t)$$

is of type (2, 1) and rank $2 + 1 = 3$, because of 2 factors dz and 1 factor $d\bar{z}$ in its general expression. Of special importance are the harmonic forms of type $(p, 0)$, $p = 0, 1, \ldots, n$. If V_n is compact, there exists only a finite number g_p of linearly independent such forms. (By the way, for a form f of type $(p, 0)$ on a Kähler variety the equation $df = 0$ implies $\partial f = 0$.)

As a last preparation I speak of fiber spaces F with the basis V_n, the fiber of which is the "line" of all complex numbers ζ with the understanding that the automorphic mappings of the individual fiber consist in the multiplications of ζ by an arbitrary complex constant $\neq 0$. I call these special fiber spaces *complex line bundles*. Thus if U, U' are two overlapping neighborhoods and the points of the part of the fiber space with basis U are represented by $z_1, \ldots, z_n; \zeta$, those with basis U' by $z'_1, \ldots, z'_n; \zeta'$, then in $U \cap U'$ the coordinates $z'_1, \ldots, z'_n$ are related to $z_1, \ldots, z_n$ by an invertible holomorphic transformation, and ζ and ζ' by a relation $\zeta' = f \,.\, \zeta$ where f is a nowhere-vanishing holomorphic function of the z (or the z') in $U \cap U'$. If V_n is covered by neighborhoods U^ϱ we have such transition factors $f = f_{\varrho\sigma}$ for each $U^{\varrho\sigma}$ and it is

$$f_{\varrho\sigma} f_{\sigma\tau} f_{\tau\varrho} = 1 \qquad \text{in } U^{\varrho\sigma\tau}. \tag{4}$$

These $f_{\varrho\sigma}$ determine the fiber bundle. But because of the relativity of the fiber coordinate ζ the factor system $f_{\varrho\sigma}$ may be replaced without changing the bundle by any equivalent factor system $f^*_{\varrho\sigma} = g_\varrho f_{\varrho\sigma} g_\sigma^{-1}$ where g_ϱ is any nowhere-vanishing holomorphic function in U^ϱ. (Equation (4) is a condition of multiplicative closure. Hence we may speak of the line bundles as elements of the multiplicative 1-dimensional cohomology group in the sheaf of non-vanishing holomorphic germs.) A *divisor* D on a complex variety V_n is locally, in a neighborhood U^ϱ, defined by a meromorphic function Φ_ϱ in U^ϱ, but in such manner that replacement of Φ_ϱ by $g_\varrho \cdot \Phi_\varrho$ does not change the divisor. By $\Phi_\sigma = f_{\varrho\sigma} \,.\, \Phi_\varrho$ we obtain transition functions $f_{\varrho\sigma}$ in $U^{\varrho\sigma}$ of the nature described before. Thus a divisor determines a line bundle, and two divisors the same line bundle if and only if they are equivalent. It is therefore convenient, with Spencer and Serre, to replace the notion of a class of equivalent divisors by that of a complex line bundle F over V_n. Kodaira introduced forms φ with coefficients in F. By this he means that form-germs ζ, ζ' defined in two overlapping neighborhoods $U = U^\varrho$, $U' = U^\sigma$ are not related by the equation $\zeta' = \zeta$, but by the equation $\zeta' = f_{\varrho\sigma} \cdot \zeta$ which describes the fiber coincidence in F.

I am sorry for this long introduction. But it will enable me to speak a little more intelligently, or intelligibly, on the most important of Kodaira's and Serre's achievements than I otherwise could have done.

Our Committee did not adopt a strict definition of what is meant by a 'young mathematician'. But if the limit is put at forty, then Kodaira is young — while Dr. Serre, born in **1926**, is young by any definition. Kunihiko Kodaira is a native of Japan and grew up in Tokyo. He studied mathematics and later also theoretical physics at Tokyo University and became in due time lecturer, assistant professor and finally full professor at that university, a position still held by him today. In 1949 he came to the Institute for Advanced Study in Princeton and has remained in America since then, at the Institute, at Johns

Hopkins and at Princeton University, to which he now belongs as a Research Associate. In a short biography written at my request he mentioned as the two main influences in his mathematical development: the teaching of Prof. S. Iyanaga and my book on Riemann surfaces. Of that I am very proud.

Kodaira's early papers deal with a number of questions in algebra, topology, group theory, almost periodic functions. They show good craftsmanship and originality, testifying also to the width of his mathematical interests; but I shall put here all the emphasis on his later work. There is first his fine treatment of the eigenfunction expansions for ordinary differential equations of order 2 and more generally of any even order, with singular ends. This work he began in Japan while still cut off from Western mathematics. After the distinction between the limit point and the limit circle case for a singular end, established as early as 1909, the main problem was how to determine explicitly the non-negative weight differential as function of the spectral parameter. This was accomplished independently by Titchmarsh and Kodaira; however I find Kodaira's approach more direct. The generalization to arbitrary even order, especially of the basic distinction between limit point and limit circle, is far from trivial.

But Kodaira's outstanding achievement lies in the theory of harmonic integrals and the numerous profound applications he made of it to Kählerian and more specially to algebraic varieties.

Riemann had based the theory of algebraic functions of one variable and their integrals on topology and construction of potentials on his Riemann surfaces by means of the transcendental Dirichlet principle. Later on, algebraic methods, Dedekind-Weber, Hensel-Landsberg, came to the fore. Weierstrass' function-theoretic treatment and the geometric approach by Clebsch and Max Noether occupy an intermediary position, but their constructions are also fundamentally algebraic. These methods carried the day when in the splendid development of the Italian school one passed from algebraic curves to algebraic varieties of higher dimensions. Algebra naturally will ask how much of the results originally stated for the coefficient field of all complex numbers survive if this field is replaced by other fields, in particular by fields of prime characteristic. Riemann's transcendental method, however, was not entirely forgotten, it suffices to mention the names of E. Picard and S. Lefschetz. But only with Hodge's book on Harmonic Integrals, in my opinion one of the great landmarks in the history of our science in the present century, its vast possibilities became evident; and Kodaira, beside Hodge himself, was the first to exploit them.

The central existence theorem in Hodge's book states that there is a uniquely determined harmonic form f of rank p with pre-assigned periods, or as I prefer to say, corresponding to a given element of the cohomology group

$\mathfrak{C}_p(R)$. Hodge's proof by means of the parametrix method, which E. E. Levi and D. Hilbert had applied to elliptic partial differential equations of the second order, contained a gap which at the time looked pretty serious. The gap was filled in 1942/43 by two people: Kodaira and the speaker; of course, independently of each other: for, as you may remember, our two nations, Japan and the USA, were at war in these years. While I stuck to the parametrix method, which had been suggested to Hodge by H. Kneser — to this proof de Rham soon gave the most elegant form —, Kodaira returned to the method of orthogonal projection, tried before by Hodge and essentially equivalent to the Dirichlet principle. It has the great advantage of being also applicable to non-compact Riemannian manifolds. In an impressive paper "Harmonic fields in Riemannian manifolds (generalized potential theory)" published in the Annals of Mathematics 1949 immediately after his arrival in the United States, Kodaira proves the existence of harmonic forms with prescribed singularities (and periods). A certain fundamental lemma needed for the justification of the principle of orthogonal projection is based here on the existence in the small of a fundamental solution; de Rham simplified this part in constructing that solution by the parametrix method. In the same paper Kodaira also gave the analog of Riemann-Roch's theorem for harmonic forms on a compact Riemannian manifold.

However, the relationship between harmonic and analytic functions is not as simple for several variables as it is for one. When turning from real Riemannian to complex Kählerian manifolds, Kodaira attacked the problem from its most difficult end, that of establishing the true Riemann-Roch theorem on algebraic varieties. From de Rham he had learnt that generalization of forms in the direction of Laurent Schwartz's *distributions* which de Rham called *currents*, and making use of a formalism developed by B. Eckmann and H. Guggenheimer, he derived a general existence theorem concerning complex currents. Then, in that wise limitation which, to use a word of Goethe's, shows the master, he first turned to the lowest case of complex dimension $n = 2$. The Riemann-Roch theorem which he obtained for a curve on the surface without multiple irreducible components, coincided with one derived by O. Goldman in algebraic fashion. It shortly afterwards enabled Kodaira and W. L. Chow to prove that a compact Kähler surface possessing two algebraically independent meromorphic functions is an algebraic surface without singularities. In '52 Kodaira passed on to the next higher case $n = 3$, — hard pioneering work that had to be done before one could hope for mastering algebraic varieties V_n of arbitrary dimension n.

The arithmetic genus γ of a compact Kähler variety V_n may be defined by means of the numbers g_p of the linearly independent harmonic forms of type

$(p, 0)$ (differentials of the first kind) as the alternating sum

$$\gamma(V_n) = g_0 - g_1 + g_2 - \cdots \pm g_n.$$

F. Severi had, by a so-called postulation formula, introduced two genera p_a and P_a for *algebraic* varieties V_n. (I tread here on hot soil, where I do not feel at home at all.) Let V_n be imbedded without singularities into an ambient complex projective space S. The hyperplanes in S define by their intersection with the variety V_n in S a class of equivalent divisors E. The genus $p_a(V_n)$ is obtained from studying the number of linearly independent meromorphic functions on V_n which are multiples of the power E^s of a divisor E of this class in its dependence on the variable positive integral exponent s. In a similar fashion $P_a(V_n)$ is obtained by means of the linear manifold of all multiples of E^s among the meromorphic n-fold differentials on V_n. (What is such a differential? A quantity represented in terms of local complex coordinates z on V_n by a meromorphic function Φ of z in such a way that, under the influence of an analytic coordinate transformation, Φ gets multiplied by the Jacobian.) — It is possible to generalize the genera γ, p_a, P_a from the underlying algebraic variety V_n to any divisor D on V_n.

By clever combination of algebraic computations with existential theorems ultimately derived from his theory of harmonic forms, Kodaira succeeded in proving one of Severi's conjectures, namely the identity of γ and P_a. I omit another famous conjecture settled by him in the same paper, concerning the so-called continuous systems on algebraic varieties. The identity of γ and p_a, also surmised by Severi, was proved not long afterwards by Kodaira and D. C. Spencer, after they had first investigated the structure of complex line bundles F over a Kähler variety V_n in several joint papers and Spencer and Serre had introduced the instrument of sheaves into this field of research. Replacing the divisor D by such a bundle F, one considers $\mathfrak{C}_q(\Omega_p(F))$, i.e. the q-dimensional cohomology group ($\mathfrak{C}_q$) in the sheaf of germs of holomorphic p-forms (Ω_p) with coefficients in F. In generalization of results due to P. Dolbeault, Kodaira showed that this group is isomorphic to the linear manifold $\Gamma_{p,q}(F)$ consisting of all harmonic forms of type (p, q) with coefficients in F and therefore is of finite dimension $d_{p,q}(F)$. The proof of the equation $\gamma = p_a$ operates with the Euler characteristic $\chi_p(F)$ defined as the alternating sum over q of these dimensions $d_{p,q}(F)$.

Finally I come to one of the most exciting results of Kodaira's investigations, published in the last number of the Ann. of Math. Hodge had remarked that algebraic varieties are not only Kähler varieties but that, in the general isomorphism between harmonic forms and cocycles modulo bounding cocycles, the harmonic ω of rank 2 connected with the Kähler metric corresponds to a

2-dimensional cocycle with *integral* coefficients. He called such Kähler varieties *of restricted type* and proved several properties for them that were known to hold for algebraic varieties. But now Kodaira has succeeded in showing that any Hodge variety, i.e. Kähler variety of restricted type, is *algebraic*, thus arriving at a very satisfactory intrinsic characterization of algebraic varieties. One can approach the same question also in opposite direction. You remember that a complex line bundle over a complex-analytic variety V_n was characterized by non-vanishing holomorphic transition factors $f_{\varrho\sigma}$ defined in $U^{\varrho\sigma}$. The automorphisms of a fiber consisted of the multiplications of the fiber variable ζ by an arbitrary non-vanishing complex constant. Multiplication can be changed into addition by passing to the logarithm. But the logarithm is not single-valued, and this has the consequence that the sum

$$\log f_{\varrho\sigma} + \log f_{\sigma\tau} + \log f_{\tau\varrho}$$

in $U^{\varrho\sigma\tau}$ is not necessarily zero but equals $2\pi i$ times an integer $n^{\varrho\sigma\tau}$; these integers are not uniquely determined, but just to that extent in which a 2-dimensional cocycle with integral coefficients modulo the bounding cocycles is determined; in other words, the bundle determines a "characteristic" element of the cohomology group $\mathfrak{C}_2(I)$. To it there corresponds a harmonic form ω of rank 2. It turns out that it is of type (1, 1) and real. But of course, the corresponding Hermitean form α with the coefficients $\alpha_{rs} = \omega_{rs}/i$ need not be positive-definite. However *if it is*, then we have exactly the situation considered by Hodge. Under these circumstances (I simplify his results a bit) Kodaira shows by an ingenious argument that all the cohomology groups $\mathfrak{C}_q(\Omega_p(F))$ reduce to zero for $q \geqq 1$. Making use of this fact for $p = 0$, he then succeeds in constructing a sufficient number of meromorphic functions on the variety, by means of which he effects its non-singular embedment in a projective space. His theorem is a profound generalization of the well-known fact that every compact Riemann surface belongs to an algebraic function field.

Kodaira has travelled a long and arduous road since he first developed the general theory of harmonic forms on a Riemannian manifold. But only if someone has the courage of attacking the primary concrete problems in all their complexity, will the general concepts gradually emerge which resolve the difficulties and ease the further progress; now the promised land seems to lie before him and his collaborators like an open plain. The last phase is characterized by a remarkable confluence of ideas to which, next to Kodaira, D. C. Spencer, F. Hirzebruch, R. Thom and also, last but not least, Serre contributed.

With this salto mortale I turn to the other of our two laureates of today, Dr. Jean-Pierre Serre. He is an ancien élève de l'Ecole Normale Supérieure, Docteur ès Sciences and at present Chargé de cours at Nancy University. In spring '52 he made his pilgrimage to Princeton. Considering the impressive line of first rank mathematicians France has in recent times given to our science, one need not look far for the formative influences that determined the beginnings of Dr. Serre's career. If one name is to be mentioned, I think, it would be that of Henri Cartan.

I have done Dr. Serre a real injustice, by passing over in silence in my report on Kodaira several results of Serre's which bore on the last phase of Kodaira's and Spencer's work. Let me try to make up for it! During 51/52 Serre in collaboration with H. Cartan succeeded in giving a brilliant reformulation and extension of some of the main results of complex variable theory in terms of cohomology in a complex-analytic sheaf. One of Serre's theorems in this line carries over de Rham's fundamental proposition referred to at the beginning of my address from real differential forms to complex holomorphic forms on a complex variety — provided this is of the type investigated by K. Stein. This shows at once that the Stein varieties have quite special topological properties, e.g. their cohomology groups $\mathfrak{C}_p(R)$ are zero for $n < p \leqq 2n$. Serre moreover proved a duality theorem concerning the cohomologies $\mathfrak{C}_q(\Omega_p(F))$ mentioned before — even in the more general case where the line of one complex variable ζ is replaced by a multi-dimensional vector space whose automorphisms are the non-singular linear transformations with complex coefficients. Serre also proved a theorem of finiteness of Kodaira's type, in one respect more limited, but in another respect much wider than his. By means of it Serre formulated a generalization of the Riemann-Roch theorem for algebraic varieties of arbitrary dimension. There is, beside the genera γ, p_a, P_a, a fourth, the Todd genus $T(D)$. Since '52, Hirzebruch had made an intensive investigation of the Todd genus after reducing its definition to a much more satisfactory form than originally given for it. By making use of Kodaira's results, including his most recent that every Hodge variety is algebraic, and Thom's theory of 'cobordisme', Hirzebruch was able, by ingenious calculations, to prove Serre's conjecture and to establish the identity of the Todd genus T with the arithmetic genus γ.

Hearing this, you may get the impression that our Committee did wrong in awarding the Fields Medals to two men whose research runs on such closely neighboring lines. This contact, however, has been established only during the last year and may well be a transient phenomenon. Serre's work before, which above all fascinated our Committee by the wealth of its surprising numerical results, is concerned with quite a different problem, the *homotopy theory of spheres*. Two mappings of the n-dimensional sphere S_n into a simply connected space M are *homotop* or equivalent to each other, and belong to the same *class*, if one can be continuously deformed into the other. In 1935 W. Hurewicz had given topology a new impetus by showing how homotopy classes may be composed; the resulting group of homotopy classes, the homotopy group, is abelian (except in the lowest case $n = 1$). Let us assume M also to be a sphere S_{n+k}, and denote then the homotopy group by $\pi(n+k, n)$. Since the case $k = 0$ has been settled long ago, we assume $k > 0$. At the Harvard Congress in 1950 Heinz Hopf chose this problem to illustrate the type of questions with which at the time topologists were mainly concerned. Said he: "Die gegenwärtige Situation in diesem Gebiet ist sehr unübersichtlich, und es ist verständlich und erfreulich, dass sich viele Geometer bemühen, hier Klarheit zu schaffen und ein Gesetz zu erkennen." Well, the man who has been most successful in carrying out this program during the intervening years is Serre, partly in collaboration with H. Cartan. Hopf had obtained quite early the result that $\pi(2n-1, n)$ is infinite for even n. Let Z_r denote the cyclic group of order r. H. Freudenthal found that the structure of the group $\pi(n+k, n)$ depends but on the difference k of the two dimensions, I denote it therefore by $\pi'(k)$, as soon as $n \geqq k+2$. Some isolated results had been obtained by various disparate methods, as e.g $\pi'(1) \simeq \pi'(2) \simeq Z_2$, Hurewicz' isomorphism $\pi(m, 2) \simeq \pi(m, 3)$ for $m \geqq 3$ and such strange ones as Eckmann's $\pi(m, 4) \simeq \pi(m-1, 3) + \pi(m, 7)$; $\pi(m, 8) \simeq \pi(m-1, 7) + \pi(m, 15)$. Serre's Comptes Rendus note of January '51, soon followed by his thesis, approached the problem by a new universal method which at once bore manifold fruit. The mystery began to be pierced. One now knows that $\pi(2n-1, n)$ for even n are the *only* infinite homotopy groups, and that each of them is the direct sum of Z_∞ and a finite group. The groups $\pi'(4)$, $\pi'(5)$, $\pi'(12)$ are zero, and one knows explicitly all $\pi(n+k, n)$ for $k \leqq 8$ and arbitrary n. Some strange laws in which the prime numbers p play a role have emerged; e.g. the p-primary component of the abelian group $\pi(n+k, n)$ is zero for all $k < 4p-6$, except $k = 2p-3$ where it is Z_p. Serre has not remained the only one working in this field, but his intervention has been decisive, both with respect to methods as well as results.

The crux of his method is the reduction of the *homotopy* group of a space M to the *homology* group of suitably constructed auxiliary spaces. Since for the investigation of homology groups strong algebraic methods have recently been developed, the homotopy problem thus comes within the range of algebraic processes, and this is probably the reason for the greater success of Serre's method in comparison to those previously applied.

Let M be a connected space. Over it as a base we construct a sort of fiber space by associating with each point a of M, as the fiber corresponding to a, the totality of all continuous curves starting from a and returning to a. Unfortunately these fibers are not homeomorphic to each other as in the classical fiber spaces. Nevertheless Eckmann's fundamental construction for fiber spaces which the French call *relèvement des homotopies* carries over to them, and this is the essential point. (For an unbounded, unramified covering manifold $\overline{M}$ over M where the fiber consists of isolated points this fundamental construction is simply the unique determination of a line $\bar{\gamma}$ on $\overline{M}$ of which initial point $\bar{a}$ and trace γ are given.) It remained tc study the relationship between the homologies of this generalized fiber space, its base and its fiber. Here Serre makes use of Leray's notion of "suite spectrale". The technical difficulties in carrying out this program are considerable, but Serre overcomes them all. Thus he proved for instance that for a simply connected space M of which all homology groups are finite or have a finite set of generators, the same statements hold for their homotopy groups. He shows that one can treat their p-primary components independently of each other for each prime p; etc.

Serre applied his results to the theory of Marston Morse concerning geodesics on Riemannian manifolds; joining hands with Armand Borel and another time with G. P. Hochschild, he won interesting results about Lie groups and algebras, thus e.g. settling a problem posed by Hopf to the effect that no sphere with the exception of S_2 and perhaps S_6 can be endowed with a complex-analytic structure. There is no time to discuss these things, and I prefer to let all the light of my lamp fall on Serre's investigations in the homotopy theory and the theory of sheaves. They are rich enough to show the enormous fecundity of his mathematical mind.

Here ends my report. If I omitted essential parts or misrepresented others, I ask for your pardon, Dr. Serre and Dr. Kodaira; it is not easy for an older man to follow your striding paces. Dear Kodaira: Your work has more than one connection with what I tried to do in my younger years; but you reached heights of which I never dreamt. Since you came to Princeton in 1949 it has been one of the greatest joys of my life to watch your mathematical development. I have no such close personal relation to you, Dr. Serre, and your research; but let me say this that never before have I witnessed such a brilliant ascension of a star in the mathematical sky as yours. The mathematical community is proud of the work you both have done. It shows that the old gnarled tree of mathematics is still full of sap and life. Carry on as you began!

165.

Address on the Unity of Knowledge delivered at the Bicentennial Conference of Columbia University

Columbia University in the City of New York Bicentennial Celebration, 1954

The present solemn occasion on which I am given the honor to address you on our general theme "The Unity of Knowledge" reminds me, you will presently see why, of another Bicentennial Conference, held 14 years ago by our neighborly university in the city of Brotherly Love. The words with which I started there a talk on "The Mathematical Way of Thinking" sound like an anticipation of today's topic; I repeat them: "By the mental process of thinking we try to ascertain truth; it is our mind's effort to bring about its own enlightenment by evidence. Hence, just as truth itself and the experience of evidence, it is something fairly uniform and universal in character. Appealing to the light in our innermost self, it is neither reducible to a set of mechanically applicable rules, nor is it divided into water-tight compartments like historical, philosophical, mathematical thinking, etc. True, nearer the surface there are certain techniques and differences; for instance, the procedures of fact-finding in a courtroom and in a physical laboratory are conspicuously different." The same conviction was more forcibly expressed by the father of our Western philosophy, Descartes, who said: "The sciences taken all together are identical with human wisdom, which always remains one and the same, however applied to different subjects, and suffers no more differentiation proceeding from them than the light of the Sun experiences from the variety of the things it illumines."

But it is easier to state this thesis in general terms than to defend it in detail when one begins to survey the various branches of human knowledge. Ernst Cassirer, whose last years were so intimately connected with this university, set out to dig for the root of unity in man by a method of his own, first developed in his great work "Philosophie der symbolischen Formen". The lucid "Essay on man" written much later in this country and published by the Yale University Press in 1944, is a revised and condensed version. In it he tries to answer the question "What is man?" by a penetrating analysis of man's cultural activities and creations: language, myth, religion, art, history, science. As a common feature of all of them he finds: the symbol, symbolic representation. He sees in them "the threads which weave the symbolic net, the tangled net of human experience." "Man", he says, "no longer lives in a merely physical universe, he lives in a symbolic universe." Since "reason is a very inadequate term with which to comprehend the forms of man's cultural life in all their richness and variety", the definition of man as the *animal rationale* had better be replaced by defining him as an *animal symbolicum.* Investigation of these symbolic forms on the basis of appropriate structural categories should ultimately tend towards displaying them as "an organic whole tied together not by a *vinculum substantiale,* but a *vinculum functionale.*" Cassirer invites us to look upon them "as so many variations on a common theme", and sets as the philosopher's task "to make this theme audible and understandable." Yet much as I admire Cassirer's analyses, which betray a mind of rare universality, culture and intellectual experience, their sequence, as one follows them in his book, resembles more a suite of bourrées, sarabands, menuets and gigues than variations on a single theme. In the concluding paragraph he himself emphasizes "the tensions and frictions, the strong contrasts and deep conflicts between the various powers of man, that cannot be reduced to a common denominator." He then finds consolation in the thought that "this multiplicity and disparateness does not denote discord or disharmony", and his last word is that of Heraclitus: "Harmony in contrariety, as in the case of the bow and the lyre." Maybe, man cannot hope to be more than that; but am I wrong when I feel that Cassirer quits with a promise unfulfilled?

In this dilemma let me now first take cover behind the shield of that special knowledge in which I have experience through my own research: the naturai sciences including mathematics. Even here doubts about their methodical unity have been raised. This, however, seems unjustified to me. Following Galileo, one may describe the method of science in general terms as a combination of passive observation refined by active experiment with that symbolic construction to which theories ultimately reduce. Physics is the paragon. Hans Driesch and the holistic school have claimed for biology a methodical approach different from, and transcending, that of physics. However, nobody doubts that the laws of physics hold for the body of an animal or myself as well as for a stone. Driesch's attempts to prove that the organic processes are incapable of mechanical explanation rest on a much too narrow notion of mechanical or physical explanation of nature. Here quantum physics has opened up new possibilities. On the other side, wholeness is not a feature limited to the organic world. Every atom is already a whole of quite definite structure; its organization is the foundation of possible organizations and structures of the utmost complexity. I do not suggest that we are safe against surprises in the future development of science. Not so long ago we had a pretty startling one in the transition from classical to quantum physics. Similar future breaks may greatly affect the epistemological interpretation, as this one did with the notion of causality; but there are no signs that the basic method itself, symbolic construction combined with experience, will change.

It is to be admitted that on the way to their goal of symbolic construction scientific theories pass preliminary stages, in particular the classifying or morphological stage. Linnaeus' classification of plants, Cuvier's comparative anatomy are early examples; comparative linguistics or jurisprudence are analogues in the historical sciences. The features which natural science determines by experiments repeatable at any place and any time are *universal;* they have that empirical necessity which is possessed by the laws of nature. But beside this domain of the necessary there remains a domain of the *contingent.* The one cosmos of stars and diffuse matter, Sun and Earth, the plants and animals living on earth, are accidental or singular phenomena. We are interested in their evolution. Primitive thinking even puts the question "How did it come about" before the question "How is it." All history in the proper sense is concerned with the development of one singular phenomenon: human civilization on earth. Yet if the experience of natural science accumulated in her own history has taught one thing, it is this, that in its field knowledge of the laws and of the inner constitution of things must be far advanced before one may hope to understand or hypothetically reconstruct their genesis. For want of such knowledge as is now slowly gathered by genetics the speculations on pedigrees and phylogeny let loose by Darwinism in the last decades of the 19th century were mostly premature. Kant and Laplace had the firm basis of Newton's gravitational law when they advanced their hypotheses about the origin of the planetary system.

After this brief glance at the methods of natural science, which are the same in all its branches, it is time now to point out the limits of science. The riddle posed by the double nature of the ego certainly lies beyond those limits. On the one hand, I am a real individual man; born by a mother and destined to die, carrying out real physical and psychical acts, one among many (far too many, I may think, if boarding a subway during rush hours). On the other hand, I am "vision" open to reason, a self-penetrating light, immanent sense-giving consciousness, or however you may call it, and as such unique. Therefore I can say to myself both: "I think, I am real and conditioned" as well as "I think, and in my thinking I am free." More clearly than in the acts of volition the decisive point in the problem of freedom comes out, as Descartes remarked, in the theoretical acts. Take for instance the statement $2+2=4$: not by blind natural causality, but because I *see* that $2+2=4$ does this judgement as a real psychic act form itself in me, and do my lips form these words: two and two make four. Reality or the realm of Being is not closed, but open toward Meaning in the ego, where Meaning and Being are merged in indissoluble union — though science will never tell us how. We do not see through the *real* origin of freedom.

And yet, nothing is more familiar and disclosed to me than this mysterious "marriage of light and darkness", of self-transparent consciousness and real being that I am myself. The access is my knowledge of myself from within, by which I am aware of my own acts of perception, thought, volition, feeling and doing, in a manner entirely different from the theoretical knowledge that represents the "parallel" cerebral processes in symbols. This inner awareness of myself is the basis for the more or less intimate understanding of my fellow-men, whom I acknowledge as beings of my own kind. Granted that I do not know of their consciousness in the same manner as of my own, nevertheless my "interpretative" understanding of it is apprehension of indisputable adequacy. As hermeneutic interpretation it is as characteristic for the historical, as symbolic construction is for the natural, sciences. Its illumining light not only falls on my fellow-men; it also reaches, though with ever increasing dimness and incertitude, deep into the animal kingdom. Kant's narrow opinion that we can feel compassion, but cannot share joy with other living creatures, is justly ridiculed by Albert Schweitzer who asks: "Did he never see a thirsty ox coming home from the fields drink?" It is idle to disparage this hold on nature "from within" as anthropomorphic and elevate the objectivity of theoretical construction, though one must admit that understanding, for the very reason that it is *concrete* and *full*, lacks the freedom of the "hollow symbol". Both roads run, as it were, in opposite directions: what is darkest for theory, man, is the most luminous for the understanding from within; and to the elementary inorganic processes, that are most easily approachable by theory, interpretation finds no access whatever. In biology the latter may serve as a guide to important problems, although it will not provide an objective theory as their solution. Such teleological statements as "The hand is there to grasp, the eye to see" drive us to find out what internal material organization enables hand and eye, according to the physical laws (that hold for them as for any inanimate object), to perform these tasks.

I will not succumb to the temptation of foisting Professor Bohr's idea of complementarity upon the two opposite modes of approach we are discussing here. However, before progressing further, I feel the need to say a little more about the constructive procedures of mathematics and physics.

Democritus, realizing that the sensuous qualities are but effects of external agents on our sense organs and hence mere apparitions, said: "Sweet and bitter, cold and warm, as well as the colors, all these exist but in opinion and not in reality (*νόμῳ, οὐ φύσει*); what really exist are unchangeable particles, atoms, which move in empty space." Following his lead, the founders of modern science, Kepler, Galileo, Newton, Huygens, with the approval of the philosophers, Descartes, Hobbes, Locke, discarded the sense qualities, on account of their subjectivity, as building material of the objective world which our perceptions reflect. But they clung to the objectivity of space, time, matter, and hence of motion and the corresponding geometric and kinematic concepts. Thus Huygens, for instance, who developed the undulatory theory of light, can say with the best of conscience that colored light beams are *in reality* oscillations of an ether consisting of tiny particles. But soon the objectivity of space and time also became suspect. Today we find it hard to realize why their intuition was thought particularly trustworthy. Fortunately Descartes' analytic geometry had provided the tool to get rid of them and to replace them by numbers, i. e. mere symbols. At the same time one learned how to introduce such concealed characters, as e. g. the inertial mass of a body, not by defining them explicitly, but by postulating certain simple laws to which one subjects the observation of reacting bodies. The upshot of it all is a purely symbolic construction that uses as its material nothing but mind's free creations: symbols. The monochromatic beam of light, which for Huygens was in reality an ether wave, has now become a formula expressing a certain undefined symbol F, called electromagnetic field, as a mathematically defined function of four other symbols x, y, z, t, called space-time coordinates. It is evident that now the words "in reality" must be put between quotation marks; who could seriously pretend that the symbolic construct *is* the true real world? Objective Being, reality, becomes elusive; and science no longer claims to erect a sublime, truly objective world above the Slough of Despond in which our daily life moves. Of course, in some way one must establish the connection between the symbols and our perceptions. Here, on the one hand, the symbolically expressed laws of nature (rather than any explicit "intuitive" definitions of the significance of the symbols) play a fundamental role, on the other hand the concretely described procedures of observation and measurement.

In this manner a theory of nature emerges which only as a whole can be confronted with experience, while the individual laws of which it consists, when taken in isolation, have no verifiable content. This discords with the traditional idea of truth, which looks at the relation between Being and Knowing from the side of Being, and may perhaps be formulated as follows: "A statement points to a fact, and it is true if the fact to which it points is so as it states." The truth of physical theory is of a different brand.

Quantum theory has gone even a step further. It has shown that observation always amounts to an uncontrollable intervention, since measurement of one quantity irretrievably destroys the possibility of measuring certain other quantities. Thereby the objective Being which we hoped to construct as one big piece of cloth each time tears off; what is left in our hands are — rags.

The notorious man-in-the-street with his common sense will undoubtedly feel a little dizzy when he sees what thus becomes of that reality which seems to surround him in such firm, reliable and unquestionable shape in his daily life. But we must point out to him that the constructions of physics are only a natural prolongation of operations his own mind performs (though mainly unconsciously) in perception, when e. g. the solid shape of a body constitutes itself as the common source of its various perspective views. These views are conceived as appearances, for a subject with its continuum of possible positions, of an entity on the next higher level of objectivity: the three-dimensional body. Carry on this "constitutive" process in which one rises from level to level, and one will land at the symbolic constructs of physics. Moreover, the whole edifice rests on a foundation which makes it binding for all reasonable thinking: of our complete experience it uses only that which is unmistakably *aufweisbar*.

Excuse me for using here the German word. I explain it by reference to the foundations of mathematics. We have come to realize that isolated statements of classical mathematics in most cases make as little sense as do the statements of physics. Thus it has become necessary to change mathematics from a system of meaningful propositions into a game of formulas which is played according to certain rules. The formulas are composed of certain clearly distinguishable symbols, as concrete as the men on a chess board. Intuitive reasoning is required and used merely for establishing the consistency of the game — a task which so far has only partially been accomplished and which we may never succeed in finishing. The visible tokens employed as symbols must be, to repeat Hilbert's words, "recognizable with certainty, independently of time and place, and independently of minor differences and the material conditions of their execution (e. g. whether written by pencil on paper or by chalk on blackboard)." It is also essential that they should be reproducible where- and whenever needed. Now here is the prototype of what we consider as *aufweisbar*, as something to which we can point *in concreto*. The inexactitude which is inseparable from continuity and thus clings inevitably to any spatial configurations is overcome here in principle, since only clearly distinguishable marks are used and slight modifications are ignored "as not affecting their identity." (Of course, even so errors are not excluded.) When putting such symbols one behind the other in a formula, like letters in a printed word, one obviously employs space and spatial intuition in a way quite different from a procedure that makes space in the sense of Euclidean geometry with its exact straight lines etc. one of the bases on which knowledge rests, as Kant does. The *Aufweisbare* we start with is not such a pure distillate, it is much more concrete.

Also the physicist's measurements, e. g. reading of a pointer, are operations performed in the *Aufweisbaren* — although here one has to take the approximate character of all measurements into account. Physical theory sets the mathematical formulas consisting of symbols into relation with the results of concrete measurements.

At this juncture I wish to mention a collection of essays by the mathematician and philosopher Kurt Reidemeister published by Springer in 1953 and 1954 under the titles "Geist und Wirklichkeit" and "Die Unsachlichkeit der Existenzphilosophie". The most important is the essay "Prolegomena einer kritischen Philosophie" in the first volume. Reidemeister is positivist in as much as he maintains the irremissible nature of the factual which science determines; he ridicules (rightly, I think) such profound sounding but hollow evocations as Heidegger indulges in, especially in his last publications. On the other hand, by his insistence that science does not make use of our full experience, but selects from it that which is *aufweisbar*, Reidemeister makes room for such other types of experience as are claimed by the windbags of profundity as their proper territory: the experience of the indisposable significant in contrast to the disposable factual. Here belongs the intuition through and in which the beautiful, whether incorporated in a vase, a piece of music or a poem, appears and becomes transparent, and the reasonable experience governing our dealings and communications with other people; an instance: the ease with which we recognize and answer a smile. Of course, the physical and the aesthetic properties of a sculpture are related to each other; it is not in vain that the sculptor is so exacting with respect to the geometric properties of his work, because the desired aesthetic effect depends on them. The same connection is perhaps even more obvious in the acoustic field. Reidemeister, however, urges us to admit our *Nicht-Wissen*, our not knowing how to combine these two sides by theory into one unified realm of Being — just as we cannot see through the union of I, the conditioned individual, and I who thinking am free. This *Nicht-Wissen* is the protecting wall behind which he wants to save the indisposable significant from the grasp of hollow profundity and restore our inner freedom for a genuine apprehension of ideas. Maybe, I overrate Reidemeister's attempt, which no doubt is still in a pretty sketchy state, when I say that, just as Kant's philosophy was based on, and made to fit, Newton's physics, so his attempt takes the present status of the foundations of mathematics as its lead. And as Kant supplements his Critique of Pure Reason by one of practical reason and of aesthetic judgement, so leaves Reidemeister's analysis room for other experiences than science makes use of, in particular for the hermeneutic understanding and interpretation on which history is based.

Let me for the few remarks I still want to make adopt the brief terms science and history for natural and historical sciences *(Natur- und Geistes-Wissenschaften)*. The first philosopher who fully realized the significance of hermeneutics as the basic method of history was Wilhelm Dilthey. He traced it back to the exegesis of the Holy Script. The chapter on history in Cassirer's "Essay on man" is one of the most successful. He rejects the assumption of a special historical logic or reason as advanced by Windelband or more recently and much more impetuously by Ortega y Gasset. According to him the essential difference between history and such branches of science as e. g. palaeontology dealing with singular phenomena lies in the necessity for the historian to interpret his "petrefacts", his monuments and documents, as having a symbolic content.

Summarizing our discussion I come to this conclusion. At the basis of all knowledge there lies: (1) *Intuition*, mind's originary act of "seeing" what is given to him; limited in science to the *Aufweisbare*, but in fact extending far beyond these boundaries. How far one should go in including here the *Wesensschau* of Husserl's phenomenology, I prefer to leave in the dark. (2) *Understanding and expression.* Even in Hilbert's formalized mathematics I must understand the directions given me by communication in words for how to handle the symbols and formulas. Expression is the active counterpart of passive understanding. (3) *Thinking the possible.* In science a very stringent form of it is exercised when, by thinking out the possibilities of the mathematical game, we try to make sure that the game will never lead to a contradiction; a much freer form is the imagination by which theories are conceived. Here, of course, lies a source of subjectivity for the direction in which science develops. As Einstein once admitted, there is no logical way leading from experience to theory, and yet the decision as to which theories are adopted turns out ultimately to be unambiguous. Imagination of the possible is of equal importance for the historian who tries to re-enliven the past. (4) On the basis of intuition, understanding and thinking of the possible, we have in science: certain practical actions, namely the *construction* of symbols and formulas on the mathematical side, the construction of the measuring devices on the empirical side. There is no analogue for this in history. Here its place is taken by *hermeneutic interpretation,* which ultimately springs from the inner awareness and knowledge of myself. Therefore the work of a great historian depends on the richness and depth of his own inner experience. Cassirer finds wonderful words for Ranke's intellectual and imaginative, not emotional, sympathy, the universality of which enabled him to write the history of the Popes and of the Reformation, of the Ottomans and the Spanish Monarchy.

Being and Knowing, where should we look for unity? I tried to make clear that the shield of Being is broken beyond repair. We need not shed too many tears about it. Even the world of our daily life is not *one,* to the extent people are inclined to assume; it would not be difficult to show up some of its cracks. Only on the side of Knowing there may be unity. Indeed, mind in the fullness of its experience has unity. Who says "I" points to it. But just because it is unity, I am unable to describe it otherwise than by such characteristic actions of the mind mutually supporting each other as I just finished enumerating. Here, I feel, I am closer to the unity of the luminous center than where Cassirer hoped to catch it: in the complex symbolic creations which this lumen built up in the history of mankind. For these, and in particular myth, religion, and alas!, also philosophy, are rather turbid filters for the light of truth, by virtue, or should I say, by vice of man's infinite capacity for self-deception.

What else than turbidity could you then have expected from a philosophical talk like this? If you found it particulary aimless, please let me make a confession before asking for your pardon. The reading of Reidemeister's essays has caused me to think over the old epistemological problems with which my own writings had dealt in the past; and I have not yet won through to a new clarity. Indecision of mind does not make for coherence in speaking one's mind. But then, would one not cease to be a philosopher, if one ceased to live in a state of wonder and mental suspense?

166.

Erkenntnis und Besinnung (Ein Lebensrückblick)

Studia Philosophica, Jahrbuch der Schweizerischen Philosophischen Gesellschaft
Annuaire de la Société Suisse de Philosophie (1954)

Die Universität Lausanne hat mir durch die Zuerkennung des Preises Arnold Reymond der fondation Charles-Eugène Guye eine große Ehre erwiesen, für die ich der Universität und ihrem Rektor Herrn Professor Bridel sowie den an der Entscheidung über die Verleihung des Preises beteiligten Fachgenossen meinen tiefgefühlten Dank abstatten möchte. Verpflichtungen, die mich während der ersten vier Monate dieses Jahres in Amerika festhielten, an der schönsten Forschungsstätte, die es für die Mathematik in der Welt gibt, an dem Institute for Advanced Study in Princeton, New Jersey, haben es mir verunmöglicht, früher vor Ihnen zu erscheinen.

Da es sich um eine Anerkennung meiner Bemühungen um die Philosophie der Wissenschaften handelt, möchte ich diese Gelegenheit benutzen, um in einem Rückblick auf mein Leben den Anteil zu schildern, den darin neben der wissenschaftlichen Erkenntnis die philosophische Besinnung gespielt hat. Stand auch die mathematische Forschung, mit gelegentlichen Ausschweifungen in die theoretische Physik, im Zentrum, so hat es mich doch immer zugleich gedrängt, reflektierend mir über Sinn und Ziel dieser Forschung Rechenschaft zu geben. In einem Vortrag über «Die Stufen des Unendlichen» habe ich einmal, im Anschluß an eine Diskussion über konstruktive Mathematik und reflektive Metamathematik, die gegenseitige Beziehung so gekennzeichnet: «Im geistigen Leben des Menschen sondern sich deutlich voneinander ein Bereich des *Handelns*, der Gestaltung, der Konstruktion auf der einen Seite, dem der tätige Künstler, Wissenschaftler, Techniker, Staatsmann hingegeben ist und der im Gebiete der Wissenschaft unter der Norm der Objektivität steht, – und ein Bereich der *Besinnung* auf der andern Seite, die in Einsichten sich vollzieht und die, als Ringen um den *Sinn* unseres Handelns, als die eigentliche Domäne des Philosophen anzusehen ist. Die Gefahr des schöpfe-

* Vortrag, gehalten an der Universität Lausanne im Mai 1954.

rischen Tuns, wenn es nicht durch Besinnung überwacht wird, ist, daß es dem Sinne entläuft, abwegig wird, in Routine erstarrt; – die Gefahr der Besinnung, daß sie zu einem die schöpferische Kraft des Menschen lähmenden unverbindlichen ‚Reden darüber' wird.»

Indem ich mich nun anschicke, von meinem Leben zu erzählen, soweit in ihm philosophische Antriebe eine Rolle gespielt haben, muß ich mich zunächst dafür entschuldigen, daß ich dies nicht in der Sprache tun kann, die den meisten von Ihnen die geläufigste ist. Seit mir vor 20 Jahren das Schicksal das Englische neben meiner Muttersprache als Sprache des Umgangs aufzwang, ist mir das Französische so fremd geworden, daß ich darauf verzichten mußte. In chronologischer Reihenfolge von meiner philosophischen Entwicklung berichtend, werde ich wie von selbst genötigt sein, der Reihe nach die großen Themen «Raum und Zeit», «dingliche Welt», «Ich und der Mensch», «Gott» zu berühren.

Für immer ist mir in Erinnerung geblieben, wie ich noch als Schüler im vorletzten Schuljahr auf dem Bodenraum meines elterlichen Hauses ein stockfleckiges, aus dem Jahre 1790 stammendes Exemplar eines kurzen Kommentars zu Kants «Kritik der reinen Vernunft» aufstöberte. Hieraus lernte ich Kants Lehre von der *Idealität von Raum und Zeit* kennen, die mich sofort aufs mächtigste ergriff: mit einem Ruck war ich aus dem «dogmatischen Schlummer» erwacht, war dem Geist des Knaben auf radikale Weise die Welt in Frage gestellt. Ist es nötig, die Quintessenz der Lehre Kants hier zu wiederholen? Er erkannte, daß Raum und Zeit nicht etwas den Dingen einer an sich, unabhängig vom Bewußtsein existierenden, Welt Einwohnendes, sondern in unserem Geiste gegründete *Formen der Anschauung* sind. Er stellte sie als solche der hyletischen Grundschicht der Wahrnehmung, den Empfindungen, gegenüber. Ich zitiere: «Da das, worin sich die Empfindungen allein ordnen und in gewisse Formen gestellt werden können, nicht selbst wiederum Empfindung sein kann, so ist uns zwar die Materie aller Erscheinung nur a posteriori gegeben, die Form derselben aber muß zu ihnen insgesamt im Gemüte a priori bereitliegen und daher abgesondert von aller Empfindung können betrachtet werden.» Oder, wie Fichte in seiner kräftigen, immer etwas verstiegenen Sprache sagt: «Der durchsichtige, durchgreifbare und durchdringliche Raum, das reinste Bild meines Wissens, wird nicht gesehen, sondern angeschaut, und in ihm wird mein Sehen selbst angeschaut. Das Licht ist nicht außer mir, sondern in mir.» Diese Lehre schien mit einem Schlage ein fast allgemein akzeptiertes Faktum zu erklären, den Umstand nämlich, daß die Grundtatsachen der Geometrie uns in unmittelbarer Evidenz einleuchten, ohne daß wir auf die Erfahrung zu rekurrieren brauchen. Kant unterscheidet *analytische* Urteile, die nichts weiter tun als aussprechen, was in den Begriffen liegt, wie etwa: «Ein rundes Ding ist rund» oder «Wenn Sokrates ein Mensch ist und alle Menschen sterblich, so ist Sokrates sterblich», von den *synthetischen*, für die zum Beispiel das Newtonsche Gravitationsgesetz ein Beispiel ist. Daß analytische Urteile a priori, unabhängig von der Erfahrung, gewiß sind, ist kein Wunder. Die Aussagen der Geometrie aber geben gemäß dem eben Gesagten ein Beispiel für synthetische Urteile, die trotz ihres synthetischen Charakters a priori gewiß, auf keine Erfahrung gegründet und von einer durch keine Erfahrung zu erschütternden Notwendigkeit sind. Kants zentrale Frage war: *Wie sind synthetische Urteile a priori möglich*, und hierauf lieferte seine Einsicht von der Natur des Raumes eine Antwort, soweit die Sätze der Geometrie in Frage stehen.

Während ich keine Schwierigkeiten hatte, diesen Teil der Kantischen Lehre mir zu eigen zu machen, quälte ich mich noch mit dem «Schematismus der reinen Verstandesbegriffe» ab, als ich 1904 die Universität Göttingen bezog. Hier lehrte David Hilbert, der kurz vorher sein epochemachendes Werk «Grundlagen der Geometrie» veröffentlicht hatte. Daraus wehte der Geist der modernen Axiomatik mich an. In einer Vollständigkeit, hinter der Euklid noch weit zurückbleibt, werden hier die Axiome der Geometrie aufgestellt. Um ihre gegenseitige logische Abhängigkeit zu untersuchen, wird nicht nur die damals schon fast ein Jahrhundert alte, sogenannte nicht-euklidische Geometrie herangezogen, sondern wird eine Fülle anderer fremdartiger Geometrien, meist auf arithmetischer Basis, konstruiert. Kants Bindung an die euklidische Geometrie erschien nun als naiv. Unter diesem überwältigenden Anstoß stürzte mir das Gebäude der Kantischen Philosophie, der ich mit gläubigem Herzen ergeben gewesen war, zusammen.

Hier unterbreche ich meine Erzählung, um kurz die mir heute vernünftig erscheinende Einstellung zum Raumproblem anzudeuten. Erstens wurden durch die spezielle Relativitätstheorie Raum und Zeit im Kosmos zu einem einzigen vierdimensionalen Kontinuum verschmolzen. Zweitens erwies es sich als wesentlich, zwischen dem amorphen Kontinuum, von welchem die heute so genannte Disziplin der Topologie handelt, und seiner Struktur, insbesondere seiner metrischen Struktur, zu scheiden. Die auf einen physikalisch nachprüfbaren Kongruenzbegriff sich stützende physische Geometrie wurde schon von Newton als ein auf Erfahrung gegründeter Teil der Mechanik angesehen. Er sagt: «Geometrie hat ihre Begründung in mechanischer Praxis und ist in der Tat nichts anderes als derjenige Teil der gesamten Mechanik, welcher die Kunst des Messens genau feststellt und begründet.» Helmholtz zeigte, daß die beiden Teile der Kantischen Lehre: 1) der Raum ist reine Form der Anschauung, 2) die Wissenschaft vom Raum, die euklidische Geometrie, gilt a priori, nicht so eng miteinander verbunden sind, daß 2) aus 1) folgt. Er ist bereit, 1) als einen richtigen Ausdruck des Sachverhalts zu akzeptieren; doch könne daraus nicht *mehr* geschlossen werden, als daß alle Dinge der Außenwelt notwendig räumlich ausgedehnt seien. Im Einklang mit Newton und Riemann weist er dann den empirisch-physikalischen Sinn der Geometrie nach. Riemanns Bemerkung, daß «die empirischen Begriffe, in welchen die räumlichen Maßbestimmungen gegründet sind, der Begriff des starren Körpers und des Lichtstrahls, im Unendlichkleinen ihre Gültigkeit verlieren», hat später den Quantenphysikern zu denken gegeben. Anderseits kam die von Riemann begründete Infinitesimalgeometrie einer Mannigfaltigkeit von beliebiger Dimensionszahl in Einsteins allgemeiner Relativitätstheorie zur Geltung, die über Riemann hinaus zeigte, daß die in der wirklichen vierdimensionalen Welt herrschende Maßbestimmung keine fest vorgegebene Entität ist, sondern sowohl von den physischen Vorgängen beeinflußt wird wie auch auf sie Wirkungen ausübt: es sind die Erscheinungen der Gravitation, in welcher sich die Fluidität des metrischen Feldes kundtut.

Erkennt man neben dem physischen einen Anschauungsraum an und behauptet von ihm, daß seine Maßstruktur aus Wesensgründen die euklidischen Gesetze erfüllt, so steht das mit unsern physikalischen Erkenntnissen nicht notwendig in Widerspruch, sofern auch diese an der euklidischen Beschaffenheit – grob gesagt an der Gültigkeit des Pythagoreischen Lehrsatzes – in der unendlich kleinen Umgebung eines Punktes O (in dem sich das Ich momentan befindet) festhält. Aber man muß dann zugeben, daß die Beziehung des Anschauungsraumes auf den physischen um so vager wird, je weiter man sich vom Ich-Zentrum O entfernt. Er ist einer Tangentenebene zu vergleichen, die im Punkte O an eine krumme Fläche, den physischen Raum, gelegt ist: in der unmittelbaren Umgebung von O decken sich beide, aber je weiter man sich von O entfernt, um so willkürlicher wird die Fortsetzung dieser Deckbeziehung zu einer ein-eindeutigen Korrespondenz zwischen Ebene und Fläche.

In der physischen Welt ist, wie gesagt, die Zeit mit dem Raum zu einem einheitlichen vierdimensionalen Kontinuum verschmolzen. In Bestätigung der Leibniz'schen These, daß die Scheidung von Vergangenheit und Zukunft auf der Kausalstruktur der vierdimensionalen Welt beruht, führte die Relativitätstheorie zu einer von der hergekommenen abweichenden Beschreibung dieser Struktur, der zufolge die Gleichzeitigkeit von Ereignissen, so gut wie ihre Gleichortigkeit, ihren objektiven Sinn einbüßt. In der Welt beschreibt mein Leib, wenn ich ihn als punktförmig betrachte, eine eindimensionale Weltlinie, längs deren sich eine physikalische Eigenzeit definieren läßt. Auf dieser Linie gibt es natürlich die durch die Worte Vergangenheit, Gegenwart, Zukunft gekennzeichnete Anordnung. Die den Bewußtseinsakten des Ich als ihre allgemeine Form inhärente phänomenale Zeit ist nicht mit der Zeitkoordinate des vierdimensionalen Kontinuums Welt in Parallele zu stellen, sondern hat ihr physisches Gegenbild an der eben erwähnten, der Weltlinie des Ich-Leibes zukommenden Eigenzeit.

Auch im Rahmen der allgemeinen Relativitätstheorie ist es möglich, am physischen Raum in einem gewissen objektiven Sinne, der nicht wie die Kantische Unterscheidung die Erkenntnisquelle und Erkenntnisart betrifft, Momenten a priori solche a posteriori entgegenzusetzen: Es steht hier der einen absolut gegebenen euklidisch-pythagoreischen *Natur* der Metrik, die nicht Teil hat an der unaufhebbaren Vagheit dessen, was eine veränderliche Stelle in einer kontinuierlichen Skala einnimmt, die gegenseitige *Orientierung* der Metriken in den verschiedenen Punkten gegenüber, der zufällige, von der Materie abhängige, wechselvolle, nur approximativ und unter Zuhülfenahme unmittelbarer anschaulicher Hinweise auf die Wirklichkeit zu gebende, quantitative Verlauf des Maßfeldes. Ich habe es einmal unternommen, gerade aus dieser Scheidung heraus die spezielle pythagoreische Natur der Metrik in ihrer mathematischen Eigenart zu erklären. Einer Aufgabe ähnlicher Art steht man gegenüber, wenn man zu verstehen sucht, warum die Welt gerade *vier*dimensional ist und nicht irgendeine andere Zahl von Dimensionen hat. Man muß nämlich wissen, daß alle unsere bisher bekannten physikalischen Gesetze (und die für sie relevanten geometrischen) sich völlig zwingend auf eine beliebige Anzahl von Dimensionen übertragen lassen, so daß in ihnen nichts liegt, was die Dimensionszahl 4 irgendwie auszeichnet. Die Mathematik aber lehrt uns, namentlich in der Gruppentheorie, Gebilde kennen, deren Struktur je nach der Dimensionszahl eine völlig verschiedene ist. Offenbar ist die Physik mit ihren uns bekannten Gesetzen noch nicht in eine Tiefe vorgestoßen, wo sie diese Art Mathematik benötigt. Darum haben wir vorläufig keine wirklich überzeugende Antwort auf die Frage nach dem Grund für die Dimensionszahl 4; und ich will es auch dahingestellt sein lassen, ob mein Versuch, die pythagoreische Natur des metrischen Feldes zu erklären, das Richtige trifft.

Soviel über Raum und Zeit. Ich fahre jetzt in meiner Erzählung fort. Nachdem durch die Berührung mit der modernen Mathematik mein Glaube an Kant zusammengebrochen war, wandte ich mich mit Inbrunst dem Studium der Mathematik zu. Was an erkenntnistheoretischen Interessen übriggeblieben war, befriedigten solche Schriften wie «La science et l'hypothèse» von Henri Poincaré, die Schriften von Ernst Mach oder die bekannte «Geschichte des Materialismus» von Friedrich Albert Lange.

Das nächste epochemachende Ereignis für mich war, daß ich eine bedeutsame mathematische Entdeckung machte. Sie betraf die Gesetzmäßigkeit in der Verteilung der Eigenfrequenzen der Schwingungen eines kontinuierlich ausgedehnten Mediums, wie es eine Membran, ein elastischer Körper oder der elektromagnetische Äther ist. Der Einfall war einer von vielen, wie sie wohl jeder, der in jugendlichem Alter sich mit Wissenschaft abgibt, hat; aber während die andern bald zerplatzende Seifenblasen waren, führte dieser, wie eine kurze Prüfung zeigte, zum Ziel. Ich war selber darüber baß erstaunt, da ich mir dergleichen durchaus nicht zugetraut hatte. Es kam hinzu, daß das Resultat, obschon von den Physikern längst vermutet, den meisten Mathematikern als etwas erschien, dessen Beweis noch in weiter Ferne liege. Während ich fieberhaft mit der Durchführung der Beweisidee beschäftigt war, hatte meine Petroleumlampe zu blaken begonnen, und als ich glücklich fertig war, regnete es von der Decke in dichten schwarzen Flocken auf Papier, Hände und Gesicht herunter.

Gottfried Keller war unbefangen genug, zu gestehen, daß die Erschütterung des Unsterblichkeitsglaubens bei ihm durch die Liebe zu einer Frau herbeigeführt wurde, zu Johanna Kapp, deren Vater dem materialistischen Philosophen Ludwig Feuerbach nahestand und in dessen Anschauungen Johanna aufgewachsen war. Ähnlich erging es auch mir: meine Beruhigung im Positivismus wurde erschüttert, als ich mich in eine junge Sängerin verliebte, deren Lebensgrund im Religiösen lag und die einem Kreis angehörte, dessen philosophischer Wortführer ein bekannter Hegelianer war. Teils wegen meiner menschlichen Unreife, aber zum Teil auch infolge dieses schwer zu überbrückenden weltanschaulichen Gegensatzes wurde nichts daraus. Die Erschütterung jedoch wirkte fort. Es war nicht lange danach, daß ich mich mit einer Philosophiestudentin verheiratete, die Schülerin des damals in Göttingen wirkenden Begründers der Phänomenologie Edmund Husserl war. So wurde es denn Husserl, der mich aus dem Positivismus wieder zu einer freieren Weltansicht herausführte. Gleichzeitig hatte ich die Mutation von einem Göttinger Privatdozenten zum Professor für Geometrie an der Eidg. Technischen Hochschule in Zürich zu vollziehen. In Zürich gerieten meine Frau und ich durch die Vermittlung von Medicus, dessen Seminar meine Frau besuchte, an die Fichtesche Wissenschaftslehre. Hier fand der metaphysische Idealismus, zu dem die Husserlsche Phänomenologie sich damals schüchtern hinzutasten begonnen hatte, den unverhohlensten und kräftigsten Ausdruck. Er packte mich, auch wenn ich meiner Frau, der die sorgfältige Methodik Husserls mehr lag als Fichtes Draufgängertum, zugestehen mußte, daß Fichte durch seine natur- und tatsachenblinde Hartnäckigkeit im Verfolgen einer Idee zu immer abstruseren Konstruktionen sich fortreißen ließ.

Husserl kam ursprünglich von der Mathematik her. Im Verfolg seiner «Logischen Untersuchungen», und zum Teil unter dem Einfluß des Philosophen Franz Brentano, war er zum Widersacher des um die Jahrhundertwende herrschenden Psychologismus geworden und hatte die Methode der Phänomenologie entwickelt, die es sich zum Ziele setzte, in der Wesensschau die dem Bewußtsein begegnenden Phänomene rein so, wie sie sich selbst geben, unabhängig von allen genetischen und andern Theorien, zu erfassen. In der Wesensschau entfaltet sich für ihn ein viel reicheres Feld evidenter apriorischer Erkenntnisse, als es die zwölf Prinzipien gewesen waren, die Kant als konstitutiv für die Welt der Erfahrung aufgestellt hatte. Ich führe ein paar Sätze an aus der systematischen Darstellung, die er 1922 in seinen «Ideen zu einer reinen Phänomenologie und phänomenologischen Philosophie» gab. «Das Was der Dinge, ‚in Idee gesetzt', ist Wesen. Erfassung von reinem Wesen in der Wesensanschauung impliziert nicht das Mindeste von Setzung irgendeines individuellen Daseins. Reine Wesenswahrheiten enthalten nicht die mindeste Behauptung über Tatsachen, also ist auch aus ihnen *allein* nicht die geringfügigste Tatsachenwahrheit zu erschließen.» Auf der andern Seite heißt es aber auch: «Jede auf Erlebnisarten bezügliche Wesensbeschreibung drückt eine unbedingt gültige Norm für mögliches empirisches Dasein aus.» Für die phänomenologische Methode typisch ist diese Feststellung: «Das unmittelbare ‚Sehen', nicht bloß das sinnliche, erfahrende Sehen, sondern das Sehen überhaupt als originär gebendes Bewußtsein welcher Art auch immer, ist die letzte Rechtsquelle aller vernünftigen Behauptungen. Was sich uns in der ‚Intuition' originär darbietet, ist einfach hinzunehmen, als was es sich gibt, aber auch nur in den Schranken, in denen es sich da gibt.» Für den Gegensatz von zufälligem faktischem Naturgesetz und notwendigem Wesensgesetz führt Husserl die beiden Aussagen an: «Alle Körper sind schwer» und «Alle Körper sind ausgedehnt.» Er hat vielleicht recht, aber man fühlt doch schon an diesem ersten Beispiel, wie unsicher in Allgemeinheit statuierte erkenntnistheoretische Unterschiede werden, sobald man vom Allgemeinen zu besonderen konkreten Anwendungen herabsteigt. Ich habe in meinen unter dem Titel «Raum, Zeit, Materie» zuerst 1918 erschienenen Vorlesungen über allgemeine Relativitätstheorie dazu dies bemerkt: «Die hier angestellten Untersuchungen über den Raum scheinen mir ein gutes Beispiel für die von der phänomenologischen Philosophie angestrebte Wesensanalyse zu sein; ein Beispiel, das typisch ist für solche Fälle, wo es sich um nicht-immanente Wesen handelt. Wir sehen da an der historischen Entwicklung des Raumproblems, wie schwer es uns in der Wirklichkeit befangenen Menschen wird, das Entscheidende zu treffen. Eine lange mathematische Entwicklung, die große Entfaltung der geometrischen Studien von Euklid bis Riemann, die physikalische Durchdringung der Natur und ihrer Gesetze seit Galilei mit all ihren immer erneuerten Anstößen aus der Empirie, endlich das Genie einzelner großer Geister – Newton, Gauss, Riemann, Einstein – waren erforderlich, uns von den zufälligen nicht wesenhaften Merkmalen loszureißen, an denen wir zunächst hängen bleiben. Freilich: ist einmal der neue, umfassendere Standpunkt gewonnen, so geht der Vernunft ein Licht auf, und sie erkennt und anerkennt das ihr Aus-sich-selbst-Verständliche; dennoch hatte sie (wenn sie natürlich auch in der ganzen Entwicklung des Problems immer «dabei war») nicht die Kraft, es mit einem Schlage zu durchschauen. Das muß der Ungeduld der Philosophen entgegengehalten werden, die da glauben, auf Grund eines einzigen Aktes exemplarischer Vergegenwärtigung das Wesen adäquat beschreiben zu können. Das Beispiel des Raumes ist zugleich sehr lehrreich für diejenige Frage der Phänomenologie, die mir die eigentlich entscheidende zu sein scheint: inwieweit die Abgrenzung der dem Bewußtsein aufgehenden Wesenheiten eine dem Reich des Gegebenen selbst eigentümliche Struktur zum Ausdruck bringt und inwieweit an ihr bloße Konvention beteiligt ist.» An dieser Auffassung des Verhältnisses von Erkenntnis und Besinnung halte ich noch jetzt im wesentlichen fest. Einsteins Aufstellung der allgemeinen Relativitätstheorie und des in ihrem Rahmen gültigen Gravitationsgesetzes durch eine aus experimentell gestützter Erfahrung, Wesensanalyse und mathematischer Konstruktion kombinierte Methode ist einer der schlagendsten und großartigsten Belege. Besinnung auf den Sinn des Bewegungsbegriffes war wichtig für Einstein, aber nur in solcher Kombination erwies sie sich als fruchtbar.

Was aber in der großen Husserlschen Arbeit aus dem Jahre 1922 zum Hauptgegenstand der Betrachtung wird, ist das Verhältnis des *immanenten Bewußtseins* und des reinen Ich, von dem seine Akte ausstrahlen, zur realen psychophysischen Welt, auf dessen Gegenstände diese Akte intentional gerichtet sind. Der Terminus intentional war von Franz Brentano aus der Scholastik übernommen worden, Husserl machte ihn sich zu eigen. Die Bewußtseinserlebnisse selbst können in der Reflektion zum intentionalen Gegenstand von darauf gerichteten immanenten Wahrnehmungen werden. Das intentionale Objekt einer äußern Wahrnehmung, dieser Baum etwa, ist das Ding, wie es sich in der Wahrnehmung selbst gibt, ohne daß die Frage erhoben wird, ob und in welchem Sinne ihm ein so oder ähnlich beschaffener *wirklicher Baum* entspricht. Umständlich beschreibt Husserl die phänomenologische ἐποχή, durch welche die zum Wesen der natürlichen Einstellung zur Welt gehörige Generalthesis des realen Daseins der Welt außer Aktion gesetzt, «eingeklammert» wird. «Bewußtsein», heißt es, «hat ein Eigensein in sich selbst, das in seinem absoluten Eigenwesen durch diese Ausschaltung nicht betroffen wird; es bleibt somit ein ‚reines' Bewußtsein als phänomenologisches Residuum zurück». Vom Raumding sagt Husserl, daß es bei all seiner Transzendenz *Wahrgenommenes*, in seiner Leibhaftigkeit bewußtseinsmäßig Gegebenes ist, indem die Empfindungsdaten in der konkreten Einheit der Wahrnehmung sich mannigfach «abschatten», durch «Auffassungen» beseelt sind und in dieser Beseelung die darstellende «Funktion» üben bzw. in eins mit ihr das ausmachen, was wir «Erscheinen von» Farbe, Gestalt usw. nennen. Ich finde es nicht leicht, dem zuzustimmen. Jedenfalls darf man nicht leugnen wollen, daß die bestimmte Art, in der sich durch jene beseelende Funktion ein leibhaftes Ding vor mich hinstellt, gerichtet wird von einer Unsumme früherer Erfahrungen, mag man auch gegen die Helmholtzsche Redewendung von den «unbewußten Schlüssen» sich sträuben. Die theoretisch-symbolische Konstruktion, durch welche die Physik das hinter dem Wahrgenommenen stehende Transzendente zu erfassen sucht, ist weit davon entfernt, bei dieser Leibhaftigkeit Halt zu machen. Ich würde darum sagen, daß Husserl lediglich eine der Stufen schildert, durch die hindurch sich die Konstitution der Außenwelt vollzieht. Im Bewußtsein unterscheidet er eine hyletische und eine noëtische Schicht, die sensuelle *ὕλη* von der intentionalen *μορφή*, und spricht von der Art, «wie (zum Beispiel hinsichtlich der Natur) Noësen, das Stoffliche beseelend und sich zu mannigfaltig-einheitlichen Kontinuen und Synthesen verflechtend, Bewußtsein von Etwas so zustande bringen, daß objektive Einheit der Gegenständlichkeit sich darin einstimmig ‚bekunden', ‚ausweisen' und ‚vernünftig' bestimmen lassen kann.» Er fährt emphatisch fort: «Bewußtsein ist eben Bewußtsein ‚von' etwas, es ist sein Wesen, ‚Sinn', sozusagen die Quintessenz von Seele, Geist, Vernunft, in sich zu bergen. Es ist nicht ein Titel für ‚psychische Komplexe', für zusammengeschmolzene ‚Inhalte', für ‚Bündel' oder Ströme von Empfindungen, die, in sich sinnlos, auch in beliebigem Gemenge keinen ‚Sinn' hergeben könnten, sondern es ist durch und durch ‚Bewußtsein', Quelle aller Vernunft und Unvernunft, alles Rechtes und Unrechtes, aller Realität und Fiktion, alles Wertes und Unwertes, aller Tat und Untat.»

Betreffs des Gegensatzes von Erlebnis und Ding behauptet Husserl das bloß phänomenale, in Abschattungen sich gebende Sein des Transzendenten, hingegen *das absolute Sein des Immanenten*, die Zweifellosigkeit der immanenten, die Zweifelhaftigkeit der transzendenten Wahrnehmung. Der Thesis der Welt in ihrer Zufälligkeit steht gegenüber die Thesis meines reinen Ich und Ich-Lebens, die eine notwendige, schlechthin zweifellose ist. «Zwischen Bewußtsein und Realität gähnt ein wahrer Abgrund des Sinnes», sagt er. «Das immanente

Sein ist in dem Sinne absolutes Sein, daß es prinzipiell nulla ‚re' indiget ad existendum; anderseits ist die Welt der transzendenten ‚res' durchaus auf Bewußtsein, und zwar nicht auf logisch erdachtes, sondern aktuelles, angewiesen.»

Hier erhebt sich in ihrem ganzen Ernst die metaphysische Frage nach der Beziehung des einen reinen Ich des immanenten Bewußtseins zu dem einzelnen verlorenen Menschen, als den ich mich unter vielen meinesgleichen in der Welt finde (zum Beispiel während der rush hour an einem Nachmittag auf der Fifth Avenue in New York). Husserl sagt nicht viel mehr darüber, als daß «nur durch die Erfahrungsbeziehung zum *Leibe* Bewußtsein zum real-psychischen des Menschen oder Tieres wird». Aber sofort pocht er wieder auf die selbstherrliche Natur des reinen Bewußtseins: es büße in diesen apperzeptiven Verflechtungen bzw. in dieser psychophysischen Beziehung auf Körperliches nichts von seinem eigenen Wesen ein. «Alle realen Einheiten sind Einheiten des Sinnes. Sinneseinheiten setzen ein sinngebendes Bewußtsein voraus, das seinerseits absolut und nicht selbst wieder durch Sinngebung ist.» Es sei darum klar, «daß trotz aller in ihrem Sinne sicherlich wohlgegründeten Rede von einem realen Sein des menschlichen Ich und seiner Bewußtseinserlebnisse in der Welt, Bewußtsein, in Reinheit betrachtet, als ein für sich geschlossener Seinszusammenhang zu gelten hat, als ein Zusammenhang absoluten Seins, in den nichts hineintreten und aus dem nichts entschlüpfen kann, der von keinem Dinge Kausalität erfahren und auf kein Ding Kausalität üben kann. Anderseits ist die ganze räumlich-zeitliche Welt, der sich Mensch und menschliches Ich als untergeordnete Einzelrealitäten zurechnen, ihrem Sinne nach bloßes intentionales Sein, also ein solches, das den bloß sekundären, relativen Sinn eines Seins *für* ein Bewußtsein hat. Es ist ein Sein, das das Bewußtsein in seinen Erfahrungen setzt, das prinzipiell nur als Identisches von einstimmig motivierten Erscheinungsmannigfaltigkeiten anschaubar und bestimmbar – *darüber hinaus aber nichts ist.*»

Radikaler noch als Husserl hat Fichte in seiner Wissenschaftslehre die Grundposition des erkenntnistheoretischen Idealismus ausgesprochen. Er ist alles andere als ein Phänomenologe, er ist ein Konstruktivist reinsten Wassers, der, ohne rechts und links zu schauen, seinen eigenwilligen Weg der Konstruktion geht. Er erinnert mich in vieler Hinsicht an Paulus. Dieselbe – wie soll ich es nennen – Klobigkeit des Denkens, das aber hinreißend ist durch seine feste Bestimmtheit. Dieselbe völlige Gleichgültigkeit gegen die Erfahrung – bei Paulus insbesondere gegenüber den Zeugnissen über das wirkliche Leben Christi. Derselbe halsstarrige, keinen Widerspruch duldende Glaube an eine verstiegene Konstruktion, der sich bei Fichte zum Beispiel in solchen Redewendungen äußert wie: «So muß es sein, und es kann nicht anders sein; darum ist es also», oder in dem Titel einer Schrift: «Sonnenklarer Bericht an das größere Publikum über das eigentlichste Wesen der neuesten Philosophie; ein Versuch, die Leser zum Verstehen zu zwingen.» Gemeinsam ist beiden auch ihr Zelotentum, das bisweilen maßlose Beschimpfen der Andersdenkenden. Dogmatismus und Idealismus als die beiden einzig möglichen Philosophien einander gegenüberstellend, tut Fichte eine Äußerung, die wie ein Vorwegnehmen des Existentialismus klingt: «Was für eine Philosophie man wählt, hängt davon ab, was man für ein Mensch ist»; aber sofort folgt der zelotische Kommentar: «Ein von Natur schlaffer oder durch Geistesknechtschaft, gelehrten Luxus und Eitelkeit erschlaffter und gekrümmter Charakter wird sich nie zum Idealismus erheben.»

Es gebricht mir hier an Zeit, eine wirkliche Darstellung der Wissenschaftslehre zu geben. Seine Methode schildert Fichte folgendermaßen: «Es ergeht die Aufforderung, einen bestimmten Begriff oder Sachverhalt zu denken. Die *notwendige* Weise, wie dieser Akt zu vollziehen ist, ist in der Natur der Intelligenz gegründet und hängt, im Gegensatz zu dem bestimmten Denkakt selbst, nicht ab von irgendwelcher Willkür. Sie ist etwas *Notwendiges*, das aber nur in und bei einer freien Handlung vorkommt; etwas *Gefundenes*, dessen Finden aber durch Freiheit bedingt ist. Insoweit weist der Idealismus im unmittelbaren Bewußtsein nach, was er behauptet. Bloße Voraussetzung aber ist, daß jenes Notwendige Grundgesetz der ganzen Vernunft sei, daß sich aus ihm das ganze System unserer notwendigen Vorstellungen, nicht nur von einer Welt, wie deren Objekte durch subsumierende und reflektierende Urteilskraft bestimmt werden, sondern auch von uns selbst, als freien und praktischen Wesen unter Gesetzen, sich ableiten lassen. Diese Voraussetzung hat er zu erweisen durch die wirkliche Ableitung, indem er zeigt, daß das zuerst als Grundsatz Aufgestellte und unmittelbar im Bewußtsein Nachgewiesene nicht möglich ist, ohne daß zugleich noch etwas Anderes geschehe, und dieses Andere nicht, ohne daß zugleich etwas Drittes geschehe, und so fort.» Das System der so abgeleiteten notwendigen Vorstellungen wird mit der gesamten Erfahrung gleichgesetzt; sie finden in der Erfahrung ihre Bestätigung; das a priori fällt darum schließlich mit dem a posteriori zusammen. Dies klingt so, als sollte nicht nur die Welt nach den in ihrer Struktur gegründeten *Möglichkeiten*, sondern die Welt in ihrer *einmaligen Faktizität* deduziert werden. Die wirkliche Ausführung dieses Vorhabens durch Fichte kann ich nur als hanebüchen bezeichnen. In dem Gegensatz von Konstruktivismus und Phänomenologie liegt meine Sympathie im ganzen auf seiner Seite. Wie aber ein konstruktives Verfahren wirklich durchgeführt werden kann, das schließlich zur Repräsentation der Welt im Symbol führt, nicht a priori, sondern mit immerwährender Bezugnahme auf die Erfahrung, das zeigt uns die Physik, vor allem in ihren beiden fortgeschrittensten Etappen der Relativitäts- und der Quanten-Theorie.

Vom Ich heißt es bei Fichte: «Das Ich fordert, daß es alle Realität in sich fasse und die Unendlichkeit erfülle. Dieser Forderung liegt notwendig zum Grunde die Idee des schlechthin durch sich selbst gesetzten unendlichen Ich; dieses ist das *absolute* Ich (welches nicht das im wirklichen Bewußtsein gegebene Ich ist). Das Ich muß über sich reflektieren; das liegt gleichfalls in seinem Begriff.» Daraus ergibt sich nach Fichte, rein aus dem Ich heraus, das hier zum *praktischen* wird, die Reihe dessen, was sein *soll*, die Reihe des *Idealen*. Einschränkung dieses unendlichen Strebens durch ein entgegengesetztes Prinzip, ein *Nicht-Ich*, führt zu der Reihe des *Wirklichen;* hier wird das Ich zur erkennenden *Intelligenz*. Doch heißt es von dieser entgegengesetzten Kraft des Nicht-Ich, daß sie von dem endlichen Wesen bloß gefühlt, aber nicht erkannt wird. «Alle möglichen Bestimmungen dieser Kraft des Nicht-Ich, die in die Unendlichkeit in unserem Bewußtsein vorkommen können, macht die Wissenschaftslehre sich anheischig, aus dem bestimmenden Vermögen des Ich abzuleiten.»

Ein Analogon aus der Geometrie kann uns nach meiner Meinung hilfreich sein zur Klärung des Problems, um das Fichte und Husserl ringen, nämlich die Brücke zu schlagen vom immanenten Bewußtsein, das nach einem Heideggerschen Ausdruck je-meiniges ist, zu dem konkreten Menschen, der ich bin, der von einer Mutter geboren ward und sterben wird. Die Objekte, die Subjekte (oder Iche) und die Erscheinung eines Objekts für ein Subjekt stelle ich in Parallele zu den Punkten, den Koordinatensystemen und den Koordinaten eines Punktes mit Bezug auf ein Koordinatensystem in der Geometrie. Jedem Punkt p einer Ebene kommen mit Bezug auf ein aus drei nicht in einer Geraden liegenden Punkten bestehendes Koordinatensystem S

drei Zahlen x_1, x_2, x_3 von der Summe 1 als seine Koordinaten (Schwerpunktskoordinaten) zu. Hier gehören Objekte (Punkte) und Subjekte (Koordinatensysteme = Punkttripel) derselben Realitätssphäre an, die Erscheinung aber liegt in einem andern Bezirk, im Reiche der Zahlen. Der naive Realismus (oder der Dogmatismus, wie Fichte diesen philosophischen Standpunkt nennt) nimmt die Punkte als etwas an sich Existierendes hin. Es ist aber ein algebraischer Aufbau der Geometrie möglich, der nur die Zahl-Erscheinungen (die Erlebnisse eines reinen Bewußtseins) benutzt. Ein Punkt, so definiert man hier, ist nichts anderes als ein Tripel von Zahlen x der Summe 1; ein Koordinatensystem besteht aus drei solchen Tripeln; es wird algebraisch erklärt, wie ein solcher Punkt p und ein solches Koordinatensystem S drei Zahlen ξ als die Koordinaten von p mit Bezug auf S bestimmen. Dieses Zahlentripel ξ stimmt mit dem den Punkt p *definierenden* Zahlentripel x überein, wenn das Koordinatensystem S das *absolute* ist, welches aus den drei Tripeln (1, 0, 0), (0, 1, 0), (0, 0, 1) besteht. Dieses korrespondiert also dem *absoluten Ich*, für welches Ding und Erscheinung zusammenfallen. Hier treten wir aus der Sphäre der Zahlen – oder, in der Analogie: des immanenten Bewußtseins – gar nicht heraus. Der im Namen der Objektivität geforderten Gleichberechtigung aller Iche kann man hier nachträglich gerecht werden, indem man erklärt, daß man sich nur für solche Zahlbeziehungen interessiert, die beim Übergang vom absoluten zu irgendeinem Koordinatensystem ungeändert bleiben oder, was das Gleiche besagt, die gegenüber beliebigen linearen Transformationen der drei Koordinaten *invariant* sind. Diese Analogie macht es verständlich, wieso das *eine* sinngebende Ich bei objektiver Einstellung, das heißt unter dem Gesichtspunkt der Invarianz, als einzelnes Subjekt unter Vielen seinesgleichen erscheinen kann. (Übrigens werden bei Übertragung in diese Analogie einige der Husserlschen Thesen erweisbar falsch, was mir als ein ernstes Verdachtsmoment gegen sie erscheint.)

Es ist aber von mir die Anerkennung des andern Ich, des Du, nicht nur so gefordert, daß ich mich im Denken einer abstrakten Norm der Invarianz oder Objektivität füge, sondern *absolut: Du* bist *für dich* noch einmal, was ich für mich bin: nicht *seiender*, sondern *bewußtseiender* Träger der Erscheinungswelt. Diesen Schritt können wir in unserer geometrischen Analogie nur tun, wenn wir von dem zahlenmäßigen Modell der Punktgeometrie übergehen zur axiomatischen Beschreibung. Hier werden die Punkte weder als vorliegende Realitäten behandelt, noch wird durch ihre Identifizierung mit Zahltripeln von vornherein ein absolutes Koordinatensystem ausgezeichnet. Sondern Punkt und die geometrischen Grundbeziehungen, vermöge deren ein Punkt p und ein Koordinatensystem = Punkttripel S ein Zahlentripel ξ bestimmen, werden als undefinierte Grundbegriffe eingeführt, für welche bestimmte Axiome gelten. So zeigt es sich, daß doch über den naiven Realismus und den Idealismus hinaus ein dritter Standpunkt, der des *Transzendentismus*, möglich ist, der ein transzendentes Sein setzt, sich aber mit seiner Nachbildung im Symbol begnügt; ihm entspricht der axiomatische Aufbau der Geometrie.

Ich will nicht sagen, daß das Rätsel der Ichheit damit gelöst sei. Leibniz glaubte den Widerstreit von menschlicher Freiheit und göttlicher Prädestination dadurch zu lösen, daß er Gott unter den unendlich vielen Möglichkeiten (aus zureichenden Gründen) gewisse, zum Beispiel die Wesen Judas und Petrus, zum Dasein erwählen läßt, deren substantiale Natur ihr ganzes Schicksal bestimmt. Die Lösung mag objektiv zureichend sein, sie zerbricht aber vor dem Verzweiflungsschrei des Judas: Warum mußte *ich* Judas sein! Die Unmöglichkeit einer objektiven Fassung dieser Frage leuchtet ein; darum kann auch keine Antwort in Form einer objektiven Erkenntnis erfolgen. Das Wissen vermag das Licht-Ich mit dem dunklen, irrenden, in ein individuelles Schicksal ausgestoßenen Menschen nicht zur Deckung zu bringen. Hier wird vielleicht auch offenbar, daß das ganze Problem bisher, namentlich von Husserl, zu einseitig theoretisch gefaßt wurde. Um sich als Intelligenz zu finden, muß das Ich nach Descartes durch den radikalen *Zweifel* hindurch; um sich als Existenz zu finden, nach Kierkegaard durch die radikale *Verzweiflung*. Durch den Zweifel hindurch stoßen wir vor zu dem Wissen um die dem immanenten Bewußtsein transzendente reale Welt; in umgekehrter Richtung aber, nicht des Erzeugnisses, sondern des Ursprungs, liegt die Transzendenz *Gottes*, aus dem herfließend das Licht des Bewußtseins, dem der Ursprung selber verdeckt ist, in seiner Selbstdurchdringung sich ergreift, gespalten und gespannt zwischen Subjekt und Objekt, zwischen *Sinn* und *Sein*.

Fichte ging in einer späteren Epoche seines Philosophierens vom Idealismus zu einem theologischen Transzendentismus über, wie ihn zum Beispiel seine Schrift «Anweisung zum seligen Leben» entwickelt. An Stelle des absoluten Ich tritt *Gott*. Ich zitiere seine Worte: «Das *Sein* ist durchaus einfach, nicht mannigfaltig, sich selbst gleich, unwandelbar und unveränderlich; es ist in ihm kein Entstehen noch Vergehen, kein Wandel und Spiel der Gestaltungen, sondern immer nur das gleiche ruhige Sein und Bestehen.» Da-sein, die Offenbarung und Äußerung des in sich selber verschlossenen Seins, ist notwendig *Bewußtsein* oder *die Vorstellung des Seins*. Davon heißt es: «Gott *ist* also nicht nur, innerlich und in sich verborgen, sondern er ist auch *da* und äußert sich; sein Dasein aber unmittelbar ist notwendig *Wissen*, welche letztere Notwendigkeit im Wissen selber sich einsehen läßt . . . Er ist da, wie er schlechthin in sich selber ist; ohne sich irgend zu verwandeln in dem Übergange vom Sein zum Dasein, ohne eine zwischenliegende Kluft oder Trennung . . . Und da das Wissen, oder *Wir*, dieses göttliche Dasein selbst sind, so kann auch in Uns . . . keine Trennung, Unterscheidung noch Zerspaltung stattfinden. So muß es sein, und es kann nicht anders sein; darum ist es also.» Aber dann muß Fichte zu Sophismen und den gewagtesten Konstruktionen seine Zuflucht nehmen, um nun doch aus dieser Einheit göttlichen Seins, die auch dem göttlichen Dasein in Uns zukommt, zur Mannigfaltigkeit der Bewußtseinsinhalte und der Welt zu gelangen.

Ich habe hier von den Philosophen gesprochen und den philosophischen Gedanken, die mich etwa in der Zeit von 1913 bis 1922 bewegt haben. Im Anschluß an Fichte bin ich dann selber monatelang metaphysischen Spekulationen über Gott, Ich und die Welt nachgegangen, in denen sich mir die letzte Wahrheit aufzuschließen schien. Ich muß Ihnen aber gestehen, daß von ihnen jede Spur in meiner Erinnerung verflogen ist. Daneben lief natürlich die eine zentralere Stelle in meinem Leben einnehmende mathematische Forschung. Das will ich hier übergehen, obschon so ein falsches Gesamtbild von der Rolle entsteht, die Erkenntnisarbeit und Besinnung in meinem Leben spielten. Nur eines mag erwähnt sein: daß ich 1918 die erste einheitliche Feldtheorie von Gravitation und Elektromagnetismus aufstellte. Obwohl ihr Grundprinzip, die «Eichinvarianz», heute in veränderter Form in die Quantentheorie aufgenommen ist, ist die Theorie selbst durch die moderne Entwicklung der Physik, die neben das elektromagnetische Feld das Wellenfeld des Elektrons und der andern Elementarteilchen gestellt hat, längst überholt. Daneben beschäftigten mich die Grundlagen der Mathematik, die so eng mit dem Problem des Unendlichen zusammenhängen.

Vom späten Fichte kam ich auf Meister Eckehart, den tiefsten der abendländischen Mystiker. Trotz Verwandtschaft mit Plotin und dem

Begriffsapparat der christlich-thomistischen Philosophie, der ihm zur Verfügung steht, kann man nicht an der Ursprünglichkeit seines religiösen Grunderlebnisses zweifeln: es ist die Eingießung der Gottheit in den Seelengrund, die er unter dem Bilde der Geburt des «Sohnes» oder des «Worts» durch Gottvater beschreibt. Im Rückgang aus der Mannigfaltigkeit des Daseins muß die Seele aber nicht nur zu diesem ihrem Urbild zurückfinden, sondern auch durch das Urbild durchbrechen zur einigen, in undurchdringlichem Schweigen wohnenden Gottheit. Wie souverän Eckehart mit dem Bibelwort umgeht, möge der Beginn einer Weihnachtspredigt illustrieren, die an Matthäus Kap. 2, Vers 2, anknüpft: «Wo ist, der nun geboren ist, der König der Juden?» «Bemerkt zunächst», sagt er, «von dieser Geburt, wo sie geschehe? Ich behaupte aber, wie schon des öfteren, daß diese ewige Geburt sich in der Seele genau in der Weise vollzieht wie in der Ewigkeit, gar nicht anders; denn es ist ein und dieselbe Geburt. Und zwar vollzieht sie sich in dem Wesen und Grunde der Seele». Die Predigt schließt mit den Worten: «In diese Geburt helfe uns der Gott, der heute von neuem als *Mensch* geboren ist, damit wir armen Erdenkinder in ihm als *Gott* geboren werden; dazu helfe er uns ewiglich! Amen.» Hier spricht, der Ton verrät es, ein Mensch von hoher Verantwortung und von ungleich größerem Adel als Fichte. Von allen geistigen Erlebnissen waren für mich die beglückendsten: als junger Student, 1905, das Studium von Hilberts großartigem «Bericht über die Theorie der algebraischen Zahlen» und 1922 die Lektüre von Eckehart, die mich während eines herrlichen Engadiner Winters gefangen hielt. Hier fand ich für mich nun auch den Zugang zur religiösen Welt, an dessen Mangel 10 Jahre früher eine sich anknüpfende Lebensbeziehung gescheitert war.

Aber mit meinen durch Fichte und Eckehart angeregten metaphysisch-religiösen Spekulationen kam ich nie ins Reine; das liegt wohl auch in der Natur der Sache. In den folgenden Jahren war ich (unter anderem) damit beschäftigt, auf Grund meiner wissenschaftlichen und philosophischen Erfahrungen die Methodologie der Wissenschaft kritisch durchzudenken. Hier wurde die Auseinandersetzung mit Leibniz von erheblicher Bedeutung. Auf den metaphysischen Hochflug folgte die Ernüchterung. Was ich von Philosophen gelernt und selber ergrübelt hatte, fand seinen Niederschlag in der 1926 veröffentlichten «Philosophie der Mathematik und Naturwissenschaft». Die Niederschrift erfolgte in wenigen Ferienwochen; aber vorher hatte ich exzerpierend ein Jahr lang philosophischer Lektüre gefrönt, wie ein Schmetterling von einer Blüte zur andern fliegend, bemüht, aus jeder etwas Honig zu saugen. Das durch die Arbeit in den exakten Wissenschaften geschärfte Erkenntnisgewissen macht es Unsereinem nicht leicht, den Mut zur philosophischen Aussage zu finden. Ganz ohne Kompromiß kommt man da nicht durch. Lassen Sie mich davon schweigen. Das Produkt dieses Ringens ist ja dem, der daran interessiert ist, im Druck zugänglich. Was ich hier allein schildern wollte, ist der philosophische Wurzelboden, dem es entsproß.

Um die gleiche Zeit erreichte ich auch in der mathematischen Forschung einen gewissen Höhepunkt in meinen Untersuchungen über halbeinfache, kontinuierliche Gruppen. Damit war meine Entwicklung im wesentlichen abgeschlossen. Ich weiß nicht, ob es andern Menschen ebenso geht: Wenn ich auf mein Leben zurückschaue, so finde ich, daß die Zeit der Jugend bis etwa zum Alter von 35–40 Jahren, in welcher die Entwicklung ständig zu neuen, noch nicht durchfühlten und durchdachten Inhalten vorstößt, unvergleichlich reicher ist als die nachfolgende Zeit der Reife und des Alterns. Natürlich bin ich in späteren Jahren weder an der Umwälzung vorübergegangen, welche bezüglich unseres Naturwissens die Quantenphysik herbeiführte, noch an der in der grausigen Zerrissenheit unseres Zeitalters emporgewachsenen Existenzphilosophie. Die erstere warf neues Licht auf das Verhältnis des erkennenden Subjekts zum Objekt, im Mittelpunkt der letzteren steht nicht ein reines Ich noch Gott, sondern der Mensch in geschichtlicher Existenz, der sich aus eigener Existenz entscheidet.

1930 war ich als Nachfolger David Hilberts von Zürich nach Göttingen zurückgekehrt. Als dann 1933 über Deutschland das Nazitum hereinbrach, wanderte ich, aufs tiefste empört über die Schande, mit der dieses Regime den deutschen Namen befleckte, nach Amerika aus. Als dort die Aufgabe an mich herantrat, eine englische Ausgabe meines alten Philosophiebuches zu veranstalten, hatte ich nicht mehr den Mut, es unter Berücksichtigung der inzwischen eingetretenen wissenschaftlichen und philosophischen Wandlungen neu zu schreiben. Ich begnügte mich damit, an den alten Text die bessernde Hand anzulegen, einige Abschnitte umzuarbeiten und eine Reihe von Anhängen hinzuzufügen, deren Abfassung mir größere Mühe bereitete als das ursprüngliche Buch. Wie oft erwog ich nicht, die Arbeit ganz aufzugeben oder, als das Manuskript der Anhänge fertiggestellt war, es ins Feuer zu werfen! Woher diese Mühen und diese Bedenken, das erläutern vielleicht die Zeilen aus T.S.Eliot's «Four Quartets», die ich als Motto der Vorrede voranstellte:

«Home is where one starts from. As one grows older
The world becomes stranger, the pattern more complicated
Of dead and living.»

Um so dankbarer bin ich dafür, daß mir nun für diese englische Ausgabe in Verbindung mit dem einstweilen nur englisch erschienenen späteren kleinen Buch über Symmetrie der Arnold-Reymond-Preis zuteil wird. Die an der Universität Princeton gehaltenen Vorlesungen über Symmetrie haben mir Freude gemacht. Mir war's dabei zumut, wie wenn ein Mann, nachdem er sich einen langen Werktag über gemüht hat, im Widerstreit der Ideen und der menschlichen Ansprüche das Seine zu tun, so gut er's vermag, nun, da die Sonne sinkt und die versöhnende Nacht hereinbricht, sich ein stilles Abendlied auf der Flöte bläst.

Hiermit ende ich meinen Rechenschaftsbericht.

167.

Rückblick auf Zürich aus dem Jahre 1930

Schweizerische Hochschulzeitung 28, 180–189 (1955)

Als Nachfolger des Geometers *C. F. Geiser* war ich 1913 an die ETH berufen worden. Meine Lehrtätigkeit erstreckte sich auf alle Gebiete der Mathematik, ja gelegentlich habe ich auch, wenn es die eigenen Forschungen mit sich brachten, theoretische Physik (Relativitäts- und Quanten-Theorie) und Philosophie der Wissenschaften eingeschlossen. Zweimal wurde meine Zürcher Wirksamkeit auf ein Jahr unterbrochen, das erste Mal, als ich während des ersten Weltkrieges als gemeiner Landsturmrekrut in die deutsche Armee einrücken mußte, ein ander Mal, als ich der Einladung zu einer Gastprofessur für mathematische Physik an der amerikanischen Universität Princeton folgte. Die schlimmste Plage während meiner Zürcher Jahre waren für mich Berufungen nach auswärts; denn Entscheidungen dieser Art fielen mir schwer. Einmal, zu Beginn der berüchtigten Inflationszeit, geschah es, daß ich gleichzeitig nach Berlin und Göttingen berufen war. Verhältnismäßig rasch entschloß ich mich zur Ablehnung von Berlin. Aber den Lehrstuhl von *Felix Klein* an der Göttinger Universität ausschlagen — das war eine härtere Nuß. Galt doch damals das mir von meiner Studenten- und Privatdozenten-Zeit wohlvertraute Göttingen neben Paris als das Zentrum der Mathematik. Als sich die Entscheidung nicht länger aufschieben ließ, lief ich im Ringen darum mit meiner Frau stundenlang um einen Häuserblock herum und sprang schließlich auf ein spätes Tram, das zum See und Telegraphenamt hinunterfuhr, ihr zurufend: „Es bleibt doch nichts anderes übrig als annehmen." Aber dann muß es mir das fröhliche Treiben, das sich an diesem schönen Sommerabend um und auf dem See entfaltete, angetan haben: ich ging zum Schalter und telegraphierte eine Ablehnung. Meine Frau war natürlich baß erstaunt, als ich heimkam. Wie weit diese Geschichte Wahrheit oder Dichtung ist, vermag ich heute nicht mehr zu sagen. Bedauert haben wir jedenfalls *diese* Entscheidung niemals.

Nach einem Zykel von 7 Jahren wiederholte sich dieselbe Situation: Berufung nach Göttingen als Nachfolger des größten Mathematikers dieser Zeit, *David Hilbert.* Diesmal widerstand ich nicht mehr — und hatte es bald bitter zu bereuen. Aber selbst die politisch Einsichtigsten meiner Schweizer Freunde sahen damals, Anfang 1930, eine Gefahr in Deutschland nur von seiten der Kommunisten, nicht der Nazis voraus.

Was ich, der immer ein schwaches Gedächtnis für Vergangenes gehabt hat, an Erinnerungen an meine dem Dienste der ETH gewidmeten Jahre bewahrt habe, ist größtenteils persönlicher Art und, wo sie sich auf Wissenschaft beziehen, drehen sie sich fast ausschließlich um die forschende, nicht die lehrende Tätigkeit. Dies ist für ein weiteres Publikum von geringem Interesse. Auch hat das inzwischen verflossene stürmische Vierteljahrhundert viele Spuren verlöscht. So ziehe ich denn vor, etwas gekürzt, eine Ansprache hier zu reproduzieren, die ich im Herbst 1930 in Göttingen an die dortige Mathematische Verbindung richtete, der ich einst als Student angehört hatte. Darin ist aus noch frischer lebendiger Erinnerung einiges ausgesagt über die Bedeutung, welche die Schweiz für mich gehabt und über die Weise, wie ich an der ETH zu wirken versucht hatte. Diese Ansprache lautete:

„Liebe Kommilitonen!“

„Als Jacob Burckhardt vierzigjährig von der Technischen Hochschule in Zürich dem Ruf an die Universität seiner Vaterstadt Basel folgte, meinte er: mit 40 Jahren müsse man dorthin gehen, wo man zu sterben wünsche. Ich will damit nicht sagen, daß ich aus der Schweiz in meine norddeutsche Heimat zurückgekehrt bin – wobei ich, der ich Holsteiner und nicht Hannoveraner bin, die Grenzen der Heimat etwas weiter ziehe als Burckhardt, für den das 2 Bahnstunden von Basel entfernte Zürich schon wilde Fremde war –, ich will nicht sagen, daß ich zurückgekehrt bin, nur um unter Ihnen zu sterben. Aber das Burckhardtsche Wort trifft mich doch. Auch ich habe erfahren, daß die Annahme einer Berufung mit 45 Jahren eine ernstere, besinnlichere, weniger enthusiastische Angelegenheit ist als mit 27. Um Ihnen das verständlich zu machen, muß ich ein wenig von mir und ein wenig von der Schweiz sprechen.

Im Jahre 1913, ein Jahr vor Ausbruch des Krieges, kam ich, jung und jung verheiratet, nach Zürich. Das Bild, das ich mir von der Schweiz machte, war, wie es damals für einen deutschen Jüngling typisch war, hauptsächlich aus den Dichtungen Gottfried Kellers und C. F. Meyers geschöpft. Ich war daher zunächst überrascht, mich in Zürich im Auslande zu finden. Ich hatte zu lernen, zu Beginn des Krieges mit vermehrter Schärfe, daß die deutsche Kultur die Schweiz nicht so selbstverständlich umschließt, wie es bei Keller geschienen hatte, daß durch die divergente Entwicklung namentlich seit der Reichsgründung 1870 ein tiefer Graben enstanden war; vor allem aber, daß die Schweiz eben nicht die deutsche Schweiz ist, sondern das Dach Europas, unter dem sich germanische und romanische Kultur treffen. Langsam wurde ich durch sie gewandelt. Wenn Sie bedenken, daß ich von den Jahren meines selbständigen Lebens, wenn ich sie mit der Studentenzeit beginnen lasse, mehr als die Hälfte in der Schweiz verbrachte – und das waren wohl die wichtigsten und produktivsten Jahre meines Lebens überhaupt –, so werden Sie sich nicht darüber wundern, daß ich mich der Schweiz nicht viel weniger verbunden fühle als Deutschland. Denn ich habe versucht, geöffnet zu leben; und das Nationale habe ich niemals als Besitz und Vorzug betrachtet, auf den man pocht, oder als Scheuklappe, mit der man sich vor dem Verständnis fremder Lebensform schützt, sondern ich habe darauf vertraut, daß das Gute, das man seiner Volksart verdankt, im eigenen Wesen und Wirken von selber zum Durchbruch kommen und, wo man auch wirke, Dank und Anerkennung finden werde. So ist mir insbesondere der romanische Einschlag, mit seiner selbstverständlicheren Schätzung und Genuß der einfachen sinnlichen Lebensgüter, seiner größeren Heiterkeit, Menschlichkeit und Form zu einem Lebenselement geworden, das ich schwer entbehre. In der Schweiz sind meine Kinder geboren und aufgewachsen. Dort habe ich – trotz der Sprödigkeit, mit welcher der Schweizer im allgemeinen dem Fremden entgegentritt – die tiefsten Freundschaften meines Lebens geschlossen. So kann ich nicht verhehlen, daß mein Herz es fast als Treulosigkeit empfand, daß ich Zürich verließ.

Und endlich hat die Schweiz, diese älteste, durch die Jahrhunderte geschulte Demokratie, ganz wesentlich meine politische Einstellung determiniert. Nur mit einiger Beklemmung finde ich mich aus ihrer freieren und entspannteren Atmosphäre zurück in das gähnende, umdüsterte und verkrampfte Deutschland der Gegenwart. Der Deutsche nimmt an geistigen Dingen lebendigeren Anteil und ist darin leidenschaftlicher als der etwas nüchterne Schweizer. Das ist ein Zug, um dessentwillen der Deutsche zu lieben ist. Und ich verkenne nicht, daß die gegenwärtige politische Erregung zum Teil von diesem edlen Feuer geistiger Leidenschaft angefacht ist. Aber daneben zeigt sich ein geradezu erschreckender Mangel an Sachlichkeit und Wirklichkeitssinn, der durchaus eine Schwäche ist und mit dem Mäntelchen Idealismus gar nicht zugedeckt werden kann – übertönt von lärmenden Explosionen rein negativer Seelenregungen: Klage über die Not, Anklage wider das Unrecht, Empörung, Haß und Schmähsucht. Der Psychologe möchte daraus erraten, daß wir, daß das deutsche Volk sich insgeheim einer inneren Schicksalsschuld bewußt ist, die nicht eingestanden werden darf und mit wilder Geste auf die andern, nur auf die andern abgewälzt werden muß. Ich hoffe, daß gerade der Teil der deutschen Studentenschaft, welcher den mathematisch-naturwissenschaftlichen Fakultäten angehört, sich von diesem Strom nicht so mitreißen läßt – da ja unsere Wissenschaften den menschlichen Dingen und Leidenschaften ferner stehen und ohne die Schulung zu strenger Objektivität nicht gedeihen können. Ich möchte jedem unter Ihnen, der es irgend vermag, anraten, womöglich 1 oder 2 Semester an einer ausländischen Hochschule zu verbringen, in der Schweiz, in England, in Frankreich, um Distanz und ein rechtes Maß für die Dinge der quälenden und brennenden deutschen Gegenwart zu gewinnen.

‚Politisch Lied – ein garstig Lied.' Lassen Sie mich jetzt lieber einige Worte über mich selber sagen und wie ich meine Aufgabe als Lehrer und Forscher auffasse und, eingedenk der Goetheschen Warnung, ‚Was euch das Innre stört, dürft ihr nicht leiden', auf Grund meiner Natur und ihrer Beschränkungen bestimmen muß. – Eine Hochschule ist meiner Überzeugung nach nicht nur Schule, oberste jener Institutionen, durch welche die Gesellschaft den Gehalt ihrer Kultur, insbesondere auch die gewonnenen wissenschaftlichen Erkenntnisse, technischen Erfahrungen und das theoretische Weltbild der heranwachsenden Generation tradiert, sondern sie dient außerdem der Forschung. Der Erkennende, der theoretisch Gestaltende ist so gut wie der Künstler ein Grundtypus des Menschen, der in der gesellschaftlichen Organisation seinen Platz finden muß, und er findet ihn heute nur an der Hochschule. Wer erkennt, den ‚verlangt nach Rede'; so mögen denn die Jungen zu seinen Füßen sitzen und ihm zuhören, wenn Rede aus ihm bricht. Dies betrachte ich als das Grundverhältnis. Ich glaube nicht daran, daß das System der Erziehung von unten aufgebaut werden müsse; die Gegenbewegung darf nicht fehlen. Was der Natur und der Notwendigkeit angehört, wächst von unten her, der Geist und seine Freiheit aber brechen von oben herein. In dieser Weise, hoffe ich, werde ich Ihnen, liebe Kommilitonen, Lehrer sein können: den Samen in den Wind streuend; fasse, wer es fassen kann. Ich bin nicht so gut dafür geschaffen, den Einzelnen an konkreten Problemen mit sicherer Hand in die wissenschaftliche Arbeit einzuführen. Meine eigenen mathematischen Arbeiten waren immer ganz unsystematisch, ohne Methode und Konsequenz. Fast liegt mir mehr an Ausdruck und Gestaltung als an der Erkenntnis selbst. Aber ich glaube auch, daß, unabhängig von meinem eigenen Wesen, die Mathematik selber, im Vergleich etwa zu den experimentellen Disziplinen, einen Zug in sich trägt, der sie der frei schaffenden Kunst nähert und daß darum die moderne wissenschaftliche Betriebsamkeit, bei der die naturwissenschaftlichen Institute gedeihen, ihr nicht so zum Heile gereicht. Das Verhältnis zwischen Lehrer und Schüler wird in der Mathematik daher immer zarter und lockerer bleiben müssen; wie wir denn ja auf dem Gebiete der Kunst normalerweise dem schaffenden Künstler überhaupt nicht die Anleitung von Schülern zu eigenem Schaffen zumuten.

Bei solchem Wesen wäre es für mich vielleicht das richtige gewesen, in Zürich zu bleiben, wo meine Stellung, mit einer verhältnismäßig geringen äußeren Wirksamkeit verbunden, der vita contemplativa besonders günstig war. Es war nicht leicht für mich, darüber zur Klarheit zu kommen. Ich habe mich darüber im Geist hauptsächlich mit zwei Männern auseinandergesetzt, mit Jacob Burckhardt und Hermann Hesse. Des Letzteren Wesen und Geschick steht mir besonders nahe. Er ging in den ‚Süden', nicht nur im räumlichen Sinne. Sein Beispiel ist lockend, aber auch warnend. Denn nur tiefere Einsamkeit ward sein Teil und eine alles Glück der Sinne, des Auges und der Dichtung seltsam beschwingende, durchdringende und auflockernde milde Verzweiflung. Burckhardt lehnte von Basel aus mit selbstverständlicher Sicherheit einen Ruf nach Berlin als Nachfolger seines großen Lehrers Ranke ab. Aber er schöpfte die Kraft zur Beschauung aus seiner ganz starken humanistisch-männlichen Verbundenheit mit der Kultur, der er als Historiker diente, und mit seiner heimatlichen Stadtrepublik Basel und ihrer Gesellschaft. — Ich bin zurückgekehrt, weil ich die Verbindung mit der Jugend nicht missen möchte, ohne die das Alter in starre Einsamkeit versinkt, weil ich sehe, daß die Wissenschaft auf die geistige Gemeinschaft als den Ort ihres Wachstums angewiesen ist und ich es darum als eine ernste Pflicht anerkenne, die Wissenschaft zu tradieren, und schließlich, um mich wieder in die eigene Volksgemeinschaft einzubetten.

An der Vergangenheit hänge ich nicht, ich hasse es sogar, sie in Zeugnissen aufzubewahren, die ihr eine Art Selbständigkeit verleihen wollen über das hinaus, wodurch sie ins Gegenwärtige fortwirkt. Was ich verrichtete, sah ich an als ein Stück am laufenden Band des Lebens. So will ich mich denn der Gegenwart, in persönlichem Kontakt mit euch, den Studierenden, und im geistigen Kontakt mit der sich fortentwickelnden Wissenschaft so offen halten, wie ich es vermag.

Dies war keine enthusiastische Rede. Die Rückkehr dessen, der als jugendlicher Individualist auszog, auf eigene Faust die Welt zu erobern, in den Hafen der Gemeinschaft, wohin der Mann gehört, schließt Entsagung ein.

> ‚Größer's wolltest auch du, aber die Liebe zwingt
> all uns nieder, das Leid beuget gewaltiger,
> doch es kehret umsonst nicht
> unser Bogen, woher er kommt'

heißt es in einem Hölderlinschen Vers, den junge Herzen so schwer verstehen ..."

So sprach ich damals, vor 25 Jahren. Aber schon fraß an der Volksgemeinschaft, in die ich mich hatte zurückbetten wollen, das von dem Rattenfänger Hitler gestreute Gift. Ich ertrug es nicht, unter der Herrschaft dieses Dämonen und Schänders des deutschen Namens zu leben, und obschon die Losreißung mir so hart fiel, daß ich darüber einen schweren seelischen Zusammenbruch erlitt, schüttelte ich den Staub des Vaterlandes von den Füßen. Ich war so glücklich, in Amerika am neugegründeten Institute for Advanced Study in Princeton, New Jersey, eine neue fruchtbare Wirkungsstätte zu finden. Der Wissenschaftler konnte sich keine schönere wünschen; was aber *Heimat* ist, habe ich verlernen müssen. Ich muß zufrieden sein, daß Kinder und Kindeskinder fest in dem neuen Land verwurzelt sind.

Hermann Weyl (1885—1955) par C. Chevalley et A. Weil

Extrait de L'Enseignement Mathématique, tome III, fasc. 3 (1957)

« Quand Hermann Weyl et Hella annoncèrent leurs fiançailles, l'étonnement fut général que ce jeune homme timide et peu loquace, étranger aux cliques qui faisaient la loi dans le monde mathématique de Göttingen, eût remporté le prix convoité par tant d'autres. Ce n'est que peu à peu que l'on comprit à quel point Hella avait eu raison dans son choix... [1] »

Peut-être les vérités mathématiques, comme les femmes, font-elles leur choix entre ceux qu'elles attirent. Est-ce le mieux doué qu'elles choisissent, ou le plus séduisant ? celui qui les désire le plus ardemment, ou celui qui les a le mieux méritées ? Elles semblent se tromper parfois; souvent il faut du temps pour s'apercevoir qu'elles ont eu raison. Timide, peu loquace, étranger aux cliques, tel apparaissait donc Hermann Weyl à ses débuts; tel il devait rester au fond de lui-même, en dépit des succès d'une brillante carrière. Comme beaucoup de timides une fois rompues les barrières de leur timidité, il était capable d'enthousiasme et d'éloquence: « Ce soir-là, dit-il en racontant sa première rencontre avec celle qu'il devait épouser [2], je décrivis l'incendie d'une grange auquel je venais d'assister; elle me dit plus tard qu'à m'écouter elle s'était éprise de moi aussitôt. » Ses propres confidences nous le montrent profondément influençable aussi, jusque dans sa pensée la plus intime: « Mon tranquille positivisme

[1] Extrait des paroles prononcées par Courant aux obsèques de Hella Weyl le 9 septembre 1948.

[2] Cette citation, comme plusieurs autres par la suite, est tirée d'une notice inédite consacrée par Hermann Weyl à la mémoire de Hella Weyl. Nos autres citations proviennent des publications de Weyl.

fut ébranlé quand je m'épris d'une jeune musicienne d'esprit très religieux, membre d'un groupe qui s'était formé autour d'un hégélien connu... Peu après, j'épousai une élève de Husserl; ainsi, ce fut Husserl qui, me dégageant du positivisme, m'ouvrit l'esprit à une conception plus libre du monde. » Il avait alors vingt-sept ans.

C'est ainsi qu'on voit se dessiner, vers l'époque de son mariage, quelques-uns des principaux traits d'une des personnalités mathématiques les plus marquantes et attachantes de la première moitié de ce siècle, mais aussi de l'une des plus difficiles à serrer de près. « A country lad of eighteen », un gars de campagne de dix-huit ans, ainsi se décrit-il lui-même à son arrivée à Göttingen. « J'avais choisi cette université, dit-il, principalement parce que le directeur de mon lycée était un cousin de Hilbert et m'avait donné pour celui-ci une lettre de recommandation. Mais il ne me fallut pas longtemps pour prendre la résolution de lire et étudier tout ce que cet homme avait écrit. Dès la fin de ma première année, j'emportai son *Zahlbericht* sous mon bras et passai les vacances à le lire d'un bout à l'autre, sans aucune notion préalable de théorie des nombres ni de théorie de Galois. Ce furent les mois les plus heureux de ma vie... [3] »

Un peu plus tard, ce sont les joies de la découverte: « Un nouvel événement fut décisif pour moi: je fis une découverte mathématique importante. Elle concernait la loi de répartition des fréquences propres d'un système continu, membrane, corps élastique ou éther électromagnétique. Le résultat, conjecturé depuis longtemps par les physiciens, semblait encore bien loin alors d'une démonstration mathématique. Tandis que j'étais fiévreusement occupé à mettre mon idée au point, ma lampe à pétrole avait commencé à fumer. Quand je terminai, une épaisse pluie de flocons noirs s'était abattue sur mon papier, sur mes mains, sur mon visage. » A ce moment, il est déjà privatdozent à Göttingen. Bientôt c'est le mariage, la chaire à Zurich, la

[3] « De toute mes expériences spirituelles, écrit-il ailleurs, celles qui m'ont comblé de la plus grande joie furent, en 1905, quand j'étais étudiant, l'étude du *Zahlbericht* et, en 1922, la lecture de maître Eckhart qui me retint fasciné pendant un splendide hiver en Engadine. »

guerre. Au bout d'un an de garnison à Sarrebruck (comme simple soldat, précise-t-il), le gouvernement suisse obtient qu'il soit rendu à son enseignement à l'Ecole polytechnique fédérale. « Je ne puis guère me souvenir d'un instant de joie plus intense que le beau jour de printemps, en mai 1916, où Hella et moi franchîmes la frontière suisse, puis, arrivés chez nous, descendîmes de nouveau jusqu'au lac à travers la belle ville paisible. »

Il reprend ses travaux. Un cours professé à Zurich sur la relativité paraît en volume, en 1918; c'est le célèbre *Raum, Zeit, Materie*, qui connaît cinq éditions en cinq ans et, profitant de la vogue extraordinaire du sujet jusque parmi les profanes, répand le nom de Weyl bien au-delà du monde des mathématiciens où sa réputation n'était plus à faire. Les offres de chaires viennent d'un peu partout; celle de Göttingen en 1922, où il s'agissait de la succession de Klein, fut l'occasion pour lui d'un débat de conscience particulièrement difficile. Ayant retardé sa décision tant qu'il pouvait, ayant encore au dernier moment parcouru avec sa femme les rues de Zurich en pesant sa réponse, il partit enfin au bureau de poste pour télégraphier son acceptation. Arrivé devant le guichet, ce fut un refus qu'il télégraphia; il n'avait pu se résoudre à échanger sa tranquillité zurichoise contre les incertitudes de l'Allemagne d'après guerre. « L'étonnement de Hella, dit-il, fut sans bornes; les événements ne tardèrent pas à me donner raison. »

Mais en 1929, quand Göttingen lui offre la succession de Hilbert, il se laisse tenter. « Les trois années qui suivirent, dit-il, furent les plus pénibles que Hella et moi ayons connues. » C'est le nazisme, d'abord imperceptible nuage à l'horizon, qui grandit à vue d'œil, s'abat en trombe sur l'Allemagne en désarroi, y recouvre tout de boue sanglante. Par bonheur pour Hermann Weyl, l'Institute for Advanced Study de Princeton, nouvellement créé, offre de le sauver du désastre. Il hésite. Il accepte, il refuse. Il accepte de nouveau l'année suivante; c'est de Zurich qu'il envoie sa démission à Göttingen en 1933 et qu'il part pour l'Amérique.

Il n'eut donc pas à subir ce stage souvent long, parfois humiliant et pénible, que les circonstances ont imposé à beaucoup de savants réfugiés aux Etats-Unis. La chaire de l'Institute lui assura d'emblée le confort matériel et la situation de premier plan dans le monde scientifique américain auxquels tout, certes, lui donnait droit. Ce furent, dit-il, des années heureuses que celles qu'il passa à Princeton. Sans doute ne s'accoutuma-t-il jamais à porter aisément ce qu'il appelle « le joug d'une langue étrangère ». Mais, grâce au respect et à l'affection qui l'entourèrent dès l'abord, il se sentait enfin chez lui; et on sent percer à nouveau le gars de campagne des premiers jours de Göttingen lorsqu'il dépeint le plaisir qu'il éprouva, en 1938, à posséder son lopin de terre et à y bâtir sa maison. Si la mort de sa femme, en 1948, le déchira cruellement, un second mariage, quelque temps après, lui fit retrouver son équilibre. Ayant pris sa retraite à l'Institute, il partagea désormais son temps entre Princeton et Zurich. Une attaque cardiaque l'emporta à l'improviste, peu après les fêtes de son soixante-dixième anniversaire.

A son arrivée en Amérique, il avait déjà donné en mathématique le meilleur de lui-même, et il le savait. Pour tout autre que lui, la tentation eût été grande de se reposer sur ses lauriers, de s'abandonner à un rôle de « pontife ». Combien n'en est-il pas dont toute l'activité, passé un certain âge, consiste à aller de commission en commission, pour y discuter gravement des mérites de travaux de « jeunes » qu'ils n'ont pas lus, qu'ils ne connaissent que par ouï-dire ! Hermann Weyl se faisait une bien autre et bien plus haute idée de son métier de professeur. Il vit que Princeton seul, à notre époque, peut être ce qu'ont été autrefois Paris, puis Göttingen: un centre d'échanges, un « clearing-house » des idées mathématiques qui circulent de par le monde. Rappelant l'intense vie mathématique qui s'était développée autrefois à Göttingen sous l'influence dominante de Hilbert, il a écrit: « Les idées font boule de neige en un pareil point de condensation de la recherche »; et il ajoute: « Nous avons assisté à quelque chose de semblable ici à Princeton pendant les premières années d'existence de l'Institute for Advanced Study. » S'il en a été ainsi, c'est en grande partie à lui qu'en revient le mérite. Il se donna pour tâche principale de se maintenir au courant de l'actualité, de renseigner et éclairer les chercheurs, de leur servir d'interprète, de comprendre mieux qu'eux ce qu'ils faisaient ou essayaient de faire; il s'y consacra en toute modestie, conscient de faire œuvre utile, conscient d'y être irremplaçable. Dans sa production, qui, pendant toute cette période, reste abondante et d'une extraordinaire variété, on retrouve la trace de ses lectures, des séminaires et discussions auxquels il prenait part, des problèmes sur lesquels de tous côtés on sollicitait ses avis. Parmi ces travaux, il n'en est guère qui n'élucide un point difficile ou ne comble une lacune fâcheuse. Cette activité s'est poursuivie jusque dans ses dernières années. Par une suprême coquetterie peut-être, sa dernière publication aura été une édition rajeunie, complètement refondue, de son premier livre, livre toujours utile, encore actuel, auquel par cette révision il a donné une vitalité nouvelle. Qui de nous ne serait satisfait de voir sa carrière scientifique se terminer de même ?

Un Protée, qui se transforme sans cesse pour se dérober aux prises de l'adversaire, et ne redevient lui-même qu'après le triomphe final: telle est l'impression que nous laisse souvent Hermann Weyl. N'est-il pas allé, poussé par le milieu sans doute, par l'occasion, mais aussi par « l'inquiétude de son génie », jusqu'à se muer en logicien, en physicien, en philosophe ? L'axiome *Ne sutor ultra crepidam* nous interdit de le suivre si loin en ses métamorphoses. Mais, dans son œuvre mathématique même, il n'est que trop fréquent qu'il vous glisse entre les mains lorsqu'on croit le mieux le saisir; et il faut avouer que la tâche de ses lecteurs n'en est pas facilitée. Il est vrai qu'il appartient à une période de transition dans l'histoire des mathématiques et qu'il s'en est trouvé profondément marqué. Souvent il a pu prendre un plaisir grisant à se laisser entraîner ou ballotter par les courants opposés qui ont agité cette époque, sûr d'ailleurs au fond de lui-même (comme lorsqu'il s'abandonna un moment à l'intuitionnisme brouwérien) que son bon sens foncier le garantirait du naufrage. Son œuvre a grandement contribué à ce changement de vision qui a fait passer de la mathématique classique, fondée sur le nombre réel, à la mathématique moderne, fondée sur la notion de structure. L'emploi systématique et tout abstrait du revêtement universel, la notion de variété analytique complexe, l'emploi courant et la popularisation, jusque parmi les physiciens, de l'algèbre vectorielle et du concept d'espace de représentation d'un groupe, tout cela vient avant tout de lui. Mais, s'il était trop élève de Hilbert pour ne pas inclure parmi ses outils la méthode axiomatique, s'il était trop mathématicien aussi pour en dédaigner les succès (son chaleureux éloge de l'œuvre d'Emmy Noether serait là, si besoin était, pour en faire foi), ce n'était pas à elle qu'allaient ses sympathies. Il y voyait « le filet dans lequel nous nous efforçons d'attraper la simple, la grande, la divine Idée »; mais, dans ce filet, il semble avoir toujours craint que l'on n'attrapât que des cadavres. A la dissection impitoyable sous le jour cru des projecteurs, il préférait, en bon romantique, le jeu troublant des analogies, auquel se prête si bien le langage de la métaphysique allemande qu'il affectionnait. Plutôt que de saisir l'idée brutalement au risque de la meurtrir, il aimait bien mieux la guetter dans la pénombre, l'accompagner dans ses évolutions, la décrire sous ses multiples aspects, dans sa vivante complexité. Etait-ce de sa faute si ses lecteurs, moins agiles que lui, éprouvaient parfois quelque peine à le suivre ?

« Le véritable principe de Dirichlet, a dit Minkowski dans un passage que Weyl citait volontiers, ce fut d'attaquer les problèmes au moyen d'un minimum de calcul aveugle, d'un maximum de réflexion lucide. » Et Weyl a écrit de son maître Hilbert: « Un trait caractéristique de son œuvre, c'est sa méthode d'attaque directe; s'affranchissant de tout algorithme, il revient toujours au problème tel qu'il se présente dans sa pureté originelle. » En deux ou trois occasions, il a atteint pleinement lui-même à cet idéal de perfection classique, par exemple dans son travail de 1916 sur l'égale répartition modulo 1, et encore dans ses mémoires jumeaux sur les fonctions presque périodiques et sur les représentations des groupes compacts. Comme il est naturel, ce sont là, parmi ses travaux, ceux qu'on relit avec le plus de plaisir, ceux dont il est le plus facile aussi de rendre compte. Aussi est-ce par eux que nous commencerons, renonçant à un ordre logique impossible à suivre lorsqu'il s'agit d'analyser une œuvre aussi riche. L'origine du premier, nous dit-il, se trouve dans un travail sur le phénomène de Gibbs, où s'était présentée incidemment une question d'approximation diophantienne; il s'était agi de faire voir que tout nombre irrationnel α peut être approché par une suite de fractions p_n/q_n satisfaisant aux conditions $q_n = o(n)$, $| \alpha - p_n/q_n | = o(1/n)$. Un peu plus tard l'attention de Weyl fut attirée par F. Bernstein sur le problème du mouvement moyen en mécanique céleste, problème qui remontait à Lagrange, et dont Bohl s'occupait alors; il s'agit là de déterminer le comportement asymptotique, pour $t \to \infty$, de l'argument d'une somme finie d'exponentielles $\Sigma a_\nu e(\lambda_\nu t)$, les λ_ν étant réels [4]. Ce fut l'occasion pour lui d'observer d'abord que son lemme diophantien entraînait aisément l'égale répartition modulo 1 de la suite $(n\alpha)$ pour α irrationnel, résultat qui fut obtenu en même temps par Bohl et par Sierpinski. Mais Weyl, à l'école de Hilbert et surtout par ses propres recherches sur les valeurs et fonctions propres, avait acquis un sens trop juste de l'analyse harmonique pour s'en tenir là.

Convenons de désigner par $\mathrm{M}(x_n)$, pour toute suite (x_n), la limite pour $n \to \infty$, si elle existe, de la moyenne des nombres $x_1, \ldots, x_n$. Dire que la suite (α_n) est également répartie modulo 1 équivaut à dire qu'on a $\mathrm{M}[f(\alpha_n)] = \int_0^1 f(x)\, dx$ pour certaines fonctions périodiques particulières, à savoir pour les fonctions de période 1 qui coïncident dans l'intervalle $[0, 1]$ avec une fonction caractéristique d'intervalle. Weyl s'aperçut que, si cette propriété est vérifiée pour les fonctions en question, elle l'est nécessairement aussi pour toute fonction périodique de période 1, intégrable au sens de Riemann, et en particulier pour les caractères $e(nx)$ du groupe additif des réels modulo 1; réciproquement, si elle l'est pour ces dernières fonctions, elle l'est aussi, en vertu des théorèmes classiques sur la série de Fourier, pour toute fonction périodique intégrable au sens de Riemann, de sorte que la suite (α_n) est également répartie modulo 1; la démonstration de ces assertions est immédiate. Le résultat sur l'égale répartition modulo 1 de la suite $(n\alpha)$ pour α irrationnel

[4] Ici, comme dans tout ce qui suit, on pose $e(t) = e^{2\pi i t}$.

découle de là aussitôt, sans aucun lemme diophantien; en remplaçant le groupe des réels modulo 1 par un tore de dimension quelconque, on obtient de même, et sans calcul, la forme quantitative des célèbres théorèmes d'approximation de Kronecker. Tout cela, si neuf à l'époque du travail de Weyl, nous paraît à présent bien simple, presque trivial. Mais aujourd'hui encore le lecteur reste étonné de voir comme Weyl, sans reprendre haleine, passe de là à l'égale répartition d'une suite (P (n)), où P est un polynôme quelconque. Cela revient naturellement, d'après ce qui précède, à l'évaluation des sommes d'exponentielles $\Sigma e(\mathrm{P}(n))$, problème qui avait été déjà l'objet des recherches de Hardy et Littlewood. Plus précisément, il s'agit de démontrer la relation

$$\sum_{n=0}^{\mathrm{N}} e\,(\mathrm{P}\,(n)) = o\,(\mathrm{N})$$

lorsque P est un polynôme où le coefficient du terme de plus haut degré est irrationnel. Pour donner une idée de la méthode de Weyl, qui (avec les perfectionnements qu'y ont apportés Vinogradov et son école) est restée fondamentale en théorie analytique des nombres, considérons le cas où P est du second degré. Posons donc:

$$s_{\mathrm{N}} = \sum_{n=0}^{\mathrm{N}} e\,(\alpha\, n^2 + \beta\, n)\,,$$

α étant irrationnel. On écrira alors, comme dans l'évaluation classique des sommes de Gauss:

$$|\, s_{\mathrm{N}}\,|^2 = s_{\mathrm{N}}\, \bar{s}_{\mathrm{N}} = \sum_{m,n=0}^{\mathrm{N}} e\,(\alpha\,(m^2 - n^2) + \beta\,(m - n))$$

$$= \sum_{r=-\mathrm{N}}^{+\mathrm{N}} e\,(\alpha\, r^2 + \beta\, r) \sum_{n\in \mathrm{I}_r} e\,(2\,\alpha\, rn)\,,$$

où on a substitué $n + r$ à m, et où I_r désigne l'intersection des deux intervalles [0, N] et $[-r, \mathrm{N} - r]$. Si on désigne par σ_r la dernière somme (celle qui est étendue à l'intervalle I_r), on a donc $|\, s_r\,|^2 \leqslant \Sigma\, |\, \sigma_r\,|$. Comme σ_r est une somme de $\mathrm{N} + 1$ termes au plus, on a $|\sigma_r| \leqslant N + 1$ quel que soit r; comme d'autre part σ_r est somme d'une progression géométrique de raison $e(2\alpha r)$, on a aussi:

$$|\sigma_r| \leqslant |\sin(2\pi\alpha r)|^{-1} \cdot$$

Soit $0 \leqslant \varepsilon < 1/2$; en vertu de l'égale répartition des nombres $2\alpha r$ modulo 1, le nombre des entiers r de l'intervalle $[-N, +N]$ qui sont tels que $2\alpha r$ soit congru modulo 1 à un nombre de l'intervalle $[-\varepsilon, +\varepsilon]$ est de la forme $4\varepsilon N + o(N)$, et est donc $\leqslant 5\varepsilon N$ dès que N est assez grand. Pour chacun de ces entiers, on a $|\sigma_r| \leqslant N + 1$; pour tous les autres, on a $|\sigma_r| \leqslant 1/\sin(\pi\varepsilon)$. On a donc, pour N assez grand:

$$|s_N|^2 \leqslant 5\varepsilon N(N+1) + \frac{2N+1}{\sin(\pi\varepsilon)} \cdot$$

Pour N assez grand, le second membre sera $\leqslant 6\varepsilon N^2$; comme il en est ainsi quel que soit ε, on a bien $s_N = o(N)$. Si le degré du polynôme P est $d + 1$ avec $d > 1$, la démonstration se fait de même (et non par récurrence sur d) au moyen d'un lemme sur l'égale répartition modulo 1 d'une fonction multilinéaire de d variables. Le résultat s'étend aux fonctions de p variables par récurrence sur p.

Avec cet admirable mémoire, Weyl était déjà très près des fonctions presque périodiques. Il s'y agissait, en effet, en premier lieu, des sous-groupes cycliques et des sous-groupes à un paramètre d'un tore de dimension finie, tandis que la théorie des fonctions presque périodiques traite, dirions-nous, des sous-groupes à un paramètre d'un tore de dimension infinie. On peut même dire que cette théorie, qui suscita tant d'intérêt pendant une dizaine d'années à la suite des publications de H. Bohr en 1924, eut pour principale utilité de ménager la transition entre le point de vue classique et le point de vue moderne au sujet des groupes compacts et localement compacts. Au temps même où Weyl s'occupait à Göttingen d'égale répartition modulo 1, vers 1913, les premières idées sur les fonctions presque périodiques y étaient « dans l'air ». Le problème du mouvement moyen portait sur les sommes d'exponentielles, finies il est vrai, et Weyl en avait traité des cas particuliers, sans d'ailleurs approfondir la question, qu'il ne devait résoudre complètement, toujours par la même méthode, que lorsqu'il s'y trouva ramené vingt-cinq ans plus tard. Mais H. Bohr, alors élève de Landau, s'occupait de séries d'exponentielles en vue de l'étude de $\zeta(s)$ dans le plan complexe, problème auquel Weyl va bientôt s'intéresser en passant, déterminant même par sa méthode le comportement asymptotique de $\zeta(1 + it)$. D'autre part, les élèves de Hilbert étaient accoutumés à considérer les termes de la série de Fourier comme fonctions propres, et les coefficients de cette série comme valeurs propres, d'opérateurs convenablement définis. Il semble donc que les voies fussent toutes préparées dans l'esprit de Weyl, lorsque apparurent les premiers travaux de H. Bohr sur les fonctions presque périodiques, pour reprendre la question du point de vue des équations intégrales.

Mais il est rare qu'un mathématicien, qu'il s'agisse du plus grand ou du plus humble, parcoure le plus court chemin d'un point à un autre de sa trajectoire. Avant de revenir aux fonctions presque périodiques à l'occasion d'une conférence de H. Bohr à Zurich, Weyl avait mené à bien ses mémorables recherches sur les groupes de Lie et leurs représentations, et avait conçu l'idée, d'une audace extraordinaire pour l'époque, de « construire » les représentations des groupes de Lie compacts par la complète décomposition d'une représentation de degré infini. Blasés que nous sommes par l'expérience des trente dernières années, cette idée ne nous étonne plus; mais son succès semble avoir fait l'effet d'un vrai miracle à son auteur; « c'est là, répète-t-il à maintes reprises, l'une des plus surprenantes applications de la méthode des équations intégrales ». Déjà I. Schur avait étendu au groupe orthogonal, au moyen de l'élément de volume invariant dans l'espace de groupe, les relations d'orthogonalité entre coefficients des représentations que Frobenius avait découvertes pour les groupes finis; mais il y avait loin de là à un théorème d'existence. Weyl n'hésite pas à introduire, sur un groupe de Lie compact, l'algèbre de groupe, toujours conçue chez lui comme algèbre des fonctions continues par rapport au produit de convolution

$$h = f * g, \quad h(s) = \int f(st^{-1})\, g(t)\, dt,$$

et dont il fait un espace préhilbertien au moyen de la norme $(\int |f|^2 ds)^{1/2}$; les intégrales, naturellement, sont prises au moyen de l'élément de volume invariant que fournit la théorie de Lie, et que Hurwitz avait sans doute été le premier à utiliser systématiquement. Dans cet espace, l'opérateur $f \to \varphi * \tilde{\varphi} * f$, où $\tilde{\varphi}$ désigne la fonction $\tilde{\varphi}(s) = \varphi(s^{-1})$, est hermitien et complètement continu; d'après la théorie de E. Schmidt, ses valeurs propres forment donc un spectre discret, et à chacune correspond un espace de fonctions propres de dimension finie, dont on constate immédiatement qu'il est invariant par le groupe; c'est donc un espace de représentation de celui-ci. Les théorèmes de Schmidt fournissent alors le développement de φ suivant les coefficients des représentations ainsi obtenues, développement qui converge au sens de la norme. C'est là une généralisation directe de la méthode de Frobenius basée sur la réduction de la représentation régulière d'un groupe fini; la seule différence, comme l'observe Weyl, c'est l'absence d'un élément unité dans l'algèbre d'un groupe compact; Weyl y supplée par un artifice tiré de la théorie des séries de Fourier, à savoir l'approximation de la masse unité placée à l'origine par une distribution de masses à densité continue, concentrée dans un voisinage de l'origine; la convolution avec celle-ci constitue un « opérateur régularisant », d'emploi courant aujourd'hui, mais dont c'était sans doute la première apparition dans le cadre de la théorie des groupes de Lie; Weyl s'en sert pour démontrer que toute fonction continue peut être approchée, non seulement au sens de la norme, mais même uniformément, par des combinaisons linéaires de coefficients de représentations.

Bien que le mémoire de Weyl se limitât nécessairement aux groupes de Lie, il avait atteint en réalité, du premier coup, à des résultats définitifs sur les représentations des groupes compacts; après la découverte de la mesure de Haar, il n'y eut pas un mot à changer à son exposé, et, chose rare en mathématique, il ne vint même à personne l'idée de le récrire. Si, comme nous le faisons aujourd'hui, on considère une fonction presque périodique comme déterminant une représentation du groupe additif des réels dans un groupe compact, et qu'on suppose acquise pour celui-ci la notion de mesure de Haar, on déduit immédiatement des résultats de Weyl exposés ci-dessus le développement de la fonction en série d'exponentielles. Les outils manquaient à Weyl, en 1926, pour adopter ce point de vue; il y supplée en remplaçant l'intégrale par une moyenne sur la droite, définie comme limite pour $T \to +\infty$ de la valeur moyenne sur l'intervalle $[t, t + T]$ lorsque cette limite est atteinte uniformément par rapport au paramètre t. En 1926, il n'allait pas de soi que la théorie des équations intégrales s'appliquât à cette moyenne; Weyl est obligé de consacrer une bonne partie de son travail à justifier cette application. Il convient d'observer d'autre part que, sur un groupe compact, la manière la plus simple de construire la mesure de Haar consiste justement à attacher à chaque fonction continue une valeur moyenne, par un procédé directement inspiré de la théorie des fonctions presque périodiques. Que Weyl, en revanche, ait cru voir dans la théorie de Bohr « le premier exemple d'une théorie des représentations d'un groupe vraiment non compact » (par opposition apparemment avec les groupes de Lie semi-simples dont les représentations, dans son esprit, se ramenaient, par la « restriction unitaire », à celles de groupes compacts), cela montre qu'il se faisait encore quelque illusion sur le degré de difficulté des problèmes qui restaient à résoudre. Ce n'en est pas moins lui qui a ouvert la voie à tous les progrès ultérieurs dans cette direction.

Sur le reste de son œuvre d'analyste, nous serons beaucoup plus brefs, d'autant plus qu'il a lui-même excellemment rendu compte d'une bonne partie de cette œuvre dans sa *Gibbs Lecture* de 1948. Débutant, il participa activement au courant de recherches qui se proposait d'approfondir et d'appliquer à des problèmes variés d'analyse la théorie spectrale des opérateurs symétriques. Citons particulièrement, dans cet ordre d'idées, sa *Habilitationsschrift* de 1910, où il étudie un opérateur différentiel autoadjoint L sur la demi-droite $[0, +\infty]$:

$$L(u) = \frac{d}{dt}\left(p(t)\frac{du}{dt}\right) - q(t)\,u\,,$$

où p, q sont à valeurs réelles et $p(t) > 0$. Sur tout intervalle fini $[0, l]$, cet opérateur, soumis aux conditions aux limites du type habituel, $(du/dt)_0 = hu(0)$, $(du/dt)_l = h' u(l)$, relève de la théorie de Sturm-Liouville ou, en termes modernes, de la théorie des opérateurs complètement continus; le spectre est réel et discret et se compose des λ pour lesquels l'équation $\mathrm{L}u = \lambda u$ a une solution satisfaisant aux conditions aux limites imposées. Le passage à la limite $l \to +\infty$ fait apparaître, non seulement un spectre continu qui peut couvrir tout l'axe réel, mais encore des phénomènes imprévus dont la découverte est due à Weyl. Les plus intéressants concernent le comportement des solutions pour $l \to +\infty$ lorsqu'on donne à λ une valeur imaginaire fixe; chose remarquable, ils sont indépendants du choix de la valeur donnée à λ. C'est ainsi que Weyl est amené en particulier à la distinction fondamentale entre le cas du « point limite » et le cas du « cercle limite »: l'une des propriétés caractéristiques du premier, c'est que l'équation $\mathrm{L}u = \lambda u$ y possède, quel que soit λ imaginaire, une solution et une seule de carré sommable sur $[0, +\infty]$, tandis que toutes ses solutions le sont, pour λ imaginaire, dans le cas du cercle limite. Weyl étudie aussi le passage à la limite $l \to +\infty$ pour les développements de Sturm-Liouville sur $[0, l]$; il en tire des formules intégrales où apparaissent en général des intégrales de Stieltjes, comme on pouvait s'y attendre. Le problème des moments de Stieltjes n'est d'ailleurs pas autre chose que le problème aux différences finies, analogue à l'équation $\mathrm{L}u = \lambda u$ sur la demi-droite, et Hellinger fit voir par la suite que la méthode de Weyl s'y transporte presque telle quelle. Mais Weyl put aussi la transposer plus tard à un problème différentiel où le paramètre spectral intervient non linéairement, ainsi qu'au problème aux différences finies correspondant (auquel il a donné le nom de problème de Pick-Nevanlinna); il apporta même à cette occasion quelques améliorations notables à son premier exposé. Si celui-ci a donné lieu depuis lors à des généralisations assez variées, il ne semble pas que la signification véritable des résultats de Weyl sur les problèmes à paramètre non linéaire ait jamais été tirée au clair.

Une autre série de travaux traite de la répartition des valeurs propres, dans divers problèmes de type elliptique. Ils reposent principalement sur un principe qui plus tard fut popularisé par Courant sous la forme suivante: si A est un opérateur symétrique complètement continu dans un espace de Hilbert H, sa n-ième valeur propre est la plus petite des valeurs (« minimum maximorum ») que peut prendre la norme de A, c'est-à-dire le nombre max $(Ax, x)/(x, x)$, sur un sous-espace de H de codimension $n - 1$. Une fois acquise la théorie des opérateurs complètement continus, la vérification de ce principe est d'ailleurs immédiate. Mais Weyl l'adapte en virtuose à toutes sortes de situations de physique mathématique. Quant au comportement asymptotique des fonctions propres, il avait, nous dit-il en 1948, certaines conjectures: « mais, n'ayant fait pendant plus de trente-cinq ans aucune tentative sérieuse pour les démontrer, je préfère, ajoute-t-il, les garder pour moi »; il aura donc laissé ce problème plus difficile en héritage à ses successeurs.

C'est en élève de Hilbert encore, et en analyste, que Weyl dut aborder le sujet d'un des premiers cours qu'il professa à Göttingen comme jeune privatdozent, la théorie des fonctions selon Riemann. Le cours terminé et rédigé, il se retrouva géomètre, et auteur d'un livre qui devait exercer une profonde influence sur la pensée mathématique de son siècle. Peut-être s'était-il proposé seulement de remettre au goût du jour, en faisant usage des idées de Hilbert sur le principe de Dirichlet, les exposés traditionnels dont l'ouvrage classique de C. Neumann fournissait le modèle. Mais il dut lui apparaître bientôt que, pour substituer aux constants appels à l'intuition de ses prédécesseurs des raisonnements corrects et, comme on disait alors, « rigoureux » (et dans l'entourage de Hilbert on n'admettait pas qu'on trichât là-dessus), c'étaient avant tout les fondements topologiques qu'il fallait renouveler. Weyl n'y semblait guère préparé par ses travaux antérieurs. Il pouvait, dans cette tâche, s'appuyer sur l'œuvre de Poincaré, mais il en parle à peine. Il mentionne, comme l'ayant profondément influencé, les recherches de Brouwer, alors dans leur première nouveauté; en réalité, il n'en fait aucun usage. De fréquents contacts avec Koebe, qui dès lors s'était consacré tout entier à l'uniformisation des fonctions d'une variable complexe, durent lui être d'une grande utilité, particulièrement dans la mise au point de ses propres idées. La première édition du livre est dédiée à Félix Klein, qui bien entendu, comme Weyl le dit dans sa préface, ne pouvait manquer de s'intéresser à un travail si voisin des préoccupations de sa jeunesse ni de donner à l'auteur des conseils inspirés de son tempérament intuitif et de sa profonde connaissance de l'œuvre de Riemann. Bien qu'il n'eût jamais connu celui-ci, c'était Klein qui, à Göttingen, incarnait la tradition riemannienne. Enfin, dans l'un de ses mémoires sur les fondements de la géométrie, Hilbert avait formulé un système d'axiomes fondé sur la notion de voisinage, en soulignant qu'on trouverait là le meilleur point de départ pour « un traitement axiomatique rigoureux de l'analysis situs ». De tous ces éléments si divers que lui fournissaient la tradition et le milieu, Weyl tira un livre profondément original et qui devait faire époque.

Le livre est divisé en deux chapitres, dont le premier contient la partie qualitative de la théorie. Les notions de « surface » (variété topologique de dimension 2 à base dénombrable) et de « surface de Riemann » (variété analytique complexe à base dénombrable, de dimension complexe 1) y sont définies au moyen de systèmes d'axiomes, inspirés naturellement de celui de Hilbert, mais qui cette fois (sauf une légère omission dans la première édition) étaient destinés à subsister sans retouches, et devaient servir de modèle à Hausdorff pour son axiomatisation de la topologie générale. Dans la première et la deuxième édition, la condition de base dénombrable apparaît sous forme de condition de triangulabilité; et la triangulation joue un grand rôle dans la suite du volume; elle devait être éliminée entièrement de la troisième édition. Les questions touchant au groupe fondamental, au revêtement universel, à l'orientation, sont élucidées avec soin dans un esprit tout moderne, ainsi que les rapports entre propriétés homologiques et périodes des intégrales simples sur la surface. Dans la première et la deuxième édition, l'auteur va jusqu'à la construction, pour les surfaces orientables compactes, d'un système de « rétrosections », c'est-à-dire essentiellement d'une base privilégiée pour le premier groupe d'homologie;

comme il le dit lui-même, il aurait pu, au prix d'un léger effort supplémentaire, aller jusqu'à la représentation de la surface au moyen d'un « polygone canonique » à $4g$ côtés (g désignant le genre), et à la détermination explicite du groupe fondamental, et on peut regretter qu'il ne l'ait pas fait. Mais la construction même des rétrosections, nécessairement basée sur la triangulation, disparaît dans la troisième édition, au profit d'un traitement plus purement homologique où n'interviennent que des recouvrements. En tout cas, pour tout l'essentiel, ce chapitre constitue une mise au point à peu près définitive des questions qu'il traite.

Les théorèmes d'existence font l'objet du deuxième chapitre. Weyl y donne du principe de Dirichlet une démonstration simplifiée, basée naturellement sur l'idée de Hilbert qui consiste, comme on sait, à opérer dans l'espace préhilbertien des fonctions différentiables avec la norme de Dirichlet; même dans la troisième édition, il n'a pas cru devoir suivre la variante qu'il avait pourtant contribué à créer lui-même, et qui consiste à opérer par projection orthogonale dans le complété de l'espace en question, puis à montrer après coup que la solution obtenue est différentiable. Une fois acquis le principe de Dirichlet, l'auteur en tire les principales propriétés des intégrales abéliennes et des fonctions multiplicatives, le théorème de Riemann-Roch, puis le théorème de l'uniformisation, c'est-à-dire la représentation conforme du revêtement universel de la surface de Riemann sur une sphère, un plan ou un disque. Si on laisse de côté les cas de genre 0 ou 1, le résultat peut s'exprimer en disant que toute surface de Riemann compacte, de genre > 1, peut se définir comme quotient du plan non-euclidien par un groupe discret de déplacements sans point fixe. « Ainsi, dit Weyl dans la préface de la première édition, ainsi nous pénétrons dans le temple où la divinité est rendue à elle-même, délivrée de ses incarnations terrestres: le *cristal non euclidien*, où l'archétype de la surface de Riemann se laisse voir dans sa pureté première... » C'est en songeant sans doute à ce passage que Weyl dit plus tard de sa préface que « plus encore que le livre lui-même, elle trahissait la jeunesse de son auteur ». Nous dirions aujourd'hui qu'on a construit pour la surface de Riemann un modèle qui est canonique à un déplacement près dans le plan non euclidien; autrement dit, on a associé canoniquement une structure à une autre. Mais qui saurait mauvais gré à Weyl, après avoir achevé un livre de cette valeur, d'avoir exprimé d'une manière peut-être un peu trop romantique son enthousiasme juvénile ?

C'est en 1916, pendant la guerre, que Weyl fit paraître en Suisse son premier mémoire de géométrie, sur le célèbre problème de la rigidité des surfaces convexes. Ici encore, Göttingen lui avait fourni son point de départ. Sous la direction de Hilbert, Weyl avait collaboré à la publication des œuvres complètes de Minkowski, où la théorie des corps convexes tient tant de place. D'autre part, Hilbert avait montré comment on peut faire dépendre les inégalités de Brunn-Minkowski de la théorie des opérateurs différentiels elliptiques. L'espace R^3 étant considéré comme espace euclidien, et $\langle x, y\rangle$ désignant le produit scalaire dans R^3, soit V un corps convexe dans cet espace, défini au moyen de la fonction d'appui H; cela veut dire que H satisfait aux conditions

$$H(x + x') \leqslant H(x) + H(x'), \quad H(\lambda x) = \lambda H(x) \quad \text{pour} \quad \lambda \geqslant 0,$$

et que V est l'ensemble des points y satisfaisant à $\langle x, y\rangle \leqslant H(x)$ quel que soit x. Si on suppose H différentiable en dehors de 0, le volume de V est alors donné par une formule

$$\mathrm{vol}(V) = \int H . Q(H)\, d\omega,$$

où l'intégrale est étendue à la sphère unité S_0 définie par $\langle x, x\rangle = 1$, où $d\omega$ désigne l'élément d'aire sur S_0, et où $Q(H)$ est une forme quadratique par rapport aux dérivées partielles secondes de H. Soient F, F′ deux fonctions, différentiables en dehors de 0, satisfaisant toutes deux à la condition d'homogénéité $F(\lambda x) = \lambda F(x)$ pour $\lambda \geqslant 0$; soit $B(F, F')$ la forme bilinéaire symétrique par rapport aux dérivées partielles secondes de F et à celles de F′ qui se déduit de la forme quadratique $Q(H)$ par linéarisation, c'est-à-dire qui est telle que $Q(H) = B(H, H)$;

posons aussi $L_F(F') = B(F, F')$. On vérifie facilement, au moyen de la formule de Stokes, que l'intégrale

$$I(F, F', F'') = \int_{S_0} F''.B(F, F')\,d\omega,$$

où F'' désigne une troisième fonction satisfaisant aux mêmes conditions que F et F', dépend symétriquement de F, F' et F''. Cela revient à dire que L_F, considéré comme opérateur différentiel sur les fonctions sur S_0 prolongées à R^3 par homogénéité, est un opérateur autoadjoint. Si V', V'' sont deux corps convexes définis par des fonctions d'appui H', H'', les formules ci-dessus montrent que les « volumes mixtes » associés par Minkowski à V, V', V'' ne sont autres que les nombres $I(H, H, H')$ et $I(H, H', H'')$; de plus, un calcul simple montre que L_H est elliptique. Dans ces conditions, comme le fait voir Hilbert dans ses *Grundzüge*, l'application à L_H de la théorie des opérateurs autoadjoints elliptiques conduit, pour le cas différentiable, à l'inégalité de Brunn-Minkowski. Mais il se trouve que L_H n'est autre qu'un opérateur qui se présente dans la théorie de la déformation infinitésimale de la surface Σ frontière de V; jointe aux résultats de Hilbert, cette observation, due à Blaschke, entraînait l'impossibilité d'une telle déformation pour Σ. Enfin, Hilbert, à propos des fondements de la géométrie, avait démontré l'impossibilité d'appliquer isométriquement une sphère sur une surface convexe non sphérique. D'ailleurs, des résultats analogues sur les polyèdres convexes avaient été obtenus jadis par Cauchy: non seulement un polyèdre convexe n'admet aucune déformation infinitésimale, mais encore, si P et P' sont deux polyèdres convexes admettant même schéma combinatoire et ayant leurs côtés correspondants égaux, ils ne peuvent différer l'un de l'autre que par un déplacement ou une symétrie. Tout cela mettait à l'ordre du jour l'extension aux surfaces convexes du second théorème de Cauchy.

Mais Weyl ne s'arrête pas là. Il considère en même temps un problème d'existence que, faute d'une conception claire de la notion de variété riemannienne abstraite, personne n'avait encore même formulé. Il s'agit de savoir si « toute surface convexe fermée, donnée *in abstracto*, est réalisable » ou, comme nous dirions maintenant, si toute variété riemannienne compacte, simplement connexe, de dimension 2, à courbure partout positive, admet un plongement isométrique dans l'espace euclidien R^3; la question d'unicité, pour ce problème d'existence, est alors celle même dont Weyl était parti. Interrompu dans son travail par sa mobilisation en 1915, il se contenta d'esquisser son idée de démonstration, et ne la mena jamais à terme. Il part du fait que toute « surface convexe *in abstracto* » peut être représentée conformément sur la sphère S_0, donc définie par un ds^2 donné sur S_0 sous la forme $ds^2 = e^{2\Phi} d\sigma^2$, où $d\sigma$ est la longueur d'arc « naturelle » et Φ une fonction différentiable sur S_0; soit $\Sigma(\Phi)$ la « surface abstraite » ainsi définie. La condition que $\Sigma(\Phi)$ soit à courbure partout positive s'exprime par une inégalité différentielle $K(\Phi) > 0$; on constate aussitôt que l'ensemble des Φ qui y satisfont est convexe; il s'ensuit que l'ensemble des surfaces convexes abstraites est connexe. L'idée de Weyl est alors d'appliquer au problème une méthode de continuité. Tout revient, I désignant l'intervalle [0, 1], à déterminer une application φ de $S_0 \times I$ dans R^3 de telle sorte que l'application $x \to \varphi_\tau(x) = \varphi(x, \tau)$ de S_0 dans R^3 applique isométriquement $\Sigma(\tau\Phi)$ sur une surface convexe $S_\tau = \varphi_\tau(S_0)$, et cela pour tout $\tau \in I$. Pour cela, Weyl considère $\partial\varphi/\partial\tau$ comme une déformation infinitésimale de S_τ, dont la détermination se ramène à la solution d'une équation $\Lambda u = f$, où f est une fonction sur S_0, dépendant de S_τ, et Λ est essentiellement l'opérateur elliptique L_H relatif à S_τ. L'application de la méthode de Hilbert à cette équation donne donc, en principe, une équation différentielle fonctionnelle pour φ_τ; il s'agit d'en trouver une solution sur l'intervalle I qui se réduise pour $\tau = 0$ à l'application identique de S_0 dans R^3; et on peut espérer y parvenir au moyen de l'une quelconque des méthodes classiques de résolution des équations différentielles. Une démonstration complète a été obtenue récemment par Nirenberg en suivant cette voie; les brèves indications données ici suffiront tout au moins à faire apparaître l'extrême hardiesse de l'idée de Hermann Weyl.

Rentré à Zurich en 1916, Weyl eut, semble-t-il, quelque velléité de revenir aux surfaces convexes; un mémoire où il reprend les résultats de Cauchy sur les polyèdres présageait peut-être un mode d'attaque basé sur des méthodes moins infinitésimales et plus directes. Mais c'est bientôt la relativité qui attire et accapare son attention.

Là encore, il était dans la tradition. Minkowski avait participé activement au courant de recherches qui s'était développé autour de la relativité restreinte. Hilbert suivait de près les travaux d'Einstein et cherchait, sans grand succès d'ailleurs, à éclaircir les problèmes de la physique par la méthode axiomatique. « Il faut en physique un autre type d'imagination que celle du mathématicien », constate plus tard Hermann Weyl, non sans quelque mélancolie, dans sa notice sur Hilbert. Sans doute, en écrivant ces mots, songeait-il aussi à sa propre expérience et à cette « théorie de Weyl » à laquelle, disait-il vers la même époque, il ne croyait plus depuis longtemps. Mais à partir de 1917, et pendant plusieurs années, son enthousiasme est débordant. En 1918, il publie son cours de l'année précédente sur la relativité sous le titre *Raum, Zeit, Materie.* « A l'occasion de ce grand sujet, écrit-il dans la préface de la première édition, j'ai voulu donner un exemple de cette interpénétration, qui me tient tant à cœur, de la pensée philosophique, de la pensée mathématique, de la pensée physique... »; mais, ajoute-t-il avec une modestie non exempte de naïveté, « le mathématicien en moi a pris le pas sur le philosophe »; et ce ne sont pas les mathématiciens qui s'en plaindront. Son ouvrage, dans ses cinq éditions successives, fit beaucoup pour répandre parmi les mathématiciens et les physiciens les connaissances géométriques et les notions essentielles de l'algèbre et de l'analyse tensorielles. A partir de la troisième édition, on y trouve aussi un exposé de la « théorie de Weyl », premier essai d'une « théorie unitaire » englobant dans un même schéma géométrique les phénomènes électromagnétiques et la gravitation. Elle était fondée, dirions-nous à présent, sur une connexion liée au groupe des similitudes (défini au moyen d'une forme quadratique de signature (1,3)), au lieu qu'Einstein s'était borné à des connexions liées au groupe de Lorentz (groupe orthogonal pour une forme de signature

(1,3)), et plus précisément à la connexion sans torsion déduite canoniquement (par transport parallèle) d'un ds^2 de signature (1,3). Cette théorie eut du moins le mérite d'élargir le cadre de la géométrie riemannienne traditionnelle et de préparer les voies aux « géométries généralisées » de Cartan, c'est-à-dire à la théorie générale des connexions liées à un groupe de Lie arbitraire.

Quant aux préoccupations philosophiques de Weyl pendant cette période d'intense fermentation, elles ne tardèrent pas (heureusement, serions-nous tentés de dire) à se couler dans un moule plus étroitement mathématique, l'amenant à chercher une base axiomatique aussi simple que possible aux structures géométriques sous-jacentes à la théorie d'Einstein et à la sienne; c'est là ce qu'il appelle le « Raumproblem », le problème de l'espace; il y consacre plusieurs articles, un cours professé à Barcelone et à Madrid, et un opuscule qui reproduit ces leçons. Il s'agit là en réalité de caractériser le groupe orthogonal (attaché à une forme quadratique, soit complexe, soit réelle et de signature quelconque) en tant que groupe linéaire, par quelques conditions simples au moyen desquelles on puisse rendre plausible que la géométrie de l'« univers » est définie localement par un tel groupe. Bien entendu, c'est la théorie des groupes de Lie et de leurs représentations qui domine la question; Weyl en donne une esquisse dans un appendice de son livre. De son côté, Cartan ne tarda pas à donner, du principal résultat mathématique de Weyl sur ce sujet, une démonstration basée sur ses propres méthodes.

Il n'était pas dans le tempérament de Hermann Weyl, une fois parvenu ainsi au seuil de l'œuvre de Cartan, de se contenter d'y jeter un coup d'œil rapide. D'autre part, à la suite peut-être d'une remarque de Study qui l'avait blessé au vif, il avait commencé à s'intéresser aux invariants des groupes classiques. Study, dans une préface de 1923, lui avait reproché, ainsi qu'aux autres relativistes, d'avoir, par leur négligence à l'égard de ce sujet, contribué à « la mise en jachère d'un riche domaine culturel »; il entendait surtout par là la théorie des invariants du groupe projectif, dans laquelle il était d'usage de faire rentrer tant bien que mal les autres groupes à l'occasion de l'étude des covariants simultanés de plusieurs formes. Par une réaction bien caractéristique, Weyl répondit à Study, avec une promptitude extraordinaire, par un mémoire où il reprend à la base la théorie classique au moyen d'identités algébriques dues à Capelli et indique aussi comment elle s'étend aux groupes orthogonaux et symplectiques; ce qui ne l'empêche pas de protester que, « si même il avait connu aussi bien que Study lui-même la théorie des invariants, il n'aurait eu nulle occasion d'en faire usage dans son livre sur la relativité: chaque chose en son lieu ! ».

La synthèse entre ces deux courants de pensée — groupes de Lie et invariants — s'opère dans son grand mémoire de 1926, mémoire divisé en quatre parties, dont il dit lui-même vers la fin de sa vie qu'il représente « en quelque sorte le sommet de sa production mathématique ». L'étude qu'avait faite Young, vers 1900, de la décomposition des tenseurs en tenseurs irréductibles définis par des conditions de symétrie avait abouti en substance à la détermination de toutes les représentations « simples », c'est-à-dire irréductibles, du groupe linéaire spécial; mais, enfermées qu'étaient ces recherches dans le cadre de la théorie traditionnelle, il leur était impossible, par définition, d'obtenir ce résultat sous la forme que nous venons de lui donner. De son côté, Cartan, parti de la théorie générale des groupes de Lie, avait déterminé toutes les représentations en question, sans d'ailleurs, semble-t-il, faire le lien entre ses résultats et ceux d'Young. Désignons par G le groupe linéaire spécial, et par $\mathfrak{g}$ son algèbre de Lie, qui se compose de toutes les matrices de trace 0; soit $\mathfrak{h}$ l'ensemble des matrices diagonales contenues dans $\mathfrak{g}$. Une représentation simple de G détermine une représentation simple ρ de $\mathfrak{g}$, donc une représentation de $\mathfrak{h}$. Cartan montre que l'espace V de la représentation ρ est engendré par des vecteurs qui sont vecteurs propres de toutes les opérations $\rho(\mathrm{H})$, pour $\mathrm{H} \in \mathfrak{h}$. Soit e l'un de ces vecteurs propres; on a $\rho(\mathrm{H}) \,.\, e = \lambda(\mathrm{H})\, e$, où λ est une forme linéaire sur $\mathfrak{h}$, qu'on appelle le *poids* de e; si H est la matrice diagonale de coefficients $a_1, \ldots, a_n$, il est facile de voir que $\lambda(\mathrm{H})$ est de la forme $m_1 a_1 + \ldots + m_n a_n$, où les m_i sont des entiers déterminés à l'addition près d'un même entier. Si on ordonne lexicographiquement l'ensemble des systèmes $(m_1, \ldots, m_n)$ de n entiers, on obtient donc une relation d'ordre dans l'ensemble des poids des représentations de G. On appelle *poids fondamental* d'une représentation simple le plus grand des poids de cette représentation pour la relation d'ordre qu'on vient de définir. Cartan avait montré que ce poids détermine complètement la représentation (à une équivalence près), qu'il correspond à un système d'entiers (m_i) tel que $m_1 \leqslant \ldots \leqslant m_n$, et que réciproquement tout système d'entiers satisfaisant à ces inégalités appartient au poids fondamental d'une représentation simple de G. Soient de plus ρ, ρ' deux représentations simples de G, opérant respectivement sur des espaces vectoriels V, V'; soient λ, λ' leurs poids fondamentaux; soient e, e' des vecteurs de V, V', de poids respectifs λ, λ'. Le produit tensoriel $\rho \otimes \rho'$ de ρ et ρ' (dit parfois encore « produit kroneckérien », et noté le plus souvent $\rho \times \rho'$ par Weyl) est une représentation opérant sur un espace $V \otimes V'$ de dimension égale au produit de celles de V et V', qui est formé de combinaisons linéaires d'éléments se transformant par G comme les produits formels xx', avec $x \in V$, $x' \in V'$; et, pour cette représentation, le vecteur $e \otimes e'$ est de poids $\lambda + \lambda'$. Soit W le sous-espace de $V \otimes V'$ engendré par $e \otimes e'$ et ses transformés par G; il découle facilement des résultats de Cartan que W ne peut pas se décomposer en somme directe de sous-espaces invariants par les opérations de G; et Cartan avait cru pouvoir déduire de là que W fournit la représentation simple de poids dominant $\lambda + \lambda'$. Weyl observa que cette conclusion est illégitime tant qu'on ne sait pas à priori que les représentations de G sont toutes semi-simples (c'est-à-dire complètement réductibles). A vrai dire, ce dernier résultat n'était pas indispensable pour se convaincre du fait que la décomposition d'Young de l'espace des tenseurs fournit toutes les représentations simples de G; Young avait en effet établi l'irréductibilité des représentations qu'il avait construites, et il suffisait d'établir par un calcul facile que leurs poids dominants sont tous ceux prévus par la théorie de Cartan. Mais on n'eût obtenu ainsi que la classification des représentations simples. Au contraire, en démontrant la complète réductibilité de toutes les représentations de G, Weyl en obtint du même coup (compte tenu des résultats de Young et Cartan) la classification définitive, qui s'exprime par le fait que toute « grandeur linéaire », comme il dit, se décompose en tenseurs irréductibles.

On sait aujourd'hui démontrer le théorème de complète réductibilité par des méthodes algébriques; c'est là le point de départ de la théorie cohomologique des algèbres de Lie. Mais c'est de considérations tout autres que Weyl tire sa démonstration. Il observe, comme l'avait déjà fait Hurwitz dans son mémoire sur la construction d'invariants par la méthode d'intégration, que la théorie des représentations du groupe linéaire spécial *complexe* G est équivalente à celle des représentations du groupe G_u formé des matrices *unitaires* appartenant à G; en dernière analyse, cela tient à ce que toute identité algébrique entre coefficients d'une matrice unitaire reste vraie pour une matrice quelconque. Or G_u possède une propriété importante qui n'appartient pas à G: il est *compact*, ce qui permet, comme l'avait fait voir Hurwitz, de construire des invariants pour G_u *et par suite pour* G par intégration dans l'espace du groupe G_u au moyen de l'élément de volume invariant fourni par la théorie de Lie. La méthode classique qui permet d'établir la complète réductibilité des représentations des groupes finis par construction d'une forme hermitienne, définie positive, invariante par les opérations du groupe, s'étend alors d'elle-même au groupe G_u.

Ce n'est pas seulement le théorème de complète réductibilité pour G que Weyl tire de la restriction au groupe unitaire G_u; il s'en sert aussi pour calculer explicitement les caractères et les degrés des représentations simples de G. On voit tout de suite, en effet, que si χ est le caractère d'une représentation de G_u, et si s est une matrice diagonale unitaire de déterminant 1 et de coefficients diagonaux $e(x_1), \dots, e(x_n)$, la valeur de $\chi(s)$ s'exprime comme somme de Fourier finie en $x_1, \dots, x_n$ et ne change pas par une permutation quelconque des x_i. Weyl montre que ces propriétés, jointes aux relations d'orthogonalité fournies, elles aussi, par la méthode d'intégration, suffisent déjà à déterminer complètement les caractères et à en obtenir des expressions explicites.

La suite du mémoire de Weyl est consacrée à l'extension des méthodes ci-dessus aux groupes orthogonaux et symplectiques, puis aux groupes semi-simples les plus généraux. Soit cette fois $\mathfrak{g}$ une algèbre de Lie semi-simple complexe; pour en étudier les représentations, Weyl va appliquer la méthode de restriction unitaire au groupe adjoint G de $\mathfrak{g}$, mis sous forme matricielle relativement à une base convenable de $\mathfrak{g}$. Pour qu'il y ait dans G « assez » d'opérations unitaires, il est nécessaire que $\mathfrak{g}$ admette ce qu'on appelle aujourd'hui une forme compacte, ou pour mieux dire une base telle que les combinaisons linéaires réelles des éléments de cette base forment l'algèbre de Lie d'un groupe compact. En examinant chaque groupe simple séparément, Cartan avait vérifié dans chaque cas l'existence d'une forme compacte; Weyl en donne une démonstration a priori basée sur les propriétés des constantes de structure de $\mathfrak{g}$. Cela fait, il introduit le groupe G_u des opérations unitaires de G, et son algèbre de Lie $\mathfrak{g}_u$. Le groupe G_u est compact, et la théorie des représentations de $\mathfrak{g}_u$ est équivalente à celle des représentations de $\mathfrak{g}$. Mais ici se présente une difficulté nouvelle; du fait que G_u peut n'être pas simplement connexe, la théorie des représentations de $\mathfrak{g}_u$ n'est plus entièrement équivalente à celle des représentations de G_u. Si on cherche à rétablir l'équivalence en remplaçant G_u par son revêtement universel G_u^*, qui, lui, est simplement connexe, il devient nécessaire de s'assurer que celui-ci est compact, et aussi d'en faire un groupe, localement isomorphe à G_u. Ce dernier point, qui devait peu après être élucidé par Schreier, est complètement laissé de côté dans le mémoire de Weyl. Mais c'est dans le premier que résidait la véritable difficulté. La question revient naturellement à faire voir que G_u a un groupe fondamental fini. Pour cela, Weyl introduit un sous-groupe A_u de G_u, qui joue le même rôle que le groupe des matrices diagonales dans la théorie du groupe unitaire spécial. Tout élément s de G_u est conjugué à un élément de A_u; excluant certains éléments s, dits singuliers, qui forment un ensemble ayant trois dimensions de moins que G_u, s n'est conjugué qu'à un nombre fini d'éléments de A_u; de plus, les éléments de A_u qui ne sont pas singuliers forment dans A_u un domaine simplement connexe Δ. Supposons que s décrive dans G_u une courbe fermée Γ qui ne rencontre pas l'ensemble des éléments singuliers. Si $s(t)$ est le point de paramètre t sur Γ, on peut déterminer par continuité une courbe $a(t)$ dans A_u telle que, pour tout t, $a(t)$ soit conjugué à $s(t)$.

Quand le point $s(t)$ revient à sa position initiale $s(1) = s(0)$, le point $a(t)$ vient en un point $a(1)$ qui est un élément de A_u conjugué de $a(0)$, ce qui ne laisse pour ce point qu'un nombre fini de possibilités. Si on a $a(1) = a(0)$, la courbe décrite par $a(t)$ est fermée, et par suite réductible à un point dans Δ; Weyl montre que Γ est alors elle-même réductible à un point. Il en résulte facilement que le groupe fondamental de l'ensemble des éléments non singuliers de G_u est fini. De cela, et du fait que les éléments singuliers se répartissent sur des sous-variétés ayant au moins trois dimensions de moins que G_u, Weyl conclut (à vrai dire sans démonstration) que G_u lui-même a un groupe fondamental fini.

Ce point établi, la voie est ouverte à la généralisation complète au cas semi-simple des résultats obtenus pour le groupe linéaire spécial. Weyl démontre la complète réductibilité des représentations de $\mathfrak{g}$, et détermine explicitement le caractère et le degré d'une représentation simple de poids dominant donné. Ici encore, cette détermination résulte des relations d'orthogonalité entre caractères et des propriétés formelles de la restriction χ d'un caractère au groupe A_u^* qui recouvre A_u dans le revêtement simplement connexe G_u^* de G_u. Ce groupe est un tore; χ est une combinaison linéaire finie de caractères de ce tore, invariante par les opérations d'un certain groupe fini S d'automorphismes du tore qui généralise le groupe des permutations de $x_1, \ldots, x_n$ dont il a été question plus haut à propos du groupe unitaire spécial. Le groupe S, dont les développements ultérieurs de la théorie ont montré qu'il y joue un rôle fondamental, s'appelle maintenant le groupe de Weyl.

Enfin la théorie s'achève par la démonstration de l'existence des représentations simples de poids fondamental donné. Pour les algèbres simples, cette existence avait été établie par Cartan par des constructions directes dans chaque cas particulier. Weyl, lui, applique au groupe compact G_u^* la méthode de décomposition de la « représentation régulière », obtenue au moyen de la théorie des équations intégrales suivant l'idée que nous avons exposée plus haut. Pour conclure à partir de là, il lui faut encore un lemme de nature plus technique, énoncé seulement dans le mémoire de 1926, et dont Weyl n'a publié la démonstration que dans son cours de 1934-35 (paru à Princeton sous forme de notes miméographiées, *The structure and representations of continuous groups*).

Beaucoup plus tard, Weyl revint sur la détermination des représentations des groupes semi-simples dans son ouvrage *The classical groups, their invariants and representations.* L'esprit de ce livre est assez différent de celui du mémoire de 1926. L'objet de l'auteur est maintenant d'une part de démontrer par des méthodes purement algébriques les résultats déjà obtenus au sujet des représentations des groupes classiques (groupe linéaire général, groupe linéaire spécial, groupe orthogonal et groupe symplectique), et d'autre part de faire la synthèse entre ces résultats et la théorie formelle des invariants qui s'était développée sous l'influence de Cayley et Sylvester au cours du XIXe siècle. Espérait-il cette fois se laver définitivement du reproche de Study en ramenant à la vie cette théorie qui était sur le point de sombrer dans l'oubli ? Il nous dit lui-même que la démonstration par Hilbert du théorème général de finitude avait « presque tué le sujet »; on peut se demander si Weyl ne lui aura pas, en réalité, porté le coup de grâce.

La situation dans laquelle on se trouve en théorie des invariants est la suivante. On a une ou plusieurs représentations linéaires ρ, ρ', ..., d'un groupe G, opérant sur des espaces vectoriels V, V$'$, ... On considère des fonctions $F(x, x', \ldots)$ dépendant d'un argument x dans V, d'un argument x' dans V$'$, etc. et s'exprimant comme polynômes par rapport aux coordonnées de ces arguments, homogènes par rapport aux coordonnées de chacun d'eux. Une telle fonction s'appelle un invariant si, pour tout s dans G, on a

$$F(s.x\,,\, s.x'\,,\, \ldots) = F(x\,,\, x'\,,\, \ldots)\,.$$

Si J_1, ..., J_h sont des invariants, tout polynôme en J_1, ..., J_h en est un aussi pourvu qu'il satisfasse aux conditions d'homogénéité imposées. Le premier problème de la théorie est de trouver des invariants J_1, ..., J_h tels que tout autre invariant puisse s'écrire comme polynôme en les J_i; cela fait, on se propose également de déterminer les relations algébriques $F(J_1, \ldots, J_h) = 0$, dites « syzygies », qui lient entre eux les invariants qu'on a construits.

Plaçons-nous plus particulièrement dans le cas où G a été identifié, au moyen d'une certaine représentation ρ_1, avec un sous-groupe du groupe linéaire à n variables, opérant sur l'espace vectoriel $V_1 = k^n$, où k est un corps de base qu'on suppose de caractéristique 0. Considérons d'abord le cas où les représentations ρ, ρ', ..., coïncident toutes avec ρ_1; on dit alors qu'on cherche les invariants d'un certain nombre de « vecteurs » (on entend par là des vecteurs de V_1). Reprenant sans grand changement son travail de 1924 par lequel il avait répondu à Study, Weyl montre alors, pour un groupe G unimodulaire, que la détermination des invariants de vecteurs en nombre quelconque peut se ramener, au moyen des identités de Capelli, au problème analogue pour $n - 1$ vecteurs. Si G n'est pas unimodulaire, ce résultat reste vrai pour les « invariants relatifs » (polynômes se multipliant par une puissance du déterminant de s quand on transforme tous les vecteurs par s). Weyl déduit de là la solution des deux problèmes ci-dessus pour le groupe unimodulaire et pour le groupe orthogonal; et il étend cette solution au cas des invariants dépendant, non seulement d'un certain nombre de vecteurs « cogrédients » (se transformant suivant ρ_1), mais aussi d'un certain nombre de vecteurs « contragrédients » (se transformant comme les formes linéaires sur V_1). Ensuite il passe aux invariants dépendant de « quantités » x, x', ... appartenant à des espaces de représentation quelconques du groupe étudié; le cas où x, x', ... sont des formes homogènes par rapport aux coordonnées d'un vecteur « contragrédient » est celui dont traitait plus particulièrement la théorie classique. Pour pouvoir aborder la question dans ce cadre général, il faut avant tout connaître les représentations simples du groupe; aussi une partie importante du livre est-elle consacrée à la détermination algébrique des représentations « tensorielles » des groupes classiques. Cela fait, Weyl montre que les invariants dépendant de plusieurs « quantités » d'espèce quelconque s'expriment comme polynômes en un nombre fini d'entre eux; il étend ce résultat, dans une certaine mesure, au groupe affine. Enfin, il emploie la méthode d'intégration pour démontrer le résultat correspondant pour les représentations quelconques d'un groupe compact, le corps de base étant cette fois le corps des réels.

Pour fêter son soixante-dixième anniversaire, les amis et élèves de Hermann Weyl publièrent un volume de *Selecta* extraits de son œuvre. Il n'y a peut-être pas lieu de se féliciter de cette mode des morceaux choisis destinés à célébrer la mise à la retraite de mathématiciens éminents. C'est trop pour les uns; ce n'est pas assez pour les autres. Du moins le volume en question contient-il une bibliographie complète de l'œuvre de Hermann Weyl, établie par ordre chronologique [5], et dont nous avons naturellement fait grand usage pour rédiger la présente notice. Pour remédier en quelque mesure aux inévitables lacunes de celle-ci, nous donnons ci-dessous une liste des mémoires de Weyl, classés par sujet; rien ne peut mieux, croyons-nous, en faire ressortir l'étonnante variété. Les numéros, bien entendu, renvoient à la liste des *Selecta*.

I. *Analyse.*

a) Equations intégrales singulières: 1, 3.

b) Problèmes de valeurs propres et développements fonctionnels associés à des équations différentielles ou aux différences finies: 6, 7, 8, 12, 103.

c) Répartition des valeurs propres d'opérateurs complètement continus en physique mathématique: 13, 16, 17, 18, 19, 22.

d) Espace de Hilbert: 4, 5.

e) Phénomène de Gibbs et analogues: 10, 11, 14.

f) Equations différentielles liées à des problèmes physiques: 36-37 (développements asymptotiques, apparentés au phénomène de Gibbs, au voisinage d'une discontinuité dans un

[5] Il convient de signaler qu'on n'a pas fait figurer dans cette bibliographie les notes de cours, publiées sous forme miméographiée par l'Institute for Advanced Study de Princeton, et qui reproduisent plusieurs des cours qu'il y professa.

II. *Géométrie.*

III. *Invariants et groupes de Lie.*

IV. *Relativité.*

V. *Théorie des quanta.*

75, 83, 84, 85, 86, 87, 90, 91, 100, 101, 140, 141.

VI. *Théorie des algèbres.*

a) Matrices de Riemann: 99, 107, 108.

b) Questions diverses: 96, 105 (spineurs, en commun avec R. Brauer), 109, 110, 143.

VII. *Théorie géométrique des nombres* (d'après Minkowski et Siegel).

120, 126, 127, 136.

VIII. *Logique.*

9, 32, 41, 60 (2e partie), 67, 77, 78.

IX. *Philosophie.*

111, 118, 119, 138, 142, 156, 163.

X. *Articles historiques et biographiques.*

15, 88, 94, 95, 102, 131, 132, 137, 147, 149, 150, 152, 157, 160, 161, 162; et la conférence « Erkenntnis und Besinnung », *Studia Philos.*, 15 (Basel, 1955) (traduction française dans *Rev. de Théol. et Philos.*, Lausanne, 1955).

XI. *Varia.*

2, 28, 38, 76, 92, 115, 128, 133, 139, 144, 146, 148, 151, 158.

Vollständige Liste aller Titel

Band I

Band II

Band III

Band IV